Applied Mathematics and Fractional Calculus

Applied Mathematics and Fractional Calculus

Editors

Francisco Martínez González
Mohammed K. A. Kaabar

MDPI • Basel • Beijing • Wuhan • Barcelona • Belgrade • Manchester • Tokyo • Cluj • Tianjin

Editors
Francisco Martínez González
Departamento de Matemática
Aplicada y Estadística
Universidad Politécnica de
Cartagena
Cartagena
Spain

Mohammed K. A. Kaabar
Institute of Mathematical
Sciences, Faculty of Science
Universiti Malaya,
Kuala Lumpur 50603
Malaysia

Editorial Office
MDPI
St. Alban-Anlage 66
4052 Basel, Switzerland

This is a reprint of articles from the Special Issue published online in the open access journal *Symmetry* (ISSN 2073-8994) (available at: www.mdpi.com/journal/symmetry/special_issues/ Applied_Mathematics_Fractional_Calculus).

For citation purposes, cite each article independently as indicated on the article page online and as indicated below:

LastName, A.A.; LastName, B.B.; LastName, C.C. Article Title. *Journal Name* **Year**, *Volume Number*, Page Range.

ISBN 978-3-0365-5148-7 (Hbk)
ISBN 978-3-0365-5147-0 (PDF)

© 2022 by the authors. Articles in this book are Open Access and distributed under the Creative Commons Attribution (CC BY) license, which allows users to download, copy and build upon published articles, as long as the author and publisher are properly credited, which ensures maximum dissemination and a wider impact of our publications.

The book as a whole is distributed by MDPI under the terms and conditions of the Creative Commons license CC BY-NC-ND.

Contents

About the Editors . . . . **ix**

Preface to "Applied Mathematics and Fractional Calculus" . . . . **xi**

Mohammad Esmael Samei, Rezvan Ghaffari, Shao-Wen Yao, Mohammed K. A. Kaabar, Francisco Martínez and Mustafa Inc
Existence of Solutions for a Singular Fractional q-Differential [-25]Equations under Riemann–Liouville Integral Boundary Condition
Reprinted from: *Symmetry* **2021**, *13*, 1235, doi:10.3390/sym13071235 . . . . **1**

Malgorzata Klimek
Spectrum of Fractional and Fractional Prabhakar Sturm–Liouville Problems with Homogeneous Dirichlet Boundary Conditions
Reprinted from: *Symmetry* **2021**, *13*, 2265, doi:10.3390/sym13122265 . . . . **21**

Yuri Luchko
General Fractional Integrals and Derivatives of Arbitrary Order
Reprinted from: *Symmetry* **2021**, *13*, 755, doi:10.3390/sym13050755 . . . . **43**

Jehad Alzabut, A. George Maria Selvam, R. Dhineshbabu and Mohammed K. A. Kaabar
The Existence, Uniqueness, and Stability Analysis of the Discrete Fractional Three-Point Boundary Value Problem for the Elastic Beam Equation
Reprinted from: *Symmetry* **2021**, *13*, 789, doi:10.3390/sym13050789 . . . . **57**

Chenkuan Li and Joshua Beaudin
Uniqueness of Abel's Integral Equations of the Second Kind with Variable Coefficients
Reprinted from: *Symmetry* **2021**, *13*, 1064, doi:10.3390/sym13061064 . . . . **75**

Shahram Rezapour, Atika Imran, Azhar Hussain, Francisco Martínez, Sina Etemad and Mohammed K. A. Kaabar
Condensing Functions and Approximate Endpoint Criterion for the Existence Analysis of Quantum Integro-Difference FBVPs
Reprinted from: *Symmetry* **2021**, *13*, 469, doi:10.3390/sym13030469 . . . . **87**

Alberto Cabada, Nikolay D. Dimitrov and Jagan Mohan Jonnalagadda
Non-Trivial Solutions of Non-Autonomous Nabla Fractional Difference Boundary Value Problems
Reprinted from: *Symmetry* **2021**, *13*, 1101, doi:10.3390/sym13061101 . . . . **109**

Maria Alessandra Ragusa and Fan Wu
Regularity Criteria for the 3D Magneto-Hydrodynamics Equations in Anisotropic Lorentz Spaces
Reprinted from: *Symmetry* **2021**, *13*, 625, doi:10.3390/sym13040625 . . . . **125**

Michael A. Awuya and Dervis Subasi
Aboodh Transform Iterative Method for Solving Fractional Partial Differential Equation with Mittag–Leffler Kernel
Reprinted from: *Symmetry* **2021**, *13*, 2055, doi:10.3390/sym13112055 . . . . **135**

Sarra Guechi, Rajesh Dhayal, Amar Debbouche and Muslim Malik
Analysis and Optimal Control of φ-Hilfer Fractional Semilinear Equations Involving Nonlocal Impulsive Conditions
Reprinted from: *Symmetry* **2021**, *13*, 2084, doi:10.3390/sym13112084 . . . . **155**

Manuel Duarte Ortigueira and Gabriel Bengochea
Bilateral Tempered Fractional Derivatives
Reprinted from: *Symmetry* **2021**, *13*, 823, doi:10.3390/sym13050823 . . . . **173**

Shugui Kang, Youmin Lu and Wenying Feng
λ-Interval of Triple Positive Solutions for the Perturbed Gelfand Problem
Reprinted from: *Symmetry* **2021**, *13*, 1606, doi:10.3390/sym13091606 . . . . **187**

Surang Sitho, Sina Etemad, Brahim Tellab, Shahram Rezapour, Sotiris K. Ntouyas and Jessada Tariboon
Approximate Solutions of an Extended Multi-Order Boundary Value Problem by Implementing Two Numerical Algorithms
Reprinted from: *Symmetry* **2021**, *13*, 1341, doi:10.3390/sym13081341 . . . . **199**

Vladimir E. Fedorov, Marina V. Plekhanova and Elizaveta M. Izhberdeeva
Initial Value Problems of Linear Equations with the Dzhrbashyan–Nersesyan Derivative in Banach Spaces
Reprinted from: *Symmetry* **2021**, *13*, 1058, doi:10.3390/sym13061058 . . . . **225**

Shazad Shawki Ahmed and Shabaz Jalil MohammedFaeq
Bessel Collocation Method for Solving Fredholm–Volterra Integro-Fractional Differential Equations of Multi-High Order in the Caputo Sense
Reprinted from: *Symmetry* **2021**, *13*, 2354, doi:10.3390/sym13122354 . . . . **239**

Tinggang Zhao and Yujiang Wu
Hermite Cubic Spline Collocation Method for Nonlinear Fractional Differential Equations with Variable-Order
Reprinted from: *Symmetry* **2021**, *13*, 872, doi:10.3390/sym13050872 . . . . **267**

Pongsakorn Sunthrayuth, Ahmed M. Zidan, Shao-Wen Yao, Rasool Shah and Mustafa Inc
The Comparative Study for Solving Fractional-Order Fornberg–Whitham Equation via ρ-Laplace Transform
Reprinted from: *Symmetry* **2021**, *13*, 784, doi:10.3390/sym13050784 . . . . **297**

Ricardo Almeida and Natália Martins
New Variational Problems with an Action Depending on Generalized Fractional Derivatives, the Free Endpoint Conditions, and a Real Parameter
Reprinted from: *Symmetry* **2021**, *13*, 592, doi:10.3390/sym13040592 . . . . **313**

Sachin Kumar, Baljinder Kour, Shao-Wen Yao, Mustafa Inc and Mohamed S. Osman
Invariance Analysis, Exact Solution and Conservation Laws of (2 + 1) Dim Fractional Kadomtsev-Petviashvili (KP) System
Reprinted from: *Symmetry* **2021**, *13*, 477, doi:10.3390/sym13030477 . . . . **331**

Nehad Ali Shah, Yasser S. Hamed, Khadijah M. Abualnaja, Jae-Dong Chung, Rasool Shah and Adnan Khan
A Comparative Analysis of Fractional-Order Kaup–Kupershmidt Equation within Different Operators
Reprinted from: *Symmetry* **2022**, *14*, 986, doi:10.3390/sym14050986 . . . . **347**

Pshtiwan Othman Mohammed, Hassen Aydi, Artion Kashuri, Y. S. Hamed and Khadijah M. Abualnaja
Midpoint Inequalities in Fractional Calculus Defined Using Positive Weighted Symmetry Function Kernels
Reprinted from: *Symmetry* **2021**, *13*, 550, doi:10.3390/sym13040550 . . . . **371**

Saima Rashid, Aasma Khalid, Sobia Sultana, Zakia Hammouch, Rasool Shah and Abdullah M. Alsharif
A Novel Analytical View of Time-Fractional Korteweg-De Vries Equations via a New Integral Transform
Reprinted from: *Symmetry* **2021**, *13*, 1254, doi:10.3390/sym13071254 . . . . **393**

About the Editors

Francisco Martínez González

Francisco Martínez is a tenured associate professor at the Universidad Politécnica de Cartagena, Spain. He received his PhD degree in Physics from Universidad de Murcia in 1992. His research interests include nonlinear dynamics methods and their applications, fractional calculus, fractional differential equations, multivariate calculus or special functions, and the divulgation of mathematics.

Mohammed K. A. Kaabar

Mohammed K. A. Kaabar received all his undergraduate and graduate degrees in Applied and Theoretical Mathematics from Washington State University (WSU), Pullman, WA, USA. Prof. Kaabar has diverse experience in teaching, globally, and has worked as Adjunct Full Professor of Mathematics, Math Lab Instructor, and Lecturer at various US institutions such as Moreno Valley College, California, USA, Washington State University, Washington, USA, and Colorado Early Colleges, Colorado, USA. Prof. Kaabar is an Elected Foreign Member of the Academy of Engineering Sciences of Ukraine and Ukrainian School of Mining Engineering, Senior Member of the Hong Kong Chemical, Biological & Environmental Engineering Society (HKCBEES), and Council Member of the International Engineering and Technology Institute (IETI). He has published more than 100 research papers indexed by Scopus and Web of Science. He has authored two math textbooks, and he served as an invited referee for more than 300 Science, Technology, Engineering, and Mathematics (STEM) international conferences and journals all over the world. He served as an editor for the American Mathematical Society (AMS) Graduate Student Blog and full editor for an educational program (Mathematics and Statistics Section) at California State University, Long Beach, CA, USA. Prof. Kaabar is currently serving as an editor for 21 international scientific journals in applied mathematics and engineering. He is an invited keynote speaker in scientific conferences in Hong Kong, France, Ukraine, Turkey, China, Malaysia, India, Romania, USA, Singapore, and Italy. His research interests are fractional calculus, applied analysis, fractal calculus, applied mathematics, mathematical physics, mathematical modelling of infectious diseases, deep learning, and nonlinear optimization.

Preface to "Applied Mathematics and Fractional Calculus"

In the last three decades, fractional calculus has broken into the field of mathematical analysis, both at the theoretical level and at the level of its applications. In essence, the fractional calculus theory is a mathematical analysis tool applied to the study of integrals and derivatives of arbitrary order, which unifies and generalizes the classical notions of differentiation and integration. These fractional and derivative integrals, which until not many years ago had been used in purely mathematical contexts, have been revealed as instruments with great potential to model problems in various scientific fields, such as : fluid mechanics, viscoelasticity, physics, biology, chemistry, dynamical systems, signal processing or entropy theory. Since the differential and integral operators of fractional order are nonlinear operators, fractional calculus theory provides a tool for modeling physical processes, which in many cases is more useful than classical formulations. This is why the application of fractional calculus theory has become a focus of international academic research. This Special Issue *"Applied Mathematics and Fractional Calculus"* has published excellent research studies in the field of applied mathematics and fractional calculus, authored by many well-known mathematicians and scientists from diverse countries worldwide such as China, USA, Canada, Germany, Mexico, Spain, Poland, Portugal, Iran, Tunisia, South Africa, Albania, Thailand, Iraq, Egypt, Italy, India, Russia, Pakistan, Taiwan, Korea, Turkey, and Saudi Arabia.

Francisco Martínez González and Mohammed K. A. Kaabar
Editors

Article

Existence of Solutions for a Singular Fractional q-Differential Equations under Riemann–Liouville Integral Boundary Condition

Mohammad Esmael Samei [1,2], Rezvan Ghaffari [1], Shao-Wen Yao [3,*], Mohammed K. A. Kaabar [4,5], Francisco Martínez [6] and Mustafa Inc [7,8]

1 Department of Mathematics, Faculty of Basic Science, Bu-Ali Sina University, Hamedan 65178, Iran; mesamei@basu.ac.ir (M.E.S.); rghaffari68@yahoo.com (R.G.)
2 Department of Medical Research, China Medical University Hospital, China Medical University, Taichung 40402, Taiwan
3 School of Mathematics and Information Science, Henan Polytechnic University, Jiaozuo 454000, China
4 Department of Mathematics and Statistics, Washington State University, Pullman, WA 99163, USA; mohammed.kaabar@wsu.edu
5 Institute of Mathematical Sciences, University of Malaya, Kuala Lumpur 50603, Malaysia
6 Department of Applied Mathematics and Statistics, Technological University of Cartagena, 30203 Cartagena, Spain; f.martinez@upct.es
7 Department of Computer Engineering, Biruni University, Istanbul 34025, Turkey; minc@firat.edu.tr
8 Department of Mathematics, Science Faculty, Firat University, Elazig 23119, Turkey
* Correspondence: yaoshaowen@hpu.edu.cn

Abstract: We investigate the existence of solutions for a system of m-singular sum fractional q-differential equations in this work under some integral boundary conditions in the sense of Caputo fractional q-derivatives. By means of a fixed point Arzelá–Ascoli theorem, the existence of positive solutions is obtained. By providing examples involving graphs, tables, and algorithms, our fundamental result about the endpoint is illustrated with some given computational results. In general, symmetry and q-difference equations have a common correlation between each other. In Lie algebra, q-deformations can be constructed with the help of the symmetry concept.

Keywords: Caputo q-derivative; singular sum fractional q-differential; fixed point; equations; Riemann–Liouville q-integral

MSC: 34A08; 34B16; 39A13

Citation: Samei, M.E.; Ghaffari, R.; Yao, S.-W.; Kaabar, M.K.A.; Martínez, F.; Inc, M. Existence of Solutions for a Singular Fractional q-Differential Equations under Riemann–Liouville Integral Boundary Condition. *Symmetry* **2021**, *13*, 1235. https://doi.org/10.3390/sym13071235

Academic Editors: Sergei D. Odintsov and Sun Young Cho

Received: 2 April 2021
Accepted: 26 May 2021
Published: 9 July 2021

Publisher's Note: MDPI stays neutral with regard to jurisdictional claims in published maps and institutional affiliations.

Copyright: © 2021 by the authors. Licensee MDPI, Basel, Switzerland. This article is an open access article distributed under the terms and conditions of the Creative Commons Attribution (CC BY) license (https://creativecommons.org/licenses/by/4.0/).

1. Introduction

There are many definitions of fractional derivatives that have been formulated according to two basic conceptions: one of a global (classical) nature and the other of a local nature. Under the first formulation, the fractional derivative is defined as an integral, Fourier, or Mellin transformation, which provides its non-local property with memory. The second conception is based on a local definition through certain incremental ratios. This global conception is associated with the appearance of the fractional calculus itself and dates back to the pioneering works of important mathematicians, such as Euler, Laplace, Lacroix, Fourier, Abel, and Liouville, until the establishment of the classical definitions of Riemann–Liouville and Caputo.

Until relatively recently, the study of these fractional integrals and derivatives was limited to a purely mathematical context; however, in recent decades, their applications in various fields of natural Sciences and technology, such as fluid mechanics, biology, physics, image processing, or entropy theory, have revealed the great potential of these fractional integrals and derivatives [1–9]. Furthermore, the study from the theoretical and practical point of view of the elements of fractional differential equations has become a focus for interested researchers [10–15].

The q-difference equations (qDifEqs) were first proposed by Jackson in 1910 [16]. After that, qDifEqs were investigated in various studies [17–24]. On the contrary, integro-differential equations (InDifEqs) have been recently studied via various fractional derivatives and formulations based on the original idea of qDifEqs (see [25–32]). The concept of symmetry and q-difference equations are connected to each other while theoretically investigating the differential equation symmetries.

The solution existence and uniqueness for the fractional qDifEqs were investigated in 2012 by Ahmad et al. as: $^c\mathcal{D}_q^\alpha[u](t) = T(t, u(t))$ with boundary conditions (B.Cs):

$$\alpha_1 u(0) - \beta_1 \mathcal{D}_q[u](0) = \gamma_1 u(\eta_1), \quad \alpha_2 u(1) - \beta_2 \mathcal{D}_q[u](1) = \gamma_2 u(\eta_2),$$

where $\alpha \in (1,2]$, $\alpha_i, \beta_i, \gamma_i, \eta_i$ are real numbers, for $i = 1, 2$ and $T \in C(J \times \mathbb{R}, \mathbb{R})$ [20]. The q-integral problem was studied in in 2013 by Zhao et al. as:

$$\mathcal{D}_q^\alpha[u](t) + f(t, u(t)) = 0,$$

with B.Cs: $u(1) = \mu \mathcal{I}_q^\beta[u](\eta)$ and $u(0) = 0$ almost $\forall\, t \in (0,1)$, where $q \in (0,1)$, $\alpha \in (1,2]$, $\beta \in (0,2]$, $\eta \in (0,1)$, μ is positive real number, and $\mathcal{D}_q^\alpha$ is the q-derivative of Riemann–Liouville (RL) and the real values continuous map u defined on $I \times [0, \infty)$ [24]. The problem:

$$^c D_q^\beta (^c D_q^\gamma + \lambda)[u](t) = pf(t, u(t)) + k\mathcal{I}_q^\zeta[g](t, u(t))$$

was investigated in 2014 by Ahmad et al. with B.Cs:

$$\alpha_1 u(0) - \beta_1 (t^{(1-\gamma)} \mathcal{D}_q[u](0))\big|_{t=0} = \sigma_1 u(\eta_1)$$

and

$$\alpha_2 u(1) + \beta_2 \mathcal{D}_q[u](1) = \sigma_2 u(\eta_2),$$

where $t, q \in [0,1]$, $^c\mathcal{D}_q^\beta$ is the Caputo fractional q-derivative (CpFqDr), $0 < \beta, \gamma \leq 1$, $\mathcal{I}_q^\zeta(.)$ represents the RL integral with $\zeta \in (0,1)$, f and g are given continuous functions, λ and p, k are real constants, $\alpha_i, \beta_i, \sigma_i \in \mathbb{R}$ and $\eta_i \in (0,1)$ for $i = 1, 2$ [19]. The solutions' existence was studied in 2019 by Samei et al. for some multi-term q-integro-differential equations with non-separated and initial B.Cs ([23]).

Inspired by all previous works, we investigate in this work the positive solutions for the singular fractional q-differential equation (SFqDEqs) as follows:

$$^c\mathcal{D}_q^\alpha[u](t) + h(t, u(t)) = 0, \tag{1}$$

with the B.Cs: $u(0) = 0$, $cu(1) = \mathcal{I}_q^\gamma[u](1)$ and $u''(0) = \cdots = u^{(n-1)}(0) = 0$, where $t \in J = (0,1)$, $\mathcal{I}_q^\gamma[u]$ is the RL q-integral of order γ for the given function: u, here $q \in J$, $c \geq 1$, $n = [\alpha] + 1$, $\alpha \geq 3$, $\gamma \in [1, \infty)$, $2\Gamma_q(\gamma) \geq \Gamma_q(\alpha)$, $h : (0,1] \times [0, \infty) \to [0, \infty)$ is continuous, $\lim_{t \to 0^+} h(t, .) = +\infty$ that is, h is singular at $t = 0$, and $^c\mathcal{D}_q^\alpha$ represents the CpFqDr of order α, $q \in J$.

This work is divided into the following: some essential notions and basic results of q-calculus are reviewed in Section 2. Our original important results are stated in Section 3. In Section 4, illustrative numerical examples are provided to validate the applicability of our main results.

2. Essential Preliminaries

Assume that $q \in (0,1)$ and $a \in \mathbb{R}$. Define $[a]_q = \frac{1-q^a}{1-q}$ [16]. The power function: $(x - y)_q^n$ with $n \in \mathbb{N}_0$ is written as:

$$(x - y)_q^{(n)} = \prod_{k=0}^{n-1}(x - yq^k)$$

for $n \geq 1$ and $(x-y)_q^{(0)} = 1$, where x and y are real numbers and $\mathbb{N}_0 := \{0\} \cup \mathbb{N}$ ([17]). In addition, for $\sigma \in \mathbb{R}$ and $a \neq 0$, we obtain:

$$(x-y)_q^{(\sigma)} = x^\sigma \prod_{k=0}^\infty \frac{x - yq^k}{x - yq^{\sigma+k}}.$$

If $y = 0$, then it is obvious that $x^{(\sigma)} = x^\sigma$. The q-Gamma function is expressed by

$$\Gamma_q(z) = \frac{(1-q)^{(z-1)}}{(1-q)^{z-1}},$$

where $z \in \mathbb{R} \setminus \{0, -1, -2, \cdots\}$ ([16]). We know that $\Gamma_q(z+1) = [z]_q \Gamma_q(z)$. The value of the q-Gamma function, $\Gamma_q(z)$, for input values q and z with counting the sentences' number n in summation by simplification analysis. A pseudo-code is constructed for estimating q-Gamma function of order n. The q-derivative of function w, is expressed as:

$$\mathcal{D}_q[w](x) = \left(\frac{d}{dx}\right)_q w(x) = \frac{w(x) - w(qx)}{(1-q)x}$$

and $\mathcal{D}_q[w](0) = \lim_{x \to 0} \mathcal{D}_q[w](x)$ ([17]). In addition, the higher order q-derivative of a function w is defined by $\mathcal{D}_q^n[w](x) = \mathcal{D}_q \mathcal{D}_q^{n-1}[w](x)$ for all $n \geq 1$, where $\mathcal{D}_q^0[w](x) = w(x)$ ([17,18]). The q-integral of a function f defined on $[0, b]$ is expressed as:

$$\mathcal{I}_q[w](x) = \int_0^x w(s) \, d_q s = x(1-q) \sum_{k=0}^\infty q^k w(xq^k),$$

for $0 \leq x \leq b$, provided that the series is absolutely convergent ([17,18]). If a in $[0, b]$, then we have:

$$\int_a^b w(u) \, d_q u = \mathcal{I}_q[w](b) - \mathcal{I}_q[w](a) = (1-q) \sum_{k=0}^\infty q^k \left[bw(bq^k) - aw(aq^k) \right],$$

if the series exists. The operator $\mathcal{I}_q^n$ is given by $\mathcal{I}_q^0[w](x) = w(x)$ and $\mathcal{I}_q^n[w](x) = \mathcal{I}_q \mathcal{I}_q^{n-1}[w](x)$ for $n \geq 1$ and $g \in C([0, b])$ ([17,18]). It is proven that $\mathcal{D}_q \mathcal{I}_q[w](x) = w(x)$ and $\mathcal{I}_q \mathcal{D}_q[w](x) = w(x) - w(0)$ whenever w is continuous at $x = 0$ ([17,18]). The fractional RL type q-integral of the function w on J for $\sigma \geq 0$ is defined by $\mathcal{I}_q^0[w](t) = w(t)$, and

$$\mathcal{I}_q^\sigma[w](t) = \frac{1}{\Gamma_q(\sigma)} \int_0^t (t - qs)^{(\sigma-1)} w(s) \, d_q s$$

$$= t^\sigma (1-q)^\sigma \sum_{k=0}^\infty q^k \frac{\prod_{i=1}^{k-1}(1 - q^{\sigma+i})}{\prod_{i=1}^{k-1}(1 - q^{i+1})} w(tq^k),$$

for $t \in J$ and $\sigma > 0$ ([22,33]). In addition, the CpFqDr of a function w is expressed as:

$$^c\mathcal{D}_q^\sigma[w](t) = \mathcal{I}_q^{[\sigma]-\sigma} \left[^c\mathcal{D}_q^{[\sigma]}[w] \right](t)$$

$$= \frac{1}{\Gamma_q([\sigma] - \alpha)} \int_0^t (t - qs)^{([\sigma]-\sigma-1)} \, ^c\mathcal{D}_q^{[\sigma]}[w](s) \, d_q s$$

$$= \frac{1}{t^\sigma (1-q)^\sigma} \sum_{k=0}^\infty q^k \frac{\prod_{i=1}^{k-1}(1 - q^{i-\sigma})}{\prod_{i=1}^{k-1}(1 - q^{i+1})} w(tq^k), \qquad (2)$$

where $t \in J$ and $\sigma > 0$ ([22]). It is proven that

$$\mathcal{I}_q^\beta \left[\mathcal{I}_q^\sigma[w] \right](x) = \mathcal{I}_q^{\sigma+\beta}[w](x) \text{ and } ^c\mathcal{D}_q^\sigma \left[\mathcal{I}_q^\sigma[w] \right](x) = w(x),$$

where $\sigma, \beta \geq 0$ ([22]).

Some essential notions and lemmas are now presented as follows: In our work, $L^1(\overline{J})$ and $C_{\mathbb{R}}(\overline{J})$ are denoted by $\overline{\mathcal{L}}$ and $\overline{\mathcal{B}}$, respectively, where $\overline{J} = [0,1]$.

Lemma 1 ([34]). *If $x \in \overline{\mathcal{B}} \cap \overline{\mathcal{L}}$ with $\mathcal{D}_q^\alpha x \in \mathcal{B} \cap \mathcal{L}$, then*

$$\mathcal{I}_q^\alpha \mathcal{D}_q^\alpha x(t) = x(t) + \sum_{i=1}^{n} c_i t^{\alpha-i},$$

where n is the smallest integer $\geq \alpha$, and c_i is some real number.

Here, we restate the well-known Arzelá–Ascoli theorem. Assume that $S = \{s_n\}_{n \geq 1}$ is a sequence of bounded and equicontinuous real valued functions on $[a,b]$. Then, S has a uniformly convergent subsequence. We need the following fixed point theorem in our main result:

Lemma 2 ([35]). *Assume that $\mathcal{A}$ is a Banach space, $P \subseteq \mathcal{A}$ is a cone, and $\mathcal{O}_1, \mathcal{O}_2$ are two bounded open balls of $\mathcal{A}$ centered at the origin with $\overline{\mathcal{O}}_1 \subset \mathcal{O}_2$. Assume that $\Omega : P \cap (\overline{\mathcal{O}_2} \backslash \mathcal{O}_1) \to P$ is a completely continuous operator such that either $\|\Omega(a)\| \leq \|a\|$ for all $a \in P \cap \partial\mathcal{O}_1$ and $\|\Omega(a)\| \geq \|a\|$ for all $a \in P \cap \partial\mathcal{O}_2$, or $\|\Omega(a)\| \geq \|a\|$ for each $a \in P \cap \partial\mathcal{O}_1$ and $\|\Omega a\| \leq \|a\|$ for $a \in P \cap \partial\mathcal{O}_2$. Then, Ω has a fixed point in $P \cap (\mathcal{O}_2 \backslash \mathcal{O}_1)$.*

3. Main Results
Differential Equation

Let us now present our fundamental lemma as follows:

Lemma 3. *The u_0 is a solution for the q-differential equation $\mathcal{D}_q^\alpha[u](t) + g(t) = 0$ with the B.Cs: $u(0) = 0$, $cu(1) = \mathcal{I}_q^\gamma u(1)$ and $u''(0) = \cdots = u^{(n-1)}(0) = 0$ if u_0 is a solution for the q-integral equation*

$$u(t) = \int_0^1 G_q(t,s) f(s) \, d_q s,$$

where

$$G_q(t,s) = \begin{cases} \dfrac{-(t-qs)^{(\alpha-1)}}{\Gamma_q(\alpha)} \\ \quad + t^2 \dfrac{\Gamma_q(\gamma+3)\left[a\Gamma_q(\alpha+\gamma)(1-qs)^{(\alpha-1)} - \Gamma_q(\alpha)(1-qs)^{(c+\gamma-1)}\right]}{\Gamma_q(\alpha)\Gamma_q(\alpha+\gamma)\left[c\Gamma_q(\gamma+3) - 2\Gamma_q(\gamma)\right]}, & s \leq t, \\[2ex] t^2 \dfrac{\Gamma_q(\gamma+3)\left[c\Gamma_q(\alpha+\gamma)(1-qs)^{(\alpha-1)} - \Gamma_q(\alpha)(1-qs)^{(c+\gamma-1)}\right]}{\Gamma_q(\alpha)\Gamma_q(\alpha+\gamma)\left[c\Gamma_q(\gamma+3) - 2\Gamma_q(\gamma)\right]}, & t \leq s, \end{cases} \quad (3)$$

for $s, t \in \overline{J}$, $n = [\alpha] + 1$, the function $g \in \overline{\mathcal{B}}$, $\alpha \geq 3$ and $\gamma \in [1, \infty)$ with $2\Gamma_q(\gamma) \geq \Gamma_q(\alpha)$.

Proof. Let us first assume that u_0 is a solution for the equation $\mathcal{D}_q^\alpha u(t) + g(t) = 0$ with the B.Cs. By using Lemma 1, we obtain:

$$u_0(t) = -\mathcal{I}_q^\alpha[g](t) + c_0 + c_1 t + c_2 t^2 + \ldots c_{n-1} t^{n-1}$$

and by using the condition $u_0(0) = u_0''(0) = \cdots = u_0^{(n-1)}(0) = 0$, we have

$$u_0(t) = -\mathcal{I}_q^\alpha[g](t) + c_2 t^2.$$

Indeed,
$$\mathcal{I}_q^\gamma[u_0](t) = -\mathcal{I}_q^{\alpha+\gamma}[g](t) + c_2 \frac{2\Gamma_q(\gamma)}{\Gamma_q(\gamma+3)} t^{\gamma+2},$$

and thus
$$\mathcal{I}_q^\gamma[u_0](1) = -\mathcal{I}_q^{(\alpha+\gamma)}[g](t) + c_2 \frac{2\Gamma_q(\gamma)}{\Gamma_q(\gamma+3)}.$$

Note that $cu_0(1) = -c\mathcal{I}_q^\alpha[g](1) + cc_2$ and

$$c_2 \left(c - \frac{2\Gamma_q(\gamma)}{\Gamma_q(\gamma+3)}\right) = c\mathcal{I}_q^\alpha g(1) - \mathcal{I}_q^{\alpha+\gamma} g(1)$$
$$= \frac{c\Gamma_q(\alpha+\gamma)}{\Gamma_q(\alpha+\gamma)} \mathcal{I}_q^\alpha[g](1) - \frac{\Gamma_q(\alpha)}{\Gamma_q(\alpha)} \mathcal{I}_q^{\alpha+\gamma}[g](1)$$
$$= \int_0^1 \frac{c\Gamma_q(\alpha+\gamma)(1-qs)^{(\alpha-1)} - \Gamma_q(\alpha)(1-qs)^{(\alpha+\gamma-1)}}{\Gamma_q(\alpha)\Gamma_q(\alpha+\gamma)} g(s)\,d_q s.$$

On the other hand,
$$c - \frac{2\Gamma_q(\gamma)}{\Gamma_q(\gamma+3)} = \frac{c\Gamma_q(\gamma+3) - 2\Gamma_q(\gamma)}{\Gamma_q(\gamma+3)}.$$

Hence,
$$c_2 = \int_0^1 \frac{\Gamma_q(\gamma+3)\left[c\Gamma_q(\alpha+\gamma)(1-qs)^{(\alpha-1)} - \Gamma_q(\alpha)(1-qs)^{(\alpha+\gamma-1)}\right]}{\Gamma_q(\alpha)\Gamma_q(\alpha+\gamma)\left[c\Gamma_q(\gamma+3) - 2\Gamma_q(\gamma)\right]} g(s)\,d_q s.$$

Therefore, we have
$$u_0(t) = -\mathcal{I}_q^\alpha[g](t)$$
$$+ t^2 \int_0^1 \frac{\Gamma(\gamma+3)\left[c\Gamma_q(\alpha+\gamma)(1-qs)^{(\alpha-1)} - \Gamma_q(\alpha)(1-qs)^{(\alpha+\gamma-1)}\right]}{\Gamma_q(\alpha)\Gamma_q(\alpha+\gamma)\left[c\Gamma_q(\gamma+3) - 2\Gamma_q(\gamma)\right]} g(s)\,d_q s$$
$$= \int_0^1 G_q(s,t) g(s)\,d_q s,$$

where
$$G_q(t,s) = \frac{-(t-qs)^{(\alpha-1)}}{\Gamma_q(\alpha)}$$
$$+ t^2 \frac{\Gamma_q(\gamma+3)\left[c\Gamma_q(\alpha+\gamma)(1-qs)^{(\alpha-1)} - \Gamma_q(\alpha)(1-qs)^{c+\gamma-1}\right]}{\Gamma_q(\alpha)\Gamma_q(\alpha+\gamma)\left[c\Gamma_q(\gamma+3) - 2\Gamma_q(\gamma)\right]},$$

whenever $0 \leq s \leq t \leq 1$ and

$$t^2 \frac{\Gamma_q(\gamma+3)\left[c\Gamma_q(\alpha+\gamma)(1-qs)^{(\alpha-1)} - \Gamma_q(\alpha)(1-qs)^{(c+\gamma-1)}\right]}{\Gamma_q(\alpha)\Gamma_q(\alpha+\gamma)\left[c\Gamma_q(\gamma+3) - 2\Gamma_q(\gamma)\right]}$$

whenever $0 \leq t \leq s \leq 1$. Hence, u_0 is an integral equation's solution. By simple review, we can see that u_0 is a solution for the equation $\mathcal{D}_q^\alpha u(t) + g(t) = 0$ with the B.Cs whenever u_0 is an integral equation's solution. □

Remark 1. *By applying some simple calculations, one can show that $G_q(t,s) \geq 0$ for each $s, t \in \overline{J}$. Now, let us define the operator Ω on the Banach space $\overline{B}$ by*

$$\Omega(u(t)) = \int_0^1 G_q(t,s)h(s,u(s))\,d_q s.$$

It is easy to check that u_0 is a fixed point of the operator Ω if u_0 is a solution for Equation (1).

Consider $\overline{B}$ together the supremum norm and cone, P is the set of all $u \in \overline{B}$ such that $u(t) \geq 0 \,\forall\, t \in \overline{J}$. Suppose that $h : (0,1] \times [0,\infty) \to [0,\infty)$ is the singular function at $t = 0$ in the Equation (1) and $G_q(t,s)$ is the q-Green function in Lemma 3. Now, define the self operator Ω on P by

$$\Omega(u(t)) = \int_0^1 G_q(t,s)h(s,u(s))\,d_q s,$$

for all $t \in \overline{J}$. At present, we can provide our first main result on the solution's existence for problem (1) under some assumptions.

Theorem 1. *Problem (1) has a unique solution if the following conditions hold.*

I. *There exists a continuous function $h : (0,1] \times [0,\infty) \to [0,\infty)$ such that*

$$\lim_{t \to 0^+} h(t,s) = \infty,$$

for $s \in [0,\infty)$.

II. *There exists $L > 0$, $\beta \in J$ and positive constant k such that*

$$kc\Gamma_q(\gamma+3) < (c\Gamma_q(\gamma+3) - 2\Gamma_q(\gamma)),$$

$|t^\beta h(t,0)| \leq L$ *for each $t \in \overline{J}$ and*

$$|t^\beta h(t,u(t)) - t^\beta h(t,v(t))| \leq k\|u - v\|,$$

for each u, v belang to P.

Proof. Note that,

$$|\Omega(u(t))| \leq t^2 \frac{c\Gamma_q(\gamma+3)}{c\Gamma_q(\gamma+3) - 2\Gamma_q(\gamma)} \mathcal{I}_q^\alpha[h](1,u(1))$$

for all $t \in \overline{J}$. Now, put

$$\ell = L \frac{c\Gamma_q(\gamma+3)\Gamma_q(1-\beta)}{c\Gamma_q(\gamma+3) - 2\Gamma_q(\gamma)}$$

and define $B = \{u \in P : \|u\| \leq \ell\}$. Clearly, B is a bounded and closed subset of $\mathcal{A}$, and thus B is complete. If $u \in B$, then we obtain:

$$|\Omega(u(t))| \leq \frac{c\Gamma_q(\gamma+3)}{\Gamma_q(\alpha)[c\Gamma(\gamma+3) - 2\Gamma_q(\gamma)]} \int_0^1 (1-qs)^{(\alpha-1)} s^{-\beta} s^\beta h(s,u(s))\,d_q s$$

$\forall\, t \in \overline{J}$ and thus

$$|F(x(t))| \leq \frac{c\Gamma_q(\gamma+3)}{\Gamma_q(\alpha)[c\Gamma_q(\gamma+3) - 2\Gamma_q(\gamma)]}$$
$$\times \int_0^1 (1-qs)^{(\alpha-1)} s^{-\beta} s^\beta (|h(s,u(s) - h(s,0)| + |h(s,0)|)\,d_q s$$
$$\leq (k\ell + L) \frac{c\Gamma_q(\gamma+3)}{\Gamma_q(\alpha)[c\Gamma_q(\gamma+3) - 2\Gamma_q(\gamma)]} B_q(1-\beta,\alpha)$$

$$= (k\ell + L)\frac{c\Gamma(\gamma+3)\Gamma_q(1-\beta)}{[c\Gamma_q(\gamma+3) - 2\Gamma_q(\gamma)]\Gamma_q(\alpha-\beta+1)}$$

$$\leq \frac{[c\Gamma_q(\gamma+3) - 2\Gamma_q(\gamma)]\ell}{c\Gamma_q(\gamma+s)\Gamma_q(1-\beta)}\left[\frac{c\Gamma_q(\gamma+3)\Gamma_q(1-\beta)}{(c\Gamma_q(\gamma+3) - 2\Gamma_q(\gamma))\Gamma_q(\alpha-\beta+1)}\right]$$

$$+ L\frac{c\Gamma_q(\gamma+3)\Gamma_q(1-\beta)}{[c\Gamma_q(\gamma+3) - 2\Gamma_q(\gamma)]\Gamma_q(\alpha-\beta+1)}$$

$$= \frac{\ell}{\Gamma_q(\alpha-\beta+1)} + \frac{\ell}{\Gamma_q(\alpha-\beta+1)}$$

$$< \frac{\ell}{\Gamma_q(\alpha)} + \frac{\ell}{\Gamma_q(\alpha)} \leq \frac{\ell}{2} + \frac{\ell}{2} = \ell.$$

Indeed, $\Omega(B) \subseteq B$, and therefore a restriction of Ω on B is an operator on B. Let $u, v \in B$. Then, we obtain

$$\|\Omega(u(t)) - \Omega(v(t))\| \leq \frac{1}{\Gamma_q(\alpha)}\int_0^1 (t-qs)^{(\alpha-1)}|h(s,u(s)) - h(s,v(s))|\,d_qs$$

$$+ \frac{ct^2\Gamma_q(\gamma+3)}{\Gamma_q(\alpha)[c\Gamma_q(\gamma+3) - 2\Gamma_q(\gamma)]}$$

$$\times \int_0^1 (1-qs)^{(\alpha-1)}s^{-\beta}s^\beta\|h(s,u(s)) - h(s,v(s))\|\,d_qs$$

$$\leq k\|u-v\|$$

$$\times \left[\frac{\Gamma_q(1-\beta)}{\Gamma_q(\alpha-\beta+1)} + \frac{c\Gamma_q(\gamma+3)\Gamma_q(1-\beta)}{[c\Gamma_q(\gamma+3) - 2\Gamma_q(\gamma)]\Gamma_q(\alpha-\beta+1)}\right]$$

$$\leq \left[\frac{c\Gamma_q(\gamma+3) - 2\Gamma_q(\gamma)}{c\Gamma_q(\gamma+3)\Gamma_q(\alpha-\beta+1)} + \frac{1}{\Gamma_q(\alpha-\beta+1)}\right]\|u-v\|$$

$$< \left[\frac{c\Gamma_q(\gamma+3) - 2\Gamma_q(\gamma)}{c\Gamma_q(\gamma+3)\Gamma_q(\alpha)} + \frac{1}{\Gamma_q(\alpha)}\right]\|u-v\|$$

for all $t \in \bar{J}$. Take

$$\lambda = \frac{c\Gamma_q(\omega+3) - 2\Gamma_q(\omega)}{c\Gamma_q(\omega+3)\Gamma_q(\alpha)} + \frac{1}{\Gamma_q(\alpha)}.$$

Since $\alpha \geq 3$, we obtain $\lambda \in J$, and therefore $\Omega : B \to B$ is a contraction. Thus, Ω has a unique fixed point in B. By employing Lemma 3, the problem (1) has a unique solution in B. □

Lemma 4. *Suppose that there exists $\beta \in J$ such that the map $t^\beta g(t)$ is a continuous map on J. If $G_q(t,s)$ is the q-Green function (3) in Lemma 3, then*

$$\Omega(t) = \int_0^1 G_q(t,s)g(s)\,d_qs,$$

is also a continuous map on J. The self-operator Ω is completely continuous whenever there exists $\beta \in J$ such that the map $t^\beta g(t)$ is a continuous map on $\bar{J}$.

Proof. Since the map $t^\beta g(t)$ is continuous and $\Omega(t) = \int_0^t G_q(t,s)s^{-\beta}s^\beta g(s)\,d_qs$, we obtain

$$|\Omega(t)| \leq \sup_{s\in\delta}\left|G_q(t,s)s^\beta g(s)\right|\int_0^t s^{-\beta}\,ds = \frac{mt^{1-\beta}}{1-\beta},$$

where $\delta = [0, t]$,

$$m = \sup_{s\in\delta}\left|G_q(t,s)s^\beta g(s)\right| < \infty.$$

Indeed, $\Omega(0) = 0$. Note that, $G_q(t,s)$ is continuous in $\bar{J}^2$. First, suppose that $t_1 = 0$ and $t_2 \in (0,1]$. By continuity $t^\beta g(t)$, there exists $L > 0$ such that

$$\sup_{t \in \bar{J}} \left| t^\beta g(t) \right| \leq L.$$

Thus, we have:

$$|\Omega(t_2) - \Omega(t_1)| = |\Omega(t_2)| \leq \int_0^{t_2} \frac{(1-qs)^{(\alpha-1)}}{\Gamma_q(\alpha)} s^{-\beta} s^\beta g(s) \, d_q s$$

$$+ t_2^2 \int_0^1 \frac{\Gamma_q(\gamma+3)[c\Gamma_q(\alpha+\gamma) + \Gamma_q(\alpha)]}{\Gamma_q(\alpha)[c\Gamma_q(\gamma+3) - 2\Gamma_q(\gamma)]} (1-qs)^{(\alpha-1)} s^{-\beta} s^\beta g(s) \, d_q s$$

$$\leq \frac{L}{\Gamma_q(\alpha)} B_q(1-\beta, \alpha) t_2^{\alpha-\beta}$$

$$+ L t_2^2 \frac{\Gamma_q(\gamma+3)[c\Gamma_q(\alpha+\gamma) + \Gamma_q(\alpha)]}{\Gamma_q(\alpha)[c\Gamma_q(\gamma+3) - 2\Gamma_q(\gamma)]} B_q(1-\beta, \alpha)$$

$$= \frac{L \Gamma_q(1-\beta)}{\Gamma_q(\alpha-\beta+1)} t_2^{\alpha-\beta}$$

$$+ L \frac{\Gamma_q(\gamma+3) \Gamma_q(1-\beta)[c\Gamma_q(\alpha+\gamma) + \Gamma_q(\alpha)]}{[c\Gamma_q(\gamma+3) - 2\Gamma_q(\gamma)]\Gamma_q(\alpha-\beta+1)} t_2^2.$$

This implies that $\lim_{t_2 \to t_1} |\Omega(t_2) - \Omega(t_1)| = 0$. At present, in the next case, we assume that $t_1 \in J$ and $t_2 \in (t_1, 1]$. Thus, we obtain:

$$|\Omega(t_2) - \Omega(t_1)| \leq \frac{1}{\Gamma_q(\alpha)} \left| - \int_0^{t_2} (t_2-qs)^{(\alpha-1)} s^{-\beta} s^\beta g(s) \, d_q s \right.$$

$$\left. + \int_0^{t_1} (t_1-qs)^{(\alpha-1)} s^{-\beta} s^\beta g(s) \, d_q s \right|$$

$$+ |t_2^2 - t_1^2| \frac{\Gamma_q(\gamma+3)[c\Gamma_q(\gamma+3) + \Gamma_q(\alpha)]}{\Gamma_q(\alpha)[c\Gamma(\gamma+3) - 2\Gamma_q(\gamma)]}$$

$$\times \int_0^1 (1-qs)^{(\alpha-1)} s^{-\beta} s^\beta g(s) \, d_q s.$$

On the other hand,

$$\frac{1}{\Gamma_q(\alpha)} \left| - \int_0^{t_2} (t_2-qs)^{(\alpha-1)} s^{-\beta} s^\beta g(s) \, d_q s + \int_0^1 (t_1-qs)^{(\alpha-1)} s^{-\beta} s^\beta g(s) \, d_q s \right|$$

$$\leq \frac{1}{\Gamma_q(\alpha)} \left| \int_0^{t_1} (t_2-qs)^{\alpha-1} s^{-\beta} s^\beta g(s) \, d_q s \right.$$

$$\left. - \int_0^{t_2} (t_2-qs)^{(\alpha-1)} s^{-\beta} s^\beta g(s) \, d_q s \right|$$

$$= \frac{1}{\Gamma_q(\alpha)} \left| \int_{t_2}^{t_1} (t_2-qs)^{(\alpha-1)} s^{-\beta} s^\beta g(s) \, d_q s \right|$$

$$\leq \frac{L}{\Gamma_q(\alpha)} \int_{t_1}^{t_2} (t_2-qs)^{(\alpha-1)} s^{-\beta} \, d_q s$$

$$\leq \frac{L}{\Gamma_q(\alpha)} \sup_{s \in [t_2, t_2]} (t_2-qs)^{(\alpha-1)} \int_{t_1}^{t_2} s^{-\beta} \, d_q s$$

$$= \frac{L}{\Gamma_q(\alpha)} (t_2-t_1)^{\alpha-1} \frac{t_2^{1-\beta} - t_1^{1-\beta}}{1-\beta}$$

and therefore $\lim_{t_2 \to t_1} |\Omega(t_2) - \Omega(t_1)| = 0$. By applying in a similar way, we conclude that

$$\lim_{t_2 \to t_1} |\Omega(t_2) - \Omega(t_1)| = 0,$$

whenever $t_1 \in \overline{J}$ and $t_2 \in [0, t_1)$. Now, we prove that the self-operator Ω is completely continuous. Assume that $\varepsilon > 0$. Since the function $t^\beta h(t, u(t))$ is continuous, there exist $\delta > 0$ such that

$$|t^\beta h(t, u(t)) - t^\beta h(t, v(t))| < \varepsilon,$$

for each $u, v \in P$ with $\|u - v\| < \delta$. Thus, we obtain

$$\|\Omega(u) - \Omega(v)\| = \sup_{t \in \overline{J}} |\Omega(u(t)) - \Omega(v(t))|$$

$$= \sup_{t \in \overline{J}} \left| \int_0^t \frac{-(t - qs)^{(\alpha-1)}}{\Gamma_q(\alpha)} s^{-\beta} (s^\beta h(s, u(s)) - s^\alpha h(s, v(s))) \, d_q s \right.$$

$$+ t^2 \int_0^1 \frac{\Gamma_q(\gamma + 3) \left[c\Gamma_q(\gamma + \alpha)(1 - qs)^{(\alpha-1)} - \Gamma_q(\alpha)(1 - qs)^{(\alpha+\gamma-1)} \right]}{\Gamma_q(\alpha) \Gamma_q(\alpha + \gamma) \left[c\Gamma_q(\gamma + 3) - 2\Gamma_q(\gamma) \right]}$$

$$\left. \times s^{-\beta} \left[s^\beta h(s, u(s)) - s^\beta h(s, u(s)) \right] d_q s \right|$$

$$\leq \sup_{t \in \overline{J}} \left[\varepsilon \int_0^t \frac{(t - qs)^{(\alpha-1)}}{\Gamma_q(\alpha)} d_q s \right.$$

$$\left. + \varepsilon t^2 \int_0^1 \frac{\Gamma_q(\gamma + 3) \left[c\Gamma_q(\gamma + \alpha) + \Gamma_q(\alpha) \right]}{\Gamma_q(\alpha) \Gamma_q(\alpha + \gamma) \left[c\Gamma_q(\gamma + 3) - 2\Gamma_q(\gamma) \right]} (1 - qs)^{(\alpha-1)} s^{-\beta} d_q s \right]$$

$$\leq \sup_{t \in \overline{J}} \varepsilon t^{\alpha - \beta} \frac{\Gamma_q(1 - \beta)}{\Gamma_q(\alpha - \beta + 1)}$$

$$+ \sup_{t \in \overline{J}} \varepsilon t^2 \frac{\Gamma_q(\gamma + 3) \Gamma_q(1 - \beta) \left[c\Gamma_q(\gamma + \alpha) + \Gamma_q(\alpha) \right]}{\Gamma_q(\alpha + \gamma) \Gamma_q(\alpha - \beta + 1) \left[c\Gamma_q(\gamma + 3) - 2\Gamma_q(\gamma) \right]}$$

$$= \left[\frac{\Gamma_q(1 - \beta)}{\Gamma_q(\alpha - \beta + 1)} + \frac{\Gamma_q(\gamma + 3) \Gamma_q(1 - \beta) \left[c\Gamma_q(\alpha + \gamma) + \Gamma_q(\alpha) \right]}{\Gamma_q(\alpha + \gamma) \Gamma_q(\alpha - \beta + 1) \left[c\Gamma_q(\gamma + 3) - 2\Gamma_q(\gamma) \right]} \right] \varepsilon.$$

Therefore, Ω is continuous. Let $Q \subset P$ be bounded. Choose $k > 0$ such that $\|u\| \leq k$ for each $u \in Q$. Since the function $t^\beta h(t, u)$ is continuous on $\overline{J} \times [0, \infty)$, the function: $t^\beta h(t, u)$ is also continuous on $\overline{J} \times [0, k]$. Select $r \geq 0$ such that $|t^\beta h(t, u)| \leq r$ for all $u \in Q$, and t belongs to $\overline{J}$. Thus,

$$|\Omega(u(t))| \leq \int_0^1 G_q(t, s) s^{-\beta} |s^\beta h(s, u(s))| \, d_q s$$

$$\leq r \left[\int_0^t \frac{(t - qs)^{(\alpha-1)}}{\Gamma_q(\alpha)} s^{-\beta} d_q s \right.$$

$$\left. + t^2 \frac{\Gamma_q(\gamma + 3) \left[c\Gamma_q(\alpha + \gamma) + \Gamma_q(\alpha) \right]}{\Gamma_q(\alpha + \gamma) \left[c\Gamma_q(\gamma + 3) - 2\Gamma_q(\gamma) \right]} \int_0^1 (1 - qs)^{(\alpha-1)} s^{-\beta} d_q s \right],$$

for each $t \in \overline{J}$, and thus

$$\|\Omega(x(t))\| = \sup_{t \in \overline{J}} |\Omega(x(t))|$$

$$\leq \frac{\Gamma_q(1 - \beta)}{\Gamma_q(\alpha - \beta + 1)} + \frac{\Gamma_q(\gamma + 3) \Gamma_q(1 - \beta) \left[c\Gamma_q(\alpha + \gamma) - \Gamma_q(\alpha) \right]}{\Gamma_q(\alpha + \gamma) \Gamma_q(\alpha - \beta + 1) \left[c\Gamma_q(\gamma + 3) - 2\Gamma_q(\gamma) \right]}$$

$$< \infty.$$

This implies that $\Omega(Q)$ is bounded. Assume that $u \in Q$ and $t_1, t_2 \in \bar{J}$ with $t_1 < t_2$. Then, we obtain

$$|\Omega(u(t_2)) - \Omega(u(t_1))| \leq \left| \int_0^{t_2} \frac{(t_2 - qs)^{(\alpha-1)}}{\Gamma_q(\alpha)} h(s, u(s)) \, d_q s \right.$$

$$\left. - \int_0^{t_1} \frac{(t_1 - qs)^{(\alpha-1)}}{\Gamma_q(\alpha)} h(s, u(s)) \, d_q s \right|$$

$$+ |t_2^2 - t_1^2| \frac{\Gamma_q(\gamma + 3)[c\Gamma_q(\alpha + \gamma) + \Gamma_q(\alpha)]}{\Gamma_q(\alpha)[c\Gamma_q(\gamma + 3) - 2\Gamma_q(\gamma)]} \int_0^1 h(s, u(s)) d_q s$$

$$\leq r \int_{t_1}^{t_2} \frac{(t_2 - qs)^{(\alpha-1)}}{\Gamma_q(\alpha)} s^{-\beta} d_q s$$

$$+ r|t_2^2 - t_1^2| \int_0^1 \frac{\Gamma_q(\gamma + 3)[c\Gamma_q(\alpha + \gamma) + \Gamma_q(\alpha)]}{\Gamma_q(\alpha)[c\Gamma_q(\gamma + 3) - 2\Gamma_q(\gamma)]} s^{-\beta} d_q s$$

$$\leq \frac{r}{\Gamma_q(\alpha)} \sup_{s \in [t_1, t_2]} (t_2 - qs)^{(\alpha-1)} \frac{t_2^{1-\beta} - t_1^{1-\beta}}{1 - \beta}$$

$$+ r(t_2^2 - t_1^2) \frac{\Gamma_q(\gamma + 3)[c\Gamma_q(\alpha + \gamma) + \Gamma_q(\alpha)] \Gamma_q(1 - \beta)}{\Gamma_q(\alpha) \Gamma_q(\alpha - \gamma + 1)[c\Gamma_q(\gamma + 3) - 2\Gamma_q(\gamma)]}.$$

Thus,

$$\lim_{t_2 \to t_1} |\Omega(u(t_2)) - \Omega(u(t_1))| = 0.$$

In other cases, one can prove a similar result. Hence, $\Omega(Q)$ is equicontinuous. Now, by applying the Arzelà–Ascoli theorem, $\overline{\Omega(Q)}$ is compact, and therefore Ω is completely continuous. □

Theorem 2. *The problem (1) has at least one positive solution whenever the hypothesis as follows holds:*

I. *There exists $\beta \in J$ such that the map $t^\beta g(t)$ is a continuous map on J.*
II. *There exists $r'_1 > 0$ and $r'_2 > 0$ with $r'_2 < r'_1$ such that $t^\beta h(t, u) \leq r'_1$ and $t^\beta h(t, u) \leq r'_2$ for each $(t, u) \in \bar{J} \times [0, r_1]$ and $(t, u) \in \bar{J} \times [0, r_2]$, respectively, where*

$$r_1 > \frac{\Gamma_q(\gamma + 3)\Gamma_q(1 - \beta)[c\Gamma_q(\alpha + \gamma) + \Gamma_q(\alpha)]}{\Gamma_q(\alpha + \gamma)\Gamma(\alpha - \sigma + 1)[c\Gamma_q(\gamma + 3) - 2\Gamma_q(\gamma)]} r'_1$$

$$> r_2$$

$$> \frac{[2\Gamma_q(\gamma)\Gamma_q(\alpha + \gamma) - \Gamma_q(\gamma + 3)\Gamma_q(\alpha)]\Gamma_q(1 - \beta)}{\Gamma_q(\alpha + \gamma)[c\Gamma_q(\gamma + 3) - 2\Gamma_q(\gamma)]\Gamma_q(\alpha - \gamma + 1)} r'_2.$$

Proof. We take the set $\mathcal{X}_1$ and $\mathcal{X}_2$ of all $u \in P$ such that

$$\|u\| < \frac{[2\Gamma_q(\gamma)\Gamma_q(\alpha + \gamma) - \Gamma_q(\gamma + 3)\Gamma_q(\alpha)]\Gamma_q(1 - \beta)}{\Gamma_q(\alpha + \gamma)[c\Gamma_q(\gamma + 3) - 2\Gamma_q(\gamma)]\Gamma_q(\alpha - \beta + 1)} r'_2$$

and

$$\|u\| < \frac{\Gamma_q(\gamma + 3)\Gamma_q(1 - \beta)[c\Gamma_q(\alpha + \gamma) + \Gamma_q(\alpha)]}{\Gamma_q(\alpha + \gamma)\Gamma_q(\alpha - \beta + 1)[c\Gamma_q(\gamma + 3) - 2\Gamma_q(\gamma)]} r'_1,$$

respectively. Since $2\Gamma_q(\gamma) > \Gamma_q(\alpha)$ and $\Gamma_q(\alpha + \gamma) > \Gamma_q(\gamma + 3)$, we have:

$$\frac{2\Gamma_q(\gamma)\Gamma_q(\alpha + \gamma) - \Gamma_q(\gamma + 3)\Gamma_q(\alpha)}{\Gamma_q(\alpha + \gamma)[c\Gamma_q(\gamma + 3) - 2\Gamma_q(\gamma)]} > 0.$$

Since $\gamma \in [1, \infty)$ and $r_1' > r_2'$, $2\Gamma_q(\gamma) < \Gamma_q(\gamma + 3)$ and

$$\frac{\Gamma_q(\gamma+3)\big[c\Gamma_q(\alpha+\gamma)+\Gamma_q(\alpha)\big]r_1'}{\Gamma_q(\alpha+\gamma)\big[c\Gamma_q(\gamma+3)-2\Gamma_q(\gamma)\big]} > \frac{2\Gamma_q(\gamma)\Gamma_q(\alpha+\gamma) - \Gamma_q(\gamma+3)\Gamma_q(\alpha)r_2'}{\Gamma_q(c+\gamma)\big[c\Gamma_q(\gamma+3)-2\Gamma_q(\gamma)\big]},$$

therefore, $\mathcal{X}_1 \subset \overline{\mathcal{X}_2}$. If $u \in P \cap \partial \mathcal{X}_1$, then

$$0 \leq u(t) \leq \frac{[2\Gamma_q(\gamma)\Gamma_q(\alpha+\gamma) - \Gamma_q(\gamma+3)\Gamma_q(\alpha)]\Gamma_q(1-\beta)}{\Gamma_q(\alpha+\gamma)\Gamma_q(\alpha-\beta+1)[c\Gamma_q(\gamma+3)-2\Gamma_q(\gamma)]} r_2',$$

$\forall\, t \in \overline{J}$, and also

$$\Omega(u(1)) = -\int_0^1 \frac{(1-qs)^{(\alpha-1)}}{\Gamma_q(\alpha)} h(s, u(s))\, d_q s$$

$$+ \int_0^1 \frac{\Gamma_q(\gamma+3)\big[c\Gamma_q(\alpha+\gamma)(1-qs)^{(\alpha-1)} - \Gamma_q(\alpha)(1-qs)^{(\alpha+\gamma-1)}\big]}{\Gamma_q(\alpha)\Gamma_q(\alpha+\gamma)\big[c\Gamma_q(\gamma+3) - 2\Gamma_q(\gamma)\big]}$$

$$\times h(s, u(s))\, d_q s$$

$$\geq \int_0^1 \frac{\Gamma_q(\gamma+3)\big[c\Gamma_q(\alpha+\gamma) - \Gamma_q(\alpha)\big] - \Gamma_q(\alpha+\gamma)\big[c\Gamma_q(\gamma+3) - 2\Gamma_q(\gamma)\big]}{\Gamma_q(\alpha)\Gamma_q(\alpha+\gamma)\big[c\Gamma_q(\gamma+3) - 2\Gamma_q(\gamma)\big]}$$

$$\times (1-qs)^{(\alpha-1)} s^{-\beta} s^{\beta} h(s, u(s))\, d_q s$$

$$\geq r_2' \int_0^1 \frac{2\Gamma_q(\gamma)\Gamma_q(\alpha+\gamma) - \Gamma_q(\gamma+3)\Gamma_q(\alpha)}{\Gamma_q(\alpha)\Gamma_q(\alpha+\gamma)\big[c\Gamma_q(\gamma+3) - 2\Gamma_q(\gamma)\big]} (1-qs)^{(\alpha-1)} s^{-\beta}\, d_q s$$

$$= A_2 \frac{[2\Gamma_q(\gamma)\Gamma_q(\alpha+\gamma) - \Gamma_q(\gamma+3)\Gamma_q(\alpha)]\Gamma_q(1-\beta)}{\Gamma_q(\alpha+\gamma)\big[c\Gamma_q(\gamma+3) - 2\Gamma_q(\gamma)\big]\Gamma_q(\alpha-\beta+1)} = \|u\|.$$

Hence, $\|\Omega(u)\| \geq \|u\|$ on $P \cap \partial \mathcal{X}_1$. If $u \in P \cap \partial \mathcal{X}_2$, then

$$\Omega(u(t)) = \int_0^t \frac{-(t-qs)^{(\alpha-1)}}{\Gamma_q(\alpha)} h(s, u(s))\, d_q s$$

$$+ t^2 \int_0^1 \frac{\Gamma_q(\gamma+3)\big[c\Gamma_q(\alpha+\gamma)(1-qs)^{(\alpha-1)} - \Gamma_q(\alpha)(1-qs)^{(\alpha+\gamma-1)}\big]}{\Gamma_q(\alpha)\Gamma_q(\alpha+\gamma)\big[c\Gamma_q(\gamma+3) - 2\Gamma_q(\gamma)\big]}$$

$$\times h(s, u(s))\, d_q s$$

$$\leq \int_0^1 \frac{\Gamma_q(p+3)\big[c\Gamma_q(\alpha+\gamma) + \Gamma_q(\alpha)\big](1-qs)^{(\alpha-1)}}{\Gamma_q(\alpha)\Gamma_q(\alpha+\gamma)\big[c\Gamma_q(\gamma+3) - 2\Gamma_q(\gamma)\big]} s^{-\beta} s^{\beta} h(s, u(s))\, d_q s$$

$$\leq r_1' \frac{\Gamma_q(\gamma+3)\big[c\Gamma_q(\alpha+\gamma) + \Gamma_q(\alpha)\big]}{\Gamma_q(\alpha)\Gamma_q(\alpha+\gamma)\big[c\Gamma_q(\gamma+3) - 2\Gamma_q(\gamma)\big]} \int_0^1 (1-qs)^{(\alpha-1)} s^{-\beta}\, d_q s$$

$$= r_0'{}_1 \frac{\Gamma_q(\gamma+3)\Gamma_q(1-\beta)\big[c\Gamma_q(\alpha+\gamma) + \Gamma_q(\alpha)\big]}{\Gamma_q(\alpha+\gamma)\Gamma_q(\alpha-\sigma+1)\big[c\Gamma_q(\gamma+3) - 2\Gamma_q(\gamma)\big]} = \|u\|$$

for $t \in \overline{J}$. Thus, $\|\Omega(u)\| \leq \|u\|$ on $P \cap \partial \mathcal{X}_2$. Since the self-operator Ω defined on P is completely continuous and $P \cap (\overline{\mathcal{X}_2} | \mathcal{X}_1)$ is a closed subset of P, the restriction $\Omega : P \cap (\overline{\mathcal{X}_2} | \mathcal{X}_1) \to P$ is completely continuous. At present, by employing Lemma 2, Ω has a fixed point in $P \cap (\overline{\mathcal{X}_2} | \mathcal{X}_1)$. By simple review, we can see that the fixed point of Ω is a positive solution for problem (1). □

4. Illustrative Examples with Application

Some illustrative examples are provided in this section to validate our original results. At the same time, a computational technique is constructed for testing the problem (1) and (2). A simplified analysis is also studied for executing the q-Gamma function's values. As a result, a pseudo-code that describes our simplified method is presented for calculating the q-Gamma function of order n in Algorithm A1 (for more details, see the following online resources: https://en.wikipedia.org/wiki/Q-gamma_function and https://www.dm.uniba.it/members/garrappa/software, accessed on 10 March 2021).

When the analytical solution is impossible to find for certain problems, we need to find the numerical approximation with a tiny step h via the implicit trapezoidal PI rule, which usually shows excellent accuracy [36]. Our numerical experiments were performed with the help of MATLAB software. Some additional supporting information are provided in Appendix A of this paper including some algorithms of the proposed method (see Algorithms A1–A5), and Tables A1–A3 present various numerical experiments to provide additional support to the validity of our results in this work.

Example 1. *Consider the SFqDEq with the B.C:*

$$\begin{cases} {}^c\mathcal{D}_q^{\frac{17}{5}}[u](t) + \frac{|\cos t|}{t^2}[1 + (u(t))^3] = 0, \\ \\ \frac{15}{7}u(1) = \mathcal{I}_q^{\frac{29}{7}}[u](1), \\ u(0) = u''(0) = u'''(0) = (0) = 0, \end{cases} \quad (4)$$

for all $t \in J = (0, 1)$ and $q \in J$.
In Problem (1), define

$$\alpha = \frac{17}{5} \geq 3, \quad n = [\frac{17}{5}] + 1 = 4, c = \frac{15}{7} \geq 1, \quad \gamma = \frac{29}{7} \in [1, \infty).$$

Define the continuous map:

$$h(t, u(t)) = \frac{|\cos t|}{t^2}\Big[1 + (u(t))^3\Big],$$

such that

$$\lim_{t \to 0^+} h(t, .) = +\infty,$$

that is, h is singular at $t = 0$. In addition to, Table 1 shows that

$$2\Gamma_q(\gamma) \geq \Gamma_q(\alpha),$$

holds for each q.

Table 1. Numerical experiment for calculating $\Gamma_q(\alpha)$, $\Gamma_q(\gamma)$ in Example 1 for $q = \frac{1}{10}, \frac{1}{2}, \frac{8}{9}$.

n	$q = \frac{1}{10}$		$q = \frac{1}{2}$		$q = \frac{8}{9}$	
	$\Gamma_q(\alpha)$	$2\Gamma_q(\gamma)$	$\Gamma_q(\alpha)$	$2\Gamma_q(\gamma)$	$\Gamma_q(\alpha)$	$2\Gamma_q(\gamma)$
1	1.1479	2.4817	2.2951	7.2266	34.0843	265.2795
2	1.1467	2.4792	2.0569	6.414	21.5589	153.3424
3	1.1466	2.479	1.9515	6.056	15.299	101.2765
4	1.1466	2.479	1.9018	5.8876	11.7053	73.0841
⋮	⋮	⋮	⋮	⋮	⋮	⋮
17	1.1466	2.479	1.8539	5.7258	3.4748	16.2557
18	1.1466	2.479	1.8539	5.7258	3.3755	15.6765
19	1.1466	2.479	1.8539	5.7257	3.2907	15.1843
20	1.1466	2.479	1.8539	5.7257	3.2177	14.7638
⋮	⋮	⋮	⋮	⋮	⋮	⋮
106	1.1466	2.479	1.8539	5.7257	2.709	11.8963
107	1.1466	2.479	1.8539	5.7257	2.709	11.8963
108	1.1466	2.479	1.8539	5.7257	2.709	11.8963
109	1.1466	2.479	1.8539	5.7257	2.709	11.8962
110	1.1466	2.479	1.8539	5.7257	2.709	11.8962

To numerically show our results, we consider the problem (2) as follows:

$$\mathcal{D}_q^{\frac{10}{3}}[u](t) + \Gamma_q(5)t^{-\frac{1}{9}}|u|^{\frac{1}{3}} + \Gamma_q(4)t^{-\frac{1}{9}}|u'|^{\frac{2}{5}}$$
$$+ \Gamma_q(6)t^{-\frac{1}{9}}|\mathcal{D}_q^{\frac{4}{15}}[u](t)|^{\frac{3}{4}} + \Gamma_q(3)t^{-\frac{1}{9}}|v_u|^{\frac{7}{9}}$$
$$+ \frac{1}{1+u^2(t)} + \frac{1}{1+(u')^2} + \frac{1}{1+(\mathcal{D}_q^{\frac{4}{15}}[u])^2} + \frac{1}{1+(v_u)^2}$$
$$\leq \mathcal{D}_q^{\frac{10}{3}}[u](t) + \Gamma_q(5)t^{-\frac{1}{9}}|u|^{\frac{1}{3}} + \Gamma_q(4)t^{-\frac{1}{9}}|u'|^{\frac{2}{5}}$$
$$+ \Gamma_q(6)t^{-\frac{1}{9}}|\mathcal{D}_q^{\frac{4}{15}}[u](t)|^{\frac{3}{4}} + \Gamma_q(3)t^{-\frac{1}{9}}|v_u|^{\frac{7}{9}}$$
$$+ (u(t))^{-2} + (u')^{-2} + (\mathcal{D}_q^{\frac{4}{15}}[u])^{-2} + (v_u)^{-2} = 0.$$

Thus,

$$\mathcal{D}_q^{\frac{10}{3}}[u](t) + \Gamma_q(5)t^{-\frac{1}{9}}|u|^{\frac{1}{3}} + \Gamma_q(4)t^{-\frac{1}{9}}|u'|^{\frac{2}{5}}$$
$$+ \Gamma_q(6)t^{-\frac{1}{9}}|\mathcal{D}_q^{\frac{4}{15}}[u](t)|^{\frac{3}{4}} + \Gamma_q(3)t^{-\frac{1}{9}}|v_u|^{\frac{7}{9}}$$
$$+ (u(t))^{-2} + (u')^{-2} + (\mathcal{D}_q^{\frac{4}{15}}[u])^{-2} + (v_u)^{-2} = 0. \quad (5)$$

Table 2 shows numerically the values of $x(t)$ in Equation (5). In addition, the curve of $x(t)$ w.r.t t in Figures 1–3 for $q = \frac{1}{10}, \frac{1}{2}$, and $\frac{6}{7}$, respectively (Algorithm A1).

Table 2. Numerical experiment of Equation (5) in Example 1 for $q \in \left\{\frac{1}{10}, \frac{1}{2}, \frac{6}{7}\right\}$ and $n = 1, \cdots 20$ (Algorithm A1).

n	$q = \frac{1}{10}$		$q = \frac{1}{2}$		$q = \frac{6}{7}$	
	t	u(t)	t	u(t)	t	u(t)
1	\multicolumn{6}{c}{$n = 1$}						
1	0	0	0	0	0	0
1	0.25	0.00172	0.25	0.00806	0.25	0.38812
1	0.5	0.01733	0.5	0.08187	0.5	4.1244
1	0.75	0.06744	0.75	0.32299	0.75	17.97576
1	1	0.17909	1	0.87607	1	56.89764
2			$n = 2$			
2	0	0	0	0	0	0
2	0.25	0.00171	0.25	0.0071	0.25	0.21494
2	0.5	0.01731	0.5	0.07216	0.5	2.26527
2	0.75	0.06737	0.75	0.2846	0.75	9.69401
2	1	0.17891	1	0.77148	1	29.82949
$\vdots$						
20			$n = 20$			
	0	0	0	0	0	0
	0.25	Inf	0.25	Inf	0.25	Inf
	0.5	Inf	0.5	Inf	0.5	Inf
	0.75	Inf	0.75	Inf	0.75	Inf
	1	Inf	1	Inf	1	Inf
	1.25	Inf	1.25	Inf	1.25	Inf
	1.5	Inf	1.5	Inf	1.5	Inf
	1.75	Inf	1.75	Inf	1.75	Inf
	$\vdots$	$\vdots$	$\vdots$	$\vdots$	$\vdots$	$\vdots$

We can see that all conditions of Theorem 2 hold. Thus, the fixed point of Ω is a positive solution for problem (4).

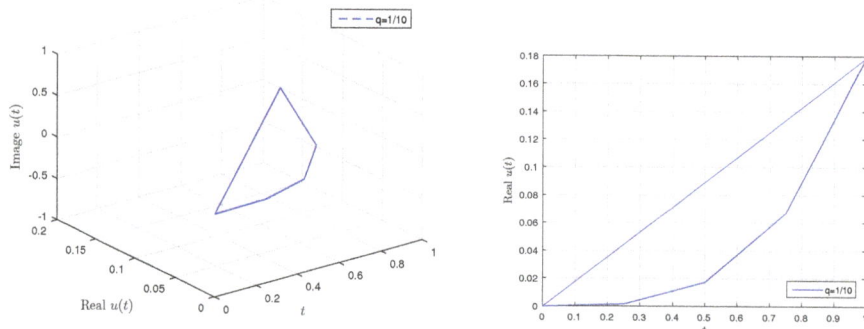

Figure 1. $u(t)$ with respect to t in Equation (5) in Example 1 for $q = \frac{1}{10}$ according to Table 2.

Linear motion is the most basic of all motion. According to Newton's first law of motion, objects that do not experience any net force will continue to move in a straight line with a constant velocity until they are subjected to a net force. In the next example, we consider an application to examine the validity of our theoretical results on the fractional order representation of the motion of a particle along a straight line.

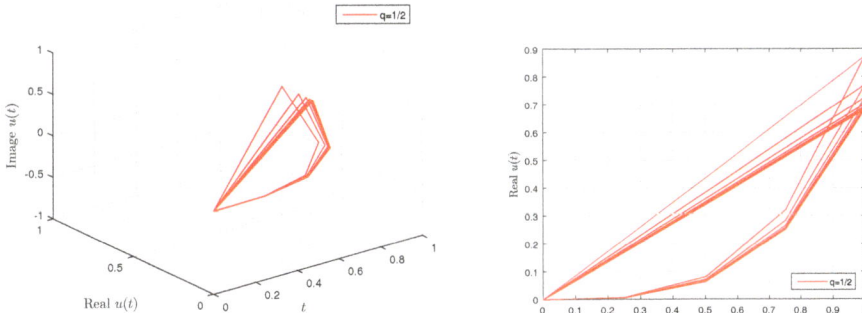

Figure 2. $u(t)$ with respect to t in Equation (5) in Example 1 for $q = \frac{1}{2}$ according to Table 2.

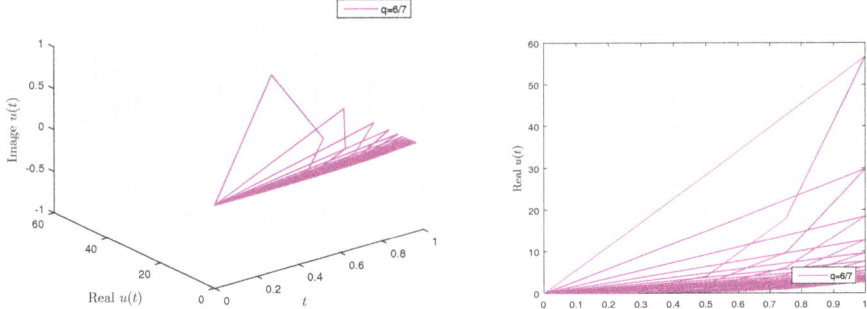

Figure 3. $u(t)$ with respect to t in Equation (5) in Example 1 for $\frac{6}{7}$ according to Table 2.

Example 2. *We consider a constrained motion of a particle along a straight line restrained by two linear springs with equal spring constants (stiffness coefficient) under an external force and fractional damping along the t-axis (Figure 4).*

The springs, unless subjected to force, are assumed to have free length (unstretched length) and resist a change in length. The motion of the system along the t-axis is independent of the initial spring tension. The springs are anchored on the t-axis at $t = -1$ and $t = 1$, and the vibration of the particle in this example is restricted to the t-axis only.

The vibration of the system is represented by a system of equations with the first equation having similar form of a simple harmonic oscillator, which cannot produce instability. Hence, the existence solution of the system depends on the following equation represented as the SFqDEq with the B.C:

$$\begin{cases} {}^c\mathcal{D}_q^{\frac{10}{3}}[u](t) + \frac{1}{8}\left[2 - 2L - \theta^2 L - \theta^2 L \cos t\right]u(t) = \nu \sin(u(t)), \\ \\ \frac{16}{9}u(1) = \mathcal{I}_q^{\frac{23}{6}}[u](1), \\ u(0) = u''(0) = u'''(0) = (0) = 0, \end{cases} \qquad (6)$$

for all $t \in J = (0,1)$, $q \in J$. Here, θ and ν are constants, and L is the unstretched length of the spring. In Problem (1),

$$\alpha = \frac{10}{3} \geq 3, \quad n = \left[\frac{10}{3}\right] + 1 = 4, c = \frac{16}{9} \geq 1, \quad \gamma = \frac{23}{6} \in [1, \infty).$$

Define the continuous map:

$$h(t, u(t)) = \frac{1}{8}\left[2 - 2L - \theta^2 L - \theta^2 L \cos t\right] u(t) - \nu \sin(u(t))$$

for $t \in (0, 1)$, such that

$$\lim_{t \to 0^+} h(t, .) = +\infty,$$

that is, h is singular at $t = 0$. Consider particular values of the parameters $L = 1.5\ m$, $\theta = 0.5$. We consider particular values of the parameter $\nu = 7.25$. Therefore, all conditions of Theorem 2 hold. Thus, the SFqDEq (6) has a solution.

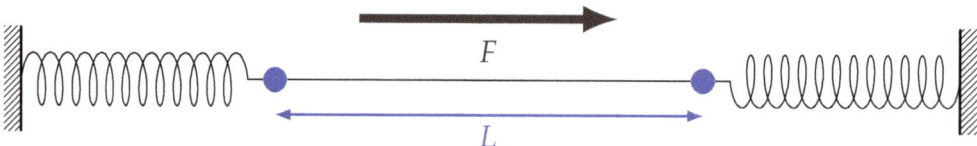

Figure 4. A particle along a straight line restrained by two linear springs with equal spring constants.

5. Conclusions

The existence of solutions was successfully investigated for a system of m-singular sum fractional q-differential equations under some integral B.Cs in the sense of CpFqDr. The positive solutions' existence was also studied with the help of a fixed point Arzelà–Ascoli theorem. Illustrative examples and numerical experiments were provided to validate our theoretical results.

Author Contributions: Conceptualization, M.E.S., R.G., M.K.A.K. and F.M.; methodology, M.E.S., M.I. and F.M.; software, M.E.S. and R.G.; validation, M.K.A.K., F.M. and S.-W.Y.; formal analysis, R.G.; investigation, M.E.S. and M.K.A.K.; resources, F.M., S.-W.Y. and M.I.; data curation, R.G.; writing—original draft preparation, M.E.S., R.G., M.K.A.K. and F.M.; writing—review and editing, M.K.A.K., F.M. and M.I.; visualization, M.E.S.; supervision, M.K.A.K., M.I. and S.-W.Y.; project administration, M.K.A.K.; funding acquisition, S.-W.Y. All authors have read and agreed to the published version of the manuscript.

Funding: Not applicable.

Institutional Review Board Statement: Not applicable.

Informed Consent Statement: Not applicable.

Data Availability Statement: Not applicable.

Acknowledgments: The first and second authors were supported by Bu-Ali Sina University.

Conflicts of Interest: The authors declare that they have no competing interest.

Appendix A. Supporting Information

Algorithm A1 The proposed method for calculating $\Gamma_q(x)$.

1: function g = qGamma(q, x, n)
2: %q-Gamma Function
3: p=1;
4: for k=0:n
5:   p=p*(1-q^(k+1))/(1- q^(x+k));
6: end;
7: g=p/(1-q)^(x-1);
8: end

Algorithm A2 The proposed method for calculating $(x-y)_q^{(\alpha)}$.

1: function p = qfunction1(x, y, q, sigma, n)
2: s=1;
3: if n==0
4:   p=1
5: else
6:   for k=1:n-1
7:     s = s*(x-y*q^k)/(x-y*q^(sigma+k));
8:   end;
9:   p=x^sigma * s;
10: end;
11: end

Algorithm A3 The proposed method for calculating $(D_q f)(x)$.

1: function g = Dq(q, x, n, fun)
2: if x==0
3:   g=limit ((fun(x)-fun(q*x))/((1-q)*x),x,0);
4: else
5:   g=(fun(x)-fun(q*x))/((1-q)*x);
6: end;
7: end

Algorithm A4 The proposed method for calculating $(D_q f)(x)$.

1: function g = Iq(q, x, n, fun)
2: p=1;
3: for k=0:n
4:   p=p+ q^k*fun(x*q^k);
5: end;
6: g=x* (1-q) * p;
7: end

Algorithm A5 The proposed method for calculating $I_q^\alpha[x]$.

1: function g = Iq_alpha(q, alpha, x, n, fun)
2: p=0;
3: for k=0:n
4:    s1=1;
5:    for i=0:k-1
6:       s1=s1*(1-q̂(alpha+i));
7:    end
8:    s2=1;
9:    for i=0:k-1
10:      s2=s2*(1-q̂(i+1));
11:    end
12:    p=p + qk̂*s1*eval(subs(fun, t*qk̂))/s2;
13: end;
14: g=round((t̂alpha)* ((1-q)âlpha)* p, 6);
15: end

Table A1. Some numerical results for the calculation of $\Gamma_q(x)$ with $q = \frac{1}{3}$ that is constant, $x = 4.5, 8.4, 12.7$ and $n = 1, 2, \ldots, 15$ of Algorithm A1.

n	x = 4.5	x = 8.4	x = 12.7	n	x = 4.5	x = 8.4	x = 12.7
1	2.472950	11.909360	68.080769	9	2.340263	11.257158	64.351366
2	2.383247	11.468397	65.559266	10	2.340250	11.257095	64.351003
3	2.354446	11.326853	64.749894	11	2.340245	11.257074	64.350881
4	2.344963	11.280255	64.483434	12	2.340244	11.257066	64.350841
5	2.341815	11.264786	64.394980	13	2.340243	11.257064	64.350828
6	2.340767	11.259636	64.365536	14	2.340243	11.257063	64.350823
7	2.340418	11.257921	64.355725	15	2.340243	11.257063	64.350822
8	2.340301	11.257349	64.352456				

Table A2. Some numerical results for the calculation of $\Gamma_q(x)$ with $q = \frac{1}{3}, \frac{1}{2}, \frac{2}{3}$, $x = 5$ and $n = 1, 2, \ldots, 35$ of Algorithm A1.

n	$q = \frac{1}{3}$	$q = \frac{1}{2}$	$q = \frac{2}{3}$	n	$q = \frac{1}{3}$	$q = \frac{1}{2}$	$q = \frac{2}{3}$
1	3.016535	6.291859	18.937427	18	2.853224	4.921884	8.476643
2	2.906140	5.548726	14.154784	19	2.853224	4.921879	8.474597
3	2.870699	5.222330	11.819974	20	2.853224	4.921877	8.473234
4	2.859031	5.069033	10.537540	21	2.853224	4.921876	8.472325
5	2.855157	4.994707	9.782069	22	2.853224	4.921876	8.471719
6	2.853868	4.958107	9.317265	23	2.853224	4.921875	8.471315
7	2.853438	4.939945	9.023265	24	2.853224	4.921875	8.471046
8	2.853295	4.930899	8.833940	25	2.853224	4.921875	8.470866
9	2.853247	4.926384	8.710584	26	2.853224	4.921875	8.470747
10	2.853232	4.924129	8.629588	27	2.853224	4.921875	8.470667
11	2.853226	4.923002	8.576133	28	2.853224	4.921875	8.470614
12	2.853224	4.922438	8.540736	29	2.853224	4.921875	8.470578
13	2.853224	4.922157	8.517243	30	2.853224	4.921875	8.470555
14	2.853224	4.922016	8.501627	31	2.853224	4.921875	8.470539
15	2.853224	4.921945	8.491237	32	2.853224	4.921875	8.470529
16	2.853224	4.921910	8.484320	33	2.853224	4.921875	8.470522
17	2.853224	4.921893	8.479713	34	2.853224	4.921875	8.470517

Table A3. Some numerical results for the calculation of $\Gamma_q(x)$ with $x = 8.4$, $q = \frac{1}{3}, \frac{1}{2}, \frac{2}{3}$ and $n = 1, 2, \ldots, 40$ of Algorithm A1.

n	$q = \frac{1}{3}$	$q = \frac{1}{2}$	$q = \frac{2}{3}$	n	$q = \frac{1}{3}$	$q = \frac{1}{2}$	$q = \frac{2}{3}$
1	11.909360	63.618604	664.767669	21	11.257063	49.065390	260.033372
2	11.468397	55.707508	474.800503	22	11.257063	49.065384	260.011354
3	11.326853	52.245122	384.795341	23	11.257063	49.065381	259.996678
4	11.280255	50.621828	336.326796	24	11.257063	49.065380	259.986893
5	11.264786	49.835472	308.146441	25	11.257063	49.065379	259.980371
6	11.259636	49.448420	290.958806	26	11.257063	49.065379	259.976023
7	11.257921	49.256401	280.150029	27	11.257063	49.065379	259.973124
8	11.257349	49.160766	273.216364	28	11.257063	49.065378	259.971192
9	11.257158	49.113041	268.710272	29	11.257063	49.065378	259.969903
10	11.257095	49.089202	265.756606	30	11.257063	49.065378	259.969044
11	11.257074	49.077288	263.809514	31	11.257063	49.065378	259.968472
12	11.257066	49.071333	262.521127	32	11.257063	49.065378	259.968090
13	11.257064	49.068355	261.666471	33	11.257063	49.065378	259.967836
14	11.257063	49.066867	261.098587	34	11.257063	49.065378	259.967666
15	11.257063	49.066123	260.720833	35	11.257063	49.065378	259.967553
16	11.257063	49.065751	260.469369	36	11.257063	49.065378	259.967478
17	11.257063	49.065564	260.301890	37	11.257063	49.065378	259.967427
18	11.257063	49.065471	260.190310	38	11.257063	49.065378	259.967394
19	11.257063	49.065425	260.115957	39	11.257063	49.065378	259.967371
20	11.257063	49.065402	260.066402	40	11.257063	49.065378	259.967357

References

1. Beyer, H.; Kempfle, S. Definition of physical consistent damping laws with fractional derivatives. *Z. Angew. Math. Mech.* **1995**, *75*, 623–635. [CrossRef]
2. He, J.H. Some applications of nonlinear fractional differential equations and their approximations. *Sci. Technol. Soc.* **1999**, *15*, 86–90.
3. He, J.H. Approximate analytic solution for seepage flow with fractional derivatives in porous media. *Comput. Methods Appl. Mech. Eng.* **1998**, *167*, 57–68. [CrossRef]
4. Caputo, M. Linear models of dissipation whose Q is almost frequency independent-II. *Geophys. J. Int.* **1967**, *13*, 529–539. [CrossRef]
5. Yan, J.P.; Li, C.P. On chaos synchronization of fractional differential equations. *Chaos Solitons Fractals* **2007**, *32*, 725–735. [CrossRef]
6. Sommacal, L.; Melchior, P.; Dossat, A.; Petit, J.; Cabelguen, J.M.; Oustaloup, A.; Ijspeert, A.J. Improvement of the muscle fractional multimodel for low-rate stimulation. *Biomed. Signal Process. Control* **2007**, *2*, 226–233. [CrossRef]
7. Rezapour, S.; Mojammadi, H.; Samei, M.E. SEIR epidemic model for COVID-19 transmission by Caputo derivative of fractional order. *Adv. Differ. Equ.* **2020**, *2020*, 490. [CrossRef] [PubMed]
8. Silva, M.F.; Machado, J.A.T.; Lopes, A.M. Fractional order control of a hexapod robot. *Nonlinear Dyn.* **2004**, *38*, 417–433. [CrossRef]
9. Mathieu, B.; Melchior, P.; Oustaloup, A.; Ceyral, C. Fractional differentiation for edge detection. *Signal Process* **2003**, *83*, 2421–2432. [CrossRef]
10. Diethelm, K.; Ford, N.J. Analysis of Fractional Differential Equations. *J. Math. Anal. Appl.* **2002**, *265*, 229–248. [CrossRef]
11. Mishra, S.K.; Samei, M.E.; Chakraborty, S.K.; Ram, B. On q-variant of Dai–Yuan conjugate gradient algorithm for unconstrained optimization problems. *Nonlinear Dyn.* **2021**, *2021*, 35. [CrossRef]
12. Ntouyas, S.K.; Samei, M.E. Existence and uniqueness of solutions for multi-term fractional q–integro-differential equations via quantum calculus. *Adv. Differ. Equ.* **2019**, *2019*, 475. [CrossRef]
13. Lakshmikantham, V.; Vatsala, A.S. Basic theory of fractional differential equations. *Nonlinear Anal. Theory Methods Appl.* **2008**, *69*, 2677–2682. [CrossRef]
14. Samei, M.E. Existence of solutions for a system of singular sum fractional q–differential equations via quantum calculus. *Adv. Differ. Equ.* **2020**, *2020*, 23. [CrossRef]
15. Matar, M.M.; Abbas, M.I.; Alzabut, J.; Kaabar, M.K.A.; Etemad, S.; Rezapour, S. Investigation of the p-Laplacian nonperiodic nonlinear boundary value problem via generalized Caputo fractional derivatives. *Adv. Differ. Equ.* **2021**, *2021*, 68. [CrossRef]
16. Jackson, F. q–difference equations. *Am. J. Math.* **1910**, *32*, 305–314. [CrossRef]
17. Adams, C. The general theory of a class of linear partial q–difference equations. *Trans. Am. Math. Soc.* **1924**, *26*, 283–312.
18. Adams, C. Note on the integro-q–difference equations. *Trans. Am. Math. Soc.* **1929**, *31*, 861–867.
19. Ahmad, B.; Nieto, J.J.; Alsaedi, A.; Al-Hutami, H. Existence of solutions for nonlinear fractional q–difference integral equations with two fractional orders and nonlocal four-point boundary conditions. *J. Franklin Inst.* **2014**, *351*, 2890–2909. [CrossRef]

20. Ahmad, B.; Ntouyas, S.; Purnaras, I. Existence results for nonlocal boundary value problems of nonlinear fractional q–difference equations. *Adv. Differ. Equ.* **2012**, *2012*, 140. [CrossRef]
21. Balkani, N.; Rezapour, S.; Haghi, R.H. Approximate solutions for a fractional q–integro-difference equation. *J. Math. Ext.* **2019**, *13*, 201–214.
22. Ferreira, R. Nontrivials solutions for fractional q–difference boundary value problems. *Electron. J. Qual. Theory Differ. Equ.* **2010**, *70*, 1–101. [CrossRef]
23. Samei, M.E.; Ranjbar, G.K.; Hedayati, V. Existence of solutions for equations and inclusions of multi-term fractional q–integro-differential with non-separated and initial boundary conditions. *J. Inequalities Appl.* **2019**, *2019*, 273. [CrossRef]
24. Zhao, Y.; Chen, H.; Zhang, Q. Existence results for fractional q–difference equations with nonlocal q–integral boundary conditions. *Adv. Differ. Equ.* **2013**, *2013*, 48. [CrossRef]
25. Agarwal, R.P.; Baleanu, D.; Hedayati, V.; Rezapour, S. Two fractional derivative inclusion problems via integral boundary condition. *Appl. Math. Comput.* **2015**, *257*, 205–212. [CrossRef]
26. Akbari Kojabad, E.; Rezapour, S. Approximate solutions of a sum-type fractional integro-differential equation by using Chebyshev and Legendre polynomials. *Adv. Differ. Equ.* **2017**, *2017*, 351. [CrossRef]
27. Liang, S.; Samei, M.E. New approach to solutions of a class of singular fractional q–differential problem via quantum calculus. *Adv. Differ. Equ.* **2020**, *2020*, 14. [CrossRef]
28. Aydogan, S.M.; Baleanu, D.; Mousalou, A.; Rezapour, S. On high order fractional integro-differential equations including the Caputo-Fabrizio derivative. *Bound. Value Probl.* **2018**, *2018*, 90. [CrossRef]
29. Baleanu, D.; Rezapour, S.; Etemad, S.; Alsaedi, A. On a time-fractional integro-differential equation via three-point boundary value conditions. *Math. Probl. Eng.* **2015**, *2015*, 12. [CrossRef]
30. He, C.Y. *Almost Periodic Differential Equations*; Higher Education Press: Beijing, China, 1992. (In Chinese)
31. Barbǎalat, I. Systems d'equations differential d'oscillations nonlinearies. *Rev. Roumaine Math. Pure Appl.* **1959**, *4*, 267–270.
32. Samei, M.E.; Hedayati, V.; Rezapour, S. Existence results for a fractional hybrid differential inclusion with Caputo-Hadamard type fractional derivative. *Adv. Differ. Equ.* **2019**, *2019*, 163. [CrossRef]
33. Annaby, M.; Mansour, Z. *q–Fractional Calculus and Equations*; Springer: Heidelberg, Germany; Cambridge, UK, 2012; [CrossRef]
34. Samko, S.; Kilbas, A.; Marichev, O. *Fractional Integrals and Derivatives: Theory and Applications*; Gordon and Breach Science Publishers: Lausanne, Switzerland; Philadelphia, PA, USA, 1993.
35. Krasnoselskii, M.A. *Positive Solution of Operator Equation*; Noordhoff: Groningen, The Netherlands, 1964.
36. Garrappa, R. Numerical Solution of Fractional Differential Equations: A Survey and a Software Tutorial. *Mathematics* **2018**, *6*, 16. [CrossRef]

Article

Spectrum of Fractional and Fractional Prabhakar Sturm–Liouville Problems with Homogeneous Dirichlet Boundary Conditions

Malgorzata Klimek

Department of Mathematics, Faculty of Mechanical Engineering and Computer Science, Czestochowa University of Technology, 42-201 Czestochowa, Poland; malgorzata.klimek@pcz.pl

Abstract: In this study, we consider regular eigenvalue problems formulated by using the left and right standard fractional derivatives and extend the notion of a fractional Sturm–Liouville problem to the regular Prabhakar eigenvalue problem, which includes the left and right Prabhakar derivatives. In both cases, we study the spectral properties of Sturm–Liouville operators on function space restricted by homogeneous Dirichlet boundary conditions. Fractional and fractional Prabhakar Sturm–Liouville problems are converted into the equivalent integral ones. Afterwards, the integral Sturm–Liouville operators are rewritten as Hilbert–Schmidt operators determined by kernels, which are continuous under the corresponding assumptions. In particular, the range of fractional order is here restricted to interval $(1/2, 1]$. Applying the spectral Hilbert–Schmidt theorem, we prove that the spectrum of integral Sturm–Liouville operators is discrete and the system of eigenfunctions forms a basis in the corresponding Hilbert space. Then, equivalence results for integral and differential versions of respective eigenvalue problems lead to the main theorems on the discrete spectrum of differential fractional and fractional Prabhakar Sturm–Liouville operators.

Keywords: fractional derivatives; fractional Prabhakar derivatives; fractional differential equations; fractional Sturm–Liouville problems; eigenfunctions and eigenvalues

1. Introduction

The aim of this paper is to study the fundamental properties of fractional eigenvalue problems developed by the construction of the Sturm–Liouville operator (SLO) with left and right fractional derivatives. In classical differential equations theory, this is a linear differential operator of the second order and yields an eigenvalue problem of the form (here, $x \in [0, b]$ in the case when we consider the problem on a finite interval):

$$\mathcal{L}_q y(x) = -\frac{d}{dx} p(x) \frac{dy(x)}{dx} + q(x) y(x) = \lambda w(x) y(x)$$

with boundary conditions appearing as follows:

$$c_1 y(0) + c_2 \frac{dy(0)}{dx} = 0, \quad d_1 y(b) + d_2 \frac{dy(b)}{dx} = 0. \tag{1}$$

Let us point out that, depending on the choice of coefficient functions and boundary conditions, such problems provide various systems of orthogonal eigenfunctions, orthogonal polynomials and families of special functions. Orthogonal systems of the solutions of classical Sturm–Liouville problems are widely applied in the analysis and solving of fundamental differential equations of mathematics, physics, mechanics , and economics.

In most of the FSLPs presented at the beginning of fractional Sturm–Liouville theory, first-order derivatives in a standard Sturm–Liouville problem were replaced with fractional order derivatives. The resulting equations were solved using some numerical schemes [1–4]. However, in these works, the essential properties, such as the orthogonality of the eigenfunctions of the fractional operator, were not investigated. In addition, the

question of whether the associated eigenvalues are real or not is not addressed. Some results concerning these properties have been obtained in papers [5,6], where the discussed equations contain a classical SLO extended by including a sum of the left and the right derivatives. Then, in paper [7], we proposed the construction of a fractional Sturm–Liouville operator which preserves the orthogonality of the eigenfunctions corresponding to distinct eigenvalues and provides real eigenvalues. The FSLO contains both the left and right derivatives and is a symmetric operator on function space restricted by fractional boundary conditions which generalize conditions (1).

A fractional version of Bessel SLO has been developed and applied to anomalous diffusion in [8], where the space-fractional differential operator has a form analogous to the FSLO proposed in a general form in [7]. Some special cases of singular fractional Sturm–Liouville problems were also studied in [9,10], where exact solutions and eigenvalues were calculated.

In our earlier works [7,11–14], we focused on the construction of a fractional version of operator $\mathcal{L}_q$, which includes standard fractional derivatives. The characteristic feature of the proposed approach is the mixture of the left and right fractional derivatives in the fractional Sturm–Liouville operator (FSLO). This construction provides eigenvalue problems with orthogonal eigenfunctions and discrete spectra under the appropriate homogeneous boundary conditions.

In recent years, fractional eigenvalue problems have also been discussed within the framework of tempered and conformable fractional calculus. In the papers [15,16], a fractional Sturm–Liouville operator is built by using the left and right tempered derivatives. Next, in [17,18], an FSLO is constructed as a composition of conformable fractional derivatives. In addition, in paper [19], the authors show how to build an FSLO with composite fractional derivatives.

Here, we add the generalization of fractional eigenvalue problems to problems with operators, including Prabhakar derivatives. The regular fractional and fractional Prabhakar Sturm–Liouville operators considered here include the left and the right derivatives, and the derived equations are in fact of a variational nature; i.e., they are Euler–Lagrange equations for respective actions (compare [11,20] and the references therein for FSLE). The properties of the spectra and eigenfunctions' systems of FSLP can be studied by applying the variational method [12,21]. Here, we shall develop the transformation method for FSLP and PSLP with Dirichlet boundary conditions, which means that we rewrite the FSLP/PSLP as the equivalent integral eigenvalue problem.

The paper is organized as follows. In the next section, we present the necessary definitions and properties of fractional and fractional Prabhakar operators, as well as the formulation of a regular fractional Sturm–Liouville problem with its generalization to the Prabhakar Sturm–Liouville problem. In Section 3, we define the problems with homogeneous Dirichlet boundary conditions and derive equivalence results for both types of fractional eigenvalue problems. It appears that by applying composition rules for derivatives and integrals, they can be converted into the equivalent integral ones. Spectral properties of integral versions of fractional and fractional Prabhakar Sturm–Liouville operators are discussed in Section 4. We shall prove that these operators are Hilbert–Schmidt integral operators, which are compact and self-adjoint on the $L_w^2(0, b)$ space. Applying the spectral Hilbert–Schmidt theorem, we derive results on discrete spectra both for fractional and fractional Prabhakar Sturm–Liouville operators. The equivalence of differential and integral versions of eigenvalue problems leads to the corresponding spectral results for differential operators.

The paper closes with a brief discussion of results and future investigations. The Appendix A contains two parts. First, we present results on Hölder continuity of kernels defining integral Sturm–Liouville operators. Then, we prove a useful theorem on the convergence of convolutions' series in a general case, which is applied in the construction of integral Sturm–Liouville operators.

2. Preliminaries

We start with a summary of definitions and properties of fractional integrals and derivatives which shall be applied in the construction of fractional and fractional Prabhakar eigenvalue problems. First, we recall the left and right Riemann–Liouville fractional derivatives of order $\alpha \in (0,1)$ [22,23]:

$$D_{0+}^\alpha y(x) := \frac{d}{dx} I_{0+}^{1-\alpha} y(x), \quad D_{b-}^\alpha y(x) := -\frac{d}{dx} I_{b-}^{1-\alpha} y(x), \tag{2}$$

where the operators I_{0+}^α and I_{b-}^α are respectively the left and the right fractional Riemann–Liouville integrals of order $\alpha > 0$ defined by the following formulas

$$I_{0+}^\alpha y(x) := \int_0^x \frac{(x-t)^{\alpha-1} y(t)}{\Gamma(\alpha)} dt, \quad x > 0, \tag{3}$$

$$I_{b-}^\alpha y(x) := \int_x^b \frac{(t-x)^{\alpha-1} y(t)}{\Gamma(\alpha)} dt, \quad x < b. \tag{4}$$

Next, we have Caputo fractional derivatives:

$$^c D_{0+}^\alpha y(x) = D_{0+}^\alpha (y(x) - y(0)), \quad ^c D_{b-}^\alpha y(x) = D_{b-}^\alpha (y(x) - y(b)) \tag{5}$$

and we note that when $y(0) = y(b) = 0$, both types of derivatives coincide, i.e.,

$$^c D_{0+}^\alpha y(x) = D_{0+}^\alpha y(x), \quad ^c D_{b-}^\alpha y(x) = D_{b-}^\alpha y(x).$$

We also recall some of the composition rules of fractional operators for the case of order $\alpha \in (0,1]$; namely, for the left-sided Caputo derivative and left-sided fractional integral, we have

$$I_{0+}^\alpha {}^c D_{0+}^\alpha y(x) = y(x) - y(0), \tag{6}$$

$$^c D_{0+}^\alpha I_{0+}^\alpha y(x) = y(x), \tag{7}$$

while for the right-sided Riemann–Liouville derivatives, the following relations are valid

$$I_{b-}^\alpha D_{b-}^\alpha y(x) = y(x) - I_{b-}^{1-\alpha} y(b) \cdot \frac{(b-x)^{\alpha-1}}{\Gamma(\alpha)}, \tag{8}$$

$$D_{b-}^\alpha I_{b-}^\alpha y(x) = y(x). \tag{9}$$

All of the above rules are fulfilled for all points $x \in [0,b]$ when function y is a continuous one. Let us note that for the continuous function fulfilling condition $y(0) = 0$, rules (6) and (8) look as follows:

$$I_{0+}^\alpha {}^c D_{0+}^\alpha y(x) = y(x), \quad I_{b-}^\alpha D_{b-}^\alpha y(x) = y(x). \tag{10}$$

The fractional operators, described above, are generalized to Prabhakar integrals and derivatives. They are defined using a three-parameter Mittag–Leffler function [22,24]:

$$E_{\rho,\mu}^\gamma(z) := \frac{1}{\Gamma(\gamma)} \sum_{k=0}^\infty \frac{\Gamma(\gamma+k)}{\Gamma(\rho k + \mu)} \cdot \frac{z^k}{k!} \tag{11}$$

and Prabhakar function [24,25]:

$$e_{\rho,\mu}^\gamma(\omega z^\rho) := z^{\mu-1} E_{\rho,\mu}^\gamma(\omega z^\rho), \tag{12}$$

both defined on the complex space when $Re(\rho) > 0$ and $Re(\mu) > 0$.

These functions lead to the left and right Prabhakar derivatives [24]:

$$D^\alpha_{\rho,\gamma,\omega,0+} y(x) := \frac{d}{dx} E^{1-\alpha}_{\rho,-\gamma,\omega,0+} y(x), \quad D^\alpha_{\rho,\gamma,\omega,b-} y(x) := -\frac{d}{dx} E^{1-\alpha}_{\rho,-\gamma,\omega,b-} y(x), \quad (13)$$

where operators $E^\alpha_{\rho,-\gamma,\omega,0+}$ and $E^\alpha_{\rho,-\gamma,\omega,b-}$ are respectively the left and the right fractional Prabhakar integrals:

$$E^\alpha_{\rho,-\gamma,\omega,0+} y(x) := \int_0^x e^{-\gamma}_{\rho,\alpha}(\omega(x-t)^\rho) y(t) dt, \quad x > 0, \quad (14)$$

$$E^\alpha_{\rho,-\gamma,\omega,b-} y(x) := \int_x^b e^{-\gamma}_{\rho,\alpha}(\omega(x-t)^\rho) y(t) dt, \quad x < b. \quad (15)$$

Similar to Caputo derivatives, given in (5), we have Caputo-type Prabhakar derivatives defined as follows

$${}^c D^\alpha_{\rho,\gamma,\omega,0+} y(x) = D^\alpha_{\rho,\gamma,\omega,0+}(y(x) - y(0)), \quad (16)$$

$${}^c D^\alpha_{\rho,\gamma,\omega,b-} y(x) = D^\alpha_{\rho,\gamma,\omega,b-}(y(x) - y(b)) \quad (17)$$

coinciding with Prabhakar derivatives (13) when $y(0) = 0$ or $y(b) = 0$, respectively. Restricting function space to continuous functions fulfilling condition $y(0) = 0$, we arrive at composition rules of Prabhakar operators analogous to (7), (9), and (10):

$${}^c D^\alpha_{\rho,\gamma,\omega,0+} E^\alpha_{\rho,\gamma,\omega,0+} y(x) = y(x), \quad (18)$$

$$E^\alpha_{\rho,\gamma,\omega,0+} {}^c D^\alpha_{\rho,\gamma,\omega,0+} y(x) = y(x), \quad (19)$$

$$D^\alpha_{\rho,\gamma,\omega,b-} E^\alpha_{\rho,\gamma,\omega,b-} y(x) = y(x), \quad (20)$$

$$E^\alpha_{\rho,\gamma,\omega,b-} D^\alpha_{\rho,\gamma,\omega,b-} y(x) = y(x). \quad (21)$$

Now, we shall quote the general formulation of the fractional eigenvalue problem, introduced and investigated in papers [7,11–14,21].

Definition 1 (compare Definition 5 in [7]). *Let $\alpha \in (0,1]$. With the notation*

$$\mathcal{L}_q := D^\alpha_{b-} p(x) \, {}^c D^\alpha_{0+} + q(x), \quad (22)$$

consider the fractional Sturm–Liouville equation (FSLE)

$$\mathcal{L}_q y_\lambda(x) = \lambda w(x) y_\lambda(x), \quad (23)$$

where $p(x) \neq 0, w(x) > 0 \ \forall x \in [0,b]$, functions p, q, w are real-valued continuous functions in $[0,b]$ and boundary conditions are:

$$c_1 y_\lambda(0) + c_2 I^{1-\alpha}_{b-} p(x) D^\alpha_{0+} y_\lambda(x)\,|_{x=0} = 0, \quad (24)$$

$$d_1 y_\lambda(b) + d_2 I^{1-\alpha}_{b-} p(x) D^\alpha_{0+} y_\lambda(x)\,|_{x=b} = 0 \quad (25)$$

with $c_1^2 + c_2^2 \neq 0$ and $d_1^2 + d_2^2 \neq 0$. The problem of finding number λ (eigenvalue) such that the BVP has a non-trivial solution, y_λ (eigenfunction) will be called the regular fractional Sturm–Liouville eigenvalue problem (FSLP).

We include Prabhakar derivatives into the construction of FSLO and formulate below the Prabhakar Sturm–Liouville problem.

Definition 2. Let $\alpha \in (0,1]$. With the notation

$$\mathcal{L}'_q := D^\alpha_{\rho,\gamma,\omega,b-} p(x) \, {}^c D^\alpha_{\rho,\gamma,\omega,0+} + q(x), \tag{26}$$

consider the fractional Prabhakar Sturm–Liouville equation (PSLE)

$$\mathcal{L}'_q y_\lambda(x) = \lambda w(x) y_\lambda(x), \tag{27}$$

where $p(x) \neq 0, w(x) > 0 \;\; \forall x \in [0,b]$, functions p, q, w are real-valued continuous functions in $[0,b]$ and boundary conditions are:

$$c_1 y_\lambda(0) + c_2 E^{1-\alpha}_{\rho,-\gamma,\omega,b-} p(x) D^\alpha_{\rho,\gamma,\omega,0+} y_\lambda(x)\,|_{x=0} = 0, \tag{28}$$

$$d_1 y_\lambda(b) + d_2 E^{1-\alpha}_{\rho,-\gamma,\omega,b-} p(x) D^\alpha_{\rho,\gamma,\omega,0+} y_\lambda(x)\,|_{x=b} = 0 \tag{29}$$

with $c_1^2 + c_2^2 \neq 0$ and $d_1^2 + d_2^2 \neq 0$. The problem of finding number λ (eigenvalue) such that the BVP has a non-trivial solution, y_λ (eigenfunction) will be called the regular fractional Prabhakar Sturm–Liouville eigenvalue problem (PSLP).

3. Formulation of the Problem and Methods

In this section, we shall focus on fractional eigenvalue problems subjected to the homogeneous Dirichlet boundary conditions. We choose values $c_2 = d_2 = 0$ in Definitions 1 and 2 and formulate the corresponding definitions of FSLP and PSLP. First, we have the fractional Sturm–Liouville problem with Dirichlet boundary conditions.

Definition 3. Let $\alpha \in (0,1]$. With the notation

$$\mathcal{L}_q := D^\alpha_{b-} p(x) \, {}^c D^\alpha_{0+} + q(x), \tag{30}$$

consider the fractional Sturm–Liouville Equation (23), where $p(x) \neq 0, w(x) > 0 \;\; \forall x \in [0,b]$, functions p, q, w are real-valued continuous functions in $[0,b]$ and the boundary conditions are:

$$y_\lambda(0) = y_\lambda(b) = 0.$$

The problem of finding number λ (eigenvalue) such that the BVP has a non-trivial solution, y_λ (eigenfunction) will be called the regular fractional Sturm–Liouville eigenvalue problem (FSLP) with homogeneous Dirichlet boundary conditions.

Next, we formulate the definition of the Prabhakar Sturm–Liouville problem with Dirichlet boundary conditions.

Definition 4. Let $\alpha \in (0,1]$. With the notation

$$\mathcal{L}'_q := D^\alpha_{\rho,\gamma,\omega,b-} p(x) \, {}^c D^\alpha_{\rho,\gamma,\omega,0+} + q(x), \tag{31}$$

consider the fractional Prabhakar Sturm–Liouville Equation (27), where $p(x) \neq 0$, $w(x) > 0 \;\; \forall x \in [0,b]$, functions p, q, w are real-valued continuous functions in $[0,b]$ and the boundary conditions are:

$$y_\lambda(0) = y_\lambda(b) = 0.$$

The problem of finding number λ (eigenvalue) such that the BVP has a non-trivial solution, y_λ (eigenfunction) is the regular fractional Prabhakar Sturm–Liouville eigenvalue problem (PSLP) with homogeneous Dirichlet boundary conditions.

We shall study the spectral properties of the eigenvalue problems described in the above definitions. Let us point out that an FSLP with a Dirichlet boundary condition spectrum was investigated in papers [12,21] using variational methods. Here, we extend the study to the Prabhakar Sturm–Liouville problem and develop the results by transforming both differential fractional problems into the respective equivalent integral ones. Then, we analyse properties of the integral versions of fractional Sturm–Liouville operators (22) and (26) and apply the Hilbert–Schmidt spectral theorem to prove that their spectrum is purely discrete. Equivalence of the respective differential and integral fractional eigenvalue problems yields the theorems on spectra of the differential fractional and fractional Prabhakar eigenvalue problems given by Definitions 3 and 4. We begin our considerations with the case when $q = 0$.

3.1. Equivalence Results for Differential and Integral FSLP, PSLP: Case $q = 0$

Here, we shall prove equivalence results for the FSLP/PSLP with an equation containing the fractional differential operators (22) and (26) and investigate the properties of the integral eigenvalue problem connected to the FSLE/PSLE in the case of order α fulfilling condition $1 \geq \alpha > 1/2$ and solutions' space restricted by the homogeneous Dirichlet boundary conditions.

In the first part, we transformed the differential fractional Sturm–Liouville problem (Definition 3) into the integral one on the subspace of the continuous functions defined below:

$$C_D[0,b] := \{y \in C[0,b]; \ y(0) = y(b) = 0\}. \tag{32}$$

Let us note that the composition rules of fractional operators (7) and (9) allow us a to write a fractional Sturm–Liouville Equation (23) on the $C_D[0,b]$ space in the case of $q = 0$ as follows:

$$\mathcal{L}_0\left(1 - \lambda I_{0+}^\alpha \frac{1}{p} I_{b-}^\alpha w(x)\right) y(x) = 0$$

which leads to the integral equation

$$\left(1 - \lambda I_{0+}^\alpha \frac{1}{p} I_{b-}^\alpha w(x)\right) y(x) = C_1^w + C_2^w I_{0+}^\alpha \frac{(b-x)^{\alpha-1}}{p(x)}.$$

Constants C_1^w and C_2^w are determined by the homogeneous Dirichlet boundary conditions

$$C_1^w = 0, \quad C_2^w = -\lambda \frac{I_{0+}^\alpha \frac{1}{p} I_{b-}^\alpha w(x) y(x)|_{x=b}}{I_{0+}^\alpha \frac{(b-x)^{\alpha-1}}{p(x)}|_{x=b}}. \tag{33}$$

The above calculations lead to the integral form of FSLE (23) with $q = 0$

$$\frac{1}{\lambda} y(x) = T_w y(x), \tag{34}$$

where linear integral operator T_w is built using the left and right Riemann–Liouville integrals and acts as follows:

$$T_w y(x) = I_{0+}^\alpha \frac{1}{p} I_{b-}^\alpha w(x) y(x) - \frac{I_{0+}^\alpha \frac{1}{p} I_{b-}^\alpha w(x) y(x)|_{x=b}}{I_{0+}^\alpha \frac{(b-x)^{\alpha-1}}{p(x)}|_{x=b}} \cdot I_{0+}^\alpha \frac{(b-x)^{\alpha-1}}{p(x)}. \tag{35}$$

Similar considerations yield the integral form of PSLE (27) when $q = 0$

$$\frac{1}{\lambda} y(x) = T_w y(x), \tag{36}$$

where linear integral operator T_w is constructed using the left and right Prabhakar integrals and acts as follows

$$T_w y(x) = E_{\rho,\gamma,w,0+}^\alpha \frac{1}{p} E_{\rho,\gamma,w,b-}^\alpha w(x) y(x) \tag{37}$$

$$-\frac{E^{\alpha}_{\rho,\gamma,\omega,0+}\frac{1}{p}E^{\alpha}_{\rho,\gamma,\omega,b-}w(x)y(x)|_{x=b}}{E^{\alpha}_{\rho,\gamma,\omega,0+}\frac{e^{-\gamma}_{\rho,\alpha}(\omega(b-x)^{\rho})}{p(x)}|_{x=b}} \cdot E^{\alpha}_{\rho,\gamma,\omega,0+}\frac{e^{-\gamma}_{\rho,\alpha}(\omega(b-x)^{\rho})}{p(x)}.$$

We note that the above integral operators (35) and (37) can be rewritten as operators indexed by the arbitrary continuous function r (here, $r = w$) and determined by the corresponding kernels—G^1 for FSLP and G^2 for PSLP:

$$T_r y(x) := \int_0^b G^j(x,s) r(s) y(s) dx, \quad j = 1,2, \tag{38}$$

where kernels are of the form:

$$G^1(x,s) := K_1(x,s) - \frac{K_1(b,x) K_1(b,s)}{K_1(b,b)}, \tag{39}$$

$$G^2(x,s) := K_1^P(x,s) - \frac{K_1^P(b,x) K_1^P(b,s)}{K_1^P(b,b)}, \tag{40}$$

$$K_1(x,s) = \int_0^{\min\{x,s\}} \frac{(x-t)^{\alpha-1}}{\Gamma(\alpha)} \cdot \frac{(s-t)^{\alpha-1}}{\Gamma(\alpha)} \frac{1}{p(t)} dt, \tag{41}$$

$$K_1^P(x,s) = \int_0^{\min\{x,s\}} \frac{e^{-\gamma}_{\rho,\alpha}(\omega(x-t)^{\rho}) e^{-\gamma}_{\rho,\alpha}(\omega(s-t)^{\rho})}{p(t)} dt. \tag{42}$$

It is easy to check the following properties of kernels. First, they are symmetric functions on square $\Delta = [0,b] \times [0,b]$

$$K_1(x,s) = K_1(s,x), \quad K_1^P(x,s) = K_1^P(s,x), \quad G^j(x,s) = G^j(s,x) \tag{43}$$

and, in addition, we have

$$K_1(0,s) = K_1(b,0) = 0, \quad K_1^P(0,s) = K_1^P(b,0) = 0, \quad G^j(0,s) = G^j(b,s) = 0. \tag{44}$$

In our results developed in this paper, we apply two types of assumptions.

Hypothesis 1 (H1). $1 \geq \alpha > 1/2$, $\frac{1}{p} \in C[0,b]$ and function $\frac{1}{p}$ be positive on $[0,b]$ or negative.

Hypothesis 2 (H2). $1 \geq \alpha > 1/2$, $\frac{1}{p} \in C[0,b]$ and function $\frac{1}{p}$ be positive on $[0,b]$ or negative. In addition, let the real parameters $\alpha, \rho, \gamma, \omega$ fulfil the conditions:

$$\min\{\rho, \gamma\} > 0, \quad \omega < 0, \quad \alpha \geq \rho\gamma, \quad \rho < 1.$$

Proposition 1. *If (H1) is fulfilled and function $y \in L^2(0,b)$, then its image $T_r y \in C_D[0,b]$ for any function $r \in C[0,b]$ and operator defined by kernel (39).*

If (H2) is fulfilled and function $y \in L^2(0,b)$, then its image $T_r y \in C_D[0,b]$ for any function $r \in C[0,b]$ and operator defined by kernel (40).

Proof. We sketch here the proof of the first part of the discussed proposition and omit the proof of the second one as it is analogous. By Corollary A1, kernel G^1 fulfills the Hölder condition; therefore, we find

$$|T_r y(x') - T_r y(x)| \leq \int_0^b |G^1(x',s) - G^1(x,s)| \cdot |r(s)y(s)| ds$$

$$\leq M_1 |x' - x|^{\beta} \int_0^b |r(s)y(s)| ds \leq M_1 \sqrt{b} \cdot ||r|| \cdot ||y||_{L^2} \cdot |x' - x|^{\beta}$$

and we infer that image $T_r y$ is a continuous function and is even uniformly continuous on interval $[0,b]$.

We check that it obeys the homogeneous Dirichlet boundary conditions as well, because kernel G^1 fulfils the conditions (44):

$$T_r y(0) = \int_0^b G^1(0,s) y(s) r(s) ds = 0,$$

$$T_r y(b) = \int_0^b G^1(b,s) y(s) r(s) ds = 0.$$

□

For functions belonging to the $C_D[0, b]$ space, we can prove the equivalence of the differential and integral form of the FSLP and PSLP, respectively. That is, the following two propositions are valid when $q = 0$. The first one concerns differential and integral fractional Sturm–Liouville problems.

Proposition 2. *If (H1) is fulfilled and $w \in C[0,b]$, then the following equivalence is valid on the $C_D[0,b]$ space*

$$\mathcal{L}_0 y(x) = \lambda w(x) y(x) \iff T_w y(x) = \frac{1}{\lambda} y(x), \tag{45}$$

where operator $\mathcal{L}_0$ is defined in (22) and operator T_w contains kernel (39).

Proof. Assuming that $y \in C_D[0,b]$ is an eigenfunction corresponding to eigenvalue λ:

$$\frac{1}{w(x)} \mathcal{L}_0 y(x) = \lambda y(x)$$

we act with the T_w operator on both sides of this equation:

$$T_w \frac{1}{w(x)} \mathcal{L}_0 y(x) = \lambda T_w y(x) \tag{46}$$

and by applying composition rules (10), we obtain the integral eigenvalue equation

$$\frac{1}{\lambda} y(x) = T_w y(x). \tag{47}$$

Next, we assume that function $y \in L^2(0,b)$ is an eigenfunction of the integral FSLP, i.e., Equation (47) is fulfilled. According to Proposition 1, eigenfunction y is a continuous one and belongs to the $C_D[0,b]$ space. Then, we calculate composition $\mathcal{L}_0 T_w$ using the composition rules (7) and (9)

$$\mathcal{L}_0 T_w y(x) = w(x) y(x) \tag{48}$$

and by applying Equation (47), we arrive at the implication

$$\mathcal{L}_0 T_w y(x) = w(x) y(x) = \frac{1}{\lambda} \mathcal{L}_0 y(x) \implies \mathcal{L}_0 y(x) = \lambda w(x) y(x).$$

Therefore, we conclude that on the $C_D[0,b]$ space, the equivalence of the differential and integral FSLP is valid. □

Below, we formulate the extended version of Proposition 2, where we describe the appropriate equivalence for Prabhakar Sturm–Liouville operators. Its proof is analogous to that presented above.

Proposition 3. *If (H2) is fulfilled and $w \in C[0,b]$, then the following equivalence*

$$\mathcal{L}_0' f(x) = \lambda w(x) f(x) \iff T_w f(x) = \frac{1}{\lambda} f(x), \tag{49}$$

is valid on the $C_D[0,b]$ space, where the $\mathcal{L}'_0$ operator is defined in (26) and the T_w operator contains kernel (40).

Equivalence of the integral and differential fractional and fractional Prabhakar eigenvalue problems is an important step in deriving results on the spectrum for the problems described in Definitions 3 and 4. In the next section, we shall extend the equivalence results to the case where $q \neq 0$.

3.2. Equivalence Results for Differential and Integral FSLP, PSLP: General Case $q \neq 0$

We begin our discussion with the fractional Sturm–Liouville problem. We write Equation (23) in the following form

$$\left(\frac{1}{w}\mathcal{L}_q - \lambda\right)y(x) = 0$$

and apply composition rules for fractional operators (7) and (9)

$$\frac{1}{w}\mathcal{L}_0\left(1 + I_{0+}^\alpha \frac{1}{p}I_{b-}^\alpha q(x) - \lambda I_{0+}^\alpha \frac{1}{p}I_{b-}^\alpha w(x)\right)y(x) = 0.$$

The fractional differential Sturm–Liouville Equation (23) now takes the form of integral equation

$$y(x) + I_{0+}^\alpha \frac{1}{p}I_{b-}^\alpha q(x) + C_1^q + C_2^q I_{0+}^\alpha \frac{(b-x)^{\alpha-1}}{p(x)}$$
$$= \lambda I_{0+}^\alpha \frac{1}{p}I_{b-}^\alpha w(x)y(x) + C_1^w + C_2^w I_{0+}^\alpha \frac{(b-x)^{\alpha-1}}{p(x)},$$

where constants are determined by the homogeneous Dirichlet boundary conditions; namely, C_1^w, C_2^w are given by (33) and for C_1^q, C_2^q, we have

$$C_1^q = 0, \quad C_2^q = -\frac{I_{0+}^\alpha \frac{1}{p}I_{b-}^\alpha q(x)y(x)|_{x=b}}{I_{0+}^\alpha \frac{(b-x)^{\alpha-1}}{p(x)}|_{x=b}}.$$

To conclude, Equation (23) is now an integral equation

$$(1 + T_q)y(x) = \lambda T_w y(x), \tag{50}$$

where the T_w operator is given in (35) and the T_q operator is given by the formula below

$$T_q y(x) = I_{0+}^\alpha \frac{1}{p}I_{b-}^\alpha q(x)y(x) - \frac{I_{0+}^\alpha \frac{1}{p}I_{b-}^\alpha q(x)y(x)|_{x=b}}{I_{0+}^\alpha \frac{(b-x)^{\alpha-1}}{p(x)}|_{x=b}} \cdot I_{0+}^\alpha \frac{(b-x)^{\alpha-1}}{p(x)}. \tag{51}$$

Let us point out that, similar to the calculations presented in the previous part, both of the above integral operators can also be rewritten as integral operators (38) with kernel (39) for $r = w$ and $r = q$, respectively.

Our aim is to reformulate the intermediate integral Equation (50) to the form of an eigenvalue equation. We apply Theorem A1 to invert the operator on the left-hand side. First, we check the assumption of Theorem A1, particularly when condition (H1) is fulfilled and $w \in C[0,b]$. We then apply Corollary A1, denoting $K(x,s) = G^1(x,s)$, and obtain:

$$\|G_w(\cdot,s))\| = \sup_{v \in [0,b]} |G_w(v,s)|$$

$$= \sup_{v \in [0,b]} |G^1(v,s)w(s)| \leq \|w\| \sup_{v \in [0,b]} |G^1(v,s)|$$

$$\leq ||w|| \sup_{v \in [0,b]} \left(|G^1(v,s) - G^1(0,s)| + |G^1(0,s)| \right)$$

$$\leq ||w|| \cdot M_1 \sup_{v \in [0,b]} v^{\alpha-1/2} = ||w|| \cdot M_1 \cdot b^{\alpha-1/2} < \infty.$$

Next, we write condition (A8) in the explicit form:

$$\xi = \sup_{x \in [0,b]} \int_0^b |q(s) G^1(x,s)| ds \qquad (52)$$

$$= \sup_{x \in [0,b]} \int_0^b |q(s)| \cdot \left| K_1(x,s) - \frac{K_1(b,x) K_1(b,s)}{K_1(b,b)} \right| ds < 1.$$

All the above considerations lead to the proposition on convergence of the series associated with the intermediate fractional integral eigenvalue problem given in (50) and (A5). Analogous convolutions' series were also studied on the $C[a,b]$ and $L^2(a,b)$ function spaces for FSLPs with homogeneous mixed and Robin boundary conditions, respectively [13,14].

Proposition 4. *Let (H1) be fulfilled, $w, q \in C[0,b]$ and function w be positive. If condition (52) is fulfilled, then for any function $y \in L^2(0,b)$ series on the right-hand side of the formula below is uniformly convergent on interval $[0,b]$:*

$$T y(x) := (1 + T_q)^{-1} T_w y(x) = T_w y(x) + \sum_{n=1}^{\infty} (-T_q)^n T_w y(x), \qquad (53)$$

where operators T_q, T_w are defined in (A6) and (A7) with $K(x,s) = G^1(x,s)$. In addition, series (A9) determining the kernel of integral operator T in (53) is uniformly convergent on square Δ and kernel G is continuous on Δ.

Proof. Let us observe that the composition of operators $T_q T_w$ is an integral operator

$$T_q T_w y(x) = \int_0^b ds \left(G_q(x,s) \int_0^b G_w(s,u) y(u) du \right)$$

$$= \int_0^b du\, y(u) \left(\int_0^b G_q(x,s) G_w(s,u) ds \right) = \int_0^b G_q * G_w(x,u) y(u) du,$$

where the kernel is defined by the following convolution:

$$A * B(x,u) := \int_0^b A(x,s) B(s,u) ds.$$

We shall prove that the compositions $(T_q)^n T_w$ are also defined by convolutions of kernels G_q and G_w. We start with the induction hypothesis:

$$(T_q)^n T_w y(x) = \int_0^b (G_q^{*n}) * G_w(x,u) y(u) du \qquad (54)$$

and we prove that this formula is valid for the next step $n+1$ as well:

$$(T_q)^{n+1} T_w y(x) = \int_0^b (G_q^{*(n+1)}) * G_w(x,u) y(u) du.$$

We begin with the left-hand side, applying the induction hypothesis and associativity property of the convolutions of continuous functions:

$$(T_q)^{n+1} T_w y(x) = \int_0^b ds G_q(x,s)(T_q)^n T_w y(s)$$

$$= \int_0^b ds G_q(x,s) \left(\int_0^b G_q^{*n} * G_w(s,u) y(u) du \right)$$

$$= \int_0^b du\, y(u) \left(G_q * G_q^{*n} * G_w(x,u) \right)$$

$$= \int_0^b G_q^{*(n+1)} * G_w(x,u) y(u) du.$$

As inductive hypothesis (54) leads to the validity of the next step $n+1$; we infer that formula (54) holds for any natural number $n \geq 1$.

Now, we apply Theorem A1 and calculate kernel G for integral operator $T := (1+T_q)^{-1} T_w$:

$$Ty(x) = T_w y(x) + \sum_{n=1}^{\infty} (-T_q)^n T_w y(x)$$

$$= \int_0^b G_w(x,s) y(s) ds + \sum_{n=1}^{\infty} (-1)^n \int_0^b G_q^{*n} * G_w(x,s) y(s) ds$$

$$= \int_0^b \left(G_w(x,s) + \sum_{n=1}^{\infty} (-1)^n G_q^{*n} * G_w(x,s) \right) y(s) ds = \int_0^b G(x,s) y(s) ds.$$

The above calculations lead to the thesis of Proposition 4; namely, operator T, defined by series (53), is correctly defined on space $L^2_w(0,b) = L^2(0,b)$ as an integral operator with a continuous kernel G:

$$Ty(x) = \int_0^b G(x,s) y(s) ds.$$

□

Having constructed operator T, we now prove the equivalence result, connecting the differential and integral fractional Sturm–Liouville problems in the general case.

Proposition 5. *If (H1) and condition (52) are fulfilled, $w, q \in C[0,b]$ and function w is positive, then the following equivalence is valid on the $C_D[0,b]$ space*

$$\mathcal{L}_q y(x) = \lambda w(x) y(x) \iff Ty(x) = \frac{1}{\lambda} y(x), \tag{55}$$

where the $\mathcal{L}_q$ operator is defined in (22) and the T operator is given in (53) with a kernel determined by series (A9) with $K(x,s) = G^1(x,s)$.

Proof. We recall that for any function $y \in C_D[0,b]$, we have (proof of Proposition 2)

$$T_w \frac{1}{w(x)} \mathcal{L}_0 y(x) = y(x),$$

and we extend this equality to the analogous formula for operators T and $\mathcal{L}_q$

$$T \frac{1}{w(x)} \mathcal{L}_q y(x) = T \frac{1}{w(x)} \mathcal{L}_0 y(x) + T \frac{q(x)}{w(x)} y(x)$$

$$= \left(T_w + \sum_{n=1}^{\infty} (-T_q)^n T_w \right) \frac{1}{w(x)} \mathcal{L}_0 y(x) + T_q y(x) + \sum_{n=1}^{\infty} (-T_q)^n T_q y(x)$$

$$= y(x) + \sum_{n=1}^{\infty}(-T_q)^n y(x) + T_q y(x) + \sum_{n=1}^{\infty}(-T_q)^n T_q y(x)$$
$$= y(x),$$

where we calculate the corresponding formulas for series by using the fact that operator T is a uniformly convergent series (Proposition 4) when acting on the $C_D[0,b]$ space. For differential FSLE,

$$\frac{1}{w(x)}\mathcal{L}_q y(x) = \lambda y(x)$$

after calculating the image of the T operator of functions on both sides of FSLE

$$T\frac{1}{w(x)}\mathcal{L}_q y(x) = y(x) = \lambda T y(x),$$

we obtain the integral fractional Sturm–Liouville equation in the form of

$$Ty(x) = \frac{1}{\lambda} y(x).$$

In the next step, we assume that the above integral FSLE is fulfilled. Then, function $y \in C_D[0,b]$. We apply the differential operator $\mathcal{L}_q$ to both sides of the integral FSLE

$$\mathcal{L}_q T y(x) = \frac{1}{\lambda} \mathcal{L}_q y(x).$$

For the composition of operators on the left-hand side, we get for continuous functions $f, y \in C_D[0,b]$

$$\mathcal{L}_0 T_w f(x) = w(x) f(x), \quad \mathcal{L}_0 T_q f(x) = q(x) f(x),$$
$$\mathcal{L}_0 (-T_q)^n T_w y(x) = -q(x)(-T_q)^{n-1} T_w y(x).$$

Applying Proposition 4 again, we obtain the following result for the composition of the $\mathcal{L}_q$ and T operators

$$\mathcal{L}_q T y(x) = (q(x) + \mathcal{L}_0)\left(T_w y(x) + \sum_{n=1}^{\infty}(-T_q)^n T_w y(x)\right)$$

$$= q(x) T_w y(x) + q(x) \sum_{n=1}^{\infty}(-T_q)^n T_w y(x) + w(x) y(x) - q(x) \sum_{n=1}^{\infty}(-T_q)^{n-1} T_w y(x)$$

$$= w(x) y(x).$$

From this relation, we derive the differential fractional eigenvalue equation

$$w(x) y(x) = \frac{1}{\lambda} \mathcal{L}_q y(x)$$

which leads to the differential fractional Sturm–Liouville equation:

$$\mathcal{L}_q y(x) = \lambda w(x) y(x)$$

and this ends the proof of equivalence (55). □

Now, we generalize the Sturm–Liouville operator $\mathcal{L}_q$ by introducing Prabhakar derivatives and we move on to the Prabhakar Sturm–Liouville problem (PSLP) determined in

Definitions 2 and 4 and discussed in [26] in the case when the solutions' space is restricted by the mixed homogeneous boundary conditions.

$$\left(\frac{1}{w}\mathcal{L}'_q - \lambda\right)y(x) = 0$$

We obtain the intermediate form of the integral fractional Prabhakar eigenvalue equation applying composition rules (18)–(21)

$$(1 + T_q)y(x) = T_w y(x), \qquad (56)$$

where integral operator T_w is given in Formula (37) and operator T_q looks as follows:

$$T_q y(x) = E^\alpha_{\rho,\gamma,\omega,0+} \frac{1}{p} E^\alpha_{\rho,\gamma,\omega,b-} q(x) y(x) \qquad (57)$$

$$- \frac{E^\alpha_{\rho,\gamma,\omega,0+} \frac{1}{p} E^\alpha_{\rho,\gamma,\omega,h-} q(x) y(x)|_{\tau=h}}{E^\alpha_{\rho,\gamma,\omega,0+} \frac{e^{-\gamma}_{\rho,\alpha}(\omega(b-x)^\rho)}{p(x)}\big|_{x=b}} \cdot E^\mu_{\rho,\gamma,\omega,0+} \frac{e^{-\gamma}_{\rho,\mu}((\omega(h-r)^\rho)}{p(x)}.$$

Similar to the previous calculations for FSLP, operators (37) and (57) can be rewritten as integral operators (38), with kernel G^2 given in (40) for $r = w$ and $r = q$, respectively. Again, we apply Theorem A1 to invert operator $1 + T_q$. First, we check the assumption of Theorem A1, assuming that (H2) is fulfilled and applying Corollary A1 with $K(x,s) = G^2(x,s)$:

$$||G_w(\cdot,s))|| = \sup_{v\in[0,b]} |G_w(v,s)|$$

$$= \sup_{v\in[0,b]} |G^2(v,s)w(s)| \le ||w|| \sup_{v\in[0,b]} |G^2(v,s)|$$

$$\le ||w|| \sup_{v\in[0,b]} \left(|G^2(v,s) - G^2(0,s)| + |G^2(0,s)|\right)$$

$$\le ||w|| \cdot M_2 \sup_{v\in[0,b]} |v|^\beta = ||w|| \cdot M_2 \cdot b^\beta < \infty.$$

Next, we write condition (A8) in the explicit form:

$$\zeta = \sup_{x\in[0,b]} \int_0^b |q(s) G^2(x,s)| ds \qquad (58)$$

$$= \sup_{x\in[0,b]} \int_0^b |q(s)| \cdot \left|K_1^P(x,s) - \frac{K_1^P(b,x) K_1^P(b,s)}{K_1^P(b,b)}\right| ds < 1.$$

In the proposition below, we describe the inverse operator $(1 + T_q)^{-1}$ connected to the intermediate Equation (56). We omit the proof as it is a straightforward corollary of Theorem A1, and the full proof is analogous to that of Proposition 4.

Proposition 6. *Let (H2) be fulfilled, $w, q \in C[0,b]$ and function w be positive. If condition (58) is fulfilled, then for any function $y \in L^2(0,b)$ series on the right-hand side of the formula below is uniformly convergent on interval $[0,b]$:*

$$Ty(x) := (1 + T_q)^{-1} T_w y(x) = T_w y(x) + \sum_{n=1}^\infty (-T_q)^n T_w y(x), \qquad (59)$$

where operators T_q, T_w are defined in (A6) and (A7) with $K(x,s) = G^2(x,s)$. In addition, series (A9) determining kernel of integral operator T in (59) is uniformly convergent on square Δ and kernel G is continuous on Δ.

Similar to Proposition 5, we formulate the equivalence result for integral and differential version of eigenvalue equations corresponding to PSLP. The proof is based on the composition rules (18) and (19) and on Proposition 6, which describes inverse integral operator $(1 + T_q)^{-1}$. We omit the proof as it is analogous to the proof of Proposition 5.

Proposition 7. *If (H2) and condition (58) are fulfilled, $w, q \in C[0, b]$ and function w is positive, then the following equivalence is valid on the $C_D[0, b]$ space*

$$\mathcal{L}'_q y(x) = \lambda w(x) y(x) \iff T y(x) = \frac{1}{\lambda} y(x), \tag{60}$$

where the $\mathcal{L}'_q$ operator is defined in (26), T operator is given in (59) with the kernel determined by the series (A9) and $K(x, s) = G^2(x, s)$.

4. Results on the Spectrum of Integral and Differential Fractional and Fractional Prabhakar Sturm–Liouville Problems

In the previous section, we discussed and proved the results on the equivalence of differential and integral forms of fractional eigenvalue problems. First, Propositions 2 and 3 describe the equivalence for fractional and fractional Prabhakar Sturm–Liouville problems when fractional differential operators are respectively $\mathcal{L}_0$ and $\mathcal{L}'_0$, i.e., $q = 0$. In this case, the corresponding integral operators are T_w with kernels G^1 and G^2. We prove the spectral results for these operators by applying the Hilbert–Schmidt theorem.

4.1. Case: $q = 0$

Theorem 1. *If (H1) is fulfilled and $w \in C[0, b]$ is a positive function, then the spectrum of operator T_w defined by (38) and (39) is a discrete one, enclosed in the interval $(-1, 1)$, with 0 being its only limit point. Eigenfunctions belong to the $C_D[0, b]$ space and form an orthogonal basis in the $L^2_w(0, b)$ space.*

If (H2) is fulfilled and $w \in C[0, b]$ is a positive function, then the spectrum of operator T_w defined by (38) and (40) is a discrete one, enclosed in the interval $(-1, 1)$, with 0 being its only limit point. Eigenfunctions belong to the $C_D[0, b]$ space and form an orthogonal basis in the $L^2_w(0, b)$ space.

Proof. Let us observe that when weight function fulfils the assumptions of the theorem, we have for functions spaces

$$L^2(0, b) = L^2_w(0, b), \quad L^2(\Delta) = L^2_{w \otimes w}(\Delta).$$

The integral Hilbert–Schmidt operator T_w, defined by kernel G^1, is a compact one, as this kernel is a function continuous on square Δ and $G^1 \in L^2_{w \otimes w}(\Delta)$.

It is also a self-adjoint operator on $L^2_w(0, b)$, because kernel G^1 is a symmetric function on square Δ, and for an arbitrary pair of functions $f, g \in L^2_w(0, b)$, we obtain:

$$\langle g, T_w f \rangle_w = \int_0^b dx \left(w(x) \overline{g(x)} \int_0^b G^1(x, s) f(s) w(s) ds \right)$$

$$= \int_0^b ds \left(w(s) f(s) \int_0^b \overline{G^1(s, x) g(x)} w(x) dx \right)$$

$$= \overline{\langle f, T_w g \rangle_w} = \langle T_w g, f \rangle_w.$$

The thesis is a straightforward result of the Hilbert–Schmidt spectral theorem. We omit the proof of the second part as it is analogous to the one presented above. □

The spectral theorem for integral fractional and Prabhakar Sturm–Liouville operators together with the equivalence results, included in Propositions 2 and 3, lead to the theorem on the spectrum of differential fractional eigenvalue problems subjected to the homogeneous Dirichlet boundary conditions in the case when $q = 0$.

Theorem 2. *If (H1) is fulfilled and $w \in C[0,b]$ is a positive function, then the spectrum of operator $\mathcal{L}_0$ defined by (22) and considered on the $C_D[0,b]$ space is a discrete one, and $|\lambda_n| \to \infty$. Eigenfunctions belonging to the $C_D[0,b]$ space form an orthogonal basis in the $L^2_w(0,b)$ space.*

If (H2) is fulfilled and $w \in C[0,b]$ is a positive function, then the spectrum of operator $\mathcal{L}'_0$, defined by (26) and considered on the $C_D[0,b]$ space is a discrete one and $|\lambda_n| \to \infty$. Eigenfunctions belonging to the $C_D[0,b]$ space form an orthogonal basis in the $L^2_w(0,b)$ space.

4.2. General Case $q \neq 0$

We observe that the analogous equivalence of differential and integral FSLP holds in the general case $q \neq 0$ as well. This result is given by Proposition 5. Analogously, Proposition 7 gives the equivalence relation of both versions of the fractional Prabhakar Sturm–Liouville problem. The results, included in the mentioned propositions, allow us to rewrite eigenvalue equations, replacing the differential FSLO and PSLO with the corresponding integral operators T. These operators, first determined as operator series with convergence described in Propositions 4 and 6, are in fact integral Hilbert–Schmidt operators. Their kernels—sums of a uniformly convergent series of convolutions—are continuous functions on square Δ. The theorem below describes the spectrum of fractional integral operators T with kernel G, determined by kernels G^1 and G^2, respectively.

Theorem 3. *If (H1) and condition (52) are fulfilled, $w, q \in C[0,b]$ and w is a positive function; then the spectrum of operator T defined by (53) with kernel G given in (A9) with $K(x,s) = G^1(x,s)$ is a discrete one, enclosed in interval $(-1,1)$, with 0 being its only limit point. Eigenfunctions belong to the $C_D[0,b]$ space and form an orthogonal basis in the $L^2_w(0,b)$ space.*

If (H2) and condition (58) are fulfilled, $w, q \in C[0,b]$ and w is a positive function, then the spectrum of operator T is defined by (59), with kernel G given in (A9) and with $K(x,s) = G^2(x,s)$ is a discrete one, enclosed in interval $(-1,1)$, with 0 being its only limit point. Eigenfunctions belong to the $C_D[0,b]$ space and form an orthogonal basis in the $L^2_w(0,b)$ space.

Proof. Let us again observe that when the weight function fulfils assumptions of the theorem; we have for spaces considered as sets of functions

$$L^2(0,b) = L^2_w(0,b), \quad L^2(\Delta) = L^2_{w \otimes w}(\Delta).$$

Integral Hilbert–Schmidt operator T, defined by kernel G, is a compact one as this kernel is a continuous function on square Δ and $G \in L^2_{w \otimes w}(\Delta)$.

We recall (proof of Theorem 1) that on the $L^2_w(0,b)$ space, the following equality holds for the arbitrary pair of functions $f, g \in L^2_w(0,b)$:

$$\langle g, T_w f \rangle_w = \langle T_w g, f \rangle_w$$

because kernel G^1 is a symmetric function on square Δ. Next, for the composition of operators $T_q T_w$, we obtain the relation

$$\langle g, T_q T_w f \rangle_w = \langle g, T_w \frac{q}{w} T_w f \rangle_w = \langle \frac{q}{w} T_w g, T_w f \rangle_w = \langle T_q T_w g, f \rangle_w.$$

Now, we apply the mathematical induction principle to prove that such relations hold for arbitrary $n > 1$ natural. We formulate an induction hypothesis in the form of

$$\langle g, (T_q)^n T_w f \rangle_w = \langle (T_q)^n T_w g, f \rangle_w \tag{61}$$

and for step $n + 1$, we achieve

$$\langle g, (T_q)^{n+1} T_w f \rangle_w = \langle g, T_q(T_q)^n T_w f \rangle_w = \langle \frac{q}{w} T_w g, (T_q)^n T_w f \rangle_w$$

$$= \langle (T_q)^n T_w \frac{q}{w} T_w g, f \rangle_w = \langle (T_q)^{n+1} T_w g, f \rangle_w.$$

Applying the mathematical induction principle, we infer that Formula (61) is valid for all natural numbers $n \geq 1$. We use this formula in the proof of the fact that integral operator T is a self-adjoint one. Remembering that it is represented by a series, uniformly convergent on the Hilbert space (Proposition 4), we calculate the scalar product term by term

$$\langle g, Tf \rangle_w = \left\langle g, \left(T_w + \sum_{n=1}^{\infty} (-T_q)^n T_w \right) f \right\rangle_w$$

$$= \langle T_w g, f \rangle_w + \sum_{n=1}^{\infty} (-1)^n \langle (T_q)^n T_w g, f \rangle_w$$

$$= \left\langle \left(T_w + \sum_{n=1}^{\infty} (-T_q)^n T_w \right) g, f \right\rangle_w$$

$$= \langle Tg, f \rangle_w.$$

To conclude, the integral operator T with a kernel G given in (A9) with $K(x, s) = G^1(x, s)$ is a compact and self-adjoint operator on Hilbert space $L^2_w(0, b)$. Therefore, the thesis of the first part of the theorem holds by the Hilbert–Schmidt spectral theorem.

Proof of the second part for operator T, associated with the integral PSLP with homogeneous Dirichlet boundary conditions, is analogous. □

Now, we apply the above spectral theorem for integral fractional eigenvalue problems, with equivalence results enclosed in Propositions 5 and 7 to formulate a theorem on discrete spectra for differential fractional and fractional Prabhakar Sturm–Liouville problems.

Theorem 4. *If (H1) and condition (52) are fulfilled, $w, q \in C[0, b]$ and w is a positive function, then the spectrum of operator $\mathcal{L}_q$ defined by (22) and considered on the $C_D[0, b]$ space is a discrete one, and $|\lambda_n| \to \infty$. Eigenfunctions, belonging to the $C_D[0, b]$ space, form an orthogonal basis in the $L^2_w(0, b)$ space.*

If (H2) and condition (58) are fulfilled, $w, q \in C[0, b]$ and w is a positive function; then the spectrum of operator $\mathcal{L}'_q$ defined by (26) and considered on the $C_D[0, b]$ space, is a discrete one and $|\lambda_n| \to \infty$. Eigenfunctions, belonging to the $C_D[0, b]$ space, form an orthogonal basis in the $L^2_w(0, b)$ space.

5. Discussion

In this paper, we presented results on the discrete spectrum of fractional and fractional Prabhakar Sturm–Liouville problems in a case when eigenfunctions' space is subjected to the homogenous Dirichlet boundary conditions. First, we extended the idea of the fractional to the fractional Prabhakar eigenvalue problem, where the Sturm–Liouville operator was constructed by using the left and right Prabhakar derivatives.

Prabhakar derivatives, with respect to time, were recently applied in anomalous diffusion models [27,28]. The derived spectral results for regular PSLP with Dirichlet boundary conditions will be used in developing equations with fractional partial derivatives with respect to the space-variable.

It appears that the method of converting the differential eigenvalue problem into the equivalent integral one can be applied to both types of Sturm–Liouville operator. This approach, developed in [13,14] for fractional eigenvalue problems subject to the homogeneous mixed and Robin boundary conditions, is extended to the case of FSLP with Dirichlet boundary conditions and generalized to PSLP with the same type of conditions.

Let us point out that the spectrum and eigenfunctions of fractional eigenvalue problems with Dirichlet boundary conditions were also studied in [11,16] by applying variational methods. The first of these papers describes the spectrum of FSLP for a fractional order in the range $(1/2, 1]$, and the spectral result was extended to range $(0, 1/2]$ in [16]. Comparing both of the methods—the variational one and the transformation into integral FSLP/PSLP—we observe that in the case of Dirichlet boundary conditions, the range of order is wider in the variational method. Nevertheless, the approach proposed here has an advantage of providing the spectral results for regular PSLP as well. Simultaneously, we obtain eigenfunctions' systems for both types of eigenvalue problems, which provide orthogonal bases in the corresponding Hilbert spaces. Such bases are a meaningful tool in applications in constructing and solving partial differential fractional equations, for example, space-fractional diffusion equations in the finite domain, as well as fractional equations governing control systems (compare references and examples in [29]).

6. Conclusions

The results developed in this paper describe the spectrum and eigenfunctions properties for FSLP and PSLP subjected to homogeneous Dirichlet boundary conditions. It seems that the conversion method can also be easily applied to other Prabhakar Sturm–Liouville problems; in particular, we shall construct the corresponding mixed, Robin, and Neumann boundary conditions and develop the equivalence results. Then, we will construct the integral PSLO with kernels analogous to those from the papers [13,14] and study the spectral properties, both for the integral and differential PSLPs.

Regarding the extension of the range of fractional order for the conversion method, we observe that so far we proved equivalence results on the space of continuous solutions. This restriction is connected to the version of Hölder condition for kernels, as discussed in Lemma A1 and Corollary A1. Thus, the aim of our future work will be to weaken this condition and to extend the range of fractional order.

Further, our investigations will include numerical simulations in order to derive approximate values of eigenvalues and eigenfunctions. As was shown in the papers [13,14], the integral form of the fractional Sturm–Liouville eigenvalue equation is particularly useful as a first step of the numerical method of solving FSLP. Thus, our aim will be to discretize integral eigenvalue problems and apply the equivalence results, enclosed in Propositions 2 and 3 for the case $q = 0$, and in Propositions 5 and 7, when $q \neq 0$. In this way, we shall arrive at numerical solutions of differential FSLP and PSLP with Dirichlet boundary conditions.

Funding: This research received no external funding.

Institutional Review Board Statement: Not applicable.

Informed Consent Statement: Not applicable.

Data Availability Statement: Not applicable.

Conflicts of Interest: The author declares no conflict of interest.

Appendix A

Appendix A.1

Let us point out that the three-parameter Mittag–Leffler function (11), which appears in the definition of the Prabhakar function (12), is a completely monotone function [30], and this property leads to the following two inequalities. First, when parameters $\alpha, \rho, \gamma, \omega$ are real and obey conditions

$$\alpha \in (0,1], \quad \min\{\rho, \gamma\} > 0, \quad \omega < 0, \quad \alpha \geq \rho\gamma, \quad \rho \leq 1,$$

the three-parameter Mittag–Leffler function is bounded on any interval $[0, b]$ (M_e is a constant)

$$|E^{\gamma}_{\rho,\alpha}(\omega x^{\rho})| \leq M_e$$

and it fulfills the Lipschitz condition on this interval (M_L is a constant)

$$|E^{\gamma}_{\rho,\alpha}(\omega (x')^{\rho}) - E^{\gamma}_{\rho,\alpha}(\omega x^{\rho})| \leq M_L |(x')^{\rho} - x^{\rho}|.$$

In addition, we remember that the power function obeys the Hölder condition on interval $[0, 1]$ when $\rho \leq 1$ (M_ρ is a constant)

$$|(y')^{\rho} - y^{\rho}| \leq M_\rho |y' - y|^{\rho}, \quad y', y \in [0, 1].$$

All the above inequalities will be applied to derive properties of a fractional integral operator associated to the differential Prabhakar Sturm–Liouville operator (PSLO). In particular, they are important in the study of Hölder continuity and the continuity of kernels of integral versions of Prabhakar Sturm–Liouville operators. The lemma below summarizes the Hölder continuity properties of kernels K_1, K_1^P and was proven in [26] (compare Properties 3.2 and 3.3).

Lemma A1. *If (H1) is fulfilled, then kernel K_1, given by (41), obeys the Hölder-type condition, i.e., there exists coefficient $\beta \in (0, 1]$ and function $m \in L^2(0, b)$ such that*

$$|K_1(x', s) - K_1(x, s)| \leq m(s)|x' - x|^{\beta}, \tag{A1}$$

where $\beta = \alpha - 1/2$ and

$$m(s) = \frac{2 b^{\alpha - 1/2} \|1/p\|}{(\Gamma(\alpha))^2 (\alpha - 1/2)}$$

is a constant function.

If (H2) is fulfilled, then kernel K_1^P, given by (42), obeys the Hölder-type condition, i.e., there exists coefficient $\beta \in (0, 1]$ and function $m \in L^2(0, b)$ such that

$$|K_1^P(x', s) - K_1^P(x, s)| \leq m(s)|x' - x|^{\beta}, \tag{A2}$$

where $\beta = \min\{\alpha - 1/2, \rho\}$ and

$$m(s) = \max\{b^{\alpha - 1/2}, b^{2\alpha - 1 - \rho}\} \cdot \frac{\|1/p\| \cdot M_e}{\alpha - 1/2} \cdot (2 M_e + M_L M_\rho b^{\rho})$$

is a constant function.

Analyzing the construction of kernels G^1 and G^2, we obtain the following corollary.

Corollary A1. *If (H1) is fulfilled, then kernel G^1, defined by Formulas (39) and (41), obeys the Hölder-type condition, i.e., there exists coefficient $\beta \in (0, 1]$ and constant M_1 such that*

$$|G^1(x', s) - G^1(x, s)| \leq M_1 |x' - x|^{\beta}, \tag{A3}$$

where $\beta = \alpha - 1/2$ and

$$M_1 = \frac{2 b^{\alpha - 1/2} \|1/p\| (1 + \|1/p\| \cdot \|p\|)}{(\Gamma(\alpha))^2 (\alpha - 1/2)}.$$

If (H2) is fulfilled, then kernel G^2, defined by Formulas (40) and (42), obeys the Hölder-type condition, i.e., there exists coefficient $\beta \in (0, 1]$ and constant M_2 such that

$$|G^2(x', s) - G^2(x, s)| \leq M_2 |x' - x|^{\beta}, \tag{A4}$$

where $\beta = \min\{\alpha - 1/2, \rho\}$ and

$$M_2 = \max\{b^{\alpha-1/2}, b^{2\alpha-1-\rho}\} \cdot \frac{||1/p||(1+||1/p|| \cdot ||p||) \cdot M_e}{\alpha - 1/2} \cdot (2M_e + M_L M_\rho b^\rho).$$

Proof. We prove the Hölder-type condition for kernel G^1 by applying Lemma A1 and the symmetry property of kernel K_1 given in (43). We begin by estimating values $K_1(b,b)$ and $K_1(b,s)$:

$$|K_1(b,b)| = \left|\int_a^b \frac{(b-t)^{2\alpha-2}}{(\Gamma(\alpha))^2 p(t)} dt\right| = \int_a^b \frac{(b-t)^{2\alpha-2}}{(\Gamma(\alpha))^2 |p(t)|} dt$$

$$\geq \frac{(b-a)^{2\alpha-1}}{(\Gamma(\alpha))^2 (2\alpha-1)||p||'}$$

$$|K_1(b,s)| = \left|\int_a^s \frac{(b-t)^{\alpha-1}}{\Gamma(\alpha)} \cdot \frac{(s-t)^{\alpha-1}}{\Gamma(\alpha)} \frac{1}{p(t)} dt\right|$$

$$\leq \left\|\frac{1}{p}\right\| \cdot \frac{(b-a)^{2\alpha-1}}{(\Gamma(\alpha))^2 (2\alpha-1)}.$$

Now, we apply the derived inequalities and condition (A1)

$$|G^1(x',s) - G^1(x,s)| =$$

$$= \left|K_1(x',s) - \frac{K_1(b,x')K_1(b,s)}{K_1(b,b)} - K_1(x,s) + \frac{K_1(b,x)K_1(b,s)}{K_1(b,b)}\right|$$

$$\leq |K_1(x',s) - K_1(x,s)| + |K_1(x',b) - K_1(x,b)| \cdot \left|\frac{K_1(b,s)}{K_1(b,b)}\right|$$

$$\leq |x' - x|^\beta m(s) \left(1 + \left|\frac{K_1(b,s)}{K_1(b,b)}\right|\right)$$

$$\leq m(s)(1 + ||1/p|| \cdot ||p||)|x' - x|^\beta$$

$$= M_1 |x' - x|^\beta,$$

where

$$M_1 = \frac{2b^{\alpha-1/2}||1/p||(1+||1/p|| \cdot ||p||)}{(\Gamma(\alpha))^2(\alpha-1/2)}.$$

The proof of the Hölder condition for kernel G^2 is analogous. □

The next corollary results from the Hölder conditions (A3) and (A4) and symmetry properties of kernels G^1, G^2 given in (43) and yields continuity of both kernels on square $\Delta = [0,b] \times [0,b]$.

Corollary A2. *If (H1) is fulfilled, then kernel G^1, defined by Formulas (39) and (41), is continuous on square $\Delta = [0,b] \times [0,b]$.*

If (H2) is fulfilled, then kernel G^2, defined by Formulas (40) and (42) is continuous on square $\Delta = [0,b] \times [0,b]$.

Proof. Let us note that the symmetry of kernel G^1 allows us to write condition (A3) in the following form

$$|G^1(x',s') - G^1(x,s)| \leq M_1 \left(|x'-x|^\beta + |s'-s|^\beta\right).$$

To prove continuity of the kernel, we apply the Cauchy definition of continuous function, i.e., we take arbitrary $\epsilon > 0$ and assume that the distance between points (x',s'), (x,s) is smaller than $\delta(\epsilon) = \left(\frac{\epsilon}{2M_1}\right)^{1/\beta}$, which means

$$|(x',s') - (x,s)| = \sqrt{(x'-x)^2 + (s'-s)^2} < \delta(\epsilon).$$

We observe that the following inequalities are then valid

$$|x' - x| < \delta(\epsilon), \quad |s' - s| < \delta(\epsilon).$$

Now, we check the distance between the values of function G^1:

$$|G^1(x',s') - G^1(x,s)| \leq M_1 \left(|x' - x|^\beta + |s' - s|^\beta \right)$$

$$\leq M_1 \cdot 2(\delta(\epsilon))^\beta = \epsilon.$$

We see that for arbitrary $\epsilon > 0$, bound $\delta(\epsilon)$ for the distance of points exists, such that the implication below is valid

$$|(x',s') - (x,s)| < \delta(\epsilon) \implies |G^1(x',s') - G^1(x,s)| < \epsilon.$$

Thus, kernel G^1 is a continuous function on square Δ by the Cauchy definition of continuity.

Proof for kernel G^2 is analogous. □

Appendix A.2

We shall study properties of integral equations of the form:

$$(1 + T_q)y(x) = \lambda T_w y(x) \tag{A5}$$

determined on the $L^2_w(a,b)$ function space. Such an equation is the intermediate stage of transformation of the fractional differential eigenvalue problems into the equivalent integral ones (see examples in papers [13,14]). In cases where the integral operator on the left-hand side of (A5) is invertible, we can convert fractional differential Sturm–Liouville operator into an integral one. Then, we can study spectral properties of the integral operator and derive results for the spectrum and eigenfunctions of the fractional differential Sturm–Liouville problems connected to various homogeneous boundary conditions.

Operators T_q and T_w are integral ones, with kernels given in the form of

$$T_q y(x) := \int_a^b G_q(x,s)y(s)ds, \quad G_q(x,s) = K(x,s)q(s), \tag{A6}$$

$$T_w y(x) := \int_a^b G_w(x,s)y(s)ds, \quad G_w(x,s) = K(x,s)w(s). \tag{A7}$$

We formulate below a theorem which we shall apply to analyse integral eigenvalue problems associated with the fractional differential ones.

Theorem A1. *Let function $q \in C[a,b]$ and function $\|G_w(\cdot,s)\| := \sup_{v \in [a,b]} |G_w(v,s)|$ be bounded on interval $[a,b]$. If condition*

$$\xi := \sup_{x \in [a,b]} \int_a^b |G_q(x,v)| dv < 1 \tag{A8}$$

is fulfilled, then the series

$$G(x,s) := G_w(x,s) + \sum_{n=1}^\infty (-1)^n G_q^{*n} * G_w(x,s) \tag{A9}$$

is uniformly convergent on square Δ; i.e., the sum of this series G is determined for all points $(x,s) \in \Delta$.

If, in addition, kernels $G_q, G_w \in C(\Delta)$, then sum $G \in C(\Delta)$.

Proof. We shall apply the mathematical induction principle to estimate all terms of series (A9). First, we estimate the absolute value of the first convolution term:

$$|G_q * G_w(x,s)| = \left| \int_a^b G_q(x,v) G_w(v,s) dv \right| \tag{A10}$$

$$\leq \int_a^b |G_q(x,v) G_w(v,s)| dv \leq ||G_w(\cdot,s)|| \cdot \sup_{x \in [a,b]} \int_a^b |G_q(x,v)| dv = \xi \cdot ||G_w(\cdot,s)||$$

and for the second term, we obtain

$$|G_q * G_q * G_w(x,s)| \leq \xi \sup_{v \in [a,b]} |G_q * G_w(v,s)| \tag{A11}$$

$$\leq \xi^2 \cdot ||G_w(\cdot,s)||.$$

Now, we formulate the induction hypothesis (here, $n > 2$ is a natural number):

$$|G_q^{*n} * G_w(x,s)| \leq \xi^n \cdot ||G_w(\cdot,s)|| \tag{A12}$$

and we shall prove that it holds for the next step $n+1$

$$|G_q^{*(n+1)} * G_w(x,s)| \leq \xi^{n+1} \cdot ||G_w(\cdot,s)||.$$

We begin from the left-hand side of the above inequality and we find

$$|G_q^{*(n+1)} * G_w(x,s)| = |G_q * \left(G_q^{*n} * G_w\right)(x,s)|$$

$$\leq \xi \sup_{v \in [a,b]} |\left(G_q^{*n} * G_w\right)(v,s)|$$

$$\leq \xi^{n+1} \sup_{v \in [a,b]} |G_w(v,s)| \leq \xi^{n+1} \cdot ||G_w(\cdot,s)||.$$

The induction hypothesis (A12) implies the validity of the next step for $n+1$; therefore, we infer that estimation (A12) is valid for all terms indexed by $n \geq 1$. Now, we are ready to consider the convergence of the function series (A9) by using the Weierstrass convergence test and inequality (A12). We observe that the majorant number series (a geometric one) is absolutely convergent under the assumption (A8). Thence, the function series (A9) is absolutely and uniformly convergent, as we achieve for any point $(x,s) \in \Delta$

$$\left| G_w(x,s) + \sum_{n=1}^{\infty} (-1)^n G_q^{*n} * G_w(x,s) \right|$$

$$\leq |G_w(x,s)| + \sum_{n=1}^{\infty} \xi^n ||G_w(\cdot,s)|| = |G_w(x,s)| + \frac{||G_w(\cdot,s)|| \cdot \xi}{1 - \xi}.$$

In the second part of Theorem 2, we note that continuity of kernels G_q, G_w implies that all terms of the series (A9) are continuous as convolutions of continuous functions. The absolutely and uniformly convergent series (A9) leads to sum G, which is also continuous on Δ. □

References

1. Al-Mdallal, Q.M. An efficient method for solving fractional Sturm-Liouville problems. *Chaos Solitons Fractals* **2009**, *40*, 183–189. [CrossRef]
2. Al-Mdallal, Q.M. On the numerical solution of fractional Sturm-Liouville problems. *Int. J. Comput. Math.* **2010**, *87*, 2837–2845. [CrossRef]

3. Erturk, V.S. Computing eigenelements of Sturm-Liouville Problems of fractional order via fractional differential transform method. *Math. Comput. Appl.* **2011**, *16*, 712–720. [CrossRef]
4. Duan, J.-S.; Wang, Z.; Liu, Y.-L.; Qiu, X. Eigenvalue problems for fractional ordinary differential equations. *Chaos Solitons Fractals* **2013**, *46*, 46–53. [CrossRef]
5. Qi, J.; Chen, S. Eigenvalue problems of the model from nonlocal continuum mechanics. *J. Math. Phys.* **2011**, *52*, 073516. [CrossRef]
6. Atanackovic, T.M.; Stankovic, S. Generalized wave equation in nonlocal elasticity. *Acta Mech.* **2009**, *208*, 1–10. [CrossRef]
7. Klimek, M.; Agrawal, O.P. On a Regular Fractional Sturm-Liouville Problem with Derivatives of Order in (0,1). In Proceedings of the 13th International Carpathian Control Conference, (ICCC 20212), High Tatras, Slovakia, 28–21 May 201; pp. 284–289. [CrossRef]
8. d'Ovidio, M. From Sturm-Liouville problems to to fractional and anomalous diffusions. *Stoch. Process. Appl.* **2012** *122*, 3513–3544. [CrossRef]
9. Rivero, M.; Trujillo, J.J.; Velasco, M.P. A fractional approach to the Sturm-Liouville problem. *Cent. Eur. J. Phys.* **2013**, *11*, 1246–1254. [CrossRef]
10. Zayernouri, M.; Karniadakis, G.E. Fractional Sturm–Liouville eigen-problems: Theory and numerical approximation. *J. Comput. Phys.* **2013**, *252*, 495–517. [CrossRef]
11. Klimek, M.; Agrawal, O.P. Fractional Sturm-Liouville Problem. *Comput. Math. Appl.* **2013**, *66*, 795–812. [CrossRef]
12. Klimek, M.; Odzijewicz, T.; Malinowska, A.B. Variational methods for the fractional Sturm-Liouville problem. *J. Math. Anal. Appl.* **2014**, *416*, 402–426. [CrossRef]
13. Klimek, M.; Ciesielski, M.; Blaszczyk, T. Exact and numerical solutions of the fractional Sturm-Liouville problem. *Fract. Calc. Appl. Anal.* **2018**, *21*, 45–71. [CrossRef]
14. Klimek, M. Homogeneous Robin boundary conditions and discrete spectrum of fractional eigenvalue problem. *Fract. Calc. Appl. Anal.* **2019**, *22*, 78–94. [CrossRef]
15. Zayernouri, M.; Ainsworth, M.; Karniadakis, G.E. Tempered fractional Sturm–Liouville eigenproblems. *SIAM J. Sci. Comput.* **2015**, *37*, A1777–A1800. [CrossRef]
16. Pandey, P.; Pandey, R.; Yadav, S.; Agrawal, O.P. Variational Approach for Tempered Fractional Sturm–Liouville Problem. *Int. J. Appl. Comput. Math.* **2021**, *7*, 51. [CrossRef]
17. Al-Refai, M.; Abdeljawad, T. Fundamental results of conformable Sturm-Liouville eigenvalue problems. *Complexity* **2017**, *2017*, 3720471. [CrossRef]
18. Mortazaasl, H.; Jodayree Akbarfam, A.A. Two classes of conformable fractional Sturm-Liouville problems: Theory and applications. *Math. Meth. Appl. Sci.* **2021**, *44*, 166–195. [CrossRef]
19. Li, J.; Qi, J. On a nonlocal Sturm–Liouville problem with composite fractional derivatives. *Math. Meth. Appl. Sci.* **2021**, *44*, 1931–1941. [CrossRef]
20. Malinowska, A.B.; Odzijewicz, T.; Torres, D.F.M. *Advanced Methods in the Fractional Calculus of Variations*; Springer International Publishing: London, UK, 2015.
21. Pandey, P.K.; Pandey, R.K.; Agrawal, O.P. Variational approximation for fractional Sturm–Liouville problem. *Fract. Calc. Appl. Anal.* **2020**, *23*, 861–874. [CrossRef]
22. Kilbas, A.A.; Srivastava, H.M.; Trujillo, J.J. *Theory and Applications of Fractional Differential Equations*; Elsevier: Amsterdam, The Netherlands, 2006.
23. Podlubny, I. *Fractional Differential Equations*; Academic Press: San Diego, CA, USA, 1999.
24. Giusti, A.; Colombaro, I.; Garra, R.; Garappa, R.; Polito, F.; Popolizio, M.; Mainardi, F. A practical guide to Prabhakar fractional calculus. *Fract. Calc. Appl. Anal.* **2020**, *23*, 9–54. [CrossRef]
25. Prabhakar, T.R. A singular integral equation with a generalized Mittag Leffler function in the kernel. *Yokohama Math. J.* **1971**, *19*, 1–11.
26. Klimek, M. On properties of exact and numerical solutions to integral eigenvalue problems associated to fractional differential ones. In *Selected Topics in Contemporary Mathematical Modeling*; Czestochowa University of Technology Press: Czestochowa, Poland, 2021; pp. 49–65.
27. Stanislavsky, A.; Weron, A. Transient anomalous diffusion with Prabhakar-type memory. *J. Chem. Phys.* **2018**, *149*, 044107. [CrossRef] [PubMed]
28. Dos Santos, M.A.F. Fractional Prabhakar derivative in diffusion equation with non-static stochastic resetting. *Physics* **2019**, *1*, 40–58. [CrossRef]
29. Shukla, A.; Sukavanam, N.; Pandey, D.N. Approximate controllabillity of semilinear fractional control systems of order $\alpha \in (1,2]$ with infinite delay. *Mediterr. J. Math.* **2016**, *13*, 2539–2550. [CrossRef]
30. Górska, K.; Horzela, A.; Lattanzi, A.; Pogany, T.K. On complete monotonicity of three parameter Mittag-Leffler function. *Appl. Anal. Discret. Math.* **2021**, *15*, 118–128. [CrossRef]

Article

General Fractional Integrals and Derivatives of Arbitrary Order

Yuri Luchko

Department of Mathematics, Physics, and Chemistry, Beuth Technical University of Applied Sciences Berlin, Luxemburger Str. 10, 13353 Berlin, Germany; luchko@beuth-hochschule.de

Abstract: In this paper, we introduce the general fractional integrals and derivatives of arbitrary order and study some of their basic properties and particular cases. First, a suitable generalization of the Sonine condition is presented, and some important classes of the kernels that satisfy this condition are introduced. Whereas the kernels of the general fractional derivatives of arbitrary order possess integrable singularities at the point zero, the kernels of the general fractional integrals can—depending on their order—be both singular and continuous at the origin. For the general fractional integrals and derivatives of arbitrary order with the kernels introduced in this paper, two fundamental theorems of fractional calculus are formulated and proved.

Keywords: Sonine kernel; general fractional derivative of arbitrary order; general fractional integral of arbitrary order; first fundamental theorem of fractional calculus; second fundamental theorem of fractional calculus

MSC: 26A33; 26B30; 44A10; 45E10

1. Introduction

In his papers [1,2], Abel derived and studied a mathematical model for the tautochrone problem in the form of the following integral equation (with slightly different notations):

$$f(t) = \frac{1}{\sqrt{\pi}} \int_0^t \frac{\phi'(\tau)\, d\tau}{\sqrt{t-\tau}}. \tag{1}$$

In fact, he considered the even more general integral equation

$$f(t) = \frac{1}{\Gamma(1-\alpha)} \int_0^t \frac{\phi'(\tau)\, d\tau}{(t-\tau)^\alpha} \tag{2}$$

under an implicit restriction $0 < \alpha < 1$. It is easy to see that the right-hand side of (2) is the operator that is currently referred to as the Caputo fractional derivative ${}_*D_{0+}^\alpha$ of the order α, $0 < \alpha < 1$. Abel's solution formula to Equation (2) is nothing else than the operator now called the Riemann–Liouville fractional integral I_{0+}^α of the order $\alpha > 0$:

$$\phi(t) = \frac{1}{\Gamma(\alpha)} \int_0^t (t-\tau)^{\alpha-1} f(\tau)\, d\tau =: (I_{0+}^\alpha f)(t),\ t > 0. \tag{3}$$

In modern notation, Formulas (2) and (3) correspond to the second fundamental theorem of FC for the Caputo fractional derivative of a function that takes the value zero at the point zero:

$$(I_{0+}^\alpha f)(t) = (I_{0+}^\alpha {}_*D_{0+}^\alpha \phi)(t) = \phi(t) - \phi(0) = \phi(t), \tag{4}$$

where the validity of the condition $\phi(0) = 0$ follows from the construction of Abel's mathematical model for the tautochrone problem. For more details regarding Abel's results and derivations presented in [1,2], see the recent paper presented in [3].

To solve the integral Equation (2), in [2], published in 1826, Abel employed the relation

$$(h_\alpha * h_{1-\alpha})(t) = \{1\}, \ t > 0, \ h_\alpha(t) := \frac{t^{\alpha-1}}{\Gamma(\alpha)}, \ \alpha > 0, \tag{5}$$

where the operation $*$ stands for the Laplace convolution,

$$(f * g)(t) = \int_0^t f(t-\tau)g(\tau)\,d\tau \tag{6}$$

and $\{1\}$ is the function that is identically equal to 1 for $t \geq 0$.

In [4], published in 1884, Sonine recognized that the relation (5) is the most crucial ingredient of Abel's solution method that can be generalized and applied to the analytical treatment of a larger class of integral equations. In place of (5), Sonine considered a pair of functions κ, k (Sonine kernels) that satisfy the relation

$$(\kappa * k)(t) = \{1\}, \ t > 0. \tag{7}$$

In what follows, we denote the set of the Sonine kernels by $\mathcal{S}$. For a given Sonine kernel κ, the kernel k that satisfies the Sonine condition (7) is called its associate Sonine kernel. Following Abel's solution method, Sonine showed that the integral equation

$$f(t) = \int_0^t \kappa(t-\tau)\phi(\tau)\,d\tau = (\kappa * \phi)(t) \tag{8}$$

has a solution in the form

$$\phi(t) = \frac{d}{dt}\int_0^t k(t-\tau)f(\tau)\,d\tau = \frac{d}{dt}(k * f)(t), \tag{9}$$

provided that the kernels κ, k satisfy the Sonine condition (7). Indeed, we obtain

$$(k * f)(t) = (k * \kappa * \phi)(t) = (\{1\} * \phi)(t) = \int_0^t \phi(\tau)\,d\tau$$

which immediately leads to the Formula (9). Of course, any concrete realization of the Sonine schema requires a precise characterization of the Sonine kernels and the spaces of functions where the operators from the right-hand sides of (8) and (9) are well defined. In [4], Sonine introduced a large class of the Sonine kernels in the form

$$\kappa(t) = h_\alpha(t) \cdot \kappa_1(t), \ \kappa_1(t) = \sum_{k=0}^{+\infty} a_k t^k, \ a_0 \neq 0, \ 0 < \alpha < 1, \tag{10}$$

$$k(t) = h_{1-\alpha}(t) \cdot k_1(t), \ k_1(t) = \sum_{k=0}^{+\infty} b_k t^k, \tag{11}$$

where the functions $\kappa_1 = \kappa_1(t), \ k_1 = k_1(t)$ are analytical on $\mathbb{R}$ and their coefficients are connected by the relations

$$a_0 b_0 = 1, \ \sum_{k=0}^n \Gamma(k+1-\alpha)\Gamma(\alpha+n-k)a_{n-k}b_k = 0, \ n \geq 1. \tag{12}$$

The most prominent pair of the Sonine kernels from this class is given by the formulas

$$\kappa(t) = (\sqrt{t})^{\alpha-1}J_{\alpha-1}(2\sqrt{t}), \ k(t) = (\sqrt{t})^{-\alpha}I_{-\alpha}(2\sqrt{t}), \ 0 < \alpha < 1, \tag{13}$$

where

$$J_\nu(t) = \sum_{k=0}^{+\infty} \frac{(-1)^k (t/2)^{2k+\nu}}{k!\Gamma(k+\nu+1)}, \quad I_\nu(t) = \sum_{k=0}^{+\infty} \frac{(t/2)^{2k+\nu}}{k!\Gamma(k+\nu+1)}, \quad \Re(\nu) > -1, \; t \in \mathbb{C} \quad (14)$$

are the Bessel and the modified Bessel functions, respectively.

Later, the evolution equations with the integro-differential operators of the convolution type (compare them with the solution by Sonine in Formula (9)),

$$(\mathbb{D}_{(k)} f)(t) = \frac{d}{dt} \int_0^t k(t-\tau) f(\tau) \, d\tau, \; t > 0, \quad (15)$$

were actively studied in the framework of the abstract Volterra integral equations on the Banach spaces (see [5] and references therein). For example, in [6], the case of operators with the completely positive kernels $k \in L^1(0, +\infty)$ was considered. The kernels from this class satisfy the condition (compare it with the Sonine condition (7))

$$a\,k(t) + \int_0^t k(t-\tau) l(\tau) \, d\tau = \{1\}, \; t > 0, \quad (16)$$

where $a \geq 0$ and $l \in L^1(0, +\infty)$ is a non-negative and non-increasing function.

However, until recently, no interpretation of these general results in the framework of fractional calculus (FC) had been suggested. The situation changed with the publication of the paper presented in [7] (see also [8–10]). In [7], Kochubei introduced a class $\mathcal{K}$ of kernels that satisfy the following conditions:

(K1) The Laplace transform $\tilde{k}$ of k,

$$\tilde{k}(p) = (\mathcal{L} k)(p) = \int_0^{+\infty} k(t) e^{-pt} \, dt, \quad (17)$$

exists for all $p > 0$;
(K2) $\tilde{k}(p)$ is a Stieltjes function (see [11] for details regarding the Stieltjes functions);
(K3) $\tilde{k}(p) \to 0$ and $p\tilde{k}(p) \to +\infty$ as $p \to +\infty$;
(K4) $\tilde{k}(p) \to +\infty$ and $p\tilde{k}(p) \to 0$ as $p \to 0$.

Using the technique of the complete Bernstein functions, Kochubei investigated the integro-differential operators in the form of (15) and their Caputo type modifications

$$({}_*\mathbb{D}_{(k)} f)(t) = (\mathbb{D}_{(k)} f)(t) - f(0) k(t) \quad (18)$$

with the kernels from $\mathcal{K}$. In [7], Kochubei showed the inclusion $\mathcal{K} \subset \mathcal{S}$, introduced the corresponding integral operator

$$(\mathbb{I}_{(\kappa)} f)(t) = (\kappa * f)(t) = \int_0^t \kappa(t-\tau) f(\tau) \, d\tau, \quad (19)$$

and proved the validity of the first fundamental theorem of FC; i.e., that the operators (15) and (18) are left-inverse to the integral operator (19) on the suitable spaces of functions.

Moreover, Kochubei treated some basic ordinary and partial fractional differential equations with the time-derivative in the form of (18) and proved that the solution to the Cauchy problem for the relaxation equation with the operator (18) and a positive initial condition is completely monotonic and that the fundamental solution to the Cauchy problem for the fractional diffusion equation with the time-derivative in the form of (18) can be interpreted as a probability density function. These results justified calling the operators (15) and (18) the general fractional derivatives (GFDs) in the Riemann–Liouville and Caputo sense, respectively. The integral operator (19) was called the general fractional integral (GFI).

The GFDs (15) and (18) with the kernels $k \in \mathcal{K} \subset \mathcal{S}$ possess a series of important properties. However, the conditions (K1)–(K4) are very strong (especially the condition (K2)), and thus in subsequent publications, the operators (15) and (18) with the Sonine kernels from some larger classes were considered from the viewpoint of FC and its applications. In [12], a class of the kernels was introduced that ensures the validity of a maximum principle for the general time-fractional diffusion equations with the operators of type (18). Another important class of the Sonine kernels was described in [8] in terms of the completely monotone functions. As shown in [8], any singular (unbounded in a neighborhood of the point zero) locally integrable completely monotone function κ is a Sonine kernel, and its associate kernel k is also a locally integrable completely monotone function.

In the recent publications presented in [9,13], the operators (15) and (18) with the Sonine kernels from the class $\mathcal{S}_{-1} \subset \mathcal{S}$ that satisfy only some minimal restrictions were studied from the viewpoint of FC. The Sonine kernels κ, $k \in \mathcal{S}_{-1}$ are continuous on $\mathbb{R}_+$ and possess the integrable singularities of the power function type at the point zero. In particular, in [9], the first and second fundamental theorems of FC for the operators (15) and (18) with the kernels $k \in \mathcal{S}_{-1}$ were formulated and proved. In [13], an operational calculus of the Mikusiński type for the operators (18) with the Sonine kernels $k \in \mathcal{S}_{-1}$ was constructed and applied for the analytical treatment of some initial value problems for the fractional differential equations with these operators.

It is clear that weakening the Kochubei conditions (K1)–(K4) on the Sonine kernels from $\mathcal{K}$ leads to the abandonment of some properties that were derived in [7] for the GFDs (15) and (18). However, it was shown in [9,13] that the operators (15) and (18) with the Sonine kernels $k \in \mathcal{S}_{-1}$ and the corresponding integral operator (19) still satisfy the main properties that the fractional derivatives and integrals should fulfill (see [14] and the references therein). Thus, these operators can also be interpreted as the GFDs and GFIs.

Another important point concerns the "generalized order" of the GFDs (15) and (18) with the Sonine kernels from the classes mentioned above. While projecting these operators to the conventional Riemann–Liouville and Caputo fractional derivatives (the case of the kernel $k(t) = h_{1-\alpha}(t)$), the derivatives' order is restricted only to the case of $\alpha \in (0, 1)$. The reason is that the Sonine condition (5) for the power functions h_α and $h_{1-\alpha}$ holds true only in the case $0 < \alpha < 1$. Moreover, even in the definition of the Caputo type general fractional derivative (18), only one initial condition is contained, which again indicates that the "generalized order" of this operator does not exceed one.

Because the Riemann–Liouville fractional integral and the Riemann–Liouville and Caputo fractional derivatives are defined for arbitrary order $\alpha \geq 0$, an extension of the GFDs (15) and (18) to the case of arbitrary order is worthy of investigation.

In a recent paper [9], the n-fold GFIs and GFDs were introduced as an attempt to extend their order behind the interval $(0, 1)$. For example, the two-fold general fractional derivative constructed for the operator (15) with the kernel $\kappa(t) = h_{1-\alpha}(t)$, $0 < \alpha < 1$ is the Riemann–Liouville fractional derivative of the order 2α:

$$(D_{0+}^{2\alpha} f)(t) = \begin{cases} \frac{d^2}{dt^2}(I_{0+}^{2-2\alpha} f)(t), & \frac{1}{2} < \alpha < 1, \ t > 0, \\ \frac{d}{dt}(I_{0+}^{1-2\alpha} f)(t), & 0 < \alpha \leq \frac{1}{2}, \ t > 0. \end{cases} \quad (20)$$

Thus, we cannot ensure that the order of this two-fold GFD is always greater than one. Depending on the values of α and n, the "generalized order" of the n-fold GFD can be any number in the interval $(0, n)$.

The main objective of this paper is to introduce the GFIs and GFDs of an arbitrary order in analogy to the Riemann–Liouville fractional integral and the Riemann–Liouville and Caputo fractional derivatives. This is done by a suitable generalization of the Sonine condition (7) and by the corresponding adjustment of Formulas (15) and (18), which define the GFDs in the Riemann–Liouville and Caputo senses.

The rest of the paper is organized as follows. In Section 2, following [9,13], we provide some basic definitions and properties of the GFDs (15) and (18) with the Sonine kernels $k \in \mathcal{S}_{-1}$. Section 3 presents our main results. First, a suitable generalization of the Sonine condition (7) is introduced and some examples of the kernels that satisfy this condition are discussed. Then, the GFDs of an arbitrary order with these kernels are defined and their properties are studied. The conventional Riemann–Liouville and Caputo fractional derivatives of arbitrary order are particular cases of these GFDs. Another important example is the integro-differential operators of convolution type with the Bessel and the modified Bessel functions in the kernels. The constructions introduced in this section allow the formulation of the fractional differential equations with the GFDs with a generalized order greater than one with several initial conditions.

2. General Fractional Integrals and Derivatives with the Sonine Kernels

In this section, we provide some basic definitions and results regarding the GFIs and GFDs with the Sonine kernels from the class S_{-1} introduced in [9]. For more details, other relevant results and the proofs, see [9,13].

In what follows, we employ the space of functions $C_{-1}(0, +\infty)$ and its sub-spaces. A family of the spaces $C_\alpha(0, +\infty)$, $\alpha \geq -1$ was first introduced in [15] as follows:

$$C_\alpha(0, +\infty) := \{f : f(t) = t^p f_1(t), t > 0, p > \alpha, f_1 \in C[0, +\infty)\}. \tag{21}$$

Evidently, the spaces $C_\alpha(0, +\infty)$ are ordered by the inclusion $\alpha_1 \geq \alpha_2 \Rightarrow C_{\alpha_1}(0, +\infty) \subseteq C_{\alpha_2}(0, +\infty)$, and thus the inclusion $C_\alpha(0, +\infty) \subseteq C_{-1}(0, +\infty)$, $\alpha \geq -1$ holds true.

In the further discussions, we also use the sub-spaces $C_{-1}^m(0, +\infty)$, $m \in \mathbb{N}_0 = \mathbb{N} \cup \{0\}$ of the space $C_{-1}(0, +\infty)$, which are defined as follows:

$$C_{-1}^m(0, +\infty) := \{f : f^{(m)} \in C_{-1}(0, +\infty)\}. \tag{22}$$

The spaces $C_{-1}^m(0, +\infty)$ were first introduced and studied in [16]. In particular, we have the following properties:

(1) $C_{-1}^0(0, +\infty) \equiv C_{-1}(0, +\infty)$;
(2) $C_{-1}^m(0, +\infty)$, $m \in \mathbb{N}_0$ is a vector space over the field $\mathbb{R}$ (or $\mathbb{C}$);
(3) If $f \in C_{-1}^m(0, +\infty)$ with $m \geq 1$, then $f^{(k)}(0+) := \lim_{t \to 0+} f^{(k)}(t) < +\infty$, $0 \leq k \leq m-1$, and the function

$$\tilde{f}(t) = \begin{cases} f(t), & t > 0, \\ f(0+), & t = 0 \end{cases}$$

belongs to the space $C^{m-1}[0, +\infty)$;
(4) If $f \in C_{-1}^m(0, +\infty)$ with $m \geq 1$, then $f \in C^m(0, +\infty) \cap C^{m-1}[0, +\infty)$.
(5) For $m \geq 1$, the following representation holds true:

$$f \in C_{-1}^m(0, +\infty) \Leftrightarrow f(t) = (I_{0+}^m \phi)(t) + \sum_{k=0}^{m-1} f^{(k)}(0) \frac{t^k}{k!}, \ t \geq 0, \ \phi \in C_{-1}(0, +\infty);$$

(6) Let $f \in C_{-1}^m(0, +\infty)$, $m \in \mathbb{N}_0$, $f(0) = \cdots = f^{(m-1)}(0) = 0$ and $g \in C_{-1}^1(0, +\infty)$. Then, the Laplace convolution $h(t) = (f * g)(t)$ belongs to the space $C_{-1}^{m+1}(0, +\infty)$ and $h(0) = \cdots = h^{(m)}(0) = 0$.

For our aims, we also need another two-parameter family of sub-spaces of $C_\alpha(0, +\infty)$ that allows us to better control the behavior of the functions at the origin:

$$C_{\alpha,\beta}(0, +\infty) = \{f : f(t) = t^p f_1(t), t > 0, \alpha < p < \beta, f_1 \in C[0, +\infty)\}. \tag{23}$$

In particular, the sub-space $C_{-1,0}(0, +\infty)$ contains the functions that are continuous on $\mathbb{R}_+$ and possess the integrable singularities of the power function type at the origin.

As mentioned in [17] (see also [8]), any Sonine kernel has an integrable singularity at the point zero. On the other hand, the kernels of the fractional integrals and derivatives should be singular [18]. Thus, the fractional integrals and derivatives with the Sonine kernels are worthy of investigation. In what follows, we consider the GFI (19) and the GFDs (15) and (18) of the Riemann–Liouville and Caputo types, respectively, with the Sonine kernels κ and k that belong to the sub-space $C_{-1,0}(0,+\infty)$ of the space $C_{-1}(0,+\infty)$.

Definition 1. *Let $\kappa, k \in C_{-1,0}(0,+\infty)$ be a pair of the Sonine kernels; i.e., let the Sonine condition (7) be fulfilled. The set of such Sonine kernels is denoted by $\mathcal{S}_{-1}$:*

$$(\kappa, k \in \mathcal{S}_{-1}) \Leftrightarrow (\kappa, k \in C_{-1,0}(0,+\infty)) \wedge ((\kappa * k)(t) = \{1\}). \tag{24}$$

Several important features of the GFI (19) on the space $C_{-1}(0,+\infty)$ follow from the well-known properties of the Laplace convolution. In particular, we mention the mapping property

$$\mathbb{I}_{(\kappa)} : C_{-1}(0,+\infty) \to C_{-1}(0,+\infty), \tag{25}$$

the commutativity law

$$\mathbb{I}_{(\kappa_1)} \mathbb{I}_{(\kappa_2)} = \mathbb{I}_{(\kappa_2)} \mathbb{I}_{(\kappa_1)}, \ \kappa_1, \kappa_2 \in \mathcal{S}_{-1}, \tag{26}$$

and the index law

$$\mathbb{I}_{(\kappa_1)} \mathbb{I}_{(\kappa_2)} = \mathbb{I}_{(\kappa_1 * \kappa_2)}, \ \kappa_1, \kappa_2 \in \mathcal{S}_{-1} \tag{27}$$

that are valid on the space $C_{-1}(0,+\infty)$.

Let $\kappa \in \mathcal{S}_{-1}$ and k be its associate Sonine kernel. The GFDs of the Riemann–Liouville and the Caputo types associated to the GFI (19) are given by the Formulas (15) and (18), respectively. It is easy to see that the GFD (18) in the Caputo sense can be rewritten as a regularized GFD (15) in the Riemann–Liouville sense:

$$(_*\mathbb{D}_{(k)} f)(t) = (\mathbb{D}_{(k)} [f(\cdot) - f(0)])(t), \ t > 0. \tag{28}$$

For the functions from $C^1_{-1}(0,+\infty)$, the Riemann–Liouville GFD (15) can be represented as

$$(\mathbb{D}_{(k)} f)(t) = (k * f')(t) + f(0)k(t), \ t > 0, \tag{29}$$

which immediately leads to the useful representation

$$(_*\mathbb{D}_{(k)} f)(t) = (k * f')(t), \ t > 0 \tag{30}$$

of the Caputo type GFD (18) that is valid on the space $C^1_{-1}(0,+\infty)$.

In the rest of this section, we formulate the first and second fundamental theorems of FC for the GFDs in the Riemann–Liouville and Caputo senses.

Theorem 1 (First Fundamental Theorem for the GFD). *Let $\kappa \in \mathcal{S}_{-1}$ and k be its associate Sonine kernel.*

Then, the GFD (15) is a left-inverse operator to the GFI (19) on the space $C_{-1}(0,+\infty)$,

$$(\mathbb{D}_{(k)} \mathbb{I}_{(\kappa)} f)(t) = f(t), \ f \in C_{-1}(0,+\infty), \ t > 0, \tag{31}$$

and the GFD (18) is a left inverse operator to the GFI (19) on the space $C_{-1,(k)}(0,+\infty)$:

$$(_*\mathbb{D}_{(k)} \mathbb{I}_{(\kappa)} f)(t) = f(t), \ f \in C_{-1,(k)}(0,+\infty), \ t > 0, \tag{32}$$

where $C_{-1,(k)}(0,+\infty) := \{f : f(t) = (\mathbb{I}_{(k)} \phi)(t), \ \phi \in C_{-1}(0,+\infty)\}$.

As shown in [9], the space $C_{-1,(k)}(0,+\infty)$ can be also characterized as follows:

$$C_{-1,(k)}(0,+\infty) = \{f : \mathbb{I}_{(\kappa)} f \in C^1_{-1}(0,+\infty) \wedge (\mathbb{I}_{(\kappa)} f)(0) = 0\}.$$

Now, we proceed with the second fundamental theorem of FC for the GFDs in the Riemann–Liouville and Caputo senses.

Theorem 2 (Second Fundamental Theorem for the GFD). *Let $\kappa \in \mathcal{S}_{-1}$ and k be its associate Sonine kernel.*

Then, the relations

$$(\mathbb{I}_{(\kappa)} \, {}_*\mathbb{D}_{(k)} f)(t) = f(t) - f(0), \; t > 0, \tag{33}$$

$$(\mathbb{I}_{(\kappa)} \mathbb{D}_{(k)} f)(t) = f(t), \; t > 0 \tag{34}$$

hold valid for the functions $f \in C_{-1}^1(0, +\infty)$.

In [9,13], the n-fold GFIs and GFDs with the Sonine kernels from $\mathcal{S}_{-1}$ were introduced and studied. For more details, we refer interested readers to these publications.

3. General Fractional Integrals and Derivatives of Arbitrary Order

As already mentioned in the Introduction, the "generalized order" of the GFIs and GFDs introduced so far is restricted to the interval $(0, 1)$. The order of the n-fold GFIs and GFDs recently introduced in [9] belongs to the interval $(0, n)$. However, it is hardly possible to fix their order between two neighboring natural numbers as in the case of the conventional Riemann–Liouville and Caputo fractional derivatives and thus to study, for example, the fractional oscillator equations or the time-fractional diffusion-wave equations with the GFDs of the order from the interval $(1, 2)$.

In this section, we define the GFIs and GFDs of arbitrary order and study their basic properties. As in the case of the conventional Riemann–Liouville and Caputo fractional derivatives, for the GFDs, we also have to distinguish between two completely different cases; namely, between the case of the integer order and the case of non-integer order. In the first case, the conventional Riemann–Liouville and Caputo fractional derivatives are defined as the integer-order derivatives, while in the second case, they are non-local integro-differential operators. Because the conventional Riemann–Liouville and Caputo fractional derivatives are important particular cases of the GFDs, we have no other choice but to follow the same strategy; namely, to separately define the GFDs of integer order as the integer-order derivatives and the GFDs of non-integer order as some integro-differential operators. In what follows, we focus on the case of the GFDs of non-integer order (the integer-order GFDs are simply the integer-order derivatives).

To introduce the GFIs and the GFDs of arbitrary non-integer order, we first formulate a condition on their kernels that generalizes the Sonine condition (7):

$$(\kappa * k)(t) = \{1\}^n(t), \; n \in \mathbb{N}, \; t > 0, \tag{35}$$

where

$$\{1\}^n(t) := (\underbrace{\{1\} * \ldots * \{1\}}_{n \text{ times}})(t) = h_n(t) = \frac{t^{n-1}}{(n-1)!}.$$

Evidently, the Sonine condition corresponds to the case $n = 1$ of the more general condition (35).

Another important ingredient of our definitions is a set of the kernels that satisfy the condition (35) and belong to the suitable spaces of functions.

Definition 2. *Let the functions κ and k satisfy the condition (35) and the inclusions $\kappa \in C_{-1}(0, +\infty)$ and $k \in C_{-1,0}(0, +\infty)$ hold true.*

The set of pairs (κ, k) of such kernels is denoted by $\mathcal{L}_n$.

Remark 1. *The set $\mathcal{L}_1$ coincides with the set of the Sonine kernels $\mathcal{S}_{-1}$ discussed in the previous section (see Definition 1). Indeed, in this case, the kernel $\kappa \in C_{-1}(0, +\infty)$ is a Sonine kernel,* and therefore it has an integrable singularity at the point zero. Thus, it belongs to the subspace $C_{-1,0}(0,+\infty)$ as required in Definition 1.

Remark 2. *For $n > 1$, Definition 2 is not symmetrical with respect to the kernels κ and k because of the non-symmetrical inclusions $\kappa \in C_{-1}(0,+\infty)$ and $k \in C_{-1,0}(0,+\infty)$ (in the case $n = 1$, Definition 1 is symmetrical and one can interchange the kernels κ and k).*

However, the same statement is valid for the kernel $\kappa(t) = h_\alpha(t)$, $\alpha > 0$ of the Riemann–Liouville integral I_{0+}^α and the kernel $k(t) = h_{n-\alpha}(t)$ of the Riemann–Liouville and Caputo fractional derivatives of order α, $n-1 < \alpha < n$, $n \in \mathbb{N}$, defined as follows:

$$(D_{0+}^\alpha f)(t) := \frac{d^n}{dt^n}(I_{0+}^{n-\alpha} f)(t), \ t > 0, \tag{36}$$

$$({}_*D_{0+}^\alpha f)(t) := \left(D_{0+}^\alpha \left(f(\cdot) - \sum_{j=0}^{n-1} f^{(j)}(0)h_{j+1}(\cdot)\right)\right)(t), \ t > 0, \tag{37}$$

with I_{0+}^α being the Riemann-Liouville fractional integral of order α:

$$(I_{0+}^\alpha f)(t) := \frac{1}{\Gamma(\alpha)} \int_0^t (t-\tau)^{\alpha-1} f(\tau)\, d\tau, \ t > 0, \ \alpha > 0. \tag{38}$$

The solution to defining the integer-order Riemann–Liouville and Caputo fractional derivatives consists of a separate definition of the Riemann–Liouville fractional integral of the order $\alpha = 0$:

$$(I_{0+}^0 f)(t) := f(t). \tag{39}$$

Of course, the definition (39) is not arbitrary and is justified inter alia by the formula

$$\lim_{\alpha \to 0+} \|(I_{0+}^\alpha f)(t) - f(t)\|_{L_1(0,T)} = 0 \tag{40}$$

that is valid for $f \in L^1(0,T)$ in every Lebesgue point of f; i.e., almost everywhere on the interval $(0, T)$, $T > 0$ (see, e.g., [19]).

Example 1. *The kernels $\kappa(t) = h_\alpha(t)$, $\alpha > 0$ and $k(t) = h_{n-\alpha}(t)$, $n-1 < \alpha < n$, $n \in \mathbb{N}$ provide a first example of the kernels from $\mathcal{L}_n$. Please note that the power functions h_α and $h_{n-\alpha}$ build a pair of the Sonine kernels only in the case $n = 1$; i.e., only in the case when the fractional derivatives' order is less than one.*

Because both the Sonine condition (7) and its generalization (35) contain the Laplace convolution of two kernels, it is very natural to transform them into the Laplace domain. Providing that the Laplace transforms $\tilde{\kappa}$, $\tilde{k}$ of the functions κ and k exist, the convolution theorem for the Laplace transform leads to the relation

$$\tilde{\kappa}(p) \cdot \tilde{k}(p) = \frac{1}{p}, \ \Re(p) > p_{\kappa,k} \in \mathbb{R} \tag{41}$$

for the Laplace transforms of the Sonine kernels and to a more general relation

$$\tilde{\kappa}(p) \cdot \tilde{k}(p) = \frac{1}{p^n}, \ \Re(p) > p_{\kappa,k} \in \mathbb{R}, \ n \in \mathbb{N} \tag{42}$$

for the kernels from the set $\mathcal{L}_n$ introduced in Definition 2.

Example 2. Formula (42) along with the works in [20,21] for the direct and inverse Laplace transforms, respectively, can be used to deduce other nontrivial examples of the kernels from $\mathcal{L}_n$. For instance, we employ the Laplace transform formulas (see [20])

$$\left(\mathcal{L}\, t^{\nu/2} J_\nu(2\sqrt{t})\right)(p) = p^{-\nu-1} \exp(-1/p), \ \Re(\nu) > -1, \ \Re(p) > 0,$$

$$\left(\mathcal{L}\, t^{\nu/2} I_\nu(2\sqrt{t})\right)(p) = p^{-\nu-1} \exp(1/p), \ \Re(\nu) > -1, \ \Re(p) > 0$$

for the Bessel function J_ν and the modified Bessel function I_ν defined by the power series (14) to introduce the kernels

$$\kappa(t) = t^{\nu/2} J_\nu(2\sqrt{t}), \ k(t) = t^{n/2-\nu/2-1} I_{n-\nu-2}(2\sqrt{t}), \ n-2 < \nu < n-1, \ n \in \mathbb{N}. \quad (43)$$

These kernels satisfy the condition (42). Moreover, for $n-2 < \nu < n-1$, $n \in \mathbb{N}$, the inclusions $\kappa \in C_{-1}(0, +\infty)$ and $k \in C_{-1,0}(0, +\infty)$ hold true, and thus the pair of the kernels (κ, k) given by (43) is from $\mathcal{L}_n$.

Now let us consider a pair of the Sonine kernels (κ, k) from $\mathcal{L}_1$ (in [4,8,9,13,17] and other related publications, many pairs of such kernels were presented). There are at least two reasonable possibilities to construct a pair (κ_n, k_n) of the kernels from $\mathcal{L}_n$, $n > 1$ based on the Sonine kernels κ, k from $\mathcal{L}_1$.

The first strategy consists of building the kernels $\kappa_n = \kappa^n$ and $k_n = k^n$. Evidently, the kernels κ_n and k_n satisfy the relation (35) because κ and k are the Sonine kernels:

$$(\kappa_n * k_n)(t) = (\kappa^n * k^n)(t) = (\kappa * k)^n(t) = \{1\}^n(t). \quad (44)$$

However, the pair (κ_n, k_n) does not always belong to the set $\mathcal{L}_n$. This is the case only under an additional condition; namely, only when the inclusion $k^n \in C_{-1,0}(0, +\infty)$ holds true (of course, $\kappa^n \in C_{-1}(0, +\infty)$ for any $n \in \mathbb{N}$). This is a very strong and restrictive condition. For example, in the case of the Riemann–Liouville fractional integral I_{0+}^α with the kernel $\kappa(t) = h_\alpha(t), 0 < \alpha < 1$ and the Riemann–Liouville fractional derivative D_{0+}^α with the kernel $k(t) = h_{1-\alpha}$, the kernel k^n takes the form $k^n(t) = h_{n(1-\alpha)}(t)$. It belongs to the space $C_{-1,0}(0, +\infty)$ only under the condition $0 < n(1-\alpha) < 1$; i.e., if $1 - \frac{1}{n} < \alpha < 1$, which is very restrictive. Moreover, the example of the kernels (43) shows that not every pair of the kernels from $\mathcal{L}_n$ can be represented in the form (κ^n, k^n) with the kernels $(\kappa, k) \in \mathcal{L}_1$.

Another and even more general and important possibility for the construction of a pair (κ_n, k_n) of the kernels from $\mathcal{L}_n$, $n > 1$ based on the Sonine kernels κ, k from $\mathcal{L}_1$ is presented in the following theorem:

Theorem 3. *Let (κ, k) be a pair of the Sonine kernels from $\mathcal{L}_1$. Then, the pair (κ_n, k_n) of the kernels given by the formula*

$$\kappa_n(t) = (\{1\}^{n-1} * \kappa)(t), \ k_n(t) = k(t) \quad (45)$$

belongs to the set $\mathcal{L}_n$.

Proof. First, we check that the kernels (45) satisfy the condition (35):

$$(\kappa_n * k_n)(t) = (\{1\}^{n-1} * \kappa * k)(t) = (\{1\}^{n-1} * \{1\})(t) = \{1\}^n(t). \quad (46)$$

Moreover, because of the inclusions $\kappa, k \in \mathcal{L}_1$, the inclusions $\kappa_n \in C_{-1}(0, +\infty)$ and $k_n = k \in C_{-1,0}(0, +\infty)$ are satisfied, and thus the kernels κ_n and k_n defined by (45) belong to the set $\mathcal{L}_n$. □

In the rest of this section, we introduce the general fractional integrals and derivatives of an arbitrary (non-integer) order and discuss their basic properties and examples.

Definition 3. Let (κ, k) be a pair of the kernels from $\mathcal{L}_n$. The GFI with the kernel κ is specified by the standard formula

$$(\mathbb{I}_{(\kappa)} f)(t) := \int_0^t \kappa(t-\tau) f(\tau) \, d\tau, \ t > 0, \tag{47}$$

whereas the GFDs of the Riemann–Liouville and Caputo types with the kernel k are defined as follows:

$$(\mathbb{D}_{(k)} f)(t) := \frac{d^n}{dt^n} \int_0^t k(t-\tau) f(\tau) \, d\tau, \ t > 0, \tag{48}$$

$$({}_*\mathbb{D}_{(k)} f)(t) := \left(\mathbb{D}_{(k)} \left(f(\cdot) - \sum_{j=0}^{n-1} f^{(j)}(0) h_{j+1}(\cdot) \right) \right)(t), \ t > 0. \tag{49}$$

Example 3. Evidently, the GFI (47) with the kernel $\kappa(t) = h_\alpha(t)$, $\alpha > 0$ is reduced to the Riemann–Liouville fractional integral (38), and the Riemann–Liouville and Caputo fractional derivatives of the order α, $n-1 < \alpha < n$, $n \in \mathbb{N}$ defined by (36) and (37), respectively, are particular cases of the GFDs (48) and (49) with the kernel $k(t) = h_{n-\alpha}(t)$. As mentioned in Example 1, the inclusion $(h_\alpha, h_{n-\alpha}) \in \mathcal{L}_n$ holds valid if and only if $n-1 < \alpha < n$, $n \in \mathbb{N}$.

It is worth mentioning that the Riemann–Liouville fractional integral (38) and the Riemann–Liouville and Caputo fractional derivatives of an arbitrary order α, $n-1 < \alpha < n$, $n \in \mathbb{N}$ can be introduced based on the Sonine pair $\kappa = h_\beta$, $k = h_{1-\beta}$, $0 < \beta < 1$ and using the construction (45) presented in Theorem 3. Indeed, in this case, we have the relations

$$\kappa_n(t) = (\{1\}^{n-1} * \kappa)(t) = (\{1\}^{n-1} * h_\beta)(t) = h_{n-1+\beta}(t), \ k_n(t) = k(t) = h_{1-\beta}(t). \tag{50}$$

Thus, the GFI (47) and the GFDs (48) and (49) with the kernels $(\kappa_n, k_n) \in \mathcal{L}_n$ take the form

$$(\mathbb{I}_{(\kappa)} f)(t) = (h_{n-1+\beta} * f)(t) = (I_{0+}^{n-1+\beta} f)(t), \ t > 0, \tag{51}$$

$$(\mathbb{D}_{(k)} f)(t) = \frac{d^n}{dt^n} (h_{1-\beta} * f)(t) = \frac{d^n}{dt^n} (I_{0+}^{1-\beta} f)(t), \ t > 0, \tag{52}$$

$$({}_*\mathbb{D}_{(k)} f)(t) = \frac{d^n}{dt^n} \left(I_{0+}^{1-\beta} \left(f(\cdot) - \sum_{j=0}^{n-1} f^{(j)}(0) h_{j+1}(\cdot) \right) \right)(t), \ t > 0. \tag{53}$$

Now we introduce a new variable $\alpha := n - 1 + \beta$. Then, $1 - \beta = n - \alpha$ and the inequalities $n - 1 < \alpha < n$ are fulfilled because of the condition $0 < \beta < 1$. Thus, the operator (51) is the Riemann–Liouville fractional integral (38) of the order α, and the operators (52) and (53) coincide with the Riemann–Liouville and Caputo fractional derivatives of the order α, $n - 1 < \alpha < n$, $n \in \mathbb{N}$.

Example 4. Another interesting and nontrivial particular case of the GFI (47) and the GFDs (48) and (49) is constructed for the pair $(\kappa, k) \in \mathcal{L}_n$ of the kernels defined by Formula (43) with $n - 2 < \nu < n - 1$, $n \in \mathbb{N}$:

$$(\mathbb{I}_{(\kappa)} f)(t) = \int_0^t (t-\tau)^{\nu/2} J_\nu(2\sqrt{t-\tau}) f(\tau) \, d\tau, \ t > 0, \tag{54}$$

$$(\mathbb{D}_{(k)} f)(t) = \frac{d^n}{dt^n} \int_0^t (t-\tau)^{n/2 - \nu/2 - 1} I_{n-\nu-2}(2\sqrt{t-\tau}) f(\tau) \, d\tau, \ t > 0, \tag{55}$$

$$({}_*\mathbb{D}_{(k)} f)(t) := \left(\mathbb{D}_{(k)} \left(f(\cdot) - \sum_{j=0}^{n-1} f^{(j)}(0) h_{j+1}(\cdot) \right) \right)(t), \ t > 0. \tag{56}$$

It is worth mentioning that the Caputo type GFD (49) can be represented in a slightly different form:

$$({}_*\mathbb{D}_{(k)}f)(t) = \left(\mathbb{D}_{(k)}\left(f(\cdot) - \sum_{j=0}^{n-1} f^{(j)}(0)h_{j+1}(\cdot)\right)\right)(t) =$$

$$(\mathbb{D}_{(k)}f)(t) - \sum_{j=0}^{n-1} f^{(j)}(0)(\mathbb{D}_{(k)}h_{j+1})(t) = (\mathbb{D}_{(k)}f)(t) - \sum_{j=0}^{n-1} f^{(j)}(0)\frac{d^n}{dt^n}(k*h_{j+1})(t) =$$

$$(\mathbb{D}_{(k)}f)(t) - \sum_{j=0}^{n-1} f^{(j)}(0)\frac{d^n}{dt^n}(I_{0+}^{j+1}k)(t) = (\mathbb{D}_{(k)}f)(t) - \sum_{j=0}^{n-1} f^{(j)}(0)\frac{d^n}{dt^n}\frac{j}{dt^{n-j-1}}k(t),\ t>0. \quad (57)$$

As regards the basic properties of the GFI (47) of an arbitrary order on $C_{-1}(0,+\infty)$, they follow from the well-known properties of the Laplace convolution (compare these to the properties of the GFI (19) of the order less than one):

$$\mathbb{I}_{(\kappa)} : C_{-1}(0,+\infty) \to C_{-1}(0,+\infty) \text{ (mapping property)}, \quad (58)$$

$$\mathbb{I}_{(\kappa_1)}\mathbb{I}_{(\kappa_2)} = \mathbb{I}_{(\kappa_2)}\mathbb{I}_{(\kappa_1)} \text{ (commutativity law)}, \quad (59)$$

$$\mathbb{I}_{(\kappa_1)}\mathbb{I}_{(\kappa_2)} = \mathbb{I}_{(\kappa_1*\kappa_2)} \text{ (index law)}. \quad (60)$$

To justify this denotation of GFIs and GFDs, in the rest of this section, we formulate and prove the two fundamental theorems of FC for the GFDs (48) and (49) of the Riemann–Liouville and Caputo types.

Theorem 4 (First Fundamental Theorem for the GFD of an Arbitrary Order). *Let (κ, k) be a pair of the kernels from $\mathcal{L}_n$.*
Then, the GFD (48) is a left-inverse operator to the GFI (47) on the space $C_{-1}(0,+\infty)$,

$$(\mathbb{D}_{(k)}\mathbb{I}_{(\kappa)}f)(t) = f(t),\ f \in C_{-1}(0,+\infty),\ t>0, \quad (61)$$

and the GFD (49) is a left-inverse operator to the GFI (47) on the space $C_{-1,(k)}(0,+\infty)$:

$$({}_*\mathbb{D}_{(k)}\mathbb{I}_{(\kappa)}f)(t) = f(t),\ f \in C_{-1,(k)}(0,+\infty),\ t>0, \quad (62)$$

where the space $C_{-1,(k)}(0,+\infty)$ is defined as in Theorem 1.

Proof. We start with a proof of the Formula (61):

$$(\mathbb{D}_{(k)}\mathbb{I}_{(\kappa)}f)(t) = \frac{d^n}{dt^n}(k*(\kappa*f))(t) = \frac{d^n}{dt^n}((k*\kappa)*f)(t) =$$

$$\frac{d^n}{dt^n}(\{1\}^n * f)(t) = \frac{d^n}{dt^n}(I_{0+}^n f)(t) = f(t).$$

A function $f \in C_{-1,(k)}(0,+\infty)$ can be represented in the form $f(t) = (\mathbb{I}_{(\kappa)}\phi)(t)$, $\phi \in C_{-1}(0,+\infty)$, and thus the following chain of equations is valid:

$$(\mathbb{I}_{(\kappa)}f)(t) = (\mathbb{I}_{(\kappa)}\mathbb{I}_{(k)}\phi)(t) = ((\kappa*k)*f)(t) = (\{1\}^n*\phi)(t) = (I_{0+}^n\phi)(t).$$

The last relation implicates the inclusion $\mathbb{I}_{(\kappa)}f \in C_{-1}^n(0,+\infty)$ and the relations

$$\frac{d^j}{dt^j}(\mathbb{I}_{(\kappa)}f)(t)\bigg|_{t=0} = (I_{0+}^{n-j}\phi)(t)\bigg|_{t=0} = 0,\ j=0,\ldots,n-1. \quad (63)$$

To derive Formula (62), we employ the representation (57) of the GFD of the Caputo type, Formula (63) and the relation (61) that we already proved:

$$({}_*\mathbb{D}_{(k)} \, \mathbb{I}_{(\kappa)} f)(t) = (\mathbb{D}_{(k)} \, \mathbb{I}_{(\kappa)} f)(t) - \sum_{j=0}^{n-1} \frac{d^j}{dt^j}(\mathbb{I}_{(\kappa)} f)(t)\Big|_{t=0} \frac{d^{n-j-1}}{dt^{n-j-1}} k(t) = f(t).$$

□

Theorem 5 (Second Fundamental Theorem for the GFD of an Abitrary Order). *Let (κ, k) be a pair of the kernels from $\mathcal{L}_n$.*
Then, the relation

$$(\mathbb{I}_{(\kappa)} \, {}_*\mathbb{D}_{(k)} f)(t) = f(t) - \sum_{j=0}^{n-1} f^{(j)}(0) \, h_{j+1}(t) \tag{64}$$

holds true on the space $C^n_{-1}(0, +\infty)$ and the formula

$$(\mathbb{I}_{(\kappa)} \, \mathbb{D}_{(k)} f)(t) = f(t), \; t > 0 \tag{65}$$

is valid for the functions $f \in C^n_{-1,(\kappa)}(0, +\infty)$.

Proof. As already mentioned in Section 2, any function f from $C^n_{-1}(0, +\infty)$ can be represented as follows (see [16]):

$$f(t) = (I^n_{0+}\phi)(t) + \sum_{j=0}^{n-1} f^{(j)}(0) \, h_{j+1}(t), \; t \geq 0, \; \phi \in C_{-1}(0, +\infty). \tag{66}$$

Then, we employ this representation and Formula (49) and arrive at the following chain of relations:

$$({}_*\mathbb{D}_{(k)} f)(t) = (\mathbb{D}_{(k)}\left(f(\cdot) - \sum_{j=0}^{n-1} f^{(j)}(0) \, h_{j+1}(\cdot)\right))(t) = (\mathbb{D}_{(k)} \, I^n_{0+}\phi)(t) =$$

$$\frac{d^n}{dt^n}(k * \{1\}^n * \phi)(t) = \frac{d^n}{dt^n}(\{1\}^n * (k * \phi))(t) = (k * \phi)(t).$$

Finally, we take into account the representation (66) and obtain Formula (64):

$$(\mathbb{I}_{(\kappa)} \, {}_*\mathbb{D}_{(k)} f)(t) = (\mathbb{I}_{(\kappa)}(k * \phi))(t) = ((\kappa * k) * \phi)(t) =$$

$$(\{1\}^n * \phi)(t) = (I^n_{0+}\phi)(t) = f(t) - \sum_{j=0}^{n-1} f^{(j)}(0) \, h_{j+1}(t).$$

To prove Formula (65), we first mention that a function $f \in C_{-1,(\kappa)}(0, +\infty)$ can be represented in the form $f(t) = (\mathbb{I}_{(\kappa)} \phi)(t)$, $\phi \in C_{-1}(0, +\infty)$, and thus the following chain of equations is valid:

$$(\mathbb{I}_{(\kappa)} \, \mathbb{D}_{(k)} f)(t) = (\mathbb{I}_{(\kappa)} \frac{d^n}{dt^n}(k * f)(t) = (\mathbb{I}_{(\kappa)} \frac{d^n}{dt^n}(k * (\kappa * \phi))(t) =$$

$$(\mathbb{I}_{(\kappa)} \frac{d^n}{dt^n}(\{1\}^n * \phi))(t) = (\mathbb{I}_{(\kappa)} \phi)(t) = f(t).$$

□

In conclusion, we emphasize once again the result of Theorem 3 and its implications on the definitions of the GFIs and the GFDs of an arbitrary order. If (κ, k) is a pair of the

Sonine kernels from $\mathcal{L}_1$, the pair (κ_n, k_n) of the kernels given by the Formula (45) belongs to the set $\mathcal{L}_n$, $n > 1$. The GFI (47) with the kernel $\kappa_n = (\{1\}^{n-1} * \kappa)(t)$ takes the form

$$(\mathbb{I}_{(\kappa_n)} f)(t) = (I_{0+}^{n-1} \mathbb{I}_{(\kappa)} f)(t), \ t > 0, \qquad (67)$$

whereas the GFDs of the Riemann–Liouville and Caputo types with the kernel $k_n = k$ can be represented as follows:

$$(\mathbb{D}_{(k_n)} f)(t) = \frac{d^n}{dt^n} (\mathbb{I}_{(k)} f)(t), \ t > 0, \qquad (68)$$

$$(_*\mathbb{D}_{(k_n)} f)(t) = \frac{d^n}{dt^n} \left(\mathbb{I}_{(k)} \left(f(\cdot) - \sum_{j=0}^{n-1} f^{(j)}(0) h_{j+1}(\cdot) \right) \right)(t), \ t > 0. \qquad (69)$$

As we see, these constructions are completely analogical to the definitions of the Riemann–Liouville fractional integral and the Riemann–Liouville and Caputo fractional derivatives of an arbitrary order.

Another point that is worth mentioning is that the kernel $\kappa_n = (\{1\}^{n-1} * \kappa)(t)$ of the GFI (67) possesses an integrable singularity of the power function type at the origin in the case $n = 1$; i.e., in the case that its order is less than one ($\kappa_1 = \kappa \in C_{-1,0}(0, +\infty)$). If the order of the GFI (67) is greater than one ($n = 2, 3 \ldots$), κ_n is continuous at the origin and $\kappa_n(0) = 0$ as in the case of the Riemann–Liouville fractional integral of the order $\alpha > 1$. Indeed, as mentioned in [16], the inclusion $g * f \in C_{\alpha_1 + \alpha_2 + 1}(0, +\infty)$ holds true for the Laplace convolution of the functions $f \in C_{\alpha_1}(0, +\infty)$, $g \in C_{\alpha_2}(0, +\infty)$, $\alpha_1, \alpha_2 \geq -1$. Thus, the function $\kappa_n = (\{1\}^{n-1} * \kappa)(t)$ with $\kappa \in C_{-1,0}(0, +\infty)$ belongs to the space $C_{n-2}(0, +\infty)$ and thus can be represented in the form $\kappa_n(t) = t^p f(t)$, $p > n - 2 \geq 0$, $f \in C[0, +\infty)$.

4. Conclusions

Starting from the work presented in [7], the so-called GFDs of the Riemann–Liouville and Caputo types have become a topic of active research in FC. In particular, both the ordinary and the partial fractional differential equations with these derivatives have been considered (see [10] for a survey of some recent results). However, the GFDs introduced to date have been based on the classical Sonine condition, and thus their "generalized order" was restricted to the interval $(0, 1)$. In particular, the initial value problems for the fractional differential equations with these derivatives permitted only one initial condition, and thus no models for the intermediate processes between diffusion and wave propagation could be formulated in terms of these GFDs.

The main contribution of this paper is an extension of the definitions of the GFIs and GFDs to the case of arbitrary order. To achieve this aim, a suitable generalization of the Sonine condition was introduced, and some important classes of the kernels that satisfy this generalized condition were described. The kernels of the GFDs of an arbitrary order possess integrable singularities at the point zero. However, the kernels of the GFIs can be both singular (in the case of an order less than one) and continuous (in the case of an order greater or equal to one) at the origin. The conventional Riemann–Liouville and Caputo fractional derivatives of arbitrary order are particular cases of these GFDs. Another important example is the integro-differential operators of the convolution type with the Bessel and the modified Bessel functions in the kernels.

To justify the denotation of GFIs and GFDs of arbitrary order, in this paper, two fundamental theorems of fractional calculus for these operators were formulated and proved. The constructions introduced in this paper allow the formulation of the initial-value problems for the fractional differential equations with GFDs of a generalized order greater than one with several initial conditions. Thus, further research regarding the properties of the GFIs and GFDs of an arbitrary order introduced in this paper as well as applications of the fractional differential equations with the GFDs of arbitrary order to model, for instance, the processes intermediate between diffusion and wave propagation is needed.

Funding: This research received no external funding.

Institutional Review Board Statement: Not applicable.

Informed Consent Statement: Not applicable.

Data Availability Statement: Not applicable.

Conflicts of Interest: The author declares no conflict of interest.

References

1. Abel, N.H. Oplösning af et par opgaver ved hjelp af bestemte integraler. *Mag. Naturvidenskaberne* **1823**, *2*, 2.
2. Abel, N.H. Auflösung einer mechanischen Aufgabe. *J. Die Reine Angew. Math.* **1826**, *1*, 153–157.
3. Podlubny, I.; Magin, R.L.; Trymorush, I. Niels Henrik Abel and the birth of Fractional Calculus. *Fract. Calc. Appl. Anal.* **2017**, *20*, 1068–1075. [CrossRef]
4. Sonine, N. Sur la généralisation d'une formule d'Abel. *Acta Math.* **1884**, *4*, 171–176. [CrossRef]
5. Prüss, J. *Evolutionary Integral Equations and Applications*; Springer: Basel, Switzerland, 1993.
6. Clément, P. On abstract Volterra equations in Banach spaces with completely positive kernels. In *Infinite-Dimensional Systems*; Kappel, F., Schappacher, W., Eds.; Lecture Notes in Math; Springer: Berlin, Germany, 1984; Volume 1076, pp. 32–40.
7. Kochubei, A.N. General fractional calculus, evolution equations, and renewal processes. *Integr. Equ. Oper. Theory* **2011**, *71*, 583–600. [CrossRef]
8. Hanyga, A. A comment on a controversial issue: A Generalized Fractional Derivative cannot have a regular kernel. *Fract. Calc. Appl. Anal.* **2020**, *23*, 211–223. [CrossRef]
9. Luchko, Y. General Fractional Integrals and Derivatives with the Sonine Kernels. *Mathematics* **2021**, *9*, 594. [CrossRef]
10. Luchko, Y.; Yamamoto, M. The General Fractional Derivative and Related Fractional Differential Equations. *Mathematics* **2020**, *8*, 2115. [CrossRef]
11. Schilling, R.L.; Song, R.; Vondracek, Z. *Bernstein Functions. Theory and Application*; De Gruyter: Berlin, Germany, 2010.
12. Luchko, Y.; Yamamoto, M. General time-fractional diffusion equation: Some uniqueness and existence results for the initial-boundary-value problems. *Fract. Calc. Appl. Anal.* **2016**, *19*, 675–695. [CrossRef]
13. Luchko, Y. Operational Calculus for the general fractional derivatives with the Sonine kernels. *Fract. Calc. Appl. Anal.* **2021**, *24*, 338–375.
14. Hilfer, R.; Luchko, Y. Desiderata for Fractional Derivatives and Integrals. *Mathematics* **2019**, *7*, 149. [CrossRef]
15. Dimovski, I.H. Operational calculus for a class of differential operators. *C. R. Acad. Bulg. Sci.* **1966**, *19*, 1111–1114.
16. Luchko, Y.; Gorenflo, R. An operational method for solving fractional differential equations. *Acta Math.* **1999**, *24*, 207–234.
17. Samko, S.G.; Cardoso, R.P. Integral equations of the first kind of Sonine type. *Int. J. Math. Sci.* **2003**, *57*, 3609–3632. [CrossRef]
18. Diethelm, K.; Garrappa, R.; Giusti, A.; Stynes, M. Why fractional derivatives with nonsingular kernels should not be used. *Fract. Calc. Appl. Anal.* **2020**, *23*, 610–634. [CrossRef]
19. Samko, S.G.; Kilbas, A.A.; Marichev, O.I. *Fractional Integrals and Derivatives. Theory and Applications*; Gordon and Breach: New York, NY, USA, 1993.
20. Prudnikov, A.P.; Brychkov, Y.A.; Marichev, O.I. *Integrals and Series: Direct Laplace Transforms, Vol. 4*; Gordon & Breach: New York, NY, USA, 1992.
21. Prudnikov, A.P.; Brychkov, Y.A.; Marichev, O.I. *Integrals and Series: Inverse Laplace Transforms, Vol. 5*; Gordon & Breach: New York, NY, USA, 1992.

Article

The Existence, Uniqueness, and Stability Analysis of the Discrete Fractional Three-Point Boundary Value Problem for the Elastic Beam Equation

Jehad Alzabut [1,2,*,†], A. George Maria Selvam [3,†], R. Dhineshbabu [4,†] and Mohammed K. A. Kaabar [5,6,*,†]

1. Department of Mathematics and General Sciences, Prince Sultan University, Riyadh 11586, Saudi Arabia
2. Group of mathematics, Faculty of Engineering, Ostim Technical University, Ankara 06374, Turkey
3. Department of Mathematics, Sacred Heart College (Autonomous), Tirupattur 635 601, Tamil Nadu, India; agms@shctpt.edu
4. Department of Mathematics, Sri Venkateswara College of Engineering and Technology (Autonomous), Chittoor 517 127, Andhra Pradesh, India; dhineshbabur@svcetedu.org
5. Department of Mathematics and Statistics, Washington State University, Pullman, WA 99163, USA
6. Institute of Mathematical Sciences, Faculty of Science, University of Malaya, Kuala Lumpur 50603, Malaysia
* Correspondence: jalzabut@psu.edu.sa (J.A.); mohammed.kaabar@wsu.edu (M.K.A.K.); Tel.: +966-114948547 (J.A.)
† These authors contributed equally to this work.

Abstract: An elastic beam equation (EBEq) described by a fourth-order fractional difference equation is proposed in this work with three-point boundary conditions involving the Riemann–Liouville fractional difference operator. New sufficient conditions ensuring the solutions' existence and uniqueness of the proposed problem are established. The findings are obtained by employing properties of discrete fractional equations, Banach contraction, and Brouwer fixed-point theorems. Further, we discuss our problem's results concerning $\mathcal{H}yers$–$\mathcal{U}lam$ ($\mathcal{HU}$), generalized $\mathcal{H}yers$–$\mathcal{U}lam$ ($\mathcal{GHU}$), $\mathcal{H}yers$–$\mathcal{U}lam$–$\mathcal{R}assias$ ($\mathcal{HUR}$), and generalized $\mathcal{H}yers$–$\mathcal{U}lam$–$\mathcal{R}assias$ ($\mathcal{GHUR}$) stability. Specific examples with graphs and numerical experiment are presented to demonstrate the effectiveness of our results.

Keywords: Riemann–Liouville fractional difference operator; boundary value problem; discrete fractional calculus; existence and uniqueness; Ulam stability; elastic beam problem

MSC: 34A12; 34B10; 34B15; 39A12; 47H10; 74B20

1. Introduction

Elastic beam (EB) deflections are commonly known phenomena in science and engineering. Based on the significance of their applications such as for aircraft design, chemical sensors, micro-electromechanical systems, material mechanics, medical diagnostics, and physics, two-point boundary value problems (BVPs) for EBEqs have received considerable attention. Recently, many researchers have investigated EBEqs with various boundary conditions (BCs) (refer to [1–6]). Gupta in [6] studied a fourth-order EBEq with two-point BCs:

$$\begin{cases} w^{(4)}(\kappa) = G(\kappa, w(\kappa)), \; \kappa \in (0,1), \\ w(0) = 0, \; w''(0) = 0, \; w'(1) = 0, \; w'''(1) = 0. \end{cases} \quad (1)$$

Equation (1) describes an elastic beam model of length 1, which is clamped with a displacement and a bending moment that are equal to zero at the left end, and this model is free to travel with disappearing angular attitude and shear force at the right end.

In addition, Cianciaruso et al. [1] studied the model of the cantilever beam equation with three-point BCs:

$$\begin{cases} w^{(4)}(\kappa) = G(\kappa, w(\kappa)), \ \kappa \in (0,1), \\ w(0) = w'(0) = w''(1) = 0, \ w'''(1) = h(w(\zeta)), \end{cases}$$

where $\zeta \in (0,1)$ is a real constant. The above is a feedback mechanism model where the shearing force at the beam's right end responds to the displacement experienced at a point ζ.

Fractional calculus (FC) is a generalized form of classical integer-order calculus. Fractional calculus examines the properties of fractional-order derivatives and integrals. Due to its numerous applications in various scientific fields, this research area has gained considerable attention over the past few years. FC can be applicable in several fields of science and engineering, along with aerodynamics, electrical circuits, fluid dynamics, heat conduction, and physics. We refer to the comprehensive works in [7–10] for a detailed analysis of its applications, and we refer to [11–15] for the latest trends in the area of FC.

Researchers have explored various aspects of fractional difference equations (FDEs). Obviously, the solutions' existence, uniqueness, and stability analysis are some important features of FDEs. Various analytical approaches and fixed-point theory have been used to examine the solutions' existence and stability for FDEs. Several researchers have contributed a number of books and papers in this regard [16]. However, finding the exact solution of nonlinear FDEs is often too difficult; therefore, the stability analysis of solutions plays a crucial role in such investigations. Various kinds of stabilities described in the past are discussed in the literature, such as Lyapunov stability [17], Mittag–Leffler stability [18], and exponential stability [19]. Presumably, the most dependable stabilities are called $\mathcal{HU}$ stability. The discussed stability was modified to $\mathcal{GHU}$ stability (refer to [20–22]). In 1970, Rassias further generalized the aforesaid stability. For FDEs with different BCs concerning Riemann–Liouville and Caputo operators, the addressed fields of existence and stability analysis are well-equipped (see [23–28]).

A new interesting research field, named discrete fractional calculus (DFC), is attracting the interest of mathematicians and researchers. With discrete fractional operators, several real-world problems are being investigated [29–32]. The fractional difference equations have recently become an interesting field for scientists because of their applications in biology, ecology, and applied sciences [33]. However, a few research studies that have been conducted on discrete fractional-order BVPs can be found in [34–47].

The above findings inspired us in this study concerning the solutions' existence and uniqueness with various types of Ulam stability results for the proposed discrete fractional elastic beam equation (FEBE) that is subject to the three-point BCs as follows:

$$\begin{cases} \Delta_{\beta-4}^{\beta} w(\kappa) = G(\kappa + \beta - 1, w(\kappa + \beta - 1)), \ \kappa \in \mathbb{N}_0^{n+3}, \\ w(\beta-4) = 0, \ \Delta^2 w(\beta-4) = 0, \ \Delta w(\beta+n) = 0, \ \Delta^3 w(\beta+n) + w(\zeta) = 0, \end{cases} \quad (2)$$

where $\beta \in (3,4]$ is a fractional order and $\zeta \in \mathbb{N}_{\beta-1}^{\beta+n+2}$ is constant. Here, we have that $G : \mathbb{N}_{\beta-4}^{\beta+n+3} \times \mathbb{R} \to \mathbb{R}$ is continuous, $w : \mathbb{N}_{\beta-4}^{\beta+n+3} \to \mathbb{R}$, $\Delta_{\beta-4}^{\beta}$ is the Riemann–Liouville discrete fractional operator, and $n \in \mathbb{N}_0$.

The rest of this research work is structured as follows. Basic background knowledge on DFC is stated in Section 2. The result for a linear version of the BVP Equation (2) is discussed in Section 3. Further, by using this solution, the existence and uniqueness conditions for the proposed discrete FEBE with three-point BCs (Equation (2)) are derived with the help of contraction mapping and the Brouwer fixed-point theorems. Different types of stability results are extensively obtained in Section 4 via the findings of nonlinear analysis. Some illustrative examples with graphs and numerical experiment are presented in Section 5 as applications to provide a better understanding of our findings. Finally, Section 6 concludes our research work.

2. Essential Preliminaries

Some important notions and preliminary lemmas are stated in this section, which are needed for discussion of our results.

Definition 1 ([30]). *For $\beta > 0$, the βth order fractional sum of G can be defined as*

$$\Delta^{-\beta} G(\kappa) = \frac{1}{\Gamma(\beta)} \sum_{i=a}^{\kappa-\beta} (\kappa - \sigma(i))^{(\beta-1)} G(i),$$

for $\kappa \in \mathbb{N}_{a+\beta}$ and $\sigma(i) = i+1$. Define the βth fractional difference for $\beta > 0$ by $\Delta^{\beta} G(\kappa) := \Delta^{M} \Delta^{\beta-M} G(\kappa)$, for $\kappa \in \mathbb{N}_{a+M-\beta}$, $M \in \mathbb{N}$ satisfies $0 \leq M - 1 < \beta \leq M$, and $\kappa^{(\beta)} := \frac{\Gamma(\kappa+1)}{\Gamma(\kappa+1-\beta)}$.

Lemma 1 ([30]). *Assume that κ and β are any numbers such that $\kappa^{(\beta)}$ and $\kappa^{(\beta-1)}$ are defined. Then we have $\Delta \kappa^{(\beta)} = \beta \kappa^{(\beta-1)}$.*

Lemma 2 (see [34,44]). *Let $0 \leq M - 1 < \beta \leq M$. Then,*

$$\Delta^{-\beta} \Delta^{\beta} G(\kappa) = G(\kappa) + C_1 \kappa^{(\beta-1)} + C_2 \kappa^{(\beta-2)} + \ldots + C_M \kappa^{(\beta-M)},$$

for some $C_j \in \mathbb{R}$, $1 \leq j \leq M$.

Lemma 3 (see [42]). *For κ and i, for which both $(\kappa - \sigma(i))^{(\beta)}$ and $(\kappa - 1 - \sigma(i))^{(\beta)}$ are defined, we obtain that $\Delta_i \left[(\kappa - \sigma(i))^{(\beta)} \right] = -\beta (\kappa - 1 - \sigma(i))^{(\beta-1)}$.*

Lemma 4 (see [43,46]). *Let $\beta, \nu > 0$. Then,*

$$\Delta^{-\beta} \kappa^{(\nu)} = \frac{\Gamma(\nu+1)}{\Gamma(\nu+\beta+1)} \kappa^{(\nu+\beta)} \text{ and } \Delta^{\beta} \kappa^{(\nu)} = \frac{\Gamma(\nu+1)}{\Gamma(\nu-\beta+1)} \kappa^{(\nu-\beta)}.$$

3. EB Existence and Uniqueness

The existence and uniqueness of EB is established in this section to the three-point BCs for the proposed discrete FEBE Equation (2). We now introduce the following theorem that deals with a linear variant solution of our proposed BVP Equation (2).

Theorem 1. *Let $H : \mathbb{N}_{\beta-4}^{\beta+n+3} \to \mathbb{R}$ be given. Then, the linear discrete FEBE with three-point BCs:*

$$\begin{cases} \Delta_{\beta-4}^{\beta} w(\kappa) = H(\kappa + \beta - 1), \kappa \in \mathbb{N}_0^{n+3}, \\ w(\beta - 4) = 0, \Delta^2 w(\beta - 4) = 0, \Delta w(\beta + n) = 0, \Delta^3 w(\beta + n) + w(\zeta) = 0, \end{cases} \quad (3)$$

has the unique solution, for $\kappa \in \mathbb{N}_{\beta-4}^{\beta+n+3}$,

$$w(\kappa) = \frac{1}{\Gamma(\beta)} \sum_{i=0}^{\kappa-\beta} (\kappa - \sigma(i))^{(\beta-1)} H(i + \beta - 1)$$

$$+ E_1(\kappa) \left[\frac{1}{\Gamma(\beta)} \sum_{i=0}^{\zeta-\beta} (\zeta - \sigma(i))^{(\beta-1)} + \frac{1}{\Gamma(\beta-3)} \sum_{i=0}^{n+3} (\beta + n - \sigma(i))^{(\beta-4)} \right] H(i + \beta - 1) \quad (4)$$

$$+ \frac{E_2(\kappa)}{\Gamma(\beta-1)} \sum_{i=0}^{n+1} (\beta + n - \sigma(i))^{(\beta-2)} H(i + \beta - 1),$$

where

$$E_1(\kappa) = \frac{\left[\frac{\kappa^{(\beta-1)}}{e_1}h_1 + \kappa^{(\beta-2)}f_1f_4 - \kappa^{(\beta-3)}f_1\right]}{K}; \quad E_2(\kappa) = \frac{\left[\kappa^{(\beta-1)}h_2 - \kappa^{(\beta-2)}e_1f_4 + \kappa^{(\beta-3)}e_1\right]}{K} \quad (5)$$

such that $h_1 = f_1(e_3 - e_2f_4) - K$, $h_2 = e_2f_4 - e_3$, $K = [e_3f_1 - e_1f_3] - f_4[e_2f_1 - e_1f_2]$,
$e_1 = (\beta-1)^{(3)}(\beta+n)^{(\beta-4)} + \zeta^{(\beta-1)}$, $e_2 = (\beta-2)^{(3)}(\beta+n)^{(\beta-5)} + \zeta^{(\beta-2)}$,
$e_3 = (\beta-3)^{(3)}(\beta+n)^{(\beta-6)} + \zeta^{(\beta-3)}$, $f_1 = (\beta-1)(\beta+n)^{(\beta-2)}$,
$f_2 = (\beta-2)(\beta+n)^{(\beta-3)}$, $f_3 = (\beta-3)(\beta+n)^{(\beta-4)}$ and $f_4 = \frac{(\beta-4)}{(\beta-2)}$.

Proof. By applying the fractional sum $\Delta^{-\beta}$ of order $\beta \in (3,4]$ along with Lemma 2 to Equation (3), we have

$$w(\kappa) = \frac{1}{\Gamma(\beta)} \sum_{i=0}^{\kappa-\beta} (\kappa - \sigma(i))^{(\beta-1)} H(i+\beta-1) + C_1\kappa^{(\beta-1)} + C_2\kappa^{(\beta-2)} + C_3\kappa^{(\beta-3)} + C_4\kappa^{(\beta-4)}, \quad (6)$$

for $\kappa \in \mathbb{N}_{\beta-4}^{\beta+n+3}$ and some constants $C_j \in \mathbb{R}$, where $j = 1,2,3,4$. By applying the first BC $w(\beta - 4) = 0$ in Equation (6), we obtain

$$w(\beta-4) = C_1(\beta-4)^{(\beta-1)} + C_2(\beta-4)^{(\beta-2)} + C_3(\beta-4)^{(\beta-3)} + C_4(\beta-4)^{(\beta-4)} = 0. \quad (7)$$

By using Definition 1, we obtain

$$(\beta-4)^{(\beta-1)} = (\beta-4)^{(\beta-2)} = (\beta-4)^{(\beta-3)} = 0 \text{ and } (\beta-4)^{(\beta-4)} = \Gamma(\beta-3). \quad (8)$$

Equations (7) and (8) imply $C_4 = 0$. Using C_4 in Equation (6) provides

$$w(\kappa) = \frac{1}{\Gamma(\beta)} \sum_{i=0}^{\kappa-\beta} (\kappa - \sigma(i))^{(\beta-1)} H(i+\beta-1) + C_1\kappa^{(\beta-1)} + C_2\kappa^{(\beta-2)} + C_3\kappa^{(\beta-3)}. \quad (9)$$

Using Lemma 1 and taking the operator Δ on both sides of Equation (9), we obtain

$$\Delta w(\kappa) = \frac{1}{\Gamma(\beta-1)} \sum_{i=0}^{\kappa-\beta+1} (\kappa - \sigma(i))^{(\beta-2)} H(i+\beta-1)$$
$$+ C_1(\beta-1)\kappa^{(\beta-2)} + C_2(\beta-2)\kappa^{(\beta-3)} + C_3(\beta-3)\kappa^{(\beta-4)}. \quad (10)$$

From the third BC $\Delta w(\beta + n) = 0$ in Equation (10), we obtain

$$\frac{1}{\Gamma(\beta-1)} \sum_{i=0}^{n+1} (\beta+n-\sigma(i))^{(\beta-2)} H(i+\beta-1) + C_1 f_1 + C_2 f_2 + C_3 f_3 = 0. \quad (11)$$

The operator Δ is applied on both sides of Equation (10) with the aid of Lemma 1, and we obtain

$$\Delta^2 w(\kappa) = \frac{1}{\Gamma(\beta-2)} \sum_{i=0}^{\kappa-\beta+2} (\kappa - \sigma(i))^{(\beta-3)} H(i+\beta-1) + C_1(\beta-1)^{(2)}\kappa^{(\beta-3)}$$
$$+ C_2(\beta-2)^{(2)}\kappa^{(\beta-4)} + C_3(\beta-3)^{(2)}\kappa^{(\beta-5)}. \quad (12)$$

The second BC of Equation (3) implies

$$C_2(\beta-2) + C_3(\beta-4) = 0. \quad (13)$$

Again, using Lemma 1 and taking the operator Δ on both sides of Equation (12), we obtain

$$\Delta^3 w(\kappa) = \frac{1}{\Gamma(\beta-3)} \sum_{i=0}^{\kappa-\beta+3} (\kappa - \sigma(i))^{(\beta-4)} H(i+\beta-1) + C_1(\beta-1)^{(3)} \kappa^{(\beta-4)}$$
$$+ C_2(\beta-2)^{(3)} \kappa^{(\beta-5)} + C_3(\beta-3)^{(3)} \kappa^{(\beta-6)}. \qquad (14)$$

Using the last BC $\Delta^3 w(\beta + n) + w(\zeta) = 0$ in Equations (9) and (14) yields

$$w(\zeta) = \frac{1}{\Gamma(\beta)} \sum_{i=0}^{\zeta-\beta} (\zeta - \sigma(i))^{(\beta-1)} H(i+\beta-1) + C_1 \zeta^{(\beta-1)} + C_2 \zeta^{(\beta-2)} + C_3 \zeta^{(\beta-3)} \qquad (15)$$

and

$$\Delta^3 w(\beta + n) = \frac{1}{\Gamma(\beta-3)} \sum_{i=0}^{n+3} (\beta + n - \sigma(i))^{(\beta-4)} H(i+\beta-1) + C_1(\beta-1)^{(3)}(\beta+n)^{(\beta-4)}$$
$$+ C_2(\beta-2)^{(3)}(\beta+n)^{(\beta-5)} + C_3(\beta-3)^{(3)}(\beta+n)^{(\beta-6)}. \qquad (16)$$

From Equations (15) and (16), and by employing the last BC Equation (3), we obtain

$$\frac{1}{\Gamma(\beta-3)} \sum_{i=0}^{n+3} (\beta + n - \sigma(i))^{(\beta-4)} H(i+\beta-1)$$
$$+ \frac{1}{\Gamma(\beta)} \sum_{i=0}^{\zeta-\beta} (\zeta - \sigma(i))^{(\beta-1)} H(i+\beta-1) + C_1 e_1 + C_2 e_2 + C_3 e_3 = 0. \qquad (17)$$

Solving Equations (11) and (17), we obtain

$$f_1 \left(\frac{1}{\Gamma(\beta-3)} \sum_{i=0}^{n+3} (\beta+n-\sigma(i))^{(\beta-4)} + \frac{1}{\Gamma(\beta)} \sum_{i=0}^{\zeta-\beta} (\zeta-\sigma(i))^{(\beta-1)} \right) H(i+\beta-1)$$
$$+ C_2(e_2 f_1 - e_1 f_2) + C_3(e_3 f_1 - e_1 f_3) - \frac{e_1}{\Gamma(\beta-1)} \sum_{i=0}^{n+1} (\beta + n - \sigma(i))^{(\beta-2)} H(i+\beta-1) = 0. \qquad (18)$$

Now, a constant C_3 is found by solving Equations (13) and (18) as follows:

$$C_3 = \frac{1}{K} \left[\frac{e_1}{\Gamma(\beta-1)} \sum_{i=0}^{n+1} (\beta+n-\sigma(i))^{(\beta-2)} H(i+\beta-1) \right.$$
$$\left. - f_1 \left(\frac{1}{\Gamma(\beta)} \sum_{i=0}^{\zeta-\beta} (\zeta-\sigma(i))^{(\beta-1)} + \frac{1}{\Gamma(\beta-3)} \sum_{i=0}^{n+3} (\beta+n-\sigma(i))^{(\beta-4)} \right) H(i+\beta-1) \right].$$

Substituting C_3 into Equation (13), we have

$$C_2 = \frac{f_4}{K} \left[f_1 \left(\frac{1}{\Gamma(\beta)} \sum_{i=0}^{\zeta-\beta} (\zeta-\sigma(i))^{(\beta-1)} + \frac{1}{\Gamma(\beta-3)} \sum_{i=0}^{n+3} (\beta+n-\sigma(i))^{(\beta-4)} \right) H(i+\beta-1) \right.$$
$$\left. - \frac{e_1}{\Gamma(\beta-1)} \sum_{i=0}^{n+1} (\beta+n-\sigma(i))^{(\beta-2)} H(i+\beta-1) \right].$$

By using the value of C_2 and C_3 in Equation (17), we arrive at

$$C_1 = \frac{1}{e_1 K} \left\{ \frac{e_1 h_2}{\Gamma(\beta - 1)} \sum_{i=0}^{n+1} (\beta + n - \sigma(i))^{(\beta-2)} H(i + \beta - 1) \right.$$

$$+ h_1 \left(\frac{1}{\Gamma(\beta)} \sum_{i=0}^{\zeta-\beta} (\zeta - \sigma(i))^{(\beta-1)} + \frac{1}{\Gamma(\beta-3)} \sum_{i=0}^{n+3} (\beta + n - \sigma(i))^{(\beta-4)} \right) H(i + \beta - 1) \bigg\}.$$

By using the constants C_j for $j = 1, 2, 3$ in Equation (9), we obtain $w(\kappa)$ in the form

$$w(\kappa) = \frac{1}{\Gamma(\beta)} \sum_{i=0}^{\kappa-\beta} (\kappa - \sigma(i))^{(\beta-1)} H(i + \beta - 1)$$

$$+ \mathbb{E}_1(\kappa) \left[\frac{1}{\Gamma(\beta)} \sum_{i=0}^{\zeta-\beta} (\zeta - \sigma(i))^{(\beta-1)} + \frac{1}{\Gamma(\beta-3)} \sum_{i=0}^{n+3} (\beta + n - \sigma(i))^{(\beta-4)} \right] H(i + \beta - 1)$$

$$+ \frac{\mathbb{E}_2(\kappa)}{\Gamma(\beta - 1)} \sum_{i=0}^{n+1} (\beta + n - \sigma(i))^{(\beta-2)} H(i + \beta - 1),$$

for $\kappa \in \mathbb{N}_{\beta-4}^{\beta+n+3}$. Therefore, the theorem's proof is complete. □

Assume that $\mathbb{B}_* : \mathbb{C}\left(\mathbb{N}_{\beta-4}^{\beta+n+3}, \mathbb{R} \right)$ is a Banach space with a norm defined by

$$\|w\| = \max\left\{ |w(\kappa)| : \kappa \in \mathbb{N}_{\beta-4}^{\beta+n+3} \right\}.$$

To discuss the theorems' existence and uniqueness, we need the following assumptions:

(A_1) There exists a constant $\mathbb{L}_G > 0$, which satisfies $|G(\kappa, w) - G(\kappa, \hat{w})| \leq \mathbb{L}_G |w - \hat{w}|$ for all $w, \hat{w} \in \mathbb{B}_*$ and each $\kappa \in \mathbb{N}_{\beta-4}^{\beta+n+3}$.

(A_2) There exists a bounded function $\chi : \mathbb{N}_{\beta-4}^{\beta+n+3} \to \mathbb{R}$ with $|G(\kappa, w)| \leq \chi(\kappa)|w|$ for all $w \in \mathbb{B}_*$.

Theorem 2. *In view of assumption (A_1), the discrete FEBE with the three-point BCs in Equation (2) has a unique solution if*

$$\Lambda := \left[\frac{(\beta + n + 3)^{(\beta)}}{\Gamma(\beta + 1)} + \mathbb{E}_1^* \left(\frac{\zeta^{(\beta)}}{\Gamma(\beta + 1)} + \frac{(\beta + n)^{(\beta-3)}}{\Gamma(\beta - 2)} \right) + \mathbb{E}_2^* \frac{(\beta + n)^{(\beta-1)}}{\Gamma(\beta)} \right] \mathbb{L}_G < 1, \tag{19}$$

where

$$\mathbb{E}_1^* = \left| \frac{1}{K} \left[\frac{(\beta + n + 3)^{(\beta-1)}}{e_1} h_1 + (\beta + n + 3)^{(\beta-2)} f_1 f_4 - (\beta + n + 3)^{(\beta-3)} f_1 \right] \right|,$$

$$\mathbb{E}_2^* = \left| \frac{1}{K} \left[(\beta + n + 3)^{(\beta-1)} h_2 - (\beta + n + 3)^{(\beta-2)} e_1 f_4 + (\beta + n + 3)^{(\beta-3)} e_1 \right] \right|, \tag{20}$$

such that K is defined in Theorem 1.

Proof. Let the operator $\mathcal{A} : \mathbb{B}_* \to \mathbb{B}_*$ be defined as

$$(\mathcal{A}w)(\kappa) = \frac{1}{\Gamma(\beta)} \sum_{i=0}^{\kappa-\beta} (\kappa - \sigma(i))^{(\beta-1)} g_w(\kappa)$$

$$+ \mathbb{E}_1(\kappa) \left[\frac{1}{\Gamma(\beta)} \sum_{i=0}^{\zeta-\beta} (\zeta - \sigma(i))^{(\beta-1)} + \frac{1}{\Gamma(\beta-3)} \sum_{i=0}^{n+3} (\beta + n - \sigma(i))^{(\beta-4)} \right] g_w(\kappa) \tag{21}$$

$$+ \frac{\mathbb{E}_2(\kappa)}{\Gamma(\beta - 1)} \sum_{i=0}^{n+1} (\beta + n - \sigma(i))^{(\beta-2)} g_w(\kappa),$$

where $g_w(\kappa) = G(\kappa + \beta - 1, w(\kappa + \beta - 1))$. Obviously, the fixed point of $\mathcal{A}$ is a solution to Equation (2). To show that $\mathcal{A}$ is a contraction, let $w, \hat{w} \in \mathbb{B}_*$ and for each $\kappa \in \mathbb{N}_{\beta-4}^{\beta+n+3}$, one has

$$|(\mathcal{A}w)(\kappa) - (\mathcal{A}\hat{w})(\kappa)| \leq \frac{1}{\Gamma(\beta)} \sum_{i=0}^{\kappa-\beta} (\kappa - \sigma(i))^{(\beta-1)} |g_w(i) - g_{\hat{w}}(i)|$$

$$+ |\mathbb{E}_1(\kappa)| \left[\frac{1}{\Gamma(\beta)} \sum_{i=0}^{\zeta-\beta} (\zeta - \sigma(i))^{(\beta-1)} + \right.$$

$$\left. \frac{1}{\Gamma(\beta-3)} \sum_{i=0}^{n+3} (\beta + n - \sigma(i))^{(\beta-4)} \right] |g_w(i) - g_{\hat{w}}(i)|$$

$$+ \frac{|\mathbb{E}_2(\kappa)|}{\Gamma(\beta-1)} \sum_{i=0}^{n+1} (\beta + n - \sigma(i))^{(\beta-2)} |g_w(i) - g_{\hat{w}}(i)|,$$

where $g_w(\kappa), g_{\hat{w}}(\kappa) \in \mathbb{C}\left(\mathbb{N}_{\beta-4}^{\beta+n+3}, \mathbb{R}\right)$ satisfies the following functional equations:

$$g_w(\kappa) = G(\kappa + \beta - 1, w(\kappa + \beta - 1)) \text{ and } g_{\hat{w}}(\kappa) = G(\kappa + \beta - 1, \hat{w}(\kappa + \beta - 1)). \quad (22)$$

By (A_1), we have

$$|g_w(\kappa) - g_{\hat{w}}(\kappa)| = |G(\kappa + \beta - 1, w(\kappa + \beta - 1)) - G(\kappa + \beta - 1, \hat{w}(\kappa + \beta - 1))|$$
$$\leq \mathbb{L}_G |w(\kappa + \beta - 1) - \hat{w}(\kappa + \beta - 1)|$$
$$|g_w(\kappa) - g_{\hat{w}}(\kappa)| \leq \mathbb{L}_G \|w - \hat{w}\|. \quad (23)$$

From which we obtain

$$\|\mathcal{A}w - \mathcal{A}\hat{w}\| \leq \frac{\mathbb{L}_G \|w - \hat{w}\|}{\Gamma(\beta)} \sum_{i=0}^{\kappa-\beta} (\kappa - \sigma(i))^{(\beta-1)}$$

$$+ |\mathbb{E}_1(\kappa)| \mathbb{L}_G \|w - \hat{w}\| \left[\frac{1}{\Gamma(\beta)} \sum_{i=0}^{\zeta-\beta} (\zeta - \sigma(i))^{(\beta-1)} + \frac{1}{\Gamma(\beta-3)} \sum_{i=0}^{n+3} (\beta + n - \sigma(i))^{(\beta-4)} \right] \quad (24)$$

$$+ \frac{|\mathbb{E}_2(\kappa)| \mathbb{L}_G \|w - \hat{w}\|}{\Gamma(\beta-1)} \sum_{i=0}^{n+1} (\beta + n - \sigma(i))^{(\beta-2)}.$$

By the application of Lemma 3, we have

$$\frac{1}{\Gamma(\beta)} \sum_{i=0}^{\kappa-\beta} (\kappa - \sigma(i))^{(\beta-1)} = \frac{1}{\Gamma(\beta)} \left[\frac{(\kappa - i)^{(\beta)}}{-\beta} \right]_{i=0}^{\kappa-\beta+1} = \frac{\kappa^{(\beta)}}{\Gamma(\beta+1)} \leq \frac{(\beta + n + 3)^{(\beta)}}{\Gamma(\beta+1)} \quad (25)$$

and

$$\frac{1}{\Gamma(\beta)} \sum_{i=0}^{\zeta-\beta} (\zeta - \sigma(i))^{(\beta-1)} = \frac{1}{\Gamma(\beta)} \left[\frac{(\zeta - i)^{(\beta)}}{-\beta} \right]_{i=0}^{\zeta-\beta+1} = \frac{\zeta^{(\beta)}}{\Gamma(\beta+1)}. \quad (26)$$

Similarly, by using Lemma 3, we also obtain

$$\frac{1}{\Gamma(\beta-1)} \sum_{i=0}^{n+1} (\beta + n - \sigma(i))^{(\beta-2)} = \frac{1}{\Gamma(\beta-1)} \left[\frac{(\beta + n - i)^{(\beta-1)}}{-(\beta-1)} \right]_{i=0}^{n+2} = \frac{(\beta + n)^{(\beta-1)}}{\Gamma(\beta)} \quad (27)$$

and

$$\frac{1}{\Gamma(\beta-3)} \sum_{i=0}^{n+3} (\beta + n - \sigma(i))^{(\beta-4)} = \frac{1}{\Gamma(\beta-3)} \left[\frac{(\beta + n - i)^{(\beta-3)}}{-(\beta-3)} \right]_{i=0}^{n+4} = \frac{(\beta + n)^{(\beta-3)}}{\Gamma(\beta-2)}. \quad (28)$$

By substituting the relations Equations (25)–(28) into Equation (24), we obtain

$$\|\mathcal{A}w - \mathcal{A}\hat{w}\| \leq \left[\frac{(\beta+n+3)^{(\beta)}}{\Gamma(\beta+1)} + \mathbb{E}_1^*\left(\frac{\zeta^{(\beta)}}{\Gamma(\beta+1)} + \frac{(\beta+n)^{(\beta-3)}}{\Gamma(\beta-2)}\right) + \mathbb{E}_2^*\frac{(\beta+n)^{(\beta-1)}}{\Gamma(\beta)}\right] L_G \|w - \hat{w}\|.$$

By Equation (19), we obtain $\|\mathcal{A}w - \mathcal{A}\hat{w}\| < \|w - \hat{w}\|$. Hence, $\mathcal{A}$ is a contraction. As a result, according to the Banach fixed-point theorem, the three-point BCs for the discrete FEBE Equation (2) has a unique solution. □

Theorem 3. *If the assumption (A_2) holds, then the discrete FEBE with three-point BCs in Equation (2) has at least one solution provided that*

$$\chi^* \leq \frac{\Gamma(\beta+1)}{[(\beta+n+3)^{(\beta)} + \mathbb{E}_1^*(\zeta^{(\beta)} + \beta^{(3)}(\beta+n)^{(\beta-3)}) + \mathbb{E}_2^*\beta(\beta+n)^{(\beta-1)}]}, \quad (29)$$

where $\chi^* = \max\{\chi(\kappa) : \kappa \in \mathbb{N}_{\beta-4}^{\beta+n+3}\}$.

Proof. Assume that $D > 0$ and consider the set $V = \{w \in \mathbb{B}_* : \|w\| \leq D\}$. For proving this theorem, let us claim that $\mathcal{A}$ maps V in V. Now, for any $w \in V$, one has

$$|(\mathcal{A}w)(\kappa)| \leq \frac{1}{\Gamma(\beta)}\sum_{i=0}^{\kappa-\beta}(\kappa-\sigma(i))^{(\beta-1)}|g_w(i)|$$

$$+ |\mathbb{E}_1(\kappa)|\left[\frac{1}{\Gamma(\beta)}\sum_{i=0}^{\zeta-\beta}(\zeta-\sigma(i))^{(\beta-1)} + \frac{1}{\Gamma(\beta-3)}\sum_{i=0}^{n+3}(\beta+n-\sigma(i))^{(\beta-4)}\right]|g_w(i)|$$

$$+ \frac{|\mathbb{E}_2(\kappa)|}{\Gamma(\beta-1)}\sum_{i=0}^{n+1}(\beta+n-\sigma(i))^{(\beta-2)}|g_w(i)|,$$

where $g_w(\kappa)$ is given in Equation (22). Using (A_2), we obtain

$$|g_w(\kappa)| = |G(\kappa+\beta-1, w(\kappa+\beta-1))| \leq \chi(\kappa)|w(\kappa+\beta-1)| \leq \chi^*\|w\|.$$

This further implies that

$$\|\mathcal{A}w\| \leq \frac{\chi^*\|w\|}{\Gamma(\beta)}\sum_{i=0}^{\kappa-\beta}(\kappa-\sigma(i))^{(\beta-1)}$$

$$+ |\mathbb{E}_1(\kappa)|\chi^*\|w\|\left[\frac{1}{\Gamma(\beta)}\sum_{i=0}^{\zeta-\beta}(\zeta-\sigma(i))^{(\beta-1)} + \frac{1}{\Gamma(\beta-3)}\sum_{i=0}^{n+3}(\beta+n-\sigma(i))^{(\beta-4)}\right] \quad (30)$$

$$+ \frac{|\mathbb{E}_2(\kappa)|\chi^*\|w\|}{\Gamma(\beta-1)}\sum_{i=0}^{n+1}(\beta+n-\sigma(i))^{(\beta-2)}.$$

Using the relations of Equations (25)–(28) in Equation (30), we obtain

$$\|\mathcal{A}w\| \leq \left[\frac{(\beta+n+3)^{(\beta)} + \mathbb{E}_1^*\left(\zeta^{(\beta)} + \beta^{(3)}(\beta+n)^{(\beta-3)}\right) + \mathbb{E}_2^*\beta(\beta+n)^{(\beta-1)}}{\Gamma(\beta+1)}\right]\chi^*D.$$

By Equation (29), we have $\|\mathcal{A}w\| \leq D$, which implies that $\mathcal{A} : V \to V$. By using the Brouwer fixed-point theorem, let us conclude that three-point BCs for discrete FEBE Equation (2) has at least one solution. □

4. EB Stability Analysis

The Ulam-type stability for the proposed problem Equation (2) is studied in this section. Now, we present some definitions of Ulam stability, and we also assume that $g_{\hat{w}}(\kappa)$: $\mathbb{C}\left(\mathbb{N}_{\beta-4}^{\beta+n+3}, \mathbb{R}\right)$ is a continuous function that satisfies $g_{\hat{w}}(\kappa) = G(\kappa + \beta - 1, \hat{w}(\kappa + \beta - 1))$.

Definition 2 ([46]). *If for every function $\hat{w} \in \mathbb{B}_*$ of*

$$\left|\Delta_{\beta-4}^{\beta}\hat{w}(\kappa) - g_{\hat{w}}(\kappa)\right| \leq \epsilon, \quad \kappa \in \mathbb{N}_0^{n+3}, \tag{31}$$

where $\epsilon > 0$, there exists solution $w \in \mathbb{B}_*$ of Equation (2) and positive number $\delta_1 > 0$ such that

$$|\hat{w}(\kappa) - w(\kappa)| \leq \delta_1 \epsilon, \quad \kappa \in \mathbb{N}_{\beta-4}^{\beta+n+3}. \tag{32}$$

Then, the discrete FEBE Equation (2) is $\mathcal{HU}$ stable. It will be $\mathcal{GHU}$ stable if we keep $\Phi(\epsilon) = \delta_1 \epsilon$ in inequality Equation (32), where $\Phi(\epsilon) \in \mathbb{C}(\mathbb{R}^+, \mathbb{R}^+)$ and $\Phi(0) = 0$.

Definition 3 ([46]). *If for every function $\hat{w} \in \mathbb{B}_*$ of*

$$\left|\Delta_{\beta-4}^{\beta}\hat{w}(\kappa) - g_{\hat{w}}(\kappa)\right| \leq \epsilon \phi(\kappa + \beta - 1), \quad \kappa \in \mathbb{N}_0^{n+3}, \tag{33}$$

where $\epsilon > 0$, there are solutions $w \in \mathbb{B}_*$ of Equation (2) and positive number $\delta_2 > 0$ such that

$$|\hat{w}(\kappa) - w(\kappa)| \leq \delta_2 \epsilon \phi(\kappa + \beta - 1), \quad \kappa \in \mathbb{N}_{\beta-4}^{\beta+n+3}. \tag{34}$$

Then, the discrete FEBE Equation (2) is $\mathcal{HUR}$ stable. It will be $\mathcal{GHUR}$ stable if $\phi(\kappa + \beta - 1) = \epsilon \phi(\kappa + \beta - 1)$ in inequality Equations (33) and (34).

Remark 1 ([46]). *A function $\hat{w} \in \mathbb{B}_*$ is a solution to Equation (31) iff there exists $\Psi : \mathbb{N}_{\beta-4}^{\beta+n+3} \to \mathbb{R}$ that satisifies, for $\kappa \in \mathbb{N}_0^{n+3}$, the following:*

(A_3) $|\Psi(\kappa + \beta - 1)| \leq \epsilon$,
(A_4) $\Delta_{\beta-4}^{\beta}\hat{w}(\kappa) = g_{\hat{w}}(\kappa) + \Psi(\kappa + \beta - 1)$.

Similarly, a remark can be constructed for inequality Equation (33).

Lemma 5. *According to Remark 1, a function $\hat{w} \in \mathbb{B}_*$ that corresponds to the discrete FEBE with three-point BCs is expressed as:*

$$\begin{cases} \Delta_{\beta-4}^{\beta}\hat{w}(\kappa) = g_{\hat{w}}(\kappa) + \Psi(\kappa + \beta - 1), \quad \kappa \in \mathbb{N}_0^{n+3}, \\ w(\beta - 4) = 0, \Delta^2 w(\beta - 4) = 0, \Delta w(\beta + n) = 0, \Delta^3 w(\beta + n) + w(\zeta) = 0, \end{cases} \tag{35}$$

satisfying the following inequality:

$$|\hat{w}(\kappa) - (\mathcal{A}\hat{w})(\kappa)| \leq \frac{\epsilon}{\Gamma(\beta+1)}(\beta + n + 3)^{(\beta)},$$

where $(\mathcal{A}\hat{w})(\kappa)$ is defined in Equation (21).

Proof. By using Theorem 1, the corresponding BVP Equation (35) becomes

$$\hat{w}(\kappa) = \frac{1}{\Gamma(\beta)} \sum_{i=0}^{\kappa-\beta} (\kappa - \sigma(i))^{(\beta-1)} g_{\hat{w}}(i)$$

$$+ \mathbb{E}_1(\kappa) \left[\frac{1}{\Gamma(\beta)} \sum_{i=0}^{\zeta-\beta} (\zeta - \sigma(i))^{(\beta-1)} + \frac{1}{\Gamma(\beta-3)} \sum_{i=0}^{n+3} (\beta + n - \sigma(i))^{(\beta-4)} \right] g_{\hat{w}}(i)$$

$$+ \frac{\mathbb{E}_2(\kappa)}{\Gamma(\beta-1)} \sum_{i=0}^{n+1} (\beta + n - \sigma(i))^{(\beta-2)} g_{\hat{w}}(i)$$

$$+ \frac{1}{\Gamma(\beta)} \sum_{i=0}^{\kappa-\beta} (\kappa - \sigma(i))^{(\beta-1)} \Psi(i + \beta - 1).$$

Using an operator $\mathcal{A}$ and taking the modulus on both sides of the above solution along with (A_3), we obtain

$$|\hat{w}(\kappa) - (\mathcal{A}\hat{w})(\kappa)| \leq \frac{\epsilon}{\Gamma(\beta+1)} (\beta + n + 3)^{(\beta)}.$$

□

Theorem 4. *Under the assumption (A_1) with the inequality Equation (19), the discrete FEBE Equation (2) is $\mathcal{HU}$ stable.*

Proof. If $\hat{w}(\kappa)$ is any solution of the inequality Equation (31), and $w(\kappa)$ is a unique solution to Equation (2), then

$$\begin{aligned}
|\hat{w}(\kappa) - w(\kappa)| &= |\hat{w}(\kappa) - (\mathcal{A}w)(\kappa)| \\
&= |\hat{w}(\kappa) - (\mathcal{A}\hat{w})(\kappa) + (\mathcal{A}\hat{w})(\kappa) - (\mathcal{A}w)(\kappa)| \\
&\leq |\hat{w}(\kappa) - (\mathcal{A}\hat{w})(\kappa)| + |(\mathcal{A}\hat{w})(\kappa) - (\mathcal{A}w)(\kappa)|.
\end{aligned} \quad (36)$$

By using Lemma 5 in Equation (36), we have

$$|\hat{w}(\kappa) - w(\kappa)| \leq \frac{\epsilon}{\Gamma(\beta+1)} (\beta + n + 3)^{(\beta)}$$

$$+ \left[\frac{(\beta + n + 3)^{(\beta)}}{\Gamma(\beta+1)} + \mathbb{E}_1^* \left(\frac{\zeta^{(\beta)}}{\Gamma(\beta+1)} + \frac{(\beta+n)^{(\beta-3)}}{\Gamma(\beta-2)} \right) + \mathbb{E}_2^* \frac{(\beta+n)^{(\beta-1)}}{\Gamma(\beta)} \right] \mathbb{L}_G \|\hat{w} - w\|.$$

This further implies that

$$\|\hat{w} - w\| \leq \delta_1 \epsilon,$$

where

$$\delta_1 = \frac{(\beta + n + 3)^{(\beta)}}{\Gamma(\beta+1) - \mathbb{L}_G \left[(\beta + n + 3)^{(\beta)} + \mathbb{E}_1^* (\zeta^{(\beta)} + \beta^{(3)} (\beta+n)^{(\beta-3)}) + \mathbb{E}_2^* \beta (\beta+n)^{(\beta-1)} \right]}.$$

Hence, the solution of Equation (2) is $\mathcal{HU}$ stable. □

Remark 2. *If $\Phi(\epsilon) = \delta_1 \epsilon$ such that $\Phi(0) = 0$, then we have*

$$\|\hat{w} - w\| \leq \Phi(\epsilon).$$

Hence, the solution of Equation (2) is $\mathcal{GHU}$ stable.

For our next result, the following hypotheses hold:

(A5) For an increasing function $\phi \in \mathbb{N}_{\beta-4}^{\beta+n+3} \to \mathbb{R}^+$, there exists $\lambda > 0$ such that, for $\kappa \in \mathbb{N}_0^{n+3}$

(i) $\dfrac{\epsilon}{\Gamma(\beta)} \sum\limits_{i=0}^{\kappa-\beta} (\kappa - \sigma(i))^{(\beta-1)} \phi(i + \beta - 1) \leq \lambda \epsilon \phi(\kappa + \beta - 1),$

(ii) $\dfrac{1}{\Gamma(\beta)} \sum\limits_{i=0}^{\kappa-\beta} (\kappa - \sigma(i))^{(\beta-1)} \phi(i + \beta - 1) \leq \lambda \phi(\kappa + \beta - 1).$

Lemma 6. *For the three-point BCs of discrete FEBE Equation (35), the following inequality holds:*

$$|\hat{w}(\kappa) - (\mathcal{A}\hat{w})(\kappa)| \leq \lambda \epsilon \phi(\kappa + \beta - 1),$$

where $(\mathcal{A}\hat{w})(\kappa)$ is defined in Equation (21).

Proof. From inequality Equation (33), for $\kappa \in \mathbb{N}_{\beta-4}^{\beta+n+3}$, we obtain a function $\Delta_{\beta-4}^{\beta} \hat{w}(\kappa) = g_{\hat{w}}(\kappa) + \Psi(\kappa + \beta - 1)$, $|\Psi(\kappa + \beta - 1)| \leq \epsilon \phi(\kappa + \beta - 1)$ and $(A_5)(i)$ such that

$$|\hat{w}(\kappa) - (\mathcal{A}\hat{w})(\kappa)| \leq \lambda \epsilon \phi(\kappa + \beta - 1).$$

□

Theorem 5. *Under the hypothesis (A_1) with the inequality Equation (19), the discrete FEBE Equation (2) is $\mathcal{HUR}$ stable.*

Proof. By using a similar procedure of Theorem 4 together with Lemma 6 for $\kappa \in \mathbb{N}_{\beta-4}^{\beta+n+3}$, we obtain

$$|\hat{w}(\kappa) - w(\kappa)| \leq \lambda \epsilon \phi(\kappa + \beta - 1)$$
$$+ \left[\dfrac{(\beta + n + 3)^{(\beta)}}{\Gamma(\beta+1)} + \mathbb{E}_1^* \left(\dfrac{\zeta^{(\beta)}}{\Gamma(\beta+1)} + \dfrac{(\beta+n)^{(\beta-3)}}{\Gamma(\beta-2)} \right) + \mathbb{E}_2^* \dfrac{(\beta+n)^{(\beta-1)}}{\Gamma(\beta)} \right] \mathbb{L}_G \|\hat{w} - w\|.$$

This further implies that

$$\|\hat{w} - w\| \leq \delta_2 \epsilon \phi(\kappa + \beta - 1),$$

where
$$\delta_2 = \dfrac{\lambda \Gamma(\beta+1)}{\Gamma(\beta+1) - \mathbb{L}_G \left[(\beta+n+3)^{(\beta)} + \mathbb{E}_1^* (\zeta^{(\beta)} + \beta^{(3)}(\beta+n)^{(\beta-3)}) + \mathbb{E}_2^* \beta(\beta+n)^{(\beta-1)} \right]}.$$

Thus, the solution of Equation (2) is $\mathcal{HUR}$ stable. □

Remark 3. *If $\phi(\kappa + \beta - 1) = \epsilon \phi(\kappa + \beta - 1)$, then we have*

$$\|\hat{w} - w\| \leq \delta_2 \phi(\kappa + \beta - 1).$$

Hence, the solution of Equation (2) is $\mathcal{GHUR}$ stable.

5. Applications

Some illustrative examples are provided in this section to demonstrate the applicability of our results in this research work.

Example 1. *Suppose that $\beta = 3.7$, $n = 2$, and $H(\kappa) = \kappa^{(13)}$ with different values of ζ. Then, a linear discrete FEBE with the three-point BCs of Equation (3) becomes*

$$\begin{cases} \Delta_{-0.3}^{3.7} w(\kappa) = (\kappa + 2.7)^{(13)}, \ \kappa \in \mathbb{N}_0^5, \\ w(-0.3) = 0, \ \Delta^2 w(-0.3) = 0, \ \Delta w(5.7) = 0, \ \Delta^3 w(5.7) + w(\zeta) = 0. \end{cases} \quad (37)$$

We shall apply Theorem 1 to find a solution $w(\kappa)$ of Equation (37) that can be expressed as:

$$w(\kappa) = \frac{1}{\Gamma(3.7)} \sum_{i=0}^{\kappa-3.7} (\kappa - \sigma(i))^{(2.7)} (i+2.7)^{(13)}$$
$$+ \mathbb{E}_1(\kappa) \left[\frac{1}{\Gamma(3.7)} \sum_{i=0}^{\zeta-3.7} (\zeta - \sigma(i))^{(2.7)} + \frac{1}{\Gamma(0.7)} \sum_{i=0}^{5} (5.7 - \sigma(i))^{(-0.3)} \right] (i+2.7)^{(13)} \quad (38)$$
$$+ \frac{\mathbb{E}_2(\kappa)}{\Gamma(2.7)} \sum_{i=0}^{3} (5.7 - \sigma(i))^{(1.7)} (i+2.7)^{(13)},$$

where $\kappa \in \mathbb{N}_{-0.3}^{8.7}$, $\mathbb{E}_1(\kappa)$ and $\mathbb{E}_2(\kappa)$ are defined in Theorem 1. With the help of both Definition 1 and Lemma 4, we obtain the expression on right-hand side of Equation (38) as follows:

$$\frac{1}{\Gamma(3.7)} \sum_{i=0}^{\kappa-3.7} (\kappa - \sigma(i))^{(2.7)} (i+2.7)^{(13)} = \Delta^{-3.7}(\kappa + 2.7)^{(13)}$$
$$= \frac{\Gamma(14)}{\Gamma(17.7)} \cdot \frac{\Gamma(\kappa + 3.7)}{\Gamma(\kappa - 13)}. \quad (39)$$

Similarly, we find

$$\frac{1}{\Gamma(3.7)} \sum_{i=0}^{\zeta-3.7} (\zeta - \sigma(i))^{(2.7)} (i+2.7)^{(13)} = \frac{\Gamma(14)}{\Gamma(17.7)} \cdot \frac{\Gamma(\zeta + 3.7)}{\Gamma(\zeta - 13)}. \quad (40)$$

$$\frac{1}{\Gamma(2.7)} \sum_{i=0}^{3} (5.7 - \sigma(i))^{(1.7)} (i+2.7)^{(13)} = \frac{\Gamma(14)}{\Gamma(16.7)} \cdot \frac{\Gamma(9.4)}{\Gamma(-6.3)}. \quad (41)$$

$$\frac{1}{\Gamma(0.7)} \sum_{i=0}^{5} (5.7 - \sigma(i))^{(-0.3)} (i+2.7)^{(13)} = \frac{\Gamma(14)}{\Gamma(14.7)} \cdot \frac{\Gamma(9.4)}{\Gamma(-4.3)}. \quad (42)$$

By substituting the expressions Equations (39)–(42) into Equation (38), we obtain Equation (37)'s solution for $\kappa \in \mathbb{N}_{-0.3}^{8.7}$, in the form

$$w(\kappa) = \left[\frac{\Gamma(14)}{\Gamma(17.7)} \cdot \frac{\Gamma(\kappa + 3.7)}{\Gamma(\kappa - 13)} \right] + \mathbb{E}_2(\kappa) \left[\frac{\Gamma(14)}{\Gamma(16.7)} \cdot \frac{\Gamma(9.4)}{\Gamma(-6.3)} \right]$$
$$+ \mathbb{E}_1(\kappa) \left[\left(\frac{\Gamma(14)}{\Gamma(17.7)} \cdot \frac{\Gamma(\zeta + 3.7)}{\Gamma(\zeta - 13)} \right) + \left(\frac{\Gamma(14)}{\Gamma(14.7)} \cdot \frac{\Gamma(9.4)}{\Gamma(-4.3)} \right) \right]. \quad (43)$$

On one hand, by choosing different values of $\zeta = 2.7, 3.7, 4.7, 5.7$ in Equation (43), we obtain different solutions for this problem, as seen in Figure 1a. On the other hand, Figure 1b shows three-dimensional solution surface plots for various values κ and ζ. In addition, a numerical experiment for our obtained solutions in Example 1 with step size 1 is presented in Table 1.

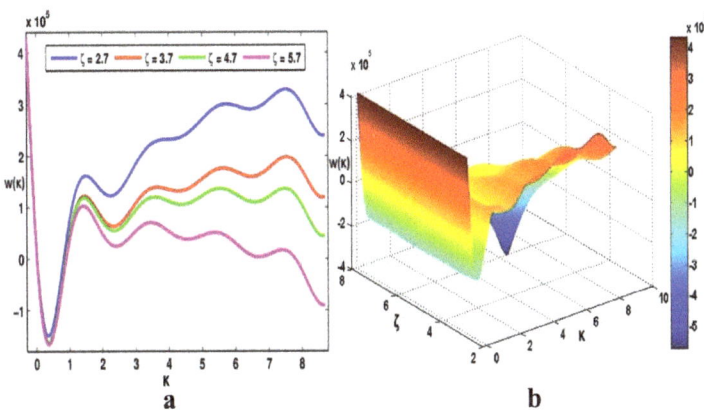

Figure 1. (a) Solution curves for various values of ζ of a discrete FEBE with the three-point BCs of Equation (37); (b) surface plots for different values of κ and ζ corresponding to Figure 1a.

Table 1. Numerical values of $w(\kappa)$ for Example 1 with step size 1.

	$w(\kappa)$			
$\kappa \zeta$	2.7	3.7	4.7	5.7
−0.3	4.3158×10^5	4.3158×10^5	4.3158×10^5	4.3158×10^5
0.7	-0.5233×10^5	-0.7356×10^5	-0.7410×10^5	-0.8078×10^5
1.7	1.5354×10^5	1.0872×10^5	1.0384×10^5	0.8468×10^5
2.7	1.4988×10^5	0.8076×10^5	0.6918×10^5	0.3413×10^5
3.7	2.3044×10^5	1.3849×10^5	1.1846×10^5	0.6559×10^5
4.7	2.5123×10^5	1.3991×10^5	1.1014×10^5	0.3865×10^5
5.7	3.0038×10^5	1.7498×10^5	1.3450×10^5	0.4457×10^5
6.7	2.9459×10^5	1.6211×10^5	1.1023×10^5	0.0290×10^5
7.7	3.2506×10^5	1.9406×10^5	1.3037×10^5	0.0745×10^5
8.7	2.3972×10^5	1.2029×10^5	0.4457×10^5	-0.9137×10^5

Example 2. Consider a discrete FEBE subject to three-point BCs:

$$\begin{cases} \Delta_{\pi-4}^{\pi} w(\kappa) = \dfrac{1}{(\kappa+\pi-1)+650}\left[\sin(w(\kappa+\pi-1)) + \dfrac{e^{-(\kappa+\pi-1)}\cos(\kappa+\pi-1)}{10\sqrt{\pi}(\kappa+\pi)}\right], \kappa \in \mathbb{N}_0^6, \\ w(\pi-4) = 0, \Delta^2 w(\pi-4) = 0, \Delta w(\pi+3) = 0, \Delta^3 w(\pi+3) + w(2.1416) = 0. \end{cases} \quad (44)$$

Clearly, $\beta = \pi, n = 3, \zeta = 2.1416$. Set $G(\kappa, w(\kappa)) = \dfrac{1}{\kappa+650}\left[\sin(w(\kappa)) + \dfrac{e^{-t}\cos(\kappa)}{10\sqrt{\pi}(1+\kappa)}\right]$ which is a continuous function for $\kappa \in \mathbb{N}_{\pi-4}^{\pi+6}$. Now, we show that Equation (44) has a unique solution.

For any $w, \hat{w} \in \mathbb{B}_*$, then

$$|G(\kappa, w(\kappa)) - G(\kappa, \hat{w}(\kappa))| = \dfrac{1}{\kappa+650}\left|\sin(w(\kappa)) + \dfrac{e^{-t}\cos(\kappa)}{10\sqrt{\pi}(1+\kappa)} - \sin(\hat{w}(\kappa)) - \dfrac{e^{-t}\cos(\kappa)}{10\sqrt{\pi}(1+\kappa)}\right|$$

$$= \dfrac{1}{\kappa+650}|\sin(w(\kappa)) - \sin(\hat{w}(\kappa))|$$

$$|G(\kappa, w(\kappa)) - G(\kappa, \hat{w}(\kappa))| \leq 0.0015|w(\kappa) - \hat{w}(\kappa)|.$$

So, we have $\mathbb{L}_G = 0.0015$, and G is Lipschitz continuous for for $\kappa \in \mathbb{N}_{\pi-4}^{\pi+6}$. Furthermore, the inequality Equation (19) is satisfied with $\Lambda \approx 0.2944 < 1$. Therefore, from Theorem 2, we conclude that problem Equation (44) has a unique solution.

Example 3. Assume that $\beta = 3.6$, $n = 4$, and $\zeta = 2.6$ with $G(\kappa, w(\kappa)) = \dfrac{\kappa}{100}e^{-\frac{w^2(\kappa)}{100}}$. Then, we obtain the following discrete FEBE Equation (2) with BCs:

$$\begin{cases} \Delta^{3.6}_{-0.4} w(\kappa) = \dfrac{1}{100}(\kappa + 2.6)e^{-\frac{1}{100}w^2(\kappa+2.6)}, \ \kappa \in \mathbb{N}_0^7, \\ w(-0.4) = 0, \ \Delta^2 w(-0.4) = 0, \ \Delta w(7.6) = 0, \ \Delta^3 w(7.6) + w(2.6) = 0. \end{cases} \quad (45)$$

Let a Banach space be $\mathbb{B}_* := \{w(\kappa) | \mathbb{N}^{10.6}_{-0.4} \to \mathbb{R}\}$. Suppose that $D = 1000$. To verify that the hypotheses of Theorem 3 hold, it is noticeable that

$$\dfrac{D\Gamma(\beta+1)}{\left[(\beta+n+3)^{(\beta)} + \mathbb{E}_1^*\left(\zeta^{(\beta)} + \beta^{(3)}(\beta+n)^{(\beta-3)}\right) + \mathbb{E}_2^*\beta(\beta+n)^{(\beta-1)}\right]} \approx 2.1790.$$

Clearly, we have $|G(\kappa, w(\kappa))| = 0.1060 \leq 2.1790$, whenever $\|w\| \leq 1000$. Thus, the problem Equation (45) has at least one solution.

Example 4. Consider the discrete FEBE with three-point BCs as follows:

$$\begin{cases} \Delta^{3.2}_{-0.8} w(\kappa) = \dfrac{1}{700}\cos(w(\kappa+2.2)) + \dfrac{1}{((\kappa+2.2)+950)}(\kappa+2.2)^{(3.2)}, \ \kappa \in \mathbb{N}_0^5, \\ w(-0.8) = 0, \ \Delta^2 w(-0.8) = 0, \ \Delta w(5.2) = 0, \ \Delta^3 w(5.2) + w(4.2) = 0. \end{cases} \quad (46)$$

Here, we have $\beta = 3.2$, $n = 2$, $\zeta = 4.2$ and $G(\kappa, w(\kappa)) = \dfrac{1}{700}\cos(w(\kappa)) + \dfrac{1}{(\kappa+950)}\kappa^{(3.2)}$ for $\kappa \in \mathbb{N}^{8.2}_{-0.8}$. Now, we prove that Equation (46) is $\mathcal{HU}$ stable. Since (A_1) holds for each $\kappa \in \mathbb{N}^{8.2}_{-0.8}$, we obtain

$$|G(\kappa, \hat{w}(\kappa)) - G(\kappa, w(\kappa))| = \left| \dfrac{1}{700}\cos(\hat{w}(\kappa)) + \dfrac{1}{(\kappa+950)}\kappa^{(3.2)} - \dfrac{1}{700}\cos(w(\kappa)) - \dfrac{1}{(\kappa+950)}\kappa^{(3.2)} \right|$$

$$= \dfrac{1}{700}|\cos(\hat{w}(\kappa)) - \cos(w(\kappa))|$$

$$|G(\kappa, \hat{w}(\kappa)) - G(\kappa, w(\kappa))| \leq 0.0014|\hat{w}(\kappa) - w(\kappa)|,$$

so $\mathbb{L}_G = 0.0014$ and G is Lipschitz continuous for $\kappa \in \mathbb{N}^{8.2}_{-0.8}$. Since

$$\dfrac{1}{\left[\dfrac{(\beta+n+3)^{(\beta)}}{\Gamma(\beta+1)} + \mathbb{E}_1^*\left(\dfrac{\zeta^{(\beta)}}{\Gamma(\beta+1)} + \dfrac{(\beta+n)^{(\beta-3)}}{\Gamma(\beta-2)}\right) + \mathbb{E}_2^*\dfrac{(\beta+n)^{(\beta-1)}}{\Gamma(\beta)}\right]} \approx 0.0080,$$

if $\mathbb{L}_G = 0.0014 < 0.0080$. Furthermore, to verify the stability results, from Theorem 4, we see that $\Lambda = 0.1758 < 1$. Hence, the solution of Equation (46) is $\mathcal{HU}$ stable with $\delta_1 = 80.8287$. In addition, it is $\mathcal{GHU}$ stable from Remark 2. For illustration, we take $\epsilon = 0.6017$ and $\hat{w}(\kappa) = \dfrac{\kappa^{(4)}}{350}$. We prove that Equation (31) holds. Indeed,

$$\left| \Delta^{3.2}_{-0.8}\hat{w}(\kappa) - G(\kappa+2.2, \hat{w}(\kappa+2.2)) \right|$$

$$= \left| \Delta^{3.2}_{-0.8}\hat{w}(\kappa) - \dfrac{\cos(\hat{w}(\kappa+2.2))}{700} - \dfrac{(\kappa+2.2)^{(3.2)}}{\kappa+952.2} \right|$$

$$= \left| \Delta^{3.2}_{-0.8}\left(\dfrac{\kappa^{(4)}}{350}\right) - 0.0014\cos\left[\dfrac{(\kappa+2.2)^{(4)}}{350}\right] - \dfrac{(\kappa+2.2)^{(3.2)}}{\kappa+952.2} \right|. \quad (47)$$

By using Lemma 4, Equation (47) becomes

$$\left|\Delta_{-0.8}^{3.2}\hat{w}(\kappa) - G(\kappa+2.2, \hat{w}(\kappa+2.2))\right|$$

$$= \left|0.0736\kappa^{(0.8)} - 0.0014\cos\left[\frac{\Gamma(\kappa+3.2)}{350\Gamma(\kappa-0.8)}\right] - \frac{\Gamma(\kappa+3.2)}{(\kappa+952.2)\Gamma(\kappa)}\right|$$

$$\leq 0.0736\left[\frac{\Gamma(\kappa+1)}{\Gamma(\kappa+0.2)}\right] + 0.0014 + \frac{1}{(\kappa+952.2)}\left[\frac{\Gamma(\kappa+3.2)}{\Gamma(\kappa)}\right]$$

$$\leq 0.6017 \leq \epsilon, \text{ for } \kappa \in \mathbb{N}_0^5.$$

Example 5. *Consider a discrete FEBE subject to the three-point BCs:*

$$\begin{cases} \Delta_{\pi-4}^{\pi} w(\kappa) = \frac{1}{700}\sin(w(\kappa+\pi-1)) + \frac{1}{310}(\kappa+\pi-1)^{(\pi)}, \ \kappa \in \mathbb{N}_0^4, \\ w(\pi-4) = 0, \ \Delta^2 w(\pi-4) = 0, \ \Delta w(\pi+1) = 0, \ \Delta^3 w(\pi+1) + w(2.1416) = 0. \end{cases} \quad (48)$$

In this example, $\beta = \pi$, $n = 1$, $\zeta = 2.1416$. Set $G(\kappa, w(\kappa)) = \frac{1}{700}\sin(w(\kappa)) + \frac{1}{310}\kappa^{(\pi)}$ for $\kappa \in \mathbb{N}_{\pi-4}^{\pi+4}$. Now, we show that Equation (48) is $\mathcal{HUR}$ stable. For any $\hat{w}, w \in \mathbb{B}_*$ and each $\kappa \in \mathbb{N}_{\pi-4}^{\pi+4}$, we obtain

$$|G(\kappa, \hat{w}(\kappa)) - G(\kappa, w(\kappa))| = \left|\frac{1}{700}\sin(\hat{w}(\kappa)) + \frac{1}{310}\kappa^{(\pi)} - \frac{1}{700}\sin(w(\kappa)) - \frac{1}{310}\kappa^{(\pi)}\right|$$

$$= \frac{1}{700}|\sin(\hat{w}(\kappa)) - \sin(w(\kappa))|$$

$$|G(\kappa, \hat{w}(\kappa)) - G(\kappa, w(\kappa))| \leq 0.0014|\hat{w}(\kappa) - w(\kappa)|.$$

This satisfies (A_1) with $\mathbb{L}_G = 0.0014$, and G is Lipschitz continuous for $\kappa \in \mathbb{N}_{\pi-4}^{\pi+4}$. Further, by assuming $\epsilon = 0.6519$ and $\phi(\kappa+\pi-1) = 1$, we have

$$\frac{0.6519}{\Gamma(\pi)}\sum_{i=0}^{\kappa-\pi}(\kappa-\sigma(i))^{(\pi-1)}(1) = \frac{(0.6519)\Gamma(\kappa+1)}{\Gamma(\pi+1)\Gamma(\kappa+1-\pi)}$$

$$\leq \frac{(0.6519)\Gamma(5)}{\Gamma(\pi+1)\Gamma(5-\pi)}$$

$$\frac{0.6519}{\Gamma(\pi)}\sum_{i=0}^{\kappa-\pi}(\kappa-\sigma(i))^{(\pi-1)}(1) \leq 2.2955, \ \kappa \in \mathbb{N}_0^4.$$

Thus, $(A_5)(i)$ holds with $\lambda = 3.5213$, $\epsilon = 0.6519$, and $\phi(\kappa+\pi-1) = 1$. Since

$$\frac{1}{\left[\frac{(\beta+n+3)^{(\beta)}}{\Gamma(\beta+1)} + \mathbb{E}_1^*\left(\frac{\zeta^{(\beta)}}{\Gamma(\beta+1)} + \frac{(\beta+n)^{(\beta-3)}}{\Gamma(\beta-2)}\right) + \mathbb{E}_2^*\frac{(\beta+n)^{(\beta-1)}}{\Gamma(\beta)}\right]} \approx 0.0137,$$

if $\mathbb{L}_G = 0.0014 < 0.0137$, from Theorem 5, we see that $\Lambda = 0.1023 < 1$. Hence, the solution to Equation (48) is $\mathcal{HUR}$ stable with $\delta_2 = 3.9224$. For illustration, we take $\epsilon = 0.6519$ and $\hat{w}(\kappa) = \frac{\kappa^{(3)}}{40}$. We prove that Equation (33) holds. Indeed,

$$\left|\Delta_{\pi-4}^{\pi}\hat{w}(\kappa) - G(\kappa+\pi-1, \hat{w}(\kappa+\pi-1))\right|$$

$$= \left|\Delta_{\pi-4}^{\pi}\hat{w}(\kappa) - \frac{1}{700}\sin(\hat{w}(\kappa+\pi-1)) - \frac{1}{310}(\kappa+\pi-1)^{(\pi)}\right|$$

$$= \left|\Delta_{\pi-4}^{\pi}\left(\frac{\kappa^{(3)}}{40}\right) - 0.0014\sin\left[\frac{(\kappa+\pi-1)^{(3)}}{40}\right] - \frac{(\kappa+\pi-1)^{(\pi)}}{310}\right|. \quad (49)$$

Using Lemma 4, Equation (49) becomes

$$\left|\Delta_{\pi-4}^{\pi}\hat{w}(\kappa) - G(\kappa+\pi-1,\hat{w}(\kappa+\pi-1))\right|$$

$$= \left|0.1358\kappa^{(3-\pi)} - 0.0014\sin\left[\frac{\Gamma(\kappa+\pi)}{40\Gamma(\kappa+\pi-3)}\right] - \frac{\Gamma(\kappa+\pi)}{310\Gamma(\kappa)}\right|$$

$$\leq 0.1358\left[\frac{\Gamma(\kappa+1)}{\Gamma(\kappa-2+\pi)}\right] + 0.0014 + \frac{\Gamma(\kappa+\pi)}{310\Gamma(\kappa)}$$

$$\leq 0.6519 \leq \epsilon\phi(\kappa+\pi-1), \text{ for } \kappa \in \mathbb{N}_0^4.$$

Furthermore, it is obviously $\mathcal{GHUR}$ stable from Remark 3.

6. Conclusions

Three-point BCs for a discrete FEBE have been investigated in this research work. For our proposed problem involving a Riemann–Liouville discrete fractional operator, some important conditions for the existence and stability theory have been developed. The required findings have been obtained with the help of fixed-point techniques such as the contraction mapping principle and Brouwer fixed-point theorem. Moreover, some new results for various types of Ulam stability of the proposed three-point BCs for a discrete FEBE have been established with the aid of nonlinear analysis. Some suitable examples have been provided and accompanied with numerical experiment for our obtained solutions for various fractional-order values in a graphical representation in order to study the effectiveness and applicability of our theoretical results. All in all, our results are new and interesting for the elastic beam problem arising from mathematical models of engineering and applied science applications.

Author Contributions: Conceptualization, J.A., R.D. and M.K.A.K.; Formal analysis, J.A., A.G.M.S., R.D. and M.K.A.K.; Investigation, J.A., A.G.M.S. and M.K.A.K.; Methodology, M.K.A.K.; Project administration, J.A.; Supervision, J.A. and M.K.A.K.; Validation, R.D. and M.K.A.K.; Visualization, A.G.M.S.; Writing—original draft, J.A., A.G.M.S., R.D. and M.K.A.K.; Writing—review and editing, J.A., A.G.M.S., R.D. and M.K.A.K. All authors have read and agreed to the published version of the manuscript.

Funding: J. Alzabut would like to thank Prince Sultan University for supporting this work.

Institutional Review Board Statement: Not applicable.

Informed Consent Statement: Not applicable.

Data Availability Statement: Not applicable.

Conflicts of Interest: The authors declare that they have no competing interest.

References

1. Cianciaruso, F.; Infante, G.; Pietramala, P. Solutions of perturbed hammerstein integral equations with applications. *arXiv* **2016**, arXiv:1602.00976v4.
2. Infante, G.; Pietramala, P. A cantilever equation with nonlinear boundary conditions. *Electron. J. Qual. Theory Differ. Equ.* **2009**, *15*, 1–14. [CrossRef]
3. Li, Y. Existence of positive solutions for the cantilever beam equations with fully nonlinear terms. *Nonlinear Anal. Real World Appl.* **2016**, *27*, 221–237. [CrossRef]
4. Song, Y. A nonlinear boundary value problem for fourth-order elastic beam equations. *Bound. Value Probl.* **2014**, *191*, 1–11. [CrossRef]
5. Li, S.; Zhai, C. New existence and uniqueness results for an elastic beam equation with nonlinear boundary conditions. *Bound. Value Probl.* **2015**, *104*, 1–12. [CrossRef]
6. Gupta, C.P. Existence and uniqueness theorems for the bending of an elastic beam equation. *Appl. Anal.* **1988**, *26*, 289–304. [CrossRef]
7. Kumar, A.; Chauhan, H.V.S.; Ravichandran, C.; Nisar, K.S.; Baleanu, D. Existence of solutions of non-autonomous fractional differential equations with integral impulse condition. *Adv. Differ. Equ.* **2020**, *434*, 1–15. [CrossRef]
8. Ismail, M.; Saeed, U.; Alzabut, J.; Rehman, M. Approximate solutions for fractional boundary value problems via green-cas wavelet method. *Mathematics* **2019** *7*, 1164. [CrossRef]

9. Kilbas, A.A.; Srivastava, H.M.; Trujillo, J.J. *Theory and Applications of Fractional Differential Equations*; Elsevier Science: Amsterdam, The Netherlands, 2006.
10. Podlubny, I. *Fractional Differential Equations*; Academic Press: New York, NY, USA, 1999.
11. Benchohra, M.; Litimein, S.; Nieto, J.J. Semilinear fractional differential equations with infinite delay and non-instantaneous impulses. *J. Fixed Point Theory Appl.* **2019**, *21*, 1–21. [CrossRef]
12. Almeida, R. What is the best fractional derivative to fit data? *Appl. Anal. Discrete Math.* **2017**, *11*, 358–368. [CrossRef]
13. Qureshi, S.; Yusuf, A. Fractional derivatives applied to MSEIR problems: Comparative study with real world data. *Eur. Phys. J. Plus* **2019**, *134*, 1–17. [CrossRef]
14. Matar, M.M.; Abbas, M.I.; Alzabut, J.; Kaabar, M.K.A.; Etemad, S.; Rezapour, S. Investigation of the p-Laplacian nonperiodic nonlinear boundary value problem via generalized Caputo fractional derivatives. *Adv. Differ. Equ.* **2021**, *68*, 1–18.
15. Pratap, A.; Raja, V.; Alzabut, J.; Dianavinnarasi, J.; Cao, J.; Rajchakit, G. Finite-time Mittag-Leffler stability of fractional-order quaternion-valued memristive neural networks with impulses. *Neural Process Lett.* **2020**, *51*, 1485–1526. [CrossRef]
16. Wu, R.C.; Hei, X.D.; Chen, L.P. Finite-time stability of fractional-order neural networks with delay. *Commun. Theor. Phys.* **2013**, *60*, 189–193. [CrossRef]
17. Trigeassou, J.C.; Maamri, N.; Sabatier, J.; Oustaloup, A. A Lyapunov approach to the stability of fractional differential equations. *Signal Process* **2011**, *91*, 437–445. [CrossRef]
18. Stamova, I. Mittag-Leffler stability of impulsive differential equations of fractional order. *Q. Appl. Math.* **2015**, *73*, 525–535. [CrossRef]
19. Lijun, G.; Wang, D.; Wang, G. Further results on exponential stability for impulsive switched nonlinear time-delay systems with delayed impulse effects. *Appl. Math. Comput.* **2015**, *268*, 186–200.
20. Hyers, D. On the stability of the linear functional equation. *Proc. Natl. Acad Sci. USA* **1941**, *27*, 222–224. [CrossRef] [PubMed]
21. Obloza, M. Hyers stability of the linear differential equation. *Rocznik Nauk-Dydakt Prace Mat.* **1993**, *13*, 259–270.
22. Ulam, S.M. *Problems in Modern Mathematics*; Wiley: New York, NY, USA, 1940.
23. Ahmad, M.; Zada, A.; Alzabut, J. Stability analysis of a nonlinear coupled implicit switched singular fractional differential system with p-Laplacian. *Adv. Differ. Equ.* **2019**, *436*, 1–22. [CrossRef]
24. Zada, A.; Alzabut, J.; Waheed, H.; Loan-Lucian, P. Ulam–Hyers stability of impulsive integrodifferential equations with Riemann-Liouville boundary conditions. *Adv. Differ. Equ.* **2020**, *1*, 1–50. [CrossRef]
25. Iswarya, M.; Raja, R.; Rajchakit, G.; Cao, J.; Alzabut, J.; Huang, C. Existence, uniqueness and exponential stability of periodic solution for discrete–time delayed BAM neural networks based on coincidence degree theory and graph theoretic method. *Mathematics* **2019**, *7*, 1055. [CrossRef]
26. Guo, Y.; Shu, X.; Li, Y.; Xu, F. The existence and Hyers–Ulam stability of solution for an impulsive Riemann–Liouville fractional neutral functional stochastic differential equation with infinite delay of order $1 < \beta < 2$. *Bound. Value Prob.* **2019**, *59*. [CrossRef]
27. Ahmad, M.; Zada, A.; Alzabut, J. Hyres–Ulam Stability of Coupled System of Fractional Differential Equations of Hilfer–Hadamard Type. *Demonstr. Math.* **2019**, *52*, 283–295. [CrossRef]
28. Salem, A.; Alzahrani, F.; Almaghamsi, L. Fractional Langevin equations with nonlocal integral boundary conditions. *Mathematics* **2019**, *7*, 402. [CrossRef]
29. Abdeljawad, T. On Riemann and Caputo fractional differences. *Comput. Math. Appl.* **2011**, *62*, 1602–1611. [CrossRef]
30. Atici, F.M.; Eloe, P.W. A transform method in discrete fractional calculus. *Int. J. Difference Equ.* **2017**, *2*, 165–176.
31. Atici, F.M.; Eloe, P.W. Initial value problems in discrete fractional calculus. *Proc. Am. Math. Soc.* **2009**, *137*, 981–989. [CrossRef]
32. Miller, K.S.; Ross, B. Fractional difference calculus. In *Proceedings of the International Symposium on Univalent Functions, Fractional Calculus and their Applications*; Nihon University: Koriyama, Japan, 1988; pp. 139–152.
33. Goodrich, C.S.; Peterson, A.C. *Discrete Fractional Calculus*; Springer: New York, NY, USA, 2015.
34. Atici, F.M.; Eloe, P.W. Two-point boundary value problems for finite fractional difference equations. *J. Differ. Equ. Appl.* **2011**, *17*, 445–456. [CrossRef]
35. Alzabut, J.; Abdeljawad, T.; Baleanu, D. Nonlinear delay fractional difference equations with applications on discrete fractional Lotka-Volterra competition model. *J. Comput. Anal. Appl.* **2018**, *25*, 889–898.
36. Chen, F.; Zhou, Y. Existence and Ulam Stability of Solutions for Discrete Fractional Boundary Value Problem. *Discret. Dyn. Nat. Soc.* **2013**, 1–7. [CrossRef]
37. Chen, H.; Jin, Z.; Kang, S. Existence of positive solution for Caputo fractional difference equation. *Adv. Differ. Equ.* **2015**, *44*, 1–12. [CrossRef]
38. Chen, C.; Bohner, M.; Jia, B. Ulam-Hyers stability of Caputo fractional difference equations. *Math. Meth. Appl. Sci.* **2019**, *42*, 7461–7470. [CrossRef]
39. Chen, C.; Bohner, M.; Jia, B. Method of upper and lower solutions for nonlinear Caputo fractional difference equations and its applications. *Fract. Calc. Appl. Anal.* **2019**, *22*, 1307–1320. [CrossRef]
40. Chen, C.; Bohner, M.; Jia, B. Existence and uniqueness of solutions for nonlinear Caputo fractional difference equations. *Turk. J. Math.* **2020**, *44*, 857–869. [CrossRef]
41. Goodrich, C.S. Solutions to a discrete right-focal fractional boundary value problem. *Int. J. Differ. Equ.* **2010**, *5*, 195–216.
42. Goodrich, C.S. Existence and uniqueness of solutions to a fractional difference equation with nonlocal conditions. *Comput. Math. Appl.* **2011**, *61*, 191–202. [CrossRef]

43. Rehman, M.; Iqbal, F.; Seemab, A. On existence of positive solutions for a class of discrete fractional boundary value problems. *Positivity* **2017**, *21*, 1173–1187. [CrossRef]
44. Selvam, A.G.M.; Dhineshbabu, R. Uniqueness of solutions of a discrete fractional order boundary value problem. *AIP Conf. Proc.* **2019**, *2095*, 1–7.
45. Selvam, A.G.M.; Alzabut, J.; Dhineshbabu, R.; Rashid, S.; Rehman, M. Discrete fractional order two-point boundary value problem with some relevant physical applications. *J. Inequal. Appl.* **2020**, *221*, 1–19.
46. Selvam, A.G.M.; Dhineshbabu, R. Ulam stability results for boundary value problem of fractional difference equations. *Adv. Math. Sci. J.* **2020**, *9*, 219–230. [CrossRef]
47. Selvam, A.G.M.; Dhineshbabu, R. Existence and uniqueness of solutions for a discrete fractional boundary value problem. *Int. J. Appl. Math.* **2020**, *33*, 283–295.

Article

Uniqueness of Abel's Integral Equations of the Second Kind with Variable Coefficients

Chenkuan Li * and Joshua Beaudin

Department of Mathematics and Computer Science, Brandon University, Brandon, MB R7A 6A9, Canada; beaudijd31@brandonu.ca
* Correspondence: lic@brandonu.ca

Abstract: This paper studies the uniqueness of the solutions of several of Abel's integral equations of the second kind with variable coefficients as well as an in-symmetry system in Banach spaces $L(\Omega)$ and $L(\Omega) \times L(\Omega)$, respectively. The results derived are new and original, and can be applied to solve the generalized Abel's integral equations and obtain convergent series as solutions. We also provide a few examples to demonstrate the use of our main theorems based on convolutions, the gamma function and the Mittag–Leffler function.

Keywords: partial Riemann–Liouville fractional integral; Babenko's approach; Banach fixed point theorem; Mittag–Leffler function; gamma function

Citation: Li, C.; Beaudin, J. Uniqueness of Abel's Integral Equations of the Second Kind with Variable Coefficients. *Symmetry* **2021**, *13*, 1064. https://doi.org/10.3390/sym13061064

Academic Editors: Francisco Martínez González and Mohammed KA Kaabar

Received: 30 May 2021
Accepted: 11 June 2021
Published: 13 June 2021

Publisher's Note: MDPI stays neutral with regard to jurisdictional claims in published maps and institutional affiliations.

Copyright: © 2021 by the authors. Licensee MDPI, Basel, Switzerland. This article is an open access article distributed under the terms and conditions of the Creative Commons Attribution (CC BY) license (https://creativecommons.org/licenses/by/4.0/).

1. Introduction

Let $0 < \Omega_i < \infty$ for $i = 1, 2, \cdots, n$, and $\Omega = [0, \Omega_1] \times [0, \Omega_2] \times \cdots \times [0, \Omega_n] \subset R^n$. Define:

$$L(\Omega) = \left\{ u \mid u \text{ is Lebesgue integrable on } \Omega \text{ and } \|u\| = \int_\Omega |u(x)| dx < \infty \right\}.$$

Furthermore, the product space $L(\Omega) \times L(\Omega)$ is given by

$$L(\Omega) \times L(\Omega) = \{ (u, v) \mid u, v \text{ are Lebesgue integrable on } \Omega \text{ and } \|(u,v)\| < \infty \},$$

where:

$$\|(u,v)\| = \int_\Omega |u(x)| dx + \int_\Omega |v(x)| dx.$$

Clearly, both $L(\Omega)$ and $L(\Omega) \times L(\Omega)$ are Banach spaces.

Let I_k^α be the partial Riemann–Liouville fractional integral of order $\alpha \in R^+$ with respect to $x_k \in [0, \Omega_k]$, with initial point zero [1]:

$$(I_k^\alpha u)(x) = \frac{1}{\Gamma(\alpha)} \int_0^{x_k} (x_k - s)^{\alpha-1} u(x_1, \cdots, x_{k-1}, s, x_{k+1}, \cdots, x_n) ds$$

for $k = 1, 2, \cdots, n$.

In particular:

$$(I_k^0 u)(x) = u(x).$$

Assume that $\lambda_{ij}(x)$ is the Lebesgue integrable and bounded on Ω for all $i = 1, 2, \cdots, n \in N$ and $j = 1, 2, \cdots, m \in N$. In this paper, we begin to construct a unique solution in the space $L(\Omega)$ using Babenko's method and properties of the gamma function for the following generalized Abel's integral equation of the second kind with variable coefficients for $f \in L(\Omega)$:

$$u(x) + \sum_{j=1}^m \left\{ \lambda_{1j}(x) I_1^{\alpha_{1j}} \right\} \left\{ \lambda_{2j}(x) I_2^{\alpha_{2j}} \right\} \cdots \left\{ \lambda_{nj}(x) I_n^{\alpha_{nj}} \right\} u(x) = f(x), \qquad (1)$$

where each fractional integral $I_k^{\alpha_{kj}}$ carries its own weight function $\lambda_{kj}(x)$, and all $\alpha_{ij} \geq 0$ satisfy a certain condition. Then, we further study the uniqueness of solutions in $L(\Omega)$ for:

$$u(x) + \sum_{j=1}^{m}\left\{\lambda_{1j}(x)I_1^{\alpha_{1j}}\right\}\left\{\lambda_{2j}(x)I_2^{\alpha_{2j}}\right\}\cdots\left\{\lambda_{nj}(x)I_n^{\alpha_{nj}}\right\}u(x) = g(x,u(x)), \quad (2)$$

where $g(x,y)$ is a mapping from $\Omega \times R$ to R. Finally, the sufficient conditions are given for the uniqueness of solutions in $L(\Omega) \times L(\Omega)$ to the symmetric system:

$$\begin{cases} u(x) + \sum_{j=1}^{m}\left\{\lambda_{1j}(x)I_1^{\alpha_{1j}}\right\}\left\{\lambda_{2j}(x)I_2^{\alpha_{2j}}\right\}\cdots\left\{\lambda_{nj}(x)I_n^{\alpha_{nj}}\right\}u(x) = g_1(x,u(x),v(x)), \\ v(x) + \sum_{j=1}^{m}\left\{\mu_{1j}(x)I_1^{\beta_{1j}}\right\}\left\{\mu_{2j}(x)I_2^{\beta_{2j}}\right\}\cdots\left\{\mu_{nj}(x)I_n^{\beta_{nj}}\right\}v(x) = g_2(x,u(x),v(x)), \end{cases} \quad (3)$$

where both $g_1(x,y_1,y_2)$ and $g_2(x,y_1,y_2)$ are mappings from $\Omega \times R \times R$ to R, and all coefficient functions $\mu_{ij}(x)$ are Lebesgue integrable and bounded on Ω. Equations (1)–(3) are new in the present studies, and have never been investigated before.

Clearly, Equation (1) turns out to be:

$$u(x) - cI^{\alpha_{11}}u(x) = f(x), \quad \alpha_{11} > 0 \quad (4)$$

if $n = m = 1$ and $\lambda_{11}(x) = -c$ (constant). Equation (4) is obviously the classical Abel's integral equation of the second kind. In 1930, Hille and Tamarkin [2] derived its solution as

$$u(x) = f(x) + c\int_0^x (x-\tau)^{\alpha_{11}-1} E_{\alpha_{11},\alpha_{11}}(c(x-\tau)^{\alpha_{11}}) f(\tau) d\tau,$$

where:

$$E_{\alpha,\beta}(z) = \sum_{j=0}^{\infty} \frac{z^j}{\Gamma(\alpha j + \beta)}, \quad \alpha, \beta > 0$$

is the Mittag–Leffler function.

There have been many analytic and numerical studies on Abel's integral equation of the second kind, including its variants and generalizations in distribution [3–11]. Cameron and McKee [12] investigated the following Abel's integral equation of the second kind, numerically based on the construction and convergence analysis of the high-order product integral:

$$u(x) + \int_0^x (x-s)^{-\alpha} k(x,s,u(s)) ds = f(x),$$

where $u(x)$ is the unknown function defined on the interval $0 \leq x \leq T < \infty$ and the kernel $k(x,s,u(s))$ is Lipschitz continuous in its third variable. Pskhu [13] constructed an explicit solution for the generalized Abel's integral equation with constant coefficients c_k for $k = 1,2,\cdots,n$:

$$u(x) - \sum_{k=1}^{n} c_k I_k^{\alpha_k} u(x) = f(x), \quad \alpha_k > 0, \quad x \in \Omega, \quad (5)$$

using the Wright function:

$$\phi(\alpha, \beta; z) = \sum_{j=0}^{\infty} \frac{z^j}{j!\Gamma(\alpha j + \beta)}, \quad \alpha > -1,$$

and convolution. Evidently, Equation (5) is a special case of our Equation (1) for particular values of m, $\lambda_{ij}(x)$ and α_{ij}. In 2019, Li and Plowman [14] derived a convergent solution for the following Abel's integral equation:

$$u(x) - (a_1(x)I_1^{\alpha_1}) \cdots (a_n(x)I_n^{\alpha_n})u(x) = f(x), \quad x \in \Omega, \tag{6}$$

based on Babenko's approach in the space $L(\Omega)$. Obviously, Equation (6) is also a particular case of Equation (1) with $m = 1$, $\lambda_{11}(x) = -a_1(x)$, $\lambda_{21}(x) = a_2(x), \cdots, \lambda_{n1}(x) = a_n(x)$, and $\alpha_{i1} = \alpha_i$ for $i = 1, 2, \cdots, n$.

In a wide range of scientific and engineering problems, the existence of a solution to an integral equation is equivalent to the existence of a fixed point for a suitable and well-defined mapping on spaces under consideration. Fixed points are therefore essential tools in studying integral equations or systems arising from the real world. Banach's contractive principle provides a general condition ensuring that, if it is satisfied, the iteration of the mapping produces a fixed point [15].

Babenko's approach [16] is a very useful method in solving differential and integral equations, which treat differential or integral operators like variables. The method itself is similar to the Laplace transform when dealing with differential or integral equations with constant coefficients, but it also works for certain equations with distributions, such as $x_+^{-1.5}$ and $\delta^{(0.5)}(x)$, whose Laplace transforms do not exist in the classical sense [6,8]. As an example, we are going to solve Equation (4) using this technique. Clearly:

$$u(x) - cI^{\alpha_{11}}u(x) = (1 - cI^{\alpha_{11}})u(x) = f(x).$$

Informally:

$$\begin{aligned}
u(x) &= (1 - cI^{\alpha_{11}})^{-1}f(x) = \sum_{k=0}^{\infty} (cI^{\alpha_{11}})^k f(x) = \sum_{k=0}^{\infty} c^k I^{\alpha_{11}k} f(x) \\
&= f(x) + \sum_{k=0}^{\infty} c^{k+1} I^{\alpha_{11}k + \alpha_{11}} f(x) \\
&= f(x) + c \sum_{k=0}^{\infty} \frac{c^k}{\Gamma(\alpha_{11}k + \alpha_{11})} \int_0^x (x - \tau)^{\alpha_{11}k + \alpha_{11} - 1} f(\tau) d\tau \\
&= f(x) + c \int_0^x (x - \tau)^{\alpha_{11} - 1} \sum_{k=0}^{\infty} \frac{c^k (x - \tau)^{\alpha_{11}k}}{\Gamma(\alpha_{11}k + \alpha_{11})} f(\tau) d\tau \\
&= f(x) + c \int_0^x (x - \tau)^{\alpha_{11} - 1} E_{\alpha_{11}, \alpha_{11}} (c(x - \tau)^{\alpha_{11}}) f(\tau) d\tau,
\end{aligned}$$

which coincides with Hille and Tamarkin's result provided above.

2. The Main Results

In this section, we are going to present our main outcomes with several examples for the illustration of the key theorems.

Theorem 1. *Assume that $f \in L(\Omega)$, $\alpha_{ij} \geq 0$, and $\lambda_{ij}(x)$ is Lebesgue integrable and bounded on Ω for all $i = 1, 2, \cdots, n$ and $j = 1, 2, \cdots, m$. In addition, there exists $1 \leq i \leq n$ such that:*

$$\alpha = \min\{\alpha_{i1}, \cdots, \alpha_{im}\} \geq 1.$$

Then, Equation (1) has a unique solution in the space $L(\Omega)$:

$$u(x) = \sum_{k=0}^{\infty} (-1)^k \sum_{k_1 + \cdots + k_m = k} \binom{k}{k_1, \cdots, k_m} \\ \left(\lambda_{11}(x)I_1^{\alpha_{11}} \cdots \lambda_{n1}(x)I_n^{\alpha_{n1}}\right)^{k_1} \cdots \left(\lambda_{1m}(x)I_1^{\alpha_{1m}} \cdots \lambda_{nm}(x)I_n^{\alpha_{nm}}\right)^{k_m} f(x). \tag{7}$$

Proof. Equation (1) turns out to be:

$$\left(1+\sum_{j=1}^{m}\{\lambda_{1j}(x)I_1^{\alpha_{1j}}\}\{\lambda_{2j}(x)I_2^{\alpha_{2j}}\}\cdots\{\lambda_{nj}(x)I_n^{\alpha_{nj}}\}\right)u(x)=f(x).$$

Thus, by Babenko's approach:

$$\begin{aligned}
u(x) &= \left(1+\sum_{j=1}^{m}\{\lambda_{1j}(x)I_1^{\alpha_{1j}}\}\{\lambda_{2j}(x)I_2^{\alpha_{2j}}\}\cdots\{\lambda_{nj}(x)I_n^{\alpha_{nj}}\}\right)^{-1}f(x) \\
&= \sum_{k=0}^{\infty}(-1)^k\left(\sum_{j=1}^{m}\{\lambda_{1j}(x)I_1^{\alpha_{1j}}\}\{\lambda_{2j}(x)I_2^{\alpha_{2j}}\}\cdots\{\lambda_{nj}(x)I_n^{\alpha_{nj}}\}\right)^k f(x) \\
&= \sum_{k=0}^{\infty}(-1)^k\sum_{k_1+\cdots+k_m=k}\binom{k}{k_1,\cdots,k_m} \\
&\quad (\lambda_{11}(x)I_1^{\alpha_{11}}\cdots\lambda_{n1}(x)I_n^{\alpha_{n1}})^{k_1}\cdots(\lambda_{1m}(x)I_1^{\alpha_{1m}}\cdots\lambda_{nm}(x)I_n^{\alpha_{nm}})^{k_m}f(x).
\end{aligned}$$

Obviously, there exists $M > 0$ such that:

$$\sup_{x\in\Omega}|\lambda_{ij}(x)| \leq M$$

for all $i = 1, 2, \cdots, n$ and $j = 1, 2, \cdots m$.

Let:

$$\omega = \max\{\Omega_1, \Omega_2, \cdots, \Omega_n\},$$

and:

$$\Phi_{i,\alpha_{ij}}(x) = \frac{(x_i)_+^{\alpha_{ij}-1}}{\Gamma(\alpha_{ij})}.$$

Then, it follows from reference [17] that:

$$\begin{aligned}
\|I_i^{\alpha_{ij}}\| &= \sup_{\|g\|\leq 1}\|I_i^{\alpha_{ij}}g\| = \sup_{\|g\|\leq 1}\|\Phi_{i,\alpha_{ij}}*g\| \leq \sup_{\|g\|\leq 1}\|\Phi_{i,\alpha_{ij}}\|\|g\| \leq \|\Phi_{i,\alpha_{ij}}\| \\
&= \int_{\Omega}\frac{(x_i)_+^{\alpha_{ij}-1}}{\Gamma(\alpha_{ij})}dx_1\cdots dx_n \\
&= \Omega_1\cdots\Omega_{i-1}\frac{\Omega_i^{\alpha_{ij}}}{\Gamma(\alpha_{ij}+1)}\Omega_{i+1}\cdots\Omega_n \leq \omega^{n-1}\frac{\omega^{\alpha_{ij}}}{\Gamma(\alpha_{ij}+1)}.
\end{aligned}$$

Therefore:

$$u(x) \leq \sum_{k=0}^{\infty}(M^n)^k\sum_{k_1+\cdots+k_m=k}\binom{k}{k_1,\cdots,k_m}\|I_1^{\alpha_{11}k_1+\cdots+\alpha_{1m}k_m}\|\cdots \\ \|I_n^{\alpha_{n1}k_1+\cdots+\alpha_{nm}k_m}\|\|f\|.$$

Clearly:

$$\|I_1^{\alpha_{11}k_1+\cdots+\alpha_{1m}k_m}\| \leq \omega^{n-1}\frac{\omega^{\alpha_{11}k_1+\cdots+\alpha_{1m}k_m}}{\Gamma(\alpha_{11}k_1+\cdots+\alpha_{1m}k_m+1)},$$

$$\cdots,$$

$$\|I_n^{\alpha_{n1}k_1+\cdots+\alpha_{nm}k_m}\| \leq \omega^{n-1}\frac{\omega^{\alpha_{n1}k_1+\cdots+\alpha_{nm}k_m}}{\Gamma(\alpha_{n1}k_1+\cdots+\alpha_{nm}k_m+1)}.$$

Since there exists $1 \leq i \leq n$ such that:
$$\alpha = \min\{\alpha_{i1}, \cdots, \alpha_{im}\} \geq 1,$$

which infers that:
$$\Gamma(\alpha_{i1}k_1 + \cdots + \alpha_{im}k_m + 1) \geq \Gamma(\alpha k + 1)$$

for all $k = 0, 1, \cdots$ by noting that $\Gamma(x+1)$ is an increasing function if $x \geq 1$. Furthermore:
$$\Gamma(\alpha_{s1}k_1 + \cdots + \alpha_{sm}k_m + 1) \geq \frac{4}{5}$$

for $s = 1, 2, \cdots, i-1, i+1, \cdots, n$ and $k = 0, 1, \cdots$, since $\Gamma(x+1) \geq 4/5$ for all $x \geq 0$. Let:
$$W = \max_{1 \leq i \leq n, 1 \leq j \leq m}\{\omega^{\alpha_{ij}}\}.$$

Applying the identity:
$$\sum_{k_1+k_2+\cdots+k_m=k} \binom{k}{k_1, k_2, \cdots, k_m} = m^k,$$

we derive that:
$$\|u(x)\| \leq \omega^{n^2-n}\left(\frac{5}{4}\right)^{n-1} \|f\| \sum_{k=0}^{\infty} \frac{(M^n m W^n)^k}{\Gamma(\alpha k + 1)}$$
$$= \omega^{n^2-n}\left(\frac{5}{4}\right)^{n-1} \|f\| E_{\alpha,1}(M^n m W^n) < \infty. \tag{8}$$

We still need to show that Equation (7) is a solution of Equation (1). Indeed:
$$\sum_{k=0}^{\infty}(-1)^k\left(\sum_{j=1}^{m}\{\lambda_{1j}(x)I_1^{\alpha_{1j}}\}\{\lambda_{2j}(x)I_2^{\alpha_{2j}}\}\cdots\{\lambda_{nj}(x)I_n^{\alpha_{nj}}\}\right)^k f(x)$$
$$= f(x) + \sum_{k=1}^{\infty}(-1)^k\left(\sum_{j=1}^{m}\{\lambda_{1j}(x)I_1^{\alpha_{1j}}\}\{\lambda_{2j}(x)I_2^{\alpha_{2j}}\}\cdots\{\lambda_{nj}(x)I_n^{\alpha_{nj}}\}\right)^k f(x),$$

and:
$$\sum_{k=1}^{\infty}(-1)^k\left(\sum_{j=1}^{m}\{\lambda_{1j}(x)I_1^{\alpha_{1j}}\}\{\lambda_{2j}(x)I_2^{\alpha_{2j}}\}\cdots\{\lambda_{nj}(x)I_n^{\alpha_{nj}}\}\right)^k f(x)$$
$$+ \sum_{k=0}^{\infty}(-1)^k\left(\sum_{j=1}^{m}\{\lambda_{1j}(x)I_1^{\alpha_{1j}}\}\{\lambda_{2j}(x)I_2^{\alpha_{2j}}\}\cdots\{\lambda_{nj}(x)I_n^{\alpha_{nj}}\}\right) \cdot$$
$$\left(\sum_{j=1}^{m}\{\lambda_{1j}(x)I_1^{\alpha_{1j}}\}\{\lambda_{2j}(x)I_2^{\alpha_{2j}}\}\cdots\{\lambda_{nj}(x)I_n^{\alpha_{nj}}\}\right)^k f(x) = 0,$$

by noting that all of the above series are uniformly and absolutely convergent in the space $L(\Omega)$ due to inequality (8).

Evidently, the uniqueness immediately follows from the fact that the homogeneous integral equation:
$$u(x) + \sum_{j=1}^{m}\{\lambda_{1j}(x)I_1^{\alpha_{1j}}\}\{\lambda_{2j}(x)I_2^{\alpha_{2j}}\}\cdots\{\lambda_{nj}(x)I_n^{\alpha_{nj}}\}u(x) = 0$$

only has solution zero by Babenko's method. This completes the proof of Theorem 2. □

Remark 1. Note that $\Gamma(x+1)$ is not a monotone increasing function on $[0,1]$ since $\Gamma(1) = 1$, $\Gamma(1.5) = \sqrt{\pi}/2$ and $\Gamma(2) = 1$.

Example 1. *Abel's integral equation:*

$$u(x_1, x_2) + x_1 I_1^{0.5} x_1^2 I_2 u(x_1, x_2) + x_2^{0.1} I_2^{1.5} u(x_1, x_2) = 1$$

has the following convergent solution in $L(\Omega)$:

$$u(x) = 1 + \sum_{k=1}^{\infty} (-1)^k \sum_{j=0}^{k} \binom{k}{j} B_j A_{k-j} \Phi_{1,1+3.5j} \Phi_{2,1+j+(k-j)1.6},$$

where the coefficients B_j and A_{k-j} are given below.

Proof. Clearly:

$$\alpha = \min\{\alpha_{21}, \alpha_{22}\} = \min\{1, 1.5\} = 1,$$

and functions x_1, x_1^2 and $x_2^{0.1}$ are Lebesgue integrable and bounded on Ω. By Theorem 1:

$$
\begin{aligned}
u(x) &= 1 + \sum_{k=1}^{\infty} (-1)^k \left(x_1 I_1^{0.5} x_1^2 I_2 + x_2^{0.1} I_2^{1.5} \right)^k 1 \\
&= 1 + \sum_{k=1}^{\infty} (-1)^k \sum_{j=0}^{k} \binom{k}{j} (x_1 I_1^{0.5} x_1^2 I_2)^j (x_2^{0.1} I_2^{1.5})^{k-j} 1.
\end{aligned}
$$

Obviously:

$$(x_2^{0.1} I_2^{1.5})^0 1 = 1,$$

$$(x_2^{0.1} I_2^{1.5}) 1 = (x_2^{0.1} I_2^{1.5}) \Phi_{2,1} = x_2^{0.1} (\Phi_{2,1.5} * \Phi_{2,1}) = x_2^{0.1} \Phi_{2,2.5} = \frac{(x_2)_+^{1.6}}{\Gamma(2.5)}$$

$$= \frac{\Gamma(2.6)}{\Gamma(2.5)} \Phi_{2,2.6},$$

$$(x_2^{0.1} I_2^{1.5})^2 1 = (x_2^{0.1} I_2^{1.5}) \frac{\Gamma(2.6)}{\Gamma(2.5)} \Phi_{2,2.6} = \frac{\Gamma(2.6)\Gamma(4.2)}{\Gamma(2.5)\Gamma(4.1)} \Phi_{2,4.2},$$

$$(x_2^{0.1} I_2^{1.5})^3 1 = (x_2^{0.1} I_2^{1.5}) \frac{\Gamma(2.6)\Gamma(4.2)}{\Gamma(2.5)\Gamma(4.1)} \Phi_{2,4.2} = \frac{\Gamma(2.6)\Gamma(4.2)\Gamma(5.8)}{\Gamma(2.5)\Gamma(4.1)\Gamma(5.7)} \Phi_{2,5.8}$$

$$\cdots,$$

$$(x_2^{0.1} I_2^{1.5})^{k-j} 1 = \frac{\Gamma(2.6)\Gamma(4.2) \cdots \Gamma(1+(k-j)1.6)}{\Gamma(2.5)\Gamma(4.1) \cdots \Gamma(0.9+(k-j)1.6)} \Phi_{2,1+(k-j)1.6},$$

$$I_2^j (x_2^{0.1} I_2^{1.5})^{k-j} 1 = \frac{\Gamma(2.6)\Gamma(4.2) \cdots \Gamma(1+(k-j)1.6)}{\Gamma(2.5)\Gamma(4.1) \cdots \Gamma(0.9+(k-j)1.6)} \Phi_{2,1+j+(k-j)1.6},$$

$$= A_{k-j} \Phi_{2,1+j+(k-j)1.6},$$

for $k - j = 1, 2, \cdots$, and:

$$A_{k-j} = \begin{cases} \dfrac{\Gamma(2.6)\Gamma(4.2) \cdots \Gamma(1+(k-j)1.6)}{\Gamma(2.5)\Gamma(4.1) \cdots \Gamma(0.9+(k-j)1.6)} & \text{if } k-j \geq 1, \\ 1 & \text{if } k-j = 0. \end{cases}$$

On the other hand:

$$(x_1 I_1^{0.5} x_1^2)^0 = 1,$$

$$x_1 I_1^{0.5} x_1^2 = x_1 \Phi_{1,0.5} * \Gamma(3) \Phi_{1,3} = \frac{\Gamma(3)\Gamma(4.5)}{\Gamma(3.5)} \Phi_{1,4.5},$$

$$(x_1 I_1^{0.5} x_1^2)^2 = \frac{\Gamma(3)\Gamma(4.5)}{\Gamma(3.5)} (x_1 I_1^{0.5} x_1^2) \Phi_{1,4.5} = \frac{\Gamma(3)\Gamma(6.5)\Gamma(8)}{\Gamma(3.5)\Gamma(7)} \Phi_{1,8},$$

$$(x_1 I_1^{0.5} x_1^2)^3 = \frac{\Gamma(3)\Gamma(6.5)\Gamma(10)\Gamma(11.5)}{\Gamma(3.5)\Gamma(7)\Gamma(10.5)} \Phi_{1,11.5},$$

$$\cdots,$$

$$(x_1 I_1^{0.5} x_1^2)^j = \frac{\Gamma(3)\Gamma(6.5) \cdots \Gamma(3+3.5(j-1))\Gamma(1+3.5j)}{\Gamma(3.5)\Gamma(7) \cdots \Gamma(3.5j)} \Phi_{1,1+3.5j},$$

$$= B_j \Phi_{1,1+3.5j},$$

for $j = 1, 2, \cdots$, and:

$$B_j = \begin{cases} \frac{\Gamma(3)\Gamma(6.5) \cdots \Gamma(3+3.5(j-1))\Gamma(1+3.5j)}{\Gamma(3.5)\Gamma(7) \cdots \Gamma(3.5j)} & \text{if } j \geq 1, \\ 1 & \text{if } j = 0. \end{cases}$$

Therefore:

$$u(x) = 1 + \sum_{k=1}^{\infty} (-1)^k \sum_{j=0}^{k} \binom{k}{j} B_j A_{k-j} \Phi_{1,1+3.5j} \Phi_{2,1+j+(k-j)1.6}.$$

This completes the proof of Example 1. □

Using Banach's fixed point theorem, we are now ready to show the uniqueness of solutions in $L(\Omega)$ for Equation (2).

Theorem 2. *Suppose that $\alpha_{ij} \geq 0$, and $\lambda_{ij}(x)$ is Lebesgue integrable and bounded on Ω for $i = 1, 2, \cdots, n$ and $j = 1, 2, \cdots, m$, and there exists $1 \leq i \leq n$ such that:*

$$\alpha = \min\{\alpha_{i1}, \cdots, \alpha_{im}\} \geq 1.$$

Let $g(x, y)$ be defined on $\Omega \times R$ satisfying:

$$|g(x, y_1) - g(x, y_2)| \leq C|y_1 - y_2|,$$

and $g(x, 0) \in L(\Omega)$. Furthermore, assume that:

$$q = C\omega^{n^2 - n} \left(\frac{5}{4}\right)^{n-1} E_{\alpha,1}(M^n m W^n) < 1,$$

where ω, M, W are given in Theorem 1 as

$$\omega = \max\{\Omega_1, \Omega_2, \cdots, \Omega_n\}, \quad \sup_{x \in \Omega} |\lambda_{ij}(x)| \leq M,$$

$$W = \max_{1 \leq i \leq n, 1 \leq j \leq m} \{\omega^{\alpha_{ij}}\}.$$

Then, Equation (2) has a unique solution in $L(\Omega)$.

Proof. Let $u \in L(\Omega)$. We first show that $g(x, u(x)) \in L(\Omega)$. Indeed:

$$\begin{aligned} |g(x, u(x))| &= |g(x, u) - g(x, 0) + g(x, 0)| \leq |g(x, u) - g(x, 0)| + |g(x, 0)| \\ &\leq C|u| + |g(x, 0)|, \end{aligned}$$

which implies that:
$$\int_\Omega |g(x,u(x))|dx \le C\int_\Omega |u|dx + \int_\Omega |g(x,0)|dx < \infty.$$

Define a nonlinear mapping T on $L(\Omega)$ by

$$T(u) = \sum_{k=0}^{\infty}(-1)^k \sum_{k_1+\cdots+k_m=k}\binom{k}{k_1,\cdots,k_m}$$
$$\left(\lambda_{11}(x)I_1^{\alpha_{11}}\cdots\lambda_{n1}(x)I_n^{\alpha_{n1}}\right)^{k_1}\cdots\left(\lambda_{1m}(x)I_1^{\alpha_{1m}}\cdots\lambda_{nm}(x)I_n^{\alpha_{nm}}\right)^{k_m}g(x,u).$$

Clearly:
$$\|T(u)\| \le \omega^{n^2-n}\left(\frac{5}{4}\right)^{n-1} E_{\alpha,1}(M^n mW^n)\int_\Omega |g(x,u(x))|dx < \infty.$$

Thus, T is a mapping from $L(\Omega)$ to $L(\Omega)$. We now need to show that T is a contractive mapping. In fact:

$$\|T(u)-T(v)\| \le \omega^{n^2-n}\left(\frac{5}{4}\right)^{n-1} E_{\alpha,1}(M^n mW^n)\int_\Omega |g(x,u)-g(x,v)|dx$$
$$\le C\omega^{n^2-n}\left(\frac{5}{4}\right)^{n-1} E_{\alpha,1}(M^n mW^n)\|u-v\| = q\|u-v\|,$$

which claims that T is contractive since $q<1$. This completes the proof of Theorem 2. □

Example 2. Let $\Omega = [0,1]\times[0,1]\times[0,1]$. Then, the generalized Abel's integral equation:

$$u(x_1,x_2,x_3) + x_1 I_1^{0.5}\sin(x_1 x_2) I_2^{1.7}\cos(x_3^2+1) I_3^{0.2} u(x_1,x_2,x_3)$$
$$- x_1^3 I_1^{1.5} x_2^{0.5} I_2 u(x_1,x_2,x_3) = \frac{1}{7\pi}\arctan(x_1^2+x_2^2)\cos(u(x_1,x_2,x_3)+1) \quad (9)$$

has a unique solution in $L(\Omega)$.

Proof. Clearly, $m=2$, $\omega = \max\{1,1,1\}=1$, $W = \max_{1\le i\le n,\, 1\le j\le m}\{\omega^{\alpha_{ij}}\} = 1$, and:

$$\alpha = \min\{1.7,1\} = 1.$$

Furthermore:
$$|x_1|\le 1, \quad |\sin(x_1 x_2)|\le 1, \quad |\cos(x_3^2+1)|\le 1,$$
$$|-x_1^3|\le 1, \quad |x_2^{0.5}|\le 1,$$

on Ω. Therefore, $M=1$. Obviously:

$$g(x_1,x_2,x_3,y) = \frac{1}{7\pi}\arctan(x_1^2+x_2^2)\cos(y+1),$$

$g(x_1,x_2,x_3,0) \in L(\Omega)$, and:

$$|g(x_1,x_2,x_3,y_1) - g(x_1,x_2,x_3,y_2)| \le \frac{1}{7\pi}\frac{\pi}{2}|y_1-y_2| = \frac{1}{14}|y_1-y_2|.$$

It remains to compute the value of q:

$$q = \frac{1}{14}\left(\frac{5}{4}\right)^{3-1} E_{1,1}(2) = \frac{25}{224}\sum_{j=0}^{\infty}\frac{2^j}{j!}$$

$$= \frac{25}{224}\left(1+2+\frac{2\cdot 2}{1\cdot 2}+\frac{2\cdot 2\cdot 2}{1\cdot 2\cdot 3}+\frac{2\cdot 2\cdot 2\cdot 2}{1\cdot 2\cdot 3\cdot 4}+\frac{2\cdot 2\cdot 2\cdot 2\cdot 2}{1\cdot 2\cdot 3\cdot 4\cdot 5}+\cdots\right)$$

$$\leq \frac{25}{224}\left(1+2+2+\left(\frac{1}{3}+\left(\frac{2}{3}\right)^0\right)+\left(\frac{2}{3}\right)^1+\left(\frac{2}{3}\right)^2+\cdots\right) = \frac{625}{672} < 1.$$

By Theorem 2, Equation (9) has a unique solution in $L(\Omega)$. This completes the proof of Example 2. □

Finally, we study the uniqueness of solutions of in-symmetry system (3) in the product space $L(\Omega) \times L(\Omega)$.

Theorem 3. *Suppose that $\alpha_{ij} \geq 0$, $\beta_{ij} \geq 0$, and $\lambda_{ij}(x)$, $\mu_{ij}(x)$ are Lebesgue integrable and bounded on Ω for $i = 1, 2, \cdots, n$ and $j = 1, 2, \cdots, m$, and there exists $1 \leq i_1, i_2 \leq n$ such that:*

$$\alpha = \min\{\alpha_{i_1 1}, \cdots, \alpha_{i_1 m}\} \geq 1,$$
$$\beta = \min\{\beta_{i_2 1}, \cdots, \beta_{i_2 m}\} \geq 1.$$

Let $g_1(x, y_1, y_2)$ and $g_2(x, y_1, y_2)$ be defined on $\Omega \times R \times R$ satisfying:

$$|g_1(x, y_1, y_2) - g_1(x, z_1, z_2)| \leq C_1|y_1 - z_1| + C_2|y_2 - z_2|,$$
$$|g_2(x, s_1, s_2) - g_2(x, t_1, t_2)| \leq C_3|s_1 - t_1| + C_4|s_2 - t_2|,$$

and $g_1(x, 0, 0), g_2(x, 0, 0) \in L(\Omega)$. Furthermore, assume that:

$$q = \max\{C_1, C_2\}\omega^{n^2-n}\left(\frac{5}{4}\right)^{n-1} E_{\alpha,1}(M_1^n m W_1^n)$$
$$+ \max\{C_3, C_4\}\omega^{n^2-n}\left(\frac{5}{4}\right)^{n-1} E_{\beta,1}(M_2^n m W_2^n) < 1,$$

where M_1, W_1 and M_2, W_2 are given as

$$\sup_{x\in\Omega}|\lambda_{ij}(x)| \leq M_1, \quad W_1 = \max_{1\leq i\leq n, 1\leq j\leq m}\{\omega^{\alpha_{ij}}\},$$
$$\sup_{x\in\Omega}|\mu_{ij}(x)| \leq M_2, \quad W_2 = \max_{1\leq i\leq n, 1\leq j\leq m}\{\omega^{\beta_{ij}}\}.$$

Then, the in-symmetry system (3) has a unique solution in $L(\Omega) \times L(\Omega)$.

Proof. Let $u, vs. \in L(\Omega)$. We first show that $g_1(x, u(x), v(x)) \in L(\Omega)$. Indeed:

$$|g_1(x, u(x), v(x))| = |g_1(x, u, v) - g_1(x, 0, 0) + g_1(x, 0, 0)|$$
$$\leq |g_1(x, u, v) - g_1(x, 0, 0)| + |g_1(x, 0, 0)|$$
$$\leq C_1|u| + C_2|v| + |g_1(x, 0, 0)|,$$

which implies that:

$$\int_\Omega |g_1(x, u(x), v(x))|dx \leq C_1\int_\Omega |u|dx + C_2\int_\Omega |v|dx + \int_\Omega |g_1(x, 0, 0)|dx < \infty.$$

Similarly, $g_2(x, u(x), v(x)) \in L(\Omega)$.

Define a mapping T on $L(\Omega) \times L(\Omega)$ as

$$T(u,v) = (T_1(u,v), T_2(u,v)),$$

where:

$$\|T(u,v)\| = \|T_1(u,v)\| + \|T_2(u,v)\|,$$

and:

$$T_1(u,v) = \sum_{k=0}^{\infty}(-1)^k \sum_{k_1+\cdots+k_m=k}\binom{k}{k_1,\cdots,k_m}$$

$$\left(\lambda_{11}(x)I_1^{\alpha_{11}}\cdots\lambda_{n1}(x)I_n^{\alpha_{n1}}\right)^{k_1}\cdots\left(\lambda_{1m}(x)I_1^{\alpha_{1m}}\cdots\lambda_{nm}(x)I_n^{\alpha_{nm}}\right)^{k_m}g_1(x,u,v),$$

and symmetrically:

$$T_2(u,v) = \sum_{k=0}^{\infty}(-1)^k \sum_{k_1+\cdots+k_m=k}\binom{k}{k_1,\cdots,k_m}$$

$$\left(\mu_{11}(x)I_1^{\beta_{11}}\cdots\mu_{n1}(x)I_n^{\beta_{n1}}\right)^{k_1}\cdots\left(\mu_{1m}(x)I_1^{\beta_{1m}}\cdots\mu_{nm}(x)I_n^{\beta_{nm}}\right)^{k_m}g_2(x,u,v).$$

By inequality (8):

$$\|T_1(u,v)\| \leq \omega^{n^2-n}\left(\frac{5}{4}\right)^{n-1} E_{\alpha,1}(M_1^n m W_1^n) \int_\Omega |g_1(x,u(x),v(x))|dx < \infty,$$

$$\|T_2(u,v)\| \leq \omega^{n^2-n}\left(\frac{5}{4}\right)^{n-1} E_{\beta,1}(M_2^n m W_2^n) \int_\Omega |g_2(x,u(x),v(x))|dx < \infty.$$

Hence, T is a mapping from $L(\Omega) \times L(\Omega)$ to $L(\Omega) \times L(\Omega)$. It remains to show that T is contractive. In fact:

$$\|T(u_1,v_1) - T(u_2,v_2)\| = \|T_1(u_1,v_1) - T_1(u_2,v_2)\| + \|T_2(u_1,v_1) - T(u_2,v_2)\|.$$

Clearly:

$$\|T_1(u_1,v_1) - T_1(u_2,v_2)\| \leq \max\{C_1,C_2\}\omega^{n^2-n}\left(\frac{5}{4}\right)^{n-1} E_{\alpha,1}(M_1^n m W_1^n) \cdot \|(u_1,v_1) - (u_2,v_2)\|,$$

and:

$$\|T_2(u_1,v_1) - T_2(u_2,v_2)\| \leq \max\{C_3,C_4\}\omega^{n^2-n}\left(\frac{5}{4}\right)^{n-1} E_{\beta,1}(M_2^n m W_2^n) \cdot \|(u_1,v_1) - (u_2,v_2)\|.$$

Thus:

$$\|T(u_1,v_1) - T(u_2,v_2)\| \leq q\|(u_1,v_1) - (u_2,v_2)\|,$$

where:

$$q = \max\{C_1,C_2\}\omega^{n^2-n}\left(\frac{5}{4}\right)^{n-1} E_{\alpha,1}(M_1^n m W_1^n)$$

$$+ \max\{C_3,C_4\}\omega^{n^2-n}\left(\frac{5}{4}\right)^{n-1} E_{\beta,1}(M_2^n m W_2^n) < 1.$$

This completes the proof of Theorem 3. □

3. Conclusions

We studied the uniqueness of the solutions of the nonlinear Abel's integral equations of the second kind with variable coefficients and the in-symmetry system based on Banach's fixed point theorem and Babenko's approach. The results are new in the current works of integral equations, which are not feasible by any integral transforms. We also presented several examples to demonstrate the use of our main theorems via some special functions and convolutions.

Author Contributions: Conceptualization, C.L. and J.B.; methodology, C.L.; software, C.L. and J.B.; validation, C.L. and J.B.; formal analysis, C.L.; investigation, C.L. and J.B.; resources, C.L.; data curation, C.L. and J.B.; writing—original draft preparation, C.L.; writing—review and editing, C.L.; visualization, C.L.; supervision, C.L.; project administration, C.L. and J.B.; funding acquisition, C.L. All authors have read and agreed to the published version of the manuscript.

Funding: This work is supported by the Natural Sciences and Engineering Research Council of Canada (Grant No. 2019-03907).

Institutional Review Board Statement: Not applicable.

Informed Consent Statement: Not applicable.

Data Availability Statement: Not applicable.

Acknowledgments: The authors are grateful to the two reviewers for their careful reading of the paper with productive comments and suggestions.

Conflicts of Interest: The authors declare no conflict of interest.

References

1. Kilbas, A.A.; Srivastava, H.M.; Trujillo, J.J. *Theory and Applications of Fractional Differential Equations*; Elsevier: Amsterdam, The Netherlands, 2006.
2. Hille, E.; Tamarkin, J.D. On the theory of linear integral equations. *Ann. Math.* **1930**, *31*, 479–528. [CrossRef]
3. Evans, G.C. Volterra's integral equation of the second kind, with discontinuous kernel. *Trans. Am. Math. Soc.* **1910**, *11*, 393–413.
4. Evans, G.C. Volterra's integral equation of the second kind, with discontinuous kernel, second paper. *Trans. Am. Math. Soc.* **1911**, *12*, 429–472. [CrossRef]
5. Gorenflo, R.; Mainardi, F. Fractional Calculus: Integral and Differential Equations of Fractional Order. In *Fractals and Fractional Calculus in Continuum Mechanics*; Springer: New York, NY, USA, 1997; pp. 223–276.
6. Podlubny, I. *Fractional Differential Equations*; Academic Press: New York, NY, USA, 1999.
7. Srivastava, H.M.; Buschman, R.G. *Theory and Applications of Convolution Integral Equations*; Kluwer Academic Publishers: Dordrecht, The Netherlands; Boston, MA, USA; London, UK, 1992.
8. Li, C.; Clarkson, K. Babenko's approach to Abel's integral equations. *Mathematics* **2018**, *6*, 32. [CrossRef]
9. Li, C. Several results of fractional derivatives in $D'(R_+)$. *Fract. Calc. Appl. Anal.* **2015**, *18*, 192–207. [CrossRef]
10. Samko, S.G.; Kilbas, A.A.; Marichev, O.I. *Fractional Integrals and Derivatives: Theory and Applications*; Gordon and Breach: New York, NY, USA, 1993.
11. Miller, K.S.; Ross, B. *An Introduction to the Fractional Calculus and Fractional Differential Equations*; Wiley: New York, NY, USA, 1993.
12. Cameron, R.F.; McKee, S. Product integration methods for second-kind Abel integral equations. *J. Comput. Appl. Math.* **1984**, *11*, 1–10. [CrossRef]
13. Pskhu, A. Solution of a multidimensional Abel integral equation of the second kind with partial fractional integrals. *Differ. Uravn.* **2017**, *53*, 1195–1199. (In Russian) [CrossRef]
14. Li, C.; Plowman, H. Solutions of the generalized Abel's integral equations of the second kind with variable coefficients. *Axioms* **2019**, *8*, 137. [CrossRef]
15. Rudin, W. *Principle of Mathematical Analysis*, 3rd ed.; McGraw-Hill: New York, NY, USA, 1976.
16. Babenkos, Y.I. *Heat and Mass Transfer*; Khimiya: Leningrad, Russia, 1986. (In Russian)
17. Barros-Neto, J. *An Introduction to the Theory of Distributions*; Marcel Dekker, Inc.: New York, NY, USA, 1973.

Article

Condensing Functions and Approximate Endpoint Criterion for the Existence Analysis of Quantum Integro-Difference FBVPs

Shahram Rezapour [1,2], Atika Imran [3], Azhar Hussain [3], Francisco Martínez [4,*], Sina Etemad [2] and Mohammed K. A. Kaabar [5]

[1] Department of Medical Research, China Medical University Hospital, China Medical University, Taichung 406040, Taiwan; sh.rezapour@azaruniv.ac.ir
[2] Department of Mathematics, Azarbaijan Shahid Madani University, Tabriz 53751-71379, Iran; sina.etemad@azaruniv.ac.ir
[3] Department of Mathematics, University of Sargodha, Sargodha 40100, Pakistan; atikaimran977@gmail.com (A.I.); azhar.hussain@uos.edu.pk (A.H.)
[4] Department of Applied Mathematics and Statistics, Technological University of Cartagena, 30203 Cartagena, Spain
[5] Department of Mathematics and Statistics, Washington State University, Pullman, WA 99163, USA; mohammed.kaabar@wsu.edu
* Correspondence: f.martinez@upct.es; Tel.: +34-968-325-586

Abstract: A nonlinear quantum boundary value problem (q-FBVP) formulated in the sense of quantum Caputo derivative, with fractional q-integro-difference conditions along with its fractional quantum-difference inclusion q-BVP are investigated in this research. To prove the solutions' existence for these quantum systems, we rely on the notions such as the condensing functions and approximate endpoint criterion (AEPC). Two numerical examples are provided to apply and validate our main results in this research work.

Keywords: condensing function; approximate endpoint criterion; quantum integro-difference BVP; existence

MSC: 34A08; 34A12

Citation: Rezapour, S.; Imran, A.; Hussain, A.; Martínez, F.; Etemad, S.; Kaabar, M.K.A. Condensing Functions and Approximate Endpoint Criterion for the Existence Analysis of Quantum Integro-Difference FBVPs. Symmetry 2021, 13, 469. https://doi.org/10.3390/sym13030469

Academic Editor: Carlo Cattani

Received: 23 February 2021
Accepted: 10 March 2021
Published: 12 March 2021

Publisher's Note: MDPI stays neutral with regard to jurisdictional claims in published maps and institutional affiliations.

Copyright: © 2021 by the authors. Licensee MDPI, Basel, Switzerland. This article is an open access article distributed under the terms and conditions of the Creative Commons Attribution (CC BY) license (https://creativecommons.org/licenses/by/4.0/).

1. Introduction

It is a fact supported by many researchers that fractional calculus (FC) establishes a flexible extension for the classical one to arbitrary orders. FC has attracted particular attention from many researchers of mathematics, applied sciences, and engineering because of the various important applications of this field in modeling certain scientific phenomena and complex physical systems. Modeling systems using fractional derivatives can provide a good interpretation of the physical behavior of the studied systems due to the nonlocality and memory effects that have been exhibited in some systems. Some studies have been conducted on the mathematical analysis of FC and its applications such as European option pricing models [1], p-Laplacian nonperiodic nonlinear boundary value problem [2], nonlocal Cauchy problem [3], economic models involving time fractal [4], complex integral [5], incompressible second-grade fluid models [6], complex-valued functions of a real variable [7], and separated homotopy method [8]. Likewise, quantum calculus is a corresponding field of the standard infinitesimal one without the concept of limits. In spite of the long history that they already have, both theories are in the field of mathematical analysis, the investigation of their properties has emerged not so long ago. The quantum fractional calculus (q-fractional calculus), considered as the fractional correspondence of the q-calculus, was initially proposed by Jackson [9–11]. Researchers such as Al-Salam [12] and Agarwal [13] gave a great boost to the fractional q-calculus and obtained important theoretical results. Based on these results, the fractional q-calculus has emerged as an instrument with great potential in the field of applications [14–17]. Even in recent years, many articles have been appeared on quantum integro-difference boundary value problems (BVPs), which are valuable abstract tools for modeling many phenomena in various fields of science [18–30].

Asawasamrit et al. [31] provided a multi-term q-integro-difference equation subject to nonlocal multi-quantum integral conditions displayed as

$$\begin{cases} {}^R_{q_1}\mathcal{D}^{\varsigma}_{0+}\hbar(r) = \phi(r,\hbar(r),{}^R_{q_2}\mathcal{I}^{\sigma_1}_{0+}\hbar(r)), & (r \in [0,K]), \\ \hbar(0) = 0, \quad \nu {}^R_{q_3}\mathcal{I}^{\sigma_2}_{0+}\hbar(\eta_1) = {}^R_{q_4}\mathcal{I}^{\sigma_3}_{0+}\hbar(\eta_2), \end{cases}$$

where $q_1, q_2, q_3, q_4 \in (0,1), \varsigma \in (1,2), \sigma_1, \sigma_2, \sigma_3 > 0, \eta_1, \eta_2 \in (0,K)$ and $\nu \in \mathbb{R}$. The approach implemented by them to arrive at the existence property of solutions for the suggested q-BVP is based on the fixed-point techniques [31]. After that in 2015, Etemad, Ettefagh and Rezapour [32] concerned the three-term q-difference FBVP

$$({}^C_q\mathcal{D}^{\varsigma}_{0+}\hbar)(r) = w(r,\hbar(r),{}^C_q\mathcal{D}^{1}_{0+}\hbar(r)),$$

with four-point q-integro-difference conditions

$$\lambda_1\hbar(0) + \zeta_1 {}^C_q\mathcal{D}^{1}_{0+}\hbar(0) = m_1 {}^R_q\mathcal{I}^{\beta}_{0+}\hbar(\xi_1) = m_1 \int_0^{\xi_1} \frac{(\xi_1 - qv)^{(\beta-1)}}{\Gamma_q(\beta)} \hbar(v) d_q v,$$

$$\lambda_2\hbar(1) + \zeta_2 {}^C_q\mathcal{D}^{1}_{0+}\hbar(1) = m_2 {}^R_q\mathcal{I}^{\beta}_{0+}\hbar(\xi_2) = m_2 \int_0^{\xi_2} \frac{(\xi_2 - qv)^{(\beta-1)}}{\Gamma_q(\beta)} \hbar(v) d_q v,$$

where $0 \leq r \leq 1, 1 < \varsigma \leq 2, q \in (0,1), \beta \in (0,2], \lambda_1, \lambda_2, \zeta_1, \zeta_2, m_1, m_2 \in \mathbb{R}$ and $\xi_1, \xi_2 \in (0,1)$ with $\xi_1 < \xi_2$. Ntouyas and Samei [33] turned to studying the solutions' existence for the q-integro-difference FBVP

$${}^C_q\mathcal{D}^{\varsigma}_{0+}\hbar(r) = w(r,h(r),(\phi_1h)(r),(\phi_2h)(r),{}^C_q\mathcal{D}^{\varsigma_1}_{0+}\hbar(r),{}^C_q\mathcal{D}^{\varsigma_2}_{0+}\hbar(r),\ldots,{}^C_q\mathcal{D}^{\varsigma_n}_{0+}\hbar(r)),$$

via boundary conditions $h(0) + ah(1) = 0$ and $h'(0) + bh'(1) = 0$, in which $r \in [0,1]$, $q \in (0,1), 1 < \varsigma < 2, \varsigma_k \in (0,1)$ with $k = 1,2,\ldots,n, a, b \neq -1, \phi_m$ are defined by the rule $(\phi_m h)(r) = \int_0^r \mu_m(r,v)h(v) d_q v$ for $m = 1,2$ and $w : [0,1] \times \mathbb{R}^{n+3} \to \mathbb{R}$ is assumed to be continuous with respect to all $(n+4)$ variables [33].

Stimulated by the above research studies, the following proposed nonlinear Caputo fractional quantum BVP is furnished with the fractional quantum integro-conditions:

$$\begin{cases} {}^C_q\mathcal{D}^{\varsigma}_{0+}\hbar(r) = \varphi_*(r,\hbar(r)), & (\varsigma \in (2,3),\ q \in (0,1)), \\ \hbar(0) + \hbar(\xi) = \ell_1 {}^R_q\mathcal{I}^{\sigma}_{0+}\hbar(1), & (\ell_1 \in \mathbb{R}^{>0}), \\ {}^C_q\mathcal{D}^{\varrho}_{0+}\hbar(0) + {}^C_q\mathcal{D}^{\varrho}_{0+}\hbar(\xi) = \ell_2 {}^R_q\mathcal{I}^{\sigma}_{0+}[{}^C_q\mathcal{D}^{\varrho}_{0+}\hbar](1), & (\ell_2 \in \mathbb{R}^{>0}), \\ {}^C_q\mathcal{D}^{1}_{0+}\hbar(0) + {}^C_q\mathcal{D}^{1}_{0+}\hbar(\xi) = \ell_3 {}^R_q\mathcal{I}^{\sigma}_{0+}[{}^C_q\mathcal{D}^{1}_{0+}\hbar](1), & (\ell_3 \in \mathbb{R}^{>0}), \end{cases} \quad (1)$$

along with its inclusion version given by

$$\begin{cases} {}^C_q\mathcal{D}^{\varsigma}_{0+}\hbar(r) \in \mathbb{T}_*(r,\hbar(r)), & (\varsigma \in (2,3),\ q \in (0,1)), \\ \hbar(0) + \hbar(\xi) = \ell_1 {}^R_q\mathcal{I}^{\sigma}_{0+}\hbar(1), & (\ell_1 \in \mathbb{R}^{>0}), \\ {}^C_q\mathcal{D}^{\varrho}_{0+}\hbar(0) + {}^C_q\mathcal{D}^{\varrho}_{0+}\hbar(\xi) = \ell_2 {}^R_q\mathcal{I}^{\sigma}_{0+}[{}^C_q\mathcal{D}^{\varrho}_{0+}\hbar](1), & (\ell_2 \in \mathbb{R}^{>0}), \\ {}^C_q\mathcal{D}^{1}_{0+}\hbar(0) + {}^C_q\mathcal{D}^{1}_{0+}\hbar(\xi) = \ell_3 {}^R_q\mathcal{I}^{\sigma}_{0+}[{}^C_q\mathcal{D}^{1}_{0+}\hbar](1), & (\ell_3 \in \mathbb{R}^{>0}), \end{cases} \quad (2)$$

where $r \in [0,1]$, $\xi \in (0,1)$, $\varrho \in (1,2)$ and $\sigma > 0$. Two operators ${}^C_q\mathfrak{D}_{0^+}^{(\cdot)}$ and ${}^R_q\mathfrak{J}_{0^+}^{(\cdot)}$ represent the Caputo quantum derivative (CpQD) and the Riemann-Liouville quantum integral (RLQI). Furthermore, continuous single-valued function $\varphi_* : [0,1] \times \mathbb{R} \to \mathbb{R}$ and multi-valued function $\mathbb{T}_* : [0,1] \times \mathbb{R} \to \mathbb{P}(\mathbb{R})$ are assumed to be arbitrary equipped with some required specifications that will be explained subsequently. In comparison to other researches on the quantum difference BVPs that were published in the literature, we here deal with two abstract and extended structures of new fractional quantum difference equations/inclusions via q-integro-difference conditions in which the existing property of the relevant solutions is derived by terms of new notions of the functional analysis such as the condensing maps and the measure of noncompactness and the approximate endpoint criterion. These procedures on the suggested q-difference-BVPs (1) and (2) have been implemented in a limited range of research studies on the quantum fractional modelings. This yields the novelty and our main motivation to finalize this manuscript.

This research scheme is outlined as follows: We present the main concepts of the quantum calculus in Section 2. Our main results caused by new fixed-point approaches about solutions' existence of quantum BVP (1) and (2) will be obtained in Section 3. In Section 4, two numerical examples will be provided to support and validate our obtained results. A conclusion about our research work will be stated in Section 5.

2. Fundamental Preliminaries

In this section, some important issues in the sense of q-calculus are discussed. We suppose that $0 < q < 1$. On the function $(m_1 - m_2)^n$ given for $n \in \mathbb{N}_0$, its q-analogue is defined by $(m_1 - m_2)^{(0)} = 1$, and

$$(m_1 - m_2)^{(n)} = \prod_{k=0}^{n-1}(m_1 - m_2 q^k),$$

so that $m_1, m_2 \in \mathbb{R}$ and $\mathbb{N}_0 := \{0,1,2,\dots\}$ [17]. Now, $n = \varsigma$ is a constant which is assumed to be contained in $\mathbb{R}$. Let us now display the follwoing q-analogue of the existing power mapping $(m_1 - m_2)^n$ in a q-fractional settings:

$$(m_1 - m_2)^{(\varsigma)} = m_1^{\varsigma} \prod_{n=0}^{\infty} \frac{1 - (\frac{m_2}{m_1})q^n}{1 - (\frac{m_2}{m_1})q^{\varsigma+n}}, \tag{3}$$

for $m_1 \neq 0$. We note that by having $m_2 = 0$, an equality $m_1^{(\varsigma)} = m_1^{\varsigma}$ is obtained immediately [17]. For the given real number $m_1 \in \mathbb{R}$, a q-number $[m_1]_q$ is expressed as:

$$[m_1]_q = \frac{1 - q^{m_1}}{1 - q} = q^{m_1-1} + \cdots + q + 1.$$

The q-Gamma function is illustrated using the following format:

$$\Gamma_q(r) = \frac{(1-q)^{(r-1)}}{(1-q)^{r-1}}, \tag{4}$$

so that $r \in \mathbb{R} \setminus \{0, -1, -2, \dots\}$ [9,17]. It is notable that $\Gamma_q(r+1) = [r]_q \Gamma_q(r)$ is valid [9]. A pseudo-code inspired by (3) and (4) is proposed in Algorithm 1 for computing various Gamma function's values in the proposed quantum settings.

Given a real-valued continuous function $\hbar$, the quantum derivative of this function can be formulated by:

$$({}_q\mathfrak{D}_{0^+}\hbar)(r) = \frac{\hbar(r) - \hbar(qr)}{(1-q)r}, \tag{5}$$

and also $({}_q\mathfrak{D}_{0^+}\hbar)(0) = \lim_{r \to 0}({}_q\mathfrak{D}_{0^+}\hbar)(r)$ [34]. Given a function $\hbar$, the quantum derivative of this function can be extended to an arbitrary higher order by $({}_q\mathfrak{D}_{0^+}^n\hbar)(r) =$

$_q\mathfrak{D}_{0^+}(\ _q\mathfrak{D}_{0^+}^{n-1}\hbar)(r)$ for any $n \in \mathbb{N}$ [34]. Obviously, we notice that $(\ _q\mathfrak{D}_{0^+}^0\hbar)(r) = \hbar(r)$. Similarly, for computing this kind of q-derivative of $\hbar$, in Algorithm 2, we propose a pseudo-code inspired by (5).

Algorithm 1 Pseudo-code for $\Gamma_q(\varsigma)$:

Require: $\varsigma \in \mathbb{R}\setminus\{0\} \cup \mathbb{Z}^-, q \in (0,1), n$
1: $w \leftarrow 1$
2: **for** $l = 0$ to n **do**
3:      $w \leftarrow w((1-q^{l+1})/(1-q^{\varsigma+l}))$
4: **end for**
5: $\Gamma_q(\varsigma) \leftarrow w/(1-q)^{\varsigma-1}$
Ensure: $\Gamma_q(\varsigma)$

Algorithm 2 Pseudo-code for $_q\mathfrak{D}_{0^+}\hbar(r)$:

Require: $q \in (0,1), \hbar(r), r$
1: syms b
2: **if** $r = 0$ **then**
3:      $\phi \leftarrow \lim((\hbar(b) - \hbar(q*b))/((1-q)b), b, 0)$
4: **else**
5:      $\phi \leftarrow (\hbar(r) - \hbar(q*r))/((1-q)*r)$
6: **end if**
Ensure: $_q\mathfrak{D}_{0^+}\hbar(r)$

Given continuous map $\hbar : [0, m_2] \to \mathbb{R}$, the quantum integral of this function can be expressed as:

$$(\ _q\mathfrak{J}_{0^+}\hbar)(r) = \int_0^r \hbar(v)\, d_q v = r(1-q)\sum_{k=0}^{\infty} \hbar(rq^k)q^k, \quad (r \in [0, m_2]) \tag{6}$$

provided the absolute convergence of the existing series holds [34]. The quantum integral of $\hbar$ can be similarly extended like quantum derivative to an arbitrary higher order using an iterative rule $(\ _q\mathfrak{J}_{0^+}^n\hbar)(r) = \ _q\mathfrak{J}_{0^+}(\ _q\mathfrak{J}_{0^+}^{n-1}\hbar)(r)$ for all $n \geq 1$ [34]. Moreover, it is clear to note that $(\ _q\mathfrak{J}_{0^+}^0\hbar)(r) = \hbar(r)$. A pseudo-code caused by (6) is proposed in in Algorithm 3. We now suppose that $m_1 \in [0, m_2]$. This time, the similar q-operator of $\hbar$ from m_1 to m_2 can be defined in this case as follows:

$$\int_{m_1}^{m_2} \hbar(v)\, d_q v = \ _q\mathfrak{J}_{0^+}\hbar(m_2) - \ _q\mathfrak{J}_{0^+}\hbar(m_1)$$

$$= \int_0^{m_2} \hbar(v)\, d_q v - \int_0^{m_1} \hbar(v)\, d_q v$$

$$= (1-q)\sum_{k=0}^{\infty}[m_2\hbar(m_2 q^k) - m_1\hbar(m_1 q^k)]q^k, \tag{7}$$

when the series exists [34]. A proposed pseudo-code caused by (7) is organized in Algorithm 4 for such a purpose.

If we assume that a function $\hbar$ is continuous at $r = 0$, then $(\ _q\mathfrak{J}_{0^+}\ _q\mathfrak{D}_{0^+}\hbar)(r) = \hbar(r) - \hbar(0)$ is obtained [34]. Moreover, the equality $(\ _q\mathfrak{D}_{0^+}\ _q\mathfrak{J}_{0^+}\hbar)(r) = \hbar(r)$ holds for each r. By considering a real number $\varsigma \geq 0$ in this case such that $n - 1 < \varsigma < n$, i.e., $n = [\varsigma] + 1$, for given function $\hbar \in \mathcal{C}_\mathbb{R}([0, +\infty))$, the RLQI of $\hbar$ is introduced by:

$$_q^R\mathfrak{J}_{0^+}^\varsigma \hbar(r) = \frac{1}{\Gamma_q(\varsigma)}\int_0^r (r - qv)^{(\varsigma-1)}\hbar(v)\, d_q v, \quad \varsigma > 0$$

provided that the above value is finite and ${}^R_q\mathfrak{J}^0_{0+}\hbar(r) = \hbar(r)$ [35,36]. Further, the semi-group specification for the mentioned q-operator occurs such that ${}^R_q\mathfrak{J}^{\varsigma_1}_{0+}({}^R_q\mathfrak{J}^{\varsigma_2}_{0+}\hbar)(r) = {}^R_q\mathfrak{J}^{\varsigma_1+\varsigma_2}_{0+}\hbar(r)$ for $\sigma_1, \sigma_2 \geq 0$ [35]. For $\theta \in (-1, \infty)$,

$$ {}^R_q\mathfrak{J}^{\varsigma}_{0+} r^\theta = \frac{\Gamma_q(\theta+1)}{\Gamma_q(\theta+\varsigma+1)} r^{\theta+\varsigma}, \quad (r > 0). $$

It is evident that if we take $\theta = 0$, then ${}^R_q\mathfrak{J}^{\varsigma}_{0+} 1(r) = \frac{1}{\Gamma_q(\varsigma+1)} r^\varsigma$ for any $r > 0$. Given a function $\hbar \in C^{(n)}_\mathbb{R}([0, +\infty))$, the CpQD for this function is formulated by:

$$ {}^C_q\mathfrak{D}^{\varsigma}_{0+}\hbar(r) = \frac{1}{\Gamma_q(n-\varsigma)} \int_0^r (r - qv)^{(n-\varsigma-1)} {}_q\mathfrak{D}^n_{0+}\hbar(v)\, d_q v, $$

if the integral exists [35,36]. The following property is valid:

$$ {}^C_q\mathfrak{D}^{\varsigma}_{0+} r^\theta = \frac{\Gamma_q(\theta+1)}{\Gamma_q(\theta-\varsigma+1)} r^{\theta-\varsigma}, \quad (r > 0). $$

It is evident that ${}^C_q\mathfrak{D}^{\varsigma}_{0+} 1(r) = 0$ for any $r > 0$. For instance, by letting $\theta = 2$, $q = 0.5$ and $\hbar(r) = r^2$, we have

$$ {}^C_{0.5}\mathfrak{D}^{\varsigma}_{0+} r^2 = \frac{\Gamma_{0.5}(3)}{\Gamma_{0.5}(3-\varsigma)} r^{2-\varsigma}. $$

In this direction, the graph of the CpQD for the function $\hbar(r) = r^2$ for $q = 0.5$ is available in Figure 1.

Algorithm 3 Pseudo-code for ${}_q\mathfrak{J}^{\varsigma}_{0+}\hbar(r)$:

Require: $\varsigma, n, \hbar(r), r, q \in (0,1)$
1: $P \leftarrow 0$
2: for $k = 0$ to n do
3: $\phi \leftarrow (1 - q^{k+1})^{\varsigma-1}$
4: $P \leftarrow P + \phi * q^k * \hbar(r * q^k)$
5: end for
6: $\psi \leftarrow (r^\varsigma * (1-q) * P)/(\Gamma_q(r))$
Ensure: ${}_q\mathfrak{J}^{\varsigma}_{0+}\hbar(r)$

Algorithm 4 Pseudo-code for $\int_{m_1}^{m_2} \hbar(v)\, d_q v$:

Require: $\hbar(r), m_1, k, m_2, q \in (0,1)$
1: $P \leftarrow 0$
2: for $l = 0 : k$ do
3: $P \leftarrow P + q^l * (m_2 * \hbar(m_2 * q^l) - m_1 * \hbar(m_1 * q^l))$
4: end for
5: $\phi \leftarrow (1-q) * P$
Ensure: $\int_{m_1}^{m_2} \hbar(v)\, d_q v$

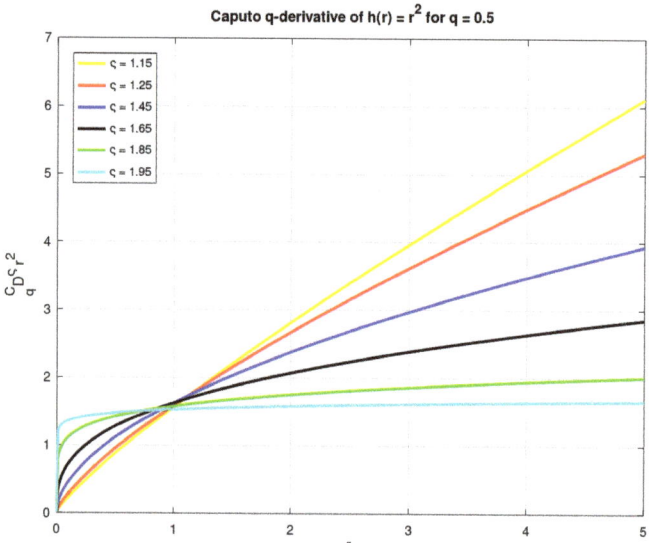

Figure 1. The graph of the Caputo q-derivative of $\hbar(r) = r^2$ for $q = 0.5$.

Lemma 1 ([37]). *Assume that $n - 1 < \varsigma < n$ and $\hbar \in C_{\mathbb{R}}^{(n)}([0, +\infty))$. Then, we have:*

$$({}_q^C\mathfrak{I}_{0+}^{\varsigma}\,{}_q^C\mathfrak{D}_{0+}^{\varsigma}\hbar)(r) = \hbar(r) - \sum_{k=0}^{n-1} \frac{r^k}{\Gamma_q(k+1)}({}_q\mathfrak{D}_{0+}^{k}\hbar)(0).$$

According to the above lemma, the given fractional quantum differential equation, ${}_q^C\mathfrak{D}_{0+}^{\varsigma}\hbar(r) = 0$, has a general solution which is obtained by $\hbar(r) = \tilde{\mu}_0 + \tilde{\mu}_1 r + \tilde{\mu}_2 r^2 + \cdots + \tilde{\mu}_{n-1} r^{n-1}$ so that $\tilde{\mu}_0, \ldots, \tilde{\mu}_{n-1} \in \mathbb{R}$, and $n = [\varsigma] + 1$ [37]. It is worth noting that for each continuous $\hbar$, according to Lemma 1, we get:

$$({}_q^R\mathfrak{I}_{0+}^{\varsigma}\,{}_q^C\mathfrak{D}_{0+}^{\varsigma}\hbar)(r) = \hbar(r) + \tilde{\mu}_0 + \tilde{\mu}_1 r + \tilde{\mu}_2 r^2 + \cdots + \tilde{\mu}_{n-1} r^{n-1},$$

where $\tilde{\mu}_0, \ldots, \tilde{\mu}_{n-1}$ illustrate constants contained in $\mathbb{R}$, and $n = [\varsigma] + 1$ [37].

Next, we recall some essential inequalities and concepts. The Kuratowski measure of noncompactness $\mathbb{O}$ is defined by

$$\mathbb{O}(\mathcal{H}) := \inf\{\varepsilon > 0 : \mathcal{H} = \bigcup_{k=1}^{n} \mathcal{H}_k \text{ and diam }(\mathcal{H}_k) \leq \varepsilon \text{ for } k = 1, \ldots, n\},$$

where $\text{diam}(\mathcal{H}_k) = \sup\{|\hbar - \hbar'| : \hbar, \hbar' \in \mathcal{H}_k\}$ and $\mathcal{H}$ is bounded subset of Banach space $\mathfrak{A}$. Moreover, it is identified that $0 \leq \mathbb{O}(\mathcal{H}) \leq \text{diam }(\mathcal{H}) < +\infty$ [38].

Lemma 2 ([38]). *Consider the bounded subsets $\mathcal{H}, \mathcal{H}_1$ and $\mathcal{H}_2$ of an arbitrary real Banach space $\mathfrak{A}$. Then, the following conditions hold:*

($\mathbb{C}_1$) $\mathbb{O}(\mathcal{H}) = 0$ iff $\mathcal{H}$ is precompact;

($\mathbb{C}_2$) $\mathbb{O}(\mathcal{H}) = \mathbb{O}(\bar{\mathcal{H}}) = \mathbb{O}(\text{cnvx}(\mathcal{H}))$, where $\bar{\mathcal{H}}$ and $\text{cnvx}(\mathcal{H})$ are the closure and convex hull of $\mathcal{H}$;

($\mathbb{C}_3$) if $\mathcal{H}_1 \subseteq \mathcal{H}_2$, then $\mathbb{O}(\mathcal{H}_1) \leq \mathbb{O}(\mathcal{H}_2)$;

($\mathbb{C}_4$) $\forall \kappa \in \mathbb{R}$, $\mathbb{O}(\kappa + \mathcal{H}) \leq \mathbb{O}(\mathcal{H})$;

(C_5) $\forall \kappa \in \mathbb{R}$, $\mathbb{O}(\kappa \mathcal{H}) = |\kappa| \mathbb{O}(\mathcal{H})$;

(C_6) $\mathbb{O}(\mathcal{H}_1 + \mathcal{H}_2) \leq \mathbb{O}(\mathcal{H}_1) + \mathbb{O}(\mathcal{H}_2)$, where $\mathcal{H}_1 + \mathcal{H}_2 = \{\hbar_1 + \hbar_2; \hbar_1 \in \mathcal{H}_1, \hbar_2 \in \mathcal{H}_2\}$;

(C_7) $\mathbb{O}(\mathcal{H}_1 \cup \mathcal{H}_2) \leq \max\{\mathbb{O}(\mathcal{H}_1) + \mathbb{O}(\mathcal{H}_2)\}$.

Lemma 3 ([39]). *Regard $\mathfrak{A}$ as a Banach space. Then, for each bounded set $\mathcal{H} \subseteq \mathfrak{A}$, a countable set $\mathcal{H}_0 \subseteq \mathcal{H}$ exists subject to $\mathbb{O}(\mathcal{H}) \leq 2\,\mathbb{O}(\mathcal{H}_0)$.*

Lemma 4 ([38]). *Regard $\mathfrak{A}$ as a Banach space. Let $\mathcal{H}$ be bounded and equi-continuous set contained in $\mathcal{C}_\mathfrak{A}([a,b])$. Then, $\mathbb{O}(\mathcal{H}(r))$ is continuous on $[a,b]$, and we have $\mathbb{O}(\mathcal{H}) = \sup_{r \in [a,b]} \mathbb{O}(\mathcal{H}(r))$.*

Lemma 5 ([38]). *Let $\mathfrak{A}$ be a Banach space. Let $\mathcal{H} = \{\hbar_n\}_{n \geq 1} \subseteq \mathcal{C}_\mathfrak{A}([a,b])$ be bounded and countable set. Then, $\mathbb{O}(\mathcal{H}(r))$ is Lebesgue integrable on $[a,b]$, and we have:*

$$\mathbb{O}\left(\left\{\int_0^r \hbar_n(v)\,dv\right\}_{n \geq 1}\right) \leq 2 \int_0^r \mathbb{O}(\{\hbar_n(v)\}_{n \geq 1})\,dv.$$

Definition 1 ([38]). *Regard $\mathfrak{A}$ as a Banach space and $\varphi_* : \mathcal{S} \subset \mathfrak{A} \to \mathfrak{A}$ as a bounded and continuous operator. Then, the map φ_* is termed condensing if for any bounded closed set $\mathcal{H} \subseteq \mathcal{S}$, the inequality $\mathbb{O}(\varphi_*(\mathcal{H})) < \mathbb{O}(\mathcal{H})$ holds.*

Theorem 1 ([38], Sadovskii's fixed point theorem). *Regard $\mathfrak{A}$ as a Banach space. Let $\mathcal{H}$ be a bounded, closed and convex set contained in $\mathfrak{A}$. Furthermore, assume that continuous mapping $\varphi_* : \mathcal{H} \to \mathcal{H}$ is condensing. Then, there exists at least one fixed point for the map φ_* in $\mathcal{H}$.*

Let us denote the normed space by $(\mathfrak{A}, \|\cdot\|_\mathfrak{A})$. Regard $\mathbb{P}(\mathfrak{A}), \mathbb{P}_{bd}(\mathfrak{A}), \mathbb{P}_{cl}(\mathfrak{A}), \mathbb{P}_{cm}(\mathfrak{A})$ and $\mathbb{P}_{cx}(\mathfrak{A})$ as a family of all non-empty, all bounded, all closed, all compact and all convex sets contained in $\mathfrak{A}$, respectively.

Definition 2 ([40]). *An element $\hbar \in \mathfrak{A}$ is termed an endpoint of a multi-valued function $\mathbb{T}_* : \mathfrak{A} \to \mathbb{P}(\mathfrak{A})$ whenever we get $\mathbb{T}_*(\hbar) = \{\hbar\}$.*

The multi-valued map $\mathbb{T}_*$ has an approximate endpoint criterion (AEPC) if

$$\inf_{\hbar_1 \in \mathfrak{A}} \sup_{\hbar_2 \in \mathbb{T}_*(\hbar_1)} d(\hbar_1, \hbar_2) = 0,$$

Ref. [40]. Next, a required theorem related to the proposed quantum boundary problem is recalled.

Theorem 2 ([40], Endpoint theorem). *Let's assume that $(\mathfrak{A}, d)$ is a complete metric space, and $\psi : [0, \infty) \to [0, \infty)$ is u.s.c subject to for each $r > 0$, $\liminf_{r \to \infty}(r - \psi(r)) > 0$, and $\psi(r) < r$. Assume that $\mathbb{T}_* : \mathfrak{A} \to \mathbb{P}_{cl,bd}(\mathfrak{A})$ is a multi-valued map such that for each $\hbar_1, \hbar_2 \in \mathfrak{A}$, the following inequality holds:*

$$\mathbb{H}_d(\mathbb{T}_* \hbar_1, \mathbb{T}_* \hbar_2) \leq \psi(d(\hbar_1, \hbar_2)).$$

Then, there is exactly one endpoint for $\mathbb{T}_*$ iff $\mathbb{T}_*$ has an approximate endpoint criterion.

3. Main Results

We regard the family of continuous functions on $[0,1]$ by $\mathfrak{A} = \mathcal{C}_\mathbb{R}([0,1])$ and the defined sup-norm $\|\hbar\|_\mathfrak{A} = \sup_{r \in [0,1]} |\hbar(r)|$, for all members $\hbar \in \mathfrak{A}$, confirms that the space $\mathfrak{A}$ becomes a Banach space. In the sequel, we will establish the existence results for quantum BVP (1) and (2). Before moving to the existence results, the following proposition will play an essential role:

Proposition 1. Let $\varphi_* \in \mathfrak{A}$, $\varsigma \in (2,3)$, $\varrho \in (1,2)$, $\xi \in (0,1)$, $\ell_1, \ell_2, \ell_3 \in \mathbb{R}^{>0}$ and $\sigma > 0$. Then, the function $\hbar^*$ satisfies as a solution for the given quantum integro-difference FBVP (CpQFP) formulated by

$$\begin{cases} {}^C_q\mathfrak{D}_{0^+}^{\varsigma} \hbar^*(r) = \varphi_*(r), & (r \in [0,1], q \in (0,1)), \\ \hbar(0) + \hbar(\xi) = \ell_1 {}^R_q\mathfrak{J}_{0^+}^{\sigma} \hbar(1), \\ {}^C_q\mathfrak{D}_{0^+}^{\varrho} \hbar(0) + {}^C_q\mathfrak{D}_{0^+}^{\varrho} \hbar(\xi) = \ell_2 {}^R_q\mathfrak{J}_{0^+}^{\sigma} [{}^C_q\mathfrak{D}_{0^+}^{\varrho} \hbar](1), \\ {}^C_q\mathfrak{D}_{0^+}^{1} \hbar(0) + {}^C_q\mathfrak{D}_{0^+}^{1} \hbar(\xi) = \ell_3 {}^R_q\mathfrak{J}_{0^+}^{\sigma} [{}^C_q\mathfrak{D}_{0^+}^{1} \hbar](1), \end{cases} \quad (8)$$

iff $\hbar^*$ is a solution for the fractional quantum integral (FQI) equation given by

$$\hbar^*(r) = \int_0^r \frac{(r-qv)^{(\varsigma-1)}}{\Gamma_q(\varsigma)} \varphi_*(v) \, d_q v + \frac{\ell_1}{\delta_1} \int_0^1 \frac{(1-qv)^{(\varsigma+\sigma-1)}}{\Gamma_q(\varsigma+\sigma)} \varphi_*(v) \, d_q v \quad (9)$$

$$- \frac{1}{\delta_1} \int_0^\xi \frac{(\xi-qv)^{(\varsigma-1)}}{\Gamma_q(\varsigma)} \varphi_*(v) \, d_q v + \ell_3 \Lambda_1(r) \int_0^1 \frac{(1-qv)^{(\varsigma+\sigma-2)}}{\Gamma_q(\varsigma+\sigma-1)} \varphi_*(v) \, d_q v$$

$$- \Lambda_1(r) \int_0^\xi \frac{(\xi-qv)^{(\varsigma-2)}}{\Gamma_q(\varsigma-1)} \varphi_*(v) \, d_q v$$

$$+ \ell_2 \Lambda_2(r) \int_0^1 \frac{(1-qv)^{(\varsigma+\sigma-\varrho-1)}}{\Gamma_q(\varsigma+\sigma-\varrho)} \varphi_*(v) \, d_q v - \Lambda_2(r) \int_0^\xi \frac{(\xi-qv)^{(\varsigma-\varrho-1)}}{\Gamma_q(\varsigma-\varrho)} \varphi_*(v) \, d_q v. \quad (10)$$

Proof. Firstly, the given function $\hbar^*$ is regarded as a solution for (8). By virtue of $\varsigma \in (2,3)$, taking the integral in the RL-settings of order ς to (8), we arrive at

$$\hbar^*(r) = \int_0^r \frac{(r-qv)^{(\varsigma-1)}}{\Gamma_q(\varsigma)} \varphi_*(v) \, d_q v + \tilde{\mu}_0 + \tilde{\mu}_1 r + \tilde{\mu}_2 r^2, \quad (11)$$

so that $\tilde{\mu}_0, \tilde{\mu}_1, \tilde{\mu}_2 \in \mathbb{R}$ are some constants that are needed to be obtained. By considering $\varrho \in (1,2)$, the following immediate results are obtained

$$^C_q\mathfrak{D}_{0^+}^{1} \hbar^*(r) = \int_0^r \frac{(r-qv)^{(\varsigma-2)}}{\Gamma_q(\varsigma-1)} \varphi_*(v) \, d_q v + \tilde{\mu}_1 + \tilde{\mu}_2(1+q)r,$$

$$^C_q\mathfrak{D}_{0^+}^{\varrho} \hbar^*(r) = \int_0^r \frac{(r-qv)^{(\varsigma-\varrho-1)}}{\Gamma_q(\varsigma-\varrho)} \varphi_*(v) \, d_q v + \tilde{\mu}_2 \frac{2r^{2-\varrho}}{\Gamma_q(3-\varrho)},$$

$$^R_q\mathfrak{J}_{0^+}^{\sigma} \hbar^*(r) = \int_0^r \frac{(r-qv)^{(\varsigma+\sigma-1)}}{\Gamma_q(\varsigma+\sigma)} \varphi_*(v) \, d_q v + \tilde{\mu}_0 \frac{r^\sigma}{\Gamma_q(\sigma+1)} + \tilde{\mu}_1 \frac{r^{\sigma+1}}{\Gamma_q(\sigma+2)}$$

$$+ \tilde{\mu}_2 \frac{(1+q)r^{\sigma+2}}{\Gamma_q(\sigma+3)},$$

$$^R_q\mathfrak{J}_{0^+}^{\sigma} \left({}^C_q\mathfrak{D}_{0^+}^{1} \hbar^*(r) \right) = \int_0^r \frac{(r-qv)^{(\varsigma+\sigma-2)}}{\Gamma_q(\varsigma+\sigma-1)} \varphi_*(v) \, d_q v + \tilde{\mu}_1 \frac{r^\sigma}{\Gamma_q(\sigma+1)} + \tilde{\mu}_2 \frac{(1+q)r^{\sigma+1}}{\Gamma_q(\sigma+2)},$$

$$^R_q\mathfrak{J}_{0^+}^{\sigma} \left({}^C_q\mathfrak{D}_{0^+}^{\varrho} \hbar^*(r) \right) = \int_0^r \frac{(r-qv)^{(\varsigma+\sigma-\varrho-1)}}{\Gamma_q(\varsigma+\sigma-\varrho)} \varphi_*(v) \, d_q v + \tilde{\mu}_2 \frac{2r^{\sigma+2-\varrho}}{\Gamma_q(\sigma+3-\varrho)}.$$

Now, by virtue of the given boundary conditions, we get

$$\tilde{\mu}_0 = \frac{\ell_1}{\delta_1} \int_0^1 \frac{(1-qv)^{(\varsigma+\sigma-1)}}{\Gamma_q(\varsigma+\sigma)} \varphi_*(v) \, d_q v - \frac{1}{\delta_1} \int_0^\xi \frac{(\xi-qv)^{(\varsigma-1)}}{\Gamma_q(\varsigma)} \varphi_*(v) \, d_q v$$

$$-\ell_3\Theta_1 \int_0^1 \frac{(1-qv)^{(\varsigma+\sigma-2)}}{\Gamma_q(\varsigma+\sigma-1)}\varphi_*(v)\,d_qv + \Theta_1 \int_0^\xi \frac{(\xi-qv)^{(\varsigma-2)}}{\Gamma_q(\varsigma-1)}\varphi_*(v)\,d_qv$$

$$+\ell_2\Theta_2 \int_0^1 \frac{(1-qv)^{(\varsigma+\sigma-\varrho-1)}}{\Gamma_q(\varsigma+\sigma-\varrho)}\varphi_*(v)\,d_qv - \Theta_2 \int_0^\xi \frac{(\xi-qv)^{(\varsigma-\varrho-1)}}{\Gamma_q(\varsigma-\varrho)}\varphi_*(v)\,d_qv,$$

$$\tilde{\mu}_1 = \frac{\ell_3}{\Delta_1}\int_0^1 \frac{(1-qv)^{(\varsigma+\sigma-2)}}{\Gamma_q(\varsigma+\sigma-1)}\varphi_*(v)\,d_qv - \frac{1}{\Delta_1}\int_0^\xi \frac{(\xi-qv)^{(\varsigma-2)}}{\Gamma_q(\varsigma-1)}\varphi_*(v)\,d_qv$$

$$-\frac{\ell_2\Delta_2}{\Delta_1\Delta_3}\int_0^1 \frac{(1-qv)^{(\varsigma+\sigma-\varrho-1)}}{\Gamma_q(\varsigma+\sigma-\varrho)}\varphi_*(v)\,d_qv + \frac{\Delta_2}{\Delta_1\Delta_3}\int_0^\xi \frac{(\xi-qv)^{(\varsigma-\varrho-1)}}{\Gamma_q(\varsigma-\varrho)}\varphi_*(v)\,d_qv,$$

and

$$\tilde{\mu}_2 = \frac{\ell_2}{\Delta_3}\int_0^1 \frac{(1-qv)^{(\varsigma+\sigma-\varrho-1)}}{\Gamma_q(\varsigma+\sigma-\varrho)}\varphi_*(v)\,d_qv - \frac{1}{\Delta_3}\int_0^\xi \frac{(\xi-qv)^{(\varsigma-\varrho-1)}}{\Gamma_q(\varsigma-\varrho)}\varphi_*(v)\,d_qv,$$

where we regard the constants

$$\delta_1 = \frac{2\Gamma_q(\sigma+1)-\ell_1}{\Gamma_q(\sigma+1)}, \quad \delta_2 = \frac{\xi\Gamma_q(\sigma+2)-\ell_1}{\Gamma_q(\sigma+2)}, \quad \delta_3 = \frac{\xi^2\Gamma_q(\sigma+3)-\ell_1(1+q)}{\Gamma_q(\sigma+3)},$$

$$\Delta_1 = \frac{2\Gamma_q(\sigma+1)-\ell_3}{\Gamma_q(\sigma+1)}, \quad \Delta_2 = \frac{(1+q)(\xi\Gamma_q(\sigma+2)-\ell_3)}{\Gamma_q(\sigma+2)},$$

$$\Delta_3 = \frac{2\xi^{2-\varrho}\Gamma_q(\sigma+3-\varrho)-2\ell_2\Gamma_q(3-\varrho)}{\Gamma_q(3-\varrho)\Gamma_q(\sigma+3-\varrho)}, \quad \Theta_1 = \frac{\delta_2}{\delta_1\Delta_1}, \quad \Theta_2 = \frac{\delta_2\Delta_2-\delta_3\Delta_1}{\delta_1\Delta_1\Delta_3},$$

along with the functions with respect to r as

$$\Lambda_1(r) = \frac{r-\Theta_1\Delta_1}{\Delta_1}, \qquad \Lambda_2(r) = \frac{r^2\Delta_1 - r\Delta_2 + \Theta_2\Delta_1\Delta_3}{\Delta_1\Delta_3}. \tag{12}$$

By substituting the values of $\tilde{\mu}_0$, $\tilde{\mu}_1$ and $\tilde{\mu}_2$ in (11), integral solution (9) is obtained. The converse part can be easily deduced. □

Remark 1. Note that for simplicity in the subsequent computations, we set the following upper bounds by virtue of the functions displayed in (12):

$$|\Lambda_1(r)| \leq \frac{1+|\Theta_1||\Delta_1|}{|\Delta_1|} := \Lambda_1^* > 0,$$

$$|\Lambda_2(r)| \leq \frac{|\Delta_1|+|\Delta_2|+|\Theta_2||\Delta_1||\Delta_3|}{|\Delta_1||\Delta_3|} := \Lambda_2^* > 0. \tag{13}$$

Theorem 3. *Let $\varphi_* : [0,1] \times \mathfrak{A} \to \mathbb{R}$ be continuous. In addition, assume that there exists a continuous $\vartheta : [0,1] \to \mathbb{R}^{>0}$ along with a nondecreasing continuous map $\wp : [0,\infty) \to (0,\infty)$ such that for each $r \in [0,1]$ and $\hbar \in \mathfrak{A}$,*

$$|\varphi_*(r,\hbar(r))| \leq \vartheta(r)\wp(\|\hbar\|_\mathfrak{A}). \tag{14}$$

We suppose that there exists a function $m_{\varphi_*} : [0,1] \to \mathbb{R}$ such that for each bounded set $\mathcal{H} \subseteq \mathfrak{A}$ and $r \in [0,1]$,

$$\mho(\varphi_*(r,\mathcal{H})) \leq m_{\varphi_*}(r)\mho(\mathcal{H}). \tag{15}$$

Then, at least one solution of the given Caputo fractional quantum BVP (1) exists on $[0,1]$ if

$$\left[\frac{\tilde{m}_{\varphi_*}}{\Gamma_q(\varsigma+1)} + \frac{\tilde{m}_{\varphi_*}}{|\delta_1|}\left(\frac{\ell_1}{\Gamma_q(\varsigma+\sigma+1)} + \frac{\xi^{(\varsigma)}}{\Gamma_q(\varsigma+1)}\right) + \tilde{m}_{\varphi_*}\Lambda_1^*\left(\frac{\ell_3}{\Gamma_q(\varsigma+\sigma)} + \frac{\xi^{(\varsigma-1)}}{\Gamma_q(\varsigma)}\right)\right.$$

$$\left. + \tilde{m}_{\varphi_*}\Lambda_2^*\left(\frac{\ell_2}{\Gamma_q(\varsigma+\sigma-\varrho+1)} + \frac{\xi^{(\varsigma-\varrho)}}{\Gamma_q(\varsigma-\varrho+1)}\right)\right] < \frac{1}{4}, \qquad (16)$$

where $\tilde{m}_{\varphi_*} = \sup_{r \in [0,1]} |m_{\varphi_*}(r)|$.

Proof. Introduce the mapping $\mathcal{G} : \mathfrak{H} \to \mathfrak{H}$ defined as:

$$\mathcal{G}(\hbar)(r) = \int_0^r \frac{(r-qv)^{(\varsigma-1)}}{\Gamma_q(\varsigma)} \varphi_*(v,\hbar(v))\, d_q v \qquad (17)$$

$$+ \frac{\ell_1}{\delta_1} \int_0^1 \frac{(1-qv)^{(\varsigma+\sigma-1)}}{\Gamma_q(\varsigma+\sigma)} \varphi_*(v,\hbar(v))\, d_q v - \frac{1}{\delta_1} \int_0^\xi \frac{(\xi-qv)^{(\varsigma-1)}}{\Gamma_q(\varsigma)} \varphi_*(v,\hbar(v))\, d_q v$$

$$+ \ell_3 \Lambda_1(r) \int_0^1 \frac{(1-qv)^{(\varsigma+\sigma-2)}}{\Gamma_q(\varsigma+\sigma-1)} \varphi_*(v,\hbar(v))\, d_q v \qquad (18)$$

$$- \Lambda_1(r) \int_0^\xi \frac{(\xi-qv)^{(\varsigma-2)}}{\Gamma_q(\varsigma-1)} \varphi_*(v,\hbar(v))\, d_q v$$

$$+ \ell_2 \Lambda_2(r) \int_0^1 \frac{(1-qv)^{(\varsigma+\sigma-\varrho-1)}}{\Gamma_q(\varsigma+\sigma-\varrho)} \varphi_*(v,\hbar(v))\, d_q v \qquad (19)$$

$$- \Lambda_2(r) \int_0^\xi \frac{(\xi-qv)^{(\varsigma-\varrho-1)}}{\Gamma_q(\varsigma-\varrho)} \varphi_*(v,\hbar(v))\, d_q v,$$

where $\mathfrak{H} = \{\hbar \in \mathfrak{A} : \|\hbar\|_\mathfrak{A} \le \varepsilon_*, \varepsilon_* \in \mathbb{R}^{>0}\} \subseteq \mathfrak{A}$ and is classified as a convex bounded closed space. Obviously, the fixed point of the proposed operator $\mathcal{G}$ is the quantum fractional BVP's solution (1).

Firstly, we verify the continuity of $\mathcal{G}$ on $\mathfrak{H}$. Take the sequence $\{\hbar_n\}_{n \ge 1}$ in $\mathfrak{H}$ such that $\hbar_n \to \hbar$ for each $\hbar \in \mathfrak{H}$. Since φ_* is continuous on $[0,1] \times \mathfrak{A}$, so we can write $\lim_{n \to \infty} \varphi_*(r, \hbar_n(r)) = \varphi_*(r, \hbar(r))$. Now, with the aid of Lebesgue dominated convergence theorem, we obtain:

$$\lim_{n \to \infty}(\mathcal{G}\hbar_n)(r) = \int_0^r \frac{(r-qv)^{(\varsigma-1)}}{\Gamma_q(\varsigma)} \lim_{n \to \infty} \varphi_*(v, \hbar_n(v))\, d_q v$$

$$+ \frac{\ell_1}{\delta_1} \int_0^1 \frac{(1-qv)^{(\varsigma+\sigma-1)}}{\Gamma_q(\varsigma+\sigma)} \lim_{n \to \infty} \varphi_*(v, \hbar_n(v))\, d_q v$$

$$- \frac{1}{\delta_1} \int_0^\xi \frac{(\xi-qv)^{(\varsigma-1)}}{\Gamma_q(\varsigma)} \lim_{n \to \infty} \varphi_*(v, \hbar_n(v))\, d_q v$$

$$+ \ell_3 \Lambda_1(r) \int_0^1 \frac{(1-qv)^{(\varsigma+\sigma-2)}}{\Gamma_q(\varsigma+\sigma-1)} \lim_{n \to \infty} \varphi_*(v, \hbar_n(v))\, d_q v$$

$$- \Lambda_1(r) \int_0^\xi \frac{(\xi-qv)^{(\varsigma-2)}}{\Gamma_q(\varsigma-1)} \lim_{n \to \infty} \varphi_*(v, \hbar_n(v))\, d_q v$$

$$+ \ell_2 \Lambda_2(r) \int_0^1 \frac{(1-qv)^{(\varsigma+\sigma-\varrho-1)}}{\Gamma_q(\varsigma+\sigma-\varrho)} \lim_{n \to \infty} \varphi_*(v, \hbar_n(v))\, d_q v$$

$$-\Lambda_2(r)\int_0^\xi \frac{(\xi-qv)^{(\varsigma-\varrho-1)}}{\Gamma_q(\varsigma-\varrho)}\lim_{n\to\infty}\varphi_*(v,\hbar_n(v))\,d_qv$$

$$=(\mathcal{G}\hbar)(r),$$

for each $r \in [0,1]$. Thus, we get $\lim_{n\to\infty}(\mathcal{G}\hbar_n)(r) = (\mathcal{G}\hbar)(r)$. Hence, the continuity of $\mathcal{G}$ on $\mathfrak{H}$ is proved. Now, we want to examine uniform boundedness of $\mathcal{G}$ on $\mathfrak{H}$. To accomplish this goal, consider $\hbar \in \mathfrak{H}$. In view of inequalities (13) and (14), we have:

$$|(\mathcal{G}\hbar)(r)| = \int_0^r \frac{(r-qv)^{(\varsigma-1)}}{\Gamma_q(\varsigma)}|\varphi_*(v,\hbar(v))|\,d_qv$$

$$+\frac{\ell_1}{|\delta_1|}\int_0^1 \frac{(1-qv)^{(\varsigma+\sigma-1)}}{\Gamma_q(\varsigma+\sigma)}|\varphi_*(v,\hbar(v))|\,d_qv$$

$$+\frac{1}{|\delta_1|}\int_0^\xi \frac{(\xi-qv)^{(\varsigma-1)}}{\Gamma_q(\varsigma)}|\varphi_*(v,\hbar(v))|\,d_qv$$

$$+\ell_3|\Lambda_1(r)|\int_0^1 \frac{(1-qv)^{(\varsigma+\sigma-2)}}{\Gamma_q(\varsigma+\sigma-1)}|\varphi_*(v,\hbar(v))|\,d_qv$$

$$+|\Lambda_1(r)|\int_0^\xi \frac{(\xi-qv)^{(\varsigma-2)}}{\Gamma_q(\varsigma-1)}|\varphi_*(v,\hbar(v))|\,d_qv$$

$$+\ell_2|\Lambda_2(r)|\int_0^1 \frac{(1-qv)^{(\varsigma+\sigma-\varrho-1)}}{\Gamma_q(\varsigma+\sigma-\varrho)}|\varphi_*(v,\hbar(v))|\,d_qv$$

$$+|\Lambda_2(r)|\int_0^\xi \frac{(\xi-qv)^{(\varsigma-\varrho-1)}}{\Gamma_q(\varsigma-\varrho)}|\varphi_*(v,\hbar(v))|\,d_qv$$

$$\leq \frac{1}{\Gamma_q(\varsigma+1)}\vartheta(r)\wp(\|\hbar\|_{\mathfrak{A}}) + \frac{\ell_1}{|\delta_1|\Gamma_q(\varsigma+\sigma+1)}\vartheta(r)\wp(\|\hbar\|_{\mathfrak{A}})$$

$$+\frac{\xi^{(\varsigma)}}{|\delta_1|\Gamma_q(\varsigma+1)}\vartheta(r)\wp(\|\hbar\|_{\mathfrak{A}})$$

$$+\frac{\ell_3\Lambda_1^*}{\Gamma_q(\varsigma+\sigma)}\vartheta(r)\wp(\|\hbar\|_{\mathfrak{A}}) + \frac{\Lambda_1^*\xi^{(\varsigma-1)}}{\Gamma_q(\varsigma)}\vartheta(r)\wp(\|\hbar\|_{\mathfrak{A}})$$

$$+\frac{\ell_2\Lambda_2^*}{\Gamma_q(\varsigma+\sigma-\varrho+1)}\vartheta(r)\wp(\|\hbar\|_{\mathfrak{A}}) + \frac{\Lambda_2^*\xi^{(\varsigma-\varrho)}}{\Gamma_q(\varsigma-\varrho+1)}\vartheta(r)\wp(\|\hbar\|_{\mathfrak{A}}).$$

Set

$$\hat{\Omega} = \frac{1}{\Gamma_q(\varsigma+1)} + \frac{1}{|\delta_1|}\left(\frac{\ell_1}{\Gamma_q(\varsigma+\sigma+1)} + \frac{\xi^{(\varsigma)}}{\Gamma_q(\varsigma+1)}\right) + \Lambda_1^*\left(\frac{\ell_3}{\Gamma_q(\varsigma+\sigma)} + \frac{\xi^{(\varsigma-1)}}{\Gamma_q(\varsigma)}\right)$$

$$+\Lambda_2^*\left(\frac{\ell_2}{\Gamma_q(\varsigma+\sigma-\varrho+1)} + \frac{\xi^{(\varsigma-\varrho)}}{\Gamma_q(\varsigma-\varrho+1)}\right). \tag{20}$$

Consequently, we can declare that $\|\mathcal{G}\hbar\|_{\mathfrak{A}} \leq \hat{\Omega}\vartheta^*\wp(\varepsilon) < \infty$, and this implies uniform boundedness of $\mathcal{G}$ on $\mathfrak{H}$. Next, we ensure the equi-continuity of $\mathcal{G}$. In order to check this, consider $r_1, r_2 \in [0, 1]$ such that $r_1 < r_2$ and $\hbar \in \mathfrak{H}$. Then, we get:

$$|(\mathcal{G}\hbar)(r_2) - (\mathcal{G}\hbar)(r_1)| \leq \int_0^{r_1} \frac{[(r_2 - qv)^{(\varsigma-1)} - (r_1 - qv)^{(\varsigma-1)}]}{\Gamma_q(\varsigma)} |\varphi_*(v, \hbar(v))| \, d_q v$$

$$+ \int_{r_1}^{r_2} \frac{(r_2 - qv)^{(\varsigma-1)}}{\Gamma_q(\varsigma)} |\varphi_*(v, \hbar(v))| \, d_q v$$

$$+ \ell_3[\Lambda_1(r_2) - \Lambda_1(r_1)] \int_0^1 \frac{(1 - qv)^{(\varsigma+\sigma-2)}}{\Gamma_q(\varsigma + \sigma - 1)} |\varphi_*(v, \hbar(v))| \, d_q v$$

$$+ [\Lambda_1(r_2) - \Lambda_1(r_1)] \int_0^{\xi} \frac{(\xi - qv)^{(\varsigma-2)}}{\Gamma_q(\varsigma - 1)} |\varphi_*(v, \hbar(v))| \, d_q v$$

$$+ \ell_2[\Lambda_2(r_2) - \Lambda_2(r_1)] \int_0^1 \frac{(1 - qv)^{(\varsigma+\sigma-\varrho-1)}}{\Gamma_q(\varsigma + \sigma - \varrho)} |\varphi_*(v, \hbar(v))| \, d_q v$$

$$+ [\Lambda_2(r_2) - \Lambda_2(r_1)] \int_0^{\xi} \frac{(\xi - qv)^{(\varsigma-\varrho-1)}}{\Gamma_q(\varsigma - \varrho)} |\varphi_*(v, \hbar(v))| \, d_q v.$$

Note that the above inequality's right hand side goes to zero as $r_1 \to r_2$ (independent of $\hbar$). Hence, it is evident that $\|(\mathcal{G}\hbar)(r_2) - (\mathcal{G}\hbar)(r_1)\|_{\mathfrak{A}} \to 0$ as $r_1 \to r_2$, and this confirms that $\mathcal{G}$ is an equi-continuous. Consequently, we conclude that $\mathcal{G}$ is a compact operator on $\mathfrak{H}$ in view of the famous Arzela–Ascoli theorem.

At this point, we will check that $\mathcal{G}$ is condensing operator on $\mathfrak{H}$. By Lemma 3, it is obvious that a countable set $\mathcal{H}_0 = \{\hbar_n\}_{n \geq 1} \subset \mathcal{H}$ exists for each bounded subset $\mathcal{H} \subset \mathfrak{H}$ such that $\mathbb{O}(\mathcal{G}(\mathcal{H})) \leq 2\mathbb{O}(\mathcal{G}(\mathcal{H}_0))$ holds. Hence, in the light of Lemmas 2, 4 and 5, the following is obtained

$$\mathbb{O}(\mathcal{G}(\mathcal{H})(r)) \leq 2\mathbb{O}(\mathcal{G}(\{\hbar_n\}_{n \geq 1}))$$

$$\leq 2 \int_0^r \frac{(r - qv)^{(\varsigma-1)}}{\Gamma_q(\varsigma)} \mathbb{O}(\varphi_*(v, \{\hbar_n(v)\}_{n \geq 1})) \, d_q v$$

$$+ \frac{2\ell_1}{|\delta_1|} \int_0^1 \frac{(1 - qv)^{(\varsigma+\sigma-1)}}{\Gamma_q(\varsigma + \sigma)} \mathbb{O}(\varphi_*(v, \{\hbar_n(v)\}_{n \geq 1})) \, d_q v$$

$$+ \frac{2}{|\delta_1|} \int_0^{\xi} \frac{(\xi - qv)^{(\varsigma-1)}}{\Gamma_q(\varsigma)} \mathbb{O}(\varphi_*(v, \{\hbar_n(v)\}_{n \geq 1})) \, d_q v$$

$$+ 2\ell_3\Lambda_1(r) \int_0^1 \frac{(1 - qv)^{(\varsigma+\sigma-2)}}{\Gamma_q(\varsigma + \sigma - 1)} \mathbb{O}(\varphi_*(v, \{\hbar_n(v)\}_{n \geq 1})) \, d_q v$$

$$+ 2\Lambda_1(r) \int_0^{\xi} \frac{(\xi - qv)^{(\varsigma-2)}}{\Gamma_q(\varsigma - 1)} \mathbb{O}(\varphi_*(v, \{\hbar_n(v)\}_{n \geq 1})) \, d_q v$$

$$+ 2\ell_2\Lambda_2(r) \int_0^1 \frac{(1 - qv)^{(\varsigma+\sigma-\varrho-1)}}{\Gamma_q(\varsigma + \sigma - \varrho)} \mathbb{O}(\varphi_*(v, \{\hbar_n(v)\}_{n \geq 1})) \, d_q v$$

$$+ 2\Lambda_2(r) \int_0^{\xi} \frac{(\xi - qv)^{(\varsigma-\varrho-1)}}{\Gamma_q(\varsigma - \varrho)} \mathbb{O}(\varphi_*(v, \{\hbar_n(v)\}_{n \geq 1})) \, d_q v$$

$$\leq 4 \int_0^r \frac{(r-qv)^{(\varsigma-1)}}{\Gamma_q(\varsigma)} m_{\varphi_*}(v) \mathbb{O}(\{\hbar_n(v)\}_{n\geq 1}) \, d_q v$$

$$+ \frac{4\ell_1}{|\delta_1|} \int_0^1 \frac{(1-qv)^{(\varsigma+\sigma-1)}}{\Gamma_q(\varsigma+\sigma)} m_{\varphi_*}(v) \mathbb{O}(\{\hbar_n(v)\}_{n\geq 1}) \, d_q v$$

$$+ \frac{4}{|\delta_1|} \int_0^\xi \frac{(\xi-qv)^{(\varsigma-1)}}{\Gamma_q(\varsigma)} m_{\varphi_*}(v) \mathbb{O}(\{\hbar_n(v)\}_{n\geq 1}) \, d_q v$$

$$+ 4\ell_3 \Lambda_1(r) \int_0^1 \frac{(1-qv)^{(\varsigma+\sigma-2)}}{\Gamma_q(\varsigma+\sigma-1)} m_{\varphi_*}(v) \mathbb{O}(\{\hbar_n(v)\}_{n\geq 1}) \, d_q v$$

$$+ 4\Lambda_1(r) \int_0^\xi \frac{(\xi-qv)^{(\varsigma-2)}}{\Gamma_q(\varsigma-1)} m_{\varphi_*}(v) \mathbb{O}(\{\hbar_n(v)\}_{n\geq 1}) \, d_q v$$

$$+ 4\ell_2 \Lambda_2(r) \int_0^1 \frac{(1-qv)^{(\varsigma+\sigma-\varrho-1)}}{\Gamma_q(\varsigma+\sigma-\varrho)} m_{\varphi_*}(v) \mathbb{O}(\{\hbar_n(v)\}_{n\geq 1}) \, d_q v$$

$$+ 4\Lambda_2(r) \int_0^\xi \frac{(\xi-qv)^{(\varsigma-\varrho-1)}}{\Gamma_q(\varsigma-\varrho)} m_{\varphi_*}(v) \mathbb{O}(\{\hbar_n(v)\}_{n\geq 1}) \, d_q v$$

$$\leq 4 \tilde{m}_{\varphi_*} \mathbb{O}(\mathcal{H}) \int_0^r \frac{(r-qv)^{(\varsigma-1)}}{\Gamma_q(\varsigma)} \, d_q v$$

$$+ \frac{4\ell_1 \tilde{m}_{\varphi_*} \mathbb{O}(\mathcal{H})}{|\delta_1|} \int_0^1 \frac{(1-qv)^{(\varsigma+\sigma-1)}}{\Gamma_q(\varsigma+\sigma)} \, d_q v + \frac{4 \tilde{m}_{\varphi_*} \mathbb{O}(\mathcal{H})}{|\delta_1|} \int_0^\xi \frac{(\xi-qv)^{(\varsigma-1)}}{\Gamma_q(\varsigma)} \, d_q v$$

$$+ 4\ell_3 \Lambda_1^* \tilde{m}_{\varphi_*} \mathbb{O}(\mathcal{H}) \int_0^1 \frac{(1-qv)^{(\varsigma+\sigma-2)}}{\Gamma_q(\varsigma+\sigma-1)} \, d_q v$$

$$+ 4\Lambda_1^* \tilde{m}_{\varphi_*} \mathbb{O}(\mathcal{H}) \int_0^\xi \frac{(\xi-qv)^{(\varsigma-2)}}{\Gamma_q(\varsigma-1)} \, d_q v$$

$$+ 4\ell_2 \Lambda_2^* \tilde{m}_{\varphi_*} \mathbb{O}(\mathcal{H}) \int_0^1 \frac{(1-qv)^{(\varsigma+\sigma-\varrho-1)}}{\Gamma_q(\varsigma+\sigma-\varrho)} \, d_q v$$

$$+ 4\Lambda_2^* \tilde{m}_{\varphi_*} \mathbb{O}(\mathcal{H}) \int_0^\xi \frac{(\xi-qv)^{(\varsigma-\varrho-1)}}{\Gamma_q(\varsigma-\varrho)} \, d_q v$$

$$\leq \frac{4 \tilde{m}_{\varphi_*} \mathbb{O}(\mathcal{H})}{\Gamma_q(\varsigma+1)} + \frac{4\ell_1 \tilde{m}_{\varphi_*} \mathbb{O}(\mathcal{H})}{|\delta_1|\Gamma_q(\varsigma+\sigma+1)} + \frac{4\xi^{(\varsigma)} \tilde{m}_{\varphi_*} \mathbb{O}(\mathcal{H})}{|\delta_1|\Gamma_q(\varsigma+1)} + \frac{4\ell_3 \Lambda_1^* \tilde{m}_{\varphi_*} \mathbb{O}(\mathcal{H})}{\Gamma_q(\varsigma+\sigma)}$$

$$+ \frac{4\xi^{(\varsigma-1)} \Lambda_1^* \tilde{m}_{\varphi_*} \mathbb{O}(\mathcal{H})}{\Gamma_q(\varsigma)} + \frac{4\ell_2 \Lambda_2^* \tilde{m}_{\varphi_*} \mathbb{O}(\mathcal{H})}{\Gamma_q(\varsigma+\sigma-\varrho+1)} + \frac{4\xi^{(\varsigma-\varrho)} \Lambda_2^* \tilde{m}_{\varphi_*} \mathbb{O}(\mathcal{H})}{\Gamma_q(\varsigma-\varrho+1)}.$$

Hence,

$$\mathbb{O}(\mathcal{G}(\mathcal{H})) \leq 4 \left[\frac{\tilde{m}_{\varphi_*}}{\Gamma_q(\varsigma+1)} + \frac{\tilde{m}_{\varphi_*}}{|\delta_1|} \left(\frac{\ell_1}{\Gamma_q(\varsigma+\sigma+1)} + \frac{\xi^{(\varsigma)}}{\Gamma_q(\varsigma+1)} \right) \right.$$

$$+ \tilde{m}_{\varphi_*} \Lambda_1^* \left(\frac{\ell_3}{\Gamma_q(\varsigma+\sigma)} + \frac{\xi^{(\varsigma-1)}}{\Gamma_q(\varsigma)} \right)$$

$$\left. + \tilde{m}_{\varphi_*} \Lambda_2^* \left(\frac{\ell_2}{\Gamma_q(\varsigma+\sigma-\varrho+1)} + \frac{\xi^{(\varsigma-\varrho)}}{\Gamma_q(\varsigma-\varrho+1)} \right) \right] \mathbb{O}(\mathcal{H}).$$

By applying condition (16), we get $\mathbb{O}(\mathcal{G}(\mathcal{H})) < \mathbb{O}(\mathcal{H})$. This clearly implies that $\mathcal{G}$ is condensing operator on $\mathfrak{H}$. Ultimately, by employing Theorem 1, we can infer that the map $\mathcal{G}$ possesses one fixed point leastwise in $\mathfrak{H}$. Thus, it is found at least one solution for the supposed quantum-integro-difference FBVP (1) and finally the proof process is terminated. □

Now, we set up an existence criterion for the given fractional quantum inclusion BVP (2). The inclusion problem's solution (2) is determined by an absolutely continuous function $\hbar : [0,1] \to \mathbb{R}$ whenever it satisfies the given fractional quantum integro-difference conditions, and a function $\mathfrak{z} \in \mathcal{L}^1([0,1],\mathbb{R})$ exists such that the inclusion $\mathfrak{z}(r) \in \mathbb{T}_*(r,\hbar(r))$ holds for almost all $r \in [0,1]$, and we have:

$$\hbar(r) = \int_0^r \frac{(r-qv)^{(\varsigma-1)}}{\Gamma_q(\varsigma)} \mathfrak{z}(v)\, d_q v + \frac{\ell_1}{\delta_1} \int_0^1 \frac{(1-qv)^{(\varsigma+\sigma-1)}}{\Gamma_q(\varsigma+\sigma)} \mathfrak{z}(v)\, d_q v$$

$$- \frac{1}{\delta_1} \int_0^{\xi} \frac{(\xi-qv)^{(\varsigma-1)}}{\Gamma_q(\varsigma)} \mathfrak{z}(v)\, d_q v$$

$$+ \ell_3 \Lambda_1(r) \int_0^1 \frac{(1-qv)^{(\varsigma+\sigma-2)}}{\Gamma_q(\varsigma+\sigma-1)} \mathfrak{z}(v)\, d_q v - \Lambda_1(r) \int_0^{\xi} \frac{(\xi-qv)^{(\varsigma-2)}}{\Gamma_q(\varsigma-1)} \mathfrak{z}(v)\, d_q v$$

$$+ \ell_2 \Lambda_2(r) \int_0^1 \frac{(1-qv)^{(\varsigma+\sigma-\varrho-1)}}{\Gamma_q(\varsigma+\sigma-\varrho)} \mathfrak{z}(v)\, d_q v - \Lambda_2(r) \int_0^{\xi} \frac{(\xi-qv)^{(\varsigma-\varrho-1)}}{\Gamma_q(\varsigma-\varrho)} \mathfrak{z}(v)\, d_q v,$$

for each $r \in [0,1]$. Let $\mathfrak{S}_{\mathbb{T}_*,\hbar}$ represents the collection of all selections of $\mathbb{T}_*$ for each $\hbar \in \mathfrak{A}$ and is defined as

$$\mathfrak{S}_{\mathbb{T}_*,\hbar} = \{\mathfrak{z} \in \mathcal{L}^1([0,1]) : \mathfrak{z}(r) \in \mathbb{T}_*(r,\hbar(r)) \text{ for almost all } r \in [0,1]\}.$$

Construct a multi-valued map $\mathcal{J} : \mathfrak{A} \to \mathbb{P}(\mathfrak{A})$ which is defined as

$$\mathcal{J}(\hbar) = \{\mathfrak{h} \in \mathfrak{A} : \mathfrak{h}(r) = \varpi(r)\}, \tag{21}$$

where

$$\varpi(r) = \int_0^r \frac{(r-qv)^{(\varsigma-1)}}{\Gamma_q(\varsigma)} \mathfrak{z}(v)\, d_q v + \frac{\ell_1}{\delta_1} \int_0^1 \frac{(1-qv)^{(\varsigma+\sigma-1)}}{\Gamma_q(\varsigma+\sigma)} \mathfrak{z}(v)\, d_q v$$

$$- \frac{1}{\delta_1} \int_0^{\xi} \frac{(\xi-qv)^{(\varsigma-1)}}{\Gamma_q(\varsigma)} \mathfrak{z}(v)\, d_q v$$

$$+ \ell_3 \Lambda_1(r) \int_0^1 \frac{(1-qv)^{(\varsigma+\sigma-2)}}{\Gamma_q(\varsigma+\sigma-1)} \mathfrak{z}(v)\, d_q v - \Lambda_1(r) \int_0^{\xi} \frac{(\xi-qv)^{(\varsigma-2)}}{\Gamma_q(\varsigma-1)} \mathfrak{z}(v)\, d_q v$$

$$+ \ell_2 \Lambda_2(r) \int_0^1 \frac{(1-qv)^{(\varsigma+\sigma-\varrho-1)}}{\Gamma_q(\varsigma+\sigma-\varrho)} \mathfrak{z}(v)\, d_q v$$

$$- \Lambda_2(r) \int_0^{\xi} \frac{(\xi-qv)^{(\varsigma-\varrho-1)}}{\Gamma_q(\varsigma-\varrho)} \mathfrak{z}(v)\, d_q v, \quad \mathfrak{z} \in \mathfrak{S}_{\mathbb{T}_*,\hbar}.$$

Theorem 4. *Let $\mathbb{T}_* : [0,1] \times \mathfrak{A} \to \mathbb{P}_{cm}(\mathfrak{A})$ be a multi-valued map. Suppose that*

($\mathcal{A}_1$) *an increasing u.s.c map $\psi : [0,\infty) \to [0,\infty)$ exists such that $\liminf_{r \to \infty}(r - \psi(r)) > 0$, and $\psi(r) < r$ for every $r > 0$;*

($\mathcal{A}_2$) $\mathbb{T}_* : [0,1] \times \mathfrak{A} \to \mathbb{P}_{cm}(\mathfrak{A})$ *is integrable and bounded and $\mathbb{T}_*(\cdot,\hbar) : [0,1] \to \mathbb{P}_{cm}(\mathfrak{A})$ is measurable for every $\hbar \in \mathfrak{A}$;*

($\mathcal{A}_3$) $\zeta \in \mathcal{C}([0,1],[0,\infty))$ exists subject to

$$\mathbb{H}_d(\mathbb{T}_*(r,\hbar_1(r)),\mathbb{T}_*(r,\hbar_2(r))) \leq \zeta(r)\psi(|\hbar_1(r) - \hbar_2(r)|)\frac{1}{\mathcal{Q}},$$

for each $r \in [0,1]$ and $\hbar_1, \hbar_2 \in \mathfrak{A}$, where $\sup_{r \in [0,1]} |\zeta(r)| = \|\zeta\|$ and

$$\mathcal{Q} = \left[\frac{1}{\Gamma_q(\varsigma+1)} + \frac{1}{|\delta_1|}\left(\frac{\ell_1}{\Gamma_q(\varsigma+\sigma+1)} + \frac{\xi^{(\varsigma)}}{\Gamma_q(\varsigma+1)}\right) + \Lambda_1^*\left(\frac{\ell_3}{\Gamma_q(\varsigma+\sigma)} + \frac{\xi^{(\varsigma-1)}}{\Gamma_q(\varsigma)}\right)\right.$$

$$\left. + \Lambda_2^*\left(\frac{\ell_2}{\Gamma_q(\varsigma+\sigma-\varrho+1)} + \frac{\xi^{(\varsigma-\varrho)}}{\Gamma_q(\varsigma-\varrho+1)}\right)\right]\|\zeta\|; \tag{22}$$

($\mathcal{A}_4$) the multi-valued map $\mathcal{J} : \mathfrak{A} \to \mathbb{P}(\mathfrak{A})$ formulated in (21) satisfies approximate endpoint criterion.

Then, a solution is found for the given quantum-difference inclusion FBVP (2).

Proof. We are going to determine that an endpoint exists for the multifunction $\mathcal{J} : \mathfrak{A} \to \mathbb{P}(\mathfrak{A})$ given by (21). Since the map $r \to \mathbb{T}_*(r,\hbar(r))$ is measurable and closed-valued set-valued mappingl therefore, it has a measurable selection. As a result, $\mathfrak{S}_{\mathbb{T}_*,\hbar} \neq \emptyset$. Firstly, we show that $\mathcal{J}(\hbar)$ is closed for every $\hbar \in \mathfrak{A}$. Consider the sequence $\{\hbar_n\}_{n\geq 1}$ in $\mathcal{J}(\hbar)$ such that $\hbar_n$ converges to $\hbar$. For each n, there exists $\mathfrak{z}_n \in \mathfrak{S}_{\mathbb{T}_*,\hbar}$ such that

$$\hbar_n(r) = \int_0^r \frac{(r-qv)^{(\varsigma-1)}}{\Gamma_q(\varsigma)}\mathfrak{z}_n(v)\,d_qv + \frac{\ell_1}{\delta_1}\int_0^1 \frac{(1-qv)^{(\varsigma+\sigma-1)}}{\Gamma_q(\varsigma+\sigma)}\mathfrak{z}_n(v)\,d_qv$$

$$- \frac{1}{\delta_1}\int_0^\xi \frac{(\xi-qv)^{(\varsigma-1)}}{\Gamma_q(\varsigma)}\mathfrak{z}_n(v)\,d_qv$$

$$+ \ell_3\Lambda_1(r)\int_0^1 \frac{(1-qv)^{(\varsigma+\sigma-2)}}{\Gamma_q(\varsigma+\sigma-1)}\mathfrak{z}_n(v)\,d_qv - \Lambda_1(r)\int_0^\xi \frac{(\xi-qv)^{(\varsigma-2)}}{\Gamma_q(\varsigma-1)}\mathfrak{z}_n(v)\,d_qv$$

$$+ \ell_2\Lambda_2(r)\int_0^1 \frac{(1-qv)^{(\varsigma+\sigma-\varrho-1)}}{\Gamma_q(\varsigma+\sigma-\varrho)}\mathfrak{z}_n(v)\,d_qv - \Lambda_2(r)\int_0^\xi \frac{(\xi-qv)^{(\varsigma-\varrho-1)}}{\Gamma_q(\varsigma-\varrho)}\mathfrak{z}_n(v)\,d_qv,$$

for almost all $r \in [0,1]$. Since the multi-valued function $\mathbb{T}_*$ is compact, we have a subsequence $\{\mathfrak{z}_n\}_{n\geq 1}$ converging to $\mathfrak{z} \in \mathcal{L}^1([0,1])$. Thus, $\mathfrak{z} \in \mathfrak{S}_{\mathbb{T}_*,\hbar}$ and

$$\lim_{n\to\infty} \hbar_n(r) = \int_0^r \frac{(r-qv)^{(\varsigma-1)}}{\Gamma_q(\varsigma)}\mathfrak{z}(v)\,d_qv$$

$$+ \frac{\ell_1}{\delta_1}\int_0^1 \frac{(1-qv)^{(\varsigma+\sigma-1)}}{\Gamma_q(\varsigma+\sigma)}\mathfrak{z}(v)\,d_qv - \frac{1}{\delta_1}\int_0^\xi \frac{(\xi-qv)^{(\varsigma-1)}}{\Gamma_q(\varsigma)}\mathfrak{z}(v)\,d_qv$$

$$+ \ell_3\Lambda_1(r)\int_0^1 \frac{(1-qv)^{(\varsigma+\sigma-2)}}{\Gamma_q(\varsigma+\sigma-1)}\mathfrak{z}(v)\,d_qv - \Lambda_1(r)\int_0^\xi \frac{(\xi-qv)^{(\varsigma-2)}}{\Gamma_q(\varsigma-1)}\mathfrak{z}(v)\,d_qv$$

$$+ \ell_2\Lambda_2(r)\int_0^1 \frac{(1-qv)^{(\varsigma+\sigma-\varrho-1)}}{\Gamma_q(\varsigma+\sigma-\varrho)}\mathfrak{z}(v)\,d_qv - \Lambda_2(r)\int_0^\xi \frac{(\xi-qv)^{(\varsigma-\varrho-1)}}{\Gamma_q(\varsigma-\varrho)}\mathfrak{z}(v)\,d_qv$$

$$= \hbar(r),$$

for almost all $r \in [0,1]$. This indicates that $\hbar \in \mathcal{J}$ and therefore, $\mathcal{J}$ is closed-valued. Since $\mathbb{T}_*$ is compact multi-valued function, it is simple to check that $\mathcal{J}(\hbar)$ is bounded for all

$\hbar \in \mathfrak{A}$. At last, we prove that $\mathbb{H}_d(\mathcal{J}(\hbar_1), \mathcal{J}(\hbar_2)) \leq \psi(\|\hbar_1 - \hbar_2\|)$ holds. Let $\hbar_1, \hbar_2 \in \mathfrak{A}$ and $\tau_1 \in \mathcal{J}(\hbar_2)$. Select $\mathfrak{z}_1 \in \mathfrak{S}_{\mathbb{T}_*,\hbar}$ such that

$$\tau_1(r) = \int_0^r \frac{(r-qv)^{(\varsigma-1)}}{\Gamma_q(\varsigma)} \mathfrak{z}_1(v) \, d_q v$$

$$+ \frac{\ell_1}{\delta_1} \int_0^1 \frac{(1-qv)^{(\varsigma+\sigma-1)}}{\Gamma_q(\varsigma+\sigma)} \mathfrak{z}_1(v) \, d_q v - \frac{1}{\delta_1} \int_0^\xi \frac{(\xi-qv)^{(\varsigma-1)}}{\Gamma_q(\varsigma)} \mathfrak{z}_1(v) \, d_q v$$

$$+ \ell_3 \Lambda_1(r) \int_0^1 \frac{(1-qv)^{(\varsigma+\sigma-2)}}{\Gamma_q(\varsigma+\sigma-1)} \mathfrak{z}_1(v) \, d_q v - \Lambda_1(r) \int_0^\xi \frac{(\xi-qv)^{(\varsigma-2)}}{\Gamma_q(\varsigma-1)} \mathfrak{z}_1(v) \, d_q v$$

$$+ \ell_2 \Lambda_2(r) \int_0^1 \frac{(1-qv)^{(\varsigma+\sigma-\varrho-1)}}{\Gamma_q(\varsigma+\sigma-\varrho)} \mathfrak{z}_1(v) \, d_q v - \Lambda_2(r) \int_0^\xi \frac{(\xi-qv)^{(\varsigma-\varrho-1)}}{\Gamma_q(\varsigma-\varrho)} \mathfrak{z}_1(v) \, d_q v,$$

for all $r \in [0,1]$. Since

$$\mathbb{H}_d(\mathbb{T}_*(r, \hbar_1(r)), \mathbb{T}_*(r, \hbar_2(r))) \leq \zeta(r)\psi(|\hbar_1(r) - \hbar_2(r)|) \frac{1}{Q}$$

for each $r \in [0,1]$, so there exists $\mathfrak{z}^* \in \mathbb{T}_*(r, \hbar_1(r))$ such that

$$|\mathfrak{z}_1(r) - \mathfrak{z}^*| \leq \zeta(r)\psi(|\hbar_1(r) - \hbar_2(r)|) \frac{1}{Q},$$

for each $r \in [0,1]$. Now, the multi-valued map $\mathfrak{X} : [0,1] \to \mathbb{P}(\mathfrak{A})$ is considered, which is characterized by

$$\mathfrak{X}(r) = \left\{ \mathfrak{z}^* \in \mathfrak{A} : |\mathfrak{z}_1(r) - \mathfrak{z}^*| \leq \zeta(r)\psi(|\hbar_1(r) - \hbar_2(r)|) \frac{1}{Q} \right\}.$$

Since $\mathfrak{z}_1$ and $\eta = \zeta(\psi(\hbar_1 - \hbar_2))\frac{1}{Q}$ are measurable, so it is obvious that the multifunction $\mathfrak{X} \cap \mathbb{T}_*(\cdot, \hbar(\cdot))$ is measurable. Now, select $\mathfrak{z}_2(r) \in \mathbb{T}_*(r, \hbar(r))$ such that

$$|\mathfrak{z}_1(r) - \mathfrak{z}_2(r)| \leq \zeta(r)(\psi(|\hbar_1(r) - \hbar_2(r)|)) \frac{1}{Q},$$

for all $r \in [0,1]$. Choose $\tau_2 \in \mathcal{J}(\hbar_1)$ such that

$$\tau_2(r) = \int_0^r \frac{(r-qv)^{(\varsigma-1)}}{\Gamma_q(\varsigma)} \mathfrak{z}_2(v) \, d_q v$$

$$+ \frac{\ell_1}{\delta_1} \int_0^1 \frac{(1-qv)^{(\varsigma+\sigma-1)}}{\Gamma_q(\varsigma+\sigma)} \mathfrak{z}_2(v) \, d_q v - \frac{1}{\delta_1} \int_0^\xi \frac{(\xi-qv)^{(\varsigma-1)}}{\Gamma_q(\varsigma)} \mathfrak{z}_2(v) \, d_q v$$

$$+ \ell_3 \Lambda_1(r) \int_0^1 \frac{(1-qv)^{(\varsigma+\sigma-2)}}{\Gamma_q(\varsigma+\sigma-1)} \mathfrak{z}_2(v) \, d_q v - \Lambda_1(r) \int_0^\xi \frac{(\xi-qv)^{(\varsigma-2)}}{\Gamma_q(\varsigma-1)} \mathfrak{z}_2(v) \, d_q v$$

$$+ \ell_2 \Lambda_2(r) \int_0^1 \frac{(1-qv)^{(\varsigma+\sigma-\varrho-1)}}{\Gamma_q(\varsigma+\sigma-\varrho)} \mathfrak{z}_2(v) \, d_q v - \Lambda_2(r) \int_0^\xi \frac{(\xi-qv)^{(\varsigma-\varrho-1)}}{\Gamma_q(\varsigma-\varrho)} \mathfrak{z}_2(v) \, d_q v,$$

for any $r \in [0,1]$. Then, we get

$$|\tau_1(r) - \tau_2(r)| \leq \int_0^r \frac{(r-qv)^{(\varsigma-1)}}{\Gamma_q(\varsigma)} |\mathfrak{z}_1(v) - \mathfrak{z}_2(v)| \, d_q v$$

$$+ \frac{\ell_1}{|\delta_1|} \int_0^1 \frac{(1-qv)^{(\varsigma+\sigma-1)}}{\Gamma_q(\varsigma+\sigma)} |\jmath_1(v) - \jmath_2(v)| \, d_q v$$

$$+ \frac{1}{|\delta_1|} \int_0^{\xi} \frac{(\xi-qv)^{(\varsigma-1)}}{\Gamma_q(\varsigma)} |\jmath_1(v) - \jmath_2(v)| \, d_q v$$

$$+ \ell_3 |\Lambda_1(r)| \int_0^1 \frac{(1-qv)^{(\varsigma+\sigma-2)}}{\Gamma_q(\varsigma+\sigma-1)} |\jmath_1(v) - \jmath_2(v)| \, d_q v$$

$$+ |\Lambda_1(r)| \int_0^{\xi} \frac{(\xi-qv)^{(\varsigma-2)}}{\Gamma_q(\varsigma-1)} |\jmath_1(v) - \jmath_2(v)| \, d_q v$$

$$+ \ell_2 |\Lambda_2(r)| \int_0^1 \frac{(1-qv)^{(\varsigma+\sigma-\varrho-1)}}{\Gamma_q(\varsigma+\sigma-\varrho)} |\jmath_1(v) - \jmath_2(v)| \, d_q v$$

$$+ |\Lambda_2(r)| \int_0^{\xi} \frac{(\xi-qv)^{(\varsigma-\varrho-1)}}{\Gamma_q(\varsigma-\varrho)} |\jmath_1(v) - \jmath_2(v)| \, d_q v$$

$$\leq \frac{1}{\Gamma_q(\varsigma+1)} \|\zeta\| \psi(\|\hbar_1 - \hbar_2\|) \frac{1}{\mathcal{Q}}$$

$$+ \frac{\ell_1}{|\delta_1|\Gamma_q(\varsigma+\sigma+1)} \|\zeta\| \psi(\|\hbar_1 - \hbar_2\|) \frac{1}{\mathcal{Q}} + \frac{\xi^{(\varsigma)}}{|\delta_1|\Gamma_q(\varsigma+1)} \|\zeta\| \psi(\|\hbar_1 - \hbar_2\|) \frac{1}{\mathcal{Q}}$$

$$+ \frac{\ell_3 \Lambda_1^*}{\Gamma_q(\varsigma+\sigma)} \|\zeta\| \psi(\|\hbar_1 - \hbar_2\|) \frac{1}{\mathcal{Q}} + \frac{\Lambda_1^* \xi^{(\varsigma-1)}}{\Gamma_q(\varsigma)} \|\zeta\| \psi(\|\hbar_1 - \hbar_2\|) \frac{1}{\mathcal{Q}}$$

$$+ \frac{\ell_2 \Lambda_2^*}{\Gamma_q(\varsigma+\sigma-\varrho+1)} \|\zeta\| \psi(\|\hbar_1 - \hbar_2\|) \frac{1}{\mathcal{Q}} + \frac{\Lambda_2^* \xi^{(\varsigma-\varrho)}}{\Gamma_q(\varsigma-\varrho+1)} \|\zeta\| \psi(\|\hbar_1 - \hbar_2\|) \frac{1}{\mathcal{Q}}$$

$$= \left[\frac{1}{\Gamma_q(\varsigma+1)} + \frac{1}{|\delta_1|} \left(\frac{\ell_1}{\Gamma_q(\varsigma+\sigma+1)} + \frac{\xi^{(\varsigma)}}{\Gamma_q(\varsigma+1)} \right) \right.$$

$$+ \Lambda_1^* \left(\frac{\ell_3}{\Gamma_q(\varsigma+\sigma)} + \frac{\xi^{(\varsigma-1)}}{\Gamma_q(\varsigma)} \right)$$

$$\left. + \Lambda_2^* \left(\frac{\ell_2}{\Gamma_q(\varsigma+\sigma-\varrho+1)} + \frac{\xi^{(\varsigma-\varrho)}}{\Gamma_q(\varsigma-\varrho+1)} \right) \right] \|\zeta\| \psi(\|\hbar_1 - \hbar_2\|) \frac{1}{\mathcal{Q}}$$

$$= \mathcal{Q} \psi(\|\hbar_1 - \hbar_2\|) \frac{1}{\mathcal{Q}}$$

$$= \psi(\|\hbar_1 - \hbar_2\|).$$

Thus, we get $\|\tau_1 - \tau_2\| \leq \psi(\|\hbar_1 - \hbar_2\|)$. Hence, $\mathbb{H}_d(\mathcal{J}(\hbar_1), \mathcal{J}(\hbar_1)) \leq \psi(\|\hbar_1 - \hbar_2\|)$ for each $\hbar_1, \hbar_2 \in \mathfrak{A}$. By utilizing $(\mathcal{A}_4)$, we realize that $\mathcal{J}$ has an approximate endpoint criterion. Now by employing Theorem 2, a member $\hbar^* \in \mathfrak{A}$ exists such that $\mathcal{J}(\hbar^*) = \{\hbar^*\}$. This indicates that $\hbar^*$ is the solution of the fractional quantum-difference inclusion problem (2), hence, our proof is finally completed. □

4. Numerical Examples

This section provides some interesting numerical examples to apply and validate our results in this research work.

Example 1. *Consider the following Caputo quantum-difference FBVP:*

$$\begin{cases} {}^C_{0.5}\mathfrak{D}^{2.5}_{0^+}\hbar(r) = \dfrac{3r+1}{8000e^{-r}}\sin(\hbar(r)), \\ \hbar(0) + \hbar(0.25) = (0.1)\,{}^R_{0.5}\mathfrak{I}^{0.75}_{0^+}\hbar(1), \\ {}^C_{0.5}\mathfrak{D}^{1.5}_{0^+}\hbar(0) + {}^C_{0.5}\mathfrak{D}^{1.5}_{0^+}\hbar(0.25) = (0.2)\,{}^R_{0.5}\mathfrak{I}^{0.75}_{0^+}\big[{}^C_{0.5}\mathfrak{D}^{1.5}_{0^+}\hbar\big](1), \\ {}^C_{0.5}\mathfrak{D}^{1}_{0^+}\hbar(0) + {}^C_{0.5}\mathfrak{D}^{1}_{0^+}\hbar(0.25) = (0.3)\,{}^R_{0.5}\mathfrak{I}^{0.75}_{0^+}\big[{}^C_{0.5}\mathfrak{D}^{1}_{0^+}\hbar\big](1), \end{cases} \qquad (23)$$

such that $q = 0.5$, $\ell_1 = 0.1$, $\varsigma = 2.5$, $\xi = 0.25$, $\ell_2 = 0.2$, $\sigma = 0.75$, $\varrho = 1.5$, $\ell_3 = 0.3$ and $r \in [0,1]$. Furthermore, we consider a continuous function $\varphi_*(r, \hbar(r)) : [0,1] \times \mathbb{R} \to \mathbb{R}$ constructed as:

$$\varphi_*(r, \hbar(r)) = \dfrac{3r+1}{8000e^{-r}}\sin(\hbar(r)).$$

The graph of this function is shown in Figure 2.

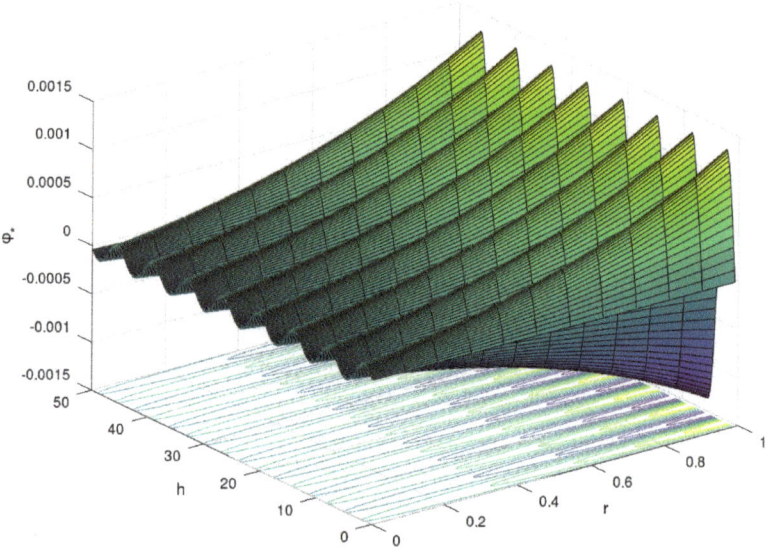

Figure 2. Graph of the function $\varphi_*(r, \hbar)$ on $[0,1] \times [0,50]$.

Then, for each $\hbar \in \mathbb{R}$, we have:

$$|\varphi_*(r, \hbar(r))| = \dfrac{3r+1}{8000e^{-r}}|\sin(\hbar(r))| \le \dfrac{3r+1}{8000e^{-r}} = \vartheta(r)\wp(\|\hbar\|_\mathbb{R}),$$

where $\vartheta : [0,1] \to \mathbb{R}^{>0}$ is a continuous function defined by $\vartheta(r) = \frac{3r+1}{8000e^{-r}}$ and $\wp : \mathbb{R}^{\ge 0} \to \mathbb{R}^{>0}$ is nondecreasing and continuous via $\wp(\|\hbar\|_\mathbb{R}) = 1$. Now, for any $\hbar_1, \hbar_2 \in \mathbb{R}$, we can write:

$$|\varphi_*(r, \hbar_1(r)) - \varphi_*(r, \hbar_2(r))| = \dfrac{3r+1}{8000e^{-r}}|\sin(\hbar_1(r)) - \sin(\hbar_2(r))|$$

$$\le \dfrac{3r+1}{8000e^{-r}}|\hbar_1(r) - \hbar_2(r)|.$$

Hence, for any bounded set $\mathcal{H}$ contained in $\mathbb{R}$, we reach

$$\mathbb{O}(\varphi_*(r,\mathcal{H})) \leq \frac{3r+1}{8000e^{-r}}\mathbb{O}(\mathcal{H}) := \tilde{m}_{\varphi_*}\mathbb{O}(\mathcal{H}).$$

We compute $\tilde{m}_{\varphi_*} = \sup_{r\in[0,1]}|m_{\varphi_*}| \simeq 0.001355$. Then, by taking into account the above calculations and the following inequality, we get

$$\left[\frac{\tilde{m}_{\varphi_*}}{\Gamma_q(\varsigma+1)} + \frac{\tilde{m}_{\varphi_*}}{|\delta_1|}\left(\frac{\ell_1}{\Gamma_q(\varsigma+\sigma+1)} + \frac{\xi^{(\varsigma)}}{\Gamma_q(\varsigma+1)}\right) + \tilde{m}_{\varphi_*}\Lambda_1^*\left(\frac{\ell_3}{\Gamma_q(\varsigma+\sigma)} + \frac{\xi^{(\varsigma-1)}}{\Gamma_q(\varsigma)}\right)\right.$$

$$\left. + \tilde{m}_{\varphi_*}\Lambda_2^*\left(\frac{\ell_2}{\Gamma_q(\varsigma+\sigma-\varrho+1)} + \frac{\xi^{(\varsigma-\varrho)}}{\Gamma_q(\varsigma-\varrho+1)}\right)\right] \simeq 0.001741 < 0.25 = \frac{1}{4}.$$

We figure out that Theorem 3 is settled. As a result, at least one solution exists for Caputo fractional quantum-difference FBVP (23).

Example 2. Consider the following Caputo fractional quantum-difference inclusion FBVP:

$$\begin{cases} {}^C_{0.8}\mathcal{D}^{2.75}_{0^+}\hbar(r) \in \left[0, \frac{5(r+1)\arctan(\hbar(r))}{256(4+3r^2)}\right], \\ \hbar(0) + \hbar(0.9) = (0.11)\,{}^R_{0.8}\mathcal{I}^{0.6}_{0^+}\hbar(1), \\ {}^C_{0.8}\mathcal{D}^{1.7}_{0^+}\hbar(0) + {}^C_{0.8}\mathcal{D}^{1.7}_{0^+}\hbar(0.9) = (0.12)\,{}^R_{0.8}\mathcal{I}^{0.6}_{0^+}[{}^C_{0.8}\mathcal{D}^{1.7}_{0^+}\hbar](1), \\ {}^C_{0.8}\mathcal{D}^{1}_{0^+}\hbar(0) + {}^C_{0.8}\mathcal{D}^{1}_{0^+}\hbar(0.9) = (0.13)\,{}^R_{0.8}\mathcal{I}^{0.6}_{0^+}[{}^C_{0.8}\mathcal{D}^{1}_{0^+}\hbar](1), \end{cases} \quad (24)$$

where $q = 0.8$, $\varsigma = 2.75$, $\xi = 0.9$, $\ell_1 = 0.11$, $\ell_2 = 0.12$, $\ell_3 = 0.13$, $\sigma = 0.6$, $\varrho = 1.7$, and $r \in [0,1]$. Now, we introduce a multi-valued function $\mathbb{T}_* : [0,1] \times \mathbb{R} \to \mathcal{P}(\mathbb{R})$ as follows:

$$\mathbb{T}_*(r,\hbar(r)) = \left[0, \frac{5(r+1)\arctan(\hbar(r))}{256(4+3r^2)}\right].$$

Next, we regard $\psi : [0,\infty) \to [0,\infty)$ as increasing upper semi-continuous function defined by $\psi(r) = \frac{r}{4}$ for any $r > 0$. It can easily be noted that $\liminf_{r\to\infty}(r - \psi(r)) > 0$ and $\psi(r) < r$ for each $r > 0$. We select $\zeta \in C([0,1], [0,\infty))$ formulated by $\zeta(r) = \frac{5(r+1)}{64(4+3r^2)}$. Thus, $\|\zeta\| \simeq 0.0390625$. For any $\hbar, \hbar^* \in \mathbb{R}$, we have:

$$\mathbb{H}_d(\mathbb{T}_*(r,\hbar(r)) - \mathbb{T}_*(r,\hbar^*(r))) = \frac{5(r+1)}{256(4+3r^2)}|\arctan(\hbar(r)) - \arctan(\hbar^*(r))|$$

$$\leq \frac{5(r+1)}{256(4+3r^2)}|\hbar(r) - \hbar^*(r)|$$

$$= \frac{5(r+1)}{64(4+3r^2)}\psi(|\hbar(r) - \hbar^*(r)|)$$

$$\leq \zeta(r)\psi(|\hbar(r) - \hbar^*(r)|)\frac{1}{Q},$$

where

$$Q = \left[\frac{1}{\Gamma_q(\varsigma+1)} + \frac{1}{|\delta_1|}\left(\frac{\ell_1}{\Gamma_q(\varsigma+\sigma+1)} + \frac{\xi^{(\varsigma)}}{\Gamma_q(\varsigma+1)}\right) + \Lambda_1^*\left(\frac{\ell_3}{\Gamma_q(\varsigma+\sigma)} + \frac{\xi^{(\varsigma-1)}}{\Gamma_q(\varsigma)}\right)\right.$$

$$+ \Lambda_2^* \left(\frac{\ell_2}{\Gamma_q(\varsigma + \sigma - \varrho + 1)} + \frac{\xi^{(\varsigma-\varrho)}}{\Gamma_q(\varsigma - \varrho + 1)} \right) \right] \|\zeta\| \simeq 0.066907.$$

The graphs of the functions: $\Lambda_1(r)$ and $\Lambda_2(r)$ for $r \in [0,1]$ are shown in Figure 3.

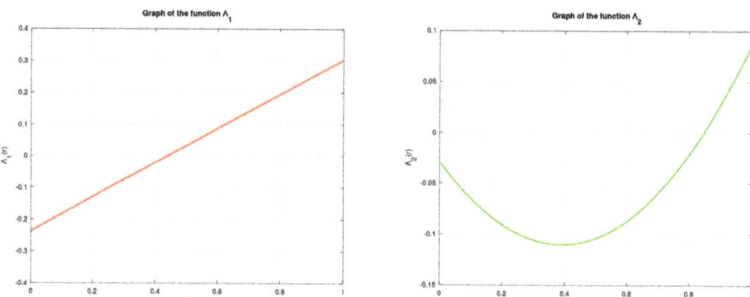

Figure 3. Graphs of functions: $\Lambda_1(r)$ and $\Lambda_2(r)$ for $r \in [0,1]$.

Next, consider the multifunction $\mathcal{J} : \mathfrak{A} \to \mathbb{P}(\mathfrak{A})$ given by:

$$\mathcal{J}(\hbar) = \{\mathfrak{h} \in \mathfrak{A} : \text{there exists } \mathfrak{z} \in \mathfrak{S}_{\mathbb{T}_*, \hbar} \text{ such that } \mathfrak{h}(r) = \varpi(r) \text{ for all } r \in [0,1]\},$$

where

$$\varpi(r) = \int_0^r \frac{(r - qv)^{(2.75-1)}}{\Gamma_q(2.75)} \mathfrak{z}(v) \, d_q v + \frac{0.11}{1.8784} \int_0^1 \frac{(1 - qv)^{(2.75+0.6-1)}}{\Gamma_q(2.75+0.6)} \mathfrak{z}(v) \, d_q v$$

$$- \frac{1}{1.8784} \int_0^{0.9} \frac{(0.9 - qv)^{(2.75-1)}}{\Gamma_q(2.75)} \mathfrak{z}(v) \, d_q v$$

$$+ (0.13)\Lambda_1(r) \int_0^1 \frac{(1 - qv)^{(2.75+0.6-2)}}{\Gamma_q(2.75 + 0.6 - 1)} \mathfrak{z}(v) \, d_q v - \Lambda_1(r) \int_0^{0.9} \frac{(0.9 - qv)^{(2.75-2)}}{\Gamma_q(2.75 - 1)} \mathfrak{z}(v) \, d_q v$$

$$+ (0.12)\Lambda_2(r) \int_0^1 \frac{(1 - qv)^{(2.75+0.6-1.7-1)}}{\Gamma_q(2.75 + 0.6 - 1.7)} \mathfrak{z}(v) \, d_q v$$

$$- \Lambda_2(r) \int_0^{0.9} \frac{(0.9 - qv)^{(2.75-1.7-1)}}{\Gamma_q(2.75 - 1.7)} \mathfrak{z}(v) \, d_q v,$$

with $\delta_1 \simeq 1.8784$ and

$$\Lambda_1(r) = 0.5387r - 0.2348 \quad \text{and} \quad \Lambda_2(r) = 0.53002r^2 - 0.4133r - 0.02959.$$

Hence, by utilizing Theorem 4, it is found a solution for the quantum-difference inclusion FBVP (24).

5. Conclusions

The proposed nonlinear Caputo quantum-difference FBVP with fractional quantum integro-conditions along with its fractional quantum-difference inclusion BVP has been studied in this work. In this direction, we proved the existence of a solution for the first quantum-difference Equation (1) with the help of some notions in topological degree theory. In other words, we defined a new operator and checked its properties and finally showed that it is a condensing function. The existence of a fixed point for this operator ensured the existence of a solution for the mentioned quantum-difference Equation (1). In the next step, we considered the inclusion version of the above FBVP which had a form as (2). To arrive at the main purpose this time for confirming the existence of solutions of (2), we used new techniques based on the approximate endpoint property and the existence of endpoints for a newly-defined multifunction. Numerical illustrative examples have been provided to display the validity and potentiality of our main results to be applied in future research works. We recommend that other researchers can study different generalizations of the proposed q-difference-FBVPs by using novel fractional difference-operators such as (p,q)-difference ones.

Author Contributions: Conceptualization, S.R., A.I., A.H. and S.E.; Formal analysis, S.R., A.I., A.H., F.M., S.E. and M.K. A.K.; Investigation, S.R., A.I., A.H., S.E. and M.K.A.K.; Methodology, S.R., A.I., S.E. and M.K.A.K.; Supervision, S.R., A.I., F.M., S.E. and M.K.A.K.; Validation, A.H., F.M., S.E. and M.K.A.K.; Writing—original draft, S.E.; Writing—review and editing, S.R., A.I., A.H., F.M., S.E. and M.K.A.K. All authors have read and agreed to the published version of the manuscript.

Funding: This research received no external funding.

Institutional Review Board Statement: Not applicable.

Informed Consent Statement: Not applicable.

Data Availability Statement: Data sharing not applicable.

Conflicts of Interest: The authors declare no conflict of interest.

References

1. Yavuz, M. European option pricing models described by fractional operators with classical and generalized Mittag-Leffler kernels. *Numer. Methods Partial Diff. Eq.* **2020**. [CrossRef]
2. Matar, M.M.; Abbas, M.I.; Alzabut, J.; Kaabar, M.K.A.; Etemad, S.; Rezapour, S. Investigation of the p-Laplacian nonperiodic nonlinear boundary value problem via generalized Caputo fractional derivatives. *Adv. Differ. Equ.* **2021**, *2021*, 68. [CrossRef]
3. Keten, A.; Yavuz, M.; Baleanu, D. Nonlocal Cauchy problem via a fractional operator involving power kernel in Banach spaces. *Fractal Fract.* **2019**, *3*, 27. [CrossRef]
4. Golmankhane, A.K.; Ali, K.K.; Yilmazer, R.; Kaabar, M.K.A. Economic models involving time fractal. *J. Math. Model. Financ.* **2020**, *1*, 181–200.
5. Martínez, F.; Martínez, I.; Kaabar, M.K.A.; Paredes, S. New results on complex conformable integral. *AIMS Math.* **2020**, *5*, 7695–7710. [CrossRef]
6. Yavuz, M.; Sene, N. Approximate solutions of the model describing fluid flow using generalized ρ-laplace transform method and heat balance integral method. *Axioms* **2020**, *9*, 123. [CrossRef]
7. Martínez, F.; Martínez, I.; Kaabar, M.K.A.; Ortíz-Munuera, R.; Paredes, S. Note on the conformable fractional derivatives and integrals of complex-valued functions of a real variable. *IAENG Int. J. Appl. Math.* **2020**, *50*, 609–615.
8. Yavuz, M. Dynamical behaviors of separated homotopy method defined by conformable operator. *Konuralp J. Math.* **2019**, *7*, 1–6.
9. Jackson, F.H. On q-definite integrals. *Quart. J. Pure Appl. Math.* **1910**, *41*, 193–203.
10. Jackson, F.H. On q-functions and a certain difference operator. *Trans. R. Soc. Edinb.* **1909**, *46*, 253–281. [CrossRef]
11. Jackson, F.H. q-difference equations. *Am. J. Math.* **1910**, *32*, 305–314. [CrossRef]
12. Al-Salam, W.A. q-analogue of the Cauchy formula. *Proc. Am. Math. Soc.* **1966**, *17*, 616–621.
13. Agarwal, R.P. Certain fractional q-integrals and q-derivatives. *Proc. Camb. Philos. Soc.* **1969**, *66*, 365–370. [CrossRef]
14. Goodrich, C.; Peterson, A. *Discrete Fractional Calculus*; Springer: Berlin, Germany, 2015.
15. Miller, W.J. Lie theory and q-difference equations. *Siam J. Math. Anal.* **1970**, *1*, 171–188. [CrossRef]
16. Purohit, S.D.; Ucar, F. An application of q-Sumudu transform for fractional q-Kinetic equation. *Turk. J. Math.* **2018**, *42*, 726–734. [CrossRef]
17. Rajkovic, P.M.; Marinkovic, S.D.; Stankovic, M.S. Fractional integrals and derivatives in q-calculus. *Appl. Anal. Disc. Math.* **2007**, *1*, 311–323.
18. Ahmad, B.; Etemad, S.; Ettefagh, M.; Rezapour, S. On the existence of solutions for fractional q-difference inclusions with q-antiperiodic boundary conditions. *Bull. Math. Soc. Sci. Math. Roum. Tome* **2016**, *59*, 119–134.
19. Ahmad, B.; Ntouyas, S.K. Existence of solutions for nonlinear fractional q-difference inclusions with nonlocal robin (separated) conditions. *Mediterr. J. Math.* **2007**, *10*, 1333–1351. [CrossRef]
20. Alzabut, J.; Mohammadaliee, B.; Samei, M.E. Solutions of two fractional q-integro-differential equations under sum and integral boundary value conditions on a time scale. *Adv. Differ. Equ.* **2020**, *2020*, 304. [CrossRef]
21. Balkani, N.; Rezapour, S.; Haghi, R.H. Approximate solutions for a fractional q-integro-difference equation. *J. Math. Ext.* **2019**, *13*, 201–214.
22. Etemad, S.; Ntouyas, S.K.; Ahmad, B. Existence theory for a fractional q-integro-difference equation with q-integral boundary conditions of different orders. *Mathematics* **2019**, *7*, 659. [CrossRef]

23. Etemad, S.; Rezapour, S.; Samei, M.E. α-ψ-contractions and solutions of a q-fractional differential inclusion with three-point boundary value conditions via computational results. *Adv. Differ. Equ.* **2020**, *2020*, 218. [CrossRef]
24. Ouncharoen, R.; Patanarapeelert, N.; Sitthiwirattham, T. Nonlocal q-symmetric integral boundary value problem for sequential q-symmetric integro-difference equations. *Mathematics* **2018**, *6*, 218. [CrossRef]
25. Samei, M.E.; Ranjbar, G.K. Some theorems of existence of solutions for fractional hybrid q-difference inclusion. *J. Adv. Math. Stud.* **2019**, *12*, 63–76.
26. Samei, M.E.; Ranjbar, G.K.; Hedayati, V. Existence of solutions for equations and inclusions of multi-term fractional q-integro-differential with non-separated and initial boundary conditions. *J. Inequal. Appl.* **2019**, *2019*, 273. [CrossRef]
27. Samei, M.E.; Rezapour, S. On a fractional q-differential inclusion on a time scale via endpoints and numerical calculations. *Adv. Differ. Equ.* **2020**, *2020*, 460. [CrossRef]
28. Sitho, S.; Sudprasert, C.; Ntouyas, S.K.; Tariboon, J. Noninstantaneous impulsive fractional quantum Hahn integro-difference boundary value problems. *Mathematics* **2020**, *8*, 671. [CrossRef]
29. Sitthiwirattham, T. On nonlocal fractional q-integral boundary value problems of fractional q-difference and fractional q-integro-difference equations involving different numbers of order and q. *Bound. Value Probl.* **2016**, *2016*, 12. [CrossRef]
30. Zhao, Y.; Chen, H.; Zhang, Q. Existence results for fractional q-difference equations with nonlocal q-integral boundary conditions. *Adv. Differ. Equ.* **2013**, *48*, 2013. [CrossRef]
31. Asawasamrit, S.; Tariboon, J.; Ntouyas, S.K. Existence of solutions for fractional q-integrodifference equations with nonlocal fractional q-integral conditions. *Abstr. Appl. Anal.* **2014**, *2014*, 474138. [CrossRef]
32. Etemad, S.; Ettefagh, M.; Rezapour, S. On the existence of solutions for nonlinear fractional q-difference equations with q-integral boundary conditions. *J. Adv. Math. Stud.* **2015**, *8*, 265–285.
33. Ntouyas, S.K.; Samei, M.E. Existence and uniqueness of solutions for multi-term fractional q-integro-differential equations via quantum calculus. *Adv. Differ. Equ.* **2019**, *475*, 2019. [CrossRef]
34. Adams, C.R. The general theory of a class of linear partial q-difference equations. *Trans. Am. Math. Soc.* **1924**, *26*, 283–312.
35. Ferreira, R.A.C. Positive solutions for a class of boundary value problems with fractional q-differences. *Comput. Math. Appl.* **2011**, *61*, 367–373. [CrossRef]
36. Graef, J.R.; Kong, L. Positive solutions for a class of higher order boundary value problems with fractional q-derivatives. *Appl. Math. Comput.* **2012**, *218*, 9682–9689. [CrossRef]
37. El-Shahed, M.; Al-Askar, F. Positive solutions for boundary value problem of nonlinear fractional q-difference equation. *ISRN Math. Anal.* **2011**, *2011*, 385459. [CrossRef]
38. Guo, D.J.; Lakshmikantham, V.; Liu, X.Z. *Nonlinear Integral Equations in Abstract Spaces*; Kluwer Academic Publishers Group: Dordrecht, The Netherland, 1996.
39. Li, Y. Existence of solutions of initial value problems for abstract semilinear evolution equations. *Acta Math. Sin. Engl. Ser. Mar.* **2005**, *48*, 1089–1094.
40. Amini-Harandi, A. Endpoints of set-valued contractions in metric spaces. *Nonlinear Anal.* **2010**, *72*, 132–134. [CrossRef]

Article

Non-Trivial Solutions of Non-Autonomous Nabla Fractional Difference Boundary Value Problems

Alberto Cabada [1,*,†], Nikolay D. Dimitrov [2,†] and Jagan Mohan Jonnalagadda [3,†]

1. Departamento de Estatística, Análise Matemática e Optimización, Instituto de Matemáticas, Facultade de Matemáticas, Universidade de Santiago de Compostela, 15782 Santiago de Compostela, Spain
2. Department of Mathematics, University of Ruse, 7017 Ruse, Bulgaria; ndimitrov@uni-ruse.bg
3. Department of Mathematics, Birla Institute of Technology and Science Pilani, Hyderabad 500078, India; jjaganmohan@hyderabad.bits-pilani.ac.in or j.jaganmohan@hotmail.com

* Correspondence: alberto.cabada@usc.gal or alberto.cabada@usc.es
† These authors contributed equally to this work.

Abstract: In this article, we present a two-point boundary value problem with separated boundary conditions for a finite nabla fractional difference equation. First, we construct an associated Green's function as a series of functions with the help of spectral theory, and obtain some of its properties. Under suitable conditions on the nonlinear part of the nabla fractional difference equation, we deduce two existence results of the considered nonlinear problem by means of two Leray–Schauder fixed point theorems. We provide a couple of examples to illustrate the applicability of the established results.

Keywords: nabla fractional difference; boundary value problem; separated boundary conditions; Green's function; existence of solutions

1. Introduction

Denote the set of all real numbers and positive real numbers by $\mathbb{R}$ and $\mathbb{R}^+$, respectively. Define by $\mathbb{N}_a = \{a, a+1, a+2, \ldots\}$ and $\mathbb{N}_a^b = \{a, a+1, a+2, \ldots, b\}$ for any $a, b \in \mathbb{R}$ such that $b - a \in \mathbb{N}_1$.

In this article, we consider the following nabla fractional difference equation associated with separated boundary conditions:

$$\begin{aligned} -\left(\nabla_a^{\nu-1}(\nabla u)\right)(t) + g(t)u(t) &= f(t, u(t)), \quad t \in \mathbb{N}_{a+2}^b, \\ \alpha u(a+1) - \beta(\nabla u)(a+1) &= 0, \\ \gamma u(b) + \delta(\nabla u)(b) &= 0. \end{aligned} \quad (1)$$

Here $a, b \in \mathbb{R}$ with $b - a \in \mathbb{N}_1$; $1 < \nu < 2$; $g : \mathbb{N}_a^b \to \mathbb{R}$; $f : \mathbb{N}_a^b \times \mathbb{R} \to \mathbb{R}$; $\nabla_a^{\nu-1}$ denotes the $(\nu - 1)$-th order Riemann–Liouville backward (nabla) difference operator; ∇ denotes the first order backward (nabla) difference operator; $\alpha, \beta, \gamma, \delta \in \mathbb{R}$ such that $\alpha^2 + \beta^2 > 0$ and $\gamma^2 + \delta^2 > 0$.

Gray and Zhang [1], Atici and Eloe [2] and Anastassiou [3] initiated the study of nabla fractional sums and differences. The combined efforts of a number of researchers has resulted in a fairly strong foundation to the basic theory of nabla fractional calculus during the past decade. For a detailed discussion on the evolution of nabla fractional calculus, we refer to the recent monograph [4] and the references therein.

We point out that problem (1) is a discrete version of the second order ordinary differential Hill's equation, which has a lot of applications in engineering and physics. We can find, among others, several problems in astronomy, circuits, electric conductivity of metals and cyclotrons. Hill's equation is named after the pioneering work of the mathematical astronomer George William Hill (1838–1914), see [5]. There is a long literature in the study of the oscillation of the solutions of such an equation and the constant sign solutions. The reader can consult the monographs [6,7] and references therein. We note that the boundary conditions cover the Sturm–Liouville conditions, which include, as particular cases, the Dirichlet, Neumann and Mixed ones.

Recently, there has been a surge of interest in the development of the theory of nabla fractional boundary value problems. Brackins [8] initiated the study of boundary value problems for linear and nonlinear nabla fractional difference equations. Following this work, several authors have studied nabla fractional boundary value problems extensively. We refer to [9–18] and the references therein to name a few.

Brackins [8] showed that for all $(t,s) \in \mathbb{N}_a^b \times \mathbb{N}_{a+1}^b$ (see Figure 1)

$$G_0(t,s) = \begin{cases} v_1(t,s), & t \in \mathbb{N}_a^{\rho(s)}, \\ v_2(t,s), & t \in \mathbb{N}_s^b \end{cases} \qquad (2)$$

is the Green's function related to the following boundary value problem:

$$\begin{cases} -\left(\nabla_a^{\nu-1}(\nabla u)\right)(t) = 0, & t \in \mathbb{N}_{a+2}^b, \\ \alpha u(a+1) - \beta(\nabla u)(a+1) = 0, \\ \gamma u(b) + \delta(\nabla u)(b) = 0. \end{cases} \qquad (3)$$

Here,

$$v_1(t,s) = \frac{1}{\xi}\Big[\alpha\gamma H_{\nu-1}(t,a)H_{\nu-1}(b,\rho(s)) + \alpha\delta H_{\nu-1}(t,a)H_{\nu-2}(b,\rho(s))$$
$$+ (\beta - \alpha)\gamma H_{\nu-1}(b,\rho(s)) + (\beta - \alpha)\delta H_{\nu-2}(b,\rho(s))\Big],$$
$$v_2(t,s) = v_1(t,s) - H_{\nu-1}(t,\rho(s)),$$
$$\xi = (\beta - \alpha)\gamma + \alpha\gamma H_{\nu-1}(b,a) + \alpha\delta H_{\nu-2}(b,a) \neq 0.$$

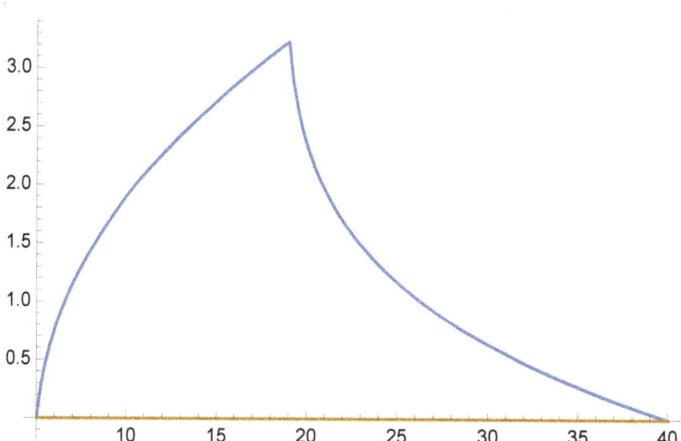

Figure 1. Graphic of $G_0(t,20)$ for $\alpha = \beta = \gamma = 1$, $\delta = 0$ (Dirichlet case), $\mu = 3/2$, $a = 5$ and $b = 40$.

This result was obtained by expressing the general solution of the nabla fractional difference equation in (3), using the method of variation of constants. Notice that, for a non-constant function g the expression of the general solution does not exist and, as a consequence, the method used in [8] is not applicable for the following boundary value problem:

$$\begin{cases} -\left(\nabla_a^{\nu-1}(\nabla u)\right)(t) + g(t)u(t) = 0, & t \in \mathbb{N}_{a+2}^b, \\ \alpha u(a+1) - \beta(\nabla u)(a+1) = 0, \\ \gamma u(b) + \delta(\nabla u)(b) = 0. \end{cases} \quad (4)$$

Due to this reason, Graef et al. [19] and Cabada et al. [20] followed a different approach. Graef et al. [19] studied the following Dirichlet problem:

$$\begin{cases} -(D_0^\mu u)(t) + g(t)u(t) = w(t)f(t,u(t)), & 0 < t < 1, \\ u(0) = u(1) = 0, \end{cases}$$

where $1 < \mu < 2$; $g : [0,1] \to \mathbb{R}$, $w : [0,1] \to \mathbb{R}^+ \cup \{0\}$, $f : [0,1] \times \mathbb{R} \to \mathbb{R}$ are continuous functions, and D_0^μ denotes the μth-th order Riemann–Liouville fractional derivative. Cabada et al. [20] studied the following Dirichlet problem:

$$-(\Delta^\mu u)(t) + g(t+\mu-1)u(t+\mu-1) = w(t)f(t+\mu-1, u(t+\mu-1)),$$

$$u(\mu-2) = u(\mu+b+1) = 0,$$

where $t \in \mathbb{N}_0^{b+1}$, $b \in \mathbb{N}_5$; $1 < \mu < 2$; $g, w : \mathbb{N}_0^{b+1} \to \mathbb{R}$ with $w \not\equiv 0$ on $\mathbb{N}_0^{b+1}$; $f : \mathbb{N}_{\mu-1}^{\mu+b} \times \mathbb{R} \to \mathbb{R}$ is a continuous function, and Δ^μ denotes the μ-th order Riemann–Liouville forward (delta) difference operator.

Similar to these works, we obtained the Green's function related to (4) as a series of functions by using the spectral theory. Then, under suitable conditions on g, w and f, we proved the existence of at least one solution of the boundary value problem (1). This work provides a new approach for constructing Green's functions for nabla fractional boundary value problems.

This article is organized as follows: In Section 2, we recall some definitions and preliminary results. In Section 3, we obtain the Green's function related to (4), and deduce some of its important properties. In Section 4, we establish a couple of existence results for the boundary value problem (1), using two different Leray–Schauder fixed point theorems and under different assumptions on the data of the problem. Finally, we give some examples to demonstrate the applicability of these results.

2. Preliminaries

In this section, we recall some elementary definitions and fundamental facts of nabla fractional calculus, which will be used throughout the article. Denote by $\mathbb{N}_a = \{a, a+1, a+2, \ldots\}$ and $\mathbb{N}_a^b = \{a, a+1, a+2, \ldots, b\}$ for any $a, b \in \mathbb{R}$ such that $b - a \in \mathbb{N}_1$. The backward jump operator $\rho : \mathbb{N}_{a+1} \to \mathbb{N}_a$ is defined by

$$\rho(t) = \max\{a, t-1\}, \quad t \in \mathbb{N}_a.$$

The Euler gamma function is defined by

$$\Gamma(z) = \int_0^\infty t^{z-1} e^{-t} dt, \quad \Re(z) > 0.$$

Using its reduction formula, the Euler gamma function can also be extended to the half-plane $\Re(z) \leq 0$ except for $z \in \{\ldots, -2, -1, 0\}$. For $t \in \mathbb{R} \setminus \{\ldots, -2, -1, 0\}$ and $r \in \mathbb{R}$ such that $(t+r) \in \mathbb{R} \setminus \{\ldots, -2, -1, 0\}$, the generalized rising function is defined by the following:

$$t^{\overline{r}} = \frac{\Gamma(t+r)}{\Gamma(t)}.$$

If $t \in \{\ldots, -2, -1, 0\}$ and $r \in \mathbb{R}$ such that $(t+r) \in \mathbb{R} \setminus \{\ldots, -2, -1, 0\}$, then we find that $t^{\overline{r}} = 0$.

Let $\mu \in \mathbb{R} \setminus \{\ldots, -2, -1\}$, define the μ-th order nabla fractional Taylor monomial by the following:

$$H_\mu(t, a) = \frac{(t-a)^{\overline{\mu}}}{\Gamma(\mu+1)},$$

provided that the right-hand side exists. Observe that $H_\mu(a, a) = 0$ and $H_\mu(t, a) = 0$ for all $\mu \in \{\ldots, -2, -1\}$ and $t \in \mathbb{N}_a$.

Let $u : \mathbb{N}_a \to \mathbb{R}$ and $N \in \mathbb{N}_1$. The first order backward (nabla) difference of u is defined by the following:

$$(\nabla u)(t) = u(t) - u(t-1), \quad t \in \mathbb{N}_{a+1},$$

and the N-th order nabla difference of u is defined recursively by

$$(\nabla^N u)(t) = \left(\nabla(\nabla^{N-1} u)\right)(t), \quad t \in \mathbb{N}_{a+N}.$$

Let $u : \mathbb{N}_{a+1} \to \mathbb{R}$ and $N \in \mathbb{N}_1$. The N-th order nabla sum of u based at a is given by the following:

$$(\nabla_a^{-N} u)(t) = \sum_{s=a+1}^{t} H_{N-1}(t, \rho(s)) u(s), \quad t \in \mathbb{N}_a,$$

where, by convention, $(\nabla_a^{-N} u)(a) = 0$.

We define $(\nabla_a^{-0} u)(t) = u(t)$ for all $t \in \mathbb{N}_{a+1}$.

Definition 1. *Let $u : \mathbb{N}_{a+1} \to \mathbb{R}$ and $\nu > 0$. The ν-th order nabla sum of u based at a is given by the following [4]:*

$$(\nabla_a^{-\nu} u)(t) = \sum_{s=a+1}^{t} H_{\nu-1}(t, \rho(s)) u(s), \quad t \in \mathbb{N}_a,$$

where, by convention, $(\nabla_a^{-\nu} u)(a) = 0$.

Definition 2. *Let $u : \mathbb{N}_{a+1} \to \mathbb{R}$, $\nu > 0$ and choose $N \in \mathbb{N}_1$ such that $N - 1 < \nu \leq N$. The ν-th order Riemann–Liouville nabla difference of u is given by the following [4]:*

$$(\nabla_a^\nu u)(t) = \left(\nabla^N (\nabla_a^{-(N-\nu)} u)\right)(t), \quad t \in \mathbb{N}_{a+N}.$$

In [21,22], Jonnalagadda obtained the following properties of the Green's function $G_0(t, s)$

Theorem 1. *Assume that the following condition holds [22]:*

(A0) $\alpha, \beta, \gamma, \delta \geq 0$, $\alpha^2 + \beta^2 > 0$, $\gamma^2 + \delta^2 > 0$ and $\beta \geq \alpha$.

Then,

1. *$G_0(t, s) \geq 0$ for all $(t, s) \in \mathbb{N}_a^b \times \mathbb{N}_{a+1}^b$;*
2. *$\max_{t \in \mathbb{N}_a^b} G_0(t, s) = G_0(\rho(s), s)$ for all $s \in \mathbb{N}_{a+1}^b$;*
3. *$G_0(\rho(s), s) < \Lambda$, where*

$$\Lambda = \frac{1}{\varsigma}\Big[\alpha\gamma H_{\nu-1}(b, a) H_{\nu-1}(b, a) + \alpha\delta H_{\nu-1}(b, a)$$
$$+ (\beta - \alpha)\gamma H_{\nu-1}(b, a) + (\beta - \alpha)\delta\Big].$$

Theorem 2. *Assume that the condition (A0) holds [21]. Then,*

$$\sum_{s=a+1}^{b} G_0(t,s) \leq \Omega,$$

for all $(t,s) \in \mathbb{N}_a^b \times \mathbb{N}_{a+1}^b$, *where*

$$\Omega = \frac{1}{\varsigma}\Big[\alpha\gamma H_{2\nu-1}(b, a+1) + \alpha\delta H_{2\nu-2}(b, a+1)$$
$$+ (\beta - \alpha)\gamma H_\nu(b, a) + (\beta - \alpha)\delta H_{\nu-1}(b, a)\Big].$$

We mention the following classical result that will be used in the next section.

Lemma 1. *Let X be a Banach space, $A : X \to X$ be a linear operator with the operator norm $\|A\|$ [23] (page 795). Then, if $\|A\| < 1$, we have that $(I - A)^{-1}$ exists and*

$$(I - A)^{-1} = \sum_{n=0}^{\infty} A^n.$$

Here, I is the identity operator.

3. Green's Function and Its Properties

In this section, we construct the Green's function related to problem (4), and deduce some significant properties.

We denote by X the set of all maps from $\mathbb{N}_a^b$ into $\mathbb{R}$. Clearly, X is a Banach space endowed with the maximum norm $\|\cdot\|$. We assume the following condition throughout the paper.

(A1) There exists $\bar{g} > 0$ such that

$$|g(t)| \leq \bar{g} < \frac{1}{\Omega}, \quad t \in \mathbb{N}_a^b.$$

We define $G : \mathbb{N}_a^b \times \mathbb{N}_{a+1}^b \to \mathbb{R}$ by the following:

$$G(t,s) = \sum_{n=0}^{\infty} (-1)^n G_n(t,s), \quad (5)$$

where $G_0(t,s)$ is given by (2), and set (see Figures 2–4).

$$G_n(t,s) = \sum_{\tau=a+1}^{b} G_0(t,\tau)G_{n-1}(\tau,s)g(\tau), \quad n \in \mathbb{N}_1. \quad (6)$$

Then, we have the following result.

Theorem 3. *Assume that conditions (A0) and (A1) are fulfilled, then function $G(t,s)$, defined in (5) as a series of functions, is convergent for $(t,s) \in \mathbb{N}_a^b \times \mathbb{N}_{a+1}^b$. Moreover, $G(t,s)$ is the Green's function for the boundary value problem (4).*

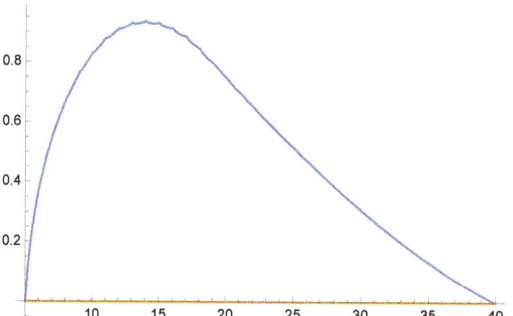

Figure 2. Graphic of $G_1(t,20)$ for $\alpha = \beta = \gamma = 1$, $\delta = 0$ (Dirichlet case), $\mu = 3/2$, $a = 5$, $b = 40$ and $g \equiv 1/100$.

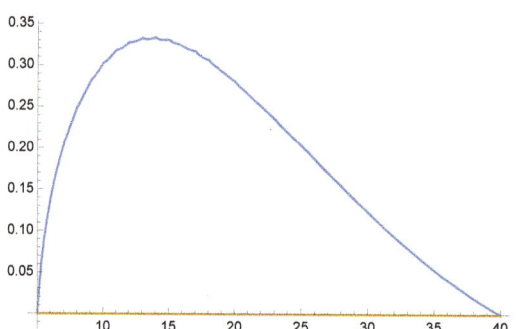

Figure 3. Graphic of $G_2(t,20)$ for $\alpha = \beta = \gamma = 1$, $\delta = 0$ (Dirichlet case), $\mu = 3/2$, $a = 5$, $b = 40$ and $g \equiv 1/100$.

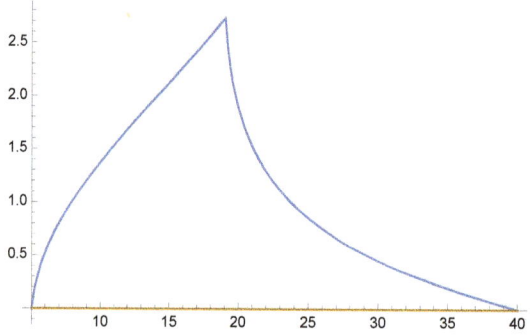

Figure 4. Graphic of the first three iterates of $G(t,20)$ for $\alpha = \beta = \gamma = 1$, $\delta = 0$ (Dirichlet case), $\mu = 3/2$, $a = 5$, $b = 40$ and $g \equiv 1/100$.

Proof. For any $h \in X$ and $t \in \mathbb{N}_a^b$, consider the following linear boundary value problem:

$$\begin{aligned}
-\left(\nabla_a^{\nu-1}(\nabla u)\right)(t) + g(t)u(t) &= h(t), \quad t \in \mathbb{N}_{a+2}^b, \\
\alpha u(a+1) - \beta(\nabla u)(a+1) &= 0, \\
\gamma u(b) + \delta(\nabla u)(b) &= 0.
\end{aligned} \quad (7)$$

114

By definition of the Green's function G_0, the solutions u of this problem satisfy the following identity:
$$u(t) = \sum_{s=a+1}^{b} G_0(t,s)[h(s) - g(s)u(s)],$$
which is the same to
$$u(t) + \sum_{s=a+1}^{b} G_0(t,s)g(s)u(s) = \sum_{s=a+1}^{b} G_0(t,s)h(s). \quad (8)$$

Now, define the operators $T_1 : X \to X$ and $T_2 : X \to X$ by the following:
$$(T_1 h)(t) = \sum_{s=a+1}^{b} G_0(t,s)h(s), \quad t \in \mathbb{N}_a^b,$$
$$(T_2 u)(t) = \sum_{s=a+1}^{b} G_0(t,s)g(s)u(s), \quad t \in \mathbb{N}_a^b.$$

Then, (8) can be expressed as the following:
$$(I + T_2)u = T_1 h.$$

Using condition (A1) and Theorem 1 the following is true:
$$\begin{aligned}
\|T_2\| = \max_{\|u\|=1} \|T_2 u\| &= \max_{\|u\|=1} \left[\max_{t \in \mathbb{N}_a^b} |(T_2 u)(t)| \right] \\
&= \max_{\|u\|=1} \left[\max_{t \in \mathbb{N}_a^b} \left| \sum_{s=a+1}^{b} G_0(t,s)g(s)u(s) \right| \right] \\
&\leq \max_{\|u\|=1} \left[\max_{t \in \mathbb{N}_a^b} \sum_{s=a+1}^{b} G_0(t,s)|g(s)||u(s)| \right] \\
&\leq \max_{\|u\|=1} \left[\bar{g}\|u\| \max_{t \in \mathbb{N}_a^b} \sum_{s=a+1}^{b} G_0(t,s) \right] \\
&< \max_{\|u\|=1} [\bar{g}\|u\|\Omega] = \bar{g}\Omega < 1.
\end{aligned}$$

Then, by Lemma 1, we have the following:
$$u = (I + T_2)^{-1} T_1 h = \sum_{n=0}^{\infty} (-T_2)^n T_1 h. \quad (9)$$

Arguing in a similar manner than in [20], we can deduce the following:
$$((-T_2)^n T_1 h)(t) = \sum_{s=a+1}^{b} (-1)^n G_n(t,s)h(s), \quad t \in \mathbb{N}_a^b, \quad n = 0, 1, 2, \ldots \quad (10)$$

Let us see now that the following inequality is fulfilled:
$$|(-1)^n G_n(t,s)| < \Lambda(\bar{g}\Omega)^n, \quad n = 0, 1, 2, \ldots \quad (11)$$

From Theorem 1, we have that (11) holds for $n = 0$. Assume now that (11) is true for some $n = k$. Then, the following is true:

$$\left|(-1)^{k+1}G_{k+1}(t,s)\right| = \left|(-1)^{k+1}\sum_{\tau=a+1}^{b}G_0(t,\tau)G_k(\tau,s)g(\tau)\right|$$

$$= \left|-\sum_{\tau=a+1}^{b}G_0(t,\tau)(-1)^k G_k(\tau,s)g(\tau)\right|$$

$$\leq \sum_{\tau=a+1}^{b}G_0(t,\tau)\left|(-1)^k G_k(\tau,s)\right||g(\tau)|$$

$$< \Lambda(\tilde{g}\Omega)^k \tilde{g}\sum_{\tau=a+1}^{b}G_0(t,\tau)$$

$$< \Lambda(\tilde{g}\Omega)^k \tilde{g}\Omega = \Lambda(\tilde{g}\Omega)^{k+1}.$$

Thus, (11) holds for $n = k+1$. By mathematical induction, (11) holds for any $n = 0, 1, 2, \ldots$.

As a direct consequence of previous inequality and condition (A1), we deduce that for all $(t,s) \in \mathbb{N}_a^b \times \mathbb{N}_{a+1}^b$ the following property is fulfilled:

$$|G(t,s)| = \left|\sum_{n=0}^{\infty}(-1)^n G_n(t,s)\right| \leq \sum_{n=0}^{\infty}|(-1)^n G_n(t,s)|$$

$$< \Lambda \sum_{n=0}^{\infty}(\tilde{g}\Omega)^n = \frac{\Lambda}{1-\tilde{g}\Omega} < \infty,$$

and, a a consequence, $G(t,s)$ converges on $\mathbb{N}_a^b \times \mathbb{N}_{a+1}^b$.

Finally, expressions (5), (9) and (10) imply that for all $t \in \mathbb{N}_a^b$ the following equality is fulfilled:

$$u(t) = \sum_{n=0}^{\infty}\left[\sum_{s=a+1}^{b}(-1)^n G_n(t,s)h(s)\right] = \sum_{s=a+1}^{b}\left[\sum_{n=0}^{\infty}(-1)^n G_n(t,s)\right]h(s)$$

$$= \sum_{s=a+1}^{b}G(t,s)h(s). \quad (12)$$

It is not difficult to verify that any function defined by (12) is a solution of the boundary value problem (7). So we conclude that problem (7) has a unique solution and, as a consequence, G is its related Green's function. □

Lemma 2. *Assume conditions (A0) and (A1). Let G be defined by (5) and the following:*

$$\tilde{G}(s) = \frac{G_0(\rho(s),s)}{1-\tilde{g}\Omega}, \quad s \in \mathbb{N}_{a+1}^b. \quad (13)$$

Then,

$$|G(t,s)| \leq \tilde{G}(s), \quad (t,s) \in \mathbb{N}_a^b \times \mathbb{N}_{a+1}^b.$$

Proof. First, we prove the following:

$$|(-1)^n G_n(t,s)| < G(s-1,s)(\tilde{g}\Omega)^n, \quad s \in \mathbb{N}_{a+1}^b, \quad n = 0,1,2,\ldots \quad (14)$$

Theorem 1 implies that inequality (14) is true for $n = 0$.

Assume now that (14) holds for some $n = k$. We will show that (14) holds for $n = k+1$. Consider the following:

$$\left|(-1)^{k+1}G_{k+1}(t,s)\right| = \left|(-1)^{k+1}\sum_{\tau=a+1}^{b}G_0(t,\tau)G_k(\tau,s)g(\tau)\right|$$

$$= \left|-\sum_{\tau=a+1}^{b}G_0(t,\tau)(-1)^k G_k(\tau,s)g(\tau)\right|$$

$$\leq \sum_{\tau=a+1}^{b}G_0(t,\tau)\left|(-1)^k G_k(\tau,s)\right||g(\tau)|$$

$$< G(s-1,s)(\bar{g}\Omega)^k \bar{g}\sum_{\tau=a+1}^{b}G_0(t,\tau)$$

$$< G(s-1,s)(\bar{g}\Omega)^k \bar{g}\Omega = G(s-1,s)(\bar{g}\Omega)^{k+1}.$$

Thus, (14) holds for $n = k+1$ and the inequalities are deduced from mathematical induction.

Now, from (5), (13) and (14), for $s \in \mathbb{N}_{a+1}^b$, we obtain the following:

$$|G(t,s)| = \left|\sum_{n=0}^{\infty}(-1)^n G_n(t,s)\right| \leq \sum_{n=0}^{\infty}|(-1)^n G_n(t,s)|$$

$$< G(s-1,s)\sum_{n=0}^{\infty}(\bar{g}\Omega)^n$$

$$= \frac{G_0(\rho(s),s)}{1-\bar{g}\Omega} = \bar{G}(s),$$

and the proof is complete. □

From the previous result, we deduce the following consequence for $g \leq 0$.

Corollary 1. *Assume that condition (A) is fulfilled and*

$$-\bar{g} < g(t) \leq 0, \quad t \in \mathbb{N}_a^b.$$

Then, $G(t,s) \geq 0$ for each $(t,s) \in \mathbb{N}_a^b \times \mathbb{N}_{a+1}^b$.

Proof. From Theorem 1, we know that $G_0(t,s) \geq 0$ for each $(t,s) \in \mathbb{N}_a^b \times \mathbb{N}_{a+1}^b$. The result follows immediately from (5) and (6). □

4. Existence of Solutions

In this section, we derive two existence results for the nonlinear problem (1). Define the operator $T: X \to X$ (X defined in previous section) by the following:

$$(Tu)(t) = \sum_{s=a+1}^{b} G(t,s) f(s, u(s)), \quad t \in \mathbb{N}_a^b. \tag{15}$$

In view of (12), it is clear that u is a fixed point of T if and only if u is a solution of (1). For any $R > 0$ given, we define the following set:

$$\mathcal{K}_R = \{u \in X : \|u\| < R\}.$$

Clearly, $\mathcal{K}_R$ is a non-empty open subset of X, $0 \in \mathcal{K}_R$ and $T: \overline{\mathcal{K}_R} \to X$.
Now, denoting by

$$\max_{t \in \mathbb{N}_a^b}|f(t,0)| = M \text{ and } \sum_{s=a+1}^{b}\bar{G}(s) = K \;(>0),$$

we enunciate the following list of assumptions:

(A2) $f : \mathbb{N}_a^b \times \mathbb{R} \to \mathbb{R}$ is continuous;

(A3) f satisfies the Lipschitz condition with respect to the second variable with the Lipschitz constant L on $\mathbb{N}_a^b \times \mathcal{K}_R$. That is, for all $(t, u), (t, v) \in \mathbb{N}_a^b \times [-R, R]$, the following inequality holds:

$$|f(t, u) - f(t, v)| \leq L|u - v|.$$

(A4) There exists a continuous function $\sigma : \mathbb{N}_a^b \to \mathbb{R}^+$ and a continuous nondecreasing function $\psi : \mathbb{R}^+ \to \mathbb{R}^+$ such that

$$|f(t, u)| \leq \sigma(t)\psi(|u|), \quad (t, u) \in \mathbb{N}_a^b \times \mathbb{R}.$$

(A5) $0 < LK < 1$.

First, we present a nonlinear alternative of Leray–Schauder for contractive maps.

Theorem 4. *(Theorem 3.2) Suppose $\mathcal{U}$ is an open subset of a Banach space X, $0 \in \mathcal{U}$ and $F : \overline{\mathcal{U}} \to X$ a contraction with $F(\overline{\mathcal{U}})$ bounded [24]. Then, either of the following is true:*

1. *F has a fixed point in $\overline{\mathcal{U}}$.*
2. *There exist $\lambda \in (0, 1)$ and $u \in \partial\mathcal{U}$ with $u = \lambda F u$,*

holds.

Now, we establish sufficient conditions on existence of solutions for (1) using Theorem 4.

Theorem 5. *Assume (A0)–(A3), (A5) hold. If we choose R such that*

$$R \geq \frac{MK}{1 - LK}, \tag{16}$$

then the boundary value problem (1) has a solution in $\overline{\mathcal{K}_R}$.

Proof. First, we show that T is a contraction. To see this, let $u, v \in \overline{\mathcal{K}_R}, t \in \mathbb{N}_a^b$, and consider the following:

$$|(Tu)(t) - (Tv)(t)| = \left| \sum_{s=a+1}^{b} G(t,s) \left[f(s, u(s)) - f(s, v(s)) \right] \right|$$

$$\leq \sum_{s=a+1}^{b} |G(t,s)||f(s, u(s)) - f(s, v(s))|$$

$$\leq L \sum_{s=a+1}^{b} \bar{G}(s)|u(s) - v(s)|$$

$$\leq LK\|u - v\|,$$

implying that

$$\|Tu - Tv\| \leq LK\|u - v\|.$$

Since

$$0 < LK < 1,$$

it follows that T is a contraction.

Next, we prove that $T(\overline{\mathcal{K}_R})$ is bounded. To see this, let $u \in \overline{\mathcal{K}_R}$ ($\|u\| \leq R$), $t \in \mathbb{N}_a^b$, and consider the following:

$$|(Tu)(t)| = \left|\sum_{s=a+1}^{b} G(t,s) f(s, u(s))\right|$$

$$\leq \sum_{s=a+1}^{b} |G(t,s)||f(s,v(s))|$$

$$= \sum_{s=a+1}^{b} |G(t,s)||f(s,u(s)) - f(s,0) + f(s,0)|$$

$$\leq \sum_{s=a+1}^{b} \tilde{G}(s)|f(s,u(s)) - f(s,0)| + \sum_{s=a+1}^{b} \tilde{G}(s)|f(s,0)|$$

$$\leq L \sum_{s=a+1}^{b} \tilde{G}(s)|u(s)| + M \sum_{s=a+1}^{b} \tilde{G}(s)$$

$$\leq LK\|u\| + MK \leq [LR + M]K,$$

implying the following:
$$\|Tu\| \leq [LR + M]K.$$

Thus, $T(\overline{\mathcal{K}_R})$ bounded.

Now, suppose there exist $v \in \partial \mathcal{K}_R$ ($\|v\| = R$) and $\lambda \in (0,1)$ such that

$$v = \lambda T v. \tag{17}$$

Using the definition of T in (17) and arguing as before, we obtain the following:

$$|v(t)| = |\lambda (Tv)(t)| \leq [LR + M]\lambda K < [LR + M]K, \quad t \in \mathbb{N}_a^b,$$

which implies the following:
$$R = \|v\| < [LR + M]K$$

or, which is the same,
$$R < \frac{MK}{1 - LK'}$$

in contradiction with (16).

Hence, by Theorem 4, we deduce that operator T has a fixed point in $\overline{\mathcal{K}_R}$ and the proof is complete. □

Remark 1. *We note that in the previous result, if $M = 0$ then we have that $u \equiv 0$ on $[0,1]$ is a solution of problem (1). On the contrary, if $M > 0$ the obtained function is non trivial on $[0,1]$*

Next, we enunciate a nonlinear alternative of Leray–Schauder for continuous and compact maps.

Theorem 6. *Let E be a Banach space, C a closed, convex subset of E, U an open subset of C and $0 \in U$ [24] (Theorem 6.6). Suppose that $F : \bar{U} \to C$ is a continuous, compact map. Then, either of the following is true:*
1. *F has a fixed point in $\bar{U}$, or*
2. *there is a $u \in \partial U$ and $\lambda \in (0,1)$ with $u = \lambda F u$.*

Now, we establish sufficient conditions on existence of solutions for (1) using Theorem 6.

Theorem 7. *Assume that conditions (A0)–(A2), and (A4) hold. If we choose R such that*

$$\frac{R}{K\|\sigma\|\psi(R)} \geq 1, \tag{18}$$

then the boundary value problem (1) has a solution in $\overline{\mathcal{K}_R}$.

Proof. Since T is a summation operator on a discrete finite set, it is trivially continuous and compact. Now, suppose that there exist $v \in \partial \mathcal{K}_R$ ($\|v\| = R$) and $\lambda \in (0,1)$ such that (17) holds. Using the definition of T in (17), we obtain the following:

$$\begin{aligned} |v(t)| &= |\lambda(Tv)(t)| \\ &= \left| \lambda \sum_{s=a+1}^{b} G(t,s) f(s,v(s)) \right| \\ &\leq \lambda \sum_{s=a+1}^{b} |G(t,s)| |f(s,v(s))| \\ &\leq \lambda \sum_{s=a+1}^{b} |G(t,s)| \sigma(s) \psi(|v(s)|) \\ &\leq \lambda \|\sigma\| \psi(\|v\|) \sum_{s=a+1}^{b} \bar{G}(s) \\ &< K \|\sigma\| \psi(\|v\|). \end{aligned}$$

So, we deduce the following:

$$R = \|v\| < K\|\sigma\|\psi(\|v\|).$$

Thus,

$$\frac{R}{K\|\sigma\|\psi(R)} < 1.$$

This is a contradiction to (18).

Hence, by Theorem 6, the boundary value problem (1) has a solution in $\overline{\mathcal{K}_R}$. The proof is complete. □

Remark 2. Note that since we have that

$$\max_{t \in \mathbb{N}_a^b} G_0(t,s) < \Lambda$$

we can set the following:

$$\overline{K} = \frac{\Lambda}{1 - \bar{g}\Omega}(b-a) > \sum_{s=a+1}^{b} \bar{G}(s) = K.$$

Thus, we can use $\overline{K}$ instead of K everywhere and we do not need to calculate the Green's function at all.

Indeed, in (A5), if we have $0 < L\overline{K} < 1$, this implies that $0 < LK < 1$.

In Theorem 4.2, if we choose $R \geq \frac{M\overline{K}}{1-L\overline{K}}$, then we will also have that $R \geq \frac{MK}{1-LK}$ since $\frac{M\overline{K}}{1-L\overline{K}} \geq \frac{MK}{1-LK}$.

Finally, in Theorem 4.4, if we choose $\frac{R}{\overline{K}\|\sigma\|\psi(R)} \geq 1$, then we will also have that $\frac{R}{K\|\sigma\|\psi(R)} \geq 1$ since $\frac{R}{K\|\sigma\|\psi(R)} \geq \frac{R}{\overline{K}\|\sigma\|\psi(R)}$.

5. Examples

In the section, we present some examples to illustrate the applicability of our main results.

Problem 1. Consider the following nabla fractional boundary value problem:

$$-\left(\nabla_0^{1/2}(\nabla u)\right)(t) + \frac{e^{-t}}{10}u(t) = \frac{1}{200}\sin(u(t)+t), \quad t \in \mathbb{N}_2^6, \tag{19}$$
$$u(0) = u(6) = 0.$$

Here, $\alpha = 1, \beta = 1, \gamma = 1$ and $\delta = 0, a = 0, b = 6$ and $\nu = 3/2$.

In addition, $g(t) = e^{-t}/10$ and $f(t,u) = (\sin(u+t))/200$. Clearly, $g: \mathbb{N}_0^6 \to \mathbb{R}$; $f: \mathbb{N}_0^6 \times \mathbb{R} \to \mathbb{R}$ is continuous and satisfies Lipschitz condition with respect to u on $\mathbb{N}_0^6 \times \mathbb{R}$ with Lipschitz constant $L = 0.005$.

We have $\zeta = H_{0.5}(6,0) \approx 2.7071, \Lambda = H_{0.5}(6,0) \approx 2.7071, \Omega = H_2(6,1) = 3$ and $\bar{g} = 0.1$ so that $|g(t)| \leq \bar{g} < \frac{1}{\Omega}$. Further,

$$\overline{K} = \frac{\Lambda}{1 - \bar{g}\Omega}(b - a) \approx 23.2037.$$

Observe that $0 < L\overline{K} \approx 0.116 < 1$. Additionally,

$$M = \max_{t \in \mathbb{N}_0^6} |f(t,0)| = \max_{t \in \mathbb{N}_0^6} \left|\frac{\sin t}{200}\right| \approx 0.00479462137.$$

If we choose

$$R \geq \frac{M\overline{K}}{1 - L\overline{K}} \approx 0.12585176,$$

then by Theorem 5 and Remark 2, the boundary value problem (19) has a solution in $\overline{\mathcal{K}_R}$.

Problem 2. Consider the following nabla fractional boundary value problem:

$$-\left(\nabla_0^{1/2}(\nabla u)\right)(t) + \frac{1}{20(t+1)}u(t) = \frac{u^2(t)}{10(t^2+10)}, \quad t \in \mathbb{N}_2^9,$$
$$u(0) + u(1) = 0, \tag{20}$$
$$u(8) + u(9) = 0.$$

Here, $\alpha = 2, \beta = 1, \gamma = 2$ and $\delta = -1$ such that $\alpha^2 + \beta^2 > 0$ and $\gamma^2 + \delta^2 > 0, a = 0, b = 9$ and $\nu = 3/2$.

In addition, $g(t) = \frac{1}{20(t+1)}$ and $f(t,u) = \frac{u^2(t)}{10(t^2+10)}$. Clearly, $g: \mathbb{N}_0^9 \to \mathbb{R}$ and $f: \mathbb{N}_0^9 \times \mathbb{R} \to \mathbb{R}$ are continuous.

We have $\zeta \approx 10.9616, \Lambda \approx 2.9403, \Omega = 15.8396$ and $\bar{g} \approx 0.05$ so that $|g(t)| \leq \bar{g} < \frac{1}{\Omega}$. Further,

$$\overline{K} = \frac{\Lambda}{1 - \bar{g}\Omega}(b - a) \approx 127.2245.$$

In addition,

$$|f(t,u)| \leq \sigma(t)\psi(|u|), \quad (t,u) \in \mathbb{N}_0^9 \times \mathbb{R},$$

where $\sigma(t) = \frac{1}{10(t^2+10)}$ and $\psi(x) = x^2$. Observe that $\sigma: \mathbb{N}_0^9 \to \mathbb{R}^+$ is continuous and $\psi: \mathbb{R}^+ \to \mathbb{R}^+$ is continuous non-decreasing with

$$\|\sigma\| = \max_{t \in \mathbb{N}_0^9} |\sigma(t)| = 0.001.$$

If we choose

$$\frac{R}{\overline{K}\|\sigma\|\psi(R)} \geq 1,$$

that is, $R \leq 7.8616$, then by Theorem 7 and Remark 2, the boundary value problem (19) has a solution in $\overline{\mathcal{K}_R}$.

Author Contributions: Conceptualization, A.C., N.D.D. and J.M.J.; methodology, A.C., N.D.D. and J.M.J.; software, A.C.; validation, A.C., N.D.D. and J.M.J.; formal analysis, A.C., N.D.D. and J.M.J.; investigation, A.C., N.D.D. and J.M.J.; resources, A.C., N.D.D. and J.M.J.; data curation, A.C.; writing—original draft preparation, J.M.J.; writing—review and editing, A.C., N.D.D. and J.M.J.; visualization, A.C., N.D.D. and J.M.J.; supervision, A.C., N.D.D. and J.M.J.; project administration, A.C., N.D.D. and J.M.J.; funding acquisition, A.C and N.D.D. All authors have read and agreed to the published version of the manuscript.

Funding: The first author is partially supported by the Agencia Estatal de Investigación (AEI) of Spain under grant MTM2016-75140-P, co-financed by the European Community fund FEDER. The second author is supported by the Bulgarian National Science Fund under Project DN 12/4 "Advanced analytical and numerical methods for nonlinear differential equations with applications in finance and environmental pollution", 2017.

Acknowledgments: The authors thanks to the anonymous referees for their useful comments that have contributed to improve this paper.

Conflicts of Interest: The authors declare no conflict of interest.

References

1. Gray, H.L.; Zhang, N.F. On a new definition of the fractional difference. *Math. Comput.* **1988**, *50*, 513–529. [CrossRef]
2. Atıcı, F.M.; Eloe, P.W. Discrete fractional calculus with the nabla operator. *Electron. J. Qual. Theory Differ. Equ.* **2009**, *12*. [CrossRef]
3. Anastassiou, G.A. Nabla discrete fractional calculus and nabla inequalities. *Math. Comput. Model.* **2010**, *51*, 562–571. [CrossRef]
4. Goodrich, C.; Peterson, A.C. *Discrete Fractional Calculus*; Springer: Cham, Switzerland, 2015.
5. Hill, G.W. On the part of the motion of the lunar perigee which is a function of the mean motions of the Sun and the Moon. *Acta Math.* **1886**, *8*, 1–36. [CrossRef]
6. Cabada, A.; Cid, J.Á.; López-Somoza, L. *Maximum Principles for the Hill's Equation*; Academic Press, London, UK, 2018; 238p.
7. Magnus, W.; Winkler, S. *Hill's Equation*; Corrected Reprint of the 1966 Edition; Dover Publications, Inc.: New York, NY, USA, 1979; 129p.
8. Brackins, A. Boundary Value Problems of Nabla Fractional Difference Equations. Ph.D. Thesis, The University of Nebraska-Lincoln, Lincoln, NE, USA, 2014; 92p.
9. Ahrendt, K.; Kissler, C. Cameron Green's function for higher-order boundary value problems involving a nabla Caputo fractional operator. *J. Differ. Equ. Appl.* **2019**, *25*, 788–800. [CrossRef]
10. Chen, C.; Bohner, M.; Jia, B. Existence and uniqueness of solutions for nonlinear Caputo fractional difference equations. *Turk. J. Math.* **2020**, *44*, 857–869. [CrossRef]
11. Gholami, Y.; Ghanbari, K. Coupled systems of fractional ∇-difference boundary value problems. *Differ. Equ. Appl.* **2016**, *8*, 459–470. [CrossRef]
12. Goodrich, C.S. Existence of a positive solution to a class of fractional differential equations. *Appl. Math. Lett.* **2010**, *23*, 1050–1055. [CrossRef]
13. Goodrich, C.S. Existence and uniqueness of solutions to a fractional difference equation with nonlocal conditions. *Comput. Math. Appl.* **2011**, *61*, 191–202. [CrossRef]
14. Ikram, A. Lyapunov inequalities for nabla Caputo boundary value problems. *J. Differ. Equ. Appl.* **2019**, *25*, 757–775. [CrossRef]
15. Baoguo, J.; Erbe, L.; Goodrich, C.; Peterson, A. The relation between nabla fractional differences and nabla integer differences. *Filomat* **2017**, *31*, 1741–1753. [CrossRef]
16. Jonnalagadda, J.M. On two-point Riemann-Liouville type nabla fractional boundary value problems. *Adv. Dyn. Syst. Appl.* **2018**, *13*, 141–166.
17. Liu, X.; Jia, B.; Gensler, S.; Erbe, L.; Peterson, A. Convergence of approximate solutions to nonlinear Caputo nabla fractional difference equations with boundary conditions. *Electron. J. Differ. Equ.* **2020**, *2020*, 1–19.
18. St. Goar, J. A Caputo Boundary Value Problem in Nabla Fractional Calculus. Ph.D. Thesis, The University of Nebraska-Lincoln, Lincoln, NE, USA, 2016; 112p.
19. Graef, J.R.; Kong, L.; Kong, Q.; Wang, M. Existence and uniqueness of solutions for a fractional boundary value problem with Dirichlet boundary condition. *Electron. J. Qual. Theory Differ. Equ.* **2013**, *2013*, 1–11. [CrossRef]
20. Cabada, A.; Dimitrov, N. Non-trivial solutions of non-autonomous Dirichlet fractional discrete problems. *Fract. Calc. Appl. Anal.* **2020**, *23*, 980–995. [CrossRef]
21. Jonnalagadda, J.M. Existence results for solutions of nabla fractional boundary value problems with general boundary conditions. *Adv. Theory Nonlinear Anal. Appl.* **2020**, *4*, 29–42. [CrossRef]
22. Jonnalagadda, J.M. On a nabla fractional boundary value problem with general boundary conditions. *AIMS Math.* **2020**, *5*, 204–215. [CrossRef]

23. Zeidler, E. *Nonlinear Functional Analysis and Its Applications. I. Fixed-Point Theorems*; Wadsack, P.R., Translator; Springer: New York, NY, USA, 1986.
24. Agarwal, R.P.; Meehan, M.; O'regan, D. *Fixed Point Theory and Applications. Cambridge Tracts in Mathematics*; Cambridge University Press: Cambridge, UK, 2001; Volume 141.

Article

Regularity Criteria for the 3D Magneto-Hydrodynamics Equations in Anisotropic Lorentz Spaces

Maria Alessandra Ragusa [1,2,*] and Fan Wu [3]

[1] Department of Mathematics, University of Catania, Viale Andrea Doria No. 6, 95128 Catania, Italy
[2] RUDN University, Miklukho-Maklay St, 117198 Moscow, Russia
[3] Key Laboratory of Computing and Stochastic Mathematics (Ministry of Education), School of Mathematics and Statistics, Hunan Normal University, Changsha 410081, China; wufan0319@smail.hunnu.edu.cn
* Correspondence: mariaalessandra.ragusa@unict.it

Abstract: In this paper, we investigate the regularity of weak solutions to the 3D incompressible MHD equations. We provide a regularity criterion for weak solutions involving any two groups functions $(\partial_1 u_1, \partial_1 b_1)$, $(\partial_2 u_2, \partial_2 b_2)$ and $(\partial_3 u_3, \partial_3 b_3)$ in anisotropic Lorentz space.

Keywords: MHD equations; weak solution; regularity criteria; anisotropic Lorentz space

MSC: 76W05; 35Q30; 35B65

1. Introduction

In this paper, we are concerned with regularity criteria for the weak solutions to the incompressible magneto-hydrodynamic (MHD) equations in $\mathbb{R}^3$ [1,2]:

$$\begin{cases} \partial_t u + (u \cdot \nabla)u - \Delta u + \nabla p = (b \cdot \nabla)b, \\ \partial_t b + (u \cdot \nabla)b - \Delta b = (b \cdot \nabla)u, \\ \nabla \cdot u = \nabla \cdot b = 0, \\ u(x,0) = u_0(x), b(x,0) = b_0(x), \end{cases} \tag{1}$$

where $u = (u_1, u_2, u_3)$ is the fluid velocity field, $b = (b_1, b_2, b_3)$ is the magnetic field, p is a scalar pressure, and u_0, b_0 is the prescribed initial data satisfying the compatibility condition $\nabla \cdot u_0 = \nabla \cdot b_0 = 0$ in the distributional sense. Physically, Equation (1) govern the dynamics of the velocity and magnetic fields in electrically conducting fluids, such as plasmas, liquid metals, and salt water.

Besides its physical applications, the MHD equations (1) have also mathematically significant. Duvaut and Lions [1] developed a global weak solution to (1) for initial data with finite energy, that is,

$$u, b \in L^\infty\left(0, T; L^2(\mathbb{R}^3)\right) \cap L^2\left(0, T; H^1(\mathbb{R}^3)\right) \quad \text{for any} \quad T > 0.$$

It is well known that the issue of regularity for weak solutions to the 3D incompressible Navier-Stokes equations has been one of the most challenging open problem in mathematical fluid mechanics [3], as well as that for the 3D incompressible magneto-hydrodynamics (MHD) equations (see Sermange and Temam [2]). Many sufficient conditions (see e.g., [4–14] and the references therein) were derived to guarantee the regularity of the weak solution. He and Xin [15] first extended the classical Prodi-Serrin conditions of Navier-Stokes equations to the MHD equations, they obtained regularity criteria involving only on velocity u, i.e.,

$$u \in L^q\left(0, T; L^p(\mathbb{R}^3)\right) \quad \text{with} \quad \frac{2}{q} + \frac{3}{p} \leq 1 \quad \text{and} \quad 3 < p \leq \infty \tag{2}$$

or
$$\nabla u \in L^q\left(0, T; L^p\left(\mathbb{R}^3\right)\right) \text{ with } \frac{2}{q}+\frac{3}{p} \leq 2 \text{ and } \frac{3}{2} < p \leq \infty. \quad (3)$$

Later, He and Wang [16] showed that a weak solution (u, b) is regular, provided only $\nabla \omega^+ = (u+b)$ or $\nabla \omega^- = (u-b)$ belongs to Beirao da Veiga's class, that is,

$$\nabla \omega^+ \text{ or } \nabla \omega^- \in L^q\left(0, T; L^{p,\infty}\left(\mathbb{R}^3\right)\right) \text{ with } \frac{2}{q}+\frac{3}{p} = 2 \text{ and } 3 \leq p \leq \infty. \quad (4)$$

Ni et al. [17] showed that one of the following conditions hold

$$\begin{cases} \nabla_h u \in L^q\left(0, T; L^p\left(\mathbb{R}^3\right)\right) \text{ with } \frac{2}{q}+\frac{3}{p} \leq 2 \text{ and } \frac{3}{2} < p \leq \infty, \\ \partial_3 b \in L^q\left(0, T; L^p\left(\mathbb{R}^3\right)\right) \text{ with } \frac{2}{q}+\frac{3}{p} \leq 2 \text{ and } \frac{3}{2} < p \leq \infty. \end{cases} \quad (5)$$

$$\begin{cases} u_3 \in L^q\left(0, T; L^p\left(\mathbb{R}^3\right)\right) \text{ with } \frac{2}{q}+\frac{3}{p} \leq 1 \text{ and } 3 < p \leq \infty, \\ \partial_3 u \in L^p\left(0, T; L^q\left(\mathbb{R}^3\right)\right) \text{ with } \frac{2}{q}+\frac{3}{p} \leq 2 \text{ and } \frac{3}{2} < p \leq \infty, \\ b_3 \in L^q\left(0, T; L^p\left(\mathbb{R}^3\right)\right) \text{ with } \frac{2}{q}+\frac{3}{p} \leq 1 \text{ and } 3 < p \leq \infty, \\ \partial_3 b \in L^q\left(0, T; L^p\left(\mathbb{R}^3\right)\right) \text{ with } \frac{2}{q}+\frac{3}{p} \leq 2 \text{ and } \frac{3}{2} < p \leq \infty, \end{cases} \quad (6)$$

$$\begin{cases} \nabla_h u \in L^q\left(0, T; L^p\left(\mathbb{R}^3\right)\right) \text{ with } \frac{2}{q}+\frac{3}{p} \leq 2 \text{ and } \frac{3}{2} < p \leq \infty, \\ \nabla_h b \in L^q\left(0, T; L^p\left(\mathbb{R}^3\right)\right) \text{ with } \frac{2}{q}+\frac{3}{p} \leq 2 \text{ and } \frac{3}{2} < p \leq \infty, \end{cases} \quad (7)$$

then the weak solution (u, b) is regular on $(0, T]$, where $\nabla_h = (\partial_1, \partial_2)$. Recently, Jia [18] showed that condition (7) can be replaced by

$$\begin{cases} \nabla_h \tilde{u} \in L^q\left(0, T; L^p\left(\mathbb{R}^3\right)\right) \text{ with } \frac{2}{q}+\frac{3}{p} \leq 2 \text{ and } \frac{3}{2} < p \leq \infty, \\ \nabla_h \tilde{b} \in L^q\left(0, T; L^p\left(\mathbb{R}^3\right)\right) \text{ with } \frac{2}{q}+\frac{3}{p} \leq 2 \text{ and } \frac{3}{2} < p \leq \infty, \end{cases} \quad (8)$$

where $\tilde{f} = (f_1, f_2)$. Regularity condition (8) was further improved by Xu et al. [19], more precisely, they proved that if any two quantities of

$$\begin{cases} A_i^{q,p}(T) := \partial_i u_i \in L^q\left(0, T; L^p\left(\mathbb{R}^3\right)\right) \text{ with } \frac{2}{q}+\frac{3}{p} = 2 \text{ and } \frac{3}{2} < p \leq \infty, \\ B_i^{q,p}(T) := \partial_i b_i \in L^q\left(0, T; L^p\left(\mathbb{R}^3\right)\right) \text{ with } \frac{2}{q}+\frac{3}{p} = 2 \text{ and } \frac{3}{2} < p \leq \infty, \end{cases} \quad (9)$$

where $i = 1, 2, 3$, then the solution is smooth on interval $(0, T]$. For readers interested in this topic for partial components, please refer to [20–26] for recent progresses.

Motivated by papers cited above, the aim of this article is to study the regularity of weak solutions for the 3D MHD equations (1) in term of the two partial derivative of the velocity components and magnetic components on framework of the anisotropic Lorentz space. Before stating our main Theorem, we shall first recall the definitions of some function spaces [27].

Lorentz Spaces

Given a measurable function $f : \mathbb{R}^n \to \mathbb{R}$ define the distribution function of f by

$$d_f(\alpha) = \mu(\{x : |f(x)| > \alpha\}),$$

where $\mu(A)$ (or $|A|$) denotes the Lebesgue measure of a set A. We now define its decreasing rearrangement $f^* : [0, \infty) \to [0, \infty]$ as

$$f^*(t) = \inf\{\alpha : d_f(\alpha) \leq t\},$$

with the convention that $\inf \emptyset = \infty$. The point of this definition is that f and f^* have the same distribution function,

$$d_{f^*}(\alpha) = d_f(\alpha),$$

but f^* is a positive non-increasing scalar function.

Definition 1. *Let $(p,q) \in [1,\infty]^2$, the Lorentz space $L^{p,q}(\mathbb{R}^3)$ consists of all measurable functions f for which the quantity*

$$\|f\|_{L^{p,q}} := \begin{cases} \left(\int_0^\infty [t^{\frac{1}{p}} f^*(t)]^q \frac{dt}{t} \right)^{\frac{1}{q}} & q < \infty, \\ \sup_{0<t<\infty} t^{\frac{1}{p}} f^*(t) & q = \infty, \end{cases}$$

is finite.

In order to give the following definition involving anisotropic Lorentz space, we denote $f = f(x_1, x_2, x_3)$ be a measurable function defined on $\mathbb{R}^3$, $f^*(t) = f^{*_1,*_2,*_3}(t_1, t_2, t_3)$. Here $f^{*_1,*_2,*_3}(t_1, t_2, t_3)$ is the multivariate decreasing rearrangement of $f(x_1, x_2, x_3)$ obtained by applying decreasing rearrangement $f^{*_1}(t_1, x_2, x_3)$ of $f(x_1, x_2, x_3)$ relating to the first variable x_1, under fixed the second, the third variables x_2, x_3, and then applying decreasing rearrangement $f^{*_1,*_2}(t_1, t_2, x_3)$ of $f^{*_1}(t_1, x_2, x_3)$ with respect to the second variable x_2 under fixed the first variable t_1 of $f^{*_1}(t_1, x_2, x_3)$ and variable x_3, finally for variable x_3, by the same trick, we obtain the multivariate decreasing rearrangement $f^{*_1,*_2,*_3}(t_1, t_2, t_3)$.

Recently, many works have been done for mixed-norm spaces. Stefanov-Torres [28] obtained the boundedness of Calderón-Zygmund operators on mixed-norm Lebesgue spaces. Georgiadis et al. [29] obtained various properties of anisotropic Triebel-Lizorkin spaces with mixed norms. In [30], Chen-Sun introduced the iterated weak and weak mixed-norm spaces and given some applications to geometric inequalities.

Definition 2. *Let multi indexes $p = (p_1, p_2, p_3), q = (q_1, q_2, q_3)$ be such that if $0 < p_i < \infty$, then $0 < q_i \leq \infty$, and if $p_i = \infty$, then $q_i = \infty$ for every $i = 1, 2, 3$ [31]. An anisotropic Lorentz space $L^{p_1,q_1}(\mathbb{R}_{x_1}; L^{p_2,q_2}(\mathbb{R}_{x_2}; L^{p_3,q_3}(\mathbb{R}_{x_3})))$ is the set of functions for which the following norm is finite:*

$$\left\| \left\| \|f\|_{L^{p_1,q_1}_{x_1}} \right\|_{L^{p_2,q_2}_{x_2}} \right\|_{L^{p_3,q_3}_{x_3}} := \left(\int_0^\infty \left(\int_0^\infty \left(\int_0^\infty [t_1^{\frac{1}{p_1}} t_2^{\frac{1}{p_2}} t_3^{\frac{1}{p_3}} f^{*_1,*_2,*_3}(t_1, t_2, t_3)]^{q_1} \frac{dt_1}{t_1} \right)^{\frac{q_2}{q_1}} \frac{dt_2}{t_2} \right)^{\frac{q_3}{q_2}} \frac{dt_3}{t_3} \right)^{\frac{1}{q_3}}.$$

Now, our main result reads:

Theorem 1. *Suppose that $(u_0, b_0) \in L^2(\mathbb{R}^3) \cap L^4(\mathbb{R}^3)$ and $\nabla \cdot u_0 = \nabla \cdot b_0 = 0$ in distributional sense. Let (u, b) be the Leray-Hopf weak solution of (1) on $(0, T]$. If any two quantities*

$$\begin{cases} A_i(T) := \int_0^T \left\| \|\partial_i u_i(t)\|_{L^{p,\infty}_{x_1}} \right\|_{L^{q,\infty}_{x_2}} \Big\|_{L^{r,\infty}_{x_3}}^{\frac{2}{2-(\frac{1}{p}+\frac{1}{q}+\frac{1}{r})}} dt, \\ B_i(T) := \int_0^T \left\| \|\partial_i b_i(t)\|_{L^{p,\infty}_{x_1}} \right\|_{L^{q,\infty}_{x_2}} \Big\|_{L^{r,\infty}_{x_3}}^{\frac{2}{2-(\frac{1}{p}+\frac{1}{q}+\frac{1}{r})}} dt, \end{cases} \quad (10)$$

are finite, where $i = 1, 2, 3$ with $2 < p, q, r \leq \infty$ and $1 - \left(\frac{1}{p} + \frac{1}{q} + \frac{1}{r}\right) \geq 0$, then the weak solution (u, b) is actually smooth on interval $(0, T]$.

Remark 1. While $L^p(\mathbb{R}^3) \hookrightarrow L^{p,\infty}(\mathbb{R}^3)$, clearly $L^{p,\infty}$ is a larger space than L^p. Therefore, from this point of view, condition (10) can be regarded as an extension of (7)–(9). In addition, our regularity criteria only depends on any two groups functions of $(\partial_1 u_1, \partial_1 b_1)$, $(\partial_2 u_2, \partial_2 b_2)$ and $(\partial_3 u_3, \partial_3 b_3)$. Hence, (10) can be as a significant improvement of condition (7) and (8). In addition, when $b = 0$, it is note that Theorem 1 is also new to the incompressible Navier-Stokes equations.

Remark 2. According to embedding relation $L^p(\mathbb{R}^3) \hookrightarrow L^{p,\infty}(\mathbb{R}^3)$, we can obtain the following regularity criteria on framework of anisotropic Lebesgue space,

$$\begin{cases} A_i(T) := \int_0^T \left\| \left\| \|\partial_i u_i(t)\|_{L^p_{x_1}} \right\|_{L^q_{x_2}} \right\|_{L^r_{x_3}}^{\frac{2}{2-(\frac{1}{p}+\frac{1}{q}+\frac{1}{r})}} dt < \infty, \\ B_i(T) := \int_0^T \left\| \left\| \|\partial_i b_i(t)\|_{L^p_{x_1}} \right\|_{L^q_{x_2}} \right\|_{L^r_{x_3}}^{\frac{2}{2-(\frac{1}{p}+\frac{1}{q}+\frac{1}{r})}} dt < \infty, \end{cases} \quad (11)$$

where we should point out that for Equation (1), the regularity criterion (11) still new.

Remark 3. Notice that when fix $p = q = r$ in condition (11), the conditions (9) naturally turn out as stated in [19]. Furthermore, if let $p = q = r$ in condition (10), it is not difficult to find that our result improves the condition (4) significantly. Hence, regularity criteria (10) or (11) is much better. In other words, Theorem 1 can be regarded as a generalization of [16,18,19,23].

Before ending this section, we state the following lemmas, which will be used in the proof of our main result.

Lemma 1. *(Young's Inequality for Lorentz Spaces [32,33]) Let $1 < p < \infty, 1 \leq q \leq \infty$ and $\frac{1}{p'} + \frac{1}{p} = 1, \frac{1}{q'} + \frac{1}{q} = 1$. Suppose as well that $1 < p_1 < p'$ and $q' \leq q \leq \infty$. If $\frac{1}{p_2} + 1 = \frac{1}{p} + \frac{1}{p_1}$ and $\frac{1}{q_2} = \frac{1}{q} + \frac{1}{q_1}$, then the convolution operator,*

$$* : L^{p,q}(\mathbb{R}^n) \times L^{p_1,q_1}(\mathbb{R}^n) \mapsto L^{p_2,q_2}(\mathbb{R}^n)$$

is a bounded bilinear operator.

Lemma 2. *(Hölder's inequality in Lorentz spaces [33]) If $1 \leq p_1, p_2, q_1, q_2 \leq \infty$, then for any $f \in L^{p_1,q_1}(\mathbb{R}^n), g \in L^{p_2,q_2}(\mathbb{R}^n)$,*

$$\|fg\|_{L^{p,q}(\mathbb{R}^n)} \leq C\|f\|_{L^{p_1,q_1}(\mathbb{R}^n)} \|g\|_{L^{p_2,q_2}(\mathbb{R}^n)},$$

where $\frac{1}{p} = \frac{1}{p_1} + \frac{1}{p_2}$ and $\frac{1}{q} = \frac{1}{q_1} + \frac{1}{q_2}$.

For any $s \geq 0$, even if s not an integer, we can define the homogeneous Sobolev space $\dot{H}^s(\mathbb{R}^n)$:

$$\dot{H}^s(\mathbb{R}^n) = \{f \in \mathcal{S}' : \hat{f} \in L^1_{loc}(\mathbb{R}^n) \text{ and } \int_{\mathbb{R}^n} |\xi|^{2s} |\widehat{f(\xi)}|^2 d\xi < \infty\}$$

with the natural norm

$$\|f\|_{\dot{H}^s} = \left(\int_{\mathbb{R}^n} |\xi|^{2s} |\widehat{f(\xi)}|^2 d\xi \right)^{\frac{1}{2}},$$

where $\mathcal{S}'$ denotes the space of the tempered distributions on $\mathbb{R}^n$.

Lemma 3. *For $2 < p < \infty$, there exists a constant $C = C(p)$ such that $f \in \dot{H}^{\frac{1}{p}}(\mathbb{R})$, then $f \in L^{\frac{2p}{p-2},2}(\mathbb{R})$ and*

$$\|f\|_{L^{\frac{2p}{p-2},2}} \leq C\|f\|_{\dot{H}^{\frac{1}{p}}}. \quad (12)$$

Proof. We first make the pointwise definition, $\gamma(\xi) = |\xi|^{\frac{1}{p}}\hat{f}(\xi)$; since $f \in \dot{H}^{\frac{1}{p}}(\mathbb{R})$, $\gamma \in L^2(\mathbb{R})$. If we set $g = \mathcal{F}^{-1}\gamma$, then $g \in L^2(\mathbb{R})$ and $\|g\|_{L^2} = \|\gamma\|_{L^2} = \|f\|_{\dot{H}^{\frac{1}{p}}}$. Now,

$$\hat{f}(\xi) = \frac{|\xi|^{\frac{1}{p}}\hat{f}(\xi)}{|\xi|^{\frac{1}{p}}} = \hat{g}(\xi)|\xi|^{-\frac{1}{p}}.$$

Combining the fact that if $P_\alpha(x) = |x|^{-\alpha}$, then $\widehat{P_\alpha(\xi)} = C_\alpha P_{1-\alpha}(\xi)$. Thus we obtain $f = g * C_{1-\frac{1}{p}}^{-1} P_{1-\frac{1}{p}}$. The function $P_{1-\frac{1}{p}} = |x|^{-\frac{p-1}{p}} \in L^{\frac{p}{p-1},\infty}(\mathbb{R})$ but not in $L^{\frac{p}{p-1}}(\mathbb{R})$. Applying Lemma 1, we find that

$$\|f\|_{L^{\frac{2p}{p-2},2}} = \left\|g * C_{1-\frac{1}{p}}^{-1} P_{1-\frac{1}{p}}\right\|_{L^{\frac{2p}{p-2},2}}$$

$$\leq C\|g\|_{L^2}\left\||x|^{-\frac{p-1}{p}}\right\|_{L^{\frac{p}{p-1},\infty}} \leq C\|f\|_{\dot{H}^{\frac{1}{p}}}. \tag{13}$$

□

Lemma 4. *There exists a positive constant C such that*

$$\left\|\left\|\|f\|_{L^{\frac{2p}{p-2},2}_{x_1}}\right\|_{L^{\frac{2q}{q-2},2}_{x_2}}\right\|_{L^{\frac{2r}{r-2},2}_{x_3}} \leq C\|\partial_1 f\|_{L^2}^{\frac{1}{p}}\|\partial_2 f\|_{L^2}^{\frac{1}{q}}\|\partial_3 f\|_{L^2}^{\frac{1}{r}}\|f\|_{L^2}^{1-\left(\frac{1}{p}+\frac{1}{q}+\frac{1}{r}\right)}, \tag{14}$$

for every $f \in C_0^\infty(\mathbb{R}^3)$ where $2 < p, q, r \leq \infty, 1 - \left(\frac{1}{p}+\frac{1}{q}+\frac{1}{r}\right) \geq 0$.

Proof. Let Λ_1^p be the Fourier multiplier defined as

$$\mathcal{F}_1\left(\Lambda_1^p f\right)(\xi_1, x_2, x_3) = |\xi_1|^p \mathcal{F}_1 f(\xi_1, x_2, x_3)$$

with

$$\mathcal{F}_1 f(\xi_1, x_2, x_3) = \int_\mathbb{R} e^{-i\xi_1 x_1} f(x_1, x_2, x_3) dx_1,$$

Λ_2^p and Λ_3^p can be defined analogously. Then by Lemma 3, Minkowski's inequality and Hölder's inequality to obtain

$$\left\|\left\|\|f\|_{L^{\frac{2p}{p-2},2}_{x_1}}\right\|_{L^{\frac{2q}{q-2},2}_{x_2}}\right\|_{L^{\frac{2r}{r-2},2}_{x_3}} \leq C\left\|\left\|\|\Lambda_1^{\frac{1}{p}} f\|_{L^2_{x_1}}\right\|_{L^{\frac{2q}{q-2},2}_{x_2}}\right\|_{L^{\frac{2r}{r-2},2}_{x_3}} \leq \left\|\left\|\|\Lambda_1^{\frac{1}{p}} f\|_{L^{\frac{2q}{q-2},2}_{x_2}}\right\|_{L^2_{x_1}}\right\|_{L^{\frac{2r}{r-2},2}_{x_3}}$$

$$\leq C\left\|\|\Lambda_2^{\frac{1}{q}}\Lambda_1^{\frac{1}{p}} f\|_{L^2_{x_1,x_2}}\right\|_{L^{\frac{2r}{r-2},2}_{x_3}} \leq C\left\|\|\Lambda_2^{\frac{1}{q}}\Lambda_1^{\frac{1}{p}} f\|_{L^{\frac{2r}{r-2},2}_{x_3}}\right\|_{L^2_{x_1,x_2}} \tag{15}$$

$$\leq C\left\|\Lambda_3^{\frac{1}{r}}\Lambda_2^{\frac{1}{q}}\Lambda_1^{\frac{1}{p}} f\right\|_{L^2}.$$

Combining the Fourier-Plancherel formula and the Hölder's inequality, we have

$$C\left\|\Lambda_3^{\frac{1}{r}}\Lambda_2^{\frac{1}{q}}\Lambda_1^{\frac{1}{p}}f\right\|_{L^2} \leq C\left(\int_{\mathbb{R}^3}|\xi_1|^{\frac{2}{p}}|\xi_2|^{\frac{2}{q}}|\xi_3|^{\frac{2}{r}}|\mathcal{F}f(\xi_1,\xi_2,\xi_3)|^2 d\xi_1 d\xi_2 d\xi_3\right)^{\frac{1}{2}}$$

$$= C\left(\int_{\mathbb{R}^3}|\xi_1|^{\frac{2}{p}}|\mathcal{F}f(\xi)|^{\frac{2}{p}}|\xi_2|^{\frac{2}{q}}|\mathcal{F}f(\xi)|^{\frac{2}{q}}|\xi_3|^{\frac{2}{r}}|\mathcal{F}f(\xi)|^{\frac{2}{r}}|\mathcal{F}f(\xi)|^{2-\left(\frac{2}{p}+\frac{2}{q}+\frac{2}{r}\right)}d\xi_1 d\xi_2 d\xi_3\right)^{\frac{1}{2}} \quad (16)$$

$$\leq C\|\mathcal{F}f\|_{L^2}^{1-\frac{1}{p}-\frac{1}{q}-\frac{1}{r}}\left(\int_{\mathbb{R}^3}|\xi_1|^2|\mathcal{F}f|^2 d\xi\right)^{\frac{1}{2p}}\left(\int_{\mathbb{R}^3}|\xi_2|^2|\mathcal{F}f|^2 d\xi\right)^{\frac{1}{2q}}\left(\int_{\mathbb{R}^3}|\xi_3|^2|\mathcal{F}f|^2 d\xi\right)^{\frac{1}{2r}}$$

$$\leq C\|\partial_1 f\|_{L^2}^{\frac{1}{p}}\|\partial_2 f\|_{L^2}^{\frac{1}{q}}\|\partial_3 f\|_{L^2}^{\frac{1}{r}}\|f\|_{L^2}^{1-\left(\frac{1}{p}+\frac{1}{q}+\frac{1}{r}\right)}.$$
$\square$

Remark 4. *In fact, since $L^{\frac{2p}{p-2},2} \hookrightarrow L^{\frac{2p}{p-2},\frac{2p}{p-2}}$ for $2 < p < \infty$, we have similar result for estimate (14) in anisotropic Lebesgue space (for more details refer to [34]). However, we should point out that Lemma 4 holds in Lorentz space mainly depends on the Sobolev's embedding in Lemma 3.*

2. Proof of Theorem 1

This section is devoted to the proof of Theorem 1. The proof is based on the establishment of a priori estimates under condition (10).

Firstly, we note that, by the energy inequality, for weak solution (u,b), we have

$$\|u\|_{L^2}^2 + \|b\|_{L^2}^2 + 2\int_0^T \|\nabla u\|_{L^2}^2 + \|\nabla b\|_{L^2}^2 dt \leq \|u_0\|_{L^2}^2 + \|b_0\|_{L^2}^2. \quad (17)$$

Next, let us convert (1) into a symmetric form. Writing

$$\omega^\pm = u \pm b,$$

we find by adding and subtracting $(1)_1$ with $(1)_2$,

$$\begin{cases} \partial_t \omega^+ + (\omega^- \cdot \nabla)\omega^+ - \Delta\omega^+ + \nabla p = 0, \\ \partial_t \omega^- + (\omega^+ \cdot \nabla)\omega^- - \Delta\omega^- + \nabla p = 0, \\ \nabla \cdot \omega^+ = \nabla \cdot \omega^- = 0, \\ \omega^+(0) = \omega_0^+ \equiv u_0 + b_0, \quad \omega^-(0) = \omega_0^- \equiv u_0 - b_0. \end{cases} \quad (18)$$

Taking the inner product of the i-th equation of $(18)_1$ with $|\omega_i^+|^2\omega_i^+$ and $(18)_2$ with $|\omega_i^-|^2\omega_i^-$ (for $i = 1,2,3$) and integrating by parts in $\mathbb{R}^3$ to get

$$\frac{1}{4}\frac{d}{dt}\left(\|\omega_i^+\|_{L^4}^4 + \|\omega_i^-\|_{L^4}^4\right) + \frac{1}{2}\left(\left\|\nabla|\omega_i^+|^2\right\|_{L^2}^2 + \left\|\nabla|\omega_i^-|^2\right\|_{L^2}^2\right)$$
$$+ \||\omega_i^+|\cdot|\nabla\omega_i^+|\|_{L^2}^2 + \||\omega_i^-|\cdot|\nabla\omega_i^-|\|_{L^2}^2 \quad (19)$$
$$= -\int_{\mathbb{R}^3}\partial_i p|\omega_i^+|^2\omega_i^+ dx - \int_{\mathbb{R}^3}\partial_i p|\omega_i^-|^2\omega_i^- dx \equiv I + J,$$

we consider the (u,b) satisfying condition (10) with any two quantities of $A_i(T)$ and $B_i(T)$ for $(i = 1,2,3)$:

$$\begin{cases} A_i(T) := \int_0^T \left\|\|\partial_i u_i(t)\|_{L_{x_1}^{p,\infty}}\right\|_{L_{x_2}^{q,\infty}}\right\|_{L_{x_3}^{r,\infty}}^{\frac{2}{2-\left(\frac{1}{p}+\frac{1}{q}+\frac{1}{r}\right)}} dt < \infty, \\ B_i(T) := \int_0^T \left\|\|\partial_i b_i(t)\|_{L_{x_1}^{p,\infty}}\right\|_{L_{x_2}^{q,\infty}}\right\|_{L_{x_3}^{r,\infty}}^{\frac{2}{2-\left(\frac{1}{p}+\frac{1}{q}+\frac{1}{r}\right)}} dt < \infty. \end{cases}$$

In order to estimate the term I and J of (19), let us first establish an estimate between the p and the w. Taking the divergence operator $\nabla\cdot$ on both sides of the first equations of (18), it follows that

$$-\Delta p = \text{div}(w^- \cdot \nabla w^+) = \text{div}\,\text{div}(w^- \otimes w^+).$$

Similarly, taking ∇div operator on both sides of the first equation of (18) to obtain

$$-\Delta(\nabla p) = \nabla\,\text{div}(w^- \cdot \nabla w^+) = \nabla\,\text{div}(w^+ \cdot \nabla w^-).$$

By using the boundedness of Riesz transformations in L^p ($1 < p < \infty$) space, so we have

$$\begin{cases} \|p\|_{L^p} \leq C\|w^+\|_{L^{2p}}\|w^-\|_{L^{2p}}, \\ \|\nabla p\|_{L^p} \leq C\|w^+ \cdot \nabla w^-\|_{L^p}, \\ \|\nabla p\|_{L^p} \leq C\|w^- \cdot \nabla w^+\|_{L^p}. \end{cases} \qquad (20)$$

Using the Hölder's inequality, Young's inequality, Lemma 4 and (20), we can deduce that

$$I = -\int_{\mathbb{R}^3} \partial_i p |\omega_i^+|^2 \omega_i^+ \, dx \leq C \left| \int_{\mathbb{R}^3} p |\omega_i^+|^2 \partial_i \omega_i^+ \, dx \right|$$

$$\leq \left\| \|\partial_i \omega_i^+\|_{L_{x_1}^{p,\infty}} \right\|_{L_{x_2}^{q,\infty}} \Big\|_{L_{x_3}^{r,\infty}} \left\| \|p\|_{L_{x_1}^{\frac{2p}{p-2},2}} \right\|_{L_{x_2}^{\frac{2q}{q-2},2}} \Big\|_{L_{x_3}^{\frac{2r}{r-2},2}} \left\| |\omega_i^+|^2 \right\|_{L^2}$$

$$\leq C \left\| \|\partial_i \omega_i^+\|_{L_{x_1}^{p,\infty}} \right\|_{L_{x_2}^{q,\infty}} \Big\|_{L_{x_3}^{r,\infty}} \|\partial_1 p\|_{L^2}^{\frac{1}{p}} \|\partial_2 p\|_{L^2}^{\frac{1}{q}} \|\partial_3 p\|_{L^2}^{\frac{1}{r}} \|p\|_{L^2}^{1-\left(\frac{1}{p}+\frac{1}{q}+\frac{1}{r}\right)} \left\| |\omega_i^+|^2 \right\|_{L^2} \qquad (21)$$

$$\leq C \left\| \|\partial_i \omega_i^+\|_{L_{x_1}^{p,\infty}} \right\|_{L_{x_2}^{q,\infty}} \Big\|_{L_{x_3}^{r,\infty}} \|\nabla p\|_{L^2}^{\frac{1}{p}+\frac{1}{q}+\frac{1}{r}} \|p\|_{L^2}^{1-\left(\frac{1}{p}+\frac{1}{q}+\frac{1}{r}\right)} \left\| |\omega_i^+|^2 \right\|_{L^2}$$

$$\leq C \left\| \|\partial_i \omega_i^+\|_{L_{x_1}^{p,\infty}} \right\|_{L_{x_2}^{q,\infty}} \Big\|_{L_{x_3}^{r,\infty}} \|\nabla p\|_{L^2}^{\frac{1}{p}+\frac{1}{q}+\frac{1}{r}} \|\omega^+\|_{L^4}^{3-\left(\frac{1}{p}+\frac{1}{q}+\frac{1}{r}\right)} \|\omega^-\|_{L^4}^{1-\left(\frac{1}{p}+\frac{1}{q}+\frac{1}{r}\right)}$$

$$\leq \epsilon \left(\|w^+ \cdot \nabla w^-\|_{L^2}^2 + \|w^- \cdot \nabla w^+\|_{L^2}^2 \right) + C \left\| \|\partial_i \omega_i^+\|_{L_{x_1}^{p,\infty}} \right\|_{L_{x_2}^{q,\infty}} \Big\|_{L_{x_3}^{r,\infty}}^{\frac{2}{2-\left(\frac{1}{p}+\frac{1}{q}+\frac{1}{r}\right)}} \left(\|\omega^+\|_{L^4}^4 + \|\omega^-\|_{L^4}^4 \right).$$

Similarly, for J, we have

$$J = -\int_{\mathbb{R}^3} \partial_i p |\omega_i^-|^2 \omega_i^+ \, dx \leq C \left| \int_{\mathbb{R}^3} p |\omega_i^-|^2 \partial_i \omega_i^- \, dx \right|$$

$$\leq \epsilon \left(\|w^+ \cdot \nabla w^-\|_{L^2}^2 + \|w^- \cdot \nabla w^+\|_{L^2}^2 \right) + C \left\| \|\partial_i \omega_i^-\|_{L_{x_1}^{p,\infty}} \right\|_{L_{x_2}^{q,\infty}} \Big\|_{L_{x_3}^{r,\infty}}^{\frac{2}{2-\left(\frac{1}{p}+\frac{1}{q}+\frac{1}{r}\right)}} \left(\|\omega^+\|_{L^4}^4 + \|\omega^-\|_{L^4}^4 \right). \qquad (22)$$

Inserting (21) and (22) into (19) and summing up with respect to the index i from 1 to 3, we get

$$\frac{1}{4}\left(\|w^+\|_{L^4}^4 + \|w^-\|_{L^4}^4\right) + \frac{1}{2}\int_0^t \left(\left\|\nabla|w^+|^2\right\|_{L^2}^2 + \left\|\nabla|w^-|^2\right\|_{L^2}^2\right)ds$$

$$+ \int_0^t \left(\||w^+| \cdot |\nabla w^+|\|_{L^2}^2 + \||w^-| \cdot |\nabla w^-|\|_{L^2}^2\right)ds$$

$$\leq C\int_0^t \sum_{i=1}^3 \left(\left\|\|\partial_i w_i^+\|_{L_{x_1}^{p,\infty}}\right\|_{L_{x_2}^{q,\infty}L_{x_3}^{r,\infty}}^{\frac{2}{2-(\frac{1}{p}+\frac{1}{q}+\frac{1}{r})}} + \left\|\|\partial_i w_i^-\|_{L_{x_1}^{p,\infty}}\right\|_{L_{x_2}^{q,\infty}L_{x_3}^{r,\infty}}^{\frac{2}{2-(\frac{1}{p}+\frac{1}{q}+\frac{1}{r})}}\right) \quad (23)$$

$$\cdot \left(\|w^+\|_{L^4}^4 + \|w^-\|_{L^4}^4\right)ds + \epsilon \int_0^t \left(\|w^+ \cdot \nabla w^-\|_{L^2}^2 + \|w^- \cdot \nabla w^+\|_{L^2}^2\right)ds + C\left(\|w_0^+\|_{L^4}^4 + \|w_0^-\|_{L^4}^4\right),$$

where we have used that for any $p \geq 1$ and some constant $C_{\gamma,p} > 0$,

$$C_{\gamma,p}^{-1}\|u\|_{L^p}^\gamma \leq \sum_{i=1}^3 \|u_i\|_{L^p}^\gamma \leq C_{\gamma,p}\|u\|_{L^p}^\gamma.$$

Due to the fact

$$\left|\nabla|w^+|^2\right| \leq 2|w^+||\nabla w^+|$$

and the inequality

$$\|u(t)\|_{L^4} \leq \frac{1}{2}(\|w^+(t)\|_{L^4} + \|w^-(t)\|_{L^4}),$$

$$\|b(t)\|_{L^4} \leq \frac{1}{2}(\|w^+(t)\|_{L^4} + \|w^-(t)\|_{L^4}).$$

We rewrite inequality (23) as follows

$$\frac{1}{4}\left(\|u(t)\|_{L^4}^4 + \|b(t)\|_{L^4}^4\right) + \frac{1}{4}\int_0^t \left(\left\|\nabla|u|^2\right\|_{L^2}^2 + \left\|\nabla|b|^2\right\|_{L^2}^2\right)ds$$

$$+ \int_0^t \left(\||u| \cdot |\nabla u|\|_{L^2}^2 + \||u| \cdot |\nabla b|\|_{L^2}^2 + \||b| \cdot |\nabla u|\|_{L^2}^2 + \||b| \cdot |\nabla b|\|_{L^2}^2\right)ds$$

$$\leq C\int_0^t \sum_{i=1}^3 \left(\left\|\|\partial_i w_i^+\|_{L_{x_1}^{p,\infty}}\right\|_{L_{x_2}^{q,\infty}L_{x_3}^{r,\infty}}^{\frac{2}{2-(\frac{1}{p}+\frac{1}{q}+\frac{1}{r})}} + \left\|\|\partial_i w_i^-\|_{L_{x_1}^{p,\infty}}\right\|_{L_{x_2}^{q,\infty}L_{x_3}^{r,\infty}}^{\frac{2}{2-(\frac{1}{p}+\frac{1}{q}+\frac{1}{r})}}\right) \quad (24)$$

$$\cdot \left(\|u\|_{L^4}^4 + \|b\|_{L^4}^4\right)ds + \epsilon \int_0^t \left(\|u \cdot \nabla u\|_{L^2}^2 + \|b \cdot \nabla u\|_{L^2}^2 + \|u \cdot \nabla b\|_{L^2}^2 + \|b \cdot \nabla b\|_{L^2}^2\right)ds$$

$$+ C\left(\|w_0^+\|_{L^4}^4 + \|w_0^-\|_{L^4}^4\right),$$

and hence we get

$$\frac{1}{4}\left(\|u(t)\|_{L^4}^4 + \|b(t)\|_{L^4}^4\right) + \frac{1}{4}\int_0^t \left(\left\|\nabla|u|^2\right\|_{L^2}^2 + \left\|\nabla|b|^2\right\|_{L^2}^2\right)ds$$

$$+ \frac{1}{4}\int_0^t \left(\||u| \cdot |\nabla u|\|_{L^2}^2 + \||u| \cdot |\nabla b|\|_{L^2}^2 + \||b| \cdot |\nabla u|\|_{L^2}^2 + \||b| \cdot |\nabla b|\|_{L^2}^2\right)ds$$

$$\leq C\int_0^t \sum_{i=1}^3 \left(\left\|\|\partial_i w_i^+\|_{L_{x_1}^{p,\infty}}\right\|_{L_{x_2}^{q,\infty}L_{x_3}^{r,\infty}}^{\frac{2}{2-(\frac{1}{p}+\frac{1}{q}+\frac{1}{r})}} + \left\|\|\partial_i w_i^-\|_{L_{x_1}^{p,\infty}}\right\|_{L_{x_2}^{q,\infty}L_{x_3}^{r,\infty}}^{\frac{2}{2-(\frac{1}{p}+\frac{1}{q}+\frac{1}{r})}}\right) \quad (25)$$

$$\cdot \left(\|u\|_{L^4}^4 + \|b\|_{L^4}^4\right)ds + C\left(\|u_0\|_{L^4}^4 + \|b_0\|_{L^4}^4\right).$$

Applying the Gronwall's inequality to obtain

$$\sup_{0\leq t\leq T}\left(\|u(t)\|_{L^4}^4+\|b(t)\|_{L^4}^4\right)$$

$$\leq C\exp C\int_0^T\sum_{i=1}^3\left(\left\|\|\|\partial_i\omega_i^+\|_{L_{x_1}^{p,\infty}}\right\|_{L_{x_2}^{q,\infty}}\right\|_{L_{x_3}^{r,\infty}}^{\overline{2-\left(\frac{1}{p}+\frac{1}{q}+\frac{1}{r}\right)}}+\left\|\|\|\partial_i\omega_i^-\|_{L_{x_1}^{p,\infty}}\right\|_{L_{x_2}^{q,\infty}}\right\|_{L_{x_3}^{r,\infty}}^{\overline{2-\left(\frac{1}{p}+\frac{1}{q}+\frac{1}{r}\right)}}\right)dt \quad (26)$$

$$\leq C\exp C\int_0^T\sum_{i=1}^3\left(\left\|\|\|\partial_i u_i\|_{L_{x_1}^{p,\infty}}\right\|_{L_{x_2}^{q,\infty}}\right\|_{L_{x_3}^{r,\infty}}^{\overline{2-\left(\frac{1}{p}+\frac{1}{q}+\frac{1}{r}\right)}}+\left\|\|\|\partial_i b_i\|_{L_{x_1}^{p,\infty}}\right\|_{L_{x_2}^{q,\infty}}\right\|_{L_{x_3}^{r,\infty}}^{\overline{2-\left(\frac{1}{p}+\frac{1}{q}+\frac{1}{r}\right)}}\right)dt$$

$<\infty.$

Since
$$u,b\in L^\infty\left(0,T;L^4\left(\mathbb{R}^3\right)\right)\subset L^8\left(0,T;L^4\left(\mathbb{R}^3\right)\right),$$
combining the classical Serrin-type regularity criterion (2), as in [15], then we complete the proof of Theorem 1.

3. Conclusions

This paper studies the MHD equations, and obtains the a regularity criterion only involving the partial components of the ∇u and ∇b. In addition, the anisotropic Lorentz space used in this article is broader than the general Lebesgue and Lorentz spaces. It seems that a slightly modified the technique in Theorem 1 can be applied to other incompressible fluid equations such as micropolar equations and the magneto-micropolar equations.

Author Contributions: Both authors contributed equally to this work. Both authors have read and agreed to the published version of the manuscript.

Data Availability Statement: All data generated or analysed during this study are included in this published article.

Acknowledgments: The first author is partially supported by I.N.D.A.M-G.N.A.M.P.A. 2019 and the "RUDN University Program 5-100".

Conflicts of Interest: The authors declare no conflict of interest.

References

1. Duvaut, G.; Lions, J.L. Inéquations en thermoélasticité et magnétohydrodynamique. *Arch. Ration. Mech. Anal.* **1972**, *46*, 241–279. [CrossRef]
2. Sermange, M.; Temam, R. Some mathematical questions related to the MHD equations. *Commun. Pure Appl. Math.* **1983**, *36*, 635–664. [CrossRef]
3. Berselli, L.C.; Spirito, S. On the Existence of Leray-Hopf Weak Solutions to the Navier-Stokes Equations. *Fluids* **2021**, *6*, 42. [CrossRef]
4. Cao, C.; Wu, J. Two regularity criteria for the 3D MHD equations. *J. Differ. Equ.* **2010**, *248*, 2263–2274. [CrossRef]
5. Da Veiga, H.B. A new regularity class for the Navier-Stokes equations in $\mathbb{R}^n$. *Chin. Ann. Math. Ser. B* **1995**, *16*, 407–412.
6. Gala, S.; Ragusa, M.A. A logarithmic regularity criterion for the two-dimensional MHD equations. *J. Math. Anal. Appl.* **2016**, *444*, 1752–1758. [CrossRef]
7. Alghamdi, A.M.; Gala, S.; Ragusa, M.A. A regularity criterion of smooth solution for the 3D viscous Hall-MHD equations. *Aims Math.* **2018**, *3*, 565–574. [CrossRef]
8. Gala, S.; Ragusa, M.A. A new regularity criterion for the 3D incompressible MHD equations via partial derivatives. *J. Math. Anal. Appl.* **2020**, *481*, 123497. [CrossRef]
9. Gala, S. Extension criterion on regularity for weak solutions to the 3D MHD equations. *Math. Methods Appl. Sci.* **2010**, *33*, 1496–1503. [CrossRef]
10. Jia, X.; Zhou, Y. Regularity criteria for the 3D MHD equations involving partial components. *Nonlinear Anal. Real World Appl.* **2012**, *13*, 410–418. [CrossRef]
11. Prodi, G. Un teorema di unicitá per le equazioni di Navier-Stokes. *Ann. Mat. Pura Appl.* **1959**, *48*, 173–182. [CrossRef]
12. Serrin, J. On the interior regularity of weak solutions of the Navier-Stokes equations. *Arch. Ration. Mech. Anal.* **1962**, *9*, 187–195. [CrossRef]

13. Xu, F. A regularity criterion for the 3D incompressible magneto-hydrodynamics equations. *J. Math. Anal. Appl.* **2018**, *460*, 634–644. [CrossRef]
14. Zhou, Y.; Gala, S. Regularity criteria for the solutions to the 3D MHD equations in the multiplier space. *Z. Angew. Math. Und Phys.* **2010**, *61*, 193–199. [CrossRef]
15. He, C.; Xin, Z. On the regularity of weak solutions to the magnetohydrodynamic equations. *J. Differ. Equ.* **2005**, *213*, 235–254. [CrossRef]
16. He, C.; Wang, Y. Remark on the regularity for weak solutions to the magnetohydrodynamic equations. *Math. Methods Appl. Sci.* **2008**, *31*, 1667–1684. [CrossRef]
17. Ni, L.; Guo, Z.; Zhou, Y. Some new regularity criteria for the 3D MHD equations. *J. Math. Anal. Appl.* **2012**, *396*, 108–118. [CrossRef]
18. Jia, X. A new scaling invariant regularity criterion for the 3D MHD equations in terms of horizontal gradient of horizontal components. *Appl. Math. Lett.* **2015**, *50*, 1–4. [CrossRef]
19. Xu, F.; Li, X.; Cui, Y.; Wu, Y. A scaling invariant regularity criterion for the 3D incompressible magneto-hydrodynamics equations. *Z. Fur Angew. Math. Und Phys.* **2017**, *68*, 125. [CrossRef]
20. Wu, F. Blow–up criterion of strong solutions to the three-dimensional double-diffusive convection system. *Bull. Malays. Math. Sci. Soc.* **2019**, *43*, 2673–2686. [CrossRef]
21. Wu, F. A refined regularity criteria of weak solutions to the magneto-micropolar fluid equations. *J. Evol. Equ.* **2020**, 1–10. [CrossRef]
22. Wu, F. Navier-Stokes Regularity Criteria in Vishik Spaces. *Appl. Math. Optim.* **2021**, 1–15. [CrossRef]
23. Zhang, X. A regularity criterion for the solutions of 3D Navier-Stokes equations. *J. Math. Anal. Appl.* **2008**, *34*, 336–339. [CrossRef]
24. Zhang, Z. A Serrin-type regularity criterion for the Navier-Stokes equations via one velocity component. *Commun. Pure Appl. Anal.* **2013**, *12*, 117–124. [CrossRef]
25. Zhang, Z.; Yang, X. Remarks on the blow-up criterion for the MHD system involving horizontal components or their horizontal gradients. *Ann. Pol. Math.* **2016**, *116*, 87–99. [CrossRef]
26. Zhang, Z.; Chen, Q. Regularity criterion via two components of vorticity on weak solutions to the Navier-Stokes equations in $\mathbb{R}^3$. *J. Differ. Equ.* **2005**, *216*, 470–481. [CrossRef]
27. Lorentz, G. Some new functional spaces. *Ann. Math.* **1950**, *51*, 37–55. [CrossRef]
28. Stefanov, A.; Torres, R.H. Calderón-Zygmund operators on mixed Lebesgue spaces and applications to null forms. *J. Lond. Math. Soc.* **2004**, *70*, 447–462. [CrossRef]
29. Georgiadis, A.G.; Johnsen, J.; Nielsen, M. Wavelet transforms for homogeneous mixed-norm Triebel?Lizorkin spaces. *Monatshefte Für Math.* **2017**, *183*, 587–624. [CrossRef]
30. Chen, T.; Sun, W. Iterated weak and weak mixed-norm spaces with applications to geometric inequalities. *J. Geom. Anal.* **2020**, *30*, 4268–4323. [CrossRef]
31. Bekmaganbetov, K.A.; Toleugazy, Y. On the Order of the trigonometric diameter of the anisotropic Nikol'skii-Besov class in the metric of anisotropic Lorentz spaces. *Anal. Math.* **2019**, *45*, 237–247.
32. Lemarie-Rieusset, P.G. *Recent Developments in the Navier-Stokes Problem*; CRC Press: Boca Raton FL, USA, 2002. [CrossRef]
33. O'Neil, R. Convolution operators and $L(p,q)$ spaces. *Duke Math. J.* **1963**, *30*, 129–142. [CrossRef]
34. Liu, Q.; Zhao, J. Blow-up criteria in terms of pressure for the 3D nonlinear dissipative system modeling electro-diffusion. *J. Evol. Equ.* **2018**, *18*, 1675–1696. [CrossRef]

Article

Aboodh Transform Iterative Method for Solving Fractional Partial Differential Equation with Mittag–Leffler Kernel

Michael A. Awuya * and Dervis Subasi

Department of Mathematics, Eastern Mediterranean University, Via Mersin 10, Famagusta TR99628, Turkey; dervis.subasi@emu.edu.tr
* Correspondence: 16600174@emu.edu.tr

Abstract: The major aim of this paper is the presentation of Aboodh transform of the Atangana–Baleanu fractional differential operator both in Caputo and Riemann–Liouville sense by using the connection between the Laplace transform and the Aboodh transform. Moreover, we aim to obtain the approximate series solutions for the time-fractional differential equations with an Atangana–Baleanu fractional differential operator in the Caputo sense using the Aboodh transform iterative method, which is the modification of the Aboodh transform by combining it with the new iterative method. The relation between the Laplace transform and the Aboodh transform is symmetrical. Some graphical illustrations are presented to describe the effect of the fractional order. The outcome reveals that Aboodh transform iterative method is easy to implement and adequately captures the behavior and the fractional effect of the fractional differential equation.

Keywords: integral transform; Atangana–Baleanu fractional derivative; fractional calculus; Aboodh transform iterative method; Mittag–Leffler function

MSC: 26A33; 34A08; 35R11

1. Introduction

The role of fractional differential operations in evaluating and simulating history dependent evolution models in physics and engineering cannot be over emphasized because of their properties [1–5]. Several definitions of fractional differential operator exist in literature. For extensive study on fractional derivative operators, refer to [6–10].

Recently, Atangana and Baleanu presented a new fractional differential operator which utilizes the Mittag–Leffler function as the kernel to replace the exponential function kernel of the Caputo-Fabrizo fractional differential operator [11,12]. This is performed with the purpose of introducing a non-local, non-singular kernel and to overcome the limitations of other fractional differential operators. For instance, the Riemman–Liouville fractional differential operator did not properly account for the initial condition, while the Caputo fractional differential operator was able to resolve this issue with the initial condition but was confronted with the limitation of singular kernel .

The use of an integral transform combined with analytical methods for the solution of fractional differential equations in the fast convergence series form is popular among researchers [13–17]. The concept of the Caputo–Fabrizio fractional derivative was extended to the model of HIV-1 infection of $CD4^+$ T-cell using the homotopy analysis transform method in [18]. The fractional Caputo–Fabrizio derivative was utilized to introduce two types of new high order derivative with their existence solutions in [19]. The authors in [20] studied the Laplace transform, Sumudu transform, Fourier transform and Mellin transform of the Atangana–Baleanu fractional differential operator. Moreover, the Shehu transform was applied on the Atangana–Baleanu fractional derivative in [21], and some new related properties are established.

The novelty of this paper is the establishment of the Aboodh transform of the Atangana–Baleanu fractional differential operator both in the Caputo and Riemman–Liouville sense using the connection between the Laplace transform and the Aboodh transform. Moreover, we validate the Aboodh transform iterative method [4] for the solution of Atangana–Baleanu fractional differential equation.

We structure this paper as follows. Section 2 consist of the fundamental concept while, in Section 3, we discuss the basic idea of Aboodh transform iterative method. In Section 4, we validate the Aboodh transform iterative method for the solution of Atangana–Baleanu fractional differential equation and provide some concluding remarks in Section 5.

2. Preliminaries

In this section, some definitions, theorems and properties that will be useful in this paper is given.

Definition 1. *The Aboodh transform of a function Q(t) with exponential order over the class of functions [4]*

$$\mathscr{C} = \{Q : |Q(t)| < Be^{p_j|t|}, if\ t \in (-1)^j \times [0,\infty), j = 1,2; (B, p_1, p_2 > 0)\} \quad (1)$$

is written as

$$\mathscr{A}[Q(t)] = \mathcal{M}(\psi), \quad (2)$$

and defined as

$$\mathscr{A}[Q(t)] = \frac{1}{\psi}\int_0^\infty Q(t)e^{-\psi t}dt = \mathcal{M}(\psi), \quad p_1 \leq \psi \leq p_2. \quad (3)$$

Obviously, The Aboodh transform is linear as the Laplace transform.

Definition 2. *The inverse Aboodh transform of a function Q(t) is defined as [4].*

$$Q(t) = \mathscr{A}^{-1}[\mathcal{M}(\psi)]. \quad (4)$$

Definition 3. *Let $Q(t) \in \mathscr{C}$, then the Laplace transform is defined by the following integral [22].*

$$\mathcal{Q}(t) = \int_0^\infty Q(t)e^{-st}dt. \quad (5)$$

The Laplace transform of Q(t) is written as follows.

$$\mathcal{L}[Q(t)] = \mathcal{Q}(s). \quad (6)$$

If ψ and s are unity, then Equations (3) and (5) are equal; hence, the relationships between the Aboodh transform and the Laplace transform are symmetrical.

Theorem 1 ([23]). *If $Q(t) \in \mathscr{C}$ with the Aboodh transform $\mathscr{A}[Q(t)]$ and Laplace transform $\mathcal{L}[Q(t)]$, then the following is the case.*

$$\mathcal{M}(\psi) = \frac{1}{\psi}\mathcal{Q}(\psi). \quad (7)$$

Definition 4. *The Mittag–Leffler function is a special function that often occurs naturally in the solution of fractional order calculus, and it is defined as follows [24].*

$$E_\beta(Z) = \sum_{\rho=0}^\infty \frac{Z^\rho}{\Gamma(\rho\beta+1)}, \quad \beta, Z \in \mathbb{C}, \operatorname{Re}(\beta) \geq 0, \quad (8)$$

In generalized form [24], it is defined as follows.

$$E_{\beta,\gamma}^{\mu} = \sum_{\rho=0}^{\infty} \frac{Z^{\rho}(\mu)_{\rho}}{\Gamma(\gamma+\rho\beta)\rho!}, \quad \beta, \gamma, Z \in \mathbb{C}, \operatorname{Re}(\beta) \geq 0, \operatorname{Re}(\gamma) \geq 0, \tag{9}$$

Moreover, we assume $(\mu)_{\rho}$ to be the Pochhammer's symbol.

Definition 5. *Let $Q \in H^1(0,1)$ and $0 < \beta < 1$, then the Atangana–Baleanu fractional derivative defined in the Caputo sense is given as follows [11].*

$$_0^{ABC}D_t^{\beta}Q(t) = \frac{N(\beta)}{1-\beta} \int_0^t Q'(x) E_{\beta}\left(\frac{-\beta(t-x)^{\beta}}{1-\beta}\right) dx. \tag{10}$$

Definition 6. *Let $Q \in H^1(0,1)$ and $0 < \beta < 1$, then the Atangana–Baleanu fractional derivative defined in the Riemann–Liouville sense is given as follows [11].*

$$_0^{ABR}D_t^{\beta}Q(t) = \frac{N(\beta)}{1-\beta} \frac{d}{dt} \int_0^t Q(x) E_{\beta}\left(\frac{-\beta(t-x)^{\beta}}{1-\beta}\right) dx, \tag{11}$$

The normalization function $N(\beta) > 0$ satisfies the condition $N(0) = N(1) = 1$.

Theorem 2 ([11]). *The Laplace transform of Atangana–Baleanu fractional derivative according to the Caputo sense is derived as follows:*

$$\mathcal{L}\left[_0^{ABC}D_t^{\beta}Q(t)\right] = \frac{N(\beta)}{1-\beta} \times \frac{s^{\beta}F(s) - s^{\beta-1}f(0)}{s^{\beta} + \frac{\beta}{1-\beta}}, \tag{12}$$

Moreover, the Laplace transform of Atangana–Baleanu fractional derivative according to the Riemann–Liouville sense is derived as follows.

$$\mathcal{L}\left[_0^{ABR}D_t^{\beta}Q(t)\right] = \frac{N(\beta)}{1-\beta} \times \frac{s^{\beta}F(s)}{s^{\beta} + \frac{\beta}{1-\beta}}. \tag{13}$$

Theorem 3. *If $\Omega, \beta \in \mathbb{C}$, with $\operatorname{Re}(\beta) > 0$, then the Aboodh transform of $E_{\beta}(\Omega t^{\beta})$ is derived as the following:*

$$\mathcal{M}(E_{\beta}(\Omega t^{\beta})) = \frac{1}{\psi^2}\left(1 - \frac{\Omega}{\psi^{\beta}}\right)^{-1}, \tag{14}$$

where $|\Omega\psi^{-\beta}| < 1$.

Proof of Theorem 3. Let us use the following Laplace transform formula:

$$\mathcal{L}[E_{\beta}(\Omega t^{\beta})] = \frac{1}{s}(1 - \Omega s^{-\beta})^{-1}, \tag{15}$$

then by using Equation (7), we have the following.

$$\mathcal{M}(\psi) = \frac{1}{\psi}\mathcal{Q}(\psi)$$

$$= \frac{1}{\psi}\left(\frac{1}{\psi}(1 - \Omega\psi^{-\beta})^{-1}\right) \tag{16}$$

$$= \psi^{-2}(1 - \Omega\psi^{-\beta})^{-1}. \tag{17}$$

□

Theorem 4. Let $\beta, \gamma \in \mathbb{C}$, with $\operatorname{Re}(\beta) > 0$, $\operatorname{Re}(\gamma) > 0$, the Aboodh transform of $t^{\gamma-1} E_{\beta,\mu}^{\mu}(\Omega t^{\beta})$ is derived as follows.

$$t^{\gamma-1} E_{\beta,\mu}^{\mu}(\Omega t^{\beta}) = \frac{1}{\psi^{\gamma+1}}(1 - \Omega\psi^{-\beta})^{-\mu}, \quad |\Omega\psi^{-\beta}| < 1. \tag{18}$$

Proof of Theorem 4. Let us use the Laplace transform formula:

$$\mathcal{L}[t^{\gamma-1} E_{\beta,\gamma}^{\mu}(\Omega t^{\beta})] = s^{-\mu}(1 - \Omega s^{-\beta})^{-\mu}, \tag{19}$$

then by using Equation (7), we have the following.

$$\mathcal{M}(t^{\gamma-1} E_{\beta,\gamma}^{\mu}(\Omega t^{\beta})) = \frac{1}{\psi} \mathcal{Q}(\psi) \tag{20}$$

$$= \frac{1}{\psi}\left(\frac{1}{\psi^{\gamma}}(1 - \Omega\psi^{-\beta})^{-\mu}\right)$$

$$= \frac{1}{\psi^{\gamma+1}}(1 - \Omega\psi^{\beta})^{\mu}. \tag{21}$$

□

Theorem 5. If $\mathcal{M}(\psi)$ is the Aboodh transform of $Q(t) \in \mathscr{C}$ and $\mathcal{Q}(s)$ is the Laplace transform of $Q(t) \in \mathscr{C}$, then the Aboodh transform of Atangana–Baleanu fractional derivative according to the Caputo sense is derived as follows.

$$\mathcal{M}(_0^{ABC}D_t^{\beta} Q(t)) = \frac{N(\beta)(\mathcal{M}(\psi) - \psi^{-2}Q(0))}{1 - \beta + \beta\psi^{-\beta}}. \tag{22}$$

Proof of Theorem 5. Using the relationship between the Aboodh transform and Laplace transform, we obtain the following.

$$\mathcal{M}(_0^{ABC}D_t^{\beta} Q(t)) = \frac{1}{\psi}\left(\frac{N(\beta)}{1-\beta} \times \frac{\psi^{\beta}\mathcal{Q}(\psi) - \psi^{\beta-1}Q(0)}{\psi^{\beta} + \frac{\beta}{1-\beta}}\right)$$

$$= \psi^{\beta}\left(N(\beta) \times \frac{\mathcal{M}(\psi) - \psi^{-2}Q(0)}{\psi^{\beta}(1 - \beta + \beta\psi^{-\beta})}\right) \tag{23}$$

$$= N(\beta) \times \frac{(\mathcal{M}(\psi) - \psi^{-2}Q(0))}{1 - \beta + \beta\psi^{-\beta}}.$$

□

Theorem 6. Assume that $\mathcal{M}(\psi)$ is the Aboodh transform of $Q(t) \in \mathscr{C}$ and $\mathcal{Q}(s)$ is the Laplace transform of $Q(t) \in \mathscr{C}$, then the Aboodh transform of Atangana–Baleanu fractional derivative according to the Riemann–Liouville sense is derived as follows.

$$\mathcal{M}(_0^{ABR}D_t^{\beta} Q(t)) = \frac{N(\beta)\mathcal{M}(\psi)}{1 - \beta + \beta\psi^{-\beta}}. \tag{24}$$

Proof of Theorem 6. By using the relationship between the Aboodh transform transform and the Laplace transform, we obtain the following.

$$\mathcal{M}(_0^{ABR}D_t^{\beta} Q(t)) = \frac{1}{\psi}\left(\frac{N(\beta)}{1-\beta} \times \frac{\psi^{\beta}\mathcal{Q}(\psi)}{\psi^{\beta} + \frac{\beta}{1-\beta}}\right)$$

$$= \frac{1}{\psi}\left(\frac{N(\beta)\mathcal{Q}(\psi)}{1-\beta+\beta\psi^{-\beta}}\right) \qquad (25)$$

$$= \frac{N(\beta)\mathcal{M}(\psi)}{1-\beta+\beta\psi^{-\beta}}.$$

□

3. Aboodh Transform Iterative Method

In this section, we consider the fundamental solution of the initial value problem using the Aboodh transform iterative method. This iterative method is a combination of the new iterative method introduced by Daftardar–Gejji and Jafari [25] with the Aboodh transform which is a modification of the Laplace transform [4].

Basic Idea of Aboodh Transform Iterative Method

Consider the fractional differential equation of the following:

$$_{0}^{ABC}D_{t}^{\beta}Q(x,t) = \mathcal{R}(Q(x,t)) + \mathcal{F}(Q(x,t)) + \Phi(x,t), \quad 0 < \beta \leq 1, \qquad (26)$$

that is subject to the following initial condition.

$$Q(x,0) = Q_0(x), \qquad (27)$$

$_{0}^{ABC}D_{t}^{\beta}$ is the Atangana–Baleanu fractional differential operator, $\Phi(x,t)$ is the source term, $\mathcal{R}$ and $\mathcal{F}$ are the linear and non-linear operators. Using the Aboodh transform on both sides of Equation (26) with the initial condition, we obtain the following.

$$\mathscr{A}[Q(x,t)] = \frac{1-\beta+\beta\psi^{-\beta}}{N(\beta)}\left(\frac{N(\beta)\psi^{-2}Q(x,0)}{1-\beta+\beta\psi^{-\beta}} + \mathscr{A}[\mathcal{R}(Q(x,t)) + \mathcal{F}(Q(x,t)) + \Phi(x,t)]\right), \qquad (28)$$

By simplifying further and taking the inverse Aboodh transform, we obtain the following.

$$Q(x,t) = \mathscr{A}^{-1}\left[\frac{1-\beta+\beta\psi^{-\beta}}{N(\beta)}\left(\frac{N(\beta)\psi^{-2}Q(x,0)}{1-\beta+\beta\psi^{-\beta}} + \mathscr{A}[\Phi(x,t)] + \mathscr{A}[\mathcal{R}(Q(x,t)) + \mathcal{F}(Q(x,t))]\right)\right]. \qquad (29)$$

The non-linear term in Equation (29) can be decompose as follows [25].

$$\mathcal{F}(Q(x,t)) = \mathcal{F}\left(\sum_{q=0}^{\infty} Q_q(x,t)\right)$$

$$= \mathcal{F}(Q_0(x,t)) + \sum_{q=1}^{\infty}\left\{\mathcal{F}\left(\sum_{j=0}^{q} Q_q(x,t)\right) - \mathcal{F}\left(\sum_{j=0}^{q-1} Q_q(x,t)\right)\right\}. \qquad (30)$$

Now, we define the k-th order approximate series as the following.

$$\mathcal{Q}^{(k)}(x,t) = \sum_{m=0}^{k} Q_m(x,t)$$

$$= Q_0(x,t) + Q_1(x,t) + Q_2(x,t) + \cdots + Q_k(x,t), \quad k \in N. \qquad (31)$$

Assume that the solution of Equation (26) is in a series form given as follow.

$$Q(x,t) = \lim_{k\to\infty} \mathcal{Q}^{(k)}(x,t) = \sum_{m=0}^{\infty} Q_m(x,t), \qquad (32)$$

Then, substituting Equations (31) and (30) into Equation (29), we obtain the following.

$$\sum_{q=0}^{\infty} Q_q(x,t) =$$

$$\mathscr{A}^{-1}\left[\frac{1-\beta+\beta\psi^{-\beta}}{N(\beta)}\left(\frac{N(\beta)\psi^{-2}Q(x,0)}{1-\beta+\beta\psi^{\beta}} + \Phi(x,t) + \mathscr{A}[\mathcal{R}(Q_0(x,t)) + \mathcal{F}(Q_0(x,t))]\right)\right] +$$

$$\mathscr{A}^{-1}\left[\frac{1-\beta+\beta\psi^{-\beta}}{N(\beta)}\left(\mathscr{A}\left[\sum_{q=1}^{\infty}\left(\mathcal{R}(Q_q(x,t)) + \left\{\mathcal{F}\left(\sum_{j=0}^{q}(x,t)\right) - \mathcal{F}\left(\sum_{j=0}^{q-1} Q_q(x,t)\right)\right\}\right)\right]\right)\right]. \quad (33)$$

From Equation (33), we define the following iterations.

$$Q_0(x,t) = \mathscr{A}^{-1}\left[\frac{1-\beta+\beta\psi^{-\beta}}{N(\beta)}\left(\frac{N(\beta)\psi^{-2}Q(x,0)}{1-\beta+\beta\psi^{\beta}} + \Phi(x,t)\right)\right], \quad (34)$$

$$Q_1(x,t) = \mathscr{A}^{-1}\left[\frac{1-\beta+\beta\psi^{-\beta}}{N(\beta)}(\mathscr{A}[\mathcal{R}(Q_0(x,t)) + \mathcal{F}(Q_0(x,t))])\right], \quad (35)$$

$$\vdots$$

$$Q_{q+1} =$$

$$\mathscr{A}^{-1}\left[\frac{1-\beta+\beta\psi^{-\beta}}{N(\beta)}\left(\mathscr{A}\left[\sum_{q=1}^{\infty}\left(\mathcal{R}(Q_q(x,t)) + \left\{\mathcal{F}\left(\sum_{j=0}^{q}(x,t)\right) - \mathcal{F}\left(\sum_{j=0}^{q-1} Q_q(x,t)\right)\right\}\right)\right]\right)\right],$$

$$q = 1,2,\ldots \quad (36)$$

Convergence Analysis

We establish the convergence analysis of the series in Equation (32) here.

Theorem 7. *Suppose that the nonlinear operator and the linear operator are from the Banach space X relative to itself, and Q(x,t) is analytic about t. Then the infinite series defined in Equation (32) computed by Equations (34), (35),..., (36) converges to the solution of Equation (26) if $0 < \rho \leq 1$, where ρ is a nonnegative real number.*

Proof. Let $\{S_q\}$ be the partial sum of the series in Equation (32). Then, we have to show that $\{S_q\}$ is Cauchy sequence in X.
Consider the following.

$$||S_{q+1}(x,t) - S_q(x,t)|| = ||Q_{q+1}(x,t)|| \leq \rho ||Q_q(x,t)|| \leq \rho^2 ||Q_{q-1}(x,t)|| \leq \cdots \leq \rho^{q+1}||Q_0(x,t)||.$$

For every $q, r \in N$ $(r \leq q)$, the following is the case.

$$||S_q - S_r|| = ||(S_q - S_{q-1}) + (S_{q-1} - S_{q-2}) + \cdots + (S_{r+1} - S_r)||$$

$$\leq ||(S_q - S_{q-1})|| + ||(S_{q-1} - S_{q-2})|| + \cdots + ||(S_{r+1} - S_r)||$$

$$\leq (\rho^q + \rho^{q+1} + \cdots + \rho^{r+1})||Q_0(x,t)||$$

$$\leq \rho^{r+1}(\rho^{q-r-1} + \rho^{q-r-2} + \rho + 1)||Q_0(x,t)||$$

$$\leq \rho^{r+1}\left(\frac{1-\rho^{q-r}}{1-\rho}\right)||Q_0(x,t)||.$$

However, $0 < \rho \leq 1$; therefore, $||S_q - S_r|| = 0$. Hence, the sequence $\{S_q\}$ is a Cauchy sequence. □

4. Applications

Here, we consider five distinct differential equations with the Atangana–Baleanu fractional derivative in order to validate the application of the scheme with different initial conditions.

Example 1. *Consider Equation (26) as the time-fractional gas dynamics equation:*

$$_0^{ABC}D_t^\beta Q + \frac{1}{2}(Q^2)_x - Q(1-Q) = 0, \quad 0 < \beta \le 1. \tag{37}$$

with the following initial condition.

$$Q_0(x) = e^{-x}. \tag{38}$$

From Equation (37) and (38), we set the following.

$$\mathcal{R}(Q(x,t)) = -Q,$$
$$\mathcal{F}(Q(x,t)) = \frac{1}{2}(Q^2)_x + Q^2,$$
$$Q_0(x,0) = e^{-x}.$$

By employing the iteration procedure described in Section 3, we obtain the following.

$$Q_0 = \mathscr{A}^{-1}\left[\frac{1-\beta+\beta\psi^{-\beta}}{N(\beta)}\left(\frac{N(\beta)\psi^{-2}Q(x,0)}{1-\beta+\beta\psi^{-\beta}}\right)\right], \quad 0 < \beta \le 1$$

$$= \mathscr{A}^{-1}[\psi^{-2}Q(x,0)] \tag{39}$$

$$= \mathscr{A}^{-1}[\psi^{-2}e^{-x}]$$

$$= e^{-x},$$

$$Q_1 = \mathscr{A}^{-1}\left[\frac{1-\beta+\beta\psi^{-\beta}}{N(\beta)}(\mathscr{A}[\mathcal{R}(Q_0(x,t))+\mathcal{F}(Q_0(x,t))])\right]$$

$$= \mathscr{A}^{-1}\left[\frac{1-\beta+\beta\psi^{-\beta}}{N(\beta)}\left(\mathscr{A}\left[Q_0 - \left(\frac{1}{2}(Q_0^2)_x + Q_0^2\right)\right]\right)\right] \tag{40}$$

$$= \mathscr{A}^{-1}\left[\left(\frac{(1-\beta)\psi^\beta+\beta}{N(\beta)}\right)\frac{e^{-x}}{\psi^{2+\beta}}\right]$$

$$= \left(\frac{(1-\beta)\psi^\beta+\beta}{N(\beta)}\right)\frac{e^{-x}}{\Gamma(\beta+1)},$$

$$Q_2 = \mathscr{A}^{-1}\left[\frac{1-\beta+\beta\psi^{-\beta}}{N(\beta)}(\mathscr{A}[\mathcal{R}(Q_1(x,t))+\{\mathcal{F}(Q_0(x,t)+Q_1(x,t))-\mathcal{F}(Q_0(x,t))\}])\right]$$

$$= \mathscr{A}^{-1}\left[\frac{1-\beta+\beta\psi^{-\beta}}{N(\beta)}\left(\mathscr{A}\left[Q_1 + \left\{\left(\frac{1}{2}((Q_0+Q_1)^2)_x + (Q_0+Q_1)^2\right) + \left(\frac{1}{2}(Q_0^2)_x + Q_0^2\right)\right\}\right]\right)\right]$$

$$= \mathscr{A}^{-1}\left[\left(\frac{(1-\beta)\psi^\beta+\beta}{N(\beta)}\right)^2 \frac{e^{-x}}{\psi^{2+2\beta}}\right] \tag{41}$$

$$= \left(\frac{(1-\beta)\psi^\beta+\beta}{N(\beta)}\right)^2 \frac{e^{-x}}{\Gamma(2\beta+1)},$$

$$\vdots$$

$$Q_k = \mathscr{A}^{-1}\left[\frac{1-\beta+\beta\psi^{-\beta}}{N(\beta)}\left(\mathscr{A}\left[\mathcal{R}(Q_{k-1}(x,t))+\left\{\mathscr{F}\left(\sum_{j=0}^{k}Q_j(x,t)\right)+\mathscr{F}\left(\sum_{j=0}^{k-1}Q_j(x,t)\right)\right\}\right]\right)\right]$$

$$= \mathscr{A}^{-1}\left[\frac{1-\beta+\beta\psi^{-\beta}}{N(\beta)}\left(\mathscr{A}\left[\left(\frac{(1-\beta)\psi^\beta+\beta}{N(\beta)}\right)^{k-1}Q_{k-1}(x,t)\right]\right)\right] \qquad (42)$$

$$= \mathscr{A}^{-1}\left[\left(\frac{(1-\beta)\psi^\beta+\beta}{N(\beta)}\right)^k \frac{e^{-x}}{\psi^{2+k\beta}}\right]$$

$$= \left(\frac{(1-\beta)\psi^\beta+\beta}{N(\beta)}\right)^k \frac{e^{-x}t^{k\beta}}{\Gamma(k\beta+1)}.$$

We derived the k-th approximate series solution as follows:

$$\mathcal{Q}^{(k)}(x,t) = \sum_{m=0}^{k} Q_m(x,t) = Q_0(x,0)+Q_1(x,t)+Q_2(x,t)+\cdots+Q_k(x,t)$$

$$= e^{-x}\left(1+\left(\frac{(1-\beta)\psi^\beta+\beta}{N(\beta)}\right)\frac{t^\beta}{\Gamma(\beta+1)}+\cdots+\left(\frac{(1-\beta)\psi^\beta+\beta}{N(\beta)}\right)^k \frac{t^{k\beta}}{\Gamma(k\beta+1)}\right)$$

$$= e^{-x}\sum_{m=0}^{k}\left(\frac{(1-\beta)\psi^\beta+\beta}{N(\beta)}\right)^m \frac{t^{m\beta}}{\Gamma(m\beta+1)}, \qquad (43)$$

when $k \to \infty$, the k-th order approximate series results in the exact solution.

$$Q(x,t) = \lim_{k\to\infty}\mathcal{Q}^{(k)}(x,t)$$

$$= e^{-x}\lim_{k\to\infty}\sum_{m=0}^{k}\left(\frac{(1-\beta)\psi^\beta+\beta}{N(\beta)}\right)^m \frac{t^{m\beta}}{\Gamma(m\beta+1)} \qquad (44)$$

$$= e^{-x}E_\beta\left(\frac{((1-\beta)\psi^\beta+\beta)t^\beta}{N(\beta)}\right).$$

When $\beta = 1$, we obtain the exact solution as follows:

$$= e^{-x}E_1(t)$$

$$= e^{t-x}, \qquad (45)$$

which is the exact solution obtained in [2]. Figure 1 reveals the effect of α and the natural behavior of the model at distinct values of α. Moreover, Figure 2a,b is the surface plot at $\alpha = 0.5$ and 1, respectively.

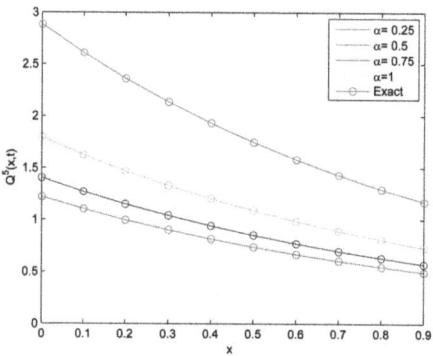

Figure 1. Comparison plot of the exact and approximate solutions for Example 1.

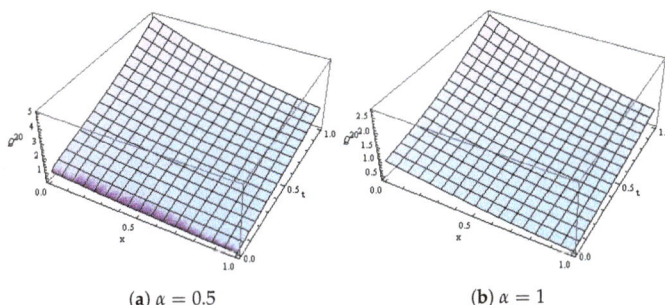

(a) α = 0.5 (b) α = 1
Figure 2. The surface plot for Example 1.

Example 2. *Consider Equation (26) as the one dimensional time-fractional biological population model according to Verhulst law [26]:*

$$^{ABC}_0 D^\beta_t Q = (Q^2)_{xx} + Q\left(1 - \frac{4}{25}Q\right), \quad t > 0, \ 0 < \beta \leq 1, \tag{46}$$

with the following initial condition.

$$Q_0(x) = e^{\frac{x}{5}}. \tag{47}$$

From Equations (46) and (47), we set the following.

$$\mathcal{R}(Q(x,t)) = Q,$$
$$\mathcal{F}(Q(x,t)) = (Q^2)_{xx} - \frac{4}{25}Q^2,$$
$$Q_0(x,0) = e^{\frac{1}{5}x}.$$

By employing the iteration procedure described in Section 3, we obtain the following.

$$Q_0 = \mathscr{A}^{-1}\left[\frac{1-\beta+\beta\psi^{-\beta}}{N(\beta)}\left(\frac{N(\beta)\psi^{-2}Q(x,0)}{1-\beta+\beta\psi^{-\beta}}\right)\right], \quad 0 < \beta \leq 1$$
$$= \mathscr{A}^{-1}\left[\psi^{-2}Q(x,0)\right] \tag{48}$$
$$= \mathscr{A}^{-1}\left[\psi^{-2}e^{\frac{1}{5}x}\right]$$
$$= e^{\frac{1}{5}x},$$

$$Q_1 = \mathscr{A}^{-1}\left[\frac{1-\beta+\beta\psi^{-\beta}}{N(\beta)}\left(\mathscr{A}[\mathcal{R}(Q_0(x,t)) + \mathcal{F}(Q_0(x,t))]\right)\right]$$
$$= \mathscr{A}^{-1}\left[\frac{1-\beta+\beta\psi^{-\beta}}{N(\beta)}\left(\mathscr{A}\left[Q_0 + (Q_0^2)_{xx} - \frac{4}{25}Q_0^2\right]\right)\right]$$
$$= \mathscr{A}^{-1}\left[\left(\frac{(1-\beta)\psi^\beta+\beta}{N(\beta)}\right)\frac{e^{\frac{x}{5}}}{\psi^{2+\beta}}\right] \tag{49}$$
$$= \left(\frac{(1-\beta)\psi^\beta+\beta}{N(\beta)}\right)\frac{e^{\frac{x}{5}}t^\beta}{\Gamma(\beta+1)},$$

$$Q_2 = \mathscr{A}^{-1}\left[\frac{1-\beta+\beta\psi^{-\beta}}{N(\beta)}\left(\mathscr{A}[\mathcal{R}(Q_1(x,t)) + \{\mathcal{F}(Q_0(x,t)+Q_1(x,t)) - \mathcal{F}(Q_0(x,t))\}]\right)\right]$$

$$= \mathscr{A}^{-1}\left[\frac{1-\beta+\beta\psi^{-\beta}}{N(\beta)}\left(\mathscr{A}\left[Q_1+\left\{((Q_0+Q_1)^2)_{xx}-\frac{4}{25}(Q_0+Q_1)^2-(Q_0^2)_{xx}+\frac{4}{25}Q_0^2\right\}\right]\right)\right]$$

$$= \mathscr{A}^{-1}\left[\left(\frac{(1-\beta)\psi^\beta+\beta}{N(\beta)}\right)^2 \frac{e^{\frac{x}{5}}}{\psi^{2+2\beta}}\right] \quad (50)$$

$$= \left(\frac{(1-\beta)\psi^\beta+\beta}{N(\beta)}\right)^2 \frac{e^{\frac{x}{5}}t^{2\beta}}{\Gamma(2\beta+1)},$$

$$\vdots$$

$$Q_k = \mathscr{A}^{-1}\left[\frac{1-\beta+\beta\psi^{-\beta}}{N(\beta)}\left(\mathscr{A}\left[\mathcal{R}(Q_{k-1}(x,t))+\left\{\mathscr{F}\left(\sum_{j=0}^{k}Q_j(x,t)\right)+\mathscr{F}\left(\sum_{j=0}^{k-1}Q_j(x,t)\right)\right\}\right]\right)\right]$$

$$= \mathscr{A}^{-1}\left[\frac{1-\beta+\beta\psi^{-\beta}}{N(\beta)}\left(\mathscr{A}\left[\left(\frac{(1-\beta)\psi^\beta+\beta}{N(\beta)}\right)^{k-1}Q_{k-1}(x,t)\right]\right)\right] \quad (51)$$

$$= \mathscr{A}^{-1}\left[\left(\frac{(1-\beta)\psi^\beta+\beta}{N(\beta)}\right)^k \frac{e^{\frac{x}{5}}}{\psi^{2+k\beta}}\right]$$

$$= \left(\frac{(1-\beta)\psi^\beta+\beta}{N(\beta)}\right)^k \frac{e^{\frac{x}{5}}t^{k\beta}}{\Gamma(k\beta+1)}.$$

We derived the k-th approximate series solution as the following:

$$\mathcal{Q}^{(k)}(x,t) = \sum_{m=0}^{k} Q_m(x,t) = Q_0(x,0)+Q_1(x,t)+Q_2(x,t)+\cdots+Q_k(x,t)$$

$$= e^{\frac{x}{5}}\left(1+\left(\frac{(1-\beta)\psi^\beta+\beta}{N(\beta)}\right)\frac{t^\beta}{\Gamma(\beta+1)}+\cdots+\left(\frac{(1-\beta)\psi^\beta+\beta}{N(\beta)}\right)^k \frac{t^{k\beta}}{\Gamma(k\beta+1)}\right) \quad (52)$$

$$= e^{\frac{x}{5}}\sum_{m=0}^{k}\left(\frac{(1-\beta)\psi^\beta+\beta}{N(\beta)}\right)^m \frac{t^{m\beta}}{\Gamma(m\beta+1)},$$

when $k \to \infty$, the k-th order approximate series results in the exact solution.

$$Q(x,t) = \lim_{k\to\infty} \mathcal{Q}^{(k)}(x,t)$$

$$= e^{\frac{x}{5}}\lim_{k\to\infty}\sum_{m=0}^{k}\left(\frac{(1-\beta)\psi^\beta+\beta}{N(\beta)}\right)^m \frac{t^{m\beta}}{\Gamma(m\beta+1)} \quad (53)$$

$$= e^{\frac{x}{5}}E_\beta\left(\frac{((1-\beta)\psi^\beta+\beta)t^\beta}{N(\beta)}\right),$$

When $\beta = 1$, we obtain the exact solution as follows.

$$= e^{\frac{x}{5}}E_1(t)$$

$$= e^{(\frac{x}{5}+t)}. \quad (54)$$

Figure 3 reveals the effect of α and the natural behavior of the model at distinct values of α. Moreover, Figure 4a,b are the surface plots at $\alpha = 0.5$ and 1, respectively.

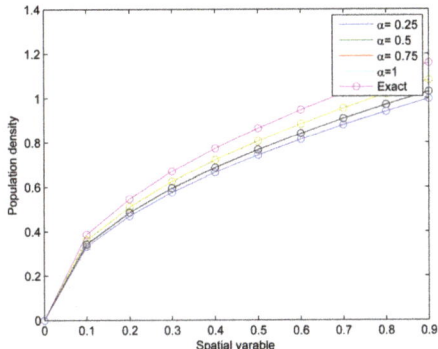

Figure 3. Comparison plot of the exact and approximate solutions for Example 2.

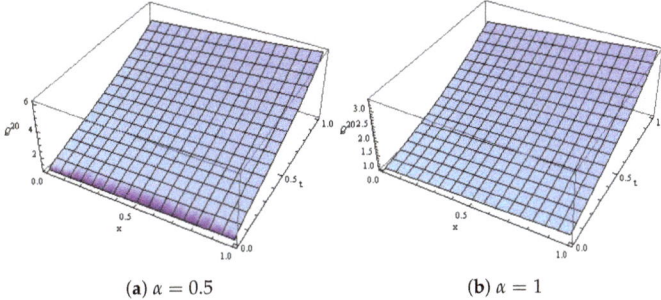

(a) $\alpha = 0.5$ (b) $\alpha = 1$

Figure 4. The surface plot for Example 2.

Example 3. *Consider Equation (26) as the time-fractional Fokker–Plane equation [2]:*

$$_0^{ABC}D_t^\beta Q + \left(\frac{4}{x}Q^2\right)_x - \left(\frac{x}{3}Q\right)_x - (Q^2)_{xx} = 0, \ t > 0, \ 0 < \beta \le 1, \tag{55}$$

with the following initial condition.

$$Q_0(x) = x^2. \tag{56}$$

From Equations (55) and (56), we set the following.

$$\mathcal{R}(Q(x,t)) = -\left(\frac{x}{3}Q\right)_x,$$

$$\mathcal{F}(Q(x,t)) = -(Q^2)_{xx} + \left(\frac{4}{x}Q^2\right)_x,$$

$$Q_0(x,0) = x^2.$$

By employing the iteration procedure described in Section 3, we obtain the following.

$$Q_0 = \mathscr{A}^{-1}\left[\frac{1-\beta+\beta\psi^{-\beta}}{N(\beta)}\left(\frac{N(\beta)\psi^{-2}Q(x,0)}{1-\beta+\beta\psi^{-\beta}}\right)\right], \ 0 < \beta \le 1$$

$$= \mathscr{A}^{-1}[\psi^{-2}Q(x,0)] \tag{57}$$

$$= \mathscr{A}^{-1}[\psi^{-2}x^2]$$

$$= x^2,$$

$$Q_1 = \mathscr{A}^{-1}\left[\frac{1-\beta+\beta\psi^{-\beta}}{N(\beta)}(\mathscr{A}[\mathcal{R}(Q_0(x,t)) + \mathcal{F}(Q_0(x,t))])\right]$$

$$= \mathscr{A}^{-1}\left[\frac{1-\beta+\beta\psi^{-\beta}}{N(\beta)}\left(\mathscr{A}\left[\left(\frac{x}{3}Q\right)_x - \left((Q_0^2)_{xx} + \left(\frac{4}{x}Q_0^2\right)_x\right)\right]\right)\right] \quad (58)$$

$$= \mathscr{A}^{-1}\left[\left(\frac{(1-\beta)\psi^\beta + \beta}{N(\beta)}\right)\frac{x^2}{\psi^{2+\beta}}\right]$$

$$= \left(\frac{(1-\beta)\psi^\beta + \beta}{N(\beta)}\right)\frac{x^2 t^\beta}{\Gamma(\beta+1)},$$

$$Q_2 = \mathscr{A}^{-1}\left[\frac{1-\beta+\beta\psi^{-\beta}}{N(\beta)}\left(\mathscr{A}[\mathcal{R}(Q_1(x,t)) + \{\mathcal{F}(Q_0(x,t)+Q_1(x,t)) - \mathcal{F}(Q_0(x,t))\}]\right)\right]$$

$$= \mathscr{A}^{-1}\left[\frac{1-\beta+\beta\psi^{-\beta}}{N(\beta)}\left(\mathscr{A}\left[\frac{x}{3}Q_1 + \left\{((Q_0+Q_1)^2)_{xx} - \left(\frac{4}{x}(Q_0+Q_1)^2\right)_x - (Q_0^2)_{xx} + \left(\frac{4}{x}Q_0^2\right)_x\right\}\right]\right)\right]$$

$$= \mathscr{A}^{-1}\left[\left(\frac{(1-\beta)\psi^\beta + \beta}{N(\beta)}\right)^2 \frac{x^2}{\psi^{2+2\beta}}\right] \quad (59)$$

$$= \left(\frac{(1-\beta)\psi^\beta + \beta}{N(\beta)}\right)^2 \frac{x^2 t^{2\beta}}{\Gamma(2\beta+1)},$$

$$\vdots$$

$$Q_k = \mathscr{A}^{-1}\left[\frac{1-\beta+\beta\psi^{-\beta}}{N(\beta)}\left(\mathscr{A}\left[\mathcal{R}(Q_{k-1}(x,t)) + \left\{\mathcal{F}\left(\sum_{j=0}^{k}Q_j(x,t)\right) + \mathcal{F}\left(\sum_{j=0}^{k-1}Q_j(x,t)\right)\right\}\right]\right)\right]$$

$$= \mathscr{A}^{-1}\left[\frac{1-\beta+\beta\psi^{-\beta}}{N(\beta)}\left(\mathscr{A}\left[\left(\frac{(1-\beta)\psi^\beta + \beta}{N(\beta)}\right)^{k-1} Q_{k-1}(x,t)\right]\right)\right] \quad (60)$$

$$= \mathscr{A}^{-1}\left[\left(\frac{(1-\beta)\psi^\beta + \beta}{N(\beta)}\right)^k \frac{x^2}{\psi^{2+k\beta}}\right]$$

$$= \left(\frac{(1-\beta)\psi^\beta + \beta}{N(\beta)}\right)^k \frac{x^2 t^{k\beta}}{\Gamma(k\beta+1)},$$

We derived the k-th approximate series solution as the following:

$$\mathscr{Q}^{(k)}(x,t) = \sum_{m=0}^{k} Q_m(x,t) = Q_0(x,0) + Q_1(x,t) + Q_2(x,t) + \cdots + Q_k(x,t)$$

$$= x^2\left(1 + \left(\frac{(1-\beta)\psi^\beta + \beta}{N(\beta)}\right)\frac{t^\beta}{\Gamma(\beta+1)} + \cdots + \left(\frac{(1-\beta)\psi^\beta + \beta}{N(\beta)}\right)^k \frac{t^{k\beta}}{\Gamma(k\beta+1)}\right) \quad (61)$$

$$= x^2 \sum_{m=0}^{k}\left(\frac{(1-\beta)\psi^\beta + \beta}{N(\beta)}\right)^m \frac{t^{m\beta}}{\Gamma(m\beta+1)},$$

when $k \to \infty$, the k-th order approximate series results in the exact solution.

$$Q(x,t) = \lim_{k\to\infty}\mathscr{Q}^{(k)}(x,t)$$

$$= x^2 \lim_{k\to\infty}\sum_{m=0}^{k}\left(\frac{(1-\beta)\psi^\beta + \beta}{N(\beta)}\right)^m \frac{t^{m\beta}}{\Gamma(m\beta+1)} \quad (62)$$

$$= x^2 E_\beta\left(\frac{((1-\beta)\psi^\beta + \beta)t^\beta}{N(\beta)}\right),$$

When $\beta = 1$, we obtain the exact solution as follows:

$$= x^2 E_1(t)$$

$$= x^2 e^t, \tag{63}$$

which is the exact solution obtained in [2]. Figure 5 reveals the effect of α and the natural behavior of the model at distinct values of α. Moreover, Figure 6a,b is the surface plot at $\alpha = 0.5$ and 1, respectively.

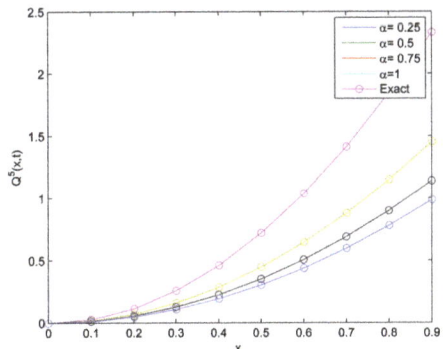

Figure 5. Comparison plot of the exact and approximate solutions for Example 3.

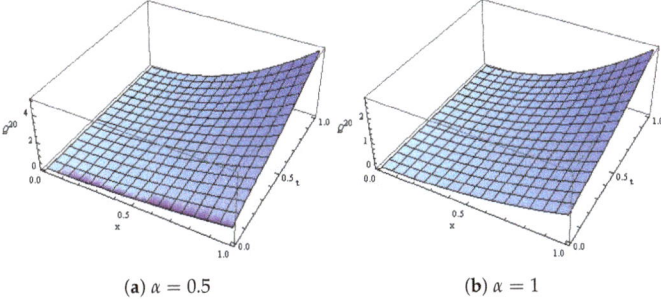

(a) $\alpha = 0.5$ (b) $\alpha = 1$

Figure 6. The surface plot for Example 3.

Example 4. *Consider Equation (26) to be the time-fractional Klomogorov equation [2]:*

$$_0^{ABC}D_t^\beta Q + x^2 e^t Q_{xx} - (x+1)Q_x = xt, \ t > 0, \ 0 < \beta \leq 1, \tag{64}$$

which is subject to the following initial condition.

$$Q_0(x) = x + 1. \tag{65}$$

From Equations (64) and (65), we set the following.

$$\mathcal{R}(Q(x,t)) = -x^2 e^t Q_{xx} + (x+1)Q_x \ ,$$
$$\mathcal{F}(Q(x,t)) = 0,$$
$$\Phi(x,t) = xt,$$
$$Q_0(x,0) = x + 1.$$

By employing the iteration procedure described in Section 3, we obtain the following.

$$Q_0 = \mathscr{A}^{-1}\left[\frac{1-\beta+\beta\psi^{-\beta}}{N(\beta)}\left(\frac{N(\beta)\psi^{-2}Q(x,0)}{1-\beta+\beta\psi^{-\beta}}+\mathscr{A}[\Phi(x,t)]\right)\right], \ 0 < \beta \leq 1$$

$$= \mathscr{A}^{-1}\left[\psi^{-2}Q(x,0) + \left(\frac{1-\beta+\beta\psi^{-\beta}}{N(\beta)}\right)\mathscr{A}[\Phi(x,t)]\right] \quad (66)$$

$$= \mathscr{A}^{-1}\left[\psi^{-2}(x+1) + \left(\frac{(1-\beta)\psi^{\beta}+\beta}{N(\beta)}\right)\frac{x}{\psi^{3+\beta}}\right]$$

$$= (x+1) + \left(\frac{(1-\beta)\psi^{\beta}+\beta}{N(\beta)}\right)\frac{xt^{\beta+1}}{\Gamma(\beta+2)},$$

$$Q_1 = \mathscr{A}^{-1}\left[\frac{1-\beta+\beta\psi^{-\beta}}{N(\beta)}\left(\mathscr{A}[\mathcal{R}(Q_0(x,t)) + \mathcal{F}(Q_0(x,t))]\right)\right]$$

$$= \mathscr{A}^{-1}\left[\frac{1-\beta+\beta\psi^{-\beta}}{N(\beta)}\left(\mathscr{A}[-x^2e^t(Q_0)_{xx} + (x+1)(Q_0)_x]\right)\right] \quad (67)$$

$$= \mathscr{A}^{-1}\left[\frac{1-\beta+\beta\psi^{-\beta}}{N(\beta)}\left(\mathscr{A}\left[(x+1)\left(1 + \left(\frac{(1-\beta)\psi^{\beta}+\beta}{N(\beta)}\right)\frac{t^{\beta+1}}{\Gamma(\beta+2)}\right)\right]\right)\right]$$

$$= (x+1)\left(\left(\frac{(1-\beta)\psi^{\beta}+\beta}{N(\beta)}\right)\frac{t^{\beta}}{\Gamma(\beta+1)} + \left(\frac{(1-\beta)\psi^{\beta}+\beta}{N(\beta)}\right)^2 \frac{t^{2\beta+1}}{\Gamma(2\beta+2)}\right),$$

$$Q_2 = \mathscr{A}^{-1}\left[\frac{1-\beta+\beta\psi^{-\beta}}{N(\beta)}\left(\mathscr{A}[\mathcal{R}(Q_1(x,t)) + \{\mathcal{F}(Q_0(x,t)+Q_1(x,t)) - \mathcal{F}(Q_0(x,t))\}]\right)\right]$$

$$= \mathscr{A}^{-1}\left[\frac{1-\beta+\beta\psi^{-\beta}}{N(\beta)}\left(\mathscr{A}[-x^2e^t(Q_1)_{xx} + (x+1)(Q_1)_x]\right)\right] \quad (68)$$

$$= \mathscr{A}^{-1}\left[\frac{1-\beta+\beta\psi^{-\beta}}{N(\beta)}\left(\mathscr{A}\left[(x+1)\left(\frac{1-\beta+\beta\psi^{-\beta}}{N(\beta)}\right)\frac{t^{\beta}}{\Gamma(\beta+1)} + \left(\frac{1-\beta+\beta\psi^{-\beta}}{N(\beta)}\right)^2 \frac{t^{2\beta+1}}{\Gamma(2\beta+2)}\right]\right)\right]$$

$$= (x+1)\left(\left(\frac{(1-\beta)\psi^{\beta}+\beta}{N(\beta)}\right)^2 \frac{t^{2\beta}}{\Gamma(2\beta+1)} + \left(\frac{(1-\beta)\psi^{\beta}+\beta}{N(\beta)}\right)^3 \frac{t^{3\beta+1}}{\Gamma(3\beta+2)}\right),$$

$$\vdots$$

$$Q_k = \mathscr{A}^{-1}\left[\frac{1-\beta+\beta\psi^{-\beta}}{N(\beta)}\left(\mathscr{A}\left[\mathcal{R}(Q_{k-1}(x,t)) + \left\{\mathcal{F}\left(\sum_{j=0}^{k}Q_j(x,t)\right) + \mathcal{F}\left(\sum_{j=0}^{k-1}Q_j(x,t)\right)\right\}\right]\right)\right]$$

$$= \mathscr{A}^{-1}\left[\frac{1-\beta+\beta\psi^{-\beta}}{N(\beta)}\left(\mathscr{A}\left[\left(\frac{(1-\beta)\psi^{\beta}+\beta}{N(\beta)}\right)^{k-1}(Q_{k-1}(x,t))_x\right]\right)\right] \quad (69)$$

$$= \mathscr{A}^{-1}\left[(x+1)\left(\frac{1}{\psi^{2+k\beta}}\left(\frac{(1-\beta)\psi^{\beta}+\beta}{N(\beta)}\right)^k + \frac{1}{\psi^{3+(k+1)\beta}}\left(\frac{(1-\beta)\psi^{\beta}+\beta}{N(\beta)}\right)^{k+1}\right)\right]$$

$$= (x+1)\left(\left(\frac{(1-\beta)\psi^{\beta}+\beta}{N(\beta)}\right)^k \frac{t^{k\beta}}{\Gamma(k\beta+1)} + \left(\frac{(1-\beta)\psi^{\beta}+\beta}{N(\beta)}\right)^{k+1} \frac{t^{(k+1)\beta+1}}{\Gamma((k+1)\beta+2)}\right),$$

We derived the k-th approximate series solution as the following.

$$\mathscr{Q}^{(k)}(x,t) = \sum_{m=0}^{k} Q_m(x,t) = Q_0(x,0) + Q_1(x,t) + Q_2(x,t) + \cdots + Q_k(x,t)$$

$$= (x+1) + ((x+1)-1)\frac{t^{\beta+1}}{\Gamma(\beta+2)}\left(\frac{(1-\beta)\psi^{\beta}+\beta}{N(\beta)}\right) +$$

$$(x+1)\left(\left(\frac{(1-\beta)\psi^{\beta}+\beta}{N(\beta)}\right)\frac{t^{\beta}}{\Gamma(\beta+1)} + \left(\frac{(1-\beta)\psi^{\beta}+\beta}{N(\beta)}\right)^2 \frac{t^{2\beta+1}}{\Gamma(2\beta+2)}\right) + \quad (70)$$

$$(x+1)\left(\left(\frac{(1-\beta)\psi^\beta+\beta}{N(\beta)}\right)^2\frac{t^{2\beta}}{\Gamma(2\beta+1)}+\left(\frac{(1-\beta)\psi^\beta+\beta}{N(\beta)}\right)^3\frac{t^{3\beta+1}}{\Gamma(3\beta+2)}\right)+$$

$$\cdots+(x+1)\left(\left(\frac{(1-\beta)\psi^\beta+\beta}{N(\beta)}\right)^k\frac{t^{k\beta}}{\Gamma(k\beta+1)}+\left(\frac{(1-\beta)\psi^\beta+\beta}{N(\beta)}\right)^{(k+1)}\frac{t^{(k+1)\beta+1}}{\Gamma((k+1)\beta+2)}\right),$$

when $k\to\infty$, the k-th order approximate series results in the exact solution:

$$Q(x,t) = \lim_{k\to\infty} \mathcal{Q}^{(k)}(x,t)$$

$$= \frac{-t^{\beta+1}}{\Gamma(\beta+2)}\left(\frac{(1-\beta)\psi^\beta+\beta}{N(\beta)}\right)+(x+1)\left(E_\beta\left(\frac{(1-\beta)\psi^\beta+\beta}{N(\beta)}\right)\right)+ \quad (71)$$

$$(x+1)\lim_{k\to\infty}\sum_{m=0}^{k}\left(\frac{(1-\beta)\psi^\beta+\beta}{N(\beta)}\right)^{m+1}\frac{t^{(m+1)\beta+1}}{\Gamma((m+1)\beta+2)},$$

When $\beta=1$, we obtain the exact solution as follows:

$$=\frac{-t^2}{2}+(x+1)\left(E_1(t)+\lim_{k\to\infty}\sum_{m=0}^{k}\frac{t^{m+2}}{\Gamma(m+3)}\right)$$

$$=\frac{-t^2}{2}+(x+1)(2e^t-t-1), \quad (72)$$

which is the exact solution obtained [2]. Figure 7 reveals the effect of α and the natural behavior of the model at distinct values of α. Moreover, Figure 8a,b is the surface plot at $\alpha = 0.5$ and 1, respectively.

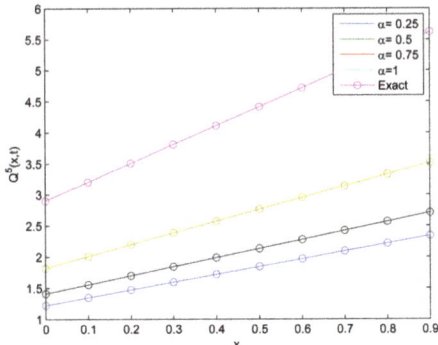

Figure 7. Comparison plot of the exact and approximate solutions for Example 4.

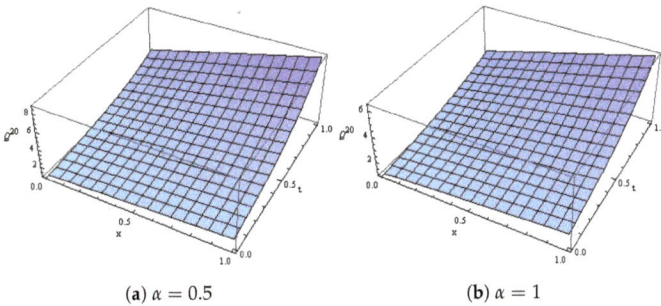

(a) $\alpha = 0.5$ (b) $\alpha = 1$

Figure 8. The surface plot for Example 4.

Example 5. Consider Equation (26) as the one dimensional time-fractional biological population model according to Verhulst law [26]:

$$_{0}^{ABC}D_{t}^{\beta}Q = (Q^2)_{xx} + \frac{1}{4}Q, \quad t > 0, \quad 0 < \beta \leq 1, \tag{73}$$

that is subject to the following initial condition.

$$Q_0(x) = x^{\frac{1}{2}}. \tag{74}$$

From Equations (73) and (74), we set the following.

$$\mathcal{R}(Q(x,t)) = \frac{1}{4}Q,$$
$$\mathcal{F}(Q(x,t)) = (Q^2)_{xx},$$
$$Q_0(x,0) = x^{\frac{1}{2}}.$$

By employing the iteration procedure described in Section 3, we obtain the following.

$$Q_0 = \mathscr{A}^{-1}\left[\frac{1-\beta+\beta\psi^{-\beta}}{N(\beta)}\left(\frac{N(\beta)\psi^{-2}Q(x,0)}{1-\beta+\beta\psi^{-\beta}}\right)\right], \quad 0 < \beta \leq 1$$
$$= \mathscr{A}^{-1}[\psi^{-2}Q(x,0)] \tag{75}$$
$$= \mathscr{A}^{-1}\left[\psi^{-2}x^{\frac{1}{2}}\right]$$
$$= x^{\frac{1}{2}}.$$

$$Q_1 = \mathscr{A}^{-1}\left[\frac{1-\beta+\beta\psi^{-\beta}}{N(\beta)}\left(\mathscr{A}[\mathcal{R}(Q_0(x,t)) + \mathcal{F}(Q_0(x,t))]\right)\right]$$
$$= \mathscr{A}^{-1}\left[\frac{1-\beta+\beta\psi^{-\beta}}{N(\beta)}\left(\mathscr{A}\left[\frac{1}{4}Q_0 + (Q_0^2)_{xx}\right]\right)\right]$$
$$= \mathscr{A}^{-1}\left[\left(\frac{(1-\beta)\psi^{\beta}+\beta}{N(\beta)}\right)\frac{\left(\frac{1}{4}\right)}{\psi^{2+\beta}}x^{\frac{1}{2}}\right] \tag{76}$$
$$= \left(\frac{(1-\beta)\psi^{\beta}+\beta}{N(\beta)}\right)\frac{\left(\frac{1}{4}\right)t^{\beta}}{\Gamma(\beta+1)}x^{\frac{1}{2}},$$

$$Q_2 = \mathscr{A}^{-1}\left[\frac{1-\beta+\beta\psi^{-\beta}}{N(\beta)}\left(\mathscr{A}[\mathcal{R}(Q_1(x,t)) + \{\mathcal{F}(Q_0(x,t) + Q_1(x,t)) - \mathcal{F}(Q_0(x,t))\}]\right)\right]$$
$$= \mathscr{A}^{-1}\left[\frac{1-\beta+\beta\psi^{-\beta}}{N(\beta)}\left(\mathscr{A}\left[\frac{1}{4}Q_1 + \left\{\left((Q_0+Q_1)^2\right)_{xx}\right\}\right]\right)\right]$$
$$= \mathscr{A}^{-1}\left[\left(\frac{(1-\beta)\psi^{\beta}+\beta}{N(\beta)}\right)^2 \frac{\left(\frac{1}{4}\right)^2}{\psi^{2+2\beta}}x^{\frac{1}{2}}\right] \tag{77}$$
$$= \left(\frac{(1-\beta)\psi^{\beta}+\beta}{N(\beta)}\right)^2 \frac{\left(\frac{1}{4}t^{\beta}\right)^2}{\Gamma(2\beta+1)}x^{\frac{1}{2}},$$

$$\vdots$$

$$Q_k = \mathscr{A}^{-1}\left[\frac{1-\beta+\beta\psi^{-\beta}}{N(\beta)}\left(\mathscr{A}\left[\mathcal{R}(Q_{k-1}(x,t)) + \left\{\mathscr{F}\left(\sum_{j=0}^{k}Q_j(x,t)\right) + \mathscr{F}\left(\sum_{j=0}^{k-1}Q_j(x,t)\right)\right\}\right]\right)\right]$$

$$= \mathscr{A}^{-1}\left[\frac{1-\beta+\beta\psi^{-\beta}}{N(\beta)}\left(\mathscr{A}\left[\left(\frac{(1-\beta)\psi^\beta+\beta}{N(\beta)}\right)^{k-1}\frac{1}{4}Q_{k-1}(x,t)\right]\right)\right] \quad (78)$$

$$= \mathscr{A}^{-1}\left[\left(\frac{(1-\beta)\psi^\beta+\beta}{N(\beta)}\right)^k \frac{e^{\frac{1}{2}x}}{\psi^{2+k\beta}}\right]$$

$$= \left(\frac{(1-\beta)\psi^\beta+\beta}{N(\beta)}\right)^k \frac{\left(\frac{1}{4}t^\beta\right)^k}{\Gamma(k\beta+1)} x^{\frac{1}{2}},$$

We derived the k-th approximate series solution as the following.

$$\mathscr{Q}^{(k)}(x,t) = \sum_{m=0}^{k} Q_m(x,t) = Q_0(x,0) + Q_1(x,t) + Q_2(x,t) + \cdots + Q_k(x,t)$$

$$= x^{\frac{1}{2}}\left(1 + \left(\frac{(1-\beta)\psi^\beta+\beta}{N(\beta)}\right)\frac{\frac{1}{4}t^\beta}{\Gamma(\beta+1)} + \cdots + \left(\frac{(1-\beta)\psi^\beta+\beta}{N(\beta)}\right)^k \frac{\left(\frac{1}{4}t^\beta\right)^k}{\Gamma(k\beta+1)}\right)$$

$$= x^{\frac{1}{2}} \sum_{m=0}^{k} \left(\frac{(1-\beta)\psi^\beta+\beta}{N(\beta)}\right)^m \frac{\left(\frac{1}{4}t^\beta\right)^m}{\Gamma(m\beta+1)}, \quad (79)$$

when $k \to \infty$, the k-th order approximate series results in the exact solution.

$$Q(x,t) = \lim_{k\to\infty} \mathscr{Q}^{(k)}(x,t)$$

$$= x^{\frac{1}{2}} \lim_{k\to\infty} \sum_{m=0}^{k} \left(\frac{(1-\beta)\psi^\beta+\beta}{N(\beta)}\right)^m \frac{\left(\frac{1}{4}t^\beta\right)^m}{\Gamma(m\beta+1)} \quad (80)$$

$$= x^{\frac{1}{2}} E_\beta\left(\frac{((1-\beta)\psi^\beta+\beta)\frac{1}{4}t^\beta}{N(\beta)}\right),$$

When $\beta = 1$, we obtain the exact solution as the following:

$$= x^{\frac{1}{2}} E_1\left(\frac{1}{4}t\right)$$

$$= x^{(\frac{1}{2})} e^{\frac{t}{4}}, \quad (81)$$

which is the exact solution obtained in [4]. Figure 9 reveals the effect of α and the natural behavior of the model at distinct values of α. Moreover, Figure 10a,b is the surface plot at $\alpha = 0.5$ and 1, respectively.

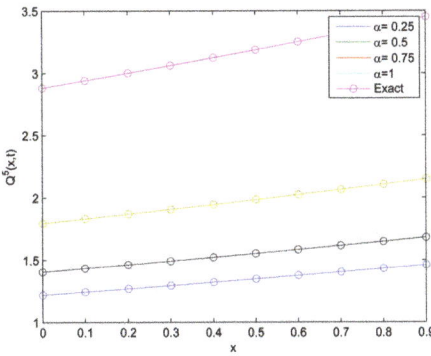

Figure 9. Comparison plot of the exact and approximate for Example 5.

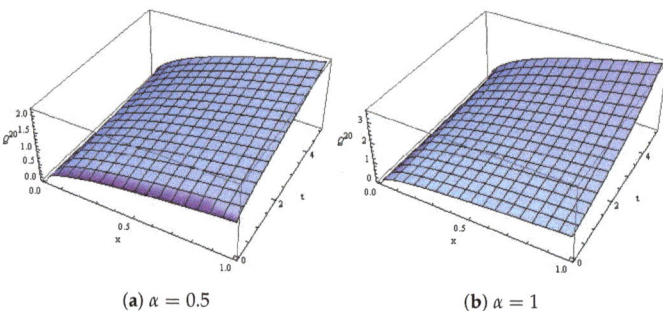

(a) $\alpha = 0.5$ (b) $\alpha = 1$

Figure 10. The surface plot for Example 5.

5. Conclusions

In this paper, we utilized the connection between the Aboodh transform and the Laplace transform to establish the Aboodh transform of Atangana–Baleanu fractional differential operator. The accuracy and validity of the Aboodh transform iterative method for fractional differential equation with Atangana–Baleanu fractional differential operator are also presented.

The graphical illustration in Figures 1–10 is presented to validate the effectiveness of the Aboodh transform iterative method and to capture the natural behavior of differential equation with the Atangana–Baleanu fractional differential operator.

Finally, we conclude that Atangana–Baleanu fractional differential operator contains a local and a singular kernel that makes the Atangana–Baleanu fractional differential operator more suitable for real life applications and that the Aboodh transform iterative method can adequately capture the effect and the behavior of fractional differential equations.

Author Contributions: Formal analysis: M.A.A.; supervision: D.S. All authors have read and agreed to the published version of the manuscript.

Funding: This research received no funding.

Institutional Review Board Statement: Not applicable.

Informed Consent Statement: Not applicable.

Data Availability Statement: Not applicable.

Conflicts of Interest: The authors declare no conflict of interest.

References

1. Iyiola, O.S.; Ojo, G.O. On the analytical solution of Fornberg–Whitham equation with the new fractional derivative. *Pramana* **2015**, *85*, 567–575. [CrossRef]
2. Akinyemi, L.; Iyiola, O.S. Exact and approximate solutions of time-fractional models arising from physics via Shehu transform. *Math. Methods Appl. Sci.* **2020**, *43*, 7442–7464. [CrossRef]
3. Iyiola, O.S.; Ojo, G.O.; Mmaduabuchi, O. The fractional Rosenau–Hyman model and its approximate solution. *Alex. Eng. J.* **2016**, *55*, 1655–1659. [CrossRef]
4. Ojo, G.O.; Mahmudov, N.I. Aboodh Transform Iterative Method for Spatial Diffusion of a Biological Population with Fractional-Order. *Mathematics* **2021**, *9*, 155. [CrossRef]
5. Iyiola, O.S.; Ojo, G.O.; Audu, J.D. A Comparison Results of Some Analytical Solutions of Model in Double Phase Flow through Porous Media. *J. Math. Syst. Sci.* **2014**, *4*, 275–284.
6. Podlubny, I. *Fractional Differential Equations: An Introduction to Fractional Derivatives, Fractional Differential Equations, to Methods of Their Solution and Some of Their Applications*; Elsevier: Amsterdam, The Netherlands, 1998.
7. Oldham, K.; Spanier, J. *The Fractional Calculus Theory and Applications of Differentiation and Integration to Arbitrary Order*; Academic Press: New York, NY, USA, 1974.
8. Miller, K.S.; Ross, B. *An Introduction to the Fractional Calculus and Fractional Differential Equations*; Wiley: New York, NY, USA, 1993.
9. Kilbas, A.A.; Srivastava, H.M.; Trujillo, J.J. *Theory and Applications of Fractional Differential Equations*; Elsevier: Amsterdam, The Netherlands, 2006; Volume 204.

10. Diethelm, K. *The Analysis of Fractional Differential Equations: An Application-Oriented Exposition Using Differential Operators of Caputo Type*; Springer Science and Business Media: Heidelberg, Germany, 2010.
11. Atangana, A.; Baleanu, D. New fractional derivatives with nonlocal and non-singular kernel: Theory and application to heat transfer model. *Therm. Sci.* **2016**, *20*, 763–769. [CrossRef]
12. Caputo, M.; Fabrizio, M. A new definition of fractional derivative without singular kernel. *Progr. Fract. Differ. Appl.* **2015**, *1*, 1–13.
13. Gondal, M.A.; Arife, A.S.; Khan, M.; Hussain, I. An efficient numerical method for solving linear and nonlinear partial differential equations by combining homotopy analysis and transform method. *World Appl. Sci. J.* **2011**, *15*, 1786–1791.
14. Arafa, A.A.; Hagag, A.M.S. Q-homotopy analysis transform method applied to fractional Kundu–Eckhaus equation and fractional massive Thirring model arising in quantum field theory. *Asian-Eur. J. Math.* **2019**, *12*, 1950045. [CrossRef]
15. Arafa, A.A.; Hagag, A.M.S. A new analytic solution of fractional coupled Ramani equation. *Chin. J. Phys.* **2019**, *60*, 388–406. [CrossRef]
16. Saad, K.M.; AL-Shareef, E.H.; Alomari, A.K.; Baleanu, D.; Gómez-Aguilar, J.F. On exact solutions for time-fractional Korteweg-de Vries and Korteweg-de Vries-Burger's equations using homotopy analysis transform method. *Chin. J. Phys.* **2020**, *63*, 149–162. [CrossRef]
17. Enyi, C.D. Efficacious Analytical Technique Applied to Fractional Fornberg–Whitham Model and Two-Dimensional Fractional Population Model. *Symmetry* **2020**, *12*, 1976. [CrossRef]
18. Baleanu, D.; Mohammadi, H.; Rezapour, S. Analysis of the model of HIV-1 infection of CD4+ T-cell with a new approach of fractional derivative. *Adv. Differ. Equ.* **2020**, *71*, 1–17. [CrossRef]
19. Aydogan, M.S.; Baleanu, D.; Mousalou, A.; Rezapour, S. On high order fractional integro-differential equations including the Caputo–Fabrizio derivative. *Bound. Value Probl.* **2018** *2018*, 90. [CrossRef]
20. Atangana, A.; Koca, I. Chaos in a simple nonlinear system with Atangana–Baleanu derivatives with fractional order. *Chaos Solitons Fractals* **2016**, *89*, 447–454. [CrossRef]
21. Bokhari, A.; Baleanu, D.; Belgacem, R. Application of Shehu transform to Atangana-Baleanu derivatives. *J. Math. Comput. Sci.* **2020**, *20*, 101–107. [CrossRef]
22. Debnath, I.; Bhatta, D. *Integral Transforms and Their Application*; CRC Press: Boca Raton, FL, USA, 2014.
23. Aboodh, K.S.; Idris, A.; Nuruddeen, R.I. On the Aboodh transform connections with some famous integral transforms. *Int. J. Eng. Inform. Syst.* **2017**, *1*, 143–151.
24. Mittag-Leffler, M.G. Sur La nonvelle Fonction $E_\alpha(x)$. *Comptes Rendus Acad. Sci. Paris* **1903**, *2*, 1003.
25. Daftardar-Gejji, V.; Jafari, H. An iterative method for solving nonlinear functional equations. *J. Math. Anal. Appl.* **2006**, *316*, 753–763. [CrossRef]
26. Gurtin, M.E.; MacCamy, R.C. On the diffusion of biological populations. *Math. Biosci.* **1977**, *33*, 35–49. [CrossRef]

Article

Analysis and Optimal Control of φ-Hilfer Fractional Semilinear Equations Involving Nonlocal Impulsive Conditions

Sarra Guechi [1], Rajesh Dhayal [2], Amar Debbouche [1,*] and Muslim Malik [3]

[1] Department of Mathematics, Guelma University, Guelma 24000, Algeria; guechi.sara@yahoo.fr
[2] School of Mathematics, Thapar Institute of Engineering and Technology, Patiala 147 004, India; rajesh.dhayal@thapar.edu
[3] School of Basic Sciences, Indian Institute of Technology Mandi, Kamand 175 005, India; muslim@iitmandi.ac.in
* Correspondence: amar_debbouche@yahoo.fr

Abstract: The goal of this paper is to consider a new class of φ-Hilfer fractional differential equations with impulses and nonlocal conditions. By using fractional calculus, semigroup theory, and with the help of the fixed point theorem, the existence and uniqueness of mild solutions are obtained for the proposed fractional system. Symmetrically, we discuss the existence of optimal controls for the φ-Hilfer fractional control system. Our main results are well supported by an illustrative example.

Keywords: φ-Hilfer fractional system with impulses; semigroup theory; nonlocal conditions; optimal controls

1. Introduction

In recent years, a lot of research attention has been paid to the study of fractional calculus, which is considered as a generalization of classical derivatives and integrals to non-integer order. Phenomena with memory and hereditary characteristics that arise in ecology, biology, medicine, electrical engineering, and mechanics, etc, may be well modelled by using fractional differential equations (FDEs for short). For more details on FDEs and its applications, see [1–5] and the references therein. In [6], Hilfer derived a new two-parameter fractional derivative $D_{a+}^{\sigma_1,\sigma_2}$ of order σ_1 and type σ_2, which is called Hilfer fractional derivative that combines the Riemann–Liouville and Caputo fractional derivatives. This kind of parameter produces more types of stationary states and gives an extra degree of freedom on the initial conditions. Systems based on Hilfer fractional derivatives are considered by many authors, see [7–11] and the references therein. Recently, Sousa and Oliveira [12] introduced a new fractional derivative with respect to another φ-function the so-called φ-Hilfer fractional derivative, and discussed their properties as well as important results of the fractional calculus. For more recent works on φ-Hilfer fractional derivative and its applications, we refer to [13–17] and the references therein.

Many real-world phenomena and processes which are subjected to external influences for a small time interval during their evolution can be represented as an impulsive differential equations. The impulsive differential equations have become the natural framework for modelling of many evolving processes and phenomena studied in the field of science and engineering such as in mechanical systems, biological systems, population dynamics, physics, economy, and control theory. Recently, based on the theory of semigroup and fixed point approach, many authors studied the qualitative properties of solutions for impulsive differential equations of order one and non-integer [18–24] and the references therein. The optimal control problem (OCP for short) plays a crucial role in biomedicine, for example, model cancer chemotherapy and recently applied to epidemiological models. When FDEs describe the system dynamics and the cost functional, an OCP reduces to a fractional optimal control problem. The fractional OCP refers to optimize the cost functional subject to dynamical constraints on the control parameter and state variables that having fractional models. For more recent works on OCP, see [25–30] and the references therein. Harrat et al. [31] investigated the existence of optimal controls for Hilfer fractional impulsive evolution inclusions with Clarke subdifferential. Moreover, optimal control problems for φ-Hilfer fractional impulsive differential equations are rarely available in the literature which serves as a motivation to our research work in this paper.

Motivated by the above facts, we consider following φ-Hilfer fractional impulsive differential system:

$$\begin{cases} {}^H D_{t_\gamma+}^{\sigma_1,\sigma_2;\varphi} z(t) = \mathcal{A}z(t) + \Delta(t,z(t)), \ t \in (0,b] - \{t_1, t_2, \ldots, t_\mathcal{H}\}, \\ I_{t_\gamma+}^{(1-\sigma_1)(1-\sigma_2);\varphi} z(t_\gamma^+) = z(t_\gamma^-) + \mathcal{I}_\gamma(z(t_\gamma^-)), \ \gamma = 1, 2, \ldots, \mathcal{H}, \\ I_{0+}^{(1-\sigma_1)(1-\sigma_2);\varphi} [z(t)]_{t=0} + \mathcal{G}(z) = z_0, \end{cases} \quad (1)$$

where ${}^H D_{t_\gamma+}^{\sigma_1,\sigma_2;\varphi}$ denotes the φ-Hilfer fractional derivative of order $1/2 < \sigma_1 < 1$, $0 < \sigma_2 < 1$ and the state $z(\cdot)$ takes values in a Hilbert space E and $\mathcal{J}_0 = [0, b]$, $0 = t_0 < t_1 < \cdots < t_\mathcal{H} < t_{\mathcal{H}+1} = b$. $\mathcal{A}$ is the generator of a C_0-semigroup $\{\mathcal{T}(t)\}_{t \geq 0}$ on E. As usual $z(t_\gamma^+)$ and $z(t_\gamma^-)$ are the right and left limits of z at the point t_γ, respectively. $\mathcal{I}_\gamma : E \to E$ are impulsive functions that characterize the jump of z at points t_γ. The functions $\Delta : \mathcal{J}_0 \times E \to E$, $\mathcal{G} : C(\mathcal{J}_0, E) \to E$ are some suitable functions that will be specified later.

The rest of the manuscript is organized as follows. In Section 2, we recall some important concepts and results. In Sections 3 and 4, we derived the mild solution by using semigroup as well as probability density function and proved the existence of mild solutions for the proposed fractional system, receptively. In Section 5, we investigated the existence of optimal controls for the φ-Hilfer fractional control system. Moreover, in Section 6, an example is presented to demonstrate the applicability of the obtained symmetry results.

2. Preliminaries

Let $\mathcal{J}_1 = [a, b]$ and $\varphi \in C^m(\mathcal{J}_1, \mathbb{R})$ an increasing function such that $\varphi'(t) \neq 0$, $\forall t \in \mathcal{J}_1$.

Definition 1. *The φ-Riemann fractional integral of order $\sigma_1 > 0$ of the function $\mathcal{R}$ is given by*

$$I_{a+}^{\sigma_1;\varphi} \mathcal{R}(t) = \frac{1}{\Gamma(\sigma_1)} \int_a^t (\varphi(t) - \varphi(s))^{\sigma_1 - 1} \mathcal{R}(s) \varphi'(s) ds.$$

Definition 2. *The φ-Riemann-Liouville fractional derivative of function $\mathcal{R}$ of order σ_1 ($m - 1 < \sigma_1 < m$, $m \in \mathbb{N}$), is defined by*

$$D_{a+}^{\sigma_1;\varphi} \mathcal{R}(t) = \left(\frac{1}{\varphi'(t)} \frac{d}{dt}\right)^m I_{a+}^{m-\sigma_1;\varphi} \mathcal{R}(t) = \frac{\left(\frac{1}{\varphi'(t)} \frac{d}{dt}\right)^m}{\Gamma(m-\sigma_1)} \int_a^t (\varphi(t) - \varphi(s))^{m-\sigma_1-1} \mathcal{R}(s) ds,$$

where $m = [\sigma_1] + 1$.

Definition 3. *The φ-Hilfer fractional derivative of function $\mathcal{R}$ of order σ_1 ($m - 1 < \sigma_1 < m$, $m \in \mathbb{N}$) and type $0 \leq \sigma_2 \leq 1$, is defined by*

$${}^H D_{a+}^{\sigma_1,\sigma_2;\varphi} \mathcal{R}(t) = I_{a+}^{\sigma_2(m-\sigma_1);\varphi} \left(\frac{1}{\varphi'(t)} \frac{d}{dt}\right)^m I_{a+}^{(1-\sigma_2)(m-\sigma_1);\varphi} \mathcal{R}(t).$$

The φ-Hilfer fractional derivative can be written as

$${}^H D_{a+}^{\sigma_1,\sigma_2;\varphi} \mathcal{R}(t) = I_{a+}^{\delta-\sigma_1;\varphi} D_{a+}^{\delta;\varphi} \mathcal{R}(t),$$

with $\delta = (\sigma_1 + \sigma_2(m - \sigma_1))$.

Lemma 1 ([12]). *If $\mathcal{R} \in C^m[a,b]$, $m-1 < \sigma_1 < m$ and $0 \leq \sigma_2 \leq 1$, then*

$$I_{a+}^{\sigma_1;\varphi}{}^H D_{a+}^{\sigma_1,\sigma_2;\varphi}\mathcal{R}(t) = \mathcal{R}(t) - \sum_{k=1}^{m} \frac{(\varphi(t)-\varphi(a))^{\delta-k}}{\Gamma(\delta-k+1)} \mathcal{R}_\varphi^{[m-k]} I_{a+}^{(1-\sigma_2)(m-\sigma_1);\varphi}\mathcal{R}(a).$$

Lemma 2 ([12]). *Let $\sigma_1 > 0$ and $\sigma_2 > 0$, then $I_{a+}^{\sigma_1,\varphi}(\varphi(t)-\varphi(a))^{\sigma_2-1} = \frac{\Gamma(\sigma_2)}{\Gamma(\sigma_2+\sigma_1)}(\varphi(t)-\varphi(a))^{\sigma_2+\sigma_1-1}$.*

Definition 4. *Let $z, \varphi : [c, \infty) \to \mathbb{R}$ be the functions such that $\varphi(t)$ is continuous and $\varphi'(t) > 0$ on $[0, \infty)$. Then the generalized Laplace transform of function $z(t)$ is given by*

$$\mathcal{L}_\varphi\{z(t)\}(s) = \int_c^\infty e^{-s(\varphi(t)-\varphi(a))} z(t) \varphi'(t) dt, \text{ for all } s.$$

For comprehensive details on φ-Hilfer fractional derivative and its properties, we refer to papers [12,14,17].

Consider the weighted space [14] defined as

$$C_{1-\rho;\varphi}(\mathcal{J}_0, E) = \{z : [0,b] \to E : (\varphi(t)-\varphi(t_\gamma))^{1-\rho} z(t) \in C(\mathcal{J}_0, E)\}.$$

Define the space of piecewise continuous functions as

$$\mathcal{PC}_{1-\rho;\varphi}(\mathcal{J}_0, E) = \{z : [0,b] \to E : z \in C_{1-\rho;\varphi}((t_\gamma, t_{\gamma+1}], E), \gamma = 1, 2, \ldots, \mathcal{H}, I_{t_\gamma^+}^{(1-\rho);\varphi} z(t_\gamma^+)$$

and $I_{t_\gamma^+}^{(1-\rho);\varphi} z(t_\gamma^-) = I_{t_\gamma^+}^{(1-\rho);\varphi} z(t_\gamma)$ exists for $\gamma = 1, 2, \ldots, \mathcal{H}, \rho = \sigma_1 + \sigma_2 - \sigma_2 \sigma_1\}$

Clearly, $\mathcal{PC}(E) = \mathcal{PC}_{1-\rho;\varphi}(\mathcal{J}_0, E)$ is a Banach space with the norm

$$\|z\|_{\mathcal{PC}} = \max_{\gamma=1,2,\ldots,\mathcal{H}} \left\{ \sup_{t \in (t_\gamma, t_{\gamma+1}]} \left\| [\varphi(t)-\varphi(t_\gamma)]^{1-\rho} z(t) \right\| \right\}.$$

3. Representation of Mild Solution

Lemma 3. *To reduce the generalized form (1), we consider the linear φ-Hilfer fractional differential system:*

$$\begin{cases} {}^H D_{0+}^{\sigma_1,\sigma_2;\varphi} z(t) = \mathcal{A} z(t) + \Delta(t), & t \in (0,b] \\ I_{0+}^{(1-\sigma_1)(1-\sigma_2);\varphi}[z(t)]_{t=0} = z_0, \end{cases} \quad (2)$$

has a mild solution, which is defined as

$$z(t) = \mathcal{S}_\varphi^{\sigma_1,\sigma_2}(t,0) z_0 + \int_0^t (\varphi(t)-\varphi(s))^{\sigma_1-1} \mathcal{T}_\varphi^{\sigma_1}(t,s) \Delta(s) \varphi'(s) ds, \quad (3)$$

where

$$\mathcal{P}_\varphi^{\sigma_1}(t,s) z = \int_0^\infty \phi_{\sigma_1}(\theta) \mathcal{T}((\varphi(t)-\varphi(s))^{\sigma_1} \theta) z d\theta,$$

$$\mathcal{S}_\varphi^{\sigma_1,\sigma_2}(t,s) z = I_{a+}^{(1-\sigma_1)(\sigma_2-1);\varphi} \mathcal{P}_\varphi^{\sigma_1}(t,s) z,$$

$$\mathcal{T}_\varphi^{\sigma_1}(t,s) z = \sigma_1 \int_0^\infty \theta \phi_{\sigma_1}(\theta) \mathcal{T}((\varphi(t)-\varphi(s))^{\sigma_1}\theta) z d\theta, \ 0 \leq s \leq t \leq b,$$

with

$$\phi_{\sigma_1}(\theta) \geq 0 \text{ for } \theta \geq 0, \quad \int_0^\infty \phi_{\sigma_1}(\theta) d\theta = 1, \text{ and } \int_0^\infty \theta \phi_{\sigma_1}(\theta) d\theta = \frac{1}{\Gamma(1+\sigma_1)}.$$

Proof. Rewrite the problem (2) in the equivalent integral equation

$$z(t) = \frac{(\varphi(t) - \varphi(0))^{(1-\sigma_1)(\sigma_2-1)}}{\Gamma(\sigma_2(1-\sigma_1) + \sigma_1)} z_0 + \frac{1}{\Gamma(\sigma_1)} \int_0^t (\varphi(t) - \varphi(s))^{\sigma_1-1} [Az(s) + \Delta(s)] \varphi'(s) ds, \quad (4)$$

provided that the integral in Equation (4) exists. Let $\beta > 0$. Applying the generalized Laplace transform

$$Z(\beta) = \frac{1}{\beta^{\sigma_2(1-\sigma_1)+\sigma_1}} z_0 + \frac{1}{\beta^{\sigma_1}} (AZ(\beta) + \hat{\Delta}(\beta)),$$

where

$$Z(\beta) = \int_0^\infty e^{-\beta(\varphi(\mu) - \varphi(0))} z(\mu) \varphi'(\mu) d\mu,$$

$$\hat{\Delta}(\beta) = \int_0^\infty e^{-\beta(\varphi(\mu) - \varphi(0))} \Delta(\mu) \varphi'(\mu) d\mu.$$

It follows that

$$Z(\beta) = \beta^{\sigma_2(\sigma_1-1)} (\beta^{\sigma_1} I - A)^{-1} z_0 + (\beta^{\sigma_1} I - A)^{-1} \hat{\Delta}(\beta)$$

$$= \beta^{\sigma_2(\sigma_1-1)} \int_0^\infty e^{-\beta^{\sigma_1} s} \mathcal{T}(s) z_0 ds + \int_0^\infty e^{-\beta^{\sigma_1} s} \mathcal{T}(s) \hat{\Delta}(\beta) ds.$$

Taking $s = \hat{t}^{\sigma_1}$, we obtain

$$Z(\beta) = \sigma_1 \beta^{(\sigma_1-1)(\sigma_2-1)} \int_0^\infty (\beta\hat{t})^{\sigma_1-1} e^{-(\beta\hat{t})^{\sigma_1}} \mathcal{T}(\hat{t}^{\sigma_1}) z_0 d\hat{t} + \sigma_1 \int_0^\infty \hat{t}^{\sigma_1-1} e^{-(\beta\hat{t})^{\sigma_1}} \mathcal{T}(\hat{t}^{\sigma_1}) \hat{\Delta}(\beta) d\hat{t}$$

$$= \beta^{(\sigma_1-1)(\sigma_2-1)} I_1 + I_2,$$

where

$$I_1 = \sigma_1 \int_0^\infty (\beta\hat{t})^{\sigma_1-1} e^{-(\beta\hat{t})^{\sigma_1}} \mathcal{T}(\hat{t}^{\sigma_1}) z_0 d\hat{t},$$

$$I_2 = \sigma_1 \int_0^\infty \hat{t}^{\sigma_1-1} e^{-(\beta\hat{t})^{\sigma_1}} \mathcal{T}(\hat{t}^{\sigma_1}) \hat{\Delta}(\beta) d\hat{t}.$$

Taking $\hat{t} = \varphi(t) - \varphi(0)$, we obtain

$$I_1 = \sigma_1 \int_0^\infty \beta^{\sigma_1-1} (\varphi(t) - \varphi(0))^{\sigma_1-1} e^{-(\beta(\varphi(t)-\varphi(0)))^{\sigma_1}} \mathcal{T}((\varphi(t) - \varphi(0))^{\sigma_1}) z_0 \varphi'(t) dt$$

$$= \int_0^\infty \frac{-1}{\beta} \frac{d}{dt} \left(e^{-(\beta(\varphi(t)-\varphi(0)))^{\sigma_1}} \right) \mathcal{T}((\varphi(t) - \varphi(0))^{\sigma_1}) z_0 dt.$$

$$I_2 = \sigma_1 \int_0^\infty (\varphi(t) - \varphi(0))^{\sigma_1-1} e^{-(\beta(\varphi(t)-\varphi(0)))^{\sigma_1}} \mathcal{T}((\varphi(t) - \varphi(0))^{\sigma_1}) \hat{\Delta}(\beta) \varphi'(t) dt$$

$$= \int_0^\infty \int_0^\infty \sigma_1 (\varphi(t) - \varphi(0))^{\sigma_1-1} e^{-(\beta(\varphi(t)-\varphi(0)))^{\sigma_1}} \mathcal{T}((\varphi(t) - \varphi(0))^{\sigma_1})$$
$$\times e^{-(\beta(\varphi(s)-\varphi(0)))} \Delta(s) \varphi'(s) \varphi'(t) ds dt.$$

We consider the following one-sided stable probability density

$$\rho_{\sigma_1}(\theta) = \frac{1}{\pi} \sum_{k=1}^\infty (-1)^{k-1} \theta^{-\sigma_1 k - 1} \frac{\Gamma(\sigma_1 k + 1)}{k!} \sin(k\pi\sigma_1), \quad \theta \in (0, \infty),$$

whose integration is given by

$$\int_0^\infty e^{-\beta\theta}\rho_{\sigma_1}(\theta)d\theta = e^{-\beta^{\sigma_1}}, \; \sigma_1 \in (0,1). \tag{5}$$

Using Equation (5), we obtain

$$\begin{aligned}
I_1 &= \int_0^\infty \frac{-1}{\beta}\frac{d}{dt}\left(\int_0^\infty e^{-(\beta(\varphi(t)-\varphi(0)))\theta}\rho_{\sigma_1}(\theta)d\theta\right)\mathcal{T}((\varphi(t)-\varphi(0))^{\sigma_1})z_0 dt \\
&= \int_0^\infty \int_0^\infty \theta\rho_{\sigma_1}(\theta)e^{-(\beta(\varphi(t)-\varphi(0)))\theta}\mathcal{T}((\varphi(t)-\varphi(0))^{\sigma_1})z_0\varphi'(t)d\theta\, dt \\
&= \int_0^\infty e^{-(\beta(\varphi(t)-\varphi(0)))}\left(\int_0^\infty \rho_{\sigma_1}(\theta)\mathcal{T}\left(\frac{(\varphi(t)-\varphi(0))^{\sigma_1}}{\theta^{\sigma_1}}\right)d\theta\right)z_0\varphi'(t)dt.
\end{aligned}$$

and

$$\begin{aligned}
I_2 &= \int_0^\infty\int_0^\infty\int_0^\infty \sigma_1(\varphi(t)-\varphi(0))^{\sigma_1-1}\rho_{\sigma_1}(\theta)e^{-(\beta(\varphi(t)-\varphi(0)))\theta}\mathcal{T}((\varphi(t)-\varphi(0))^{\sigma_1})e^{-(\beta(\varphi(s)-\varphi(0)))} \\
&\quad \times \Delta(s)\varphi'(s)\varphi'(t)d\theta\, ds dt \\
&= \int_0^\infty\int_0^\infty\int_0^\infty \sigma_1 e^{-(\beta(\varphi(t)+\varphi(s)-2\varphi(0)))}\frac{(\varphi(t)-\varphi(0))^{\sigma_1-1}}{\theta^{\sigma_1}}\rho_{\sigma_1}(\theta)\mathcal{T}\left(\frac{(\varphi(t)-\varphi(0))^{\sigma_1}}{\theta^{\sigma_1}}\right) \\
&\quad \times \Delta(s)\varphi'(s)\varphi'(t)d\theta\, ds dt \\
&= \int_0^\infty\int_0^\mu\int_0^\infty \sigma_1 e^{-(\beta(\varphi(\mu)-\varphi(0)))}\rho_{\sigma_1}(\theta)\frac{(\varphi(t)-\varphi(0))^{\sigma_1-1}}{\theta^{\sigma_1}}\mathcal{T}\left(\frac{(\varphi(t)-\varphi(0))^{\sigma_1}}{\theta^{\sigma_1}}\right) \\
&\quad \times \Delta(\varphi^{-1}(\varphi(\mu)-\varphi(t)+\varphi(0))))\varphi'(\mu)\varphi'(t)d\theta dt d\mu \\
&= \int_0^\infty e^{-(\beta(\varphi(\mu)-\varphi(0)))}\left(\int_0^\mu\int_0^\infty \sigma_1\rho_{\sigma_1}(\theta)\frac{(\varphi(\mu)-\varphi(s))^{\sigma_1-1}}{\theta^{\sigma_1}}\mathcal{T}\left(\frac{(\varphi(\mu)-\varphi(s))^{\sigma_1}}{\theta^{\sigma_1}}\right)\right. \\
&\quad \left.\times \Delta(s)\varphi'(s)d\theta ds\right)\varphi'(\mu)d\mu.
\end{aligned}$$

Hence, we obtain

$$\begin{aligned}
Z(\beta) &= \beta^{(\sigma_1-1)(\sigma_2-1)}\int_0^\infty e^{-(\beta(\varphi(t)-\varphi(0)))}\left(\int_0^\infty \rho_{\sigma_1}(\theta)\mathcal{T}\left(\frac{\varphi(t)-\varphi(0))^{\sigma_1}}{\theta^{\sigma_1}}\right)z_0 d\theta\right)\varphi'(t)dt \\
&\quad + \int_0^\infty e^{-(\beta(\varphi(\mu)-\varphi(0)))}\left(\int_0^\mu\int_0^\infty \sigma_1\rho_{\sigma_1}(\theta)\frac{(\varphi(\mu)-\varphi(s))^{\sigma_1-1}}{\theta^{\sigma_1}}\mathcal{T}\left(\frac{(\varphi(\mu)-\varphi(s))^{\sigma_1}}{\theta^{\sigma_1}}\right)\right. \\
&\quad \left.\times \Delta(s)\varphi'(s)d\theta ds\right)\varphi'(\mu)d\mu.
\end{aligned}$$

By using inverse Laplace transform, we obtain

$$\begin{aligned}
z(t) &= I_{a+}^{(1-\sigma_1)(\sigma_2-1);\varphi}\int_0^\infty \rho_{\sigma_1}(\theta)\mathcal{T}\left(\frac{\varphi(t)-\varphi(0))^{\sigma_1}}{\theta^{\sigma_1}}\right)z_0 d\theta \\
&\quad + \int_0^t\int_0^\infty \sigma_1\rho_{\sigma_1}(\theta)\frac{(\varphi(t)-\varphi(s))^{\sigma_1-1}}{\theta^{\sigma_1}}\mathcal{T}\left(\frac{(\varphi(t)-\varphi(s))^{\sigma_1}}{\theta^{\sigma_1}}\right)\Delta(s)\varphi'(s)d\theta ds.
\end{aligned}$$

Thus, we obtain

$$\begin{aligned}
z(t) &= I_{a+}^{(1-\sigma_1)(\sigma_2-1);\varphi}\int_0^\infty \phi_{\sigma_1}(\theta)\mathcal{T}(\varphi(t)-\varphi(0))^{\sigma_1}\theta)z_0 d\theta \\
&\quad + \sigma_1\int_0^t\int_0^\infty \theta\phi_{\sigma_1}(\theta)(\varphi(t)-\varphi(s))^{\sigma_1-1}\mathcal{T}((\varphi(t)-\varphi(s))^{\sigma_1}\theta)\Delta(s)\varphi'(s)d\theta ds,
\end{aligned}$$

where $\phi_{\sigma_1}(\theta) = \frac{1}{\sigma_1}\theta^{-1-\frac{1}{\sigma_1}}\rho_{\sigma_1}(\theta^{-\frac{1}{\sigma_1}})$ is the probability density function defined on $(0,\infty)$. For any $z \in E$, the operators $S_\varphi^{\sigma_1,\sigma_2}(t,s)$ and $T_\varphi^{\sigma_1}(t,s)$ defined as

$$\mathcal{P}_\varphi^{\sigma_1}(t,s)z = \int_0^\infty \phi_{\sigma_1}(\theta)\mathcal{T}((\varphi(t) - \varphi(s))^{\sigma_1}\theta)z d\theta,$$

$$S_\varphi^{\sigma_1,\sigma_2}(t,s)z = I_{a+}^{(1-\sigma_1)(\sigma_2-1);\varphi}\mathcal{P}_\varphi^{\sigma_1}(t,s)z,$$

and

$$\mathcal{T}_\varphi^{\sigma_1}(t,s)z = \sigma_1 \int_0^\infty \theta \phi_{\sigma_1}(\theta)\mathcal{T}((\varphi(t) - \varphi(s))^{\sigma_1}\theta)z d\theta, \; 0 \leq s \leq t \leq b.$$

Hence, we obtain

$$z(t) = S_\varphi^{\sigma_1,\sigma_2}(t,0)z_0 + \int_0^t (\varphi(t) - \varphi(s))^{\sigma_1-1}\mathcal{T}_\varphi^{\sigma_1}(t,s)\Delta(s)\varphi'(s)ds.$$

□

Remark 1. Let $\mathcal{A}$ be the generator of a C_0-semigroup $\{\mathcal{T}(t)\}_{t\geq 0}$ on E. Then there exists $\mathcal{M} \geq 1$ such that $\mathcal{M} = \sup_{t\in[0,b]} \mathcal{T}(t)$

Lemma 4 ([17,32]). The operators $S_\varphi^{\sigma_1,\sigma_2}$ and $\mathcal{T}_\varphi^{\sigma_1}$ have the subsequent conditions

1. $S_\varphi^{\sigma_1,\sigma_2}(t,s)$ and $\mathcal{T}_\varphi^{\sigma_1}(t,s)$ are linear and bounded operators for any fixed $t \geq s \geq 0$, and

$$\|S_\varphi^{\sigma_1,\sigma_2}(t,s)(z)\| \leq \frac{\mathcal{M}(\varphi(b) - \varphi(0))^{(1-\sigma_1)(\sigma_2-1)}}{\Gamma(\sigma_1 + \sigma_2 - \sigma_1\sigma_2)}\|z\| = \mathcal{M}_1\|z\|,$$

$$\|\mathcal{T}_\varphi^{\sigma_1}(t,s)(z)\| \leq \frac{\sigma_1 \mathcal{M}}{\Gamma(1+\sigma_1)}\|z\| = \frac{\mathcal{M}}{\Gamma(\sigma_1)}\|z\| = \mathcal{M}_2\|z\|.$$

2. If $\mathcal{T}(t)$ is compact operator for all $t > 0$, then $S_\varphi^{\sigma_1,\sigma_2}(t,s)$, $\mathcal{T}_\varphi^{\sigma_1}(t,s)$ are compact for all $t,s > 0$. Hence, $S_\varphi^{\sigma_1,\sigma_2}(t,s)$ and $\mathcal{T}_\varphi^{\sigma_1}(t,s)$ are strongly continuous.
3. The operators $S_\varphi^{\sigma_1,\sigma_2}(t,s)$ and $\mathcal{T}_\varphi^{\sigma_1}(t,s)$ are strongly continuous. For every $z \in E$ and $0 \leq s \leq t_1 < t_2 \leq b$, we have

$$\|S_\varphi^{\sigma_1,\sigma_2}(t_2,s)z - S_\varphi^{\sigma_1,\sigma_2}(t_1,s)z\| \to 0 \text{ and } \|\mathcal{T}_\varphi^{\sigma_1}(t_2,s)z - \mathcal{T}_\varphi^{\sigma_1}(t_1,s)z\| \to 0 \text{ as } t_1 \to t_2.$$

Definition 5. A function $z \in PC(E)$ is called a mild solution of problem (1) if for every $t \in \mathcal{J}_0$, $z(t)$ fulfills $I_{0+}^{(1-\sigma_1)(1-\sigma_2);\varphi}[z(t)]_{t=0} + \mathcal{G}(z) = z_0$, $I_{t_\gamma+}^{(1-\sigma_1)(1-\sigma_2);\varphi}z(t_\gamma^+) = z(t_\gamma^-) + \mathcal{I}_\gamma(z(t_\gamma^-))$, $\gamma = 1,2,\ldots,\mathcal{H}$, and

$$z(t) = S_\varphi^{\sigma_1,\sigma_2}(t,0)[z_0 - \mathcal{G}(z)] + \int_0^t (\varphi(t) - \varphi(s))^{\sigma_1-1}\mathcal{T}_\varphi^{\sigma_1}(t,s)\Delta(s,z(s))\varphi'(s)ds,$$

for every $t \in [0,t_1]$ and

$$z(t) = S_\varphi^{\sigma_1,\sigma_2}(t,t_\gamma)\left[z(t_\gamma^-) + \mathcal{I}_\gamma(z(t_\gamma^-))\right] + \int_{t_\gamma}^t (\varphi(t) - \varphi(s))^{\sigma_1-1}\mathcal{T}_\varphi^{\sigma_1}(t,s)\Delta(s,z(s))\varphi'(s)ds,$$

for every $t \in (t_\gamma, t_{\gamma+1}]$.

4. Existence and Uniqueness

In this section, we prove the existence outcomes of the proposed system (1). Let us assume the following hypotheses

[X1]: $\mathcal{T}(t)$ is compact for every $t > 0$.
[X2]: The function $\Delta : \mathcal{J}_0 \times E \to E$ satisfies

(a) For all $z \in E$, the function $t \to \Delta(t,z)$ is strongly measurable and the function $\Delta(t, \cdot) : E \to E$ is continuous for a.e $t \in \mathcal{J}_0$.
(b) There exists a continuous function $\hat{\mathcal{K}}_\Delta \in L^1(\mathcal{J}_0, \mathbb{R}^+)$ such that

$$\|\Delta(t,z)\| \leq \hat{\mathcal{K}}_\Delta(t) \|z\|, \, \forall\, (t,z) \in \mathcal{J}_0 \times E,$$

with $\mathcal{K}_\Delta = \sup_{t \in \mathcal{J}_0} \hat{\mathcal{K}}_\Delta(t)$.

[X3]: The function $\mathcal{G} : C(\mathcal{J}_0, E) \to E$ is Lipschitz continuous, i.e.; there exists a positive constant $\hat{\mathcal{K}}_\mathcal{G}$ such that

$$\|\mathcal{G}(z_1) - \mathcal{G}(z_2)\| \leq \hat{\mathcal{K}}_\mathcal{G} \|z_1 - z_2\|, \, \forall\, z_1, z_2 \in E.$$

[X4]: For every $z, z_1, z_2 \in E$ and all $t \in (t_\gamma, t_{\gamma+1}]$, $\gamma = 1, 2, \ldots, \mathcal{H}$, there exist $\mathcal{D}_\gamma, \mathcal{K}_\gamma > 0$, satisfies

$$\|\mathcal{I}_\gamma(z(t_\gamma^-))\| \leq \mathcal{K}_\gamma, \quad \|\mathcal{I}_\gamma(z_1(t_\gamma^-)) - \mathcal{I}_\gamma(z_2(t_\gamma^-))\| \leq \mathcal{D}_\gamma \|z_1(t_\gamma^-) - z_2(t_\gamma^-)\|.$$

[X5]: The following inequalities hold

$$\hat{\mathcal{O}} = \max_{1 \leq \gamma \leq \mathcal{H}} [\mathcal{M}_1 \hat{\mathcal{K}}_\mathcal{G}, \mathcal{M}_1(1 + \mathcal{D}_\gamma)] < 1.$$

[X6]: There exists a constant $\hat{\mathcal{R}}_\Delta > 0$ such that

$$\|\Delta(t, z_1) - \Delta(t, z_2)\| \leq \hat{\mathcal{R}}_\Delta \|z_1 - z_2\|, \, \forall\, z_1, z_2 \in E.$$

Theorem 1. *Suppose the hypotheses [X1]–[X5] are fulfilled. If*

$$\mathcal{M}_1 \hat{\mathcal{K}}_\mathcal{G} + \mathcal{M}_2 \mathcal{K}_\Delta \frac{\Gamma(\sigma_1) \Gamma(\rho)}{\Gamma(\rho + \sigma_1)} (\varphi(b) - \varphi(0))^{\sigma_1} < 1, \quad (6)$$

then φ-fractional system (1) has at least one mild solution on $\mathcal{J}_0$.

Proof. For any $\pi > 0$, we define

$$\Omega_\pi = \{z \in \mathcal{PC}(E) : \|z\|_{\mathcal{PC}} \leq \pi\}.$$

Clearly, Ω_π is closed convex and bounded subset of $\mathcal{PC}(E)$. Define an operator $\Pi : \Omega_\pi \to \mathcal{PC}(E)$ by

$$(\Pi z)(t) = \begin{cases} \mathcal{S}_\varphi^{\sigma_1, \sigma_2}(t, 0)[z_0 - \mathcal{G}(z)] \\ + \int_0^t (\varphi(t) - \varphi(s))^{\sigma_1 - 1} \mathcal{T}_\varphi^{\sigma_1}(t, s) \Delta(s, z(s)) \varphi'(s) ds, & t \in [0, t_1], \gamma = 0, \\ \mathcal{S}_\varphi^{\sigma_1, \sigma_2}(t, t_\gamma) [z(t_\gamma^-) + \mathcal{I}_\gamma(z(t_\gamma^-))] \\ + \int_{t_\gamma}^t (\varphi(t) - \varphi(s))^{\sigma_1 - 1} \mathcal{T}_\varphi^{\sigma_1}(t, s) \Delta(s, z(s)) \varphi'(s) ds, & t \in (t_\gamma, t_{\gamma+1}], \gamma \geq 1. \end{cases}$$

Now, we split Π as $\Pi_1 + \Pi_2$, where

$$(\Pi_1 z)(t) = \begin{cases} \mathcal{S}_\varphi^{\sigma_1, \sigma_2}(t, 0)[z_0 - \mathcal{G}(z)], & t \in [0, t_1], \gamma = 0, \\ \mathcal{S}_\varphi^{\sigma_1, \sigma_2}(t, t_\gamma)[z(t_\gamma^-) + \mathcal{I}_\gamma(z(t_\gamma^-))], & t \in (t_\gamma, t_{\gamma+1}], \gamma \geq 1, \end{cases}$$

and

$$(\Pi_2 z)(t) = \begin{cases} \int_0^t (\varphi(t) - \varphi(s))^{\sigma_1 - 1} \mathcal{T}_\varphi^{\sigma_1}(t, s) \Delta(s, z(s)) \varphi'(s) ds, & t \in [0, t_1], \gamma = 0, \\ \int_{t_\gamma}^t (\varphi(t) - \varphi(s))^{\sigma_1 - 1} \mathcal{T}_\varphi^{\sigma_1}(t, s) \Delta(s, z(s)) \varphi'(s) ds, & t \in (t_\gamma, t_{\gamma+1}], \gamma \geq 1. \end{cases}$$

Step 1. There exists $\pi > 0$ such that $\Pi(\Omega_\pi) \subset \Omega_\pi$. If we assume that the assertion is not true, then for $\pi > 0$, we take $t \in \mathcal{J}_0$ and $z^\pi \in \Omega_\pi$ such that $\|\Pi(z^\pi)\|_{\mathcal{PC}} > \pi$. For $t \in [0, t_1]$, we obtain

$$\begin{aligned}
\pi < \|\Pi(z^\pi)\|_{\mathcal{PC}} &\leq \left\|(\varphi(t) - \varphi(0))^{1-\rho} \mathcal{S}_\varphi^{\sigma_1,\sigma_2}(t,0)[z_0 - \mathcal{G}(z^\pi)]\right\| \\
&+ \left\|(\varphi(t) - \varphi(0))^{1-\rho} \int_0^t (\varphi(t) - \varphi(s))^{\sigma_1-1} \mathcal{T}_\varphi^{\sigma_1}(t,s) \Delta(s, z^\pi(s)) \varphi'(s) ds\right\| \\
&\leq \mathcal{M}_1[\|z_0\|_{\mathcal{PC}} + \hat{\mathcal{K}}_\mathcal{G}\pi + \|\mathcal{G}(0)\|_{\mathcal{PC}}] \\
&+ \mathcal{M}_2 \mathcal{K}_\Delta (\varphi(t_1) - \varphi(0))^{1-\rho} \int_0^t (\varphi(t) - \varphi(s))^{\sigma_1-1} \|z^\pi(s)\| \varphi'(s) ds \\
&\leq \mathcal{M}_1[\|z_0\|_{\mathcal{PC}} + \hat{\mathcal{K}}_\mathcal{G}\pi + \|\mathcal{G}(0)\|_{\mathcal{PC}}] \\
&+ \pi \mathcal{M}_2 \mathcal{K}_\Delta (\varphi(t_1) - \varphi(0))^{1-\rho} \int_0^t (\varphi(t) - \varphi(s))^{\sigma_1-1} (\varphi(s) - \varphi(0))^{\rho-1} \varphi'(s) ds \\
&\leq \mathcal{M}_1[\|z_0\|_{\mathcal{PC}} + \hat{\mathcal{K}}_\mathcal{G}\pi + \|\mathcal{G}(0)\|_{\mathcal{PC}}] \\
&+ \pi \mathcal{M}_2 \mathcal{K}_\Delta (\varphi(t_1) - \varphi(0))^{1-\rho} \Gamma(\sigma_1) I_{0+}^{\sigma_1;\varphi}(\varphi(s) - \varphi(0))^{\rho-1} \\
&\leq \mathcal{M}_1[\|z_0\|_{\mathcal{PC}} + \hat{\mathcal{K}}_\mathcal{G}\pi + \|\mathcal{G}(0)\|_{\mathcal{PC}}] \\
&+ \pi \mathcal{M}_2 \mathcal{K}_\Delta (\varphi(t_1) - \varphi(0))^{1-\rho} \frac{\Gamma(\sigma_1)\Gamma(\rho)}{\Gamma(\rho + \sigma_1)} (\varphi(t_1) - \varphi(0))^{\rho+\sigma_1-1} \\
&\leq \mathcal{M}_1[\|z_0\|_{\mathcal{PC}} + \hat{\mathcal{K}}_\mathcal{G}\pi + \|\mathcal{G}(0)\|_{\mathcal{PC}}] \\
&+ \pi \mathcal{M}_2 \mathcal{K}_\Delta \frac{\Gamma(\sigma_1)\Gamma(\rho)}{\Gamma(\rho + \sigma_1)} (\varphi(t_1) - \varphi(0))^{\sigma_1}.
\end{aligned}$$

For every $t \in (t_\gamma, t_{\gamma+1}]$, $\gamma = 1, 2, \ldots, \mathcal{H}$, we obtain

$$\begin{aligned}
\pi < \|\Pi(z^\pi)\|_{\mathcal{PC}} &\leq \left\|(\varphi(t) - \varphi(t_\gamma))^{1-\rho} \mathcal{S}_\varphi^{\sigma_1,\sigma_2}(t,t_\gamma)\left[z^\pi(t_\gamma^-) + \mathcal{I}_\gamma(z^\pi(t_\gamma^-))\right]\right\| \\
&+ \left\|(\varphi(t) - \varphi(t_\gamma))^{1-\rho} \int_{t_\gamma}^t (\varphi(t) - \varphi(s))^{\sigma_1-1} \mathcal{T}_\varphi^{\sigma_1}(t,s) \Delta(s, z^\pi(s)) \varphi'(s) ds\right\| \\
&\leq \mathcal{M}_1\left[\|z^\pi(t_\gamma^-)\|_{\mathcal{PC}} + (\varphi(t_{\gamma+1}) - \varphi(t_\gamma))^{1-\rho} \mathcal{K}_\gamma\right] \\
&+ \mathcal{M}_2 \mathcal{K}_\Delta (\varphi(t_{\gamma+1}) - \varphi(t_\gamma))^{1-\rho} \int_{t_\gamma}^t (\varphi(t) - \varphi(s))^{\sigma_1-1} \|z^\pi(s)\| \varphi'(s) ds \\
&\leq \mathcal{M}_1\left[\|z^\pi(t_\gamma^-)\|_{\mathcal{PC}} + (\varphi(t_{\gamma+1}) - \varphi(t_\gamma))^{1-\rho} \mathcal{K}_\gamma\right] \\
&+ \pi \mathcal{M}_2 \mathcal{K}_\Delta (\varphi(t_{\gamma+1}) - \varphi(t_\gamma))^{1-\rho} \int_{t_\gamma}^t (\varphi(t) - \varphi(s))^{\sigma_1-1} (\varphi(s) - \varphi(t_\gamma))^{\rho-1} \varphi'(s) ds \\
&\leq \mathcal{M}_1\left[\|z^\pi(t_\gamma^-)\|_{\mathcal{PC}} + (\varphi(t_{\gamma+1}) - \varphi(t_\gamma))^{1-\rho} \mathcal{K}_\gamma\right] \\
&+ \pi \mathcal{M}_2 \mathcal{K}_\Delta (\varphi(t_{\gamma+1}) - \varphi(t_\gamma))^{1-\rho} \Gamma(\sigma_1) I_{t_\gamma^+}^{\sigma_1;\varphi}(\varphi(s) - \varphi(t_\gamma))^{\rho-1} \\
&\leq \mathcal{M}_1\left[\|z^\pi(t_\gamma^-)\|_{\mathcal{PC}} + (\varphi(t_{\gamma+1}) - \varphi(t_\gamma))^{1-\rho} \mathcal{K}_\gamma\right] \\
&+ \pi \mathcal{M}_2 \mathcal{K}_\Delta (\varphi(t_{\gamma+1}) - \varphi(t_\gamma))^{1-\rho} \frac{\Gamma(\sigma_1)\Gamma(\rho)}{\Gamma(\rho + \sigma_1)} (\varphi(t_{\gamma+1}) - \varphi(t_\gamma))^{\rho+\sigma_1-1} \\
&\leq \mathcal{M}_1\left[\|z^\pi(t_\gamma^-)\|_{\mathcal{PC}} + (\varphi(t_{\gamma+1}) - \varphi(t_\gamma))^{1-\rho} \mathcal{K}_\gamma\right] \\
&+ \pi \mathcal{M}_2 \mathcal{K}_\Delta \frac{\Gamma(\sigma_1)\Gamma(\rho)}{\Gamma(\rho + \sigma_1)} (\varphi(t_{\gamma+1}) - \varphi(t_\gamma))^{\sigma_1}.
\end{aligned}$$

For every $t \in \mathcal{J}_0$, we obtain

$$\pi < \|\Pi(z^\pi)\|_{\mathcal{PC}} \leq \mathcal{W}^* + \mathcal{M}_1 \hat{\mathcal{K}}_\mathcal{G} \pi + \pi \mathcal{M}_2 \mathcal{K}_\Delta \frac{\Gamma(\sigma_1)\Gamma(\rho)}{\Gamma(\rho + \sigma_1)} (\varphi(b) - \varphi(0))^{\sigma_1}, \tag{7}$$

where

$$\mathcal{W}^* = \max_{1\leq \gamma \leq \mathcal{H}} \left\{ \mathcal{M}_1[\|z_0\|_{\mathcal{PC}} + \|\mathcal{G}(0)\|_{\mathcal{PC}}] + \mathcal{M}_1\left[\|z^\pi(t_\gamma^-)\|_{\mathcal{PC}} + (\varphi(t_{\gamma+1}) - \varphi(t_\gamma))^{1-\rho}\mathcal{K}_\gamma\right]\right\}.$$

Here, $\mathcal{W}^*$ is independent of π, both sides of Equation (7) are dividing by π and taking $\pi \to \infty$, we obtain

$$1 < \mathcal{M}_1\hat{\mathcal{K}}_\mathcal{G} + \mathcal{M}_2\mathcal{K}_\Delta \frac{\Gamma(\sigma_1)\Gamma(\rho)}{\Gamma(\rho+\sigma_1)}(\varphi(b) - \varphi(0))^{\sigma_1},$$

which contradicts to Equation (6). Hence, for some $\pi > 0$, $\Pi(\Omega_\pi) \subset \Omega_\pi$.

Step 2. We will prove that Π_1 is a contraction map.
For $z^*, z^{**} \in \Omega_\pi$, if $t \in [0, t_1]$, then we obtain

$$\begin{aligned}
\|\Pi_1 z^* - \Pi_1 z^{**}\|_{\mathcal{PC}} &= \|(\varphi(t) - \varphi(0))^{1-\rho}\mathcal{S}_\varphi^{\sigma_1,\sigma_2}(t,0)[\mathcal{G}(z^*) - \mathcal{G}(z^{**})]\| \\
&\leq \mathcal{M}_1\hat{\mathcal{K}}_\mathcal{G}\|z^* - z^{**}\|_{\mathcal{PC}}.
\end{aligned} \tag{8}$$

Similarly, if $t \in (t_\gamma, t_{\gamma+1}]$, $\gamma = 1, 2, \ldots, \mathcal{H}$, then we get

$$\begin{aligned}
\|\Pi_1 z^* - \Pi_1 z^{**}\|_{\mathcal{PC}} &= \|(\varphi(t) - \varphi(t_\gamma))^{1-\rho}\mathcal{S}_\varphi^{\sigma_1,\sigma_2}(t,t_\gamma)[z^*(t_\gamma^-) - z^{**}(t_\gamma^-)]\| \\
&+ \|(\varphi(t) - \varphi(t_\gamma))^{1-\rho}\mathcal{S}_\varphi^{\sigma_1,\sigma_2}(t,t_\gamma)[\mathcal{I}_\gamma(z^*(t_\gamma^-)) - \mathcal{I}_\gamma(z^{**}(t_\gamma^-))]\| \\
&\leq \mathcal{M}_1(1 + \mathcal{D}_\gamma)\|z^* - z^{**}\|_{\mathcal{PC}}.
\end{aligned} \tag{9}$$

From Equations (8) and (9), we obtain

$$\|\Pi_1 z^* - \Pi_1 z^{**}\|_{\mathcal{PC}} \leq \hat{\mathcal{O}}\|z^* - z^{**}\|_{\mathcal{PC}},$$

where $\hat{\mathcal{O}} = \max_{1\leq \gamma \leq \mathcal{H}}[\mathcal{M}_1\hat{\mathcal{K}}_\mathcal{G}, \mathcal{M}_1(1+\mathcal{D}_\gamma)]$. By [X5], we see that $\hat{\mathcal{O}} < 1$. Hence, Π_1 is a contraction mapping.

Step 3. We will prove that $\Pi_2 : \Omega_\pi \to \Omega_\pi$ is continuous.
Let $\{z_k\} \subset \Omega_\pi$ with $z_k \to z$ as $k \to \infty$. By [X2], we obtain

$$\Delta(t, z_k) \to \Delta(t, z) \text{ as } k \to \infty,$$

and

$$\|\Delta(t, z_k(t)) - \Delta(t, z(t))\| \leq 2\hat{\mathcal{K}}_\Delta(t)\pi.$$

For every $t \in (t_\gamma, t_{\gamma+1}]$, $\gamma = 0, 1, \ldots, \mathcal{H}$, we obtain

$$\begin{aligned}
\|\Pi_2(z_k) - \Pi_2(z)\|_{\mathcal{PC}} &\leq \left\|(\varphi(t) - \varphi(t_\gamma))^{1-\rho}\int_{t_\gamma}^t (\varphi(t) - \varphi(s))^{\sigma_1-1}\mathcal{T}_\varphi^{\sigma_1}(t,s)\right. \\
&\times \left.[\Delta(s, z_k(s)) - \Delta(s, z(s))]\varphi'(s)ds\right\| \\
&\leq \mathcal{M}_2(\varphi(t_{\gamma+1}) - \varphi(t_\gamma))^{1-\rho} \\
&\times \int_{t_\gamma}^t (\varphi(t) - \varphi(s))^{\sigma_1-1}\|\Delta(s, z_k(s)) - \Delta(s, z(s))\|\varphi'(s)ds.
\end{aligned}$$

By the Lebesgue dominated convergence theorem, we obtain

$$\|\Pi_2(z_k) - \Pi_2(z)\|_{\mathcal{PC}} \to 0 \text{ as } k \to \infty.$$

Hence, Π_2 is continuous.

Step 4. We prove that $\{\Pi_2 z : z \in \Omega_\pi\}$ is equicontinuous.

Let $\kappa_1, \kappa_2 \in (t_\gamma, t_{\gamma+1}]$, with $t_\gamma < \kappa_1 < \kappa_2 \leq t_{\gamma+1}$, then we obtain for every $t \in (t_\gamma, t_{\gamma+1}]$, $\gamma = 0, 1, \ldots, \mathcal{H}$,

$$\|(\varphi(\kappa_2) - \varphi(t_\gamma))^{1-\rho}(\Pi_2 z)(\kappa_2) - (\varphi(\kappa_1) - \varphi(t_\gamma))^{1-\rho}(\Pi_2 z)(\kappa_1)\|$$
$$\leq \int_{t_\gamma}^{\kappa_1} \|(\varphi(\kappa_2) - \varphi(t_\gamma))^{1-\rho}(\varphi(\kappa_2) - \varphi(s))^{\sigma_1-1}\mathcal{T}_\varphi^{\sigma_1}(\kappa_2, s)$$
$$- (\varphi(\kappa_1) - \varphi(t_\gamma))^{1-\rho}(\varphi(\kappa_1) - \varphi(s))^{\sigma_1-1}\mathcal{T}_\varphi^{\sigma_1}(\kappa_1, s)\| \|\Delta(s, z(s))\| \varphi'(s) ds$$
$$+ \int_{\kappa_1}^{\kappa_2} \|(\varphi(\kappa_2) - \varphi(t_\gamma))^{1-\rho}(\varphi(\kappa_2) - \varphi(s))^{\sigma_1-1}\mathcal{T}_\varphi^{\sigma_1}(\kappa_2, s)\| \|\Delta(s, z(s))\| \varphi'(s) ds. \tag{10}$$

As $\kappa_2 \to \kappa_1$, the right-hand side of Equation (10) tends to zero. Thus, the equicontinuity of $\{\Pi_2 z : z \in \Omega_\pi\}$ is obtained.

Step 5. We prove that $\delta(t) = \{(\Pi_2 z)(t) : z \in \Omega_\pi\}$ is relatively compact in E.

Obviously, $\delta(0) = \{0\}$ is relatively compact. Let $t \in (t_\gamma, t_{\gamma+1}]$ be fixed, $0 < \epsilon < t$, and ϵ is real number. For $z \in \Omega_\pi$, we define

$$(\Pi_2^\epsilon z)(t) = \begin{cases} \int_0^{t-\epsilon}(\varphi(t) - \varphi(s))^{\sigma_1-1}\mathcal{T}_\varphi^{\sigma_1}(t, s)\Delta(s, z(s))\varphi'(s) ds, & t \in [0, t_1], \gamma = 0, \\ \int_{t_\gamma}^{t-\epsilon}(\varphi(t) - \varphi(s))^{\sigma_1-1}\mathcal{T}_\varphi^{\sigma_1}(t, s)\Delta(s, z(s))\varphi'(s) ds, & t \in (t_\gamma, t_{\gamma+1}], \gamma \geq 1. \end{cases}$$

By [X1], we obtain $\delta^\epsilon(t) = \{(\Pi^\epsilon z)(t) : z \in \Omega_\pi\}$ is relatively compact in E. for every $z \in \Omega_\pi$, we get

$$\|(\varphi(t) - \varphi(t_\gamma))^{1-\rho}[(\Pi_2 z)(t) - (\Pi_2^\epsilon z)(t)]\| \leq \pi \mathcal{M}_2 \mathcal{K}_\Delta (\varphi(t_{\gamma+1}) - \varphi(t_\gamma))^{1-\rho}$$
$$\times \int_{t-\epsilon}^t (\varphi(t) - \varphi(s))^{\sigma_1-1}(\varphi(s) - \varphi(t_\gamma))^{\rho-1}\varphi'(s) ds$$
$$\to 0 \text{ as } \epsilon \to 0.$$

Then $\delta(t)$ is relatively compact in E. By steps 3–5 and Arzela-Ascoli theorem, Π_2 is completely continuous. Hence, by the fixed point theorem of Krasnoselskii's [33], there exists at least one mild solution on $\mathcal{J}_0$. □

Theorem 2. *Suppose the hypotheses [X1]–[X6] are fulfilled. Then φ-fractional system (1) has a unique mild solution on $\mathcal{J}_0$.*

Proof. Let z_1 and z_2 be the mild solutions of the φ-fractional system (1) in Ω_π. Then, for each $k \in \{1, 2\}$, the mild solutions z_k satisfies

$$(\Pi z_k)(t) = \begin{cases} \mathcal{S}_\varphi^{\sigma_1, \sigma_2}(t, 0)[z_0 - \mathcal{G}(z_k)] \\ + \int_0^t (\varphi(t) - \varphi(s))^{\sigma_1-1}\mathcal{T}_\varphi^{\sigma_1}(t, s)\Delta(s, z_k(s))\varphi'(s) ds, & t \in [0, t_1], \gamma = 0, \\ \mathcal{S}_\varphi^{\sigma_1, \sigma_2}(t, t_\gamma)\left[z_k(t_\gamma^-) + \mathcal{I}_\gamma(z_k(t_\gamma^-))\right] \\ + \int_{t_\gamma}^t (\varphi(t) - \varphi(s))^{\sigma_1-1}\mathcal{T}_\varphi^{\sigma_1}(t, s)\Delta(s, z_k(s))\varphi'(s) ds, & t \in (t_\gamma, t_{\gamma+1}], \gamma \geq 1. \end{cases}$$

For every $t \in [0, t_1]$, $\gamma = 0$, we obtain

$$\begin{aligned}
\|(\varphi(t) - \varphi(0))^{1-\rho}[z_1(t) - z_2(t)]\| &= \|(\varphi(t) - \varphi(0))^{1-\rho}[(\Pi z_1)(t) - (\Pi z_2)(t)]\| \\
&\leq \mathcal{M}_1 \hat{\mathcal{K}}_\mathcal{G} \|(\varphi(t) - \varphi(0))^{1-\rho}[z_1(t) - z_2(t)]\| \\
&\quad + \mathcal{M}_2 \hat{\mathcal{R}}_\Delta (\varphi(t_1) - \varphi(0))^{1-\rho} \int_0^t (\varphi(t) - \varphi(s))^{\sigma_1 - 1} \\
&\quad \times (\varphi(s) - \varphi(0))^{\rho-1} \|(\varphi(s) - \varphi(0))^{1-\rho}[z_1(s) - z_2(s)]\| \|\varphi'(s)\| ds \\
&\leq \mathcal{M}_1 \hat{\mathcal{K}}_\mathcal{G} \|(\varphi(t) - \varphi(0))^{1-\rho}[z_1(t) - z_2(t)]\| \\
&\quad + \mathcal{M}_2 \hat{\mathcal{R}}_\Delta K_0^* (\varphi(t_1) - \varphi(0))^{1-\rho} \int_0^t (\varphi(t) - \varphi(s))^{\sigma_1 - 1} \\
&\quad \times \|(\varphi(s) - \varphi(0))^{1-\rho}[z_1(s) - z_2(s)]\| \|\varphi'(s)\| ds,
\end{aligned}$$

where $K_0^* = \sup_{0 \leq s \leq t_1} (\varphi(s) - \varphi(0))^{\rho-1}$.

Then we obtain

$$\begin{aligned}
\|(\varphi(t) - \varphi(0))^{1-\rho}[z_1(t) - z_2(t)]\| &\leq \frac{\mathcal{M}_2 \hat{\mathcal{R}}_\Delta K_0^* (\varphi(t_1) - \varphi(0))^{1-\rho}}{(1 - \mathcal{M}_1 \hat{\mathcal{K}}_\mathcal{G})} \int_0^t (\varphi(t) - \varphi(s))^{\sigma_1 - 1} \\
&\quad \times \|(\varphi(s) - \varphi(0))^{1-\rho}[z_1(s) - z_2(s)]\| \|\varphi'(s)\| ds,
\end{aligned}$$

where $\mathcal{M}_1 \hat{\mathcal{K}}_\mathcal{G} < 1$.

For every $t \in (t_\gamma, t_{\gamma+1}]$, $\gamma = 1, 2, \ldots, \mathcal{H}$, we get

$$\begin{aligned}
\|(\varphi(t) - \varphi(t_\gamma))^{1-\rho}[z_1(t) - z_2(t)]\| &= \|(\varphi(t) - \varphi(t_\gamma))^{1-\rho}[(\Pi z_1)(t) - (\Pi z_2)(t)]\| \\
&\leq \mathcal{M}_1 (1 + \mathcal{D}_\gamma) \|(\varphi(t) - \varphi(t_\gamma))^{1-\rho}[z_1(t_\gamma^-) - z_2(t_\gamma^-)]\| \\
&\quad + \mathcal{M}_2 \hat{\mathcal{R}}_\Delta (\varphi(t_{\gamma+1}) - \varphi(t_\gamma))^{1-\rho} \int_{t_\gamma}^t (\varphi(t) - \varphi(s))^{\sigma_1 - 1} \\
&\quad \times (\varphi(s) - \varphi(t_\gamma))^{\rho-1} \|(\varphi(s) - \varphi(t_\gamma))^{1-\rho}[z_1(s) - z_2(s)]\| \|\varphi'(s)\| ds \\
&\leq \mathcal{M}_1 (1 + \mathcal{D}_\gamma) \|(\varphi(t) - \varphi(t_\gamma))^{1-\rho}[z_1(t_\gamma^-) - z_2(t_\gamma^-)]\| \\
&\quad + \mathcal{M}_2 \hat{\mathcal{R}}_\Delta K_\gamma^* (\varphi(t_{\gamma+1}) - \varphi(t_\gamma))^{1-\rho} \int_{t_\gamma}^t (\varphi(t) - \varphi(s))^{\sigma_1 - 1} \\
&\quad \times \|(\varphi(s) - \varphi(t_\gamma))^{1-\rho}[z_1(s) - z_2(s)]\| \|\varphi'(s)\| ds,
\end{aligned}$$

where $K_\gamma^* = \sup_{t_\gamma \leq s \leq t_{\gamma+1}} (\varphi(s) - \varphi(0))^{\rho-1}$, $\gamma = 1, 2, \ldots, \mathcal{H}$.

Then we obtain

$$\begin{aligned}
\|(\varphi(t) - \varphi(t_\gamma))^{1-\rho}[z_1(t) - z_2(t)]\| &\leq \frac{\mathcal{M}_2 \hat{\mathcal{R}}_\Delta K_\gamma^* (\varphi(t_{\gamma+1}) - \varphi(t_\gamma))^{1-\rho}}{(1 - \mathcal{M}_1(1 + \mathcal{D}_\gamma))} \int_{t_\gamma}^t (\varphi(t) - \varphi(s))^{\sigma_1 - 1} \\
&\quad \times \|(\varphi(s) - \varphi(t_\gamma))^{1-\rho}[z_1(s) - z_2(s)]\| \|\varphi'(s)\| ds,
\end{aligned}$$

where $\mathcal{M}_1(1 + \mathcal{D}_\gamma) < 1$.

By using the Gronwall's inequality (Theorem 2.11, [17]), we get

$$\|z_1 - z_2\|_{\mathcal{PC}} = 0,$$

which implies that $z_1 \equiv z_2$. Therefore, φ-fractional system (1) has a unique mild solution on $\mathcal{J}_0$. □

5. Existence of Optimal Controls

Let v takes the value in the separable reflexive Banach space $\mathcal{T}$ and $\mathcal{V}_f(\mathcal{T})$ is a class of subsets of $\mathcal{T}$, which is nonempty convex and closed. The multifunction $g : \mathcal{J} \to \mathcal{V}_f(\mathcal{T})$ is measurable and $g(\cdot) \subset \Delta$, the admissible control set

$$\mathcal{U}_{ad} = \{v \in L^2(\Delta) : v(t) \in g(t) \text{ a.e.}\},$$

where Δ is a bounded set of $\mathcal{T}$. Then $\mathcal{U}_{ad} \neq \phi$.

Consider following φ-Hilfer fractional impulsive differential control system:

$$\begin{cases} {}^H D_{t_\gamma^+}^{\sigma_1,\sigma_2;\varphi} z(t) = \mathcal{A}z(t) + \mathcal{D}v(t) + \Delta(t,z(t)), \ t \in (0,b] - \{t_1,t_2,\ldots,t_\mathcal{H}\}, \\ I_{t_\gamma^+}^{(1-\sigma_1)(1-\sigma_2);\varphi} z(t_\gamma^+) = z(t_\gamma^-) + \mathcal{I}_\gamma(z(t_\gamma^-)), \ \gamma = 1,2,\ldots,\mathcal{H}, \\ I_{0^+}^{(1-\sigma_1)(1-\sigma_2);\varphi} [z(t)]_{t=0} + \mathcal{G}(z) = z_0. \end{cases} \quad (11)$$

Let us assume the following hypotheses

[X7]: $\mathcal{D} \in L^\infty(\mathcal{J}_0, L(\mathcal{T}, E))$, that implies that $\mathcal{D}v \in L^2(\mathcal{J}_0, E)$ for $v \in \mathcal{U}_{ad}$.
[X8]: $\mathcal{K}_* = \sup_{t \in \mathcal{J}_0} \varphi'(t) < \infty$.

Theorem 3. *Suppose the hypotheses of Theorem 2 and [X7]–[X8] are fulfilled. Then for each $v \in \mathcal{U}_{ad}$, φ-fractional system (11) has a mild solution which is given by*

$$z^v(t) = \begin{cases} S_\varphi^{\sigma_1,\sigma_2}(t,0)[z_0 - \mathcal{G}(z)] \\ + \int_0^t (\varphi(t) - \varphi(s))^{\sigma_1-1} T_\varphi^{\sigma_1}(t,s)[\mathcal{D}v(s) + \Delta(s,z(s))]\varphi'(s)ds, & t \in [0,t_1], \ \gamma = 0, \\ S_\varphi^{\sigma_1,\sigma_2}(t,t_\gamma)[z(t_\gamma^-) + \mathcal{I}_\gamma(z(t_\gamma^-))] \\ + \int_{t_\gamma}^t (\varphi(t) - \varphi(s))^{\sigma_1-1} T_\varphi^{\sigma_1}(t,s)[\mathcal{D}v(s) + \Delta(s,z(s))]\varphi'(s)ds, & t \in (t_\gamma, t_{\gamma+1}], \ \gamma \geq 1. \end{cases}$$

Proof. Let us consider

$$\mathcal{H}(t) = \int_{t_\gamma}^t (\varphi(t) - \varphi(s))^{\sigma_1-1} T_\varphi^{\sigma_1}(t,s) \mathcal{D}v(s) \varphi'(s) ds.$$

By Hölder's inequality and [X7], we get

$$\begin{aligned} \|(\varphi(t) - \varphi(t_\gamma))^{1-\rho} \mathcal{H}(t)\| &\leq M_2 \|\mathcal{D}\|_\infty (\varphi(t_{\gamma+1}) - \varphi(t_\gamma))^{1-\rho} \int_{t_\gamma}^t (\varphi(t) - \varphi(s))^{\sigma_1-1} \|v(s)\|_\mathcal{T} \varphi'(s) ds \\ &\leq M_2 \|\mathcal{D}\|_\infty (\varphi(t_{\gamma+1}) - \varphi(t_\gamma))^{1-\rho} \\ &\quad \times \left(\int_{t_\gamma}^t (\varphi(t) - \varphi(s))^{2(\sigma_1-1)} \varphi'(s) ds\right)^{1/2} \left(\int_{t_\gamma}^t \|v(s)\|_\mathcal{T}^2 \varphi'(s) ds\right)^{1/2} \\ &\leq \frac{M_2 \|\mathcal{D}\|_\infty (\varphi(t_{\gamma+1}) - \varphi(t_\gamma))^{\sigma_1-\rho+(1/2)}}{(2\sigma_1 - 1)^{1/2}} \left(\int_{t_\gamma}^t \|v(s)\|_\mathcal{T}^2 \varphi'(s) ds\right)^{1/2} \\ &\leq \frac{M_2 \mathcal{K}_*^{1/2} \|\mathcal{D}\|_\infty (\varphi(t_{\gamma+1}) - \varphi(t_\gamma))^{\sigma_1-\rho+(1/2)}}{(2\sigma_1 - 1)^{1/2}} \|v\|_{L^2(\mathcal{J}_0,\mathcal{T})}. \end{aligned}$$

It follows that $(\varphi(t) - \varphi(s))^{\sigma_1-1} T_\varphi^{\sigma_1}(t,s) \mathcal{D}v(s) \varphi'(s) ds$ are integrable on $\mathcal{J}_0$, here, $\|\mathcal{D}\|_\infty$ is the norm of $\mathcal{D}$ in Banach space $L^\infty(\mathcal{J}_0, L(\mathcal{T}, E))$. Hence, $\mathcal{H}(\cdot) \in \Omega_\pi$. Using Theorem 2, we get the required results. □

We consider the Lagrange problem

$$(\mathcal{LP}) \begin{cases} \text{Find } (z^*, v^*) \in PC(E) \times \mathcal{U}_{ad} \\ \text{such that } \mathcal{J}(z^*, v^*) \leq \mathcal{J}(z^v, v), \ (z^v, v) \in PC(E) \times \mathcal{U}_{ad}, \end{cases}$$

where the cost functional is

$$\mathcal{J}(z^v, v) = \sum_{\gamma=0}^{\mathcal{H}} \int_{t_\gamma}^{t_{\gamma+1}} \mathcal{L}(t, z^v(t), v(t)) dt,$$

where z^v be the mild solution of (11) with respect to control $v \in \mathcal{U}_{ad}$.

Next, we assume

[X9]:
1. The functional $\mathcal{L} : \mathcal{J}_0 \times E \times \mathcal{T} \to \mathbb{R} \cup \{\infty\}$ is Borel measurable.
2. For almost all $t \in \mathcal{J}_0$, $\mathcal{L}(t, \cdot, \cdot)$ is sequentially lower semicontinuous on $E \times \mathcal{T}$.
3. For each $z^v \in E$ and almost all $t \in \mathcal{J}_0$, $\mathcal{L}(t, z^v, \cdot)$ is convex on $\mathcal{T}$.
4. There exist constants $d_1 \geq 0$, $d_2 > 0$, ϕ is non-negative function in $L^1(\mathcal{J}_0, \mathbb{R})$ such that

$$\mathcal{L}(t, z^v, v) \geq \phi(t) + d_1 \|z^v\| + d_2 \|v\|_{\mathcal{T}}^2.$$

[X10]: $\mathcal{D}$ is a strongly continuous operator.

Theorem 4. *If the assumptions [X1]–[X10] are fulfilled, then the problem $(\mathcal{LP})$ admits at least one optimal pair.*

Proof. Assume that $\inf\{\mathcal{J}(z^v, v) : v \in \mathcal{U}_{ad}\} = \epsilon < +\infty$. By using [X9], we obtain $\epsilon > -\infty$. By definition of infimum there exists a minimizing sequence feasible pair $(z^k, v^k) \subset \mathcal{P}_{ad}$, where $\mathcal{P}_{ad} = \{(z^v, v) : z^v \text{ is a solution of (11) with respect to } v \in \mathcal{U}_{ad}\}$ such that $\mathcal{J}(z^k, v^k) \to \epsilon$ as $k \to +\infty$. Since $v^k \subseteq \mathcal{U}_{ad}$, v^k is bounded in $L^2(\mathcal{J}_0, \mathcal{T})$, there exists a subsequence which is still represented by v^k and $v^* \in L^2(\mathcal{J}_0, \mathcal{T})$ such that

$$v^k \xrightarrow{w} v^*$$

in $L^2(\mathcal{J}_0, \mathcal{T})$. Since $\mathcal{U}_{ad}$ is convex and closed, by using Marzur Lemma, we get $v^* \in \mathcal{U}_{ad}$. Let z^k and z^* be the mild solution of system (11) with respect to v^k and v^*, respectively

$$z^k(t) = \begin{cases} S_\varphi^{\sigma_1,\sigma_2}(t,0)[z_0 - \mathcal{G}(z^k)] \\ + \int_0^t (\varphi(t) - \varphi(s))^{\sigma_1-1} T_\varphi^{\sigma_1}(t,s)[\mathcal{D}v^k(s) + \Delta(s, z^k(s))]\varphi'(s)ds, & t \in [0, t_1], \gamma = 0, \\ S_\varphi^{\sigma_1,\sigma_2}(t, t_\gamma)[z^k(t_\gamma^-) + \mathcal{I}_\gamma(z^k(t_\gamma^-))] \\ + \int_{t_\gamma}^t (\varphi(t) - \varphi(s))^{\sigma_1-1} T_\varphi^{\sigma_1}(t,s)[\mathcal{D}v^k(s) + \Delta(s, z^k(s))]\varphi'(s)ds, & t \in (t_\gamma, t_{\gamma+1}], \gamma \geq 1, \end{cases}$$

and

$$z^*(t) = \begin{cases} S_\varphi^{\sigma_1,\sigma_2}(t,0)[z_0 - \mathcal{G}(z^*)] \\ + \int_0^t (\varphi(t) - \varphi(s))^{\sigma_1-1} T_\varphi^{\sigma_1}(t,s)[\mathcal{D}v^*(s) + \Delta(s, z^*(s))]\varphi'(s)ds, & t \in [0, t_1], \gamma = 0, \\ S_\varphi^{\sigma_1,\sigma_2}(t, t_\gamma)[z^*(t_\gamma^-) + \mathcal{I}_\gamma(z^*(t_\gamma^-))] \\ + \int_{t_\gamma}^t (\varphi(t) - \varphi(s))^{\sigma_1-1} T_\varphi^{\sigma_1}(t,s)[\mathcal{D}v^*(s) + \Delta(s, z^*(s))]\varphi'(s)ds, & t \in (t_\gamma, t_{\gamma+1}], \gamma \geq 1. \end{cases}$$

It follows from the boundedness of $\{v^k\}$, $\{v^*\}$ and Theorem 2, we obtain there exists a constant $\Theta > 0$ such that $\|z^k\|_\infty, \|z^*\|_\infty \leq \Theta$.

For every $t \in [0, t_1]$, $\gamma = 0$, we get

$$\begin{aligned}
\|(\varphi(t)-\varphi(0))^{1-\rho}[z^k(t)-z^*(t)]\| &\leq \mathcal{M}_1\hat{\mathcal{K}}_\mathcal{G}\|(\varphi(t)-\varphi(0))^{1-\rho}[z^k(t)-z^*(t)]\| \\
&+ \mathcal{M}_2\hat{\mathcal{R}}_\Delta(\varphi(t_1)-\varphi(0))^{1-\rho}\int_0^t(\varphi(t)-\varphi(s))^{\sigma_1-1} \\
&\times (\varphi(s)-\varphi(0))^{\rho-1}\|(\varphi(s)-\varphi(0))^{1-\rho}[z^k(s)-z^*(s)]\|\varphi'(s)ds \\
&+ \mathcal{M}_2(\varphi(t_1)-\varphi(0))^{1-\rho}\int_0^t(\varphi(t)-\varphi(s))^{\sigma_1-1} \\
&\times \|\mathcal{D}v^k(s)-\mathcal{D}v^*(s)\|_{L^2(\mathcal{J}_0,E)}\varphi'(s)ds \\
&\leq \mathcal{M}_1\hat{\mathcal{K}}_\mathcal{G}\|(\varphi(t)-\varphi(0))^{1-\rho}[z^k(t)-z^*(t)]\| \\
&+ \mathcal{M}_2\hat{\mathcal{R}}_\Delta(\varphi(t_1)-\varphi(0))^{1-\rho}\int_0^t(\varphi(t)-\varphi(s))^{\sigma_1-1} \\
&\times (\varphi(s)-\varphi(0))^{\rho-1}\|(\varphi(s)-\varphi(0))^{1-\rho}[z^k(s)-z^*(s)]\|\varphi'(s)ds \\
&+ \frac{\mathcal{M}_2\mathcal{K}_*^{1/2}(\varphi(t_1)-\varphi(0))^{\sigma_1-\rho+(1/2)}}{(2\sigma_1-1)^{1/2}}\|\mathcal{D}v^k-\mathcal{D}v^*\|_{L^2(\mathcal{J}_0,E)}.
\end{aligned}$$

For every $t \in (t_\gamma, t_{\gamma+1}]$, $\gamma = 1, 2, \ldots, \mathcal{H}$, we get

$$\begin{aligned}
\|(\varphi(t)-\varphi(t_\gamma))^{1-\rho}[z^k(t)-z^*(t)]\| &\leq \mathcal{M}_1(1+\mathcal{D}_\gamma)\|(\varphi(t)-\varphi(t_\gamma))^{1-\rho}[z^k(t_\gamma^-)-z^*(t_\gamma^-)]\| \\
&+ \mathcal{M}_2\hat{\mathcal{R}}_\Delta(\varphi(t_{\gamma+1})-\varphi(t_\gamma))^{1-\rho}\int_{t_\gamma}^t(\varphi(t)-\varphi(s))^{\sigma_1-1} \\
&\times (\varphi(s)-\varphi(t_\gamma))^{\rho-1}\|(\varphi(s)-\varphi(t_\gamma))^{1-\rho}[z^k(s)-z^*(s)]\|\varphi'(s)ds \\
&+ \mathcal{M}_2(\varphi(t_{\gamma+1})-\varphi(t_\gamma))^{1-\rho}\int_{t_\gamma}^t(\varphi(t)-\varphi(s))^{\sigma_1-1} \\
&\times \|\mathcal{D}v^k(s)-\mathcal{D}v^*(s)\|_{L^2(\mathcal{J}_0,E)}\varphi'(s)ds \\
&\leq \mathcal{M}_1(1+\mathcal{D}_\gamma)\|(\varphi(t)-\varphi(t_\gamma))^{1-\rho}[z^k(t_\gamma^-)-z^*(t_\gamma^-)]\| \\
&+ \mathcal{M}_2\hat{\mathcal{R}}_\Delta(\varphi(t_{\gamma+1})-\varphi(t_\gamma))^{1-\rho}\int_{t_\gamma}^t(\varphi(t)-\varphi(s))^{\sigma_1-1} \\
&\times (\varphi(s)-\varphi(0))^{\rho-1}\|(\varphi(s)-\varphi(t_\gamma))^{1-\rho}[z^k(s)-z^*(s)]\|\varphi'(s)ds \\
&+ \frac{\mathcal{M}_2\mathcal{K}_*^{1/2}(\varphi(t_{\gamma+1})-\varphi(t_\gamma))^{\sigma_1-\rho+(1/2)}}{(2\sigma_1-1)^{1/2}}\|\mathcal{D}v^k-\mathcal{D}v^*\|_{L^2(\mathcal{J}_0,E)}.
\end{aligned}$$

For every $t \in \mathcal{J}_0$, we obtain

$$\begin{aligned}
\|z^k-z^*\|_{\mathcal{PC}} &\leq \mathcal{M}_1(1+\mathcal{D}_\gamma)\|z^k-z^*\|_{\mathcal{PC}} + \mathcal{M}_1\hat{\mathcal{K}}_\mathcal{G}\|z^k-z^*\|_{\mathcal{PC}} \\
&+ \mathcal{M}_2\hat{\mathcal{R}}_\Delta\frac{\Gamma(\sigma_1)\Gamma(\rho)}{\Gamma(\rho+\sigma_1)}(\varphi(b)-\varphi(0))^{\sigma_1}\|z^k-z^*\|_{\mathcal{PC}} \\
&+ \frac{\mathcal{M}_2\mathcal{K}_*^{1/2}(\varphi(b)-\varphi(0))^{\sigma_1-\rho+(1/2)}}{(2\sigma_1-1)^{1/2}}\|\mathcal{D}v^k-\mathcal{D}v^*\|_{L^2(\mathcal{J}_0,E)},
\end{aligned}$$

then there exists a constant $\mathcal{N}^* > 0$ such that

$$\|z^k-z^*\|_{\mathcal{PC}} \leq \mathcal{N}^*\|\mathcal{D}v^k-\mathcal{D}v^*\|_{L^2(\mathcal{J}_0,E)}, \tag{12}$$

where

$$\mathcal{N}^* = \frac{\mathcal{M}_2\mathcal{K}_*^{1/2}(\varphi(b)-\varphi(0))^{\sigma_1-\rho+(1/2)}}{(2\sigma_1-1)^{1/2}\left(1-\mathcal{M}_1(1+\mathcal{D}_\gamma)-\mathcal{M}_1\hat{\mathcal{K}}_\mathcal{G}-\mathcal{M}_2\hat{\mathcal{R}}_\Delta\frac{\Gamma(\sigma_1)\Gamma(\rho)}{\Gamma(\rho+\sigma_1)}(\varphi(b)-\varphi(0))^{\sigma_1}\right)},$$

with $\mathcal{M}_1(1+\mathcal{D}_\gamma) + \mathcal{M}_1\hat{\mathcal{K}}_{\mathcal{G}} + \mathcal{M}_2\hat{\mathcal{R}}_\Delta \frac{\Gamma(\sigma_1)\Gamma(\rho)}{\Gamma(\rho+\sigma_1)}(\varphi(b)-\varphi(0))^{\sigma_1} < 1$ for every $\gamma = 1, 2, \ldots, \mathcal{H}$.

Since $\mathcal{D}$ is strongly continuous, we obtain

$$\|\mathcal{D}v^k - \mathcal{D}v^*\|_{L^2(\mathcal{J}_0, E)} \longrightarrow 0 \text{ as } k \to \infty.$$

Thus, we have

$$\|z^k - z^*\|_{\mathcal{PC}} \longrightarrow 0 \text{ as } k \to \infty,$$

this yields that $z^k \longrightarrow z^*$ in $\mathcal{PC}(E)$ as $k \to \infty$. Since $\mathcal{PC}(E) \subset L^1(\mathcal{J}_0, E)$, by using [X9] and Balder's theorem, we obtain

$$\epsilon = \lim_{k \to \infty} \sum_{\gamma=0}^{\mathcal{H}} \int_{t_\gamma}^{t_{\gamma+1}} \mathcal{L}(t, z^k(t), v^k(t)) dt$$

$$\geq \sum_{\gamma=0}^{\mathcal{H}} \int_{t_\gamma}^{t_{\gamma+1}} \mathcal{L}(t, z^*(t), v^*(t)) dt = \mathcal{J}(z^*, v^*) \geq \epsilon, \, \gamma = 0, 1, \ldots, \mathcal{H}.$$

Thus J attains its minimum at $v^* \in \mathcal{U}_{ad}$. □

6. Example

Consider the following φ-Hilfer fractional impulsive differential control system to verify the proposed results:

$$\begin{cases} {}^H\mathbb{D}_{t_\gamma^+}^{\sigma_1, \sigma_2; \varphi} z(t, \alpha) = z_{\alpha\alpha}(t, \alpha) + v(t, \alpha) + \frac{t e^{-t} z(t, \alpha)}{18(1 + |z(t, \alpha)|)}, \, t \in (0, 1] - \{t_1\}, \, \alpha \in [0, \pi], \\ I_{t_1^+}^{(1-\sigma_1)(1-\sigma_2); \varphi} z(t_1^+, \alpha) = z(t_1^-, \alpha) + \frac{1}{100} z(t_1^-, \alpha), \, \alpha \in [0, \pi], \\ I_{0^+}^{(1-\sigma_1)(1-\sigma_2); \varphi}[z(t, \alpha)]_{t=0} + \frac{1}{15} z(t, \alpha) = z_0(\alpha), \\ z(t, 0) = 0 = z(t, \pi), \end{cases} \quad (13)$$

with cost functional as

$$\mathcal{J}(z^v, v) = \sum_{\gamma=0}^{\mathcal{H}} \left[\int_{t_\gamma}^{t_{\gamma+1}} \int_0^\pi |z^v(t, \alpha)|^2 d\alpha \, dt + \int_{t_\gamma}^{t_{\gamma+1}} \int_0^\pi |v(t, \alpha)|^2 d\alpha \, dt \right]$$

subject to the problem (13), where $\gamma = 0, 1$, $\sigma_1 = 2/3$, $\sigma_2 = 1/4$ and $0 = t_0 < t_1 < t_2 = b$ with $t_1 = 0.5$, $b = 1$. Let $\varphi(t) = t$ and $E = \mathcal{T} = L^2([0, \pi])$. Define an operator $\mathcal{A} : \mathcal{D}(\mathcal{A}) \subseteq E \to E$ by $\mathcal{A}\psi = \psi''$ with

$$\mathcal{D}(\mathcal{A}) = \{\psi \in E : \psi, \psi' \text{ are absolutely continuous and } \psi'' \in E, \psi(0) = 0 = \psi(\pi)\}.$$

$\mathcal{A}$ has a discrete spectrum, the normalized eigenvectors $e_n(\alpha) = \sqrt{2/\pi} \sin(n\alpha)$ corresponding to eigenvalue are $-n^2$, $n \in \mathbb{N}$ and $\mathcal{A}$ generates an analytic semigroup $\{\mathcal{T}(t)\}_{t \geq 0}$ in E, which uniformly bounded and defined as

$$\mathcal{T}(t)\alpha = \sum_{n=1}^\infty e^{-n^2 t} \langle \alpha, e_n \rangle e_n, \, \alpha \in E,$$

with $\|\mathcal{T}(t)\| \leq e^{-t} \, \forall \, t \geq 0$. Thus, we choose $\mathcal{M} = 1$ that implies that $\sup_{t \in [0, \infty)} \|\mathcal{T}(t)\| = 1$ and [X1] is fulfilled. We obtain $\mathcal{M}_1 = 0.8161$ and $\mathcal{M}_2 = 0.7385$. The admissible controls set

$$\mathcal{U}_{ad} = \{v \in \mathcal{T} : \|v\| \in L^2([0, 1], \mathcal{T}) \leq 1\}.$$

Let $z(t)(\alpha) = z(t,\alpha)$ and the functions Δ, $\mathcal{I}_1$ and $\mathcal{G}$ are defined as

$$\Delta(t,z)(\alpha) = \frac{te^{-t}z(t,\alpha)}{18(1+|z(t,\alpha)|)}, \quad \mathcal{I}_1 = \frac{1}{100}z(t_1^-,\alpha), \quad \mathcal{G}(z)(\alpha) = \frac{1}{15}z(t,\alpha).$$

We obtain $\mathcal{K}_\Delta = \mathcal{R}_\Delta = 1/18$, $\hat{\mathcal{K}}_\mathcal{G} = 1/15$, $\mathcal{D}_1 = 1/100$ and

1. $\hat{\mathcal{O}} = \max[\mathcal{M}_1 \hat{\mathcal{K}}_\mathcal{G}, \mathcal{M}_1(1+\mathcal{D}_1)] = \max[0.0544, 0.8243] < 1$,
2. $\mathcal{M}_1 \hat{\mathcal{K}}_\mathcal{G} + \mathcal{M}_2 \mathcal{K}_\Delta \dfrac{\Gamma(\sigma_1)\Gamma(\rho)}{\Gamma(\rho+\sigma_1)}(\varphi(b) - \varphi(0))^{\sigma_1} = 0.1312 < 1$,
3. $\mathcal{M}_1(1+\mathcal{D}_1) + \mathcal{M}_1 \hat{\mathcal{K}}_\mathcal{G} + \mathcal{M}_2 \hat{\mathcal{R}}_\Delta \dfrac{\Gamma(\sigma_1)\Gamma(\rho)}{\Gamma(\rho+\sigma_1)}(\varphi(b) - \varphi(0))^{\sigma_1} = 0.9555 < 1$.

The system (13) can be transformed into (11) with the functional

$$\mathcal{J}(z^v, v) = \sum_{\gamma=0}^{\mathcal{H}} \int_{t_\gamma}^{t_{\gamma+1}} \left[\|z^v(t)\|^2 + \|v(t)\|_{\mathcal{T}}^2\right] dt.$$

All hypotheses of Theorems 3 and 4 are satisfied. Hence, the problem (13) has at least one optimal pair.

7. Discussion

The solvability and optimal control results for a class of φ-Hilfer fractional differential equations with impulses and nonlocal conditions have been investigated. Standard techniques combined with the notion of piecewise continuous mild solutions were used for the main results. Moreover, by using the minimizing sequence concept, we proved the optimal controls for deriving the optimality conditions. At end, we presented an illustrative example to provide the obtained theoretical results. In the forthcoming papers, as new direction, we intend to investigate the relaxation in nonconvex optimal control problems for a new class of φ-Hilfer fractional stochastic differential equations driven by the Rosenblatt process with non-instantaneous impulses [34,35].

Author Contributions: Conceptualization, A.D.; methodology, M.M.; validation, A.D.; formal analysis, R.D.; investigation, S.G.; writing—original draft preparation, S.G., A.D.; writing—review and editing, A.D., M.M.; supervision, A.D.; project administration, R.D. All authors have read and agreed to the published version of the manuscript.

Funding: This research received no external funding.

Institutional Review Board Statement: Not applicable.

Informed Consent Statement: Not applicable.

Data Availability Statement: Not applicable.

Acknowledgments: We are very thankful to the anonymous reviewers and associate editor for their constructive comments and suggestions which help us to improve the manuscript.

Conflicts of Interest: The authors declare no conflict of interest. The funders had no role in the design of the study; in the collection, analyses, or interpretation of data; in the writing of the manuscript, or in the decision to publish the results.

References

1. Zhou, Y.; Wang, J.; Zhang, L. *Basic Theory of Fractional Differential Equations*; World Scientific Publishing Co. Pte. Ltd.: Hackensack, NJ, USA, 2017.
2. Miller, K.S.; Ross, B. *An Introduction to the Fractional Calculus and Fractional Differential Equations*; A Wiley-Interscience Publication; John Wiley and Sons, Inc.: New York, NY, USA, 1993.
3. Kilbas, A.A.; Srivastava, H.M.; Trujillo, J.J. Theory and applications of fractional differential equations. In *North-Holland Mathematics Studies*; Elsevier Science B.V.: Amsterdam, The Netherlands, 2006; Volume 204.

4. Podlubny, I. Fractional differential equations. In *Mathematics in Science and Engineering*; Academic Press, Inc.: San Diego, CA, USA, 1999; Volume 198.
5. Dhayal, R.; Malik, M.; Abbas, S. Solvability and optimal controls of non-instantaneous impulsive stochastic fractional differential equation of order $q \in (1,2)$. *Stochastics* **2020**, *93*, 780–802. [CrossRef]
6. Hilfer, R. *Application of Fractional Calculus in Physics*; World Scientific: Singapore, 2000.
7. Wang, J.; Zhang, Y. Nonlocal initial value problems for differential equations with Hilfer fractional derivative. *Appl. Math. Comput.* **2015**, *266*, 850–859. [CrossRef]
8. Karthikeyan, K.; Debbouche, A.; Torres, D.F.M. Analysis of Hilfer fractional integro-differential equations with almost sectorial operators. *Fractal Fract.* **2021**, *5*, 22. [CrossRef]
9. Yang, M.; Wang, Q. Approximate controllability of Hilfer fractional differential inclusions with nonlocal conditions. *Math. Methods Appl. Sci.* **2017**, *40*, 1126–1138. [CrossRef]
10. Vijayakumar, V.; Udhayakumar, R. Results on approximate controllability for non-densely defined Hilfer fractional differential system with infinite delay. *Chaos Solitons Fractals* **2020**, *139*, 110019. [CrossRef]
11. Debbouche, A.; Antonov, V. Approximate controllability of semilinear Hilfer fractional differential inclusions with impulsive control inclusion conditions in Banach spaces. *Chaos Solitons Fractals* **2017**, *102*, 140–148. [CrossRef]
12. Sousa, J.V.C.; Oliveira, E.C. On the ψ-Hilfer fractional derivative. *Commun. Nonlinear Sci. Numer. Simul.* **2018**, *60*, 72–91. [CrossRef]
13. Sousa, J.V.C.; Kucche, K.D.; Oliveira, E.C. Stability of ψ-Hilfer impulsive fractional differential equations. *Appl. Math. Lett.* **2019**, *88*, 73–80. [CrossRef]
14. Kucche, K.D.; Kharade, J.P.; Sousa, J.V.C. On the nonlinear impulsive ψ-Hilfer fractional differential equations. *Math. Model. Anal.* **2020**, *25*, 642–660. [CrossRef]
15. Sousa, J.V.C.; Oliveira, E.C. On the Ulam-Hyers-Rassias stability for nonlinear fractional differential equations using the ψ-Hilfer operator. *J. Fixed Point Theory Appl.* **2018**, *20*, 96. [CrossRef]
16. Sousa, J.V.C.; Rodrigues, F.G.; Oliveira, E.C. Stability of the fractional Volterra integro-differential equation by means of ψ-Hilfer operator. *Math. Methods Appl. Sci.* **2019**, *42*, 3033–3043. [CrossRef]
17. Suechoei, A.; Sa Ngiamsunthorn, P. Existence uniqueness and stability of mild solutions for semilinear ψ-Caputo fractional evolution equations. *Adv. Differ. Equ.* **2020**, 114. [CrossRef]
18. Malti, A.I.N.; Benchohra, M.; Graef, J.R.; Lazreg, J.E. Impulsive boundary value problems for nonlinear implicit Caputo-exponential type fractional differential equations. *Electron. J. Qual. Theory Differ. Equ.* **2020**, *2020*, 1–17. [CrossRef]
19. Vadivoo, B.S.; Ramachandran, R.; Cao, J.; Zhang, H.; Li, X. Controllability analysis of nonlinear neutral-type fractional-order differential systems with state delay and impulsive effects. *Int. J. Control. Autom. Syst.* **2018**, *16*, 659–669. [CrossRef]
20. Aimene, D.; Baleanu, D.; Seba, D. Controllability of semilinear impulsive Atangana-Baleanu fractional differential equations with delay. *Chaos Solitons Fractals* **2019**, *128*, 51–57. [CrossRef]
21. Dhayal, R.; Malik, M.; Abbas, S. Existence, stability and controllability results of stochastic differential equations with non-instantaneous impulses. *Int. J. Control* **2020**, 1–12. [CrossRef]
22. Yu, X.; Debbouche, A.; Wang, J. On the iterative learning control of fractional impulsive evolution equations in Banach spaces. *Math. Methods Appl. Sci.* **2017**, *40*, 6061–6069. [CrossRef]
23. Ahmed, H.M.; El-Borai, M.M.; El-Owaidy, H.M.; Ghanem, A.S. Impulsive Hilfer fractional differential equations. *Adv. Differ. Equ.* **2018**, 226. [CrossRef]
24. Kucche, K.D.; Kharade, J.P. Analysis of impulsive ψ-Hilfer fractional differential equations. *Mediterr. J. Math.* **2020**, *17*, 163. [CrossRef]
25. Wang, J.; Zhou, Y. A class of fractional evolution equations and optimal controls. *Nonlinear Anal. Real World Appl.* **2011**, *12*, 262–272. [CrossRef]
26. Dhayal, R.; Malik, M.; Abbas, S.; Debbouche, A. Optimal controls for second-order stochastic differential equations driven by mixed-fractional Brownian motion with impulses. *Math. Methods Appl. Sci.* **2020**, *43*, 4107–4124. [CrossRef]
27. Debbouche, A.; Nieto, J.J. Sobolev type fractional abstract evolution equations with nonlocal conditions and optimal multi-controls. *Appl. Math. Comput.* **2014**, *245*, 74–85. [CrossRef]
28. Liu, S.; Wang, J. Optimal controls of systems governed by semilinear fractional differential equations with not instantaneous impulses. *J. Optim. Theory Appl.* **2017**, *174*, 455–473. [CrossRef]
29. Balasubramaniam, P.; Tamilalagan, P. The solvability and optimal controls for impulsive fractional stochastic integro-differential equations via resolvent operators. *J. Optim. Theory Appl.* **2017**, *174*, 139–155. [CrossRef]
30. Yan, Z.; Jia, X. Optimal controls of fractional impulsive partial neutral stochastic integro-differential syste Infin. Delay Hilbert Spaces. *Int. J. Control. Autom. Syst.* **2017**, *15*, 1051–1068. [CrossRef]
31. Harrat, A.; Nieto, J.J.; Debbouche, A. Solvability and optimal controls of impulsive Hilfer fractional delay evolution inclusions with Clarke subdifferential. *J. Comput. Appl. Math.* **2018**, *344*, 725–737. [CrossRef]
32. Zhou, Y.; Jiao, F. Existence of mild solutions for fractional neutral evolution equations. *Comput. Math. Appl.* **2010**, *59*, 1063–1077. [CrossRef]
33. Sakthivel, R.; Revathi, P.; Ren, Y. Existence of solutions for nonlinear fractional stochastic differential equations. *Nonlinear Anal. Theory Methods Appl.* **2013**, *81*, 70–86. [CrossRef]

34. Debbouche, A.; Nieto, J.J.; Torres, D.F.M. Optimal solutions to relaxation in multiple control problems of Sobolev type with nonlocal nonlinear fractional differential equations. *J. Optim. Theory Appl.* **2017**, *174*, 7–31. [CrossRef]
35. Debbouche, A.; Nieto, J.J. Relaxation in controlled systems described by fractional integro-differential equations with nonlocal control conditions. *Electron. J. Differ. Equ.* **2015**, *2015*, 1–18.

Article

Bilateral Tempered Fractional Derivatives

Manuel Duarte Ortigueira [1,*] and Gabriel Bengochea [2]

1 Centre of Technology and Systems-UNINOVA, NOVA School of Science and Technology of NOVA University of Lisbon, Quinta da Torre, 2829-516 Caparica, Portugal
2 Academia de Matemática, Universidad Autónoma de la Ciudad de México, Ciudad de México 04510, Mexico; gabriel.bengochea@uacm.edu.mx
* Correspondence: mdo@fct.unl.pt

Abstract: The bilateral tempered fractional derivatives are introduced generalising previous works on the one-sided tempered fractional derivatives and the two-sided fractional derivatives. An analysis of the tempered Riesz potential is done and shows that it cannot be considered as a derivative.

Keywords: tempered fractional derivative; one-sided tempered fractional derivative; bilateral tempered fractional derivative; tempered riesz potential

MSC: Primary 26A33; Secondary 34A08; 35R11

1. Introduction

In a recent paper [1], we presented a unified formulation for the one-sided Tempered Fractional Calculus, that includes the classic, tempered, substantial, and shifted fractional operators [2–9].

Here, we continue in the same road by presenting a study on the two-sided tempered operators that generalize and include the one-sided. The most interesting is the tempered Riesz potential that was proposed in analogy with the one-sided tempered derivatives [10]. However, a two-sided tempering was introduced before, in the study of the called variance gamma processes [11,12], in Statistical Physics for modelling turbulence, under the concept of truncated Lévy flight [8,13–17], and for defining the Regular Lévy Processes of Exponential type [2,10,18]. The tempered stable Lévy motion appeared in a previous work [19]. Meanwhile, the Feynman–Kac equation used in normal diffusion was generalized for anomalous diffusion and tempered [20,21]. These studies led to the introduction of the tempered Riesz derivative [14] and some applications. Sabzikar et al. [22] described a new variation on the fractional calculus which was called tempered fractional calculus and introduced the tempered fractional diffusion equation. The solutions to this equation are tempered stable probability densities, with semi-heavy tails that state a transition from power law to Gaussian. They proposed a new stochastic process model for turbulence, based on tempered fractional Brownian motion. Li et al. [23] designed a high order difference scheme for the tempered fractional diffusion equation on bounded domain. Their approach is based in properties of the tempered fractional calculus using first order Grünwald type difference approximations. Alternatively, Arshad et al. [24] proposed another difference scheme to solve time–space fractional diffusion equation where the Riesz derivative is approximated by means of a centered difference. They obtained Volterra integral equations which were approximated using the trapezoidal rule. For solving space–time tempered fractional diffusion-wave equation in finite domain another fourth-order technique was proposed in [25,26]. D'Ovidio et al. [27] presented fractional equations governing the distribution of reflecting drifted Brownian motions. In Zhang et al. [28] approximated the tempered Riemann–Liouville and Riesz derivatives by means of second-order difference operator. In [29] new computational methods for the tempered fractional

Laplacian equation were introduced, including the cases with the homogeneous and non-homogeneous generalized Dirichlet type boundary conditions. In [30], by means of a linear combination of the left and right normalized tempered Riemann–Liouville fractional operators, tempered fractional Laplacian (tempered Riesz fractional derivative) was defined as $(\Delta + \lambda)^{\beta/2}$. This operator was used to develop finite difference schemes to solve the tempered fractional Laplacian equation that governs the probability distribution function of the positions of particles. Similarly, Duo et al. [31] presented a finite difference method to discretize the d-dimensional (for $d \geq 1$) tempered integral fractional Laplacian $(-\Delta + \lambda)^{\alpha/2}$. By means of this approximation they resolved fractional Poisson problems. Hu et al. [32] present the implicit midpoint method for solving Riesz tempered fractional diffusion equation with a nonlinear source term. The Riesz tempered fractional derivative was worked in finite domain. An interesting application of the tempered Riesz derivative in solving the fractional Schrödinger equation was described in [33].

These works suggest us that the tempered Riesz derivative (TRD) is a very important operator. However, and despite such importance, there are no significative theoretical results about such operator. Furthermore, nobody has placed the question: is *the tempered Riesz derivative* really a derivative?

In this paper, we follow the work described in our previous paper [1] where a deep study on the tempered one-sided derivative was performed. Therefore, we intend here to enlarge the results we obtained previously by combining them with the two-sided derivatives studied in [34]. This approach intends to show that the TRD is not really a fractional derivative according to the criterion introduced in [35]. Instead, we propose a formulation for general tempered two-sided derivatives defined with the help of the Tricomi function [36].

The paper is outlined as follows. In Section 2.1 two preliminary descriptions are done: the one-sided tempered fractional derivatives (TFDs) and the two-sided (non tempered) fractional derivatives (TSFDs). The Riesz–Feller tempered derivatives are introduced and studied in Section 3. Their study in frequency domain shows that they should not be considered as derivatives. The bilateral tempered fractional derivatives (BTFDs) are studied in Section 4. Both versions, continuous- and discrete-time are considered and compared with Riesz-Feller's. Finally, some conclusions are drawn.

Remark 1. *We adopt here the assumptions in [1], namely*
- *We work on $\mathbb{R}$.*
- *We use the two-sided Laplace transform (LT):*

$$F(s) = \mathcal{L}[f(t)] = \int_{\mathbb{R}} f(t) e^{-st} dt, \tag{1}$$

where $f(t)$ is any function defined on $\mathbb{R}$ and $F(s)$ is its transform, provided that it has a non empty region of convergence (ROC).
- *The Fourier transform (FT), $\mathcal{F}[f(t)]$, is obtained from the LT through the substitution $s = i\kappa$, with $\kappa \in \mathbb{R}$.*

2. Preliminaries

2.1. The Unilateral Tempered Fractional Derivatives

The one-sided (unilateral) Tempered Fractional Derivatives TFD (UTFD) were formally introduced and studied in [1]. In Table 1 we depict the most important characteristics of the most interesting derivatives, namely the transfer function and corresponding region of convergence (ROC). The tempering parameter λ is assumed to be a nonnegative real number. We present only the stable derivatives. This stability manifests in the fact that the ROC of the LT of stable TFD include the imaginary axis. Therefore, the corresponding FT exist and are obtained by setting $s = i\kappa$. The ROC abscissa is $-\lambda$ in the causal (forward) and λ in the anti-causal (backward) cases. The parameter $\alpha \in \mathbb{R}$ is the derivative order and $N = \lfloor \alpha \rfloor$.

Table 1. Stable TFD with $\lambda \geq 0$.

Derivative	$_\lambda D^\alpha_{\pm a} f(t)$	LT	ROC
Forward Grünwald-Letnikov	$\lim_{h \to 0^+} h^{-\alpha} \sum_{n=0}^{\infty} \frac{(-\alpha)_n}{n!} e^{-n\lambda h} f(t - nh)$	$(s + \lambda)^\alpha$	$Re(s) > -\lambda$
Backward Grünwald-Letnikov	$\lim_{h \to 0^+} h^{-\alpha} \sum_{n=0}^{\infty} \frac{(-\alpha)_n}{n!} e^{-n\lambda h} f(t + nh)$	$(-s + \lambda)^\alpha$	$Re(s) < \lambda$
Regularised forward Liouville	$\int_0^\infty \left[f(t - \tau) - \varepsilon(\alpha) \sum_0^N \frac{(-1)^m f^{(m)}(t)}{m!} \tau^m \right] e^{-\lambda \tau} \frac{\tau^{-\alpha-1}}{\Gamma(-\alpha)} d\tau$	$(s + \lambda)^\alpha$	$Re(s) > -\lambda$
Regularised backward Liouville	$\int_0^\infty \left[f(t + \tau) - \varepsilon(\alpha) \sum_0^N \frac{f^{(m)}(t)}{m!} \tau^m \right] e^{-\lambda \tau} \frac{\tau^{-\alpha-1}}{\Gamma(-\alpha)} d\tau$	$(-s + \lambda)^\alpha$	$Re(s) < \lambda$

Relatively to [1], a complex factor in the backward derivatives was removed to keep coherence with the mathematical developments presented below. The corresponding LT was changed accordingly. Throughout the paper, we will use the designations "Grünwald–Letnikov" (GL) and "Liouville derivative" (L) for the cases corresponding to $\lambda = 0$.

2.2. The Two-Sided Fractional Derivatives

Definition 1. *In [34], we introduced formally a general two-sided fractional derivative (TSFD), $_0 D^\beta_\theta$, through its Fourier transform*

$$\mathcal{F}\left[_0 D^\beta_\theta f(x) \right] = |\kappa|^\beta e^{i \frac{\pi}{2} \theta \cdot sgn(\kappa)} F(\kappa), \tag{2}$$

where β and θ are any real numbers that we will call derivative order and asymmetry parameter, respectively.

The inverse Fourier transform computation of (2) is not important here (see, [34]). In Table 2 we present the most interesting definitions of the two-sided derivatives together with the corresponding Fourier transform. It is important to note that we present the regularised Riesz and Feller derivatives.

Table 2. TSFD ($\lambda = 0$).

Derivative	$_0 D^\beta_\theta f(t)$	FT
TSGL symmetric	$\lim_{h \to 0^+} h^{-\beta} \sum_{n=-\infty}^{+\infty} \frac{(-1)^n \Gamma(\beta+1)}{\Gamma(\frac{\beta}{2}-n+1)\Gamma(\frac{\beta}{2}+n+1)} f(x - nh)$	$|\kappa|^\beta$
TSGL anti-symmetric	$\lim_{h \to 0^+} h^{-\beta} \sum_{n=-\infty}^{+\infty} \frac{(-1)^n \Gamma(\beta+1)}{\Gamma(\frac{\beta+1}{2}-n+1)\Gamma(\frac{\beta-1}{2}+n+1)} f(x - nh)$	$i|\kappa|^\beta sgn(\kappa)$
TSGL general	$\lim_{h \to 0^+} h^{-\beta} \sum_{n=-\infty}^{+\infty} \frac{(-1)^n \Gamma(\beta+1)}{\Gamma(\frac{\beta+\theta}{2}-n+1)\Gamma(\frac{\beta-\theta}{2}+n+1)} f(x - nh)$	$|\kappa|^\beta e^{i \frac{\pi}{2} \theta \cdot sgn(\kappa)}$
Riesz derivative	$\frac{1}{2 \cos(\beta \frac{\pi}{2}) \Gamma(-\beta)} \int_{-\infty}^{\infty} \left[f(x - y) - 2 \sum_{k=0}^{M} \frac{f^{(2k)}(x)}{(2k)!} y^{2k} \right] |y|^{-\beta-1} dy$	$|\kappa|^\beta$
Feller derivative	$\frac{1}{2 \sin(\beta \frac{\pi}{2}) \Gamma(-\beta)} \int_{-\infty}^{\infty} \left[f(x - y) - 2 \sum_{k=0}^{M} \frac{f^{(2k+1)}(x)}{(2k+1)!} y^{2k+1} \right] |y|^{-\beta-1} sgn(y) dy$	$i|\kappa|^\beta sgn(\kappa)$
Riesz-Feller potential	$\frac{1}{2 \sin(\beta \pi) \Gamma(-\beta)} \int_{\mathbb{R}} f(x - y) \sin[(\beta + \theta \cdot sgn(y)) \pi / 2] |y|^{-\beta-1} dy$	$|\kappa|^\beta e^{i \frac{\pi}{2} \theta \cdot sgn(\kappa)}$

Some properties of this definition can be drawn [34,37,38]. Here we are mainly interested in the folowing

1. **Eigenfunctions**
   Let $f(x) = e^{i\kappa x}$, $\kappa, x \in \mathbb{R}$. Then
   $$_0D_\theta^\beta e^{i\kappa x} = |\kappa|^\beta e^{i\frac{\pi}{2}\theta \cdot sgn(\kappa)} e^{i\kappa x}, \tag{3}$$
   meaning that the sinusoids are the eigenfunctions of the TSFD.

2. **The Liouville and GL derivatives as particular cases**
   With $\theta = \pm\beta$ we obtain the forward (left) (+) and backward (−) Liouville one-sided derivatives:
   $$\mathcal{F}\left[_0D_{\pm\beta}^\beta f(x)\right] = (\pm\kappa)^\beta F(\kappa). \tag{4}$$

3. **The Riesz and Feller derivatives as special cases**
   $$\mathcal{F}\left[_0D_0^\beta f(x)\right] = |\kappa|^\beta F(\kappa), \tag{5}$$
   and
   $$\mathcal{F}\left[_0D_1^\beta f(x)\right] = i|\kappa|^\beta \cdot sgn(\kappa) F(\kappa). \tag{6}$$

4. **Relations involving the sum/difference of Liouville derivatives [39]**
   Let $\kappa, \beta \in \mathbb{R}$. It is a simple task to show that
   $$|\kappa|^\beta = \frac{(i\kappa)^\beta + (-i\kappa)^\beta}{2\cos(\beta\frac{\pi}{2})}, \quad \beta \neq 1,3,5\cdots \tag{7}$$
   $$i|\kappa|^\beta sgn(\kappa) = \frac{(i\kappa)^\beta - (-i\kappa)^\beta}{2\sin(\beta\frac{\pi}{2})}, \quad \beta \neq 2,4,6\cdots \tag{8}$$
   which means that the Riesz derivative is, aside a constant, equal to the sum of the left and right Liouville derivatives. Similarly, the Feller derivative is the difference. Then,
   $$_0D_0^\beta = \frac{_0D_\beta^\beta + _0D_{-\beta}^\beta}{2\cos(\beta\frac{\pi}{2})}, \quad \beta \neq 1,3,5\cdots \tag{9}$$
   $$_0D_1^\beta = \frac{_0D_\beta^\beta - _0D_{-\beta}^\beta}{2\sin(\beta\frac{\pi}{2})}, \quad \beta \neq 2,4,6\cdots \tag{10}$$

5. **Relations involving the composition of Liouville derivatives [34]**
   The composition of the GL, or L, derivatives in (4) is defined by:
   $$\mathcal{F}\left[_0D_{\beta_1}^{\beta_1} {_0D_{-\beta_2}^{\beta_2}} f(x)\right] = (i\kappa)^{\beta_1}(-i\kappa)^{\beta_2} F(\kappa). \tag{11}$$
   Setting $\beta = \beta_1 + \beta_2$ and $\theta = \beta_1 - \beta_2$ we obtain
   $$\Psi_\theta^\beta(\kappa) = (i\kappa)^{\beta_1}(-i\kappa)^{\beta_2} = |\kappa|^\beta e^{i\frac{\pi}{2}\theta \cdot sgn(\kappa)}, \tag{12}$$
   showing that any bilateral fractional derivative can be considered as the composition of a forward and a backward GL, or L, derivatives.

6. **The TSFD as a linear combination of Riesz and Feller derivatives [34]**
   $$_0D_\theta^\beta f(x) = \cos\left(\frac{\pi}{2}\theta\right) {_0D_0^\beta} f(x) + \sin\left(\frac{\pi}{2}\theta\right) {_0D_1^\beta} f(x). \tag{13}$$
   Therefore, any TSFD can be expressed as a linear combinations of pairs: causal/anti-causal GL, or L, or Riesz/Feller derivatives.

3. Riesz–Feller Tempered Derivatives

The Riesz tempered potential has been used by several authores as referred in Section 1. Here, we will deduce its general regularised form from the TFD in Section 2.1 while using the relation (9).

Definition 2. *We define the tempered Riesz derivative by:*

$$_\lambda D_0^\beta = \frac{_\lambda D_\beta^\beta + _\lambda D_{-\beta}^\beta}{2\cos(\beta\frac{\pi}{2})} \qquad \beta \neq 1,3,5\cdots \qquad (14)$$

This definition allows us to state that

Theorem 1.

$$_\lambda D_0^\beta f(x) = \frac{1}{2\Gamma(-\beta)\cos(\beta\frac{\pi}{2})} \int_{-\infty}^\infty \left[f(x-\tau) - \sum_{m=0}^M \frac{f^{(2m)}(x)}{(2m)!} \tau^{2m} \right] e^{-\lambda|\tau|} |\tau|^{-\beta-1} d\tau, \qquad (15)$$

for $2M < \beta < 2M+2$, $M \in \mathbb{Z}^+$.

Remark 2. *The integer order case leads to a singular situation that we can solve using the relations introduced in [34]. We will not do it here.*

Proof. We only have to insert the expressions from Table 1 into (14). Let $N = \lfloor \beta \rfloor$ If we use the Liouville derivatives, we obtain:

$$_\lambda D_0^\beta f(x) = \frac{1}{2\Gamma(-\beta)\cos(\beta\frac{\pi}{2})} \int_0^\infty \left[f(x-\tau) - \varepsilon(\beta)\sum_{m=0}^N \frac{(-1)^m f^{(m)}(x)}{m!} \tau^m \right] e^{-\lambda\tau} \tau^{-\beta-1} d\tau$$

$$+ \frac{1}{2\Gamma(-\beta)\cos(\beta\frac{\pi}{2})} \int_0^\infty \left[f(x+\tau) - \varepsilon(\beta)\sum_0^N \frac{(+1)^m f^{(m)}(x)}{m!} \tau^m \right] e^{-\lambda\tau} \tau^{-\beta-1} d\tau$$

or

$$_\lambda D_0^\beta f(x) = \frac{1}{2\Gamma(-\beta)\cos(\beta\frac{\pi}{2})}$$

$$\int_0^\infty \left\{ f(x-\tau) + f(x+\tau) - \varepsilon(\beta) \left[\sum_0^N \frac{(-1)^m f^{(m)}(x)}{m!} \tau^m + \sum_{m=0}^N \frac{f^{(m)}(x)}{m!} \tau^m \right] \right\} e^{-\lambda|\tau|} |\tau|^{-\beta-1} d\tau.$$

The odd terms in the inner summation are null. Therefore,

$$_\lambda D_0^\beta f(x) = \frac{1}{2\Gamma(-\beta)\cos(\beta\frac{\pi}{2})}$$

$$\int_0^\infty \left\{ f(x-\tau) + f(x+\tau) - 2\varepsilon(\beta) \sum_{m=0}^M \frac{f^{(2m)}(x)}{(2m)!} \tau^{2m} \right\} e^{-\lambda|\tau|} |\tau|^{-\beta-1} d\tau.$$

As the integrand is an even function, we are led to (15). □

In which concerns the Laplace and Fourier transforms, we remark that

$$\mathcal{L}\left[_\lambda D_0^\beta f(x) \right] = \frac{(s+\lambda)^\beta + (-s+\lambda)^\beta}{2\cos(\beta\frac{\pi}{2})} F(s),$$

for $|Re(s)| < \lambda$, meaning that the ROC is a vertical strip that contains the imaginary axis, $s = i\kappa$. Therefore, as $(\pm i\kappa + \lambda)^\beta = |\kappa^2 + \lambda^2|^{\frac{\beta}{2}} e^{\pm i\beta \arctan(\frac{\kappa}{\lambda})}$, and using relation (7), we obtain

$$\mathcal{F}\left[_\lambda D_0^\beta f(x)\right] = \frac{|\kappa^2 + \lambda^2|^{\frac{\beta}{2}} \cos\left(\beta \arctan(\frac{\kappa}{\lambda})\right)}{\cos(\beta \frac{\pi}{2})} F(i\kappa), \qquad (16)$$

that is coherent with the usual Riesz derivative ($\lambda = 0$).

Definition 3. *Similarly to the Riesz case, we use the relation (10) to find expressions for the tempered Feller derivative that we can define through*

$$_\lambda D_0^\beta = \frac{_\lambda D_\beta^\beta - _\lambda D_{-\beta}^\beta}{2 \sin(\beta \frac{\pi}{2})}, \qquad \beta \neq 2, 4, 6 \cdots \qquad (17)$$

Theorem 2. *The tempered Feller derivative is given by:*

$$_\lambda D_0^\beta f(x) = \frac{1}{2\Gamma(-\alpha) \sin(\beta \frac{\pi}{2})} \int_{-\infty}^{\infty} \left[f(x - \tau) - \sum_{m=0}^{M} \frac{f^{(2m+1)}(x)}{(2m+1)!} \tau^{(2m+1)}\right] e^{-\lambda |\tau|} |\tau|^{-\beta-1} d\tau, \qquad (18)$$

for $2M + 1 < \beta < 2M + 3$.

The proof is similar to the Riesz derivative. Therefore we omit it.
Now, the corresponding Laplace transform is

$$\mathcal{L}\left[_\lambda D_0^\beta f(x)\right] = \frac{(s + \lambda)^\beta - (-s + \lambda)^\beta}{2 \sin(\beta \frac{\pi}{2})},$$

for $|Re(s)| < \lambda$. Therefore, using relation (8), we obtain

$$\mathcal{F}\left[_\lambda D_0^\beta f(x)\right] = i \frac{|\kappa^2 + \lambda^2|^{\frac{\beta}{2}} \sin\left(\beta \arctan(\frac{\kappa}{\lambda})\right)}{\sin(\beta \frac{\pi}{2})} F(\kappa), \qquad (19)$$

that is coherent with the usual Feller derivative ($\lambda = 0$). In fact $\lim_{\lambda \to 0^+} \sin(\beta \arctan(\frac{\kappa}{\lambda})) = \sin[\beta \frac{\pi}{2} sgn(\kappa)]$.

Remark 3. *These procedures and the TSGL derivative (3) suggest that the GL type tempered Riesz–Feller derivatives should read*

$$_\lambda D_0^\beta f(x) = \lim_{h \to 0^+} h^{-\beta} \sum_{n=-\infty}^{+\infty} \frac{(-1)^n \Gamma(\beta + 1)}{\Gamma(\frac{\beta+\theta}{2} - n + 1) \Gamma(\frac{\beta-\theta}{2} + n + 1)} e^{-\lambda |n| h} f(x - nh). \qquad (20)$$

We will not study it, since it leads to the results stated above.

The relation (13) allows us to obtain the general tempered Riesz–Feller derivatives. We only have to insert there the expressions (14) and (18). Proceeding as in [34] we obtain:

Definition 4. *Let $\beta \in \mathbb{R} \setminus \mathbb{Z}$ and $f(x)$ in $L_1(\mathbb{R})$ or in $L_2(\mathbb{R})$. The generalised TSFD is defined by*

$$_\lambda \mathcal{D}_\theta^\beta f(x) := \frac{1}{2 \sin(\beta \pi) \Gamma(-\beta)} \int_\mathbb{R} f(x - \tau) \sin[(\beta + \theta \cdot sgn(\tau))\pi/2] e^{-\lambda |\tau|} |\tau|^{-\beta-1} d\tau. \qquad (21)$$

In terms of the Fourier transform, we have from (13)

$$\mathcal{F}\left[{}_\lambda D_\theta^\beta f(x)\right] = 2\left|\kappa^2 + \lambda^2\right|^{\frac{\beta}{2}} \left[\frac{\cos\left(\theta\frac{\pi}{2}\right)\cos\left(\beta\arctan\left(\frac{\kappa}{\lambda}\right)\right)}{\cos\left(\beta\frac{\pi}{2}\right)} + i\frac{\sin\left(\theta\frac{\pi}{2}\right)\sin\left(\beta\arctan\left(\frac{\kappa}{\lambda}\right)\right)}{\sin\left(\beta\frac{\pi}{2}\right)}\right] F(\kappa). \quad (22)$$

Remark 4. *It is important to note that none of these operators, tempered Riesz and Feller, and the general Riesz–Feller, can be considered as fractional derivatives. This is easy to see, for example, from (16) that*

$$_\lambda D_0^{\alpha+\beta} f(x) \neq {}_\lambda D_0^\alpha {}_\lambda D_0^\beta f(x),$$

for any pairs $\alpha, \beta \in \mathbb{R}$, since

$$2\left|\kappa^2 + \lambda^2\right|^{\frac{\alpha+\beta}{2}} \cos\left((\alpha+\beta)\arctan\left(\frac{\kappa}{\lambda}\right)\right) \neq \qquad (23)$$
$$2\left|\kappa^2 + \lambda^2\right|^{\frac{\alpha}{2}} \cos\left[\alpha\arctan\left(\frac{\kappa}{\lambda}\right)\right] \cdot 2\left|\kappa^2 + \lambda^2\right|^{\frac{\beta}{2}} \cos\left[\beta\arctan\left(\frac{\kappa}{\lambda}\right)\right].$$

These considerations show that although appealing this way into bilateral tempered fractional derivatives is not correct, since we do not obtain effectively derivatives according to the criteria stated in [35]. In Figure 1, we observe the effect of the tempering on the spectra and on the time kernel corresponding to $\beta = -1.8$ and $\lambda = 0, 0.25, 0.5, 0.75$.

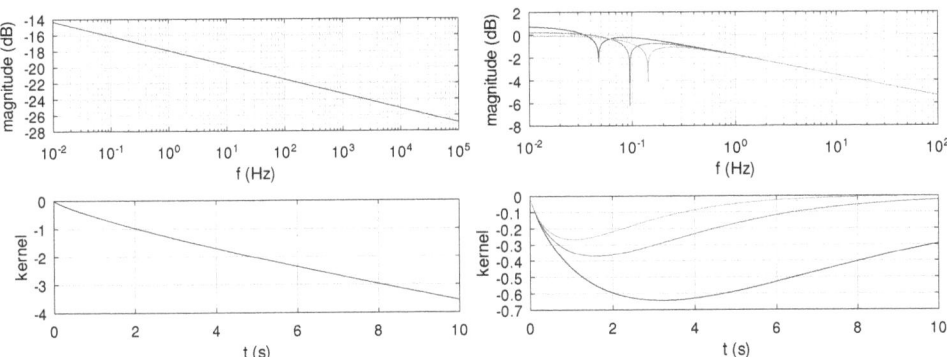

Figure 1. Frequency responses and kernels of Riesz potential ($\beta = -1.8$) without and with tempering ($\lambda = 0.25, 0.5, 0.75$).

4. Bilateral Tempered Fractional Derivatives

Above, we profit the fact that Riesz and Feller derivatives are expressed as sum and difference of one-sided derivatives. However, such approach was not successful, attending to the characteristics of the obtained operators that do not make them derivatives. Anyway, there is an alternative approach.

Definition 5. *We define the Bilateral Tempered Fractional Derivatives (BTFD), ${}_\lambda D_\theta^\alpha$, as a composition of forward and backward unilateral TFD derivatives, Liouville or Grünwald–Letnikov. Let a, b, α, and θ be real numbers, such that $\alpha = a + b$ and $\theta = a - b$. Then*

$$_\lambda D_\theta^\alpha f(x) = {}_\lambda D_a^a\left[{}_\lambda D_{-b}^b f(x)\right], \qquad (24)$$

or, using the Fourier transform:

$$\mathcal{F}[{}_\lambda D_\theta^\alpha f(x)] = (i\kappa + \lambda)^a (-i\kappa + \lambda)^b \qquad (25)$$
$$= \left|\kappa^2 + \lambda^2\right|^{\frac{\alpha}{2}} e^{i\theta \arctan\left(\frac{\kappa}{\lambda}\right)} F(\kappa).$$

It is important to note that $\lim_{\lambda \to 0^+} \arctan(\frac{\kappa}{\lambda}) = \frac{\pi}{2} \text{sgn}(\kappa)$.

Let
$$\lambda \psi_\theta^\alpha(t) = \mathcal{F}^{-1}[\lambda \Psi_\theta^\alpha(\omega)], \tag{26}$$

and
$$T(\alpha, \theta, 2\lambda|t|) = \frac{1}{\Gamma\left(-\frac{\alpha+\text{sgn}(t)\theta}{2}\right)\Gamma\left(\frac{\alpha-\text{sgn}(t)\theta}{2}\right)} \int_0^\infty e^{-2\lambda|t|u} u^{-\frac{\alpha+\text{sgn}(t)\theta}{2}-1}(u+1)^{-\frac{\alpha-\text{sgn}(t)\theta}{2}-1} du, \tag{27}$$

closely related (aside a factor) with the Tricomi function [36]. Then

Theorem 3. *For $\alpha, \beta < 0$,*
$$\lambda \psi_\theta^\alpha(t) = e^{-\lambda|t|}|t|^{-\alpha-1} T(\alpha, \theta, 2\lambda|t|). \tag{28}$$

Proof. Suppose that $a, b < 0$. As
$$\int_0^\infty f(t+\tau) e^{-\lambda \tau} \frac{\tau^{-a-1}}{\Gamma(-a)} d\tau = \int_{-\infty}^0 f(t-\tau) e^{\lambda \tau} \frac{(-\tau)^{-a-1}}{\Gamma(-a)} d\tau,$$

then
$$\lambda D_\theta^\alpha f(t) = \left[e^{-\lambda t} \frac{t^{-a-1}}{\Gamma(-a)} \varepsilon(t)\right] * \left[e^{\lambda t} \frac{(-t)^{-b-1}}{\Gamma(-b)} \varepsilon(-t)\right] * f(t), \tag{29}$$

where $*$ denotes the usual convolution. Let
$$\lambda \psi_\theta^\alpha(t) = \left[e^{-\lambda t} \frac{t^{-a-1}}{\Gamma(-a)} \varepsilon(t)\right] * \left[e^{\lambda t} \frac{(-t)^{-b-1}}{\Gamma(-b)} \varepsilon(-t)\right].$$

Hence
$$\lambda \psi_\theta^\alpha(t) = \int_0^\infty e^{-\lambda \tau} \frac{\tau^{-a-1}}{\Gamma(-a)} e^{\lambda(t-\tau)} \frac{(\tau-t)^{-b-1}}{\Gamma(-b)} \varepsilon(\tau-t) d\tau.$$

We have two possibilities

1. $t \geq 0$
$$\lambda \psi_\theta^\alpha(t) = \int_t^\infty e^{-\lambda \tau} \frac{\tau^{-a-1}}{\Gamma(-a)} e^{\lambda(t-\tau)} \frac{(\tau-t)^{-b-1}}{\Gamma(-b)} d\tau = \int_0^\infty e^{-\lambda(\tau+t)} \frac{(\tau+t)^{-a-1}}{\Gamma(-a)} e^{\lambda(-\tau)} \frac{\tau^{-b-1}}{\Gamma(-b)} d\tau$$

2. $t < 0$
$$\lambda \psi_\theta^\alpha(t) = \int_0^\infty e^{-\lambda \tau} \frac{\tau^{-a-1}}{\Gamma(-a)} e^{\lambda(t-\tau)} \frac{(\tau-t)^{-b-1}}{\Gamma(-b)} d\tau = \int_0^\infty e^{-\lambda \tau} \frac{\tau^{-a-1}}{\Gamma(-a)} e^{-\lambda(|t|+\tau)} \frac{(\tau+|t|)^{-b-1}}{\Gamma(-b)} d\tau$$

Setting $a = \frac{\alpha+\theta}{2}$ and $b = \frac{\alpha-\theta}{2}$ we can write
$$\lambda \psi_\theta^\alpha(t) = \frac{e^{-\lambda|t|}}{\Gamma(-\frac{\alpha+\text{sgn}(t)\theta}{2})\Gamma(-\frac{\alpha-\text{sgn}(t)\theta}{2})} \int_0^\infty e^{-2\lambda \tau} \tau^{-\frac{\alpha+\text{sgn}(t)\theta}{2}-1}(\tau+|t|)^{-\frac{\alpha-\text{sgn}(t)\theta}{2}-1} d\tau$$
$$= \frac{|t|^{-\alpha-1}}{\Gamma(-\frac{\alpha+\text{sgn}(t)\theta}{2})\Gamma(-\frac{\alpha-\text{sgn}(t)\theta}{2})} \int_0^\infty e^{-\lambda|t|(1+2\frac{\tau}{|t|})} \left(\frac{\tau}{|t|}\right)^{-\frac{\alpha+\text{sgn}(t)\theta}{2}-1} \left(\frac{\tau}{|t|}+1\right)^{-\frac{\alpha-\text{sgn}(t)\theta}{2}-1} \frac{d\tau}{|t|},$$

and
$$\lambda \psi_\theta^\alpha(t) = \frac{e^{-\lambda|t|}|t|^{-\alpha-1}}{\Gamma(-\frac{\alpha+\text{sgn}(t)\theta}{2})\Gamma(-\frac{\alpha-\text{sgn}(t)\theta}{2})} \int_0^\infty e^{-2\lambda|t|u} u^{-\frac{\alpha+\text{sgn}(t)\theta}{2}-1}(u+1)^{-\frac{\alpha-\text{sgn}(t)\theta}{2}-1} du. \tag{30}$$

□

Remark 5. With (29) we can write

$$_\lambda D_\theta^\alpha f(t) = \int_{-\infty}^{\infty} f(t-\tau) e^{-\lambda|\tau|} |\tau|^{-\alpha-1} T(\alpha, \theta, 2\lambda|\tau|) d\tau, \qquad (31)$$

that is valid for $\alpha \leq 0$. We can extend its validity for $\alpha > 0$, through a regularization as shown above in Section 4. It is important to note the similarity between (31) and (15).

Another version of this derivative can be obtained from the tempered unilateral GL derivatives in Table 1. It has the advantage of not needing any regularization.

Theorem 4. For any $\alpha, \theta \in \mathbb{R}$,

$$_\lambda D_\theta^\alpha f(t) = \lim_{h \to 0^+} h^{-\alpha} \sum_{m=-\infty}^{\infty} T_m(\alpha, \theta, 2\lambda h) e^{-|m|\lambda h} f(t - mh), \qquad (32)$$

where $T_m(\alpha, \beta, 2\lambda h)$ is defined below (37).

Proof. We have successively

$$g(t) = \sum_{n=0}^{\infty} \frac{(-a)_n}{n!} e^{-n\lambda h} \sum_{k=0}^{\infty} \frac{(-b)_k}{k!} e^{-k\lambda h} f(t - (n-k)h)$$

$$= \sum_{m=-\infty}^{\infty} \left[\sum_{n=\max(0,m)}^{\infty} e^{-2n\lambda h} \frac{(-a)_n}{n!} \frac{(-b)_{n-m}}{(n-m)!} e^{(m-2n)\lambda h} \right] f(t - mh).$$

Let us work out the series

$$\sum_{n=\max(m,0)}^{\infty} \frac{(-a)_n}{n!} \frac{(-b)_{n-m}}{(n-m)!} e^{(m-2n)\lambda h}.$$

For $m \geq 0$

$$\sum_{n=\max(m,0)}^{\infty} \frac{(-a)_n}{n!} \frac{(-b)_{n-m}}{(n-m)!} e^{(m-2n)\lambda h} = \sum_{n=0}^{\infty} \frac{(-a)_{n+m}}{(n+m)!} \frac{(-b)_n}{n!} e^{(-m-2n)\lambda h}. \qquad (33)$$

Therefore,

$$\sum_{n=\max(m,0)}^{\infty} \frac{(-a)_n}{n!} \frac{(-b)_{n-m}}{(n-m)!} e^{(-2n+m)\lambda h} = \begin{cases} \sum_{n=0}^{\infty} \frac{(-a)_{n+m}}{(n+m)!} \frac{(-b)_n}{n!} e^{(-m-2n)\lambda h}, & m \geq 0 \\ \sum_{n=0}^{\infty} \frac{(-a)_n}{n!} \frac{(-b)_{n-m}}{(n-m)!} e^{(m-2n)\lambda h}, & m < 0 \end{cases} \qquad (34)$$

Using the relations $(-a)_{n+|m|} = (-a)_{|m|}(-a+|m|)_n$ and $(-b)_{n+|m|} = (-b)_{|m|}(-b+|m|)_n$ and simplifying, we get

$$\begin{cases} e^{-m\lambda h} \dfrac{(-a)_m}{m!} \sum_{n=0}^{\infty} \dfrac{(-a+m)_n}{(m+1)_n} \dfrac{(-b)_n}{n!} e^{-2n\lambda h}, & m \geq 0 \\ e^{-|m|\lambda h} \dfrac{(-b)_{|m|}}{|m|!} \sum_{n=0}^{\infty} \dfrac{(-b+|m|)_n}{(|m|+1)_n} \dfrac{(-a)_n}{n!} e^{-2n\lambda h}, & m < 0. \end{cases} \qquad (35)$$

From this relation, we define a new discrete function $T_m(a,b,2\lambda h)$ by

$$T(a,b,2\lambda h) = \begin{cases} \dfrac{(-a)_m}{m!} \sum_{n=0}^{\infty} \dfrac{(-a+m)_n}{(m+1)_n} \dfrac{(-b)_n}{n!} e^{-2n\lambda h}, & m \geq 0 \\ \dfrac{(-b)_{|m|}}{|m|!} \sum_{n=0}^{\infty} \dfrac{(-b+|m|)_n}{(|m|+1)_n} \dfrac{(-a)_n}{n!} e^{-2n\lambda h}, & m < 0 \end{cases} \tag{36}$$

Therefore,

$$g(t) = \sum_{m=-\infty}^{\infty} T_m(a,b,2\lambda h) e^{-|m|\lambda h} f(t-mh).$$

It is interesting to note that $T_{-m}(a,b,2\lambda h) = T_m(b,a,2\lambda h)$. Setting $\alpha = a+b$ and $\theta = a-b$, we obtain

$$T_m(\alpha,\theta,2\lambda h) = \begin{cases} \dfrac{(-\frac{\alpha+\theta}{2})_m}{m!} \sum_{n=0}^{\infty} e^{-2n\lambda h} \dfrac{(-\frac{\alpha+\theta}{2}+m)_n}{(m+1)_n} \dfrac{(-\frac{\alpha-\theta}{2})_n}{n!} & m \geq 0 \\ \dfrac{(-\frac{\alpha-\theta}{2})_{|m|}}{|m|!} \sum_{n=0}^{\infty} e^{-2n\lambda h} \dfrac{(-\frac{\alpha-\theta}{2}+|m|)_n}{(|m|+1)_n} \dfrac{(-\frac{\alpha+\theta}{2})_n}{n!} & m < 0. \end{cases}$$

Then

$$T_m(\alpha,\theta,2\lambda h) = T_{-m}(\alpha,-\theta,2\lambda h), \quad m \in \mathbb{Z}$$

and consequently,

$$T_m(\alpha,\theta,2\lambda h) = \dfrac{(-\frac{\alpha+\theta}{2})_{|m|}}{|m|!} \sum_{n=0}^{\infty} e^{-2n\lambda h} \dfrac{(-\frac{\alpha+\theta}{2}+|m|)_n}{(|m|+1)_n} \dfrac{(-\frac{\alpha-\theta}{2})_n}{n!}, \tag{37}$$

for any integer m. □

Remark 6. *The similarity of (37) and (27) must be noted.*
We can give a more symmetric form of the summation in (37) using a Pfaff transformation, but it seems not to be of particular interest.

To verify the coherence of this result, we note that:
1. The second term in (37) is the Hypergeometric function;
2. If $\lambda = 0$, using a well-known property of the Hypergeometric function, we have

$$\sum_{n=0}^{\infty} \dfrac{(-\frac{\alpha+\theta}{2}+|m|)_n}{(|m|+1)_n} \dfrac{(-\frac{\alpha-\theta}{2})_n}{n!} = \dfrac{\Gamma(1+\alpha)|m|!}{\Gamma(\frac{\alpha+\theta}{2}+1)\Gamma(\frac{\alpha-\theta}{2}+|m|+1)},$$

and,

$$T_m(\alpha,\theta,0) = \dfrac{(-\frac{\alpha+\theta}{2})_{|m|}}{|m|!} \dfrac{\Gamma(1+\alpha)|m|!}{\Gamma(\frac{\alpha+\theta}{2}+1)\Gamma(\frac{\alpha-\theta}{2}+|m|+1)}. \tag{38}$$

3. As $(1-z)_n = (-1)^n \Gamma(z)/\Gamma(z-n)$,

$$(-\dfrac{\alpha+\theta}{2})_{|m|} = (-1)^m \dfrac{\Gamma(1+\frac{\alpha+\theta}{2})}{\Gamma(\frac{\alpha+\theta}{2}-|m|+1)},$$

and

$$T_m(\alpha,\theta,0) = (-1)^m \dfrac{\Gamma(1+\alpha)}{\Gamma(\frac{\alpha+\theta}{2}-|m|+1)\Gamma(\frac{\alpha-\theta}{2}+|m|+1)}, \tag{39}$$

in agreement with (20). Another interesting result can be obtained by dividing (37) by (38) to obtain the factor

$$Q_m(\alpha, \theta, 2\lambda h) = \frac{\Gamma(\frac{\alpha+\theta}{2}+1)\Gamma(\frac{\alpha-\theta}{2}+|m|+1)}{\Gamma(1+\alpha)|m|!} \sum_{n=0}^{\infty} e^{-2n\lambda h} \frac{(-\frac{\alpha+\theta}{2}+|m|)_n}{(|m|+1)_n} \frac{(-\frac{\alpha-\theta}{2})_n}{n!}, \quad (40)$$

that expresses the "deviation" of the BTFD from the tempered Riesz–Feller derivative (22). In Figure 2 we illustrate the behaviour of this factor for two derivative orders, $\alpha = \pm 0.5$ and three values of the tempering exponent, $\lambda = 0.25, 0.5, 1$ with $\theta = 0.4$. It is important to note that

- In the derivative case, Q_m increases slowly and monotonuously with m, contributing for an enlargement of the kernel duration;
- In the anti-derivative case, Q_m decreases slowly and monotonuously to zero with increasing m reducing the kernel duration and consequently the memory of the operator.

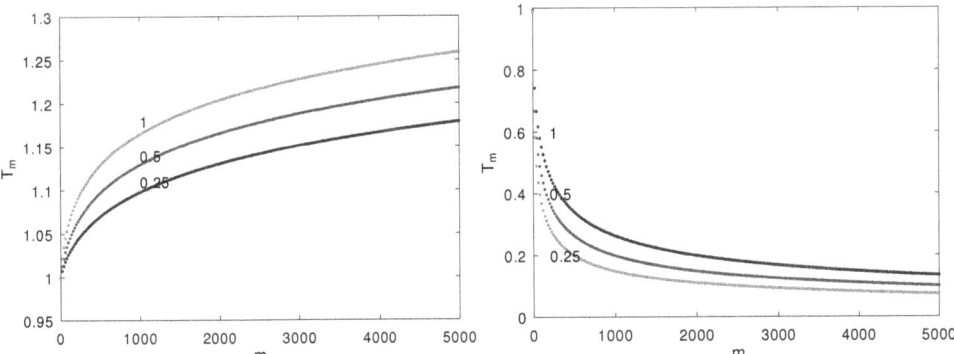

Figure 2. The Q-factor for $\beta = \pm 0.5; \theta = 0.4$, and $\lambda = 0.25, 0.5, 1$.

Knowing that the first term in (37) tends asymptotically to $\frac{1}{|m|^{\alpha+1}}$ [39], it will be interesting to study the behaviour of the summation term. In Figure 3 we examplify its variation for positive and negative derivative orders for three values of λ.

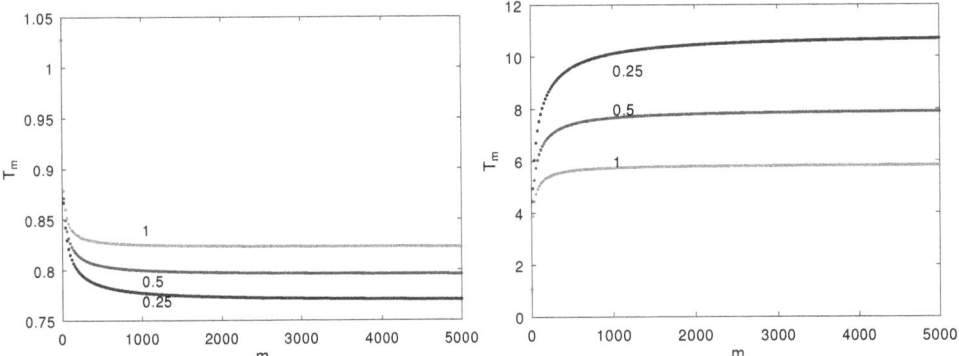

Figure 3. The summation factor in (37) for $\beta = \pm 0.5; \theta = 0.4$, and $\lambda = 0.25, 0.5, 1$.

As seen, it seems to approach a constant depending on λ.

Can We Consider the BTFD as Fractional Derivatives?

In Section 4 we noted that the tempered Riesz and Feller potentials could not be considered as fractional derivatives, since the composition property was not valid for any pairs of orders. We wonder if this is also true for the BTFD. We will base our study in the SSC as proposed in [35].

It is not a hard task to show that the BTFD verify the following properties

P1 Linearity

The BTFD we introduced in the last sub-section is linear.

P2 Identity

The zero order BTFD of a function returns the function itself, since $(i\kappa + \lambda)^0 = 1$, for any $\lambda, \kappa \in \mathbb{R}$.

P3 Backward compatibility

When the order is integer, the BTFD gives the same result as the integer order two-sided TD and recovers the ordinary bilateral derivative, for $\lambda = 0$.

P4 The index law holds

$$_\lambda D_\theta^\alpha {}_\lambda D_\eta^\beta f(t) = {}_\lambda D_{\theta+\eta}^{\alpha+\beta} f(t), \tag{41}$$

for any α and β, since

$$\left|\kappa^2 + \lambda^2\right|^{\frac{\alpha}{2}} e^{i\theta \arctan(\frac{\kappa}{\lambda})} \left|\kappa^2 + \lambda^2\right|^{\frac{\beta}{2}} e^{i\eta \arctan(\frac{\kappa}{\lambda})} = \left|\kappa^2 + \lambda^2\right|^{\frac{\alpha+\beta}{2}} e^{i(\theta+\eta) \arctan(\frac{\kappa}{\lambda})}$$

P5 The generalised Leibniz rule reads

$$_\lambda D_\theta^\alpha [f(t)g(t)] = \sum_{i=0}^{\infty} \binom{\alpha}{i} D^i f(t) {}_\lambda D_\theta^{\alpha-i} g(t), \tag{42}$$

a bit different from the usual. Its deduction is similar to the one described in [1].

We conclude that *the BTFD verifies the SSC and therefore can be considered a derivative.*

5. Conclusions

This paper addressed the study of tempered two-sided derivatives. Two versions were considered: integral and GL like. The conformity of these operators as studied in the perspective of a criterion for fractional derivatives was stated. In passing we showed that a simple tempering of the traditional Riesz and Feller potentials does not lead to fractional derivatives.

Author Contributions: These two authors contribute equally to this paper. All authors have read and agreed to the published version of the manuscript.

Funding: This work was partially funded by National Funds through the Foundation for Science and Technology of Portugal, under the projects UIDB/00066/2020.

Institutional Review Board Statement: Not applicable.

Informed Consent Statement: Not applicable.

Data Availability Statement: Not applicable.

Conflicts of Interest: The authors declare no conflict of interest.

Abbreviations

The following abbreviations are used in this manuscript:

LT	Laplace transform
FT	Fourier transform
FD	Fractional derivative
FP	Feller Potential
GL	Grünwald-Letnikov
L	Liouville
RL	Riemann-Liouville
TF	Transfer function
TFD	Tempered Fractional Derivative
BTFD	Bilateral Tempered Fractional Derivatives
RP	Riesz Potential
RD	Riesz Derivative
RFD	Riesz-Feller Derivative

References

1. Ortigueira, M.D.; Bengochea, G.; Machado, J.T. Substantial, Tempered, and Shifted Fractional Derivatives: Three Faces of a Tetrahedron. *Math. Methods Appl. Sci.* **2021**, 1–19. [CrossRef]
2. Barndorff-Nielsen, O.E.; Shephard, N. Normal modified stable processes. *Theory Probab. Math. Stat.* **2002**, *65*, 1–20.
3. Cao, J.; Li, C.; Chen, Y. On tempered and substantial fractional calculus. In Proceedings of the 2014 IEEE/ASME 10th International Conference on Mechatronic and Embedded Systems and Applications (MESA), Senigallia, Italy, 10–12 September 2014; pp. 1–6.
4. Chakrabarty, A.; Meerschaert, M.M. Tempered stable laws as random walk limits. *Stat. Probab. Lett.* **2011**, *81*, 989–997. [CrossRef]
5. Hanyga, A.; Rok, V.E. Wave propagation in micro-heterogeneous porous media: A model based on an integro-differential wave equation. *J. Acoust. Soc. Am.* **2000**, *107*, 2965–2972. [CrossRef]
6. Meerschaert, M.M. Fractional calculus, anomalous diffusion, and probability. In *Fractional Dynamics: Recent Advances*; World Scientific: Singapore, Singapore, 2012; pp. 265–284.
7. Pilipovíc, S. The α-Tempered Derivative and some spaces of exponential distributions. *Publ. L'Institut Mathématique Nouv. Série* **1983**, *34*, 183–192.
8. Rosiński, J. Tempering stable processes. *Stoch. Process. Their Appl.* **2007**, *117*, 677–707. [CrossRef]
9. Skotnik, K. On tempered integrals and derivatives of non-negative orders. *Ann. Pol. Math.* **1981**, *XL*, 47–57. [CrossRef]
10. Carr, P.; Geman, H.; Madan, D.B.; Yor, M. The fine structure of asset returns: An empirical investigation. *J. Bus. B* **2002**, *75*, 305–332. [CrossRef]
11. Madan, D.B.; Milne, F. Option pricing with vg martingale components 1. *Math. Financ.* **1991**, *1*, 39–55. [CrossRef]
12. Madan, D.B.; Carr, P.P.; Chang, E.C. The Variance Gamma Process and Option Pricing. *Rev. Financ.* **1998**, *2*, 79–105. Available online: https://engineering.nyu.edu/sites/default/files/2018-09/CarrEuropeanFinReview1998.pdf (accessed on 7 May 2021). [CrossRef]
13. Cartea, A.; del Castillo-Negrete, D. Fractional diffusion models of option prices in markets with jumps. *Phys. A Stat. Mech. Appl.* **2007**, *374*, 749–763. [CrossRef]
14. Cartea, A.; del Castillo-Negrete, D. Fluid limit of the continuous-time random walk with general Lévy jump distribution functions. *Phys. Rev. E* **2007**, *76*, 041105. [CrossRef]
15. Mantegna, R.N.; Stanley, H.E. Stochastic Process with Ultraslow Convergence to a Gaussian: The Truncated Lévy Flight. *Phys. Rev. Lett.* **1994**, *73*, 2946–2949. [CrossRef]
16. Novikov, E.A. Infinitely divisible distributions in turbulence. *Phys. Rev. E* **1994**, *50*, R3303–R3305. [CrossRef]
17. Sokolov, I.; Chechkin, A.V.; Klafter, J. Fractional diffusion equation for a power-law-truncated Lévy process. *Phys. A Stat. Mech. Appl.* **2004**, *336*, 245–251. [CrossRef]
18. Carr, P.; Geman, H.; Madan, D.B.; Yor, M. Stochastic Volatility for Lévy Processes. *Math. Financ.* **2003**, *13*, 345–382. Available online: https://onlinelibrary.wiley.com/doi/abs/10.1111/1467-9965.00020 (accessed on 7 May 2021). [CrossRef]
19. Baeumer, B.; Meerschaert, M.M. Tempered stable Lévy motion and transient super-diffusion. *J. Comput. Appl. Math.* **2010**, *233*, 2438–2448. [CrossRef]
20. Wu, X.; Deng, W.; Barkai, E. Tempered fractional Feynman-Kac equation: Theory and examples. *Phys. Rev. E* **2016**, *93*, 032151. [CrossRef] [PubMed]
21. Hou, R.; Deng, W. Feynman–Kac equations for reaction and diffusion processes. *J. Phys. A Math. Theor.* **2018**, *51*, 155001. [CrossRef]
22. Sabzikar, F.; Meerschaert, M.; Chen, J. Tempered fractional calculus. *J. Comput. Phys.* **2015**, *293*, 14–28. [CrossRef] [PubMed]
23. Li, C.; Deng, W. High order schemes for the tempered fractional diffusion equations. *Adv. Comput. Cathematics* **2016**, *42*, 543–572. [CrossRef]
24. Arshad, S.; Huang, J.; Khaliq, A.; Tang, Y. Trapezoidal scheme for time–space fractional diffusion equation with Riesz derivative. *J. Comput. Phys.* **2017**, *350*, 1–15. [CrossRef]

25. Çelik, C.; Duman, M. Crank–Nicolson method for the fractional diffusion equation with the Riesz fractional derivative. *J. Comput. Phys.* **2012**, *231*, 1743–1750. [CrossRef]
26. Dehghan, M.; Abbaszadeh, M.; Deng, W. Fourth-order numerical method for the space–time tempered fractional diffusion-wave equation. *Appl. Math. Lett.* **2017**, *73*, 120–127. [CrossRef]
27. D'Ovidio, M.; Iafrate, F.; Orsingher, E. Drifted Brownian motions governed by fractional tempered derivatives. *Mod. Stochastics Theory Appl.* **2018**, *5*, 445–456. [CrossRef]
28. Zhang, Y.; Li, Q.; Ding, H. High-order numerical approximation formulas for Riemann-Liouville (Riesz) tempered fractional derivatives: Construction and application (I). *Appl. Math. Comput.* **2018**, *329*, 432–443. [CrossRef]
29. Zhang, Z.; Deng, W.; Karniadakis, G. A Riesz basis Galerkin method for the tempered fractional Laplacian. *SIAM J. Numer. Anal.* **2018**, *56*, 3010–3039. [CrossRef]
30. Zhang, Z.; Deng, W.; Fan, H. Finite Difference Schemes for the Tempered Fractional Laplacian. *Numer. Math. Theory Methods Appl.* **2019**, *12*, 492–516. [CrossRef]
31. Duo, S.; Zhang, Y. Numerical approximations for the tempered fractional Laplacian: Error analysis and applications. *J. Sci. Comput.* **2019**, *81*, 569–593. [CrossRef]
32. Hu, D.; Cao, X. The implicit midpoint method for Riesz tempered fractional diffusion equation with a nonlinear source term. *Adv. Differ. Equ.* **2019**, *2019*, 1–14. [CrossRef]
33. Herrmann, R. Solutions of the fractional Schrödinger equation via diagonalization—A plea for the harmonic oscillator basis part 1: The one dimensional case. *arXiv* **2018**, arXiv:1805.03019.
34. Ortigueira, M.D. Two-sided and regularised Riesz-Feller derivatives. *Math. Methods Appl. Sci.* **2019**. Available online: https://onlinelibrary.wiley.com/doi/abs/10.1002/mma.5720 (accessed on 7 May 2021). [CrossRef]
35. Ortigueira, M.D.; Machado, J.A.T. What is a fractional derivative? *J. Comput. Phys.* **2015**, *293*, 4–13. [CrossRef]
36. Tricomi, F. Sulle funzioni ipergeometriche confluenti. *Ann. Mat. Pura Appl.* **1947**, *26*, 141–175. [CrossRef]
37. Ortigueira, M.D. Riesz potential operators and inverses via fractional centred derivatives. *Int. J. Math. Math. Sci.* **2006**, *2006*, 48391. [CrossRef]
38. Ortigueira, M.D. Fractional central differences and derivatives. *J. Vib. Control* **2008**, *14*, 1255–1266. [CrossRef]
39. Samko, S.G.; Kilbas, A.A.; Marichev, O.I. *Fractional Integrals and Derivatives: Theory and Applications*; Gordon and Breach Science Publishers: Amsterdam, The Netherlands, 1993.

Article

λ-Interval of Triple Positive Solutions for the Perturbed Gelfand Problem

Shugui Kang [1], Youmin Lu [2] and Wenying Feng [3,*]

[1] The Institute of Applied Mathematics, Shanxi Datong University, Datong 037009, China; kangshugui@sxdtdx.edu.cn
[2] Department of Mathematical and Digital Sciences, Bloomsburg University, Bloomsburg, PA 17815, USA; ylu@bloomu.edu
[3] Departments of Mathematics and Computer Science, Trent University, Peterborough, ON K9L 0G2, Canada
* Correspondence: wfeng@trentu.ca

Abstract: We study a two-point Boundary Value Problem depending on two parameters that represents a mathematical model arising from the combustion theory. Applying fixed point theorems for concave operators, we prove uniqueness, existence, upper, and lower bounds of positive solutions. In addition, we give an estimation for the value of λ_* such that, for the parameter $\lambda \in [\lambda_*, \lambda^*]$, there exist exactly three positive solutions. Numerical examples are presented to illustrate various cases. The results complement previous work on this problem.

Keywords: boundary value problem; concave operator; fixed point theorem; Gelfand problem; order cone

MSC: Primary 34B08; 34B18; Secondary 34C11

1. Introduction

As a mathematical model arising from the combustion theory [1,2], the following two-point Boundary Value Problem (BVP) has been well studied by a number of authors [3–10]:

$$\begin{cases} u''(t) + \lambda \exp\left(\frac{\alpha u(t)}{\alpha + u(t)}\right) = 0, & -1 < t < 1, \\ u(-1) = u(1) = 0, \end{cases} \quad (1)$$

where $\lambda > 0$ is the Frank–Kamenetskii parameter, $\alpha > 0$ is the activation energy parameter, u is the dimensionless temperature, and the reaction term $\exp\left(\frac{\alpha u}{\alpha + u}\right)$ shows the temperature dependence. Representing the steady case in the thermal explosion, BVP (1.1) is well-known as the one-dimensional perturbed Gelfand problem [1,2,5].

In the literature, bifurcation curve, existence, and multiplicity of positive solutions for BVP (1.1) have been extensively studied. In particular, Shivaji [8] first shows that, for every $\alpha > 0$, BVP (1.1) has a unique nonnegative solution when λ is small enough or large enough. Hastings and McLeod [4] and Brown et al. [3] prove that the bifurcation curve of (1.1) is S-shaped on the $(\lambda, ||u||)$ plane when α is large enough, where $||u||$ is the norm in the space $C[-1, 1]$. That is, when α is large enough, there exist λ_*, λ^* such that (1.1) has a unique nonnegative solution for $0 < \lambda < \lambda_*, \lambda > \lambda^*$, exactly three nonnegative solutions for $\lambda_* < \lambda < \lambda^*$, and exactly two nonnegative solutions for $\lambda = \lambda_*(\alpha)$ and $\lambda^*(\alpha)$. Later, it was proved that the BVP (1.1) has multiple solutions when $\alpha > 4.4967$ [11]. This lower bound was improved to 4.35 by Korman and Li [12]. Recently, it was shown in [5,6] that the number can be as close to 4 as 4.166. The problem has also been considered for general operator equations in abstract Banach spaces [10]. Most recently, a similar problem has been studied for the Neumann boundary value problem [9]. The techniques applied mostly are the quadrature method.

In this paper, we first apply a new result on a unique solution for a class of concave operators in a partially ordered Banach space [13] to prove that there exists a unique solution for BVP (1.1) when $\alpha \leq 4$. Previously, it was shown that, when $\alpha \leq 4$, the bifurcation curve for $(\lambda, ||u||)$ is monotonically increasing, which implies that the sup norm of the solutions must be unique [11]. With a totally different approach, we are able to directly prove the uniqueness of solutions. Then, we prove a general result for all parameters on the existence of a solution using a new fixed point theorem on order intervals that was recently introduced in [14]. As an advantage of this new method, we obtain upper and lower bounds of the solutions depending on the values of λ and α. Next, assuming that $\alpha > 4$, it is known that there exists an λ-interval (λ_*, λ^*) such that BVP (1.1) has at least three nonnegative solutions for $\lambda \in (\lambda_*, \lambda^*)$ [3–6,11,12]. However, nothing is known for the range of the λ-interval, or the values of λ_* and λ^*. We obtain a range of λ_* by an upper bound and a lower bound. The accuracy of the estimation is shown by the fact that the range is usually very small. From our knowledge, this is the first time to give a concrete estimation for the λ-intervals that ensure solution multiplicity. Lastly, some numerical results are given to illustrate the upper and lower bounds and multiplicity of solutions.

The rest of the paper is organized as the following: Section 2 provides some preliminary results that will be used in the sequel. Section 3 proves the uniqueness theorem. Section 4 discusses existence, upper, and lower bounds of solutions. Section 5 gives the λ-intervals for multiplicity. Numerical solutions obtained by MatLab are presented in Section 6.

2. Preliminary

Let $(E, ||\cdot||)$ be a real Banach space and θ be the zero element of E. We first introduce the concept of order cone.

Definition 1 ([15], p. 276). *A subset P of E is called an order cone iff:*

(i) P is closed, nonempty, and $P \neq \{0\}$;
(ii) $a, b \in \mathbb{R}, a, b \geq 0, x, y \in P \Rightarrow ax + by \in P$;
(iii) $x \in P$ and $-x \in P \Rightarrow x = 0$.

A Banach space E is partially ordered by an order cone P, i.e., $x \leq y$ if and only if $y - x \in P$ for any $x, y \in E$. P is normal if there exists $N > 0$ such that $||x|| \leq N||y||$ if $x, y \in E$ and $\theta \leq x \leq y$. The infimum of such constants N is called the normality constant of P. Following the notation of [13,16], for $x, y \in E$, $x \sim y$ means that there exist $\lambda > 0$ and $\mu > 0$ such that $\lambda x \leq y \leq \mu x$. It is clear that $\sim$ is an equivalence relation. For fixed $h > \theta$, $P_h = \{x \in E \mid x \sim h\}$. It is easy to see that $P_h \subset P$.

Definition 2. *An operator $A : E \to E$ is increasing if $x \leq y$ implies $Ax \leq Ay$.*

Definition 3 ([13]). *Let $e \in P$ with $\theta \leq e \leq h$. Define the set*

$$P_{h,e} = \{x \in E \mid x + e \in P_h\}.$$

An operator $A : P_{h,e} \to E$ is said to be a ϕ-(h, e)-concave operator if there exists $\phi(\lambda) > \lambda$ for $\lambda \in (0, 1)$ such that

$$A(\lambda x + (\lambda - 1)e) \geq \phi(\lambda) Ax + (\phi(\lambda) - 1)e \text{ for any } x \in P_{h,e}.$$

Theorem 1 ([16]). *Suppose that A is an increasing ϕ-(h, θ)-concave operator, P is normal, and $Ah \in P_h$. Then, A has a unique fixed point x^* in P_h. Moreover, for any given point $w_0 \in P_h$, $||w_n - x^*|| \to 0$ as $n \to \infty$ if $w_n = Aw_{n-1}$ for $n = 1, 2, \ldots$*

Theorem 2 ([14]). *Assume that X is an ordered Banach space with the order cone X_+. Let $0 \le u_0 \le \phi$ be such that $||u_0|| \le 1$ and $||\phi|| = 1$ satisfying the condition that if $x \in X_+, ||x|| \le 1$, then $x \le \phi$. If there exist positive numbers, $0 < a < b$ such that $T : P_{u_0} \cap (\overline{\Omega_b} \setminus \Omega_a) \to P_{u_0}$ is a completely continuous operator. If the conditions*

$$||T(x)||_{x \in [au_0, a\phi]} \le a, \text{ and } ||T(x)||_{x \in [bu_0, b\phi]} \ge b \qquad (2)$$

or

$$||T(x)||_{x \in [au_0, a\phi]} \ge a, \text{ and } ||T(x)||_{x \in [bu_0, b\phi]} \le b \qquad (3)$$

are satisfied, then T has a fixed point $x_0 \in [au_0, b\phi]$.

3. Uniqueness for $\alpha \le 4$

In this section, we apply Theorem 1 to prove the following theorem on existence and uniqueness of solutions for BVP (1.1) with the assumption of $\alpha \le 4$.

Let $X = C[-1, 1]$ with the standard norm $u \in X$, $||u|| = \max_{-1 \le t \le 1} |u(t)|$. Let $P = \{u | u \in X, u(t) \ge 0, t \in [-1, 1]\}$. It is clear that P is a normal cone of $C[-1, 1]$.

Theorem 3. *BVP problem (1.1) has a unique solution for all $\alpha \le 4$.*

Proof. It can be verified that $u \in X$ is a solution of BVP (1.1) if and only if $Tu = u$, where $T : X \to X$ is the Hammerstein integral operator defined as

$$(Tu)(t) = \frac{\lambda}{2} \int_{-1}^{1} G(s, t) \exp\left(\frac{\alpha u(s)}{\alpha + u(s)}\right) ds, \; t \in [-1, 1], \qquad (4)$$

and the Green's function $G(s, t)$ is calculated as

$$G(s, t) = \begin{cases} (1-t)(1+s), & -1 < s \le t < 1, \\ (1+t)(1-s), & -1 < t \le s < 1. \end{cases}$$

It is easy to see that $(1 - |s|)(1 - |t|) < G(s, t) \le 1 - s^2$ for all $-1 < s < 1$ and $-1 < t < 1$ and $\int_{-1}^{1} G(s, t) ds = 1 - t^2$.

Since both λ and G are positive and the function $f(x) = \exp\left(\frac{\alpha x}{\alpha + x}\right)$ is increasing with respect to x, the operator T is increasing. Let $h(t) = 1 - t^2$. One can easily find that

$$\frac{\lambda}{2}(1 - t^2) \le Tu(t) \le \frac{\lambda}{2} e^{\alpha} (1 - t^2).$$

Therefore, $Th(t) \in P_{h,\theta}$, where $P_{h,\theta}$ is defined by Definition 3.

To prove that $T : P_{h,\theta} \to X$ is a ϕ-(h, θ)-concave operator, denote $f(x) = \exp\left(\frac{x}{1+\epsilon x}\right)$ for $\epsilon = \frac{1}{a}$ and let $\phi(\mu) = \frac{f(\mu x)}{f(x)} = \exp\left(\frac{\mu x}{1+\epsilon \mu x} - \frac{x}{1+\epsilon x}\right)$. Then,

$$\phi'(\mu)(x) = \phi(\mu)(x) \frac{\mu(1+\epsilon x)^2 - (1+\epsilon \mu x)^2}{(1+\epsilon x)^2(1+\epsilon \mu x)^2}.$$

Since $\phi(\mu) > 0$, the numerator is the only part that may change sign. It can be verified that the numerator is less than 0 when $x \in [0, \frac{1}{\epsilon \sqrt{\mu}}]$ and greater than 0 when $x \in [\frac{1}{\epsilon \sqrt{\mu}}, \infty]$. Therefore, $\phi(\mu)$ has only one critical point at $x = \frac{1}{\epsilon \sqrt{\mu}}$ and it has its minimum value $\phi(\mu)(\frac{1}{\epsilon \sqrt{\mu}}) = \exp\left(\frac{\sqrt{\mu}-1}{(\sqrt{\mu}+1)\epsilon}\right)$. Hence, $f(\mu x) \ge \phi(\mu) f(x)$.

Next, denoting $k(\mu) = \frac{\sqrt{\mu}-1}{(\sqrt{\mu}+1)\ln \mu} < \epsilon$, we show that $k'(\mu) > 0$. Let $q(\mu) = \ln \mu - \mu^{\frac{1}{2}} + \mu^{-\frac{1}{2}}$. Then, $q'(\mu) = -\frac{1}{2}\mu^{-\frac{3}{2}}(\sqrt{\mu}-1)^2 < 0$, $q(1) = 0$ and $q(\mu) > 0$ ensure that $k'(\mu) > 0$ for all $\mu \in (0, 1)$. It follows that k is increasing and its superum over $(0, 1)$ is $\frac{1}{4}$. Hence, the inequality $\epsilon \ge \frac{1}{4}$ or $\alpha \le 4$ implies that $\phi(\mu) > \mu$ with all $\mu \in (0, 1)$. Consequently, the operator T defined (3.1) satisfies all the conditions of Theorem 1 when $\alpha \le 4$, and it has a unique fixed point in $P_{h,\theta}$. Since operator (3.1) guarantees that all solutions are in P_h, BVP (1.1) has a unique solution when $\alpha \leq 4$ for every $\lambda > 0$. □

Remark 1. *Existence of solutions for BVP (1.1) was previously shown by the S-shaped bifurcation curve on $(\lambda, \|u\|)$ [3,4,6,11]. Since the bifurcation curve depends on $\|u\|$, some qualitative properties for the maximum of solutions can be observed. For example, it was proved in [3] that the sup norm of the solutions of BVP (1.1) is unique when $\alpha \leq 4$.*

4. Upper, Lower Bounds and Order Sequence of Solutions

In this section, we prove the existence of upper and lower bounds for the general case of BVP (1.1). The approach is by Theorem 2, a new fixed point theorem on order intervals recently introduced in [14].

Let X, P and f be defined as in the proof of Theorem 3 and $g(x) = \frac{f(x)}{x}$. Then, g has the properties of

$$\lim_{x \to 0^+} g(x) = \infty, \quad \lim_{x \to \infty} g(x) \to 0. \tag{5}$$

Theorem 4. *Select positive parameters $a, b,$ and δ such that*

$$a = \frac{\lambda}{2}, \quad g(b) = \frac{2}{\lambda}, \quad \delta = \frac{\lambda}{2b}. \tag{6}$$

Then BVP (1.1) has a solution u such that

$$\frac{\lambda}{2}\delta(1-t^2) \leq \delta u(0)(1-t^2) \leq u(t) \leq b(1-t^2), \quad t \in [0,1]. \tag{7}$$

Proof. From the proof of Theorem 3, $u \in X$ is a solution of BVP (1.1) if and only if $Tu = u$, where T is defined by (4). Let $u_0 = \delta(1-t^2)$ and $\varphi = 1$. Then, u_0 and φ satisfy the conditions of Theorem 2. Define

$$P_{u_0} = \{u \in P | \ \|u\| = u(0), \ u(-t) = u(t), \ u(t) \geq \delta u(0)(1-t^2), \ t \in [-1,1]\}.$$

It can be verified that P_{u_0} is a subcone of P. To prove $T : P_{u_0} \cap (\overline{\Omega_b} \setminus \Omega_a) \to P_{u_0}$, let $u \in P_{u_0}$ with $\|u\| \leq b$. We have

$$(Tu)(t) = \frac{\lambda}{2} \int_{-1}^{1} G(s,t) \exp\left(\frac{\alpha u}{\alpha + u}\right) ds \geq \frac{\lambda}{2}(1-t^2). \tag{8}$$

On the other hand,

$$\begin{aligned}
\delta(Tu)(0) &= \frac{\lambda \delta}{2} \int_{-1}^{1} G(s,0) \exp\left(\frac{\alpha u}{\alpha + u}\right) ds \\
&\leq \frac{\lambda \delta}{2} \exp\left(\frac{\alpha b}{\alpha + b}\right) \int_{-1}^{1} G(s,0) ds \\
&= \frac{\lambda}{2}.
\end{aligned}$$

Therefore, $(Tu)(t) \geq \delta(Tu)(0)(1-t^2)$. Assume that $u(t) = u(-t)$ for $t \in [-1,1]$.

$$(Tu)(t) = \frac{\lambda}{2} \int_{-1}^{t} (1+s)(1-t) f(u(s)) ds + \frac{\lambda}{2} \int_{t}^{1} (1-s)(1+t) f(u(s)) ds. \tag{9}$$

$$(Tu)(-t) = \frac{\lambda}{2}\int_{-1}^{-t}(1+s)(1+t)f(u(s))ds + \frac{\lambda}{2}\int_{-t}^{1}(1-s)(1-t)f(u(s))ds$$
$$= \frac{\lambda}{2}\int_{t}^{1}(1-x)(1+t)f(u(x))dx + \frac{\lambda}{2}\int_{-1}^{t}(1+x)(1-t)f(u(x))dx$$
$$= (Tu)(t),$$

where $x = -s$. To show that $\|Tu\| = (Tu)(0)$, let $g(t) = (Tu)(t)$, by (4.5),

$$g'(t) = -\frac{\lambda}{2}\int_{-1}^{t}(1+s)f(u(s))ds + \frac{\lambda}{2}\int_{t}^{1}(1-s)f(u(s))ds.$$

Hence, $g'(-1) > 0$, $g'(1) < 0$ and $g''(t) = -\lambda f(u(t)) \leq 0$. This implies that g' is decreasing and only has one zero point. Since g is symmetric about zero, $g'(0) = 0$ and $\|g\| = g(0)$. This implies that $Tu \in P_{u_0}$. The Hammerstein integral operator T is completely continuous. For $u \in [au_0, a\varphi]$, we have

$$\|Tu\| = (Tu)(0) = \frac{1}{2}\lambda \int_{-1}^{1} G(s,0)\exp\left(\frac{\alpha u}{\alpha + u}\right)ds$$
$$\geq \frac{\lambda}{2} = a.$$

On the other hand, let $\delta b(1-t^2) \leq u(t) \leq b$,

$$(Tu)(t) = \frac{1}{2}\lambda \int_{-1}^{1} G(s,t)\exp\left(\frac{\alpha u(s)}{\alpha + u(s)}\right)ds$$
$$\leq \frac{\lambda}{2}\exp\left(\frac{\alpha b}{\alpha + b}\right)\int_{-1}^{1} G(s,t)ds$$
$$= \frac{\lambda}{2}\exp\left(\frac{\alpha b}{\alpha + b}\right)(1-t^2)$$
$$= b(1-t^2) \leq b.$$

By Theorem 2, BVP (1.1) has a solution u such that $u(t) \in [a\delta(1-t^2), b]$ and $u \in P_{u_0}$. From (4.5), we can see that $u(0) = (Tu)(0) \geq \frac{\lambda}{2} = a$. It follows that the solution u satisfies

$$\frac{\lambda}{2}\delta(1-t^2) \leq \delta u(0)(1-t^2) \leq u(t) \leq b. \quad (10)$$

Moreover, from $\|u\| = u(0) \leq b$, we obtain

$$u(t) = Tu(t) = \frac{\lambda}{2}\int_{-1}^{1} G(s,t)\exp\left(\frac{\alpha u}{\alpha + u}\right)ds$$
$$\leq \frac{\lambda}{2}\exp\left(\frac{\alpha b}{\alpha + b}\right)\int_{-1}^{1} G(s,t)ds$$
$$= b(1-t^2).$$

Combining it with (4.6), we have

$$\frac{\lambda}{2}\delta(1-t^2) \leq \delta u(0)(1-t^2) \leq u(t) \leq b(1-t^2).$$

The proof is complete. □

The lower bound given in Theorem 4 depends on both parameters b and λ. When $\lambda > \left(\frac{\pi}{2}\right)^2$, a uniform lower bound can be obtained for all values of λ.

Theorem 5. *Let x_0 be the smallest value satisfying $g(x_0) = 1$. BVP (1.1) has a solution $u(t) \geq x_0 \sin(\frac{\pi}{2}t + \frac{\pi}{2})$ provided that $\lambda \geq (\frac{\pi}{2})^2$.*

Proof. We will construct a bounded increasing sequence using the Hammerstein operator T defined as (3.1). Let

$$u_0(t) = x_0 \sin(\frac{\pi}{2}t + \frac{\pi}{2}) \text{ and } u_1(t) = \frac{\lambda}{2} \int_{-1}^{1} G(s,t)f(u_0(s))ds.$$

By the definition of x_0, we have $g(u_0(t)) \geq 1$ or $f(u_0(t)) \geq u_0(t)$ and

$$u_1(t) \geq \frac{\lambda}{2} \int_{-1}^{1} G(s,t) u_0(s) ds$$

$$\geq \frac{\pi^2}{8} \int_{-1}^{1} G(s,t) u_0(s) ds.$$

Since $(\frac{\pi}{2})^2$ is an eigenvalue of the linear equation $u''(t) = -\lambda u(t)$ and $\sin(\frac{\pi}{2}t + \frac{\pi}{2})$ is its corresponding eigenvector, we have

$$\frac{\pi^2}{8} \int_{-1}^{1} G(s,t) u_0(s) ds = u_0(t) \leq u_1(t), \ t \in [-1,1].$$

Construct the sequence

$$u_n(t) = \frac{\lambda}{2} \int_{-1}^{1} G(s,t) f(u_{n-1}(s)) ds, \ n = 2,3,\ldots \quad (11)$$

The fact that f is increasing ensures that u_n is increasing. Let $x_3 > x_0$ be a constant such that $\frac{\lambda}{2} g(x_3) < 1$, then $u_0(t) < x_3$ and

$$u_n(t) = \frac{\lambda}{2} \int_{-1}^{1} G(s,t) f(u_{n-1}(s)) ds$$

$$\leq \frac{\lambda}{2} \int_{-1}^{1} G(s,t) f(x_3) ds$$

$$\leq x_3 \int_{-1}^{1} G(s,t) ds$$

$$= x_3(1 - t^2) \leq x_3.$$

Therefore, the sequence u_n is bounded above and it converges to a solution u of BVP (1.1). Obviously, the solution satisfies that

$$u(t) \geq u_0(t) = x_0 \sin(\frac{\pi}{2}t + \frac{\pi}{2}).$$

□

The construction method used in the proof of Theorem 5 has the advantage to provide numerical approximation with iterations. Following the similar idea, we can show that, for the same α value, a solution sequence can be constructed according to the order of the λ values.

Theorem 6. *For each $\lambda > 0$, there exists a positive solution $u_\lambda(t)$ for BVP (1.1) such that for $\lambda_1 < \lambda_2$, $u_{\lambda_1}(t) < u_{\lambda_2}(t), t \in [-1,1]$.*

Proof. As in the proof of Theorem 4, let $b_1, b_2 > 0$ satisfy $g(b_i) = \frac{2}{\lambda_i}, i = 1, 2$. Then,

$$\frac{\lambda_1}{2}g(b_2) < \frac{\lambda_2}{2}g(b_2) = 1.$$

Letting

$$u_0(t) = b_2 \int_{-1}^{1} G(s,t)ds = b_2(1-t^2),$$

$\|u_0(t)\| = b_2$. Define $u_{\lambda_1}^{(1)}(t) = \frac{\lambda_1}{2}\int_{-1}^{1} G(s,t)f(u_0(s))ds$, we have

$$\begin{aligned}
u_{\lambda_1}^{(1)}(t) &\leq \frac{\lambda_1}{2}\int_{-1}^{1} G(s,t)\exp\left(\frac{\alpha\|u_0\|}{\alpha + \|u_0\|}\right)ds \\
&= \frac{\lambda_1}{\lambda_2}b_2 \int_{-1}^{1} G(s,t)ds \\
&\leq b_2(1-t^2) = u_0(t),
\end{aligned}$$

and

$$\begin{aligned}
u_{\lambda_1}^{(2)}(t) &= \frac{\lambda_1}{2}\int_{-1}^{1} G(s,t)\exp\left(\frac{\alpha u_{\lambda_1}^{(1)}}{\alpha + u_{\lambda_1}^{(1)}}\right)ds \\
&\leq \frac{\lambda_1}{2}\int_{-1}^{1} G(s,t)\exp\left(\frac{\alpha u_0}{\alpha + u_0}\right)ds = u_{\lambda_1}^{(1)}(t).
\end{aligned}$$

By iteration, we can obtain the sequence

$$u_0 \geq u_{\lambda_1}^{(1)} \geq u_{\lambda_1}^{(2)} \geq \cdots \geq u_{\lambda_1}^{k} \geq u_{\lambda_1}^{k+1} \geq \cdots \geq 0.$$

Let $\lim_{k\to\infty} u_{\lambda_1}^{(k)}(t) = u_{\lambda_1}(t), t \in [-1,1]$, then $u_{\lambda_1}(t)$ is a positive solution for BVP (1.1) with parameter λ_1. Similarly, we can obtain the monotonic sequence $u_{\lambda_2}^{(k)}, k = 1, 2, 3, \cdots$ and

$$\begin{aligned}
u_{\lambda_2}^{(1)}(t) &= \frac{\lambda_2}{2}\int_{-1}^{1} G(s,t)\exp\left(\frac{\alpha u_0(s)}{\alpha + u_0(s)}\right)ds \\
&> \frac{\lambda_1}{2}\int_{-1}^{1} G(s,t)\exp\left(\frac{\alpha u_0(s)}{\alpha + u_0(s)}\right)ds = u_{\lambda_1}^{(1)}(t).
\end{aligned}$$

By mathematical induction, $u_{\lambda_2}^{(k)} \geq u_{\lambda_1}^{(k)}$ for $k = 1, 2, 3, \cdots$.

Let $\lim_{k\to\infty} u_{\lambda_2}^{(k)}(t) = u_{\lambda_2}(t), t \in [-1,1]$. Then, $u_{\lambda_2}(t)$ is a positive solution for BVP (1.1) with parameter λ_2 and $u_{\lambda_1}(t) \leq u_{\lambda_2}(t)$. □

5. λ-Interval for Triple Positive Solutions

The existence of multiple solutions is always a challenge. It is known that there exists α_0 such that the bifurcation curve of $(\lambda, \|u\|)$ is S-shaped when $\alpha > \alpha_0$, and this result ensures that there exist λ^* and λ_* such that BVP (1.1) has at least three solutions when $\lambda_* < \lambda < \lambda^*$, at least two solutions for $\lambda = \lambda_*$ and $\lambda = \lambda^*$ and at least one solution otherwise. Over the last two decades, the value of α_0 has been a focus of a series of publications [3–5,11,12,14]. Consequently, the estimation for α_0 has been improved again and again. Most recently, it is shown by numerical methods that $\alpha_0 \approx 4.069$ [5,6]. However, there is no result on the range of the λ-intervals or estimations for λ_* and λ^*.

In this section, we give an estimation for the value of λ_* by obtaining both upper and lower bounds and also show that the estimation is accurate since the difference between the upper bound and lower bound is actually very small. We use the functions f and g

defined in Section 4 again. When $\alpha > 4$, the following lemma shows the different behavior of function g from the case of $\alpha \leq 4$.

Lemma 1. *Let $f(x) = \exp(\frac{\alpha x}{\alpha+x})$ and $g(x) = \frac{f(x)}{x}$. Then,*
1. *When $\alpha \leq 4$, g is decreasing over $(0, \infty)$.*
2. *When $\alpha > 4$, g has a local minimum at $x_1 = \frac{\alpha^2 - 2\alpha - \sqrt{\alpha^4 - 4\alpha^3}}{2}$ and a local maximum at $x_2 = \frac{\alpha^2 - 2\alpha + \sqrt{\alpha^4 - 4\alpha^3}}{2}$.*
3. *When $\alpha > 4$, $g(x_1)$ is increasing with respect to α and $\frac{e^2}{4} < g(x_1) < \frac{e^2}{2}$.*

Theorem 7. *If $\alpha > 4$, and $\frac{2x_1}{f(x_1)} \geq \lambda > \frac{4x_2}{f(x_2)+0.5}$. BVP (1.1) has at least two non-negative solutions.*

Proof. For $\lambda \leq \frac{2x_1}{f(x_1)}$, since $g(x)$ is decreasing for $x \in (0, x_1)$, we have $x_1 \geq b$, where b is selected for condition (6). Therefore, Theorem 4 guarantees that BVP (1.1) has a solution $u^*(t) \leq x_1(1 - t^2)$.

Next, using the idea of Brown, Ibrahin, and Shivaji [6], we construct another solution using the condition $\lambda > \frac{4x_2}{f(x_2)+0.5}$. Define

$$u_0(t) = \begin{cases} x_2, & -\frac{1}{2} \leq t \leq \frac{1}{2}, \\ 0, & -1 < t < -\frac{1}{2} \text{ or } \frac{1}{2} < t < 1, \end{cases} \quad (12)$$

and

$$u_1(t) = \frac{\lambda}{2} \int_{-1}^{1} G(s,t) f(u_0(s)) ds. \quad (13)$$

When $-1 < t < -\frac{1}{2}$ or $\frac{1}{2} < t < 1$, it is clear that $u_1(t) \geq u_0(t)$. For $-\frac{1}{2} \leq t \leq \frac{1}{2}$, we have

$$\begin{aligned} u_1(t) &= \frac{\lambda}{2} \int_{-\frac{1}{2}}^{\frac{1}{2}} G(s,t) f(x_2) ds + \frac{\lambda}{2} \int_{-1}^{-\frac{1}{2}} G(s,t) ds + \frac{\lambda}{2} \int_{\frac{1}{2}}^{1} G(s,t) ds \\ &= \frac{\lambda}{2} f(x_2)(\frac{3}{4} - t^2) + \frac{\lambda}{8} \\ &\geq \frac{\lambda}{4} f(x_2) + \frac{\lambda}{8}. \end{aligned}$$

The condition $\lambda > \frac{4x_2}{f(x_2)+0.5}$ implies $u_1(t) \geq u_0(t)$ and the sequence defined as

$$u_n(t) = \frac{\lambda}{2} \int_{-1}^{1} G(s,t) f(u_{n-1}(s)) ds, \quad n = 0, 1, 2, \cdots \quad (14)$$

is increasing. It is also clear that $u_n(t) < \frac{\lambda}{2} e^\alpha x_2, n = 0, 1, 2, \cdots$. Therefore, this sequence converges and its limit $u^{**}(t)$ is a solution of BVP (1.1). The inequality

$$u^{**}(t) \geq x_2 > x_1 \geq x_1(1 - t^2) \geq u^*(t) \quad (15)$$

shows that problem (1.1) has at least two solutions. □

Remark 2. *Theorem 7 gives the estimation of $\lambda_* \leq \frac{2x_1}{f(x_1)} = \overline{\lambda}$.*

Remark 3. *It is shown by numerical calculation that, when $\alpha > 5.758$, the condition $\frac{2x_1}{f(x_1)} > \frac{4x_2}{f(x_2)+0.5}$ is always true.*

Remark 4. *We can calculate that $f'(x) = f(x)\frac{\alpha^2}{(x+\alpha)^2}$ has an absolute maximum value $\frac{4e^{\alpha-2}}{\alpha^2}$. The fixed point problem for the Hammerstein operator T defined by (4) has a unique solution when*

$\frac{2\lambda e^{\alpha-2}}{\alpha^2} < 1$ or $\lambda < \frac{\alpha^2}{2e^{\alpha-2}} = \overline{\lambda}$ by the standard contraction mapping theorem. This implies that $\lambda_* > \overline{\lambda}$. It is reasonable to conjecture that $\lambda_* = \frac{2x_2}{f(x_2)}$. The comparison in Table 1 indicates that the interval $[\overline{\lambda}, \overline{\overline{\lambda}}]$ is in fact very small.

Table 1. Upper and lower bounds for the value of λ_*.

α	$\overline{\lambda}$	$\frac{2x_2}{f(x_2)}$	$\overline{\overline{\lambda}}$
4.01	1.0773	1.0798	1.08155
4.02	1.0719	1.07676	1.07776
5	0.6223	0.70256	0.959057
5.5	0.4567	0.5329	0.92795
6	0.3297	0.3945	0.904837
100	$1.374392504 \times 10^{-39}$	2.002116×10^{-39}	0.743229

6. Numerical Solutions

In this section, we produce some numerical solutions using Matlab to give some direct illustration for the solutions. Figure 1 shows that the order sequence of solutions follow the value of λ as proved in Theorem 6. In both cases of $\alpha < 4$ (Figure 1a) and $\alpha > 4$ (Figure 1b), the order of the solutions follows the order of the parameter λ.

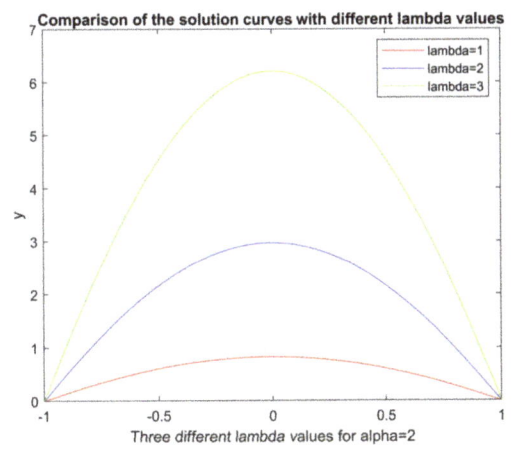

(a) $\alpha < 4$

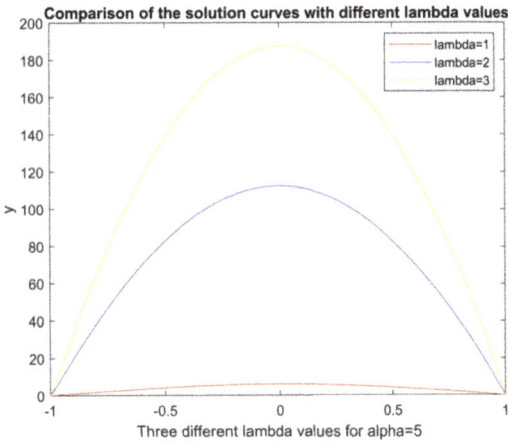

(b) $\alpha > 4$

Figure 1. Order sequences for λ values.

Lemma 2 ([5], p. 479). *If $u(t)$ is a solution of BVP (1.1), then $u(t)$ is symmetric about $t = 0$. Thus, $u(t) = u(-t)$.*

The following property on the norm and order of the solutions are new, to our knowledge.

Proposition 1. *If $u_1(t)$ and $u_2(t)$ are two solutions of BVP (1.1) for the same λ and $\|u_1\| > \|u_2\|$, then $u_1(t) > u_2(t)$ for $t \in (-1, 1)$.*

Proof. Since $u_1(t)$ and $u_2(t)$ are symmetric about $t = 0$, it is sufficient to prove that $u_1(t) > u_2(t)$ for $t \in (-1,0]$. First, we prove that $u_1(t) \geq u_2(t)$ for $t \in (-1,0]$. Let $f(x) = \exp\left(\frac{\alpha x}{\alpha+x}\right)$ for $x \geq 0$ and $F(u) = \int_0^u f(s)ds$. From (1.1), we have

$$u''u' + \lambda f(u)u' = 0.$$

Integrating both sides from 0 to $u(t)$, we obtain

$$\frac{1}{2}(u'(t))^2 + \lambda F(u) = C,$$

where C is a constant. Since $u(0) = \|u\|$ and $u'(0) = 0$, we find $C = \lambda F(\|u\|)$. Therefore,

$$\frac{1}{2}(u'(t))^2 + \lambda F(u) = \lambda F(\|u\|). \tag{16}$$

At $t = -1$, $u'(-1) = \sqrt{2\lambda F(\|u\|)}$. Thus,

$$u_1'(-1) = \sqrt{2\lambda F(\|u_1\|)} > \sqrt{2\lambda F(\|u_2\|)} = u_2'(-1).$$

There exists an interval $(-1,c)$ such that $u_1(t) > u_2(t)$ for $t \in (-1,c)$. Suppose that $-1 < r < 0$ is the first value such that $u_1(r) = u_2(r)$ and $u_1(t) < u_2(t)$ for $t > r$ in an interval. Using (6.1), we have

$$\begin{aligned} u_1'(r) &= \sqrt{2\lambda F(\|u_1\|) - 2\lambda F(u_1(r))} \\ &> \sqrt{2\lambda F(\|u_2\|) - 2\lambda F(u_2(r))} = u_2'(r). \end{aligned}$$

This is clearly a contradiction. Next, from the corresponding integral equation, we have

$$\begin{aligned} u_1(t) &= \frac{\lambda}{2}\int_{-1}^1 G(s,t)\exp\left(\frac{\alpha u_1(s)}{\alpha+u_1(s)}\right)ds \\ &> \frac{\lambda}{2}\int_{-1}^1 G(s,t)\exp\left(\frac{\alpha u_2(s)}{\alpha+u_2(s)}\right)ds = u_2(t). \end{aligned}$$

The proof is complete. □

It is interesting to see that all three solutions were found, as shown in Figure 2, where $\alpha = 6$ and $\lambda = 0.7$. In addition, $\frac{\lambda}{2} = 0.35$ and the value of b satisfying $\frac{0.7f(b)}{2b} = 1$ is 0.608. Figure 2a is consistent with Theorem 5. The value of $x_2 = 22.39$ and the solution curve in Figure 2c clearly supports the result in Theorem 7.

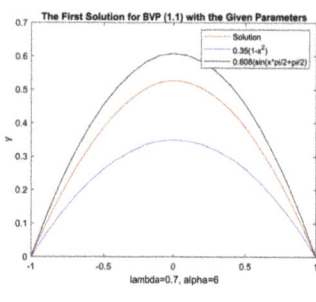

(a) The first solution

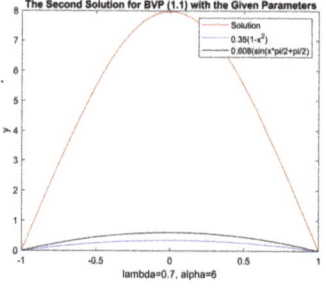

(b) The second solution

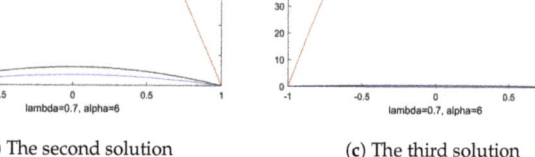

(c) The third solution

Figure 2. Three solutions.

Remark 5. When $\lambda > \left(\frac{\pi}{2}\right)^2 \approx 2.4674$, combining Theorems 4 and 5, there exist solutions u_1 and u_2 such that
$$u_1(t) \leq b(1-t^2) \text{ and } u_2(t) \geq x_0 \sin\left(\frac{\pi}{2}t + \frac{\pi}{2}\right),$$
where the constant b satisfying $g(b) = \frac{2}{\lambda}$, x_0 is the smallest value satisfying $g(x_0) = 1$. Since $\lambda \geq \left(\frac{\pi}{2}\right)^2$, $g(b) < g(x_0)$. Thus, $b > x_0$ because they must be values exceeding x_2 in Theorem 7 when $\alpha > 4$. If $\alpha \leq 4$, g is decreasing. Assuming a unique solution exists, then $u_1 = u_2 = u$, and we have
$$x_0 \sin\left(\frac{\pi}{2}t + \frac{\pi}{2}\right) \leq u(t) \leq b(1-t^2) \text{ if } \lambda \geq \left(\frac{\pi}{2}\right)^2. \tag{17}$$

Figure 3 illustrates the upper bound and lower bound given by (17). In (A), the solution of BVP (1.1) for $\lambda = 2.47 \geq \left(\frac{\pi}{2}\right)^2$ and $\alpha = 5 > 4$. In this case, $x_0 = 121.869$ and $b = 157.093$, and so $121.869 \sin\left(\frac{\pi}{2}t + \frac{\pi}{2}\right) < u(t) < 157.093(1-t^2)$. In (B), one calculated the solution of BVP (1.1) for $\lambda = 2.47$ and $\alpha = 2 < 4$, In this case, $x_0 = 3.632$ and $b = 5.26$ and so $3.632 \sin\left(\frac{\pi}{2}t + \frac{\pi}{2}\right) < u(t) < 3.26(1-t^2)$.

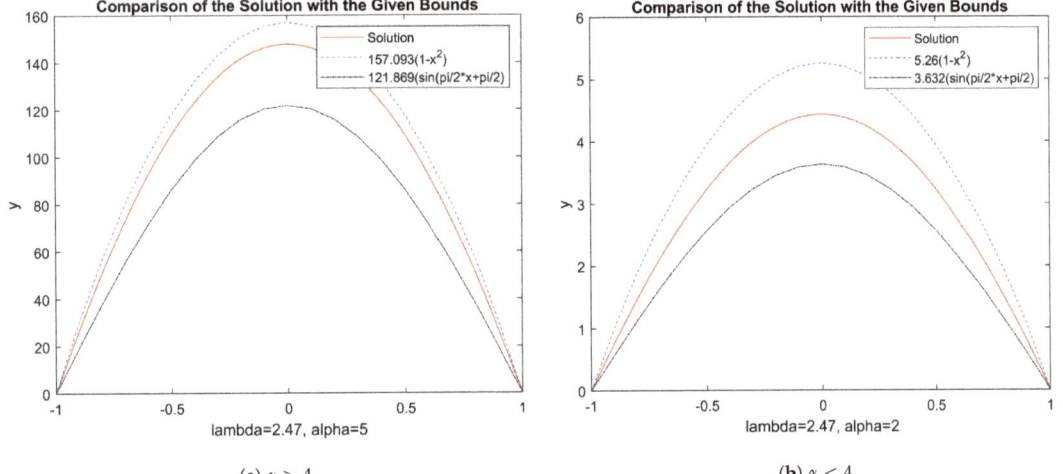

Figure 3. Upper and lower bounds for solutions.

Remark 6. With the advantages of the concrete equation (1.1), we are able to obtain more detailed quantitative properties for the solutions as given in the above sections. The results provide ideas for solving similar problems for more abstract problems. For example, similar approaches may be applied to study parameter dependent operator equations in abstract partial ordered Banach spaces.

In conclusion, we studied a two-point boundary value problem arising from the combustion theory. The second-order system of differential equations involves two positive parameters λ and α that are physically significant in the process.

Using topological methods, we proved results on uniqueness, existence, and multiplicity of positive solutions depending on the range of the two parameters. The results enriched previous work on this important application problem.

Author Contributions: Conceptualization, S.K., Y.L. and W.F.; methodology, S.K., Y.L. and W.F.; software, Y.L; formal analysis S.K., Y.L. and W.F.; investigation, S.K., Y.L. and W.F.; resources, S.K.; data curation, Y.L.; writing—original draft preparation, S.K., Y.L. and W.F.; writing—review and editing, S.K., Y.L. and W.F.; supervision, S.K., Y.L. and W.F.; project administration, S.K., Y.L. and W.F.; funding acquisition, S.K. and W.F. All authors have read and agreed to the published version of the manuscript.

Funding: This research was funded by the Natural Sciences Foundation of China (No. 11871314, 61803241) and the Natural Sciences and Engineering Research Council of Canada (NSERC).

Institutional Review Board Statement: Not applicable.

Informed Consent Statement: Not applicable.

Data Availability Statement: Not applicable.

Conflicts of Interest: The authors declare no conflict of interest. The funders had no role in the design of the study; in the collection, analyses, or interpretation of data; in the writing of the manuscript, or in the decision to publish the results.

References

1. Bebernes, J.; Eberly, D. *Mathematical Problems from Combustion Theory*; Springer: New York, NY, USA, 1989.
2. Boddington, T.; Gray, P.; Robinson, C. Thermal explosion and the disappearance of criticality at small activation energies: Exact results for the slab. *Proc. R. Soc. Lond. Ser. A Math. Phus. Eng. Sci.* **1979**, *368*, 441–461.
3. Brown, K.J.; Ibrahim, M.M.A.; Shivaji, R. S-Shaped bifurcation curves. *Nonlinear Anal.* **1981**, *5*, 475–486. [CrossRef]
4. Hastings, S.P.; McLeod, J.B. The number of solutions to an equation from catalysis. *Proc. R. Soc. Edinb. Sect. A* **1985**, *101*, 213–230. [CrossRef]
5. Hung, K.C.; Wang, S.H. A theorem on S-shaped bifurcation curve for a positone problem with convex, concave nonlinearity and its applications to the perturbed Gelfand problem. *J. Differ. Equ.* **2011**, *251*, 223–237. [CrossRef]
6. Huang, S.Y.; Wang, S.H. On S-shaped bifurcation curves for a two-point boundary value problem arising in a theory of thermal explosion. *Discret. Contin. Dyn. Syst.* **2015**, *35*, 4839–4858. [CrossRef]
7. Huang, S.Y.; Wang, S.H. Proof of a conjecture for the one-dimensional perturbed Gelfand problem from combustion theory. *Arch. Ration. Mech. Anal.* **2016**, *222*, 769–825. [CrossRef]
8. Shivaji, R. Uniqueness results for a class of positone problems. *Nonlinear Anal.* **1983**, *7*, 223–230. [CrossRef]
9. Xing, H.; Chen, H.; Yao, R. Reversed S-shaped bifurcation curve for a Neumann problem. *Discret. Dyn. Nat. Soc.* **2018**, *2018*, 5376075. [CrossRef]
10. Xu, B.; Wang, Z. On S-shaped bifurcation curves for a class of perturbed semilinear equations. *Chin. Ann. Math. B* **2008**, *29*, 641–662. [CrossRef]
11. Wang, S.H. On S-shaped bifurcation curves. *Nonlinear Anal. TMS* **1994**, *22*, 1475–1485. [CrossRef]
12. Korman, P.; Li, Y. On the exactness of an S-Shaped bifurcation curve. *Proc. Am. Math. Soc.* **1999**, *127*, 1011–1020. [CrossRef]
13. Zhai, C.; Wang, F. ϕ-(h,e)-concave operators and applications. *J. Math. Anal. Appl.* **2017**, *4543*, 571–584. [CrossRef]
14. Feng, W.; Zhang, G. New fixed point theorems on order intervals and their applications. *Fixed Point Theory Appl.* **2015**, *2015*, 1–10. [CrossRef]
15. Zeidler, E. *Nonlinear Functional Analysis and Its Application I*; Springer: New York, NY, USA, 1986.
16. Zhai, C.; Wang, F. Properties of positive solutions for the operator equation $Ax = \lambda x$ and applications to fractional differential equations with integral boundary conditions. *Adv. Differ. Equ.* **2015**, *366*, 1–10.

Article

Approximate Solutions of an Extended Multi-Order Boundary Value Problem by Implementing Two Numerical Algorithms

Surang Sitho [1,†], Sina Etemad [2,†], Brahim Tellab [3,†], Shahram Rezapour [2,*,†], Sotiris K. Ntouyas [4,5,†] and Jessada Tariboon [6,*,†]

1. Department of Social and Applied Sciences, College of Industrial Technology, King Mongkut's University of Technology North Bangkok, Bangkok 10800, Thailand; surang.s@sci.kmutnb.ac.th
2. Department of Mathematics, Azarbaijan Shahid Madani University, Tabriz 53751-71379, Iran; sina.etemad@azaruniv.ac.ir
3. Laboratory of Applied Mathematics, Kasdi Merbah University, B.P. 511, Ouargla 30000, Algeria; tellab.brahim@univ-ouargla.dz
4. Department of Mathematics, University of Ioannina, 451 10 Ioannina, Greece; sntouyas@uoi.gr
5. Nonlinear Analysis and Applied Mathematics (NAAM)-Research Group, Department of Mathematics, Faculty of Science, King Abdulaziz University, P.O. Box 80203, Jeddah 21589, Saudi Arabia
6. Intelligent and Nonlinear Dynamic Innovations Research Center, Department of Mathematics, Faculty of Applied Science, King Mongkut's University of Technology North Bangkok, Bangkok 10800, Thailand
* Correspondence: sh.rezapour@azaruniv.ac.ir (S.R.); jessada.t@sci.kmutnb.ac.th (J.T.)
† These authors contributed equally to this work.

Abstract: In this paper, we establish several necessary conditions to confirm the uniqueness-existence of solutions to an extended multi-order finite-term fractional differential equation with double-order integral boundary conditions with respect to asymmetric operators by relying on the Banach's fixed-point criterion. We validate our study by implementing two numerical schemes to handle some Riemann–Liouville fractional boundary value problems and obtain approximate series solutions that converge to the exact ones. In particular, we present several examples that illustrate the closeness of the approximate solutions to the exact solutions.

Keywords: approximate solutions; boundary value problem; existence; Riemann–Liouville derivative

1. Introduction

Fractional calculus is extending quickly, and its interesting and attractive applications are perfectly utilized in different parts of science [1–3]. It has appeared in financial models [4], optimal control [5,6], chaotic systems [7], epidemiological models [8,9], engineering [10,11], etc. Particularly, the fractional systems of boundary value problems (FBVP) of fractional differential equations usually yield other operational mathematical models for the description of special chemical, physical, and biological processes, which one can find in recently published works [12–19]. Along with these real models describing the phenomena, many mathematicians conduct research on the existence theory of solutions for different abstract structures of FBVPs with general boundary conditions including three-point, multi-point, multi-order, multi-strip, and nonlocal integral ones [20–29].

Several studies have also concentrated on the numerical techniques to obtain the analytical and approximate solutions of FBVPs. New numerical methods are introduced by researchers that have improved the convergence rate and error resulting from the approximate solutions. Examples of these methods and how to use them are Haar wavelet method [30,31], CAS wavelet method [32], homotopy analysis transform method (HATM) [33], q-HATM [34], Bernstein polynomials [35], iterative reproducing kernel Hilbert space method [36], Legendre functions with fractional orders [37], variational iteration method [38], and so on.

Since multi-term multi-order fractional differential equations have appeared in a wide range of fields, many mathematicians have started to review the properties and numerical solutions of this type of fractional differential equations. On the other side, because most of the time the exact solution cannot be found or it is very difficult to find, various numerical techniques have been applied for such FBVPs to obtain the approximate solutions. For instance, Bolandtalat, Babolian and Jafari [39] compared the convergence effects of exact and numerical solutions of multi-order fractional differential equations by means of Boubaker polynomials. In 2016, Hesameddini, Rahimi, and Asadollahifard [40] presented a new version of the reliable algorithm to solve multi-order fractional differential equations and investigated the convergence of it. Firoozjaee et al. [41] implemented a numerical approach on a multi-order fractional differential equation with mixed boundary-initial conditions. Recently, Dabiri and Butcher [42] invoked a numerical technique based on the spectral collocation methods and obtained the numerical solutions of multi-order fractional differential equations subject to multiple delays.

In recent years, many FBVPs with integral boundary conditions have been formulated by researchers of this field. Ali, Sarwar, Zada, and Shah [43] developed some conditions with the aid of topological degree results for confirming the existence of solutions to the nonlinear integral FBVP

$$\begin{cases} {}^c\mathfrak{D}_{0+}^\varrho v(z) = h(z, v(\mu z)), z \in I := [0,1], \mu \in (0,1), \\ c_1 v(0) + c_2 v(1) = \mathfrak{I}_{0+}^\varrho \varphi_1(1, v(1)), c_3 v'(0) + c_4 v'(1) = \mathfrak{I}_{0+}^\varrho \varphi_2(1, v(1)), \end{cases}$$

in which $\varrho \in (1,2]$, $c_1, c_2, c_3, c_4 \in \mathbb{R}^+$ and $h, \varphi_1, \varphi_2 \in C(I \times \mathbb{R}, \mathbb{R})$. ${}^c\mathfrak{D}_{0+}^\varrho$ denotes the Caputo fractional derivative of order ϱ and $\mathfrak{I}_{0+}^\varrho$ is the Riemann–Liouville fractional integral of order ϱ. Liu, Li, Dai, and Liu [44] implemented the fixed point techniques to establish the existence and uniqueness of solutions for the nonlocal integral FBVP

$$\begin{cases} \mathfrak{D}_{0+}^\varrho v(z) + \psi(z) h(z, v(z)) = 0, z \in (0,1), \\ v(0) = v'(0) = \cdots = v^{(k-2)}(0) = 0, \qquad v'(1) = p \mathfrak{I}_{0+}^\mu v(\xi), \end{cases}$$

where $\varrho \in (k-1, k]$, $\xi \in (0,1]$, $p, \mu > 0$, $\dfrac{p \Gamma(\varrho) \xi^{\varrho + \mu - 1}}{\Gamma(\varrho + \mu)} < 1$ and $\mathfrak{D}_{0+}^\varrho$ is the Riemann–Liouville fractional derivative of order ϱ. In 2018, Padhi, Graef, and Pati [45] studied positive solutions for the given fractional differential equation with Riemann–Stieltjes integral conditions

$$\begin{cases} \mathfrak{D}_{0+}^\varrho v(z) + \psi(z) h(z, v(z)) = 0, z \in (0,1), \\ v(0) = v'(0) = \cdots = v^{(k-2)}(0) = 0, \qquad \mathfrak{D}_{0+}^\varpi v(1) = \int_0^1 \varphi(r, v(r)) \, dA(r), \end{cases}$$

where $\varrho \in (k-1, k]$ with $k > 2$ and $1 \leq \varpi \leq \varrho - 1$.

In 2021, Thabet, Etemad, and Rezapour [46] designed and discussed the notion of the existence for possible solutions of a coupled system of the Caputo conformable FBVPs of the pantograph differential equation by

$$\begin{cases} {}^{cc}\mathfrak{D}_{z_0}^{\varrho, \sigma_1^*} v(z) = \tilde{\mathcal{P}}_1(z, m(z), m(\ell z)), z \in [z_0, \tilde{K}], z_0 \geq 0, \\ {}^{cc}\mathfrak{D}_{z_0}^{\varrho, \sigma_2^*} m(z) = \tilde{\mathcal{P}}_2(z, v(z), v(\ell z)), \end{cases}$$

with three-point RL-conformable integral conditions

$$\begin{cases} v(z_0) = 0, c_1 v(\tilde{K}) + c_2 {}^{RC}\mathfrak{J}_{z_0}^{\varrho,\theta^*} v(\delta) = w_1^*, \\ m(z_0) = 0, c_1^* m(\tilde{K}) + c_2^* {}^{RC}\mathfrak{J}_{z_0}^{\varrho,\theta^*} m(\nu) = w_2^*, \end{cases}$$

in which $\varrho \in (0,1]$, $\sigma_1^*, \sigma_2^* \in (1,2)$, $\delta, \nu \in (z_0, \tilde{K})$, $c_1, c_2, c_1^*, c_2^*, w_1^*, w_2^* \in \mathbb{R}$, $\ell \in (0,1)$ and $\tilde{\mathcal{P}}_1, \tilde{\mathcal{P}}_2 \in C([z_0, \tilde{K}] \times \mathbb{R} \times \mathbb{R}, \mathbb{R})$. In all the above fractional models with integral conditions, only the required conditions of the existence of solutions have been investigated and FBVPs have not been solved numerically. Due to the complexity of the structure of these FBVPs with integral boundary conditions and the difficulty associated with finding their exact solutions, some modern numerical algorithms have been developed to find approximate and analytical solutions.

In 2005, Daftardar-Gejji and Jafari [47] employed the Adomian decomposition method (ADM) to find solutions to a generalized initial system of multi-order fractional differential equations. One year later, they [48] presented an iterative algorithm jointly for solving a general functional equation approximately and called it the Daftardar-Gejji and Jafari method (DGJIM). Among other numerical algorithms, these two methods, i.e., DGJIM and ADM, are known as two numerical tools with high accuracy and rapid convergence to an exact solution. For more details, one can point out to some works in this regard [49–51]. We apply these two strong numerical tools to approximate possible solutions of our suggested FBVP.

In precise terms and with the help of the above ideas, in this paper, we propose a double-order integral FBVP of the multi-term multi-order differential equation in the framework of the Riemann–Liouville (RL) asymmetric derivation operators displayed as

$$\begin{cases} \mathfrak{D}_{0^+}^{\varrho} u(z) = \hbar(z, u(z), \mathfrak{D}_{0^+}^{\sigma_1} u(z), \mathfrak{D}_{0^+}^{\sigma_2} u(z), \dots, \mathfrak{D}_{0^+}^{\sigma_{n-1}} u(z), \mathfrak{D}_{0^+}^{\sigma_n} u(z)), \\ u(0) = 0, \quad u(1) = p\mathfrak{J}_{0^+}^{\mu} k_1(\xi, u(\xi)) + q\mathfrak{J}_{0^+}^{\nu} k_2(\eta, u(\eta)), \end{cases} \quad (1)$$

where $0 \le z \le 1$, $1 < \varrho < 2$, $0 < \sigma_1 < \sigma_2 < \cdots < \sigma_n < 1$, $\varrho > \sigma_n + 1$, $\hbar : [0,1] \times \mathbb{R}^{n+1} \to \mathbb{R}$, $k_j : [0,1] \times \mathbb{R} \to \mathbb{R}$, $(j = 1, 2)$ are continuous functions; $\mathfrak{D}_{0^+}^{\varrho}, \mathfrak{D}_{0^+}^{\sigma_1}, \dots, \mathfrak{D}_{0^+}^{\sigma_n}$ are RL-derivatives of order $\varrho, \sigma_1, \dots, \sigma_n$, respectively; and $\mathfrak{J}_{0^+}^{\gamma}$ denotes the RL-integral of order $\gamma \in \{\mu, \nu\}$ with $\mu, \nu, p, q > 0$ and $0 < \xi, \eta < 1$. Here, we first obtain the corresponding integral equation of the given multi-term multi-order RLFBVP (1) based on a theoretical argument and then establish the existence and uniqueness results with the aid of the fixed point tool. After that, we propose two numerical algorithms entitled DGJIM along with ADM to find approximate solutions.

Indeed, we must emphasize that the novelty and motivation of our work is that, although other papers use the ADM and DGJIM methods for solving IVPs, we here intend to compute approximate solutions for a complicated multi-order multi-term RLFBVP with boundary conditions including double-order RL-fractional integrals. In addition, note that, in the second boundary condition, the value of the unknown function at the end point $z = 1$ is proportional to a linear combination of RL-integrals with different orders $\mu, \nu > 0$ at the intermediate points $z = \xi, \eta \in (0, 1)$, respectively. Along with this, we consider the right-hand side nonlinear term $\hbar$ as a multi-variable function including multi-order RL-derivatives finitely.

The rest of this paper is organized as follows. Section 2 recalls fundamental notions on fractional calculus. Section 3 is devoted to establishing some criteria for confirming the existence of solutions. Section 4 introduces the two numerical methods named ADM and DGJIM. In Section 5, the proposed approximation techniques are described using different examples. Some concluding remarks are provided in Section 6.

2. Basic Concepts

First, for the convenience of the readers, we need some fundamental properties and lemmas on fractional calculus which are used further in this paper.

Definition 1. *[3] Let $\varrho > 0$ and $\phi : [0, \infty) \to \mathbb{R}$ be a continuous function. The following integral*

$$(\mathfrak{I}_{0+}^{\varrho} \phi)(z) = \frac{1}{\Gamma(\varrho)} \int_0^z (z-s)^{\varrho-1} \phi(s) ds,$$

is called the Riemann–Liouville integral of order ϱ such that the integral on the right-hand side exists.

Definition 2. *[3] Let $n - 1 < \varrho < n$. Then, the ϱth Riemann–Liouville derivative of a continuous function $\phi : [0, \infty) \to \mathbb{R}$ is defined as*

$$\mathfrak{D}_{0+}^{\varrho} \phi(z) = \frac{1}{\Gamma(n-\varrho)} \left(\frac{d}{dz}\right)^n \int_0^z (z-s)^{n-\varrho-1} \phi(s) ds$$

$$= \left(\frac{d}{dz}\right)^n \mathfrak{I}_{0+}^{n-\varrho} \phi(z),$$

provided that the integral on the right-hand side exists and $n = [\varrho] + 1$, where $[\varrho]$ denotes the greatest integer less than ϱ.

The following properties of the fractional operators are necessary for our paper.

Lemma 1. *[2] Let $u \in L^1(0, 1)$ and $\sigma > \varrho > 0$. Then,*

- $\mathfrak{I}_{0+}^{\sigma} \mathfrak{I}_{0+}^{\varrho} u(z) = \mathfrak{I}_{0+}^{\sigma+\varrho} u(z)$,
- $\mathfrak{D}_{0+}^{\varrho} \mathfrak{I}_{0+}^{\sigma} u(z) = \mathfrak{I}_{0+}^{\sigma-\varrho} u(z)$,
- $\mathfrak{D}_{0+}^{\sigma} \mathfrak{I}_{0+}^{\sigma} u(z) = u(z)$.

Lemma 2. *[2] If $\varrho > 0$ and $\nu > 0$, then*

- $\mathfrak{D}_{0+}^{\varrho} z^{\nu-1} = \begin{cases} \dfrac{\Gamma(\nu)}{\Gamma(\nu-\varrho)} z^{\nu-\varrho-1}, \\ \mathfrak{D}_{0+}^{\varrho} z^{\nu-1} = 0, \quad \text{if} \quad \nu - \varrho \in \{0\} \cup \mathbb{Z}^-, \end{cases}$

- $\mathfrak{I}_{0+}^{\varrho} z^{\nu} = \dfrac{\Gamma(\nu+1)}{\Gamma(\nu+\varrho+1)} z^{\nu+\varrho}$.

Lemma 3. *[2] Let $n - 1 < \varrho < n$ and $u \in C(0, 1)$ and $\mathfrak{D}_{0+}^{\varrho} u \in L^1(0, 1)$. Then,*

$$\mathfrak{I}_{0+}^{\varrho} \mathfrak{D}_{0+}^{\varrho} u(z) = u(z) - \sum_{j=1}^{n} \frac{\mathfrak{I}_{0+}^{n-\varrho} u(0)}{\Gamma(\varrho - j + 1)} z^{\varrho-j},$$

where $n = [\varrho] + 1$ and $[\varrho]$ denotes the greatest integer less than ϱ.

3. Results of the Existence Criterion

In this section, we first derive an integral equation corresponding to the given multi-term multi-order RLFBVP (1) and then establish required conditions to confirm the existence of solutions for (1).

Definition 3. *The function $u(z)$ is called a solution for the suggested multi-term multi-order RLFBVP (1) if u satisfies (1) and $\mathfrak{D}_{0+}^{\varrho} u(z) \in C[0, 1]$ and $u(z) \in C[0, 1]$.*

Theorem 1. *Let* $1 < \varrho < 2, 0 < \sigma_1 < \sigma_2 < \cdots < \sigma_n < 1, \varrho > \sigma_n + 1, \mu, \nu, p, q > 0,$ *and* $0 < \xi, \eta < 1$. *Then, the function* $u(z)$ *is a solution of the RLFBVP (1) if and only if* $m(z) = \mathfrak{D}_{0+}^{\sigma_n} u(z)$ *satisfies the integral equation*

$$m(z) = \mathfrak{I}_{0+}^{\varrho-\sigma_n} \hat{h}\left(z, \mathfrak{I}_{0+}^{\sigma_n} m(z), \mathfrak{I}_{0+}^{\sigma_n-\sigma_1} m(z), \ldots, \mathfrak{I}_{0+}^{\sigma_n-\sigma_{n-1}} m(z), m(z)\right)$$

$$+ \frac{\Gamma(\varrho)}{\Gamma(\varrho-\sigma_n)} \left[\frac{p}{\Gamma(\mu)} \int_0^\xi (\xi-s)^{\mu-1} k_1\left(s, \mathfrak{I}_{0+}^{\sigma_n} m(s)\right) ds \right.$$

$$+ \frac{q}{\Gamma(\nu)} \int_0^\eta (\eta-s)^{\nu-1} k_2\left(s, \mathfrak{I}_{0+}^{\sigma_n} m(s)\right) ds$$

$$\left. - \mathfrak{I}_{0+}^{\varrho} \hat{h}\left(z, \mathfrak{I}_{0+}^{\sigma_n} m(z), \mathfrak{I}_{0+}^{\sigma_n-\sigma_1} m(z), \ldots, \mathfrak{I}_{0+}^{\sigma_n-\sigma_{n-1}} m(z), m(z)\right) \Big|_{z=1} \right] z^{\varrho-\sigma_n-1}. \quad (2)$$

Proof. In the first step, let $u(z) \in C[0,1]$ be a solution of the multi-term multi-order RLFBVP (1) which it gives $m(z) = \mathfrak{D}_{0+}^{\sigma_n} u(z) \in C[0,1]$. Applying the RL-operator $\mathfrak{I}_{0+}^{\sigma_n}$ on both sides of equation $m(z) = \mathfrak{D}_{0+}^{\sigma_n} u(z)$, we get

$$\mathfrak{I}_{0+}^{\sigma_n} m(z) = \mathfrak{I}_{0+}^{\sigma_n} \mathfrak{D}_{0+}^{\sigma_n} u(z) = u(z) - \frac{(\mathfrak{I}_{0+}^{1-\sigma_n} u)(0)}{\Gamma(\sigma_n)} z^{\sigma_n-1}. \quad (3)$$

Since $(\mathfrak{I}_{0+}^{1-\sigma_n} u)(0) = 0$, then we have

$$u(z) = \mathfrak{I}_{0+}^{\sigma_n} m(z). \quad (4)$$

In view of the second property in Lemma 1 and by (4), it follows that

$$\mathfrak{D}_{0+}^{\sigma_{n-1}} u(z) = \mathfrak{D}_{0+}^{\sigma_{n-1}} \mathfrak{I}_{0+}^{\sigma_n} m(z) = \mathfrak{I}_{0+}^{\sigma_n-\sigma_{n-1}} m(z),$$

$$\vdots = \vdots$$

$$\mathfrak{D}_{0+}^{\sigma_1} u(z) = \mathfrak{D}_{0+}^{\sigma_1} \mathfrak{I}_{0+}^{\sigma_n} m(z) = \mathfrak{I}_{0+}^{\sigma_n-\sigma_1} m(z).$$

Since $1 < \varrho < 2$, by definition of the Riemann–Liouville fractional derivative, $\mathfrak{D}_{0+}^{\varrho} u(z) = \mathfrak{D}_{0+}^{2} \mathfrak{I}_{0+}^{2-\varrho} u(z)$. Now, by (4), we get $\mathfrak{D}_{0+}^{\varrho} u(z) = \mathfrak{D}_{0+}^{2} \mathfrak{I}_{0+}^{2-\varrho} \mathfrak{I}_{0+}^{\sigma_n} m(z)$. Now, by Lemma 1, if we use the semi-group property for Riemann–Liouville fractional integrals, we have

$$\mathfrak{I}_{0+}^{2-\varrho} \mathfrak{I}_{0+}^{\sigma_n} m(z) = \mathfrak{I}_{0+}^{2-\varrho+\sigma_n} m(z).$$

Again, by definition of the Riemann–Liouville fractional derivative, we have

$$\mathfrak{D}_{0+}^{2} \mathfrak{I}_{0+}^{2-\varrho+\sigma_n} m(z) = \mathfrak{D}_{0+}^{-(-\varrho+\sigma_n)} m(z) = \mathfrak{D}_{0+}^{\varrho-\sigma_n} m(z),$$

and so

$$\mathfrak{D}_{0+}^{\varrho} u(z) = \mathfrak{D}_{0+}^{\varrho-\sigma_n} m(z).$$

Consequently, the multi-term multi-order equation illustrated by (1), becomes

$$\mathfrak{D}_{0+}^{\varrho-\sigma_n} m(z) = \hat{h}\left(z, \mathfrak{I}_{0+}^{\sigma_n} m(z), \mathfrak{I}_{0+}^{\sigma_n-\sigma_1} m(z), \ldots, \mathfrak{I}_{0+}^{\sigma_n-\sigma_{n-1}} m(z), m(z)\right), \quad 0 \le z \le 1. \quad (5)$$

Setting $\lambda = \varrho - \sigma_n > 1, \lambda_j = \sigma_n - \sigma_j, \sigma_0 = 0$ $(j = 0, 1, \ldots n)$, then (5) can be rewritten as

$$\mathfrak{D}_{0+}^{\lambda} m(z) = \hat{h}\left(z, \mathfrak{I}_{0+}^{\lambda_0} m(z), \mathfrak{I}_{0+}^{\lambda_1} m(z), \ldots, \mathfrak{I}_{0+}^{\lambda_{n-1}} m(z), m(z)\right), \quad 0 \le z \le 1. \quad (6)$$

Hence, by (4), it follows that $u(0) = 0$, and one can determine the value of the initial condition $m(0)$. Therefore, since $m(z) \in C[0,1]$,

$$\mathfrak{I}_{0+}^{\sigma_n} m(z) = \frac{1}{\Gamma(\sigma_n)} \int_0^z (z-s)^{\sigma_n - 1} m(s) ds,$$

and so we can arbitrarily provide the initial value of $m(z)$ such that $u(0) = \mathfrak{I}_{0+}^{\sigma_n} m(z)\big|_{z=0} = 0$. We assume that

$$m(0) = 0. \tag{7}$$

Now, taking the Riemann–Liouville fractional integral $\mathfrak{I}_{0+}^{\lambda}$ on both sides of (6), we find that

$$\mathfrak{I}_{0+}^{\lambda} \mathfrak{D}_{0+}^{\lambda} m(z) = \mathfrak{I}_{0+}^{\lambda} \hat{h}\big(z, \mathfrak{I}_{0+}^{\lambda_0} m(z), \mathfrak{I}_{0+}^{\lambda_1} m(z), \ldots, \mathfrak{I}_{0+}^{\lambda_{n-1}} m(z), m(z)\big), \quad 0 \leq z \leq 1. \tag{8}$$

By the hypothesis of the theorem, we have $\lambda = \varrho - \sigma_n > 1$. Then, from Lemma 3, the left-hand side of (8) becomes

$$\mathfrak{I}_{0+}^{\lambda} \mathfrak{D}_{0+}^{\lambda} m(z) = m(z) + c_1 z^{\lambda - 1} + c_2 z^{\lambda - 2},$$

hence Equation (8) is rewritten in the following form

$$m(z) = \mathfrak{I}_{0+}^{\lambda} \hat{h}\big(z, \mathfrak{I}_{0+}^{\lambda_0} m(z), \mathfrak{I}_{0+}^{\lambda_1} m(z), \ldots, \mathfrak{I}_{0+}^{\lambda_{n-1}} m(z), m(z)\big) - c_1 z^{\lambda - 1} - c_2 z^{\lambda - 2}. \tag{9}$$

By (7), since $m(0) = 0$ and $2 > \lambda > 1$, we get $c_2 = 0$. Therefore, Equation (9) becomes

$$m(z) = \mathfrak{I}_{0+}^{\lambda} \hat{h}\big(z, \mathfrak{I}_{0+}^{\lambda_0} m(z), \mathfrak{I}_{0+}^{\lambda_1} m(z), \ldots, \mathfrak{I}_{0+}^{\lambda_{n-1}} m(z), m(z)\big) - c_1 z^{\lambda - 1}. \tag{10}$$

By using the second boundary condition given in (1) and by (4), we have

$$u(1) = \mathfrak{I}_{0+}^{\sigma_n} m(z)\big|_{z=1} = p \mathfrak{I}_{0+}^{\mu} k_1\big(\xi, \mathfrak{I}_{0+}^{\sigma_n} m(\xi)\big) + q \mathfrak{I}_{0+}^{\nu} k_2\big(\eta, \mathfrak{I}_{0+}^{\sigma_n} m(\eta)\big). \tag{11}$$

With the help of Lemma 1 and from (10) and (11), we figure out that

$$u(1) = \mathfrak{I}_{0+}^{\sigma_n} m(z)\big|_{z=1}$$

$$= \mathfrak{I}_{0+}^{\sigma_n + \lambda} \hat{h}\big(z, \mathfrak{I}_{0+}^{\lambda_0} m(z), \mathfrak{I}_{0+}^{\lambda_1} m(z), \ldots, \mathfrak{I}_{0+}^{\lambda_{n-1}} m(z), m(z)\big)\big|_{z=1} - c_1 \mathfrak{I}_{0+}^{\sigma_n} z^{\lambda - 1}\big|_{z=1}$$

$$= \mathfrak{I}_{0+}^{\sigma_n + \lambda} \hat{h}\big(z, \mathfrak{I}_{0+}^{\lambda_0} m(z), \mathfrak{I}_{0+}^{\lambda_1} m(z), \ldots, \mathfrak{I}_{0+}^{\lambda_{n-1}} m(z), m(z)\big)\big|_{z=1} - c_1 \frac{\Gamma(\lambda)}{\Gamma(\lambda + \sigma_n)} z^{\lambda + \sigma_n - 1}\big|_{z=1}$$

$$= \frac{p}{\Gamma(\mu)} \int_0^{\xi} (\xi - s)^{\mu - 1} k_1(s, \mathfrak{I}_{0+}^{\sigma_n} m(s)) ds + \frac{q}{\Gamma(\nu)} \int_0^{\eta} (\eta - s)^{\nu - 1} k_2(s, \mathfrak{I}_{0+}^{\sigma_n} m(s)) ds.$$

However, we have $\lambda + \sigma_n - 1 = \varrho - \sigma_n + \sigma_n - 1 = \varrho - 1 > 0$. Then, one can write

$$\frac{p}{\Gamma(\mu)} \int_0^{\xi} (\xi - s)^{\mu - 1} k_1(s, \mathfrak{I}_{0+}^{\sigma_n} m(s)) ds + \frac{q}{\Gamma(\nu)} \int_0^{\eta} (\eta - s)^{\nu - 1} k_2(s, \mathfrak{I}_{0+}^{\sigma_n} m(s)) ds$$

$$= \mathfrak{I}_{0+}^{\sigma_n + \lambda} \hat{h}\big(z, \mathfrak{I}_{0+}^{\lambda_0} m(z), \mathfrak{I}_{0+}^{\lambda_1} m(z), \ldots, \mathfrak{I}_{0+}^{\lambda_{n-1}} m(z), m(z)\big)\big|_{z=1} - c_1 \frac{\Gamma(\lambda)}{\Gamma(\lambda + \sigma_n)}$$

$$= \mathfrak{I}_{0+}^{\varrho} \hat{h}\big(z, \mathfrak{I}_{0+}^{\sigma_n} m(z), \mathfrak{I}_{0+}^{\sigma_n - \sigma_1} m(z), \ldots, \mathfrak{I}_{0+}^{\sigma_n - \sigma_{n-1}} m(z), m(z)\big)\big|_{z=1} - c_1 \frac{\Gamma(\varrho - \sigma_n)}{\Gamma(\varrho)}.$$

Thus, we get

$$c_1 = \frac{\Gamma(\varrho)}{\Gamma(\varrho - \sigma_n)} \left[\mathfrak{I}_{0+}^{\varrho} \hbar(z, \mathfrak{I}_{0+}^{\sigma_n} m(z), \mathfrak{I}_{0+}^{\sigma_n - \sigma_1} m(z), \ldots, \mathfrak{I}_{0+}^{\sigma_n - \sigma_{n-1}} m(z), m(z)) \Big|_{z=1} \right.$$

$$- \frac{p}{\Gamma(\mu)} \int_0^{\xi} (\xi - s)^{\mu - 1} k_1(s, \mathfrak{I}_{0+}^{\sigma_n} m(s)) ds$$

$$\left. - \frac{q}{\Gamma(\nu)} \int_0^{\eta} (\eta - s)^{\nu - 1} k_2(s, \mathfrak{I}_{0+}^{\sigma_n} m(s)) ds \right].$$

By substituting the value of c_1 into Equation (10), we obtain the following equation

$$m(z) = \mathfrak{I}_{0+}^{\varrho - \sigma_n} \hbar(z, \mathfrak{I}_{0+}^{\sigma_n} m(z), \mathfrak{I}_{0+}^{\sigma_n - \sigma_1} m(z), \ldots, \mathfrak{I}_{0+}^{\sigma_n - \sigma_{n-1}} m(z), m(z))$$

$$+ \frac{\Gamma(\varrho)}{\Gamma(\varrho - \sigma_n)} \left[\frac{p}{\Gamma(\mu)} \int_0^{\xi} (\xi - s)^{\mu - 1} k_1(s, \mathfrak{I}_{0+}^{\sigma_n} m(s)) ds \right.$$

$$+ \frac{q}{\Gamma(\nu)} \int_0^{\eta} (\eta - s)^{\nu - 1} k_2(s, \mathfrak{I}_{0+}^{\sigma_n} m(s)) ds$$

$$\left. - \mathfrak{I}_{0+}^{\varrho} \hbar(z, \mathfrak{I}_{0+}^{\sigma_n} m(z), \mathfrak{I}_{0+}^{\sigma_n - \sigma_1} m(z), \ldots, \mathfrak{I}_{0+}^{\sigma_n - \sigma_{n-1}} m(z), m(z)) \Big|_{z=1} \right] z^{\varrho - \sigma_n - 1},$$

which implies that $m(z) = \mathfrak{D}_{0+}^{\sigma_n} u(z) \in C[0, 1]$ is a solution of (2).

Conversely, suppose that $m(z) = \mathfrak{D}_{0+}^{\sigma_n} u(z) \in C[0, 1]$ is a solution of (2). By applying the Riemann–Liouville fractional integral $\mathfrak{I}_{0+}^{\sigma_n}$ on both sides of $m(z) = \mathfrak{D}_{0+}^{\sigma_n} u(z)$, we have

$$\mathfrak{I}_{0+}^{\sigma_n} m(z) = \mathfrak{I}_{0+}^{\sigma_n} \mathfrak{D}_{0+}^{\sigma_n} u(z) = u(z) - \frac{(\mathfrak{I}_{0+}^{1-\sigma_n} u)(0)}{\Gamma(\sigma_n)} z^{\sigma_n - 1}.$$

Due to $(\mathfrak{I}_{0+}^{1-\sigma_n} u)(0) = 0$, we obtain $u(z) = \mathfrak{I}_{0+}^{\sigma_n} m(z)$. In the next steps, we obtain other fractional derivatives recursively and the second property in Lemma 1 as follows

$$u(z) = \mathfrak{I}_{0+}^{\sigma_n} m(z),$$

$$\mathfrak{D}_{0+}^{\sigma_{n-1}} u(z) = \mathfrak{D}_{0+}^{\sigma_{n-1}} \mathfrak{I}_{0+}^{\sigma_n} m(z) = \mathfrak{I}_{0+}^{\sigma_n - \sigma_{n-1}} m(z),$$

$$\vdots = \vdots$$

$$\mathfrak{D}_{0+}^{\sigma_1} u(z) = \mathfrak{D}_{0+}^{\sigma_1} \mathfrak{I}_{0+}^{\sigma_n} m(z) = \mathfrak{I}_{0+}^{\sigma_n - \sigma_1} m(z). \qquad (12)$$

By taking the Riemann–Liouville operator $\mathfrak{I}_{0+}^{\sigma_n}$ on both sides of (2), it becomes

$$\mathfrak{I}_{0+}^{\sigma_n} m(z) = \mathfrak{I}_{0+}^{\varrho} \hbar(z, \mathfrak{I}_{0+}^{\sigma_n} m(z), \mathfrak{I}_{0+}^{\sigma_n - \sigma_1} m(z), \ldots, \mathfrak{I}_{0+}^{\sigma_n - \sigma_{n-1}} m(z), m(z))$$

$$+ \frac{\Gamma(\varrho)}{\Gamma(\varrho - \sigma_n)} \left[\frac{p}{\Gamma(\mu)} \int_0^{\xi} (\xi - s)^{\mu - 1} k_1(s, \mathfrak{I}_{0+}^{\sigma_n} m(s)) ds \right.$$

$$+ \frac{q}{\Gamma(\nu)} \int_0^{\eta} (\eta - s)^{\nu - 1} k_2(s, \mathfrak{I}_{0+}^{\sigma_n} m(s)) ds$$

$$\left. - \mathfrak{I}_{0+}^{\varrho} \hbar(z, \mathfrak{I}_{0+}^{\sigma_n} m(z), \mathfrak{I}_{0+}^{\sigma_n - \sigma_1} m(z), \ldots, \mathfrak{I}_{0+}^{\sigma_n - \sigma_{n-1}} m(z), m(z)) \Big|_{z=1} \right] \mathfrak{I}_{0+}^{\sigma_n} z^{\varrho - \sigma_n - 1},$$

and so

$$u(z) = \mathfrak{I}_{0+}^{\varrho}\hat{\hbar}\big(z,\mathfrak{I}_{0+}^{\sigma_n}m(z),\mathfrak{I}_{0+}^{\sigma_n-\sigma_1}m(z),\ldots,\mathfrak{I}_{0+}^{\sigma_n-\sigma_{n-1}}m(z),m(z)\big)$$

$$+\frac{\Gamma(\varrho)}{\Gamma(\varrho-\sigma_n)}\left[\frac{p}{\Gamma(\mu)}\int_0^{\xi}(\xi-s)^{\mu-1}k_1\big(s,\mathfrak{I}_{0+}^{\sigma_n}m(s)\big)ds\right.$$

$$+\frac{q}{\Gamma(\nu)}\int_0^{\eta}(\eta-s)^{\nu-1}k_2\big(s,\mathfrak{I}_{0+}^{\sigma_n}m(s)\big)ds$$

$$\left.-\mathfrak{I}_{0+}^{\varrho}\hat{\hbar}\big(z,\mathfrak{I}_{0+}^{\sigma_n}m(z),\mathfrak{I}_{0+}^{\sigma_n-\sigma_1}m(z),\ldots,\mathfrak{I}_{0+}^{\sigma_n-\sigma_{n-1}}m(z),m(z)\big)\Big|_{z=1}\right]\mathfrak{I}_{0+}^{\sigma_n}z^{\varrho-\sigma_n-1}. \quad (13)$$

In the sequel, by applying the Riemann–Liouville operator $\mathfrak{D}_{0+}^{\varrho}$ on both sides of (13), it follows

$$\mathfrak{D}_{0+}^{\varrho}u(z) = \mathfrak{D}_{0+}^{\varrho}\mathfrak{I}_{0+}^{\varrho}\hat{\hbar}\big(z,\mathfrak{I}_{0+}^{\sigma_n}m(z),\mathfrak{I}_{0+}^{\sigma_n-\sigma_1}m(z),\ldots,\mathfrak{I}_{0+}^{\sigma_n-\sigma_{n-1}}m(z),m(z)\big)$$

$$+\frac{\Gamma(\varrho)}{\Gamma(\varrho-\sigma_n)}\left[\frac{p}{\Gamma(\mu)}\int_0^{\xi}(\xi-s)^{\mu-1}k_1\big(s,\mathfrak{I}_{0+}^{\sigma_n}m(s)\big)ds\right.$$

$$+\frac{q}{\Gamma(\nu)}\int_0^{\eta}(\eta-s)^{\nu-1}k_2\big(s,\mathfrak{I}_{0+}^{\sigma_n}m(s)\big)ds$$

$$\left.-\mathfrak{I}_{0+}^{\varrho}\hat{\hbar}\big(z,\mathfrak{I}_{0+}^{\sigma_n}m(z),\mathfrak{I}_{0+}^{\sigma_n-\sigma_1}m(z),\ldots,\mathfrak{I}_{0+}^{\sigma_n-\sigma_{n-1}}m(z),m(z)\big)\Big|_{z=1}\right]\mathfrak{D}_{0+}^{\varrho}\mathfrak{I}_{0+}^{\sigma_n}z^{\varrho-\sigma_n-1}.$$

Since, by Lemma 2, $\mathfrak{I}_{0+}^{\sigma_n}z^{\varrho-\sigma_n-1} = \frac{\Gamma(\varrho-\sigma_n)}{\Gamma(\varrho)}z^{\varrho-1}$ and $\mathfrak{D}_{0+}^{\varrho}z^{\varrho-1} = 0$, we get

$$\mathfrak{D}_{0+}^{\varrho}u(z) = \hat{\hbar}\big(z,\mathfrak{I}_{0+}^{\sigma_n}m(z),\mathfrak{I}_{0+}^{\sigma_n-\sigma_1}m(z),\ldots,\mathfrak{I}_{0+}^{\sigma_n-\sigma_{n-1}}m(z),m(z)\big)$$

$$+\left[\frac{p}{\Gamma(\mu)}\int_0^{\xi}(\xi-s)^{\mu-1}k_1\big(s,\mathfrak{I}_{0+}^{\sigma_n}m(s)\big)ds\right.$$

$$+\frac{q}{\Gamma(\nu)}\int_0^{\eta}(\eta-s)^{\nu-1}k_2\big(s,\mathfrak{I}_{0+}^{\sigma_n}m(s)\big)ds$$

$$\left.-\mathfrak{I}_{0+}^{\varrho}\hat{\hbar}\big(z,\mathfrak{I}_{0+}^{\sigma_n}m(z),\mathfrak{I}_{0+}^{\sigma_n-\sigma_1}m(z),\ldots,\mathfrak{I}_{0+}^{\sigma_n-\sigma_{n-1}}m(z),m(z)\big)\Big|_{z=1}\right]\mathfrak{D}_{0+}^{\varrho}z^{\varrho-1}$$

$$= \hat{\hbar}\big(z,\mathfrak{I}_{0+}^{\sigma_n}m(z),\mathfrak{I}_{0+}^{\sigma_n-\sigma_1}m(z),\ldots,\mathfrak{I}_{0+}^{\sigma_n-\sigma_{n-1}}m(z),m(z)\big). \quad (14)$$

According to (12), the fractional differential Equation (14) reduces to

$$\mathfrak{D}_{0+}^{\varrho}u(z) = \hat{\hbar}\big(z,u(z),\mathfrak{D}_{0+}^{\sigma_1}u(z),\mathfrak{D}_{0+}^{\sigma_2}u(z),\ldots,\mathfrak{D}_{0+}^{\sigma_{n-1}}u(z),\mathfrak{D}_{0+}^{\sigma_n}u(z)\big).$$

Finally, we check both boundary conditions of problem (1). In view of Equation (2) and by definition of the Riemann–Liouville integral of the function

$$\hat{\hbar}\big(z,\mathfrak{I}_{0+}^{\sigma_n}m(z),\mathfrak{I}_{0+}^{\sigma_n-\sigma_1}m(z),\ldots,\mathfrak{I}_{0+}^{\sigma_n-\sigma_{n-1}}m(z),m(z)\big)$$

of order $\varrho - \sigma_n$ at point $z = 0$, it is immediately deduced that

$$m(0) = \mathfrak{I}_{0+}^{\varrho-\sigma_n}\hat{\hbar}\big(z,\mathfrak{I}_{0+}^{\sigma_n}m(z),\mathfrak{I}_{0+}^{\sigma_n-\sigma_1}m(z),\ldots,\mathfrak{I}_{0+}^{\sigma_n-\sigma_{n-1}}m(z),m(z)\big)\Big|_{z=0}$$

$$+ \frac{\Gamma(\varrho)}{\Gamma(\varrho-\sigma_n)}\left[\frac{p}{\Gamma(\mu)}\int_0^\xi (\xi-s)^{\mu-1} k_1(s, \mathfrak{I}_{0+}^{\sigma_n} m(s))ds\right.$$

$$+ \frac{q}{\Gamma(\nu)}\int_0^\eta (\eta-s)^{\nu-1} k_2(s, \mathfrak{I}_{0+}^{\sigma_n} m(s))ds$$

$$\left.- \mathfrak{I}_{0+}^\varrho \hat{h}(z, \mathfrak{I}_{0+}^{\sigma_n} m(z), \mathfrak{I}_{0+}^{\sigma_n-\sigma_1} m(z), \ldots, \mathfrak{I}_{0+}^{\sigma_n-\sigma_{n-1}} m(z), m(z))\Big|_{z=1}\right] z^{\varrho-\sigma_n-1}\Big|_{z=0}$$

$$= 0 + 0 = 0. \tag{15}$$

Thus, $m(0) = 0$. Hence, we have $u(z) = \mathfrak{I}_{0+}^{\sigma_n} m(z)$, and so $u(0) = \mathfrak{I}_{0+}^{\sigma_n} m(z)\big|_{z=0} = 0$. Thus, $u(0) = 0$. This means that the first boundary condition holds. Now, to check the second boundary condition, by substituting $z = 1$ into (13), we obtain

$$u(1) = \mathfrak{I}_{0+}^\varrho \hat{h}(z, \mathfrak{I}_{0+}^{\sigma_n} m(z), \mathfrak{I}_{0+}^{\sigma_n-\sigma_1} m(z), \ldots, \mathfrak{I}_{0+}^{\sigma_n-\sigma_{n-1}} m(z), m(z))\big|_{z=1}$$

$$+ \frac{\Gamma(\varrho)}{\Gamma(\varrho-\sigma_n)}\left[\frac{p}{\Gamma(\mu)}\int_0^\xi (\xi-s)^{\mu-1} k_1(s, \mathfrak{I}_{0+}^{\sigma_n} m(s))ds\right.$$

$$+ \frac{q}{\Gamma(\nu)}\int_0^\eta (\eta-s)^{\nu-1} k_2(s, \mathfrak{I}_{0+}^{\sigma_n} m(s))ds$$

$$\left.- \mathfrak{I}_{0+}^\varrho \hat{h}(z, \mathfrak{I}_{0+}^{\sigma_n} m(z), \mathfrak{I}_{0+}^{\sigma_n-\sigma_1} m(z), \ldots, \mathfrak{I}_{0+}^{\sigma_n-\sigma_{n-1}} m(z), m(z))\Big|_{z=1}\right] \mathfrak{I}_{0+}^{\sigma_n} z^{\varrho-\sigma_n-1}\Big|_{z=1}$$

$$= \mathfrak{I}_{0+}^\varrho \hat{h}(z, \mathfrak{I}_{0+}^{\sigma_n} m(z), \mathfrak{I}_{0+}^{\sigma_n-\sigma_1} m(z), \ldots, \mathfrak{I}_{0+}^{\sigma_n-\sigma_{n-1}} m(z), m(z))\big|_{z=1}$$

$$+ \frac{\Gamma(\varrho)}{\Gamma(\varrho-\sigma_n)}\left[\frac{p}{\Gamma(\mu)}\int_0^\xi (\xi-s)^{\mu-1} k_1(s, \mathfrak{I}_{0+}^{\sigma_n} m(s))ds\right.$$

$$+ \frac{q}{\Gamma(\nu)}\int_0^\eta (\eta-s)^{\nu-1} k_2(s, \mathfrak{I}_{0+}^{\sigma_n} m(s))ds$$

$$\left.- \mathfrak{I}_{0+}^\varrho \hat{h}(z, \mathfrak{I}_{0+}^{\sigma_n} m(z), \mathfrak{I}_{0+}^{\sigma_n-\sigma_1} m(z), \ldots, \mathfrak{I}_{0+}^{\sigma_n-\sigma_{n-1}} m(z), m(z))\Big|_{z=1}\right] \frac{\Gamma(\varrho-\sigma_n)}{\Gamma(\varrho)} z^{\varrho-1}\Big|_{z=1}$$

$$= p\mathfrak{I}_{0+}^\mu k_1(\xi, u(\xi)) + q\mathfrak{I}_{0+}^\nu k_2(\eta, u(\eta)).$$

Therefore, we figure out that $u(z)$ satisfies the multi-term multi-order RLFBVP (1) and so u will be a solution of the mentioned RLFBVP, and the proof is completed. □

Here, we introduce the Banach space $E = C[0, 1]$ with the norm $\|m\| = \max_{z \in [0,1]} |m(z)|$, and, along with this, by Theorem 1, we define an operator $\Psi : E \longrightarrow E$ by

$$(\Psi m)(z) = \mathfrak{I}_{0+}^{\varrho-\sigma_n} \hat{h}(z, \mathfrak{I}_{0+}^{\sigma_n} m(z), \mathfrak{I}_{0+}^{\sigma_n-\sigma_1} m(z), \ldots, \mathfrak{I}_{0+}^{\sigma_n-\sigma_{n-1}} m(z), m(z))$$

$$+ \frac{\Gamma(\varrho)}{\Gamma(\varrho-\sigma_n)}\left[\frac{p}{\Gamma(\mu)}\int_0^\xi (\xi-s)^{\mu-1} k_1(s, \mathfrak{I}_{0+}^{\sigma_n} m(s))ds\right.$$

$$+ \frac{q}{\Gamma(\nu)}\int_0^\eta (\eta-s)^{\nu-1} k_2(s, \mathfrak{I}_{0+}^{\sigma_n} m(s))ds$$

$$\left.- \mathfrak{I}_{0+}^\varrho \hat{h}(z, \mathfrak{I}_{0+}^{\sigma_n} m(z), \mathfrak{I}_{0+}^{\sigma_n-\sigma_1} m(z), \ldots, \mathfrak{I}_{0+}^{\sigma_n-\sigma_{n-1}} m(z), m(z))\Big|_{z=1}\right] z^{\varrho-\sigma_n-1}. \tag{16}$$

We clearly have the following equation

$$\Psi m = m, \quad m \in E, \qquad (17)$$

which is equivalent to Equation (2). If Ψ has a fixed point, then it will be the solution of the multi-term multi-order RLFBVP (1). On the other side, notice that the continuity of all three functions $\hat{h}, k_1$, and k_2 confirms that of the operator Ψ. In this place, we want to express the existence theorem in relation to solutions of the multi-term multi-order RLFBVP (1).

Theorem 2. *Assume that these assumptions are valid:*
($\mathcal{AS}1$) There exist real constants $M_j (j = 0, 1, \ldots, n)$ such that

$$\left| \hat{h}(z, u_0, u_1, \ldots, u_n) - \hat{h}(z, U_0, U_1, \ldots, U_n) \right| \leq \sum_{j=0}^{n} M_j |u_j - U_j|,$$

for all $z \in [0, 1]$ and $(u_0, u_1, \ldots, u_n), (U_0, U_1, \ldots, U_n) \in \mathbb{R}^{n+1}$.
($\mathcal{AS}2$) There exist two real constants $\theta_1, \theta_2 > 0$ such that

$$|k_1(z, m) - k_1(z, u)| \leq \theta_1 |m - u|, \quad m, u \in \mathbb{R},$$
$$|k_2(z, m) - k_2(z, u)| \leq \theta_2 |m - u|, \quad m, u \in \mathbb{R}.$$

($\mathcal{AS}3$) Let

$$0 < \Phi = \frac{\Gamma(\varrho) p \theta_1 \xi^{\mu}}{\Gamma(\varrho - \sigma_n)\Gamma(\mu + 1)\Gamma(\sigma_n + 1)} + \frac{\Gamma(\varrho) q \theta_2 \eta^{\nu}}{\Gamma(\varrho - \sigma_n)\Gamma(\nu + 1)\Gamma(\sigma_n + 1)}$$

$$+ \sum_{j=0}^{n} \left[\frac{M_j}{\Gamma(\varrho - \sigma_j + 1)} + \frac{M_j \Gamma(\varrho)}{\Gamma(\varrho - \sigma_n)\Gamma(\varrho + \sigma_n - \sigma_j + 1)} \right] < 1.$$

Then, the multi-term multi-order RLFBVP (1) has a unique solution.

Proof. In view of Theorem 1, it is explicit that the existence of solutions to the multi-term multi-order RLFBVP (1) is derived from the existence of solutions to Equation (16) or (17). Thus, it suffices to prove that (16) has a unique fixed point. Now, let $\lambda = \varrho - \sigma_n$, $\sigma_0 = 0$, and $\lambda_j = \sigma_n - \sigma_j$ for $j = 0, 1, \ldots, n$. Then, from ($\mathcal{AS}1$), it follows that for any $m_1, m_2 \in E$, we have

$$\left| \hat{h}(z, \mathfrak{I}_{0+}^{\lambda_0} m_1(z), \ldots, \mathfrak{I}_{0+}^{\lambda_{n-1}} m_1(z), m_1(z)) - \hat{h}(z, \mathfrak{I}_{0+}^{\lambda_0} m_2(z), \ldots, \mathfrak{I}_{0+}^{\lambda_{n-1}} m_2(z), m_2(z)) \right|$$

$$\leq \sum_{j=0}^{n} M_j \left| \mathfrak{I}_{0+}^{\lambda_j} m_1(z) - \mathfrak{I}_{0+}^{\lambda_j} m_2(z) \right|. \qquad (18)$$

Taking the Riemann–Liouville operator $\mathfrak{I}_{0+}^{\lambda}$ on both sides of inequality (18), we find that

$$\mathfrak{I}_{0+}^{\lambda} \left| \hat{h}(z, \mathfrak{I}_{0+}^{\lambda_0} m_1(z), \ldots, \mathfrak{I}_{0+}^{\lambda_{n-1}} m_1(z), m_1(z)) - \hat{h}(z, \mathfrak{I}_{0+}^{\lambda_0} m_2(z), \ldots, \mathfrak{I}_{0+}^{\lambda_{n-1}} m_2(z), m_2(z)) \right|$$

$$\leq \mathfrak{I}_{0+}^{\lambda} \sum_{j=0}^{n} M_j \left| \mathfrak{I}_{0+}^{\lambda_j} m_1(z) - \mathfrak{I}_{0+}^{\lambda_j} m_2(z) \right| \leq \sum_{j=0}^{n} M_j \mathfrak{I}_{0+}^{\lambda + \lambda_j} |m_1(z) - m_2(z)|$$

$$\leq \|m_1 - m_2\| \sum_{j=0}^{n} \frac{M_j}{\Gamma(\lambda + \lambda_j + 1)} = \|m_1 - m_2\| \sum_{j=0}^{n} \frac{M_j}{\Gamma(\varrho - \sigma_n + \sigma_n - \sigma_j + 1)}$$

$$= \|m_1 - m_2\| \sum_{j=0}^{n} \frac{M_j}{\Gamma(\varrho - \sigma_j + 1)}. \tag{19}$$

On the other side, by using ($\mathcal{AS}2$), we get

$$\left| \frac{\Gamma(\varrho)}{\Gamma(\lambda)} \left[\frac{p}{\Gamma(\mu)} \int_0^{\xi} (\xi-s)^{\mu-1} k_1(s, \mathfrak{I}_{0^+}^{\sigma_n} m_1(s)) ds + \frac{q}{\Gamma(\nu)} \int_0^{\eta} (\eta-s)^{\nu-1} k_2(s, \mathfrak{I}_{0^+}^{\sigma_n} m_1(s)) ds \right. \right.$$

$$\left. - \mathfrak{I}_{0^+}^{\sigma_n + \lambda} \hbar(z, \mathfrak{I}_{0^+}^{\lambda_0} m_1(z), \ldots, \mathfrak{I}_{0^+}^{\lambda_{n-1}} m_1(z), m_1(z)) \Big|_{z=1} \right] z^{\lambda-1}$$

$$- \frac{\Gamma(\varrho)}{\Gamma(\lambda)} \left[\frac{p}{\Gamma(\mu)} \int_0^{\xi} (\xi-s)^{\mu-1} k_1(s, \mathfrak{I}_{0^+}^{\sigma_n} m_2(s)) ds + \frac{q}{\Gamma(\nu)} \int_0^{\eta} (\eta-s)^{\nu-1} k_2(s, \mathfrak{I}_{0^+}^{\sigma_n} m_2(s)) ds \right.$$

$$\left. \left. - \mathfrak{I}_{0^+}^{\sigma_n + \lambda} \hbar(z, \mathfrak{I}_{0^+}^{\lambda_0} m_2(z), \ldots, \mathfrak{I}_{0^+}^{\lambda_{n-1}} m_2(z), m_2(z)) \Big|_{z=1} \right] z^{\lambda-1} \right|$$

$$\leq \frac{\Gamma(\varrho)}{\Gamma(\lambda)} \left[\frac{p}{\Gamma(\mu)} \int_0^{\xi} (\xi-s)^{\mu-1} \left| k_1(s, \mathfrak{I}_{0^+}^{\sigma_n} m_1(s)) - k_1(s, \mathfrak{I}_{0^+}^{\sigma_n} m_2(s)) \right| ds \right.$$

$$+ \frac{q}{\Gamma(\nu)} \int_0^{\eta} (\eta-s)^{\nu-1} \left| k_2(s, \mathfrak{I}_{0^+}^{\sigma_n} m_1(s)) - k_2(s, \mathfrak{I}_{0^+}^{\sigma_n} m_2(s)) \right| ds$$

$$\left. + \mathfrak{I}_{0^+}^{\sigma_n + \lambda} \left| \hbar(z, \mathfrak{I}_{0^+}^{\lambda_0} m_1(z), \ldots, \mathfrak{I}_{0^+}^{\lambda_{n-1}} m_1(z), m_1(z)) - \hbar(z, \mathfrak{I}_{0^+}^{\lambda_0} m_2(z), \ldots, \mathfrak{I}_{0^+}^{\lambda_{n-1}} m_2(z), m_2(z)) \right| \Big|_{z=1} \right]$$

$$\leq \frac{\Gamma(\varrho)}{\Gamma(\lambda)} \left[\frac{p\theta_1}{\Gamma(\mu)} \int_0^{\xi} (\xi-s)^{\mu-1} \left| \mathfrak{I}_{0^+}^{\sigma_n} m_1(s) - \mathfrak{I}_{0^+}^{\sigma_n} m_2(s) \right| ds \right.$$

$$\left. + \frac{q\theta_2}{\Gamma(\nu)} \int_0^{\eta} (\eta-s)^{\nu-1} \left| \mathfrak{I}_{0^+}^{\sigma_n} m_1(s) - \mathfrak{I}_{0^+}^{\sigma_n} m_2(s) \right| ds + \mathfrak{I}_{0^+}^{\sigma_n + \lambda} \sum_{j=0}^{n} M_j \mathfrak{I}_{0^+}^{\lambda_j} |m_1(z) - m_2(z)| \Big|_{z=1} \right] \tag{20}$$

$$\leq \frac{\Gamma(\varrho)}{\Gamma(\lambda)} \left[\frac{p\theta_1 \xi^{\mu}}{\Gamma(\mu+1)\Gamma(\sigma_n+1)} + \frac{q\theta_2 \eta^{\nu}}{\Gamma(\nu+1)\Gamma(\sigma_n+1)} + \sum_{j=0}^{n} \frac{M_j}{\Gamma(\sigma_n + \lambda + \lambda_j + 1)} \right] \|m_1 - m_2\|$$

$$= \frac{\Gamma(\varrho)}{\Gamma(\varrho - \sigma_n)} \left[\frac{p\theta_1 \xi^{\mu}}{\Gamma(\mu+1)\Gamma(\sigma_n+1)} + \frac{q\theta_2 \eta^{\nu}}{\Gamma(\nu+1)\Gamma(\sigma_n+1)} + \sum_{j=0}^{n} \frac{M_j}{\Gamma(\sigma_n + \varrho - \sigma_j + 1)} \right] \|m_1 - m_2\|.$$

Consequently, by adding both sides of (19) and (20) and according to the definition of Ψ in (16), we have

$$|\Psi m_1(z) - \Psi m_2(z)| \leq \left[\frac{\Gamma(\varrho) p \theta_1 \xi^{\mu}}{\Gamma(\varrho - \sigma_n)\Gamma(\mu+1)\Gamma(\sigma_n+1)} + \frac{\Gamma(\varrho) q \theta_2 \eta^{\nu}}{\Gamma(\varrho - \sigma_n)\Gamma(\nu+1)\Gamma(\sigma_n+1)} \right.$$

$$\left. + \sum_{j=0}^{n} \left(\frac{M_j}{\Gamma(\varrho - \sigma_j + 1)} + \frac{M_j \Gamma(\varrho)}{\Gamma(\varrho - \sigma_n)\Gamma(\varrho + \sigma_n - \sigma_j + 1)} \right) \right] \|m_1 - m_2\|.$$

By using ($\mathcal{AS}3$), we find

$$\|\Psi m_1 - \Psi m_2\| \leq \Phi \|m_1 - m_2\|,$$

where $\Phi \in (0, 1)$. Hence, by the Banach fixed point theorem [52], it follows that Ψ has a unique fixed point which points out that the suggested multi-term multi-order RLFBVP (1) has a unique solution. □

4. Approximation of Solutions via DGJIM and ADM Methods

This section is devoted to implementing the numerical methods named DGJIM and ADM. Indeed, we here state how we can employ these methods to our suggested multi-term multi-order RLFBVP. In both algorithms, appropriate recursion relations are formulated to approximate the solutions of (1) along with their convergence. Our techniques are inspired by [47,48].

4.1. DGJIM Numerical Method

We prove above that the solutions of Equations (1) and (2) are equivalent. Thus, we now suppose that the right-hand side of (17) is written under the following decomposition (not uniquely)

$$(\Psi m)(z) = \widetilde{L}(m(z)) + \widetilde{N}(m(z)) + \zeta(z),$$

where the operator $\widetilde{L}$ is linear, the operator $\widetilde{N}$ stands for the nonlinear terms, and ζ is a known function. Then, one can rewrite (2) in the decomposed form

$$m(z) = \widetilde{L}(m(z)) + \widetilde{N}(m(z)) + \zeta(z). \tag{21}$$

Suppose that the solution of (21) is written as a series as follows

$$m(z) = \sum_{n=0}^{+\infty} m_n(z). \tag{22}$$

By combining (22) and (21), we get

$$\sum_{n=0}^{+\infty} m_n(z) = \widetilde{L}\Big(\sum_{n=0}^{+\infty} m_n(z)\Big) + \widetilde{N}\Big(\sum_{n=0}^{+\infty} m_n(z)\Big) + \zeta(z). \tag{23}$$

Since $\widetilde{L}$ is linear, by a simple manipulation, we obtain the following algorithm known as the DGJIM numerical method:

$$\begin{cases} m_0(z) = \zeta(z), \\ m_1(z) = \widetilde{L}(m_0(z)) + \widetilde{N}(m_0(z)), \\ m_2(z) = \widetilde{L}(m_1(z)) + \widetilde{N}(m_0(z) + m_1(z)) - \widetilde{N}(m_0(z)), \\ m_3(z) = \widetilde{L}(m_2(z)) + \widetilde{N}(m_0(z) + m_1(z) + m_2(z)) - \widetilde{N}(m_0(z) + m_1(z)), \\ \vdots = \vdots \\ m_n(z) = \widetilde{L}(m_{n-1}(z)) + \widetilde{N}\Big(\sum_{i=0}^{n-1} m_i(z)\Big) - \widetilde{N}\Big(\sum_{i=0}^{n-2} m_i(z)\Big), \\ \vdots = \vdots \end{cases} \tag{24}$$

Therefore, we can obtain the n-term approximate solution of the integral Equation (2) as

$$w_n(z) = \sum_{i=0}^{n} m_i(z). \tag{25}$$

In view of (25), we simply get

$$m_n(z) = w_n(z) - w_{n-1}(z). \tag{26}$$

Thus, a combination of (24) and (26) gives

$$w_n(z) = w_{n-1}(z) + \tilde{L}(w_{n-1}(z) - w_{n-2}(z)) + \tilde{N}(w_{n-1}(z)) - \tilde{N}(w_{n-2}(z)). \tag{27}$$

Now, let

$$\|\tilde{L}m - \tilde{L}u\| \leq \mu_1 \|m - u\|, \quad 0 < \mu_1 < 1,$$

$$\|\tilde{N}m - \tilde{N}u\| \leq \mu_2 \|m - u\|, \quad 0 < \mu_2 < 1,$$

where $\mu_1 + \mu_2 < 1$. Therefore, the Banach fixed point principle guarantees the existence of a unique solution $\tilde{w}(z)$ for (21) and so for the integral Equation (2). According to the relation (27), the following iterative expression is derived

$$\|w_n - w_{n-1}\| \leq \mu_1 \|w_{n-1} - w_{n-2}\| + \mu_2 \|w_{n-1} - w_{n-2}\|$$

$$= (\mu_1 + \mu_2)\|w_{n-1} - w_{n-2}\|$$

$$\leq (\mu_1 + \mu_2)^2 \|w_{n-2} - w_{n-3}\|$$

$$\leq \vdots$$

$$\leq (\mu_1 + \mu_2)^{n-1} \|w_1 - w_0\|,$$

which implies the absolute convergence and the uniform convergence of the sequence $\{w_n\}$ to the exact solution $\tilde{w}(z)$.

4.2. ADM Numerical Method

To implement the ADM numerical method, the nonlinear term $\tilde{N}\left(\sum\limits_{n=0}^{+\infty} m_n(z)\right)$ introduced in (23) is decomposed into a series of Adomian polynomials as

$$\tilde{N}\left(\sum_{n=0}^{+\infty} m_n(z)\right) = \sum_{n=0}^{+\infty} A_n(m_0, m_1, \ldots, m_n),$$

where $A_n(m_0, m_1, \ldots, m_n)$ is produced by

$$A_n(m_0, m_1, \ldots, m_n) = \frac{1}{n!} \frac{\partial^n}{\partial z^n}\left[\tilde{N}\left(\sum_{k=0}^{+\infty} m_k z^k\right)\right]_{z=0}, \quad (n \in \mathbb{N} \cup \{0\}). \tag{28}$$

Consequently, Equation (23) reduces to

$$\sum_{n=0}^{+\infty} m_n(z) = \tilde{L}\left(\sum_{n=0}^{+\infty} m_n(z)\right) + \sum_{n=0}^{+\infty} A_n(m_0(z), m_1(z), \ldots, m_n(z)) + \zeta(z),$$

which gives us the following iterative schemes called the ADM method:

$$\begin{cases} m_0(z) = \zeta(z), \\ m_1(z) = \widetilde{L}(m_0(z)) + A_0(m_0(z), m_1(z), \ldots, m_n(z)), \\ m_2(z) = \widetilde{L}(m_1(z)) + A_1(m_0(z), m_1(z), \ldots, m_n(z)), \\ m_3(z) = \widetilde{L}(m_2(z)) + A_2(m_0(z), m_1(z), \ldots, m_n(z)), \\ \vdots \\ m_n(z) = \widetilde{L}(m_{n-1}(z)) + A_{n-1}(m_0(z), m_1(z), \ldots, m_n(z)), \\ \vdots \end{cases} \quad (29)$$

Finally, by writing M-term approximate solution of the integral Equation (2) as

$$w_M(z) = \sum_{n=0}^{M} m_n(z), \quad (30)$$

we obtain the exact solution of (2) by

$$m(z) = \lim_{M \to +\infty} w_M(z). \quad (31)$$

Lastly, we find that the approximate solutions and the exact solution of the multi-term multi-order RLFBVP (1) are extracted as $u_n(z) = \mathfrak{I}_{0^+}^{\sigma_n} w_n(z)$ and $u(z) = \mathfrak{I}_{0^+}^{\sigma_n} m(z)$, respectively.

5. Application

Here, we prepare two distinct examples. In the first, the theoretical existence results are examined, and, in the second, the approximate solutions of a given RLFBVP are obtained with the help of the DGJIM and ADM numerical methods introduced above. Note that, in the second example, we compare the approximate solutions obtained by two mentioned numerical methods with the exact ones for different given fractional orders.

Example 1. *Let us consider the following RLFBVP*

$$\begin{cases} \mathfrak{D}_{0^+}^{1.8} u(z) = z^2 + \dfrac{1}{8}\sin(2u(z)) + \dfrac{1}{4}\mathfrak{D}_{0^+}^{0.4} u(z) + \dfrac{2}{10}\arctan\left(\mathfrak{D}_{0^+}^{0.5} u(z)\right), \quad z \in (0,1), \\ u(0) = 0, \\ u(1) = 6\displaystyle\int_0^{\frac{1}{2}} \dfrac{(1-2s)^3(1+u(s))}{8\Gamma(4)(4+s^2)} ds + 24\displaystyle\int_0^{\frac{1}{4}} \dfrac{(1-4s)^4(e^{-s}+\sin(u(s)))}{\Gamma(5)1024} ds, \end{cases}$$

where we take data $\varrho = 1.8$, $n = 2$, $\sigma_0 = 0$, $\sigma_1 = 0.4$, $\sigma_2 = 0.5$, $\xi = \dfrac{1}{2}$, $\eta = \dfrac{1}{4}$, $p = 6$, $q = 24$, $\mu = 4$, and $\nu = 5$. Along with these, continuous functions

$$\hbar(z, s(z), x(z), y(z)) = z^2 + \dfrac{1}{8}\sin(2s(z)) + \dfrac{1}{4}x(z) + \dfrac{2}{10}\arctan(y(z)),$$

and

$$k_1(z, u(z)) = \dfrac{1 + u(z)}{4 + z^2}, k_2(z, u(z)) = \dfrac{e^{-z} + \sin(u(z))}{4},$$

are defined on their domain. Clearly, $M_0 = M_1 = 0.25$ and $M_2 = 0.2$. On the other side, we get

$$|k_1(z, u(z)) - k_1(z, U(z))| \leq \left|\frac{1 + u(z)}{4 + z^2} - \frac{1 + U(z)}{4 + z^2}\right| \leq \frac{1}{4 + z^2}|u(z) - U(z)|,$$

and

$$|k_2(z, u(z)) - k_2(z, U(z))| \leq \left|\frac{e^{-z} + \sin(u(z))}{4} - \frac{e^{-z} + \sin(U(z))}{4}\right| \leq \frac{1}{4}|u(z) - U(z)|.$$

Thus, $\theta_1 = \theta_2 = 0.25$. In addition,

$$\Phi = \frac{\Gamma(\varrho)p\theta_1\xi^\mu}{\Gamma(\varrho - \sigma_n)\Gamma(\mu + 1)\Gamma(\sigma_n + 1)} + \frac{\Gamma(\varrho)q\theta_2\eta^\nu}{\Gamma(\varrho - \sigma_n)\Gamma(\nu + 1)\Gamma(\sigma_n + 1)}$$

$$+ \sum_{j=0}^{n}\left[\frac{M_j}{\Gamma(\varrho - \sigma_j + 1)} + \frac{M_j\Gamma(\varrho)}{\Gamma(\varrho - \sigma_n)\Gamma(\varrho + \sigma_n - \sigma_j + 1)}\right] \approx 0.8951 < 1$$

In consequence, by Theorem 2, a unique solution exists for the multi-term multi-order RLFBVP considered above.

For the next example, we consider three different cases for the order of the proposed RLFBVP and compare obtained approximate results with exact outcomes, which shows the effectiveness of both DGJIM and ADM numerical methods together.

Example 2. *In the present example, we consider three distinct values for ϱ as $\varrho = 1.4$, $\varrho = 1.7$ and $\varrho = 1.9$.*
- **Case(I)** :$\varrho = 1.4$: *Let us consider the following RLFBVP which has a structure as*

$$\begin{cases} \mathfrak{D}_{0^+}^{1.4}u(z) = u(z) + \mathfrak{D}_{0^+}^{0.3}u(z) + \hat{\varphi}(z), & z \in (0, 1), \\ u(0) = 0, u(1) = 8\int_0^{\frac{1}{2}} u(s)\,ds + 54\int_0^{\frac{1}{3}} u(s)\,ds, \end{cases} \tag{32}$$

where

$$\hat{\varphi}(z) = \frac{2}{\Gamma(1.6)}z^{0.6} - \frac{2}{\Gamma(2.7)}z^{1.7} - z^2.$$

In this problem, we have taken data $\varrho = 1.4$, $\xi = 1/2$, $\eta = 1/3$, $\sigma_n = 0.3$, $\mu = \nu = 1$, $p = 8$ and $q = 54$. It is known that $\varrho - \sigma_n = 1.1 > 1$. In addition, $k_1(z, u(z)) = k_2(z, u(z)) = u(z)$ for $z \in [0, 1]$. By assuming $m(z) = \mathfrak{D}_{0^+}^{0.3}u(z)$, the equivalent integral equation of the problem (32) is the following

$$m(z) = \mathfrak{J}_{0^+}^{1.1}[\mathfrak{J}_{0^+}^{0.3}m(z) + m(z) + \hat{\varphi}(z)] + \frac{\Gamma(1.4)}{\Gamma(1.1)}\left(8\int_0^{\frac{1}{2}}\mathfrak{J}_{0^+}^{0.3}m(s)\,ds\right.$$

$$\left.+ 54\int_0^{\frac{1}{3}}\mathfrak{J}_{0^+}^{0.3}m(s)\,ds - \mathfrak{J}_{0^+}^{1.4}[\mathfrak{J}_{0^+}^{0.3}m(z) + m(z) + \hat{\varphi}(z)]\bigg|_{z=1}\right)z^{0.1}$$

$$= \mathfrak{J}_{0^+}^{1.4}m(z) + \mathfrak{J}_{0^+}^{1.1}m(z) + \mathfrak{J}_{0^+}^{1.1}\hat{\varphi}(z) + z^{0.1}\frac{8\Gamma(1.4)}{\Gamma(1.1)}\int_0^{\frac{1}{2}}\mathfrak{J}_{0^+}^{0.3}m(s)\,ds$$

$$+ z^{0.1}\frac{54\Gamma(1.4)}{\Gamma(1.1)}\int_0^{\frac{1}{3}}\mathfrak{J}_{0^+}^{0.3}m(s)\,ds - \frac{\Gamma(1.4)z^{0.1}}{\Gamma(1.1)}(\mathfrak{J}_{0^+}^{1.7}m(z)|_{z=1})$$

$$- \frac{\Gamma(1.4)z^{0.1}}{\Gamma(1.1)}(\mathfrak{J}_{0^+}^{1.4}m(z)|_{z=1}) - \frac{\Gamma(1.4)z^{0.1}}{\Gamma(1.1)}(\mathfrak{J}_{0^+}^{1.4}\hat{\varphi}(z)|_{z=1}). \tag{33}$$

Thus, we decompose the right-hand side of (33) as

$$m(z) = \widetilde{\mathbf{L}}(m(z)) + \widetilde{\mathbf{N}}(m(z)) + \zeta(z),$$

where

$$\widetilde{\mathbf{L}}(m(z)) = \mathfrak{I}_{0^+}^{1.4} m(z) - \mathfrak{I}_{0^+}^{1.1} m(z),$$

$$\widetilde{\mathbf{N}}(m(z)) = \frac{8\Gamma(1.4)z^{0.1}}{\Gamma(1.1)} \int_0^{\frac{1}{2}} \mathfrak{I}_{0^+}^{0.3} m(s) ds + \frac{54\Gamma(1.4)z^{0.1}}{\Gamma(1.1)} \int_0^{\frac{1}{3}} \mathfrak{I}_{0^+}^{0.3} m(s) ds$$

$$- \frac{\Gamma(1.4)z^{0.1}}{\Gamma(1.1)} \left(\mathfrak{I}_{0^+}^{1.7} m(z)\big|_{z=1}\right) - \frac{\Gamma(1.4)z^{0.1}}{\Gamma(1.1)} \left(\mathfrak{I}_{0^+}^{1.4} m(z)\big|_{z=1}\right),$$

and

$$\zeta(z) = \mathfrak{I}_{0^+}^{1.1} \hat{\varphi}(z) - \frac{\Gamma(1.4)z^{0.1}}{\Gamma(1.1)} \left(\mathfrak{I}_{0^+}^{1.4} \hat{\varphi}(z)\big|_{z=1}\right).$$

Then, the sequence of approximate solutions of (32) and (33) are obtained by means of algorithms of the DGJIM and ADM methods as follows:

- **Approximate solutions via DGJIM method for $\varrho = 1.4$:**

By using the suggested algorithm known as DGJIM numerical method in (24), we get

$$m_0(z) = 1.2948z^{1.7} - 0.4262z^{2.8} - 0.2936z^{3.1} - 0.4748z^{0.1},$$

$$m_1(z) = 0.2936z^{3.1} - 0.1228z^{4.2} - 0.0382z^{4.5} - 0.2398z^{1.5} + 0.4261z^{2.8} - 0.0968z^{3.9},$$

$$m_2(z) = 0.172z^{3.5} - 0.0165z^{5.6} - 0.0033z^{5.9} - 0.0852z^{2.9} + 0.1228z^{4.2} - 0.0297z^{5.3}$$
$$- 0.0430z^{2.6} - 0.0315z^{1.6} + 0.0968z^{3.9} - 0.0167z^5 - 0.1683z^{2.3} * 0.4086z^{1.2} + 4.4196z^{0.1}.$$

Therefore,

$$w_0(z) = 1.2948z^{1.7} - 0.4262z^{2.8} - 0.2936z^{3.1} - 0.4748z^{0.1},$$

$$w_1(z) = 1.2948z^{1.7} - 0.1228z^{4.2} - 0.0382z^{4.5} - 0.3398z^{1.5}$$
$$- 0.0968z^{3.9} - 0.4100z^{1.2} - 4.4292z^{0.1},$$

$$w_2(z) = 1.2948z^{1.7} - 0.0165z^{5.6} - 0.0033z^{5.9} - 0.0852z^{2.9}$$
$$- 0.0297z^{5.3} - 0.0430z^{2.6} - 0.0315z^{1.6} - 0.0167z^5$$
$$- 0.0683z^{2.3} - 0.0014z^{1.2} - 0.0096z^{0.1} - 0.0382z^{4.5} + 0.2398z^{1.5},$$

and

$$u_0(z) = z^2 - 0.2937z^{3.1} - 0.1973z^{3.4} - 0.5091z^{0.4},$$

$$u_1(z) = z^2 - 0.0764z^{4.5} - 0.0234z^{4.8} - 0.2694z^{1.8}$$
$$- 0.0614z^{4.2} - 0.3398z^{1.5} - 4.7491z^{0.4},$$

$$u_2(z) = z^2 - 0.0095z^{5.9} - 0.0019z^{6.9} - 0.0582z^{3.2} - 0.0174z^{5.6}$$
$$- 0.0302z^{2.9} - 0.0246z^{1.9} - 0.0099z^{5.3} - 0.0493z^{2.6} - 0.0012z^{1.5}$$
$$- 0.0103z^{0.4} - 0.0234z^{4.8} + 0.1901z^{1.8}.$$

- **Approximate solutions via ADM method for $\varrho = 1.4$:**
  By using the suggested algorithm known as ADM numerical method in (29), we get

$$m_0(z) = 1.2948z^{1.7} - 0.4262z^{2.8} - 0.2936z^{3.1} - 0.4748z^{0.1},$$

$$m_1(z) = 0.2936z^{3.1} - 0.1228z^{4.2} - 0.0382z^{4.5} - 0.3398z^{1.5} + 0.4261z^{2.8} - 0.0968z^{3.9} - 0.4100z^{1.2} - 3.9544z^{0.1},$$

$$m_2(z) = 0.1720z^{3.5} - 0.0165z^{5.6} - 0.0033z^{5.9} - 0.0852z^{2.9} + 0.1228z^{4.2} - 0.0297z^{5.3} - 0.2430z^{2.6}$$
$$- 0.0015z^{1.6} + 0.0968z^{3.9} - 0.0167z^{5} - 0.1683z^{2.3} + 0.4050z^{1.2}.$$

Therefore,

$$w_0(z) = 1.2948z^{1.7} - 0.4262z^{2.8} - 0.2936z^{3.1} - 0.4748z^{0.1},$$

$$w_1(z) = 1.2948z^{1.7} + 4.4110z^{0.1} - 0.1228z^{4.2} - 0.0382z^{4.5}$$
$$+ 0.3398z^{1.5} - 0.0968z^{3.9} - 0.4100z^{1.2},$$

$$w_2(z) = 1.2948z^{1.7} + 0.1720z^{3.5} - 0.0166z^{5.6} - 0.0033z^{5.9} - 0.0852z^{2.9}$$
$$- 0.0297z^{5.3} - 0.2430z^{2.6} - 0.0015z^{1.6} - 0.0167z^{5} - 0.1683z^{2.3}$$
$$- 0.0044z^{1.2} - 0.0282z^{0.1} - 0.0382z^{4.5} + 0.3398z^{1.5},$$

and

$$u_0(z) = z^2 - 0.0716z^{4.1} - 0.1973z^{3.4} - 0.5091z^{0.4},$$

$$u_1(z) = z^2 + 4.7296z^{0.4} - 0.0764z^{4.5} - 0.0234z^{4.8}$$
$$+ 0.2694z^{1.8} - 0.0614z^{4.2} - 0.3398z^{1.5},$$

$$u_2(z) = z^2 + 0.1122z^{3.8} - 0.0095z^{5.9} - 0.0019z^{6.2} - 0.0582z^{3.2}$$
$$- 0.0174z^{5.6} - 0.1704z^{2.9} - 0.0012z^{1.9} - 0.0099z^{5.3} - 0.1215z^{2.6}$$
$$- 0.0036z^{1.5} - 0.0302z^{0.4} - 0.0234z^{4.8} + 0.2694z^{1.8}.$$

In this case, the graphs of the three-term approximate solutions obtained by the DGJIM and ADM algorithms for the suggested RLFBVP (32) and the integral Equation (33) are plotted in Figure 1.

Note that, in view of Theorem 1, we prove that $u(z)$ is the solution of RLFBVP (1) if and only if $m(z) = \mathfrak{D}_{0+}^{\sigma_n} u(z)$ is the solution of the integral Equation (2). Now, in the case $\varrho = 1.4$, since the exact solution of RLFBVP is given by $u(z) = z^2$, the corresponding exact solution of the equivalent integral equation is

$$m(z) = \mathfrak{D}_{0+}^{\sigma_n} z^2 = \mathfrak{D}_{0+}^{0.3} z^2 = \frac{2}{\Gamma(2.7)} z^{1.7} = 1.2948z^{1.7}.$$

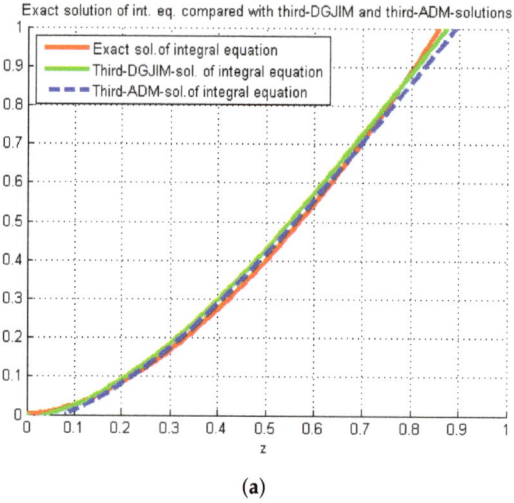

(a)

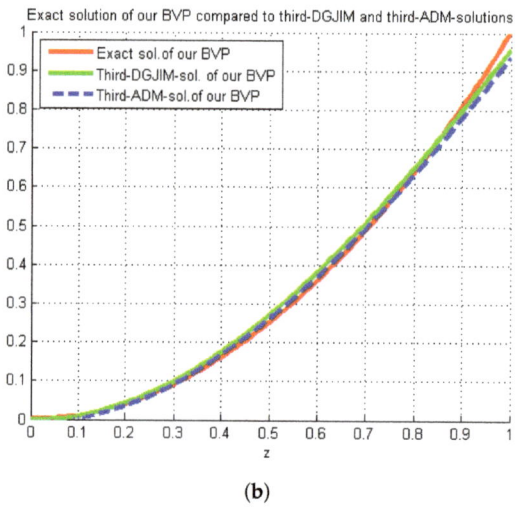

(b)

Figure 1. Graphs of the exact solutions of the (**a**) integral Equation (33) and (**b**) RLFBVP (32) compared with their third-DGJIM and third-ADM approximate solutions for $\varrho = 1.4$.

- **Case(II)** : $\varrho = 1.7$

In the next case, we consider the same problem for $\varrho = 1.7$. In fact, at this time, we consider the following RLFBVP

$$\begin{cases} \mathfrak{D}_{0^+}^{1.7} u(z) = u(z) + \mathfrak{D}_{0^+}^{0.3} u(z) + \hat{\varphi}(z), & z \in (0,1), \\ u(0) = 0, \\ u(1) = 8 \int_0^{\frac{1}{2}} u(s)\, ds + 54 \int_0^{\frac{1}{3}} u(s)\, ds, \end{cases} \quad (34)$$

where

$$\hat{\varphi}(z) = \frac{2}{\Gamma(1.3)} z^{0.3} - \frac{2}{\Gamma(2.7)} z^{1.7} - z^2,$$

such that we consider parameters $\varrho = 1.7$, $\xi = 1/2$, $\eta = 1/3$, $\sigma_n = 0.3$, $\mu = \nu = 1$, $p = 8$, and $q = 54$. Obviously, $\varrho - \sigma_n = 1.4 > 1$. In addition, $k_1(z, u(z)) = k_2(z, u(z)) = u(z)$ for $z \in [0, 1]$. By assuming $m(z) = \mathfrak{D}_{0^+}^{0.3} u(z)$, the equivalent integral equation of the problem (34) is given by

$$m(z) = \mathfrak{I}_{0^+}^{1.4}[\mathfrak{I}_{0^+}^{0.3} m(z) + m(z) + \hat{\varphi}(z)] + \frac{\Gamma(1.7)}{\Gamma(1.4)}\left(8 \int_0^{\frac{1}{2}} \mathfrak{I}_{0^+}^{0.3} m(s)\,ds\right.$$

$$\left. + 54 \int_0^{\frac{1}{3}} \mathfrak{I}_{0^+}^{0.3} m(s)\,ds - \mathfrak{I}_{0^+}^{1.7}[\mathfrak{I}_{0^+}^{0.3} m(z) + m(z) + \hat{\varphi}(z)]\Big|_{z=1}\right) z^{0.4}$$

$$= \mathfrak{I}_{0^+}^{1.7} m(z) + \mathfrak{I}_{0^+}^{1.4} m(z) + \mathfrak{I}_{0^+}^{1.4} \hat{\varphi}(z) + z^{0.4} \frac{8\Gamma(1.7)}{\Gamma(1.4)} \int_0^{\frac{1}{2}} \mathfrak{I}_{0^+}^{0.3} m(s)\,ds$$

$$+ z^{0.4} \frac{54\Gamma(1.7)}{\Gamma(1.4)} \int_0^{\frac{1}{3}} \mathfrak{I}_{0^+}^{0.3} m(s)\,ds - \frac{\Gamma(1.7)z^{0.4}}{\Gamma(1.4)} (\mathfrak{I}_{0^+}^2 m(z)|_{z=1})$$

$$- \frac{\Gamma(1.7)z^{0.4}}{\Gamma(1.4)} (\mathfrak{I}_{0^+}^{1.7} m(z)|_{z=1}) - \frac{\Gamma(1.7)z^{0.4}}{\Gamma(1.4)} (\mathfrak{I}_{0^+}^{1.7} \hat{\varphi}(z)|_{z=1}). \tag{35}$$

Then, we decompose the right-hand side of (35) as

$$m(z) = \tilde{\mathbf{L}}(m(z)) + \tilde{\mathbf{N}}(m(z)) + \zeta(z),$$

where

$$\tilde{\mathbf{L}}(m(z)) = \mathfrak{I}_{0^+}^{1.7} m(z) + \mathfrak{I}_{0^+}^{1.4} m(z),$$

$$\tilde{\mathbf{N}}(m(z)) = \frac{8\Gamma(1.7)z^{0.4}}{\Gamma(1.4)} \int_0^{\frac{1}{2}} \mathfrak{I}_{0^+}^{0.3} m(s)\,ds + \frac{54\Gamma(1.7)z^{0.4}}{\Gamma(1.4)} \int_0^{\frac{1}{3}} \mathfrak{I}_{0^+}^{0.3} m(s)\,ds$$

$$- \frac{\Gamma(1.7)z^{0.4}}{\Gamma(1.4)} (\mathfrak{I}_{0^+}^2 m(z)|_{z=1}) - \frac{\Gamma(1.7)z^{0.4}}{\Gamma(1.4)} (\mathfrak{I}_{0^+}^{1.7} m(z)|_{z=1}),$$

$$\zeta(z) = \mathfrak{I}_{0^+}^{1.4} \hat{\varphi}(z) - \frac{\Gamma(1.7)z^{0.4}}{\Gamma(1.4)} (\mathfrak{I}_{0^+}^{1.7} \hat{\varphi}(z)|_{z=1}).$$

Then, the sequence of approximate solutions of (34) and (35) are obtained by means of two DGJIM and ADM methods as follows:

- **Approximate solutions via DGJIM method for $\varrho = 1.7$:**

$$w_0(z) = 1.2948z^{1.7} - 0.3186z^{3.1} - 0.1973z^{3.4} - 0.6893z^{0.4},$$

$$w_1(z) = 1.2948z^{1.7} + 0.0169z^{3.4} - 2.7346z^{0.4} - 0.0487z^{4.8}$$

$$- 0.1040z^{5.1} - 0.2783z^{2.1} - 0.1886z^{4.5} - 0.3839z^{1.8},$$

$$w_2(z) = 1.2948z^{1.7} + 0.0169z^{3.4} - 0.0115z^{0.4} - 0.0234z^{4.8}$$

$$- 0.0888z^{5.1} - 0.0019z^{2.1} - 0.1471z^{4.5} + 0.2654z^{1.8}$$

$$- 0.0033z^{6.5} - 0.1078z^{3.5} + 0.0044z^{5.8} - 0.0165z^{5.9},$$

and

$$u_0(z) = z^2 - 0.2141z^{3.4} - 0.1296z^{3.7} - 0.6731z^{0.7},$$

$$u_1(z) = z^2 + 0.0111z^{3.7} - 2.6703z^{0.7} - 0.1493z^{5.1}$$
$$- 0.0615z^{5.4} - 0.2052z^{2.4} - 0.0982z^{4.8} - 0.2921z^{2.1},$$

$$u_2(z) = z^2 + 0.0111z^{3.7} - 0.0112z^{0.7} - 0.0141z^{5.1} - 0.0525z^{5.4}$$
$$- 0.0014z^{2.4} - 0.0002z^{7.1} - 0.0449z^{4.1} - 0.0075z^{6.5} - 0.0703z^{3.8}$$
$$+ 0.0025z^{6.1} - 0.0018z^{6.8} - 0.0094z^{6.2} - 0.0899z^{4.8} + 0.2025z^{2.1}.$$

- **Approximate solutions via ADM method for $\varrho = 1.7$:**

$$w_0(z) = 1.2948z^{1.7} - 0.3186z^{3.1} - 0.1973z^{3.4} - 0.6893z^{0.4},$$

$$w_1(z) = 1.2948z^{1.7} + 0.0169z^{3.4} - 0.0346z^{0.4} - 0.0487z^{4.8}$$
$$- 0.1040z^{5.1} - 0.2783z^{2.1} - 0.1886z^{4.5} - 0.3839z^{1.8},$$

$$w_2(z) = 1.2948z^{1.7} + 0.0169z^{3.4} - 0.0346z^{0.4} - 0.0234z^{4.8} + 0.0012z^{5.1}$$
$$+ 1.7119z^{2.1} - 0.1471z^{4.5} - 1.3654z^{1.8} - 0.0033z^{6.5} - 0.0005z^{6.8}$$
$$- 0.0703z^{3.8} - 0.0134z^{6.2} - 0.1078z^{3.5} + 0.0044z^{5.8} - 0.0165z^{5.9},$$

and

$$u_0(z) = z^2 - 0.2141z^{3.4} - 0.1296z^{3.7} - 0.6731z^{0.7},$$

$$u_1(z) = z^2 + 0.1055z^{3.7} - 0.0338z^{0.7} - 0.0111z^{3.7} - 0.0293z^{5.1}$$
$$- 0.0083z^{5.4} - 0.2052z^{2.4} - 0.1153z^{4.8} - 0.2921z^{2.1},$$

$$u_2(z) = z^2 + 0.0111z^{3.7} - 0.0338z^{0.7} + 0.0111z^{3.7} - 0.0141z^{5.1}$$
$$+ 0.0007z^{5.4} + 1.2619z^{2.4} - 0.0899z^{4.8} - 1.0416z^{2.1} - 0.0018z^{6.8}$$
$$- 0.0002z^{7.1} - 0.0449z^{4.1} - 0.0075z^{6.5} - 0.0703z^{3.8} + 0.0025z^{6.1} - 0.0094z^{6.2}.$$

In consequence, the graphs of the three-term approximate solutions obtained by the DGJIM and ADM algorithm for the suggested RLFBVP (34) and the integral Equation (35) are plotted in Figure 2.

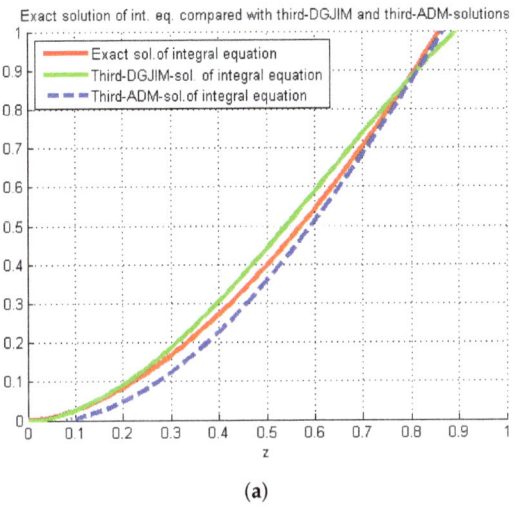

(a)

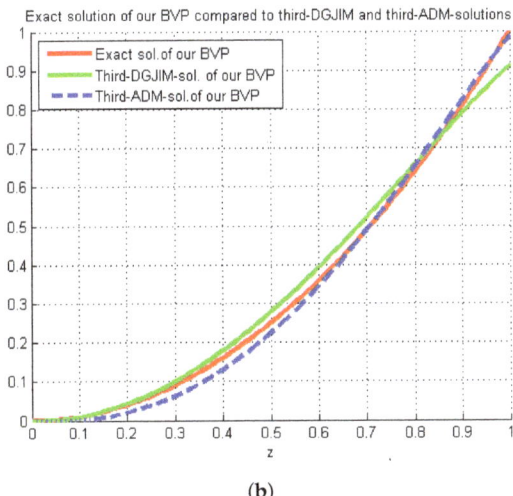

(b)

Figure 2. Graphs of the exact solutions of (**a**) the integral Equation (35) and (**b**) RLFBVP (34) compared with their third-DGJIM and third-ADM approximate solutions for $\varrho = 1.7$.

- **Case(III)** :$\varrho = 1.9$

Finally, we consider the first problem for $\varrho = 1.9$ as the third case. Consider the following RLFBVP

$$\begin{cases} \mathfrak{D}_{0^+}^{1.9} u(z) = u(z) + \mathfrak{D}_{0^+}^{0.3} u(z) + \hat{\varphi}(z), & z \in (0,1), \\ u(0) = 0, \\ u(1) = 8 \int_0^{\frac{1}{2}} u(s)\, ds + 54 \int_0^{\frac{1}{3}} u(s)\, ds, \end{cases} \qquad (36)$$

where

$$\hat{\varphi}(z) = \frac{2}{\Gamma(1.1)} z^{0.1} - \frac{2}{\Gamma(2.7)} z^{1.7} - z^2.$$

The parameters $\varrho = 1.9$, $\xi = 1/2$, $\eta = 1/3$, $\sigma_n = 0.3$, $\mu = \nu = 1$, $p = 8$, and $q = 54$ are assumed here. Evidently, $\varrho - \sigma_n = 1.6 > 1$. In addition, $k_1(z, u(z)) = k_2(z, u(z)) = u(z)$ for $z \in [0, 1]$. By assuming $m(z) = \mathfrak{D}_{0^+}^{0.3} u(z)$, the equivalent integral equation of the problem (36) is given in the following form

$$m(z) = \mathfrak{J}_{0^+}^{1.6}[\mathfrak{J}_{0^+}^{0.3} m(z) + m(z) + \hat{\varphi}(z)] + \frac{\Gamma(1.9)}{\Gamma(1.6)}\left(8\int_0^{\frac{1}{2}} \mathfrak{J}_{0^+}^{0.3} m(s)\,ds\right.$$

$$\left. + 54\int_0^{\frac{1}{3}} \mathfrak{J}_{0^+}^{0.3} m(s)\,ds - \mathfrak{J}_{0^+}^{1.9}[\mathfrak{J}_{0^+}^{0.3} m(z) + m(z) + \hat{\varphi}(z)]\Big|_{z=1}\right)z^{0.6}$$

$$= \mathfrak{J}_{0^+}^{1.9} m(z) + \mathfrak{J}_{0^+}^{1.6} m(z) + \mathfrak{J}_{0^+}^{1.6} \hat{\varphi}(z) + z^{0.6} \frac{8\Gamma(1.9)}{\Gamma(1.6)} \int_0^{\frac{1}{2}} \mathfrak{J}_{0^+}^{0.3} m(s)\,ds$$

$$+ z^{0.6} \frac{54\Gamma(1.9)}{\Gamma(1.6)} \int_0^{\frac{1}{3}} \mathfrak{J}_{0^+}^{0.3} m(s)\,ds - \frac{\Gamma(1.9) z^{0.6}}{\Gamma(1.6)} (\mathfrak{J}_{0^+}^{2.2} m(z)\big|_{z=1})$$

$$- \frac{\Gamma(1.9) z^{0.6}}{\Gamma(1.6)} (\mathfrak{J}_{0^+}^{1.9} m(z)\big|_{z=1}) - \frac{\Gamma(1.9) z^{0.6}}{\Gamma(1.6)} (\mathfrak{J}_{0^+}^{1.9} \hat{\varphi}(z)\big|_{z=1}). \quad (37)$$

By decomposing the right-hand side of (37), we get

$$m(z) = \tilde{\mathbf{L}}(m(z)) + \tilde{\mathbf{N}}(m(z)) + \zeta(z),$$

where

$$\tilde{\mathbf{L}}(m(z)) = \mathfrak{J}_{0^+}^{1.9} m(z) + \mathfrak{J}_{0^+}^{1.6} m(z),$$

$$\tilde{\mathbf{N}}(m(z)) = \frac{8\Gamma(1.9) z^{0.6}}{\Gamma(1.6)} \int_0^{\frac{1}{2}} \mathfrak{J}_{0^+}^{0.3} m(s)\,ds + \frac{54\Gamma(1.9) z^{0.6}}{\Gamma(1.6)} \int_0^{\frac{1}{3}} \mathfrak{J}_{0^+}^{0.3} m(s)\,ds$$

$$- \frac{\Gamma(1.9) z^{0.6}}{\Gamma(1.6)} (\mathfrak{J}_{0^+}^{2.2} m(z)\big|_{z=1}) - \frac{\Gamma(1.9) z^{0.6}}{\Gamma(1.6)} (\mathfrak{J}_{0^+}^{1.9} m(z)\big|_{z=1}),$$

$$\zeta(z) = \mathfrak{J}_{0^+}^{1.6} \hat{\varphi}(z) - \frac{\Gamma(1.9) z^{0.6}}{\Gamma(1.6)} (\mathfrak{J}_{0^+}^{1.9} \hat{\varphi}(z)\big|_{z=1}).$$

Then, the sequence of approximate solutions are obtained by means of two DGJIM and ADM methods illustrated as:

- **Approximate solutions via DGJIM method for $\varrho = 1.9$:**

$$w_0(z) = 1.2948 z^{1.7} - 0.2259 z^{3.3} - 0.1495 z^{3.6} - 0.8114 z^{0.6},$$

$$w_1(z) = 1.2948 z^{1.7} - 1.8726 z^{0.6} - 0.0236 z^{5.2} - 0.0069 z^{5.5}$$

$$- 0.2182 z^{2.5} - 0.0198 z^{4.9} - 0.2991 z^{2.2},$$

$$w_2(z) = 1.2948 z^{1.7} + 0.0427 z^{0.6} + 0.0017 z^{5.2} - 0.5008 z^{2.5}$$

$$+ 0.4866 z^{2.2} - 0.0009 z^{7.1} - 0.0001 z^{7.4} - 0.0163 z^{4.4}$$

$$- 0.0017 z^{6.8} - 0.0520 z^{4.1} - 0.0011 z^{6.5} - 0.0406 z^{3.8},$$

and

$$u_0(z) = z^2 - 0.1495z^{3.6} - 0.0968z^{3.9} - 0.7538z^{0.9},$$

$$u_1(z) = z^2 - 1.7397z^{0.9} - 0.0139z^{5.5} - 0.0040z^{5.8}$$
$$- 0.1545z^{2.8} - 0.0118z^{5.2} - 0.2182z^{2.5},$$

$$u_2(z) = z^2 + 0.0397z^{0.9} + 0.0010z^{5.5} - 0.3546z^{2.8}$$
$$+ 0.3549z^{25} - 0.0004z^{7.4} - 0.0005z^{7.7} - 0.0118z^{4.7}$$
$$- 0.0009z^{7.1} - 0.0326z^{4.4} - 0.0006z^{6.8} - 0.0025z^{4.1}.$$

- **Approximate solutions via ADM method for $\varrho = 1.9$:**

$$w_0(z) = 1.2948z^{1.7} - 0.2259z^{3.3} - 0.1495z^{3.6} - 0.8114z^{0.6},$$

$$w_1(z) = 1.2948z^{1.7} - 1.8626z^{0.6} - 0.0236z^{5.2} - 0.0069z^{5.5}$$
$$- 0.2182z^{2.5} - 0.0198z^{4.9} - 0.2991z^{2.2},$$

$$w_2(z) = 1.2948z^{1.7} - 0.0126z^{0.6} + 0.0017z^{5.2} - 0.5008z^{2.5}$$
$$+ 0.4866z^{2.2} - 0.0009z^{7.1} - 0.0001z^{7.4} - 0.0163z^{4.4}$$
$$- 0.0017z^{6.8} - 0.0520z^{4.1} - 0.0011z^{6.5} - 0.0406z^{3.8},$$

and

$$u_0(z) = z^2 - 0.1495z^{3.6} - 0.0968z^{3.9} - 0.7528z^{0.9},$$

$$u_1(z) = z^2 - 1.7304z^{0.9} - 0.0139z^{5.5} - 0.0040z^{5.8}$$
$$- 0.1545z^{2.8} - 0.0118z^{5.2} - 0.2182z^{2.5},$$

$$u_2(z) = z^2 - 0.0117z^{0.9} + 0.0010z^{5.5} - 0.3546z^{2.8}$$
$$+ 0.3549z^{2.5} - 0.0004z^{7.41} - 0.0005z^{7.7} - 0.0100z^{4.7}$$
$$- 0.0009z^{7.1} - 0.0626z^{4.4} - 0.0006z^{6.8} - 0.0259z^{4.1}.$$

In consequence, the graphs of the three-term approximate solutions obtained by the DGJIM and ADM algorithm for the suggested RLFBVP (36) and the integral Equation (37) are plotted in Figure 3.

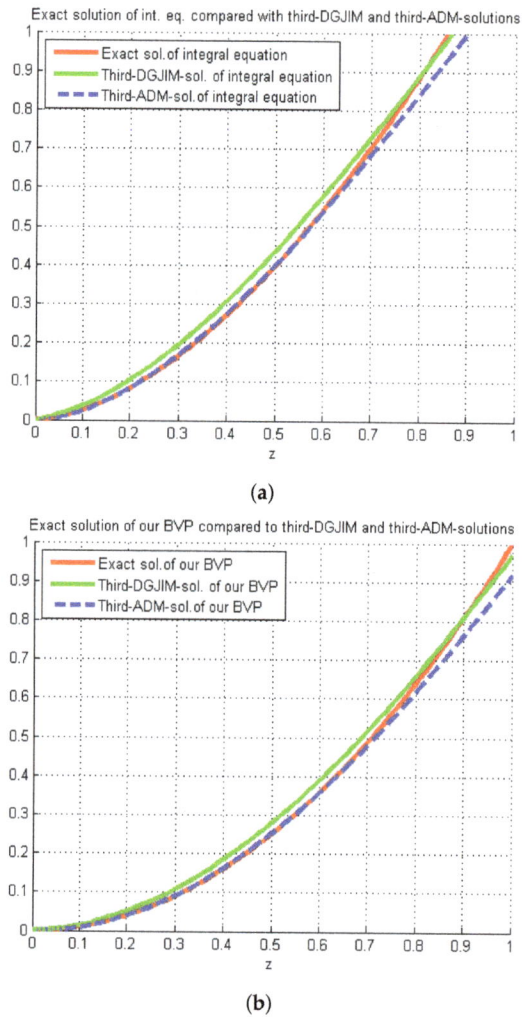

Figure 3. Graphs of the exact solutions of (a) the integral Equation (37) and (b) RLFBVP (36) compared with their third-DGJIM and third-ADM approximate solutions for $\varrho = 1.9$.

6. Conclusions

In this paper, we study the existence of solutions for a multi-term multi-order RLF-BVP with integral boundary conditions in the first step. Next, we apply two numerical methods (i.e., DGJIM and ADM algorithms) for solving the suggested multi-term fractional differential equation based on the decomposition technique. We show by an example that the approximate solutions obtained by these methods are in excellent agreement with the exact solutions. These give the solution as a series that quickly converges to the exact one if it exists. Therefore, this paper states that these two numerical methods can be utilized in many other multi-term FBVPs with different boundary value conditions by terms of some symmetric and asymmetric operators.

Author Contributions: Conceptualization, S.E. and S.R.; Formal analysis, S.S., S.E., B.T., S.R., S.K.N. and J.T.; Funding acquisition, J.T.; Methodology, S.E., B.T., S.R., S.K.N. and J.T.; Software, B.T. All authors have read and agreed to the published version of the manuscript.

Funding: This research was funded by King Mongkut's University of Technology North Bangkok (Contract No. KMUTNB-61-KNOW-030).

Institutional Review Board Statement: Not applicable.

Informed Consent Statement: Not applicable.

Data Availability Statement: Data sharing is not applicable to this article as no datasets were generated or analyzed during the current study.

Acknowledgments: The second and fourth authors would like to thank Azarbaijan Shahid Madani University. The authors also acknowledge the reviewers for their constructive remarks on our work.

Conflicts of Interest: The authors declare no conflict of interest.

References

1. Anastassiou, G.A.; Argyros, I.K.; Kumar, S. Monotone convergence of extended iterative methods and fractional calculus with applications. *Fundam. Inform.* **2017**, *151*, 241–253. [CrossRef]
2. Kilbas, A.A.; Srivastava, H.M.; Trujillo, J.J. *Theory and Applications of the Fractional Differential Equations*; Elsevier: Amsterdam, The Netherlands, 2006; Volume 204.
3. Podlubny, I. *Fractional Differential Equations*; Academic Press: New York, NY, USA, 1999.
4. Jajarmi, A.; Hajipour, M.; Baleanu, D. New aspects of the adaptive synchronization and hyperchaos suppression of a financial model. *Chaos Solitons Fractals* **2017**, *99*, 285–296. [CrossRef]
5. Baleanu, D.; Jajarmi, A.; Hajipour, M. A new formulation of the fractional optimal control problems involving Mittag-Leffler nonsingular kernel. *J. Optim. Theory Appl.* **2017**, *175*, 718–737. [CrossRef]
6. Alizadeh, S.; Baleanu, D.; Rezapour, S. Analyzing transient response of the parallel RCL circuit by using the Caputo-Fabrizio fractional derivative. *Adv. Differ. Equ.* **2020**, *2020*, 55. [CrossRef]
7. Hajipour, M.; Jajarmi, A.; Baleanu, D. An efficient nonstandard finite difference scheme for a class of fractional chaotic systems. *J. Comput. Nonlinear Dyn.* **2017**, *13*, 021013. [CrossRef]
8. Dokuyucu, M.A.; Celik, E.; Bulut, H.; Baskonus, H.M. Cancer treatment model with the Caputo-Fabrizio fractional derivative. *Eur. Phys. J. Plus* **2018**, *133*, 92. [CrossRef]
9. Singh, J.; Kumar, D.; Hammouch, Z.; Atangana, A. A fractional epidemiological model for computer viruses pertaining to a new fractional derivative. *Appl. Math. Comput.* **2018**, *316*, 504–515. [CrossRef]
10. Kosmatov, N.; Jiang, W. Resonant functional problems of fractional order. *Chaos Solitons Fractals* **2016**, *91*, 573–579. [CrossRef]
11. Lu, C.; Fu, C.; Yang, H. Time-fractional generalized Boussinesq equation for Rossby solitary waves with dissipation effect in stratified fluid and conservation laws as well as exact solutions. *Appl. Math. Comput.* **2018**, *327*, 104–116. [CrossRef]
12. Abdo, M.S.; Shah, K.; Wahash, H.A.; Panchal, S.K. On a comprehensive model of the novel coronavirus (COVID-19) under Mittag-Leffler derivative. *Chaos Solitons Fractal* **2020**, *135*, 109867. [CrossRef] [PubMed]
13. Baleanu, D.; Etemad, S.; Rezapour, S. A hybrid Caputo fractional modeling for thermostat with hybrid boundary value conditions. *Bound. Value Probl.* **2020**, *2020*, 64. [CrossRef]
14. Rezapour, S.; Etemad, S.; Mohammadi, H. A mathematical analysis of a system of Caputo-Fabrizio fractional differential equations for the anthrax disease model in animals. *Adv. Differ. Equ.* **2020**, *2020*, 481. [CrossRef]
15. Baleanu, D.; Mohammadi, H.; Rezapour, S. Analysis of the model of HIV-1 infection of CD4+ T-cell with a new approach of fractional derivative. *Adv. Differ. Equ.* **2020**, *2020*, 71. [CrossRef]
16. Rezapour, S.; Mohammadi, H.; Jajarmi, A. A new mathematical model for Zika virus transmission. *Adv. Differ. Equ.* **2020**, *2020*, 589. [CrossRef]
17. Khan, S.A.; Shah, K.; Zaman, G.; Jarad, F. Existence theory and numerical solutions to smoking model under Caputo-Fabrizio fractional derivative. *Chaos Interdiscip. J. Nonlinear Sci.* **2019**, *29*, 013128. [CrossRef]
18. Mohammadi, H.; Kumar, S.; Rezapour, S.; Etemad, S. A theoretical study of the Caputo-Fabrizio fractional modeling for hearing loss due to Mumps virus with optimal control. *Chaos Solitons Fractals* **2021**, *144*, 110668. [CrossRef]
19. Thabet, S.T.M.; Etemad, S.; Rezapour, S. On a new structure of the pantograph inclusion problem in the Caputo conformable setting. *Bound. Value Probl.* **2020**, *2020*, 171. [CrossRef]
20. Adiguzel, R.S.; Aksoy, U.; Karapinar, E.; Erhan, I.M. On the solution of a boundary value problem associated with a fractional differential equation. *Math. Methods Appl. Sci.* **2020**. [CrossRef]
21. Afshari, H.; Kalantari, S.; Karapinar, E. Solution of fractional differential equations via coupled fixed point. *Electron. J. Diff. Equ.* **2015**, *286*, 1–12.
22. Ahmad, B.; Alsaedi, A.; Salem, S.; Ntouyas, S.K. Fractional differential equation involving mixed nonlinearities with nonlocal multi-point and Riemann-Stieltjes integral-multi-strip conditions. *Fractal Fract.* **2019**, *3*, 34. [CrossRef]

23. Ahmad, B.; Ntouyas, S.K.; Alsaedi, A.; Agarwal, R.P. A study of nonlocal integro-multi-point boundary value problems of sequential fractional integro-differential inclusions. *Dyn. Contin. Disc. Impuls. Syst. Ser. A Math. Anal.* **2018**, *25*, 125–140.
24. Baitiche, Z.; Derbazi, C.; Benchohra, M. ψ-Caputo fractional differential equations with multi-point boundary conditions by topological degree theory. *Results Nonlinear Anal.* **2020**, *3*, 167–178.
25. Baleanu, D.; Etemad, S.; Rezapour, S. On a fractional hybrid integro-differential equation with mixed hybrid integral boundary value conditions by using three operators. *Alex. Eng. J.* **2020**. [CrossRef]
26. Etemad, S.; Rezapour, S. On the existence of solutions for fractional boundary value problems on the ethane graph. *Adv. Differ. Equ.* **2020**, *2020*, 276. [CrossRef]
27. Boucenna, D.; Boulfoul, A.; Chidouh, A.; Ben Makhlouf, A.; Tellab, B. Some results for initial value problem of nonlinear fractional equation in Sobolev space. *J. Appl. Math. Comput.* **2021**. [CrossRef]
28. Boulfoul, A.; Tellab, B.; Abdellouahab, N.; Zennir, K. Existence and uniqueness results for initial value problem of nonlinear fractional integro-differential equation on an unbounded domain in a weighted Banach space. *Math. Methods Appl. Sci.* **2020**. [CrossRef]
29. Si Bachir, F.; Abbas, S.; Benbachir, M.; Benchohra, M. Hilfer-Hadamard fractional differential equations; Existence and attractivity. *Adv. Theory Nonlinear Anal. Appl.* **2021**, *5*, 49–57.
30. Chen, Y.M.; Han, X.N.; Liu, L.C. Numerical solution for a class of linear system of fractional differential equations by the Haar wavelet method and the convergence analysis. *Comput. Model. Eng. Sci.* **2014**, *97*, 391–405.
31. Jong, K.; Choi, H.; Jang, Y.; Pak, S. A new approach for solving one-dimensional fractional boundary value problems via Haar wavelet collocation method. *Appl. Num. Math.* **2021**, *160*, 313–330. [CrossRef]
32. Saeed, U. CAS Picard method for fractional nonlinear differential equation. *Appl. Math. Comput.* **2017**, *307*, 102–112. [CrossRef]
33. Kumar, D.; Singh, J.; Baleanu, D. A new numerical algorithm for fractional Fitzhugh-Nagumo equation arising in transmission of nerve impulses. *Nonlinear Dyn.* **2018**, *91*, 307–317. [CrossRef]
34. Veeresha, P.; Prakasha, D.G.; Baskonus, H.M. Solving smoking epidemic model of fractional order using a modified homotopy analysis transform method. *Math. Sci.* **2019**, *13*, 115–125. [CrossRef]
35. Chen, Y.M.; Liu, L.Q.; Li, B.F.; Sun, Y.N. Numerical solution for the variable order linear cable equation with Bernstein polynomials. *Appl. Math. Comput.* **2014**, *238*, 329–341. [CrossRef]
36. Sakar, M.G.; Akgul, A.; Baleanu, D. On solutions of fractional Riccati differential equations. *Adv. Differ. Equ.* **2017**, *2017*, 39. [CrossRef]
37. Yin, F.; Song, J.; Wu, Y.; Zhang, L. Numerical solution of the fractional partial differential equations by the two-dimensional fractional-order Legendre functions. *Abstr. Appl. Anal.* **2013**, *13*, 562140. [CrossRef]
38. Odibat, Z.M.; Momani, S. Application of variational iteration method to nonlinear differential equations of fractional order. *Int. J. Nonlinear Sci. Numer. Simul.* **2016**, *7*, 7–34. [CrossRef]
39. Bolandtalat, A.; Babolian, E.; Jafari, H. Numerical solutions of multi-order fractional differential equations by Boubaker polynomials. *Open Phys.* **2016**, *14*, 226–230. [CrossRef]
40. Hesameddini, E.; Rahimi, A.; Asadollahifard, E. On the convergence of a new reliable algorithm for solving multi-order fractional differential equations. *Commun. Nonlinear Sci. Numer. Simul.* **2016**, *34*, 154–164. [CrossRef]
41. Firoozjaee, M.A.; Yousefi, S.A.; Jafari, H.; Baleanu, D. On a numerical approach to solve multi order fractional differential equations with boundary initial conditions. *J. Comput. Nonlinear Dynam.* **2015**, *10*, 061025. [CrossRef]
42. Dabiri, A.; Butcher, E.A. Stable fractional Chebyshev differentiation matrix for the numerical solution of multi-order fractional differential equations. *Nonlinear Dyn.* **2017**, *90*, 185–201. [CrossRef]
43. Ali, A.; Sarwar, M.; Zada, M.B.; Shah, K. Existence of solutions to fractional differential equation with fractional integral type boundary conditions. *Math. Methods Appl. Sci.* **2021**, *44*, 1615–1627. [CrossRef]
44. Liu, S.; Li, H.; Dai, Q.; Liu, J. Existence and uniqueness results for nonlocal integral boundary value problems for fractional differential equations. *Adv. Differ. Equ.* **2016**, *2016*, 122. [CrossRef]
45. Padhi, S.; Graef, J.R.; Pati, S. Multiple positive solutions for a boundary value problem with nonlinear nonlocal Riemann-Stieltjes integral boundary conditions. *Fract. Calc. Appl. Anal.* **2018**, *21*, 716–745. [CrossRef]
46. Thabet, S.T.M.; Etemad, S.; Rezapour, S. On a coupled Caputo conformable system of pantograph problems. *Turk. J. Math.* **2021**, *45*, 496–519. [CrossRef]
47. Daftardar-Gejji, V.; Jafari, H. Adomian decomposition: a tool for solving a system of fractional differential equations. *J. Math. Anal. Appl.* **2005**, *301*, 508–518. [CrossRef]
48. Daftardar-Gejji, V.; Jafari, H. An iterative method for solving nonlinear functional equations. *J. Math. Anal. Appl.* **2006**, *316*, 753–763. [CrossRef]
49. Babolian, E.; Vahidi, A.R.; Shoja, A. An efficient method for nonlinear fractional differential equations: combination of the Adomian decomposition method and spectral method. *Indian J. Pure Appl. Math.* **2014**, *45*, 1017–1028. [CrossRef]
50. Khodabakhshi, N.; Vaezpour, S.M.; Baleanu, D. Numerical solutions of the initial value problem for fractional differential equations by modification of the Adomian decomposition method. *Fract. Calc. Appl. Anal.* **2014**, *17*, 382–400. [CrossRef]
51. Loghmani, G.B.; Javanmardi, S. Numerical methods for sequential fractional differential equations for Caputo operator. *Bull. Malays. Math. Sci. Soc.* **2012**, *35*, 315–323.
52. Granas, A.; Dugundji, J. Elementary Fixed Point Theorems. In *Fixed Point Theory*; Springer: New York, NY, USA, 2003.

Article

Initial Value Problems of Linear Equations with the Dzhrbashyan–Nersesyan Derivative in Banach Spaces

Vladimir E. Fedorov [1,2,*], Marina V. Plekhanova [1,3] and Elizaveta M. Izhberdeeva [1]

[1] Mathematical Analysis Department, Chelyabinsk State University, 129, Kashirin Brothers St., 454001 Chelyabinsk, Russia; plekhanovamv@susu.ru (M.V.P.); iem@csu.ru (E.M.I.)
[2] Laboratory of Functional Materials, South Ural State University, 76, Lenin Av., 454080 Chelyabinsk, Russia
[3] Computational Mechanics Department, South Ural State University, 76, Lenin Av., 454080 Chelyabinsk, Russia
* Correspondence: kar@csu.ru; Tel.: +7-351-799-7106

Abstract: Among the many different definitions of the fractional derivative, the Riemann–Liouville and Gerasimov–Caputo derivatives are most commonly used. In this paper, we consider the equations with the Dzhrbashyan–Nersesyan fractional derivative, which generalizes the Riemann–Liouville and the Gerasimov–Caputo derivatives; it is transformed into such derivatives for two sets of parameters that are, in a certain sense, symmetric. The issues of the unique solvability of initial value problems for some classes of linear inhomogeneous equations of general form with the fractional Dzhrbashyan–Nersesyan derivative in Banach spaces are investigated. An inhomogeneous equation containing a bounded operator at the fractional derivative is considered, and the solution is presented using the Mittag–Leffler functions. The result obtained made it possible to study the initial value problems for a linear inhomogeneous equation with a degenerate operator at the fractional Dzhrbashyan–Nersesyan derivative in the case of relative p-boundedness of the operator pair from the equation. Abstract results were used to study a class of initial boundary value problems for equations with the time-fractional Dzhrbashyan–Nersesyan derivative and with polynomials in a self-adjoint elliptic differential operator with respect to spatial variables.

Keywords: fractional differential equation; fractional Dzhrbashyan–Nersesyan derivative; degenerate evolution equation; initial value problem; initial boundary value problem

MSC: 34G10; 35R11; 34A08

1. Introduction

One of the rapidly developing areas of modern mathematics is the theory of fractional differential equations and their applications [1–7] (also see the references therein). Among the many different definitions of the fractional derivative, the Riemann–Liouville [8] and Gerasimov–Caputo [8–10] derivatives are most commonly used. In this paper, we consider the equations with the Dzhrbashyan–Nersesyan fractional derivative [11], which generalizes the Riemann–Liouville and Gerasimov–Caputo derivatives; it is transformed into such derivatives for two sets of parameters that are, in a certain sense, symmetric. In this sense, the concepts of the Riemann–Liouville and Gerasimov–Caputo derivatives are symmetric. We investigate initial value problems with the Dzhrbashyan–Nersesyan fractional derivative, and the results obtained in these symmetric cases will be valid for the initial problems of equations with the Riemann–Liouville and the Gerasimov–Caputo derivatives, respectively. To begin, let us give the following definition.

Let $\{\alpha_k\}_0^n = \{\alpha_0, \alpha_1, \ldots, \alpha_n\}$ be the set of real numbers satisfying the condition $0 < \alpha_k \leq 1$, $k = 0, 1, \ldots, n$, $n \in \mathbb{N} \cup \{0\}$. We denote

$$D^{\sigma_0} z(t) = D_t^{\alpha_0 - 1} z(t), \tag{1}$$

$$D^{\sigma_k}z(t) = D_t^{\alpha_k-1}D_t^{\alpha_{k-1}}D_t^{\alpha_{k-2}}\ldots D_t^{\alpha_0}z(t), \quad k=1,2,\ldots,n. \tag{2}$$

The fractional Dzhrbashyan–Nersesyan derivative of the order σ_n associated with the sequence $\{\alpha_k\}$ is determined by the relations (1) and (2), and it includes the Riemann–Liouville ($\alpha_0 \in (0,1)$, $\alpha_k = 1$, $k = 1,2,\ldots,n$) and the Gerasimov–Caputo ($\alpha_k = 1$, $k = 0,1,\ldots,n-1$, $\alpha_n \in (0,1)$) fractional derivatives.

In [11], M.M. Dzhrbashyan and A.B. Nersesyan proved the existence of a unique continuous solution lying in $L_p(0,l;\mathbb{R})$ for the initial value problem

$$D^{\sigma_k}z(0) = z_k, \quad k = 0,1,\ldots,n-1 \tag{3}$$

for the equation $D^{\sigma_n}z(t) + p_0(t)D^{\sigma_{n-1}}z(t) + \cdots + p_{n-1}(t)D^{\sigma_0}z(t) + p_n(t)z(t) = f(t)$ with some functions $p_k : (0,T) \to \mathbb{R}$, $k = 0,1,\ldots n-1$, $f : (0,T) \to \mathbb{R}$. In the partial case, $p_0 \equiv p_1 \equiv \cdots \equiv p_{n-1} \equiv f(t) = 0$, $p_n \equiv a \in \mathbb{R}$, the solution is presented in the form of a linear combination of the Mittag–Leffeler functions.

Various differential equations with the Dzhrbashyan–Nersesyan derivative were considered in the works of A.V. Pskhu. For example, in [12], the fundamental solution of a diffusion-wave equation with the Dzhrbashyan–Nersesyan time-fractional derivative was obtained, and the unique solvability of the initial value problem $D^{\sigma_k}z(x,0) = z_k(x)$, $k = 0,1,\ldots,n-1$, $x \in \mathbb{R}^n$ for the equation in $\mathbb{R}^n \times (0,T]$ was studied. In [13], similar issues were researched for the case of the discretely distributed Dzhrbashyan–Nersesyan time-fractional derivative.

In this paper, we study the unique solvability issues (in the classical sense) for some classes of linear equations with operator coefficients in Banach spaces. In Section 2, the formula of the Laplace transform for the fractional Dzhrbashyan–Nersesyan derivative is obtained, and the initial value problem (3) with z_k from a Banach space $\mathcal{Z}$, $k = 0,1,\ldots,n-1$, for the class of homogeneous equations $D^{\sigma_n}z(t) = Az(t)$ with a linear bounded operator in $\mathcal{Z}$ is studied; $z : \mathbb{R}_+ \to \mathcal{Z}$. Using the Laplace transform, we obtain the resolving operators' families for this equation, which are presented in the form of the Mittag–Leffler functions with an operator argument. In Section 3, the same initial value problem for the inhomogeneous equation

$$D^{\sigma_n}z(t) = Az(t) + f(t), \tag{4}$$

with a function $f \in C([0,T];\mathcal{Z})$ is investigated.

These results are used for the proof of the unique solvability of the problem

$$D^{\sigma_k}x(0) = x_k, \quad k = 0,1,\ldots,n-1, \tag{5}$$

$$D^{\sigma_n}Lx(t) = Mx(t) + g(t). \tag{6}$$

Here, $\mathcal{X}, \mathcal{Y}$ are Banach spaces, $L \in \mathcal{L}(\mathcal{X};\mathcal{Y})$ (linear and continuous operator from $\mathcal{X}$ into $\mathcal{Y}$), and $M \in \mathcal{Cl}(\mathcal{X};\mathcal{Y})$ (linear closed operator with a dense domain D_M in the space $\mathcal{X}$ and with an image in $\mathcal{Y}$). We consider the case $\ker L \neq \{0\}$; hence, Equation (5) is called a degenerate evolution equation. For this equation, we will use the condition of (L,p)-boundedness of the operator M. It allows us to reduce this equation to a system of two equations on two mutual subspaces. One of them has the form (4), and the other has a nilpotent operator at the fractional derivative. It is shown that the initial value problem

$$D^{\sigma_k}Px(0) = x_k, \quad k = 0,1,\ldots,n-1, \tag{7}$$

is more natural for the degenerate Equation (6). Here, P is a projector on one of the above-mentioned subspaces along the other subspace. A theorem of the existence and uniqueness of a classical solution of the problem in (6) and (7) is also obtained.

Abstract results for non-degenerate and degenerate equations in Banach spaces are applied to the investigation of a class of initial boundary value problems for partial differential equations with a time-fractional derivative and with polynomials in a self-adjoint elliptic differential operator with respect to spatial variables.

This article is a continuation of the previous work of the authors, who investigated equations in Banach spaces with other fractional derivatives [14–17] with applications to initial boundary value problems for partial differential equations and systems of equations.

2. Homogeneous Equation with the Dzhrbashyan–Nersesyan Fractional Derivative

Consider the fractional Dzhrbashyan–Nersesyan derivative, which is a generalization of two well-known fractional derivatives: the Riemann–Liouville and Gerasimov–Caputo [11] derivatives. Let us present their definitions.

Let $\alpha > 0$, $z : [0, T] \to \mathcal{Z}$, for some $T > 0$ and Banach space $\mathcal{Z}$. The Riemann–Liouville fractional integral of an order $\alpha > 0$ of a function z has the form

$$J_t^\alpha z(t) := \int_0^t \frac{(t-s)^{\alpha-1}}{\Gamma(\alpha)} z(s) ds, \quad t > 0.$$

The Riemann–Liouville fractional derivative of an order $\alpha > 0$ for a function z is defined as

$$^R D_t^\alpha z(t) := D_t^m J_t^{m-\alpha} z(t),$$

where $m - 1 < \alpha \leq m \in \mathbb{N}$, and $D_t^m := \frac{d^m}{dt^m}$ is the integer-order derivative. Further, we use the notations $^R D_t^\alpha := D_t^\alpha$, $D_t^{-\alpha} := J_t^\alpha$ for $\alpha > 0$. The Gerasimov—Caputo fractional derivative of an order $\alpha > 0$ is defined as

$$^C D_t^\alpha z(t) := {^R D_t^\alpha} \left(z(t) - \sum_{k=0}^{m-1} z^{(k)}(0) \frac{t^k}{k!} \right).$$

Let $\{\alpha_k\}_0^n = \{\alpha_0, \alpha_1, \ldots, \alpha_n\}$ be the set of real numbers that satisfy the condition $0 < \alpha_k \leq 1$, $k = 0, 1, \ldots, n \in \mathbb{N}$. We denote

$$\sigma_k := \sum_{j=0}^k \alpha_j - 1, \quad k = 0, 1, \ldots, n,$$

so $-1 < \sigma_k \leq k - 1$. Further, it is assumed that the condition $\sigma_n > 0$ is met everywhere. We define the Dzhrbashyan–Nersesyan fractional derivatives, which are associated with a sequence $\{\alpha_k\}_0^n$, with the relations

$$D^{\sigma_0} z(t) := D_t^{\alpha_0 - 1} z(t), \tag{8}$$

$$D^{\sigma_k} z(t) := D_t^{\alpha_k - 1} D_t^{\alpha_{k-1}} D_t^{\alpha_{k-2}} \ldots D_t^{\alpha_0} z(t), \quad k = 1, 2, \ldots, n. \tag{9}$$

Let function $z : \mathbb{R}_+ \to \mathcal{Z}$, $\alpha > 0$, $m = \lceil \alpha \rceil$; then, the Laplace transform, which we will denote as $\hat{z}$—or when the expressions are too large for z, we denote it as $\text{Lap}[z]$—has the form

$$\widehat{D_t^\alpha z}(\lambda) = \lambda^\alpha \hat{z}(\lambda) - \sum_{k=0}^{m-1} \lambda^k D_t^{\alpha-m+k} z(0).$$

Therefore,

$$\widehat{D^{\sigma_n} z}(\lambda) = \lambda^{\alpha_n - 1} \text{Lap}[D_t^{\alpha_{n-1}} D_t^{\alpha_{n-2}} \ldots D_t^{\alpha_0} z](\lambda) =$$

$$= \lambda^{\alpha_{n-1} + \alpha_n - 1} \text{Lap}[D_t^{\alpha_{n-2}} \ldots D_t^{\alpha_0} z](\lambda) - \lambda^{\alpha_n - 1} D^{\sigma_{n-1}} z(0) = \cdots =$$

$$= \lambda^{\alpha_1 + \cdots + \alpha_n - 1} \widehat{D_t^{\alpha_0} z}(\lambda) - \lambda^{\alpha_n - 1} D^{\sigma_{n-1}} z(0) - \lambda^{\alpha_{n-1} + \alpha_n - 1} D^{\sigma_{n-2}} z(0) -$$

$$- \cdots - \lambda^{\alpha_2 + \cdots + \alpha_n - 1} D^{\sigma_1} z(0) =$$

$$= \lambda^{\alpha_0 + \cdots + \alpha_n - 1} \hat{z}(\lambda) - \lambda^{\alpha_n - 1} D^{\sigma_{n-1}} z(0) - \lambda^{\alpha_{n-1} + \alpha_n - 1} D^{\sigma_{n-2}} z(0) -$$

$$- \cdots - \lambda^{\alpha_1 + \cdots + \alpha_n - 1} D^{\sigma_0} z(0) =$$

$$= \lambda^{\sigma_n}\hat{z}(\lambda) - \lambda^{\sigma_n-\sigma_{n-1}-1}D^{\sigma_{n-1}}z(0) - \lambda^{\sigma_n-\sigma_{n-2}-1}D^{\sigma_{n-2}}z(0) - \cdots -$$
$$- \lambda^{\sigma_n-\sigma_0-1}D^{\sigma_0}z(0) = \lambda^{\sigma_n}\hat{z}(\lambda) - \sum_{k=0}^{n-1}\lambda^{\sigma_n-\sigma_k-1}D^{\sigma_k}z(0). \tag{10}$$

Let $\mathcal{L}(\mathcal{Z})$ be the Banach space of all linear bounded operators on $\mathcal{Z}$, $A \in \mathcal{L}(\mathcal{Z})$, and let D^{σ_n} be the Dzhrbashyan–Nersesyan fractional derivative, which is defined by a set of numbers $\{\alpha_k\}_0^n = \{\alpha_0, \alpha_1, \ldots, \alpha_n\}$, $0 < \alpha_k \leq 1$, $k = 0, 1, \ldots, n \in \mathbb{N}$ using Formulas (8) and (9). It is required that the inequality $\sigma_n > 0$ is satisfied. Consider the equation

$$D^{\sigma_n}z(t) = Az(t), \quad t > 0, \tag{11}$$

with the initial conditions

$$D^{\sigma_k}z(0) = z_k, \quad k = 0, 1, \ldots, n-1. \tag{12}$$

A function $z \in C(\mathbb{R}_+; \mathcal{Z})$ is called a solution to problem (11), (12), if $D_t^{\sigma_k}z \in C(\overline{\mathbb{R}}_+; \mathcal{Z})$, $k = 0, 1, \ldots, n-1$, $D_t^{\sigma_n}z \in C(\mathbb{R}_+; \mathcal{Z})$, equality (11) is fulfilled for all $t \in \mathbb{R}_+$, and conditions (12) are true. Here, $\overline{\mathbb{R}}_+ := \mathbb{R}_+ \cup \{0\}$.

Let a solution of (11) have the Laplace transform; then, Equation (11) implies that

$$\lambda^{\sigma_n}\hat{z}(\lambda) - \sum_{k=0}^{n-1}\lambda^{\sigma_n-\sigma_k-1}D^{\sigma_k}z(0) = A\hat{z}(\lambda). \tag{13}$$

For a fixed value $l \in \{0, 1, \ldots, n-1\}$, consider the problem

$$D^{\sigma_l}z(0) = z_l, \quad D^{\sigma_k}z(0) = 0, \quad k \in \{0, 1, \ldots, n-1\} \setminus \{l\}. \tag{14}$$

for Equation (11). If its solution has the Laplace transform, then the equality (13) for it has the form

$$\lambda^{\sigma_n}\hat{z}(\lambda) - \lambda^{\sigma_n-\sigma_l-1}z_l = A\hat{z}(\lambda).$$

From here, we have

$$\hat{z}(\lambda) = \lambda^{\sigma_n-\sigma_l-1}(\lambda^{\sigma_n}I - A)^{-1}z_l,$$

$$z(t) = \frac{1}{2\pi i}\int_\gamma \lambda^{\sigma_n-\sigma_l-1}(\lambda^{\sigma_n}I - A)^{-1}e^{\lambda t}d\lambda\, z_l,$$

where $\gamma = \{\lambda = re^{-i\pi} \in \mathbb{C} : r \in (\infty, a]\} \cup \{\lambda = ae^{i\varphi} \in \mathbb{C} : \varphi \in (-\pi, \pi)\} \cup \{\lambda = re^{i\pi} \in \mathbb{C} : r \in [a; \infty)\}$ with $a > \|A\|_{\mathcal{L}(\mathcal{Z})}^{1/\sigma_n}$.

So, we define the operators for $k = 0, 1, \ldots, n-1$:

$$Z_k(t) = \frac{1}{2\pi i}\int_\gamma \lambda^{\sigma_n-\sigma_k-1}(\lambda^{\sigma_n}I - A)^{-1}e^{\lambda t}d\lambda, \quad t > 0.$$

Note that due to the boundedness of the operator A,

$$Z_k(t) = \sum_{j=0}^\infty \frac{A^j}{2\pi i}\int_\gamma \lambda^{-\sigma_k-1-j\sigma_n}e^{\lambda t}d\lambda = \sum_{j=0}^\infty \frac{t^{j\sigma_n+\sigma_k}A^j}{2\pi i}\int_{\gamma_t}\frac{e^v dv}{v^{j\sigma_n+\sigma_k+1}} =$$

$$= \sum_{j=0}^\infty \frac{t^{j\sigma_n+\sigma_k}A^j}{\Gamma(j\sigma_n+\sigma_k+1)} = t^{\sigma_k}E_{\sigma_n,\sigma_k+1}(t^{\sigma_n}A).$$

The Mittag–Leffler function is used here:

$$E_{\alpha,\beta}(Z) = \sum_{j=0}^{\infty} \frac{Z^j}{\Gamma(\alpha j + \beta)}, \quad Z \in \mathcal{L}(\mathcal{Z}).$$

Lemma 1. *Let $A \in \mathcal{L}(\mathcal{Z})$, $z_l \in \mathcal{Z}$ for $l \in \{0, \ldots, n-1\}$, $\sigma_n > 0$, $\alpha_n + \sigma_l > 0$. Then, function $Z_l(t) = t^{\sigma_l} E_{\sigma_n, \sigma_l+1}(t^{\sigma_n} A)$ is the unique solution to the problem in (11) and (14).*

Proof. We have

$$D_t^{\sigma_0} t^{\sigma_l} E_{\sigma_n, \sigma_l+1}(t^{\sigma_n} A) = D_t^{\alpha_0 - 1} \sum_{j=0}^{\infty} \frac{t^{j\sigma_n + \sigma_l} A^j}{\Gamma(j\sigma_n + \sigma_l + 1)} = \sum_{j=0}^{\infty} \frac{t^{j\sigma_n + \sigma_l - \sigma_0} A^j}{\Gamma(j\sigma_n + \sigma_l - \sigma_0 + 1)},$$

$\sigma_l - \sigma_0 > 0$ for $l > 0$, so $D^{\sigma_0} Z_l(0) = 0$, $D^{\sigma_0} Z_0(0) = I$. For $k \in \{0, 1, \ldots, l\}$,

$$D^{\sigma_k} Z_l(t) = D_t^{\alpha_k - 1} D_t^{\alpha_k - 1} D_t^{\alpha_k - 2} \ldots D_t^{\alpha_0} Z_l(t) = \sum_{j=0}^{\infty} \frac{t^{j\sigma_n + \sigma_l - \sigma_k} A^j}{\Gamma(j\sigma_n + \sigma_l - \sigma_k + 1)},$$

at $k < l$ $\sigma_l - \sigma_k = \alpha_{k+1} + \alpha_{k+2} + \cdots + \alpha_l > 0$; hence, $D^{\sigma_k} Z_l(0) = 0$, $D^{\sigma_l} Z_l(0) = I$.
Further,

$$D^{\sigma_{l+1}} Z_l(t) = D_t^{\alpha_{l+1} - 1} D_t^{\alpha_l} D_t^{\alpha_{l-1}} \ldots D_t^{\alpha_0} Z_l(t) = D_t^{\alpha_{l+1} - 1} D_t^{\alpha_l} \sum_{j=0}^{\infty} \frac{t^{j\sigma_n + \alpha_l - 1} A^j}{\Gamma(j\sigma_n + \alpha_l)} =$$

$$= D_t^{\alpha_{l+1} - 1} \sum_{j=1}^{\infty} \frac{t^{j\sigma_n - 1} A^j}{\Gamma(j\sigma_n)} = \sum_{j=1}^{\infty} \frac{t^{j\sigma_n + \sigma_l - \sigma_{l+1} + 1} A^j}{\Gamma(j\sigma_n + \sigma_l - \sigma_{l+1} + 1)},$$

for $k \in \{l+1, l+2, \ldots, n-1\}$

$$D^{\sigma_k} Z_l(t) = \sum_{j=1}^{\infty} \frac{t^{j\sigma_n + \sigma_l - \sigma_k} A^j}{\Gamma(j\sigma_n + \sigma_l - \sigma_k + 1)}.$$

For $l \in \{0, 1, \ldots, n-2\}$, $k \in \{l+1, l+2, \ldots, n-1\}$, we have

$$\sigma_n + \sigma_l - \sigma_k \geq \sigma_n + \sigma_l - \sigma_{n-1} = \alpha_n + \sigma_l > 0.$$

Therefore, $D^{\sigma_k} Z_l(0) = 0$.
Finally,

$$D^{\sigma_n} Z_l(t) = \sum_{j=1}^{\infty} \frac{t^{j\sigma_n + \sigma_l - \sigma_n} A^j}{\Gamma(j\sigma_n + \sigma_l - \sigma_n + 1)} = A \sum_{j=0}^{\infty} \frac{t^{j\sigma_n + \sigma_l} A^j}{\Gamma(j\sigma_n + \sigma_l + 1)} = AZ_l(t).$$

We will prove the uniqueness of the solution. Suppose that $z_1(t)$ and $z_2(t)$ are two solutions of the problem in (11) and (14). Let us fix some $T > 0$; then, $y(t) = z_1(t) - z_2(t)$ is a solution of the problem $D^{\sigma_k} y(0) = 0$, $k = 0, 1, \ldots, n$, for Equation (11) on the interval $(0, T)$. We define the function $y(t)$ as zero on $[T, +\infty)$. Such a function is bounded and is also a solution to this problem for Equation (11) for $t > 0$, except it may be a point $t = T$. After acting with the Laplace transform on both parts of the equality $D_t^{\sigma_n} y(t) = Ay(t)$, we get $\lambda^{\sigma_n} \widehat{y}(\lambda) = A\widehat{y}(\lambda)$. Therefore, $(\lambda^{\sigma_n} - A)\widehat{y}(\lambda) \equiv 0$. If $|\lambda| > \|A\|_{\mathcal{L}(\mathcal{Z})}^{1/\sigma_n}$; then, $\widehat{y}(\lambda) = 0$. Consequently, $z_1(t) - z_2(t) = y(t) \equiv 0$ for all $t \in (0, T)$. Because $T > 0$ can be chosen at a large enough value, then $z_1(t) = z_2(t)$ for all $t > 0$. □

Theorem 1. *Let $A \in \mathcal{L}(\mathcal{Z})$, $z_k \in \mathcal{Z}$, $k = 0, 1 \ldots, n-1$, $0 < \alpha_k \leq 1$, $k = 0, 1 \ldots, n$, $\sigma_n > 0$, $\alpha_0 + \alpha_n > 1$. Then, the function*

$$z(t) = \sum_{k=0}^{n-1} t^{\sigma_k} E_{\sigma_n, \sigma_k + 1}(t^{\sigma_n} A) z_k$$

is a unique solution of the problem in (11) and (12).

Proof. For any $l \in \{0, 1, \ldots, n-1\}$, we have

$$\sigma_l + \alpha_n \geq \sigma_0 + \alpha_n = \alpha_0 + \alpha_n - 1 > 0.$$

Therefore, Lemma 1 is valid for all l. From the linearity of the problem in (11) and (12), we get what we need. □

Remark 1. *The result for $\mathcal{Z} = \mathbb{R}$ was obtained in [11].*

3. Inhomogeneous Equation

Consider the inhomogeneous equation

$$D^{\sigma_n} z(t) = Az(t) + f(t), \quad t \in (0, T], \tag{15}$$

for some $f \in C([0, T]; \mathcal{Z})$. A function $z \in C((0, T]; \mathcal{Z})$ is called a solution of the problem in (12) and (15) if $D_t^{\sigma_k} z \in C([0, T]; \mathcal{Z})$, $k = 0, 1, \ldots, n-1$, $D_t^{\sigma_n} z \in C((0, T]; \mathcal{Z})$, equality (15) is satisfied for all $t \in (0, T]$, and conditions (12) are true.

Assuming the convergence of the corresponding integrals, we denote

$$Z(t) = \frac{1}{2\pi i} \int_\gamma (\lambda^{\sigma_n} I - A)^{-1} e^{\lambda t} d\lambda = \sum_{j=0}^\infty \frac{A^j}{2\pi i} \int_\gamma \lambda^{-(j+1)\sigma_n} e^{\lambda t} d\lambda =$$

$$= \sum_{j=0}^\infty \frac{t^{(j+1)\sigma_n - 1} A^j}{2\pi i} \int_{\gamma_t} \frac{e^\nu d\nu}{\nu^{(j+1)\sigma_n}} = \sum_{j=0}^\infty \frac{t^{(j+1)\sigma_n - 1} A^j}{\Gamma((j+1)\sigma_n)} = t^{\sigma_n - 1} E_{\sigma_n, \sigma_n}(t^{\sigma_n} A),$$

for $k = 0, 1, \ldots, n-1$, and

$$Z_{\sigma_k}(t) = \frac{1}{2\pi i} \int_\gamma \lambda^{\sigma_k} (\lambda^{\sigma_n} I - A)^{-1} e^{\lambda t} d\lambda = \sum_{j=0}^\infty \frac{A^j}{2\pi i} \int_\gamma \lambda^{\sigma_k - (j+1)\sigma_n} e^{\lambda t} d\lambda =$$

$$= \sum_{j=0}^\infty \frac{t^{(j+1)\sigma_n - \sigma_k - 1} A^j}{2\pi i} \int_{\gamma_t} \frac{e^\nu d\nu}{\nu^{(j+1)\sigma_n - \sigma_k}} = \sum_{j=0}^\infty \frac{t^{(j+1)\sigma_n - \sigma_k - 1} A^j}{\Gamma((j+1)\sigma_n - \sigma_k)} = t^{\sigma_n - \sigma_k - 1} E_{\sigma_n, \sigma_n - \sigma_k}(t^{\sigma_n} A).$$

We note that $\sigma_n - \sigma_k > 0$, and by assumption, $\sigma_n > 0$; hence, as $t \to 0+$,

$$Z(t) \sim \frac{t^{\sigma_n - 1}}{\Gamma(\sigma_n)}, \quad Z_{\sigma_k}(t) \sim \frac{t^{\sigma_n - \sigma_k - 1}}{\Gamma(\sigma_n - \sigma_k)}, \quad k = 0, 1, \ldots, n-1. \tag{16}$$

Lemma 2. *Let $A \in \mathcal{L}(\mathcal{Z})$, $0 < \alpha_k \leq 1$, $k = 0, 1 \ldots, n$, $\sigma_n > 0$, $\alpha_0 + \alpha_n > 1$, $f \in C([0, T]; \mathcal{Z})$. Then, the function*

$$z_f(t) = \int_0^t (t-s)^{\sigma_n - 1} E_{\sigma_n, \sigma_n}((t-s)^{\sigma_n} A) f(s) ds$$

is a unique solution for the problem

$$D^{\sigma_k} z(0) = 0, \quad k = 0, 1, \ldots, n-1, \tag{17}$$

for Equation (15).

Proof. We have
$$\|z_f(t)\|_Z \le \max_{s\in[0,T]} \|E_{\sigma_n,\sigma_n}(s^{\sigma_n}A)\|_{\mathcal{L}(Z)} \max_{s\in[0,T]} \|f(s)\|_Z \frac{t^{\sigma_n}}{\sigma_n},$$

so $z_f(0) = 0$. For $\alpha_0 \in (0,1)$,

$$\|D^{\sigma_0}z_f(t)\|_Z = \left\|\frac{1}{\Gamma(1-\alpha_0)}\int_0^t (t-s)^{-\alpha_0}z_f(s)ds\right\|_Z \le \frac{\max_{s\in[0,T]}\|z_f(s)\|_Z}{\Gamma(1-\alpha_0)} \frac{t^{1-\alpha_0}}{1-\alpha_0}.$$

Therefore, $D^{\sigma_0}z_f(0) = 0$.

The Laplace transform is

$$\widehat{Z}(\mu) = \frac{1}{2\pi i}\int_\gamma (\lambda^{\sigma_n}I - A)^{-1}\int_0^\infty e^{(\lambda-\mu)t}dt d\lambda =$$

$$= \frac{1}{2\pi i}\int_\gamma (\lambda^{\sigma_n}I - A)^{-1}\frac{d\lambda}{\mu-\lambda} = (\mu^{\sigma_n}I - A)^{-1},$$

because

$$\left\|\frac{1}{\mu-\lambda}(\lambda^{\sigma_n}I - A)^{-1}\right\|_Z \le \frac{C}{|\lambda|^{1+\sigma_n}}.$$

We define f with zero outside the segment $[0,T]$. We have $z_f = Z * f$; consequently,

$$\widehat{z}_f(\mu) = \widehat{Z}(\mu)\widehat{f}(\mu) = (\mu^{\sigma_n}I - A)^{-1}\widehat{f}(\mu),$$

$$\widehat{D^{\sigma_0}z_f}(\mu) = \mu^{\sigma_0}(\mu^{\sigma_n}I - A)^{-1}\widehat{f}(\mu), \quad D^{\sigma_0}z_f(t) = \int_0^t Z_{\sigma_0}(t-s)f(s)ds,$$

$$\widehat{D^{\sigma_1}z_f}(\mu) = \mu^{\sigma_1}(\mu^{\sigma_n}I - A)^{-1}\widehat{f}(\mu), \quad D^{\sigma_1}z_f(t) = \int_0^t Z_{\sigma_1}(t-s)f(s)ds$$

due to (10). Then, for $k = 0,1$, by virtue of (16),

$$\|D^{\sigma_k}z_f(t)\| \le C_k \max_{s\in[0,T]}\|f(s)\|_Z \int_0^t (t-s)^{\sigma_n-\sigma_k-1}ds = \frac{C_k t^{\sigma_n-\sigma_k}}{\sigma_n-\sigma_k}\max_{s\in[0,T]}\|f(s)\|_Z.$$

Consequently, $D^{\sigma_k}z_f(0) = 0$, and

$$\widehat{D^{\sigma_2}z_f}(\mu) = \mu^{\sigma_2}(\mu^{\sigma_n}I - A)^{-1}\widehat{f}(\mu), \quad D^{\sigma_2}z_f(t) = \int_0^t Z_{\sigma_2}(t-s)f(s)ds.$$

Continuing these arguments, we get

$$D^{\sigma_k}z_f(t) = \int_0^t Z_{\sigma_k}(t-s)f(s)ds, \quad k=0,1,\ldots,n,$$

$$D^{\sigma_k}z_f(0) = 0, \quad k=0,1,\ldots,n-1.$$

Therefore, conditions (17) are valid.

Due to the boundedness of the operator A,

$$A\widehat{z}_f(\mu) = A\widehat{z}_f(\mu) = A(\mu^{\sigma_n}I - A)^{-1}\widehat{f}(\mu) = \mu^{\sigma_n}(\mu^{\sigma_n}I - A)^{-1}\widehat{f}(\mu) - \widehat{f}(\mu),$$

so

$$Az_f(t) = \int_0^t Z_{\sigma_n}(t-s)f(s)ds - f(t) = D^{\sigma_n}z_f(t) - f(t)$$

for all $t > 0$. Thus, equality (15) is satisfied for the function z_f.

The uniqueness of the solution can be proved in the same way as for the homogeneous equation above. □

From Theorem 1 and Lemma 2, we immediately get the following result.

Theorem 2. *Let $A \in \mathcal{L}(\mathcal{Z})$, $z_k \in \mathcal{Z}$, $k = 0, 1\ldots, n-1$, $0 < \alpha_k \leq 1$, $k = 0, 1\ldots, n$, $\sigma_n > 0$, $\alpha_0 + \alpha_n > 1$, $f \in C([0, T]; \mathcal{Z})$. Then, function*

$$z(t) = \sum_{k=0}^{n-1} t^{\sigma_k} E_{\sigma_n, \sigma_k+1}(t^{\sigma_n} A) z_k + \int_0^t (t-s)^{\sigma_n - 1} E_{\sigma_n, \sigma_n}((t-s)^{\sigma_n} A) f(s) ds$$

is a unique solution of the problem (12) in (15).

4. Degenerate Equation

Let $L \in \mathcal{L}(\mathcal{X}; \mathcal{Y})$ and $M \in \mathcal{C}l(\mathcal{X}; \mathcal{Y})$; D_M is a domain of an operator M. We define the L-resolvent set $\rho^L(M) = \{\mu \in \mathbb{C} : (\mu L - M)^{-1} \in \mathcal{L}(\mathcal{Y}; \mathcal{X})\}$ of an operator M and denote $R_\mu^L(M) := (\mu L - M)^{-1}L$, $L_\mu^L := L(\mu L - M)^{-1}$.

An operator M is called (L, σ)-bounded if

$$\exists a > 0 \quad \forall \mu \in \mathbb{C} \quad (|\mu| > a) \Rightarrow (\mu \in \rho^L(M)).$$

Lemma 3. *([18], pp. 89, 90). Let an operator M be (L, σ)-bounded; $\gamma = \{\mu \in \mathbb{C} : |\mu| = r > a\}$. Then, operators*

$$P = \frac{1}{2\pi i} \int_\gamma R_\mu^L(M) \, d\mu \in \mathcal{L}(\mathcal{X}), \quad Q = \frac{1}{2\pi i} \int_\gamma L_\mu^L(M) \, d\mu \in \mathcal{L}(\mathcal{Y})$$

are projections.

Set $\mathcal{X}^0 = \ker P$, $\mathcal{X}^1 = \operatorname{im} P$, $\mathcal{Y}^0 = \ker Q$, $\mathcal{Y}^1 = \operatorname{im} Q$. We denote by L_k (M_k) the restriction of the operator L (M) on $\mathcal{X}^k$ ($D_{M_k} = D_M \cap \mathcal{X}^k$), $k = 0, 1$.

Theorem 3. *([18], pp. 90, 91). Let an operator M be (L, σ)-bounded. Then,*
(i) *$M_1 \in \mathcal{L}(\mathcal{X}^1; \mathcal{Y}^1)$, $M_0 \in \mathcal{C}l(\mathcal{X}^0; \mathcal{Y}^0)$, $L_k \in \mathcal{L}(\mathcal{X}^k; \mathcal{Y}^k)$, $k = 0, 1$;*
(ii) *there exist operators $M_0^{-1} \in \mathcal{L}(\mathcal{Y}^0; \mathcal{X}^0)$, $L_1^{-1} \in \mathcal{L}(\mathcal{Y}^1; \mathcal{X}^1)$.*

We denote $G := M_0^{-1} L_0$. For $p \in \mathbb{N}_0 := \mathbb{N} \cup \{0\}$, the operator M is called (L, p)-bounded if it is (L, σ)-bounded; $G^p \neq 0$, $G^{p+1} = 0$.

Consider the initial problem

$$D^{\sigma_k} x(0) = x_k, \quad k = 0, 1, \ldots, n-1, \tag{18}$$

for a linear inhomogeneous fractional-order equation

$$D^{\sigma_n} L x(t) = M x(t) + g(t), \tag{19}$$

in which, as before, D^{σ_n} is the Dzhrbashyan–Nersesyan fractional derivative, which is defined by a set of numbers $\{\alpha_0, \alpha_1, \ldots, \alpha_n\}$, $0 < \alpha_k \leq 1$, $k = 0, 1, \ldots, n$, $g \in C([0, T]; \mathcal{Y})$.

A solution to the problem in (18) is (19) is called a function $x : (0, T] \to D_M$, for which $Mx \in C((0, T]; \mathcal{Y})$, $D^{\sigma_k}x \in C([0, T]; \mathcal{X})$, $k = 0, 1, \ldots, n-1$, $D^{\sigma_n}Lx \in C((0, T]; \mathcal{X})$, the equality (19) is valid for all $t \in (0, T]$, and conditions (18) are true.

Lemma 4. *Let $H \in \mathcal{L}(\mathcal{X})$ be a nilpotent operator with a power $p \in \mathbb{N}_0$, $h : [0, T] \to \mathcal{X}$, such that $(D^{\sigma_n}H)^l h \in C((0, T]; \mathcal{X})$ at $l = 0, 1, \ldots, p$, $D^{\sigma_k}(D^{\sigma_n}H)^l h \in C([0, T]; \mathcal{X})$ for $k = 0, 1, \ldots, n-1$, $l = 0, 1, \ldots, p$. Then, there exists a unique solution to the equation*

$$D^{\sigma_n} H x(t) = x(t) + h(t). \tag{20}$$

It has the form

$$x(t) = -\sum_{l=0}^{p}(D^{\sigma_n}H)^l h(t). \tag{21}$$

Proof. Let $z = z(t)$ be a solution of Equation (20). We act with the operator H on both parts of (20) and get the equality $HD^{\sigma_n}Hz(t) = Hz(t) + Hh(t)$. Due to the theorem's conditions, there exists a fractional derivative D^{σ_n} for the the right-hand side of this equality, as well as for its left-hand side. Acting with the operator D^{σ_n} on both parts of this equality, we will have

$$(D^{\sigma_n}H)^2 z = D^{\sigma_n}Hz + D^{\sigma_n}_t Hh = z + h + D^{\sigma_n}Hh.$$

At the p-th step, sequentially continuing this reasoning, we obtain the equality

$$(D^{\sigma_n}H)^{p+1}z = z + \sum_{l=0}^{p}(D^{\sigma_n}H)^l h.$$

By virtue of the continuity and nilpotency of the operator H, we have

$$(D^{\sigma_n}H)^{p+1}z = (D^{\sigma_n})^{p+1}H^{p+1}z \equiv 0.$$

Hence, equality (21) for is true the function z. This equality implies the existence of a solution to Equation (20) (it is checked by substituting this function into the equation) and its uniqueness. Indeed, the difference of two solutions corresponds to a solution of Equation (20) with the function $h \equiv 0$. According to Formula (21), its solution is identically equal to zero. The lemma has been proved. □

Theorem 4. *Let an operator M be (L, p)-bounded, $0 < \alpha_k \leq 1$, $k = 0, 1 \ldots, n$, $\sigma_n > 0$, $\alpha_0 + \alpha_n > 1$, $g \in C([0, T]; \mathcal{Y})$, $(D^{\sigma_n}G)^l M_0^{-1}(I - Q)g \in C((0, T]; \mathcal{X})$, $l = 0, 1, \ldots, p$, $D^{\sigma_k}(D^{\sigma_n}G)^l M_0^{-1}(I - Q)g \in C([0, T]; \mathcal{X})$ for $k = 0, 1, \ldots, n-1$, $l = 0, 1, \ldots, p$, and let $x_k \in \mathcal{X}$ satisfy the conditions*

$$(I - P)x_k = -D^{\sigma_k}\sum_{l=0}^{p}(D^{\sigma_n}G)^l M_0^{-1}(I - Q)g(t)|_{t=0}, \quad k = 0, 1, \ldots n-1. \tag{22}$$

Then, there exists a unique solution to the problem (18) in (19); it has the form

$$x(t) = \sum_{k=0}^{n-1} t^{\sigma_k} E_{\sigma_n, \sigma_k+1}(t^{\sigma_n} L_1^{-1} M) P x_k + \int_0^t (t-s)^{\sigma_n-1} E_{\sigma_n, \sigma_n}((t-s)^{\sigma_n} L_1^{-1} M) L_1^{-1} Qg(s) ds -$$

$$- \sum_{l=0}^{p}(D^{\sigma_n}G)^l M_0^{-1}(I - Q)g(t). \tag{23}$$

Proof. Acting on (19) with the operator $L_1^{-1}Q \in \mathcal{L}(\mathcal{Y}^1; \mathcal{X}^1)$, we get the equation

$$D^{\sigma_n}v(t) = L_1^{-1}Mv(t) + L_1^{-1}Qg(t), \qquad (24)$$

where $v(t) = Px(t)$. Indeed, $L_1^{-1}QD^{\sigma_n}Lx(t) = D^{\sigma_n}L_1^{-1}QLx(t) = D^{\sigma_n}L_1^{-1}L_1Px(t) = D^{\sigma_n}v(t) = L_1^{-1}Q(Mx(t)+g(t)) = L_1^{-1}MPx(t) + L_1^{-1}Qg(t) = L_1^{-1}Mv(t) + L_1^{-1}Qg(t)$. In this case, the equality $MP = QM$ is used (see Lemma 3 and Theorem 3).

If we use the operator $M_0^{-1}(I - Q) \in \mathcal{L}(\mathcal{Y}^0; \mathcal{X}^0)$ in the same way, then we get the equation

$$D^{\sigma_n}Gw(t) = w(t) + M_0^{-1}(I - Q)g(t), \qquad (25)$$

$w(t) = (I - P)x(t)$. Here, we use the equalities $MP_0 = M(I - P) = (I - Q)M = Q_0M$.

Equations (24) and (25) are endowed with the initial conditions

$$D^{\sigma_k}v(0) = Px_k, \quad k = 0, 1, \ldots, n-1, \qquad (26)$$

$$D^{\sigma_k}w(0) = (I - P)x_k, \quad k = 0, 1, \ldots, n-1. \qquad (27)$$

By Theorem 2 and with $\mathcal{Z} = \mathcal{X}^1$, $A = L_1^{-1}M_1 \in \mathcal{L}(\mathcal{X}^1)$ (see Theorem 3), $f(t) = L_1^{-1}Qg(t)$, $z_k = Px_k$, $k = 0, 1, n-1$, the problem in (24) and (26) has a unique solution, and it has the form

$$v(t) = \sum_{k=0}^{n-1} t^{\sigma_k} E_{\sigma_n,\sigma_k+1}(t^{\sigma_n}L_1^{-1}M)Px_k + \int_0^t (t-s)^{\sigma_n-1} E_{\sigma_n,\sigma_n}((t-s)^{\sigma_n}L_1^{-1}M)L_1^{-1}Qg(s)ds.$$

By virtue of Lemma 4, if conditions (22) are fulfilled, the problem in (25) and (27) has a unique solution:

$$w(t) = -\sum_{l=0}^{p}(D^{\sigma_n}G)^l M_0^{-1}(I - Q)g(t).$$

In this case, the following conditions are used: $D^{\sigma_k}(D^{\sigma_n}G)^l M_0^{-1}(I-Q)g \in C([0,T]; \mathcal{X})$ for $k = 0, \ldots, n-1$, $l = 0, 1, \ldots, p$. □

To avoid the need to satisfy the approval conditions (22), consider the problem

$$D^{\sigma_k}Px(0) = x_k, \quad k = 0, 1, \ldots, n-1, \qquad (28)$$

for Equation (19). Its solution is called a function $x : (0, T] \to D_M$, for which $x \in C((0, T]; D_M)$, $D^{\sigma_k}Px \in C([0, T]; \mathcal{X})$, $k = 0, 1, \ldots, n-1$, $D^{\sigma_n}Lx \in C((0, T]; \mathcal{X})$, equality (19) are fulfilled for all $t \in (0, T]$, and conditions (28) are valid.

Remark 2. *It is not difficult to make sure that, for $p = 0$, the initial conditions (28) are equivalent to the conditions*

$$D^{\sigma_k}Lx(0) = y_k, \quad k = 0, 1, \ldots, n-1, \qquad (29)$$

where $y_k = Lx_k$, or $x_k = L_1^{-1}y_k$, $k = 0, 1, \ldots, n-1$.

The existence and uniqueness theorem for the problem in (19) and (28) is proved similarly with help of a reduction to the system in (24) and (25) with initial conditions (26) and without conditions (27).

Theorem 5. *Let an operator M be (L, p)-bounded, $0 < \alpha_k \leq 1$, $k = 0, 1 \ldots, n$, $\sigma_n > 0$, $\alpha_0 + \alpha_n > 1$, $g \in C([0, T]; \mathcal{Y})$, $(D^{\sigma_n}G)^l M_0^{-1}(I - Q)g \in C((0, T]; \mathcal{X})$, $l = 0, 1, \ldots, p$, $x_k \in \mathcal{X}^1$, $k = 0, 1, \ldots n-1$. Then, there exists a unique solution to the problem in (19) and (28), and it has the form of (23).*

5. Application to a Class of Initial Boundary Value Problems

Let $P_\varrho(\lambda) = \sum_{j=0}^{\varrho} c_j \lambda^j$, $Q_\varrho(\lambda) = \sum_{j=0}^{\varrho} d_j \lambda^j$, $c_j, d_j \in \mathbb{C}$, $j = 0, 1, \ldots, \varrho \in \mathbb{N}_0$, $c_\varrho \neq 0$, $\Omega \subset \mathbb{R}^d$ be a bounded region with a smooth boundary $\partial\Omega$,

$$(\Lambda u)(s) := \sum_{|q| \leq 2r} a_q(s) \frac{\partial^{|q|} u(s)}{\partial s_1^{q_1} \partial s_2^{q_2} \ldots \partial s_d^{q_d}}, \quad a_q \in C^\infty(\overline{\Omega}),$$

$$(B_l u)(s) := \sum_{|q| \leq r_l} b_{lq}(s) \frac{\partial^{|q|} u(s)}{\partial s_1^{q_1} \partial s_2^{q_2} \ldots \partial s_d^{q_d}}, \quad b_{lq} \in C^\infty(\partial\Omega), \, l = 1, 2, \ldots, r,$$

$q = (q_1, q_2, \ldots, q_d) \in \mathbb{N}_0^d$, $|q| = q_1 + \cdots + q_d$, and let the operator pencil $\Lambda, B_1, B_2, \ldots, B_r$ be regularly elliptical [19]. Let an operator $\Lambda_1 \in \mathcal{Cl}(L_2(\Omega))$ with the domain

$$D_{\Lambda_1} = H_{\{B_l\}}^{2r}(\Omega) := \{v \in H^{2r}(\Omega) : B_l v(s) = 0, \, l = 1, 2, \ldots, r, \, s \in \partial\Omega\}$$

act as $\Lambda_1 u := \Lambda u$. Assume that Λ_1 is a self-adjoint operator; then, the spectrum $\sigma(\Lambda_1)$ of the operator Λ_1 is real and discrete, with finite multiplicity [19]. In addition, the spectrum $\sigma(\Lambda_1)$ is bounded from the right and does not contain zero; $\{\varphi_k : k \in \mathbb{N}\}$ is orthonormal in the $L_2(\Omega)$ system of eigenfunctions of the operator Λ_1, which is numbered in the non-increasing order of the corresponding eigenvalues $\{\lambda_k : k \in \mathbb{N}\}$, taking their multiplicity into account.

Consider the initial boundary value problem

$$D_t^{\sigma_k} u(s, 0) = u_k(s), \, k = 0, 1, \ldots, n-1, \, s \in \Omega, \tag{30}$$

$$B_l \Lambda^k u(s, t) = 0, \, k = 0, 1, \ldots, \varrho - 1, \, l = 1, 2, \ldots, r, \, (s, t) \in \partial\Omega \times (0, T], \tag{31}$$

$$D_t^{\sigma_n} P_\varrho(\Lambda) u(s, t) = Q_\varrho(\Lambda) u(s, t) + h(s, t), \, (s, t) \in \Omega \times (0, T], \tag{32}$$

where $D_t^{\sigma_k}$ are the Dzhrbashyan–Nersesyan fractional derivatives with respect to the variable t, corresponding to the set $\{\alpha_k\}_{k=0}^n$, $\alpha_k \in (0, 1]$, $k = 0, 1, \ldots, n$, $h : \Omega \times [0, T] \to \mathbb{R}$. Take

$$\mathcal{X} = \{v \in H^{2r\varrho}(\Omega) : B_l \Lambda^k v(s) = 0, \, k = 0, 1, \ldots, \varrho - 1, \, l = 1, 2, \ldots, r, \, s \in \partial\Omega\},$$

$$\mathcal{Y} = L_2(\Omega), L = P_\varrho(\Lambda), M = Q_\varrho(\Lambda) \in \mathcal{L}(\mathcal{X}; \mathcal{Y}).$$

Let $P_\varrho(\lambda_k) \neq 0$ for all $k \in \mathbb{N}$; then, there exists an inverse operator $L^{-1} \in \mathcal{L}(\mathcal{Y}; \mathcal{X})$, and the problem in (30)–(32) is representable as the problem in (12) and (15), where $\mathcal{Z} = \mathcal{X}$, $A = L^{-1}M \in \mathcal{L}(\mathcal{Z})$, $z_k = u_k(\cdot)$, $k = 0, 1, \ldots, n - 1$, $f(t) = L^{-1}h(\cdot, t)$. By Theorem 2, for $\sigma_n > 0$, $\alpha_0 + \alpha_n > 1$, there exists a unique solution to problem (30)–(32) for any $u_k \in \mathcal{X}$, $k = 0, 1, \ldots, n - 1$, and $h \in C([0, T]; L_2(\Omega))$ (in this case, $L^{-1}h \in C([0, T]; \mathcal{X})$).

Example 1. Take $\varrho = 2$, $P_2(\lambda) = \lambda^2$, $Q_2(\lambda) = a_0 + a_1\lambda$, $d = 1$, $\Omega = (0, \pi)$, $r = 1$, $\Lambda u = \frac{\partial^2 u}{\partial s^2}$, $B_1 = I$. Then, the problem in (30)–(32) has the form

$$D_t^{\sigma_n} \frac{\partial^4 u}{\partial s^4}(s, t) = a_0 u(s, t) + a_1 \frac{\partial^2 u}{\partial s^2}(s, t) + h(s, t), \, (s, t) \in (0, \pi) \times (0, T],$$

$$u(0, t) = u(\pi, t) = \frac{\partial^2 u}{\partial s^2}(0, t) = \frac{\partial^2 u}{\partial s^2}(\pi, t) = 0, \, t \in (0, T],$$

$$D_t^{\sigma_k} u(s, 0) = u_k(s), \, k = 0, 1, \ldots, n - 1, \, s \in (0, \pi).$$

Now, consider the degenerate case. Suppose that $P_\varrho(\lambda_k) = 0$ for some $k \in \mathbb{N}$. If the polynomials P_ϱ and Q_ϱ have no common roots on the set $\{\lambda_k\}$, the operator M is $(L, 0)$-bounded (see [20]), and the projectors have the form

$$P = \sum_{P_\varrho(\lambda_k)\neq 0} \langle \cdot, \varphi_k \rangle \varphi_k, \quad Q = \sum_{P_\varrho(\lambda_k)\neq 0} \langle \cdot, \varphi_k \rangle \varphi_k,$$

where $\langle \cdot, \varphi_k \rangle$ is the inner product in $L_2(\Omega)$. Considering Remark 2, the initial conditions can be given in the form

$$D_t^{\sigma_k} P_\varrho(\Lambda) u(s, 0) = y_k(s), \quad k = 0, 1, \ldots, n-1, \quad s \in \Omega. \tag{33}$$

Then, the problem in (31)–(33) is represented as (19) and (29) with the spaces $\mathcal{X}, \mathcal{Y}$ and the operators L and M selected above. Theorem 5 implies the unique solvability of the problem in (31)–(33) if $\sigma_n > 0$, $\alpha_0 + \alpha_n > 1$, $h \in C([0,T]; L_2(\Omega))$, and $y_k \in L_2(\Omega)$, $k = 0, 1, \ldots, n-1$, such that $\langle y_k, \varphi_l \rangle = 0$ for all $l \in \mathbb{N}$, for which $P_\varrho(\lambda_l) = 0$ (in other words, $y_k \in \mathcal{Y}^1$, $k = 0, 1, \ldots, n-1$).

Example 2. Let $\varrho = 2$, $P_2(\lambda) \equiv \lambda(\lambda + 9)$, $Q_2(\lambda) = 1 + \lambda$, $d = 1$, $\Omega = (0, \pi)$, $r = 1$, $\Lambda u = \frac{\partial^2 u}{\partial s^2}$, $B_1 = I$. Then, the degenerate problem in (31)–(33) has the form

$$D_t^{\sigma_n}\left(\frac{\partial^4 u}{\partial s^4} + 9\frac{\partial^2 u}{\partial s^2}\right)(s,t) = \left(u + \frac{\partial^2 u}{\partial s^2}\right)(s,t), \quad (s,t) \in (0,\pi) \times (0,T],$$

$$u(0,t) = u(\pi,t) = \frac{\partial^2 u}{\partial s^2}(0,t) = \frac{\partial^2 u}{\partial s^2}(\pi,t) = 0, \quad t \in (0,T],$$

$$D_t^{\sigma_k}\left(\frac{\partial^4 u}{\partial s^4} + 9\frac{\partial^2 u}{\partial s^2}\right)(s,0) = y_k(s), \quad k = 0, 1, \ldots, n-1, \quad s \in (0,\pi).$$

Here, $P_2(0) = P_2(-9) = 0$, $0 \notin \sigma(\Lambda_1)$, $-9 = -3^2 \in \sigma(\Lambda_1)$; therefore, $\mathcal{X}^0 = \mathcal{Y}^0 = \operatorname{span}\{\sin 3s\}$, $\mathcal{X}^1$, and $\mathcal{Y}^1$ are closures of $\operatorname{span}\{\sin ks : k \in \mathbb{N} \setminus \{3\}\}$ in $H^4(0,\pi)$ and $L_2(0,\pi)$, respectively. Thus, the conditions

$$\langle y_k, \sin 3s \rangle = \int_0^\pi y_k(s) \sin 3s\, ds = 0, \quad k = 0, 1, \ldots, n-1,$$

must be satisfied for the solvability of this initial boundary value problem.

Author Contributions: Conceptualization, V.E.F. and M.V.P.; methodology, V.E.F.; software, E.M.I.; validation, E.M.I. and M.V.P.; formal analysis, E.M.I.; investigation, E.M.I. and V.E.F.; resources, E.M.I.; data curation, E.M.I.; writing—original draft preparation, E.M.I.; writing—review and editing, V.E.F. and M.V.P.; visualization, E.M.I.; supervision, V.E.F. and M.V.P.; project administration, V.E.F.; funding acquisition, M.V.P. All authors have read and agreed to the published version of the manuscript.

Funding: This research was funded by the Russian Foundation for Basic Research, grant number 21-51-54003. The APC was funded by the Research Support Foundation of Chelyabinsk State University.

Institutional Review Board Statement: Not applicable.

Informed Consent Statement: Not applicable.

Conflicts of Interest: The authors declare no conflict of interest.

References

1. Samko, S.G.; Kilbas, A.A.; Marichev, O.I. *Fractional Integrals and Derivatives. Theory and Applications*; Gordon and Breach Science Publishers: Philadelphia, PA, USA, 1993.
2. Kiryakova, V. *Generalized Fractional Calculus and Applications*; Longman Scientific & Technical: Harlow, UK, 1994.

3. Podlubny, I. *Fractional Differential Equations*; Academic Press: Boston, MA, USA, 1999.
4. Tarasov, V.E. *Fractional Dynamics: Applications of Fractional Calculus to Dynamics of Particles, Fields and Media*; Springer: New York, NY, USA, 2011.
5. Agarwal, R.P.; Gala, S.; Ragusa, M.A. A regularity criterion of the 3D MHD equations involving one velocity and one current density component in Lorentz space. *Z. Angew. Math. Und Phys.* **2020**, *71*, 95. [CrossRef]
6. Ahmad, B.; Alsaedi, A.; Alruwaily, Y. On Riemann—Stieltjes integral boundary value problems of Caputo—Riemann—Liouville type fractional integro-differential equations. *Filomat* **2020**, *34*, 2723–2738. [CrossRef]
7. Mamchuev, M. Cauchy problem for a linear system of ordinary differential equations of the fractional order. *Mathematics* **2020**, *8*, 1475. [CrossRef]
8. Kilbas, A.A.; Srivastava, H.M.; Trujillo, J.J. *Theory and Applications of Fractional Differential Equations*; Elsevier Science Publishing: Amsterdam, The Netherlands, 2006.
9. Gerasimov, A.N. Generalization of linear laws of deformation and their application to problems of internal friction. *Appl. Math. Mech.* **1948**, *12*, 251–260. (In Russian)
10. Caputo, M. Linear model of dissipation whose Q is almost frequancy independent. II. *Geophys. J. R. Astron. Soc.* **1967**, *13*, 529–539. [CrossRef]
11. Dzhrbashyan, M.M.; Nersesyan, A.B. Fractional derivatives and the Cauchy problem for differential equations of fractional order. *Izv. Akad. Nauk. Armyanskoy Ssr. Mat.* **1968**, *3*, 3–28. (In Russian)
12. Pskhu, A.V. The fundamental solution of a diffusion-wave equation of fractional order. *Izv. Math.* **2009**, *73*, 351–392. [CrossRef]
13. Pskhu, A.V. Fractional diffusion equation with a discretely distributed differentiation operator. *Sib. Elektron. Math. Rep.* **2016**, *13*, 1078–1098.
14. Plekhanova, M.V. Nonlinear equations with degenerate operator at fractional Caputo derivative. *Math. Methods Appl. Sci.* **2016**, *40*, 41–44. [CrossRef]
15. Fedorov, V.E.; Plekhanova, M.V.; Nazhimov R.R. Degenerate linear evolution equations with the Riemann—Liouville fractional derivative. *Sib. Math. J.* **2018**, *59*, 136–146. [CrossRef]
16. Fedorov, V.E. Generators of analytic resolving families for distributed order equations and perturbations. *Mathematics* **2020**, *8*, 1306. [CrossRef]
17. Fedorov, V.E.; Phuong, T.D.; Kien, B.T.; Boyko, K.V.; Izhberdeeva, E.M. A class of semilinear distributed order equations in Banach spaces. *Chelyabinsk Phys. Math. J.* **2020**, *5*, 343–351.
18. Sviridyuk, G.A.; Fedorov, V.E. *Linear Sobolev Type Equations and Degenerate Semigroups of Operators*; VSP: Utrecht, The Netherlands, 2003.
19. Triebel, H. *Interpolation Theory. Function Spaces. Differential Operators*; North-Holland Publishing Company: Amsterdam, The Netherlands, 1978.
20. Fedorov, V.E. Strongly holomorphic groups of linear equations of Sobolev type in locally convex spaces. *Differ. Equ.* **2004**, *40*, 753–765. [CrossRef]

Article

Bessel Collocation Method for Solving Fredholm–Volterra Integro-Fractional Differential Equations of Multi-High Order in the Caputo Sense

Shazad Shawki Ahmed * and Shabaz Jalil MohammedFaeq *

Department of Mathematics, College of Science, University of Sulaimani, Sulaymaniyah 46001, Iraq
* Correspondence: shazad.ahmed@univsul.edu.iq (S.S.A.); shabaz.mohammedfaeq@univsul.edu.iq (S.J.M.)

Abstract: The approximate solutions of Fredholm–Volterra integro-differential equations of multi-fractional order within the Caputo sense (F-VIFDEs) under mixed conditions are presented in this article apply a collocation points technique based completely on Bessel polynomials of the first kind. This new approach depends particularly on transforming the linear equation and conditions into the matrix relations (some time symmetry matrix), which results in resolving a linear algebraic equation with unknown generalized Bessel coefficients. Numerical examples are given to show the technique's validity and application, and comparisons are made with existing results by applying this process in order to express these solutions, most general programs are written in Python V.3.8.8 (2021).

Keywords: Fredholm–Volterra integral Equations; fractional derivative; Bessel polynomials; Caputo derivative; collocation points

Citation: Ahmed, S.S.; MohammedFaeq, S.J. Bessel Collocation Method for Solving Fredholm–Volterra Integro-Fractional Differential Equations of Multi-High Order in the Caputo Sense. *Symmetry* **2021**, *13*, 2354. https://doi.org/10.3390/sym13122354

Academic Editors: Francisco Martínez González and Mohammed KA Kaabar

Received: 3 November 2021
Accepted: 20 November 2021
Published: 7 December 2021

Publisher's Note: MDPI stays neutral with regard to jurisdictional claims in published maps and institutional affiliations.

Copyright: © 2021 by the authors. Licensee MDPI, Basel, Switzerland. This article is an open access article distributed under the terms and conditions of the Creative Commons Attribution (CC BY) license (https://creativecommons.org/licenses/by/4.0/).

1. Introduction

Fractional calculus (FC) deals with the differentiation and integration of arbitrary order and it is used in the real world to model and analyze big problems. Fluid flow, electrical networks, fractals theory, control theory, electromagnetic theory, probability, statistics, optics, potential theory, biology, chemistry, diffusion, and viscoelasticity are just a few of the many fields where fractional calculus is used [1–4].

In recent years, fractional differential equations and integro-fractional differential equations (IFDEs) have captivated the hobby of many researchers in various fields of science and era due to the reality that realistic modeling of a bodily phenomenon with dependencies not only in the immediate time, but also in the past time history can be accomplished effectively using FC. However, in addition to modeling, the solution approaches and their dependability are crucial in detecting key points when a rapid divergence, convergence, or bifurcation begins. As a result, high-precision solutions are always required. Several strategies for solving fractional order differential equations were presented for this purpose (or integro-differential equations), [1,3,4]. The Adomian decomposition method [5], variational iteration method [6], fractional differential transform method [7], fractional difference method [8], and power series method [9] are the most commonly used ideas.

However, from the beginning of 1994, Laguerre, Legendre, Taylor, Fourier, Hermite, and Bessel polynomials have been employed in works [10–15] to solve linear differential, integral, and integro-differential difference equations and related systems. In addition, the Bessel polynomial of the first kind method has been used to find approximate solutions of differential, fractional differential equations, integro-differential equations of fractional order, LVIDEs, and LF-VIDEs [16–19].

The aim of this paper is to expand and apply the first kind of Bessel polynomial in matrix form, as well as the collocation techniques, to evaluate the approximate solution for the multi-high-order linear Fredholm–Volterra integro-fractional differential equations (FVIFDEs) of the general type:

$$_a^C D_x^{\sigma_n} u(x) \; + \sum_{l=1}^{n-1} p_l(x) {}_a^C D_x^{\sigma_{n-l}} u(x) + p_n(x) u(x)$$
$$= g(x) + \sum_{i=0}^{m_1} \lambda_i \int_a^b F_i(x,t) {}_a^C D_t^{\alpha_i} u(t) dt + \sum_{j=0}^{m_2} \overline{\lambda}_j \int_a^x V_j(x,t) {}_a^C D_t^{\beta_j} u(t) dt, \; x \in [a,b] \quad (1)$$

together with mixed conditions:

$$\sum_{\ell=0}^{\mu-1} \left\{ \hbar_{k\ell} u^{(\ell)}(a) + \overline{\hbar}_{k\ell} u^{(\ell)}(b) \right\} = C_k, \; k = 0, 1, \ldots, \mu - 1. \quad (2)$$

where the fractional orders: $\sigma_n > \sigma_{n-1} > \cdots > \sigma_1 > \sigma_0 = 0$, $\alpha_{m_1} > \alpha_{m_1-1} > \cdots > \alpha_1 > \alpha_0 = 0$, and $\beta_{m_2} > \beta_{m_2-1} > \cdots > \beta_1 > \beta_0 = 0$, and $\mu = \max\{\lceil \sigma_n \rceil, \lceil \alpha_{m_1} \rceil, \lceil \beta_{m_2} \rceil\}$. In addition, $u(x)$ is an unknown function, the functions $p_l(x)$, $g(x) \in C([a,b], \mathbb{R})$, for all $l = 1, 2, \ldots, n$, and $F_i(x,t)$, $V_j(x,t) \in C(S, \mathbb{R})$, (with $S = \{(x,t) : a \leq t \leq x \leq b\}$) are known, with constants $\hbar_{k\ell}$, $\overline{\hbar}_{k\ell}$, $\lambda_i, \overline{\lambda}_j$ and $C_k \in \mathbb{R}$ for all $k, \ell = 0, 1, \ldots, \mu - 1$, $i = 0, 1, \ldots, m_1$, $j = 0, 1, \ldots, m_2$, (n, m_1, $m_2 \in \mathbb{Z}^+$) are given.

2. Preliminary Considerations
2.1. Basic Definitions and Some Lemmas

Many mathematical definitions of fractional integration and differentiation have come to light in recent years. The most frequently used definitions of fractional calculus involves the Riemann–Liouville fractional derivative and Caputo derivative. In terms of applicability, the Caputo concept is more dependable than the Riemann–Liouville definition. In this section, we are interested some basic definitions and lemmas which are used later on in this paper [1,3,4,20,21].

Definition 1 [22]. *A real valued function u defined on closed bounded interval $[a,b] = I$ be in the space $C_\gamma(I)$, $\gamma \in \mathbb{R}$, if there exist a real number $k > \gamma$, such that $u(x) = (x-a)^k u_0(x)$, where $u_0(x) \in C(I)$, and it is said to be in the space $C_\gamma(I)$, $\gamma \in \mathbb{R}$, if there exist a real number $k > \gamma$, such that $u(x) = (x-a)^k u_0(x)$, where $u_0(x) \in C(I)$, and it is said to be in the space $C_\gamma^n(I)$ iff $u^{(n)}(x) \in C_\gamma(I)$, where $n \in \mathbb{Z}^+ \cup \{0\}$.*

Definition 2 [23]. *The Riemann–Liouville (R-L) fractional integral operator, ${}_a J_x^\alpha$, of order $\alpha > 0$ of a function $u \in C_\gamma(I), \gamma \geq -1$ is defined as:*

$$_a J_x^\alpha u(x) = \frac{1}{\Gamma(\alpha)} \int_a^x (x-t)^{\alpha-1} u(t) dt, \quad \alpha \in \mathbb{R}^+$$
$$_a J_x^0 u(x) = u(x).$$

Definition 3 [24]. *The Riemann–Liouville (R-L) fractional derivative operator, ${}_a^R D_x^\alpha$, of order $\alpha \geq 0$ of a function $u(x)$ and $u \in C_{-1}^m(I)$, $m = \alpha$ is normally defined as:*

$$_a^R D_x^\alpha u(x) = D_x^m {}_a J_x^{m-\alpha} u(x), \quad m-1 < \alpha \leq m, \quad m \in \mathbb{N}.$$

Definition 4 [23]. *The Caputo fractional derivative operator, ${}_a^C D_x^\alpha$, of a function $u \in C_{-1}^m(I)$ and $m = \lceil \alpha \rceil$, (ceiling function), is defined as:*

$$_a^C D_x^\alpha = {}_a J_x^{m-\alpha} [D_x^m u(x)]$$
$$= \begin{cases} \frac{1}{\Gamma(m-\alpha)} \int_a^x (x-t)^{m-\alpha-1} \frac{\partial^m u(t)}{\partial t^m} dt, & m-1 < \alpha < m \\ \frac{\partial^m u(x)}{\partial x^m}, & \alpha = m, m \in \mathbb{N} \end{cases}$$

where the parameter α is the order of the derivative and is allowed to be any positive real number. The operators $_aJ_x^\alpha$ and $_a^C D_x^\alpha$ are linear operators. Furthermore, we have

Lemma 1 [4]. *Let $x > a$, $a \in \mathbb{R}$ and for $u(x) = (x-a)^\beta$ for some $\beta \neq -m$ is not negative integer, then*

$$_aJ_x^\alpha (x-a)^\beta = \frac{\Gamma(\beta+1)}{\Gamma(\beta+\alpha+1)} (x-a)^{\beta+\alpha}.$$

Lemma 2 [20]. *The Caputo derivative of order $\alpha \geq 0$ with $n = \lceil \alpha \rceil$ of the power function $u(x) = (x-a)^\beta$ for some $\beta \geq 0$ is formed by:*

$$_a^C D_x^\alpha u(x) = \begin{cases} 0 & \text{if } \beta \in \{0,1,2,\cdots,n-1\} \\ \frac{\Gamma(\beta+1)}{\Gamma(\beta-\alpha+1)}(x-a)^{\beta-\alpha} & \text{if } \beta \in \mathbb{N} \text{ and } \beta \geq n \\ & \text{or } \beta \notin \mathbb{N} \text{ and } \beta > n-1 \end{cases}$$

Lemma 3 [20]. *Let $\alpha \geq 0$, $m = \lceil \alpha \rceil$. Moreover, assume that $u \in C_{-1}^m(I)$. Then the Caputo fractional derivative $_a^C D_x^\alpha u(x)$ is continuous on $I = [a,b]$ and $\lim_{x \to a} [_a^C D_x^\alpha u(x)] = 0$.*

2.2. Bessel Polynomial of the First Kind

The r-th degree N-truncated Bessel polynomials of the first kind, [25,26], $J_r(x)$, $r = 0, 1, \ldots, N$ are defined by

$$J_r(x) = \sum_{k=0}^{\lfloor \frac{N-r}{2} \rfloor} \frac{(-1)^k}{k!(k+r)!} \left(\frac{x}{2}\right)^{2k+r}, \quad r \in \mathbb{N},\ 0 \leq x < \infty.$$

Here, N is a positive integer that is selected in such a way that $N \geq r$. On the other hand, we may express the $J_r(x)$ as follows in the matrix form.

$$\mathbf{J}(x) = \mathbf{X}(x)\mathbf{D}^T \text{ or } \mathbf{J}^T(x) = \mathbf{D}\mathbf{X}^T(x). \qquad (3)$$

where $\mathbf{J}(x) = [J_0(x)\ J_1(x) \ldots J_N(x)]$ and $\mathbf{X}(x) = [1\ x\ x^2\ \ldots x^N]$

If N is odd

$$\mathbf{D} = \begin{bmatrix} \frac{1}{0!\,0!\,2^0} & 0 & \frac{-1}{1!\,1!\,2^2} & \cdots & \frac{(-1)^{\frac{N-1}{2}}}{\left(\frac{N-1}{2}\right)!\,\left(\frac{N-1}{2}\right)!\,2^{N-1}} & 0 \\ 0 & \frac{1}{0!\,1!\,2^1} & 0 & \cdots & 0 & \frac{(-1)^{\frac{N-1}{2}}}{\left(\frac{N-1}{2}\right)!\,\left(\frac{N-1}{2}\right)!\,2^N} \\ 0 & 0 & \frac{1}{0!\,2!\,2^2} & \cdots & \frac{(-1)^{\frac{N-3}{2}}}{\left(\frac{N-3}{2}\right)!\,\left(\frac{N+1}{2}\right)!\,2^{N-1}} & 0 \\ \vdots & \vdots & \vdots & \ddots & \vdots & \vdots \\ 0 & 0 & 0 & \cdots & \frac{1}{0!\,(N-1)!\,2^{N-1}} & 0 \\ 0 & 0 & 0 & \cdots & 0 & \frac{1}{0!\,N!\,2^N} \end{bmatrix}_{(N+1)\times(N+1)}$$

If N is even

$$D = \begin{bmatrix} \frac{1}{0!\,0!\,2^0} & 0 & \frac{-1}{1!\,1!\,2^2} & \cdots & 0 & \frac{(-1)^{\frac{N}{2}}}{(\frac{N}{2})!\,(\frac{N}{2})!\,2^N} \\ 0 & \frac{1}{0!\,1!\,2^1} & 0 & \cdots & \frac{(-1)^{\frac{N-2}{2}}}{(\frac{N-2}{2})!\,(\frac{N}{2})!\,2^{N-1}} & 0 \\ 0 & 0 & \frac{1}{0!\,2!\,2^2} & \cdots & 0 & \frac{(-1)^{\frac{N-2}{2}}}{(\frac{N+2}{2})!\,(\frac{N+1}{2})!\,2^N} \\ \vdots & \vdots & \vdots & \ddots & \vdots & \vdots \\ 0 & 0 & 0 & \cdots & \frac{1}{0!\,(N-1)!\,2^{N-1}} & 0 \\ 0 & 0 & 0 & \cdots & 0 & \frac{1}{0!\,N!\,2^N} \end{bmatrix}_{(N+1)\times(N+1)}$$

3. Fundamental Matrix Relations

Recall Equation (1) and rewrite it as follows:

$$D(\{\sigma_l\}_{l=1}^n, x) = g(x) + I_f\left(\{\alpha_i\}_{i=0}^{m_1}, x\right) + I_v\left(\{\beta_j\}_{j=0}^{m_2}, x\right). \tag{4}$$

where

$$D(\{\sigma_l\}_{l=1}^n, x) = {}^c_a D_x^{\sigma_n} u(x) + \sum_{l=1}^{n-1} p_l(x)\,{}^c_a D_x^{\sigma_{n-l}} u(x) + p_n(x) u(x)$$

and the integral parts:

$$I_f(\{\alpha_i\}_{i=0}^{m_1}, x) = \sum_{i=0}^{m_1} \lambda_i \int_a^b F_i(x,t)\,{}^c_a D_t^{\alpha_i} u(t)\,dt, \quad I_v\left(\{\beta_j\}_{j=0}^{m_2}, x\right) = \sum_{j=0}^{m_2} \bar{\lambda}_j \int_a^x V_j(x,t)\,{}^c_a D_t^{\beta_j} u(t)\,dt.$$

our purpose is to find a close approximation of Equation (1) in the N-truncated Bessel series arrangement

$$u(x) \cong \sum_{r=0}^{N} a_r J_\nabla(x). \tag{5}$$

So that a_∇, for all $\nabla = 0, 1, \ldots, N$ are the unknown Bessel coefficients. Before we begin the approximate solution we must convert the solution $u(x)$ and its ${}^c_a D_x^{\sigma_n} u(x)$, ${}^c_a D_x^{\sigma_{n-l}} u(x)$, ${}^c_a D_x^{\alpha_i} u(x)$ and ${}^c_a D_x^{\beta_j} u(x)$, for all $l = 1, 2, \ldots, n-1$, $i = 0, 1, \ldots, m_1$, $j = 0, 1, \ldots, m_2$ in the parts $D(\{\sigma_l\}_{l=1}^n, x)$, $I_f(\{\alpha_i\}_{i=0}^{m_1}, x)$ and $I_v\left(\{\beta_j\}_{j=0}^{m_2}, x\right)$, to matrix form, within the mixed conditions of Equation (2).

3.1. Matrix Relation for the Fractional Derivative Part D

To describe the solution $u(x)$ of Equation (1), which is specified by the N-truncated Bessel series of Equation (5). The function defined in relation (5) in a matrix form

$$[u(x)] = J(x)A\,; A = [a_0\ a_1\ a_2 \ldots a_N]^T \tag{6}$$

or from Equation (3)

$$[u(x)] = X(x) D^T A. \tag{7}$$

The relationship between the matrix $X(x)$ and its derivative $X^{(1)}(x)$ is also written as follows:

$$X^{(1)}(x) = X(x) B^T. \tag{8}$$

where

$$B^T = \begin{bmatrix} 0 & 1 & 0 & 0 & \cdots & 0 \\ 0 & 0 & 2 & 0 & \cdots & 0 \\ 0 & 0 & 0 & 3 & \cdots & 0 \\ \vdots & \vdots & \vdots & \vdots & \ddots & \vdots \\ 0 & 0 & 0 & 0 & \cdots & N \\ 0 & 0 & 0 & 0 & \cdots & 0 \end{bmatrix}_{(N+1)\times(N+1)}$$

We will also get the recurrence relations from Equation (8):

$$X^{(0)}(x) = X(x)$$
$$X^{(1)}(x) = X(x)B^T$$
$$X^{(2)}(x) = X^{(1)}(x)B^T = X(x)(B^T)^2 \qquad (9)$$
$$\vdots$$
$$X^{(\iota)}(x) = X^{(\iota-1)}(x)B^T = X(x)(B^T)^{\iota}.$$

Here, note that $(B^T)^0 = [\mathbf{I}]_{(N+1)\times(N+1)}$ is an identity matrix of dimension $(N+1)$. Using mathematical induction, we can prove that Equation (9) is correct. By applying the same concept to Equation (7) and using Equation (9), we attain matrix relation

$$\begin{aligned} u^{(l)}(x) &= X^{(l)}(x)D^T A \\ u^{(l)}(x) &= X(x)(B^T)^l D^T A, \text{ for each} |= 0,1,\ldots,\mu \text{ and } \mu = \max\{\lceil \sigma_n \rceil, \lceil \alpha_{m_1} \rceil, \lceil \beta_{m_2} \rceil\} \end{aligned} \qquad (10)$$

By using Equation (7) with (9) and applying the Caputo Definition 4, with Lemma 1 and 2, we can convert the fractional terms ${}_a^c D_x^{\sigma_{n-l}} u(x), \overline{n}(\sigma_{n-l}) - 1 < \sigma_{n-l} \le \overline{n}(\sigma_{n-l})$, that is $\overline{n}(\sigma_{n-l}) = \lceil \sigma_{n-l} \rceil$, for all $l = 0,1,\ldots,n-1$ to matrix form:

$$\begin{aligned} {}_a^c D_x^{\sigma_{n-l}} u(x) &= {}_a^c D_x^{\sigma_{n-l}} X(x) D^T A \\ &= {}_a J_x^{(\overline{n}(\sigma_{n-l})-\sigma_{n-l})} D^{\overline{n}(\sigma_{n-l})} X(x) D^T A \\ &= {}_a J_x^{(\overline{n}(\sigma_{n-l})-\sigma_{n-l})} X(x) (B^T)^{\overline{n}(\sigma_{n-l})} D^T A \\ &= x^{\overline{n}(\sigma_{n-l})-\sigma_{n-l}} X(x) C(\overline{n}(\sigma_{n-l}) - \sigma_{n-l}) (B^T)^{\overline{n}(\sigma_n)} D^T A. \end{aligned}$$

Since

$${}_a J_x^{(\overline{n}(\sigma_{n-l})-\sigma_{n-l})} X(x) = {}_a J_x^{(\overline{n}(\sigma_{n-l})-\sigma_{n-l})} \begin{bmatrix} 1 & x & x^2 & \ldots & x^N \end{bmatrix}$$

$$= \left[\frac{\Gamma(1)}{\Gamma(\overline{n}(\sigma_{n-l}) - \sigma_{n-l} + 1)} x^{\overline{n}(\sigma_{n-l})-\sigma_{n-l}+0}, \frac{\Gamma(2)}{\Gamma(\overline{n}(\sigma_{n-l}) - \sigma_{n-l} + 2)} x^{\overline{n}(\sigma_{n-l})-\sigma_{n-l}+1}, \right.$$
$$\left. \ldots, \frac{\Gamma(N+1)}{\Gamma(\overline{n}(\sigma_{n-l}) - \sigma_{n-l} + N+1)} x^{\overline{n}(\sigma_{n-l})-\sigma_{n-l}+N} \right]$$

$$= x^{\overline{n}(\sigma_{n-l})-\sigma_{n-l}} \begin{bmatrix} 1 & x & x^2 & \ldots & x^N \end{bmatrix} \begin{bmatrix} \frac{\Gamma(1)}{\Gamma(\overline{n}(\sigma_{n-l}) - \sigma_{n-l} + 1)} & 0 & \cdots & 0 \\ 0 & \frac{\Gamma(2)}{\Gamma(\overline{n}(\sigma_{n-l}) - \sigma_{n-l} + 2)} & \cdots & 0 \\ \vdots & \vdots & \ddots & \vdots \\ 0 & 0 & \cdots & \frac{\Gamma(N+1)}{\Gamma(\overline{n}(\sigma_{n-l}) - \sigma_{n-l} + N+1)} \end{bmatrix}$$

Putting

$$C(\overline{n}(\sigma_{n-l}) - \sigma_{n-l}) = \begin{bmatrix} \frac{\Gamma(1)}{\Gamma(\overline{n}(\sigma_{n-l}) - \sigma_{n-l} + 1)} & 0 & \cdots & 0 \\ 0 & \frac{\Gamma(2)}{\Gamma(\overline{n}(\sigma_{n-l}) - \sigma_{n-l} + 2)} & \cdots & 0 \\ \vdots & \vdots & \ddots & \vdots \\ 0 & 0 & \cdots & \frac{\Gamma(N+1)}{\Gamma(\overline{n}(\sigma_{n-l}) - \sigma_{n-l} + N + 1)} \end{bmatrix} \quad (11)$$

Thus, for all $l = 1, 2, \ldots, n-1$ in general we obtain

$${}_a^c D_x^{\sigma_{n-l}} u(x) = {}_a^c D_x^{\sigma_{n-l}} X(x) D^T A = x^{\overline{n}(\sigma_{n-l}) - \sigma_{n-l}} X(x) C(\overline{n}(\sigma_{n-l}) - \sigma_{n-l}) \left(B^T\right)^{\overline{n}(\sigma_{n-l})} D^T A. \quad (12)$$

and

$${}_a^c D_x^{\sigma_n} u(x) = {}_a^c D_x^{\sigma_n} X(x) D^T A = x^{\overline{n}(\sigma_n) - \sigma_n} X(x) C(\overline{n}(\sigma_n) - \sigma_n) \left(B^T\right)^{\overline{n}(\sigma_n)} D^T A. \quad (13)$$

Using mathematical induction, we can prove that Equations (12) and (13) are correct. By substituting expressions (7), (12) and (13) into (4), As well we can make this assumption $y(n, x) = x^{\overline{n}(\sigma_n) - \sigma_n}$, $y(n-l, x) = x^{\overline{n}(\sigma_{n-l}) - \sigma_{n-l}}$, for all $l = 1, 2, \ldots, n-1$, we have

$$D(\{\sigma_l\}_{l=1}^n, x) = [y(n, x) X(x) C(\overline{n}(\sigma_n) - \sigma_n) \left(B^T\right)^{\overline{n}(\sigma_n)} D^T$$
$$+ \sum_{l=1}^{n-1} p_l(x) y(n-l, x) X(x) C(\overline{n}(\sigma_{n-l}) - \sigma_{n-l}) \left(B^T\right)^{\overline{n}(\sigma_{n-l})} D^T + p_n(x) X(x) D^T] A. \quad (14)$$

3.2. Matrix Relation for the Fredholm Integral Part I_f

The N-truncated Taylor series around $(0,0)$, [27] and the N-truncated Bessel series can be used to approximate the Fredholm kernel functions $F_i(x, t)$, $i = 0, 1, \ldots, m_1$, respectively

$$F_i(x, t) = \sum_{\updownarrow=0}^{N} \sum_{\backslash=0}^{N} {}_t F_{\updownarrow\backslash}^i x^\updownarrow t^\backslash \text{ and } F_i(x, t) = \sum_{\updownarrow=0}^{N} \sum_{\backslash=0}^{N} {}_b F_{\updownarrow\backslash}^i J_\updownarrow(x) J_\backslash(t), \quad i = 0, 1, \ldots, m_1 \quad (15)$$

where

$$\left[{}_t F_{\updownarrow\backslash}^i\right] = \frac{1}{\updownarrow! \backslash!} \frac{\partial^{\updownarrow + \backslash} F_i(0,0)}{\partial x^\updownarrow \partial t^\backslash}, \quad i = 0, 1, \ldots, m_1, \; \updownarrow, \backslash = 0, 1, \ldots, N.$$

In matrix forms, the Equation (15) may be written as Equations (16) and (17), respectively

$$F_i(x, t) = X(x) F_t^i X^T(t), \quad F_t^i = \left[{}_t F_{\updownarrow\backslash}^i\right], \quad i = 0, 1, \ldots, m_1, \; \updownarrow, \backslash = 0, 1, \ldots, N. \quad (16)$$

and

$$F_i(x, t) = J(x) F_b^i J^T(t), \quad F_b^i = \left[{}_b F_{\updownarrow\backslash}^i\right], \quad i = 0, 1, \ldots, m_1, \; \updownarrow, \backslash = 0, 1, \ldots, N. \quad (17)$$

From Equations (16) and (17), it also comes out according to Equation (3), the following relation

$$X(x) F_t^i X^T(t) = J(x) F_b^i J^T(t), \quad i = 0, 1, \ldots, m_1$$
$$X(x) F_t^i X^T(t) = X(x) D^T F_b^i D X^T(t), \quad i = 0, 1, \ldots, m_1 \quad (18)$$
$$F_t^i = D^T F_b^i D \text{ or } F_b^i = (D^T)^{-1} F_t^i D^{-1}, \quad i = 0, 1, \ldots, m_1$$

In the same way from Equations (12) and (13), convert ${}_a^c D_x^{\alpha_i} u(x)$, and $\overline{n}(\alpha_i) - 1 < \alpha_i \leq \overline{n}(\alpha_i)$, i.e., $\overline{n}(\alpha_i) = \lceil \alpha_i \rceil$ for all $i = 0, 1, \ldots, m_1$, by apply the Caputo Definition 4 with Lemma 1 to the matrix form, we obtain

$$\begin{aligned}
{}_a^c D_x^{\alpha_i} u(x) &= {}_a^c D_x^{\alpha_i} X(x) D^T A \\
&= {}_a J_x^{(\overline{n}(\alpha_i)-\alpha_i)} D^{\overline{n}(\alpha_i)} X(x) D^T A \\
&= {}_a J_x^{(\overline{n}(\alpha_i)-\alpha_i)} X(x) \left(B^T\right)^{\overline{n}(\alpha_i)} D^T A \\
&= x^{\overline{n}(\alpha_i)-\alpha_i} X(x) C(\overline{n}(\alpha_i)-\alpha_i) \left(B^T\right)^{\overline{n}(\alpha_i)} D^T A, \quad i = 0, 1, \ldots, m_1
\end{aligned} \quad (19)$$

where $C(\overline{n}(\alpha_i) - \alpha_i)$ is defined at Equation (11). We obtain the matrix relation (20), put the Equation (3) into (17), and then replace the obtained matrix with matrix (19) in the Fredholm integral part I_f in Equation (4)

$$\begin{aligned}
I_f(\{\alpha_i\}_{i=0}^{m_1}, x) &= \sum_{i=0}^{m_1} \lambda_i \int_a^b F_i(x,t) {}_a^c D_t^{\alpha_i} u(t) dt \\
&= \sum_{i=0}^{m_1} \lambda_i \int_a^b J(x) F_b^i J^T(t) t^{\overline{n}(\alpha_i)-\alpha_i} X(t) C(\overline{n}(\alpha_i) - \alpha_i) \left(B^T\right)^{\overline{n}(\alpha_i)} D^T A dt \\
&= \sum_{i=0}^{m_1} \lambda_i \int_a^b X(x) D^T F_b^i D X^T(t) t^{\overline{n}(\alpha_i)-\alpha_i} X(t) C(\overline{n}(\alpha_i) - \alpha_i) \left(B^T\right)^{\overline{n}(\alpha_i)} D^T A dt \\
&= \sum_{i=0}^{m_1} \lambda_i X(x) D^T F_b^i D \left(\int_a^b X^T(t) X(t) t^{\overline{n}(\alpha_i)-\alpha_i} dt \right) C(\overline{n}(\alpha_i) - \alpha_i) \left(B^T\right)^{\overline{n}(\alpha_i)} D^T A \\
&= \sum_{i=0}^{m_1} \lambda_i X(x) D^T F_b^i D H_{f,i} C(\overline{n}(\alpha_i) - \alpha_i) \left(B^T\right)^{\overline{n}(\alpha_i)} D^T A.
\end{aligned} \quad (20)$$

where

$$H_{f,i} = \int_a^b X^T(t) X(t) t^{\overline{n}(\alpha_i)-\alpha_i} dt = \left[h_{rs}^{f,i}\right], \quad i = 0, 1, \ldots, m_1, \; r, s = 0, 1, \ldots, N.$$

$$\left[h_{rs}^{f,i}\right] = \frac{b^{\overline{n}(\alpha_i)-\alpha_i+r+s+1} - a^{\overline{n}(\alpha_i)-\alpha_i+r+s+1}}{\overline{n}(\alpha_i)-\alpha_i+r+s+1}, \quad i = 0, 1, \ldots, m_1, \; r, s = 0, 1, \ldots, N.$$

We can get the last matrix form (21) by replacing the matrix relation (18) into expression (20).

$$I_f(\{\alpha_i\}_{i=0}^{m_1}, x) = \sum_{i=0}^{m_1} \lambda_i X(x) F_t^i H_{f,i} C(\overline{n}(\alpha_i) - \alpha_i) \left(B^T\right)^{\overline{n}(\alpha_i)} D^T A. \quad (21)$$

3.3. Matrix Relation for the Volterra Integral Part I_v

The N-truncated Taylor series around (0,0), [27] and the N-truncated Bessel series can be used to approximate the Volterra kernel functions $V_j(x,t)$, $j = 0, 1, \ldots, m_2$, respectively

$$V_j(x,t) = \sum_{\updownarrow=0}^{N} \sum_{\backslash=0}^{N} {}_t V_{\updownarrow\backslash}^j x^{\updownarrow} t^{\backslash} \text{ and } V_j(x,t) = \sum_{\updownarrow=0}^{N} \sum_{\backslash=0}^{N} {}_b V_{\updownarrow\backslash}^j J_{\updownarrow}(x) J_{\backslash}(t), \; j = 0, 1, \ldots, m_2 \quad (22)$$

where

$$\left[{}_t V_{\updownarrow\backslash}^j\right] = \frac{1}{\updownarrow!\backslash!} \frac{\partial^{\updownarrow+\backslash} V_j(0,0)}{\partial x^{\updownarrow} \partial t^{\backslash}}, \; j = 0, 1, \ldots, m_2, \; \updownarrow, \backslash = 0, 1, \ldots, N.$$

The relations in Equation (22) can be transformed into matrix forms:

$$V_j(x,t) = X(x) V_t^j X^T(t), \quad V_t^j = \left[{}_t V_{\updownarrow\backslash}^j\right], \; j = 0, 1, \ldots, m_2, \; \updownarrow, \backslash = 0, 1, \ldots, N \quad (23)$$

and

$$V_j(x,t) = J(x) V_b^j J^T(t), \quad V_b^j = \left[{}_b V_{\updownarrow\backslash}^j\right], \; j = 0, 1, \ldots, m_2, \; \updownarrow, \backslash = 0, 1, \ldots, N \quad (24)$$

from Equations (23) and (24), it also comes out according to Equation (3) we obtain the following relation:

$$X(x)V_t^j X^T(t) = J(x)V_b^j J^T(t), \quad j = 0,1,\ldots,m_2$$
$$X(x)V_t^j X^T(t) = X(x)D^T V_b^j D X^T(t), \quad j = 0,1,\ldots,m_2 \qquad (25)$$
$$V_t^j = D^T V_b^j D \quad \text{or} \quad V_b^j = (D^T)^{-1} V_t^j D^{-1}, \quad j = 0,1,\ldots,m_2$$

Finally, in the same way from Equations (12) and (13), convert ${}_a^c D_x^{\beta_j} u(x)$, $\overline{n}(\beta_j) - 1 < \beta_j \leq \overline{n}(\beta_j)$, i.e., $\overline{n}(\beta_j) = \lceil \beta_j \rceil$, for all $j = 0,1,\ldots,m_2$, by applying the Caputo Definition 4 with Lemma 1, 2 and 3 to matrix form, we obtain

$$\begin{aligned}
{}_a^c D_x^{\beta_j} u(x) &= {}_a^c D_x^{\beta_j} X(x) D^T A \\
&= {}_a I_x^{(\overline{n}(\beta_j) - \beta_j)} D^{\overline{n}(\beta_j)} X(x) D^T A \\
&= {}_a I_x^{(\overline{n}(\beta_j) - \beta_j)} X(x) (B^T)^{\overline{n}(\beta_j)} D^T A \\
&= x^{\overline{n}(\beta_j) - \beta_j} X(x) C(\overline{n}(\beta_j) - \beta_j) (B^T)^{\overline{n}(\beta_j)} D^T A, \quad j = 0,1,\ldots,m_2
\end{aligned} \qquad (26)$$

where $C(\overline{n}(\beta_j) - \beta_j)$ define at Equation (11). We obtain the matrix relation (27), put the Equation (3) into (24) and then replace the obtained matrix with matrix (26) in Fredholm integral part I_v in Equation (4)

$$\begin{aligned}
I_v\left(\{\beta_j\}_{j=0}^{m_2}, x\right) &= \sum_{j=0}^{m_2} \overline{\lambda}_j \int_a^x V_j(x,t) {}_a^c D_t^{\beta_j} u(t) dt \\
&= \sum_{j=0}^{m_2} \overline{\lambda}_j \int_a^x J(x) V_b^j J^T(t) t^{\overline{n}(\beta_j) - \beta_j} X(t) C(\overline{n}(\beta_j) - \beta_j) (B^T)^{\overline{n}(\beta_j)} D^T A dt \\
&= \sum_{j=0}^{m_2} \overline{\lambda}_j \int_a^x X(x) D^T V_b^j D X^T(t) t^{\overline{n}(\beta_j) - \beta_j} X(t) C(\overline{n}(\beta_j) - \beta_j) (B^T)^{\overline{n}(\beta_j)} D^T A dt \\
&= \sum_{j=0}^{m_2} \overline{\lambda}_j X(x) D^T V_b^j D \left(\int_a^x X^T(t) X(t) t^{\overline{n}(\beta_j) - \beta_j} dt \right) C(\overline{n}(\beta_j) \\
&\quad - \beta_j) (B^T)^{\overline{n}(\beta_j)} D^T A \\
&= \sum_{j=0}^{m_2} \overline{\lambda}_j X(x) D^T V_b^j D H_{v,j}(x) C(\overline{n}(\beta_j) - \beta_j) (B^T)^{\overline{n}(\beta_j)} D^T A.
\end{aligned} \qquad (27)$$

where
$$H_{v,j}(x) = \int_a^x X^T(t) X(t) t^{\overline{n}(\beta_j) - \beta_j} dt = \left[h_{rs}^{v,j}(x) \right], \quad j = 0,1,\ldots,m_2, \ r,s = 0,1,\ldots,N.$$

$$\left[h_{rs}^{v,j}(x) \right] = \frac{x^{\overline{n}(\beta_j) - \beta_j + r + s + 1} - a^{\overline{n}(\beta_j) - \beta_j + r + s + 1}}{\overline{n}(\beta_j) - \beta_j + r + s + 1}, \quad j = 0,1,\ldots,m_2, \ r,s = 0,1,\ldots,N.$$

We can get the last matrix form (28) by replacing the matrix relation (25) into expression (27).

$$I_v\left(\{\beta_j\}_{j=0}^{m_2}, x\right) = \sum_{j=0}^{m_2} \overline{\lambda}_j X(x) V_t^j H_{v,j}(x) C(\overline{n}(\beta_j) - \beta_j) (B^T)^{\overline{n}(\beta_j)} D^T A. \qquad (28)$$

3.4. Matrix Relation for the Conditions

For each $k = 0,1,\ldots,\mu - 1$ and $\mu = \max\{\lceil \sigma_n \rceil, \lceil \alpha_{m_1} \rceil, \lceil \beta_{m_2} \rceil\}$, applying the relation (10) to each mixed condition of Equation (2), we obtain the corresponding condition matrix forms as follows.

$$\sum_{\ell=0}^{\mu-1} \left\{ \zeta_{k\ell} X(a) (B^T)^\ell D^T A + \overline{\zeta}_{k\ell} X(b) (B^T)^\ell D^T A \right\} = [C_k], \ k = 0,1,\ldots,\mu - 1$$

Thus

$$\sum_{\ell=0}^{\mu-1} \left[\overline{\langle_{k\ell}} X(a) + \overline{\langle_{k\ell}} X(b) \right] \left(B^T \right)^\ell D^T A = [C_k], \quad k = 0, 1, \ldots, \mu-1. \tag{29}$$

4. Method of Solution

To construct the fundamental matrix equation that corresponds to Equation (1), insert the matrix relations (14), (21), and (28) into Equation (4) to obtain the following matrix equation

$$\left[y(n,x) X(x) \, C(\overline{n}(\sigma_n) - \sigma_n) \, (B^T)^{\overline{n}(\sigma_n)} D^T + \sum_{l=1}^{n-1} p_l(x) y(n-l,x) X(x) \, C(\overline{n}(\sigma_{n-l}) - \sigma_{n-l}) \, (B^T)^{\overline{n}(\sigma_{n-l})} D^T \right.$$
$$\left. + p_n(x) X(x) D^T \right] A$$
$$= g(x) + \sum_{i=0}^{m_1} \lambda_i X(x) F_t^i \, H_{f,i} \, C(\overline{n}(\alpha_i) - \alpha_i) \, (B^T)^{\overline{n}(\alpha_i)} D^T A \tag{30}$$
$$+ \sum_{j=0}^{m_2} \overline{\lambda}_j X(x) V_t^j H_{v,j}(x) C(\overline{n}(\beta_j) - \beta_j) \, (B^T)^{\overline{n}(\beta_j)} D^T A.$$

We get the following system of equations by setting the collocation points, [28], described by $x_i = a + \frac{b-a}{N} i, \ i = 0, 1, \ldots, N$:

$$\left[y(n,x_i) X(x_i) \, C(\overline{n}(\sigma_n) - \sigma_n) \, (B^T)^{\overline{n}(\sigma_n)} D^T + \sum_{l=1}^{n-1} p_l(x_i) y(n-l,x_i) X(x_i) \, C(\overline{n}(\sigma_{n-l}) - \sigma_{n-l}) \, (B^T)^{\overline{n}(\sigma_{n-l})} D^T \right.$$
$$\left. + p_n(x_i) X(x_i) D^T \right] A$$
$$= g(x_i) + \sum_{i=0}^{m_1} \lambda_i X(x_i) F_t^i \, H_{f,i} \, C(\overline{n}(\alpha_i) - \alpha_i) \, (B^T)^{\overline{n}(\alpha_i)} D^T A$$
$$+ \sum_{j=0}^{m_2} \overline{\lambda}_j X(x_i) V_t^j H_{v,j}(x_i) C(\overline{n}(\beta_j) - \beta_j) \, (B^T)^{\overline{n}(\beta_j)} D^T A, \quad i = 0, 1, \ldots, N.$$

or in brief, the most important matrix equation is

$$\left[y(n) X \, C(\overline{n}(\sigma_n) - \sigma_n) \, (B^T)^{\overline{n}(\sigma_n)} D^T + \sum_{l=1}^{n-1} p_l y(n-l) X \, C(\overline{n}(\sigma_{n-l}) - \sigma_{n-l}) \, (B^T)^{\overline{n}(\sigma_{n-l})} D^T + p_n X D^T \right.$$
$$\left. - \sum_{i=0}^{m_1} \lambda_i X F_t^i \, H_{f,i} \, C(\overline{n}(\alpha_i) - \alpha_i) \, (B^T)^{\overline{n}(\alpha_i)} D^T - \sum_{j=0}^{m_2} \overline{\lambda}_j \overline{XV^j H}_j \overline{C}_j B^i D \right] A = G. \tag{31}$$

where

$$y(n-l) = \begin{bmatrix} x_0^{(\overline{n}(\sigma_{n-l})-\sigma_{n-l})} & 0 & \cdots & 0 \\ 0 & x_1^{(\overline{n}(\sigma_{n-l})-\sigma_{n-l})} & \cdots & 0 \\ \vdots & \vdots & \ddots & \vdots \\ 0 & 0 & \cdots & x_N^{(\overline{n}(\sigma_{n-l})-\sigma_{n-l})} \end{bmatrix}, \quad P_l = \begin{bmatrix} p_l(x_0) & 0 & \cdots & 0 \\ 0 & p_l(x_1) & \cdots & 0 \\ \vdots & \vdots & \ddots & \vdots \\ 0 & 0 & \cdots & p_l(x_N) \end{bmatrix}$$

for all $l = 0, 1, \ldots, n$, and $y(0) = [\mathbf{I}]_{(N+1) \times (N+1)}$, $P_0 = [\mathbf{I}]_{(N+1) \times (N+1)}$ are the unit matrix,

$$X = \begin{bmatrix} X(x_0) \\ X(x_1) \\ \vdots \\ X(x_N) \end{bmatrix} = \begin{bmatrix} 1 & x_0 & x_0^2 & \cdots & x_0^N \\ 1 & x_1 & x_1^2 & \cdots & x_1^N \\ \vdots & \vdots & \vdots & \cdots & \vdots \\ 1 & x_N & x_N^2 & \cdots & x_N^N \end{bmatrix}, \quad \overline{X} = \begin{bmatrix} X(x_0) & 0 & \cdots & 0 \\ 0 & X(x_1) & \cdots & 0 \\ \vdots & \vdots & \ddots & \vdots \\ 0 & 0 & \cdots & X(x_N) \end{bmatrix}$$

also, for each fractional order $\gamma = \{\sigma_n, \sigma_{n-1}, \alpha_i \text{ and } \beta_j, \text{ for all } l = 1, 2, \ldots, n, i = 0, 1, \ldots, m_1, j = 0, 1, \ldots, m_2\}$ we are putting

$$C(\overline{n}(\gamma)-\gamma) = \begin{bmatrix} \frac{\Gamma(1)}{\Gamma(\overline{n}(\gamma)-\gamma+1)} & 0 & \cdots & 0 \\ 0 & \frac{\Gamma(2)}{\Gamma(\overline{n}(\gamma)-\gamma+2)} & \cdots & 0 \\ \vdots & \vdots & \ddots & \vdots \\ 0 & 0 & \cdots & \frac{\Gamma(N+1)}{\Gamma(\overline{n}(\gamma)-\gamma+N+1)} \end{bmatrix}$$

$$(B^T)^{\overline{n}(\gamma)} = \begin{bmatrix} 0 & 1 & 0 & 0 & \cdots & 0 \\ 0 & 0 & 2 & 0 & \cdots & 0 \\ 0 & 0 & 0 & 3 & \cdots & 0 \\ \vdots & \vdots & \vdots & \vdots & \ddots & \vdots \\ 0 & 0 & 0 & 0 & \cdots & N \\ 0 & 0 & 0 & 0 & \cdots & 0 \end{bmatrix}^{\overline{n}(\gamma)}, \quad {_tF_t^i} = \begin{bmatrix} {_tF_{00}^i} & {_tF_{01}^i} & \cdots & {_tF_{0N}^i} \\ {_tF_{10}^i} & {_tF_{11}^i} & \cdots & {_tF_{1N}^i} \\ \vdots & \vdots & \ddots & \vdots \\ {_tF_{N0}^i} & {_tF_{N1}^i} & \cdots & {_tF_{NN}^i} \end{bmatrix}, \quad {_tV_t^j} = \begin{bmatrix} {_tV_{00}^j} & {_tV_{01}^j} & \cdots & {_tV_{0N}^j} \\ {_tV_{10}^j} & {_tV_{11}^j} & \cdots & {_tV_{1N}^j} \\ \vdots & \vdots & \ddots & \vdots \\ {_tV_{N0}^j} & {_tV_{N1}^j} & \cdots & {_tV_{NN}^j} \end{bmatrix}$$

$$i = 0,1,\ldots,m_1, \quad j = 0,1,\ldots,m_2$$

where

$$\left[{_tF_{\updownarrow\backslash}^i}\right] = \frac{1}{\updownarrow!\backslash!} \frac{\partial^{\updownarrow+\backslash} F_i(0,0)}{\partial x^{\updownarrow} \partial t^{\backslash}}, \quad i = 0,1,\ldots,m_1, \quad \updownarrow, \backslash = 0,1,\ldots,N$$

$$\left[{_tV_{\updownarrow\backslash}^j}\right] = \frac{1}{\updownarrow!\backslash!} \frac{\partial^{\updownarrow+\backslash} V_j(0,0)}{\partial x^{\updownarrow} \partial t^{\backslash}}, \quad j = 0,1,\ldots,m_2, \quad \updownarrow, \backslash = 0,1,\ldots,N$$

$$H_{f,i} = \begin{bmatrix} h_{00}^{f,i} & h_{01}^{f,i} & \cdots & h_{0N}^{f,i} \\ h_{10}^{f,i} & h_{11}^{f,i} & \cdots & h_{1N}^{f,i} \\ \vdots & \vdots & \ddots & \vdots \\ h_{N0}^{f,i} & h_{N1}^{f,i} & \cdots & h_{NN}^{f,i} \end{bmatrix}, \quad H_{v,j}(x_\iota) = \begin{bmatrix} h_{00}^{v,j}(x_\iota) & h_{01}^{v,j}(x_\iota) & \cdots & h_{0N}^{v,j}(x_\iota) \\ h_{10}^{v,j}(x_\iota) & h_{11}^{v,j}(x_\iota) & \cdots & h_{1N}^{v,j}(x_\iota) \\ \vdots & \vdots & \ddots & \vdots \\ h_{N0}^{v,j}(x_\iota) & h_{N1}^{v,j}(x_\iota) & \cdots & h_{NN}^{v,j}(x_\iota) \end{bmatrix}, \iota = 0,1,\ldots,N$$

where, respectively

$$\left[h_{rs}^{f,i}\right] = \frac{b^{\overline{n}(\alpha_i)-\alpha_i+r+s+1} - a^{\overline{n}(\alpha_i)-\alpha_i+r+s+1}}{\overline{n}(\alpha_i)-\alpha_i+r+s+1}, \quad i=0,1,\ldots,m_1, \quad r,s=0,1,\ldots,N$$

$$\left[h_{rs}^{v,j}(x_\iota)\right] = \frac{x_\iota^{\overline{n}(\beta_j)-\beta_j+r+s+1} - a^{\overline{n}(\beta_j)-\beta_j+r+s+1}}{\overline{n}(\beta_j)-\beta_j+r+s+1}, \quad j=0,1,\ldots,m_2, \quad r,s=0,1,\ldots,N$$

$$\iota = 0,1,\ldots,N$$

$$\overline{V}^j = \begin{bmatrix} V_t^j & 0 & \cdots & 0 \\ 0 & V_t^j & \cdots & 0 \\ \vdots & \vdots & \ddots & \vdots \\ 0 & 0 & \cdots & V_t^j \end{bmatrix}, \quad \overline{C}_j = \begin{bmatrix} C(\overline{n}(\beta_j)-\beta_j) & 0 & \cdots & 0 \\ 0 & C(\overline{n}(\beta_j)-\beta_j) & \cdots & 0 \\ \vdots & \ddots & \ddots & \vdots \\ 0 & 0 & \cdots & C(\overline{n}(\beta_j)-\beta_j) \end{bmatrix}$$

$$\overline{H}_j = \begin{bmatrix} H_{v,j}(x_0) & 0 & \cdots & 0 \\ 0 & H_{v,j}(x_1) & \cdots & 0 \\ \vdots & \vdots & \ddots & \vdots \\ 0 & 0 & \cdots & H_{v,j}(x_N) \end{bmatrix}, \quad \overline{B}^j = \begin{bmatrix} (B^T)^{\overline{n}(\beta_j)} & 0 & \cdots & 0 \\ 0 & (B^T)^{\overline{n}(\beta_j)} & \cdots & 0 \\ \vdots & \vdots & \ddots & \vdots \\ 0 & 0 & \cdots & (B^T)^{\overline{n}(\beta_j)} \end{bmatrix}$$

$$j = 0,1,\ldots,m_2$$

$$\overline{D} = \begin{bmatrix} D^T \\ D^T \\ \vdots \\ D^T \end{bmatrix}, \quad G = \begin{bmatrix} g(x_0) \\ g(x_1) \\ \vdots \\ g(x_N) \end{bmatrix}, \quad \text{and} \quad A = \begin{bmatrix} a_0 \\ a_1 \\ \vdots \\ a_N \end{bmatrix}$$

When the matrices

$y(n)$, $y(n-1)$, p_n, p_l, X, $C(\overline{n}(\sigma_n) - \sigma_n)$, $C(\overline{n}(\sigma_{n-1}) - \sigma_{n-1})$, $C(\overline{n}(\alpha_i) - \alpha_i)$, $C(\overline{n}(\beta_j) - \beta_j)$, $(B^T)^{\overline{n}(\sigma_n)}$,

$$(B^T)^{\overline{n}(\sigma_{n-1})}, \ (B^T)^{\overline{n}(\alpha_i)}, \ (B^T)^{\overline{n}(\beta_j)}, \ F_t^i, \ V_t^j, \ H_{f,i}, \ H_{v,j}(x_i) \text{ and } D^T$$

For all $l = 1, 2, \ldots, n-1$, $i = 0, 1, \ldots, m_1$, $j = 0, 1, \ldots, m_2$, $i = 0, 1, \ldots, N$. In Equation (31) we have explained that their dimensions are similar to those of $(N+1) \times (N+1)$. Moreover, in Equation (31), these matrices $\overline{X}$, $\overline{V}^j$, $\overline{H}_j$, $\overline{C}_j$, $\overline{B}^j$ and $\overline{D}$, for all $j = 0, 1, \ldots, m_2$ are written in full, their measured dimensions can be observed by $(N+1) \times (N+1)^2$, $(N+1)^2 \times (N+1)^2$, $(N+1)^2 \times (N+1)^2$, $(N+1)^2 \times (N+1)^2$, $(N+1)^2 \times (N+1)^2$, and $(N+1)^2 \times (N+1)$ respectively.

As a result, the fundamental matrix Equation (31) that corresponds to Equation (1) may be expressed as

$$WA = G \text{ or } [W:G]. \tag{32}$$

where

$$W = y(n)X\,C(\overline{n}(\sigma_n) - \sigma_n)\,(B^T)^{\overline{n}(\sigma_n)}D^T + \sum_{l=1}^{n-1} p_l y(n-l) X\,C(\overline{n}(\sigma_{n-1}) - \sigma_{n-1})\,(B^T)^{\overline{n}(\sigma_{n-1})}D^T$$

$$+ p_n X D^T - \sum_{i=0}^{m_1} \lambda_i X F_t^i\, H_{f,i}\, C(\overline{n}(\alpha_i) - \alpha_i)\,(B^T)^{\overline{n}(\alpha_i)}D^T - \sum_{j=0}^{m_2} \overline{\lambda}_j \overline{X}\overline{V}^j \overline{H}_j \overline{C}_j \overline{B}^j \overline{D}.$$

Note that, Equation (32) is a set of $(N+1)$ linear algebraic equations with unknown Bessel coefficients $A = [a_0, a_1, \ldots, a_N]$. The matrix form (29) for the conditions, on the other hand, may be represented as

$$U_k A = [C_k] \text{ or } [U_k : C_k]; \ k = 0, 1, \ldots, \mu - 1, \qquad \mu = \max\{\lceil \sigma_n \rceil, \lceil \alpha_{m_1} \rceil, \lceil \beta_{m_2} \rceil\}. \tag{33}$$

$$U_k = \sum_{\ell=0}^{\mu-1} \left[\langle_{k\ell} \mathcal{X}(a) + \overline{\langle}_{k\ell} \mathcal{X}(b)\right] (B^T)^\ell D^T A.$$

$$= [u_{k0}\ u_{k1}\ u_{k2}\ \cdots\ u_{kN}], \ k = 0, 1, \ldots, \mu - 1$$

Hence, we may solve Equation (1) under mixed conditions (2) by substituting the rows of the matrices W and G for the rows of the matrices U_k and C_k, respectively.

$$\widetilde{W}A = \widetilde{G}.$$

The new augmented matrix (some time may be symmetry) of the preceding system is as follows if the last μ-rows of the matrix (32) are replaced for simplicity:

$$[\widetilde{W} : \widetilde{G}] = \begin{bmatrix} w_{00} & w_{01} & w_{02} & \cdots & w_{0N} & : & g(x_0) \\ w_{10} & w_{11} & w_{12} & \cdots & w_{1N} & : & g(x_1) \\ w_{20} & w_{21} & w_{22} & \cdots & w_{2N} & : & g(x_2) \\ \vdots & \vdots & \vdots & \ddots & \vdots & : & \vdots \\ w_{N-m,0} & w_{N-m,1} & w_{N-m,2} & \cdots & w_{N-m,N} & : & g(x_{N-m}) \\ u_{00} & u_{01} & u_{02} & \cdots & u_{0N} & : & c_0 \\ u_{10} & u_{11} & u_{12} & \cdots & u_{1N} & : & c_1 \\ u_{20} & u_{21} & u_{22} & \cdots & u_{2N} & : & c_2 \\ \vdots & \vdots & \vdots & \ddots & \vdots & : & \vdots \\ u_{\mu-1,0} & u_{\mu-1,1} & u_{\mu-1,2} & \cdots & u_{\mu-1,N} & : & c_{\mu-1} \end{bmatrix} \tag{34}$$

Take note that rank $\widetilde{W}$ = rank $[\widetilde{W} : \widetilde{G}] = N + 1$. If it isn't, the suggested technique fails to offer a solution; but in this case, in this situation, the number of collocation points (or, equivalently, the dimension of the matrix $\widetilde{W}$) can be increased to get the specific or general answer. As a result, we may write $A = \left(\widetilde{W}\right)^{-1} \widetilde{G}$, and therefore the elements $a_0, a_1, \ldots, a_N$ of A are uniquely determined.

Moreover, select that N we define needs to be greater than μ, i.e., $N > \mu = max\{\lceil \sigma_n \rceil, \lceil \alpha_{m_1} \rceil, \lceil \beta_{m_2} \rceil\}$. If it is not, the proposed strategy is thus unable to give a solution, because matrix B^T becomes a zero matrix, we only get zero solution.

5. Numerical Examples

In this work, we choose several examples where the exact solution already exists to demonstrate the accuracy. They were all carried out on a computer using a Python program V3.8.8 (2021). The least square errors (L.S.E) in tables are the values of $\sum_{i=0}^{M}[u(x_i) - \widetilde{u}_N(x_i)]^2$, $M \in \mathbb{N}$ at M-selected collocation points x_i. and the running time is also provided in tabular form.

Example 1. *Consider the linear Fredholm–Volterra integro-differential equation of multi-higher fractional order, given by*

$$_0^C D_x^{1.2} u(x) + x_0^C D_x^{0.1} u(x) - xu(x) = g(x) + \int_0^1 \left(e^x u(t) - t_0^C D_t^{0.9} u(t)\right) dt + \int_0^x \left(-(x-t)u(t) + 2x_0^C D_t^{1.8} u(t)\right) dt$$

where $0 \leq x, t \leq 1$

$$g(x) = \frac{-2}{\Gamma(1.8)} x^{0.8} - \frac{2}{\Gamma(2.9)} x^{2.9} + \frac{1}{\Gamma(1.9)} x^{1.9} + \frac{4}{\Gamma(2.2)} x^{2.2} - \frac{1}{6} e^x + \frac{1.1}{\Gamma(3.1)} - \frac{4.2}{\Gamma(4.1)} - x^2 + \frac{7}{6} x^3 - \frac{1}{12} x^4$$

with the boundary conditions

$$2u^{(1)}(0) + u^{(1)}(1) = 1 \text{ and } u^{(1)}(1) = -1$$

which is the exact solution $u(x) = x(1-x)$.

Let us now determine the N-truncated Bessel series approximate solution $u_N(x)$

$$u(x) \cong u_N(x) = \sum_{\nabla=0}^{N} a_\nabla J_\nabla(x)$$

Here, from the considered, example we have:

$\sigma_0 = 0, \sigma_1 = 0.1, \sigma_2 = 1.2 \rightarrow \overline{n}(\sigma_0) = \sigma_0 = 0, \overline{n}(\sigma_1) = \sigma_1 = 1, \overline{n}(\sigma_2) = \sigma_2 = 2$

$\alpha_0 = 0, \alpha_1 = 0.9 \rightarrow \overline{n}(\alpha_0) = \alpha_0 = 0, \overline{n}(\alpha_1) = \alpha_1 = 1$

$\beta_0 = 0, \beta_1 = 1.8 \rightarrow \overline{n}(\beta_0) = \beta_0 = 0, \overline{n}(\beta_1) = \beta_1 = 2$

$\mu = max\{\lceil 1.2 \rceil, \lceil 0.9 \rceil, \lceil 1.8 \rceil \} = 2$

$p_1(x) = x, p_2(x) = -x, F_0(x,t) = e^x, F_1(x,t) = t, V_0(x,t) = x - t, V_1(x,t) = x,$

$\lambda_0 = 1, \lambda_1 = -1, \overline{\lambda}_0 = -1, \overline{\lambda}_1 = 2$

Hence $\mu = 2$ so take, the collocation point sets are $\left\{x_0 = 0, \ x_1 = \frac{1}{3}, \ x_2 = \frac{2}{3}, \ x_3 = 1\right\}$, and the fundamental matrix equation of the given (LF-VFIDEs) is derived from Equation (31), written as

$$\left[y(2)X\,C(\overline{n}(\sigma_2) - \sigma_2)\,(B^T)^{\overline{n}(\sigma_2)}D^T + p_1 y(1)X\,C(\overline{n}(\sigma_1) - \sigma_1)\,(B^T)^{\overline{n}(\sigma_1)}D^T + p_2 X D^T \right. $$
$$\left. - \sum_{i=0}^{1} \lambda_i X F_i^i\,H_{f,i}\,C(\overline{n}(\alpha_i) - \alpha_i)\,(B^T)^{\overline{n}(\alpha_i)}D^T - \sum_{j=0}^{1} \overline{\lambda}_j \overline{XV}^j \overline{H}_j \overline{C}_j B^j D \right] A = G$$

where

$$X(x) = \begin{bmatrix} 1 & x & x^2 & x^3 \end{bmatrix}, X = \begin{bmatrix} X(0) \\ X(1/3) \\ X(2/3) \\ X(1) \end{bmatrix} = \begin{bmatrix} 1 & 0 & 0 & 0 \\ 1 & 1/3 & 1/9 & 1/27 \\ 1 & 2/3 & 4/9 & 8/27 \\ 1 & 1 & 1 & 1 \end{bmatrix},$$

$$p_1 = \begin{bmatrix} 0 & 0 & 0 & 0 \\ 0 & 1/3 & 0 & 0 \\ 0 & 0 & 2/3 & 0 \\ 0 & 0 & 0 & 1 \end{bmatrix}, p_2 = \begin{bmatrix} 0 & 0 & 0 & 0 \\ 0 & -\frac{1}{3} & 0 & 0 \\ 0 & 0 & -\frac{2}{3} & 0 \\ 0 & 0 & 0 & -1 \end{bmatrix},$$

$$y(2) = \begin{bmatrix} 0 & 0 & 0 & 0 \\ 0 & (1/3)^{0.8} & 0 & 0 \\ 0 & 0 & (2/3)^{0.8} & 0 \\ 0 & 0 & 0 & 1 \end{bmatrix}, y(1) = \begin{bmatrix} 0 & 0 & 0 & 0 \\ 0 & (1/3)^{0.9} & 0 & 0 \\ 0 & 0 & (2/3)^{0.9} & 0 \\ 0 & 0 & 0 & 1 \end{bmatrix},$$

$$C(\overline{n}(\sigma_2) - \sigma_2) = \begin{bmatrix} \frac{1}{\Gamma(1.8)} & 0 & 0 & 0 \\ 0 & \frac{1}{\Gamma(2.8)} & 0 & 0 \\ 0 & 0 & \frac{2}{\Gamma(3.8)} & 0 \\ 0 & 0 & 0 & \frac{6}{\Gamma(4.8)} \end{bmatrix}, C(\overline{n}(\sigma_1) - \sigma_1) = \begin{bmatrix} \frac{1}{\Gamma(1.9)} & 0 & 0 & 0 \\ 0 & \frac{1}{\Gamma(2.9)} & 0 & 0 \\ 0 & 0 & \frac{2}{\Gamma(3.9)} & 0 \\ 0 & 0 & 0 & \frac{6}{\Gamma(4.9)} \end{bmatrix},$$

$$C(\overline{n}(\alpha_1) - \alpha_1) = \begin{bmatrix} \frac{1}{\Gamma(1.1)} & 0 & 0 & 0 \\ 0 & \frac{1}{\Gamma(2.1)} & 0 & 0 \\ 0 & 0 & \frac{2}{\Gamma(3.1)} & 0 \\ 0 & 0 & 0 & \frac{6}{\Gamma(4.1)} \end{bmatrix}, C(\overline{n}(\beta_1) - \beta_1) = \begin{bmatrix} \frac{1}{\Gamma(1.2)} & 0 & 0 & 0 \\ 0 & \frac{1}{\Gamma(2.2)} & 0 & 0 \\ 0 & 0 & \frac{2}{\Gamma(3.2)} & 0 \\ 0 & 0 & 0 & \frac{6}{\Gamma(4.2)} \end{bmatrix},$$

$$B^T = (B^T)^{\overline{n}(\sigma_1)} = (B^T)^{\overline{n}(\alpha_1)} = \begin{bmatrix} 0 & 1 & 0 & 0 \\ 0 & 0 & 2 & 0 \\ 0 & 0 & 0 & 3 \\ 0 & 0 & 0 & 0 \end{bmatrix}, (B^T)^2 = (B^T)^{\overline{n}(\sigma_2)} = (B^T)^{\overline{n}(\beta_1)} = \begin{bmatrix} 0 & 0 & 2 & 0 \\ 0 & 0 & 0 & 6 \\ 0 & 0 & 0 & 0 \\ 0 & 0 & 0 & 0 \end{bmatrix},$$

$$F_t^0 = \begin{bmatrix} 1 & 0 & 0 & 0 \\ 1 & 0 & 0 & 0 \\ \frac{1}{2} & 0 & 0 & 0 \\ \frac{1}{6} & 0 & 0 & 0 \end{bmatrix}, F_t^1 = \begin{bmatrix} 0 & 1 & 0 & 0 \\ 0 & 0 & 0 & 0 \\ 0 & 0 & 0 & 0 \\ 0 & 0 & 0 & 0 \end{bmatrix}, V_t^0 = \begin{bmatrix} 0 & -1 & 0 & 0 \\ 1 & 0 & 0 & 0 \\ 0 & 0 & 0 & 0 \\ 0 & 0 & 0 & 0 \end{bmatrix}, V_t^1 = \begin{bmatrix} 0 & 0 & 0 & 0 \\ 1 & 0 & 0 & 0 \\ 0 & 0 & 0 & 0 \\ 0 & 0 & 0 & 0 \end{bmatrix},$$

$$H_{f,0} = \begin{bmatrix} \frac{1}{1} & \frac{1}{2} & \frac{1}{3} & \frac{1}{4} \\ \frac{1}{2} & \frac{1}{3} & \frac{1}{4} & \frac{1}{5} \\ \frac{1}{3} & \frac{1}{4} & \frac{1}{5} & \frac{1}{6} \\ \frac{1}{4} & \frac{1}{5} & \frac{1}{6} & \frac{1}{7} \end{bmatrix}, H_{f,1} = \begin{bmatrix} \frac{1}{1.1} & \frac{1}{2.1} & \frac{1}{3.1} & \frac{1}{4.1} \\ \frac{1}{2.1} & \frac{1}{3.1} & \frac{1}{4.1} & \frac{1}{5.1} \\ \frac{1}{3.1} & \frac{1}{4.1} & \frac{1}{5.1} & \frac{1}{6.1} \\ \frac{1}{4.1} & \frac{1}{5.1} & \frac{1}{6.1} & \frac{1}{7.1} \end{bmatrix}, D^T = \begin{bmatrix} 1 & 0 & 0 & 0 \\ 0 & \frac{1}{2} & 0 & 0 \\ \frac{-1}{4} & 0 & \frac{1}{8} & 0 \\ 0 & \frac{-1}{16} & 0 & \frac{1}{48} \end{bmatrix},$$

$$\overline{X} = \begin{bmatrix} X(0) & 0 & 0 & 0 \\ 0 & X(1/3) & 0 & 0 \\ 0 & 0 & X(2/3) & 0 \\ 0 & 0 & 0 & X(1) \end{bmatrix}, \overline{V}^0 = \begin{bmatrix} V_t^0 & 0 & 0 & 0 \\ 0 & V_t^0 & 0 & 0 \\ 0 & 0 & V_t^0 & 0 \\ 0 & 0 & 0 & V_t^0 \end{bmatrix}, \overline{V}^1 = \begin{bmatrix} V_t^1 & 0 & 0 & 0 \\ 0 & V_t^1 & 0 & 0 \\ 0 & 0 & V_t^1 & 0 \\ 0 & 0 & 0 & V_t^1 \end{bmatrix},$$

$$\overline{H}_0 = \begin{bmatrix} H_{v,0}(0) & 0 & 0 & 0 \\ 0 & H_{v,0}(1/3) & 0 & 0 \\ 0 & 0 & H_{v,0}(2/3) & 0 \\ 0 & 0 & 0 & H_{v,0}(1) \end{bmatrix}, \overline{H}_1 = \begin{bmatrix} H_{v,1}(0) & 0 & 0 & 0 \\ 0 & H_{v,1}(1/3) & 0 & 0 \\ 0 & 0 & H_{v,1}(2/3) & 0 \\ 0 & 0 & 0 & H_{v,1}(1) \end{bmatrix},$$

$$H_{v,0}(0) = \begin{bmatrix} 0 & 0 & 0 & 0 \\ 0 & 0 & 0 & 0 \\ 0 & 0 & 0 & 0 \\ 0 & 0 & 0 & 0 \end{bmatrix}, H_{v,0}(1/3) = \begin{bmatrix} \frac{1}{3} & \frac{(\frac{1}{3})^2}{2} & \frac{(\frac{1}{3})^3}{3} & \frac{(\frac{1}{3})^4}{4} \\ \frac{(\frac{1}{3})^2}{2} & \frac{(\frac{1}{3})^3}{3} & \frac{(\frac{1}{3})^4}{4} & \frac{(\frac{1}{3})^5}{5} \\ \frac{(\frac{1}{3})^3}{3} & \frac{(\frac{1}{3})^4}{4} & \frac{(\frac{1}{3})^5}{5} & \frac{(\frac{1}{3})^6}{6} \\ \frac{(\frac{1}{3})^4}{4} & \frac{(\frac{1}{3})^5}{5} & \frac{(\frac{1}{3})^6}{6} & \frac{(\frac{1}{3})^7}{7} \end{bmatrix}, H_{v,0}(2/3) = \begin{bmatrix} \frac{2}{3} & \frac{(\frac{2}{3})^2}{2} & \frac{(\frac{2}{3})^3}{3} & \frac{(\frac{2}{3})^4}{4} \\ \frac{(\frac{2}{3})^2}{2} & \frac{(\frac{2}{3})^3}{3} & \frac{(\frac{2}{3})^4}{4} & \frac{(\frac{2}{3})^5}{5} \\ \frac{(\frac{2}{3})^3}{3} & \frac{(\frac{2}{3})^4}{4} & \frac{(\frac{2}{3})^5}{5} & \frac{(\frac{2}{3})^6}{6} \\ \frac{(\frac{2}{3})^4}{4} & \frac{(\frac{2}{3})^5}{5} & \frac{(\frac{2}{3})^6}{6} & \frac{(\frac{2}{3})^7}{7} \end{bmatrix},$$

$$H_{v,0}(1) = \begin{bmatrix} \frac{1}{1} & \frac{1}{2} & \frac{1}{3} & \frac{1}{4} \\ \frac{1}{2} & \frac{1}{3} & \frac{1}{4} & \frac{1}{5} \\ \frac{1}{3} & \frac{1}{4} & \frac{1}{5} & \frac{1}{6} \\ \frac{1}{4} & \frac{1}{5} & \frac{1}{6} & \frac{1}{7} \end{bmatrix}, H_{v,1}(0) = \begin{bmatrix} 0 & 0 & 0 & 0 \\ 0 & 0 & 0 & 0 \\ 0 & 0 & 0 & 0 \\ 0 & 0 & 0 & 0 \end{bmatrix}, H_{v,1}(1/3) = \begin{bmatrix} \frac{(\frac{1}{3})^{1.2}}{1.2} & \frac{(\frac{1}{3})^{2.2}}{2.2} & \frac{(\frac{1}{3})^{3.2}}{3.2} & \frac{(\frac{1}{3})^{4.2}}{4.2} \\ \frac{(\frac{1}{3})^{2.2}}{2.2} & \frac{(\frac{1}{3})^{3.2}}{3.2} & \frac{(\frac{1}{3})^{4.2}}{4.2} & \frac{(\frac{1}{3})^{5.2}}{5.2} \\ \frac{(\frac{1}{3})^{3.2}}{3.2} & \frac{(\frac{1}{3})^{4.2}}{4.2} & \frac{(\frac{1}{3})^{5.2}}{5.2} & \frac{(\frac{1}{3})^{6.2}}{6.2} \\ \frac{(\frac{1}{3})^{4.2}}{4.2} & \frac{(\frac{1}{3})^{5.2}}{5.2} & \frac{(\frac{1}{3})^{6.2}}{6.2} & \frac{(\frac{1}{3})^{7.2}}{7.2} \end{bmatrix},$$

$$H_{v,1}(2/3) = \begin{bmatrix} \frac{(\frac{2}{3})^{1.2}}{1.2} & \frac{(\frac{2}{3})^{2.2}}{2.2} & \frac{(\frac{2}{3})^{3.2}}{3.2} & \frac{(\frac{2}{3})^{4.2}}{4.2} \\ \frac{(\frac{2}{3})^{2.2}}{2.2} & \frac{(\frac{2}{3})^{3.2}}{3.2} & \frac{(\frac{2}{3})^{4.2}}{4.2} & \frac{(\frac{2}{3})^{5.2}}{5.2} \\ \frac{(\frac{2}{3})^{3.2}}{3.2} & \frac{(\frac{2}{3})^{4.2}}{4.2} & \frac{(\frac{2}{3})^{5.2}}{5.2} & \frac{(\frac{2}{3})^{6.2}}{6.2} \\ \frac{(\frac{2}{3})^{4.2}}{4.2} & \frac{(\frac{2}{3})^{5.2}}{5.2} & \frac{(\frac{2}{3})^{6.2}}{6.2} & \frac{(\frac{2}{3})^{7.2}}{7.2} \end{bmatrix}, H_{v,1}(1) = \begin{bmatrix} \frac{1}{1.2} & \frac{1}{2.2} & \frac{1}{3.2} & \frac{1}{4.2} \\ \frac{1}{2.2} & \frac{1}{3.2} & \frac{1}{4.2} & \frac{1}{5.2} \\ \frac{1}{3.2} & \frac{1}{4.2} & \frac{1}{5.2} & \frac{1}{6.2} \\ \frac{1}{4.2} & \frac{1}{5.2} & \frac{1}{6.2} & \frac{1}{7.2} \end{bmatrix},$$

$$\overline{C}_1 = \begin{bmatrix} C(\overline{n}(\beta_1) - \beta_1) & 0 & 0 & 0 \\ 0 & C(\overline{n}(\beta_1) - \beta_1) & 0 & 0 \\ 0 & 0 & C(\overline{n}(\beta_1) - \beta_1) & 0 \\ 0 & 0 & 0 & C(\overline{n}(\beta_1) - \beta_1) \end{bmatrix},$$

$$\overline{B}^1 = \begin{bmatrix} (B^T)^{\overline{n}(\beta_1)} & 0 & 0 & 0 \\ 0 & (B^T)^{\overline{n}(\beta_1)} & 0 & 0 \\ 0 & 0 & (B^T)^{\overline{n}(\beta_1)} & 0 \\ 0 & 0 & 0 & (B^T)^{\overline{n}(\beta_1)} \end{bmatrix}, \overline{G} = \begin{bmatrix} -0.28262788 \\ -0.90165393 \\ -0.47693449 \\ 0.94267344 \end{bmatrix} \text{ and } \overline{D} = \begin{bmatrix} D^T \\ D^T \\ D^T \\ D^T \end{bmatrix}.$$

putting all above matrices in matrix Equation (32) and calculating it, this fundamental matrix equation's augmented matrix is:

$$[W:G] = \begin{bmatrix} -1.07079233 & -0.02572358 & 0.03539616 & 0.00866477 & : & -0.28262788 \\ -1.85498405 & -0.12829141 & 0.09107227 & 0.01393314 & : & -0.90165393 \\ -2.40595015 & -0.22844205 & 0.01161705 & 0.01241051 & : & -0.47693449 \\ -2.7722549 & -0.23879802 & -0.19720588 & -0.02479584 & : & 0.94267344 \end{bmatrix}$$

For our consider example, the boundary conditions from Equation (33) have the following matrix forms:

$$U_k A = [C_k] \text{ or } [U_k : C_k]; \; k = 0, 1$$

or clearly

$$\begin{matrix} [U_0 : C_0] = [& -0.5 & 1.3125 & 0.25 & 0.0625 & : & 1 \;] \\ [U_1 : C_1] = [& -0.5 & 0.3125 & 0.25 & 0.0625 & : & -1 \;] \end{matrix}$$

The new augmented matrix depending on conditions is constructed as follows from the system (34):

$$[\widetilde{W} : \widetilde{G}] = \begin{bmatrix} -1.07079233 & -0.02572358 & 0.03539616 & 0.00866477 & : & -0.28262788 \\ -1.85498405 & -0.12829141 & 0.09107227 & 0.01393314 & : & -0.90165393 \\ -0.5 & 1.3125 & 0.25 & 0.0625 & : & 1 \\ -0.5 & 0.3125 & 0.25 & 0.0625 & : & -1 \end{bmatrix}$$

The Bessel coefficient matrix A is obtained by solving this system.

$$A = \begin{bmatrix} -1.98357176 \times 10^{-06} & 2.00000000 & -8.00269502 & 6.01076421 \end{bmatrix}^T$$

hence, for $N = 3$ the approximate solution of the problem is formed as

$$u_3(x) = 0.0002242544 x^3 - 1.0003363816 x^2 + 1.0 x - 0.0000019836$$

for $N = 6$ and $N = 10$, similarly as steps above and running the general python program which are written for this purpose we obtain the approximate solution of the problem, respectively.

$$u_6(x) = -0.0003537963x^6 + 0.0007932088x^5 - 0.0006209581x^4 + 0.000229553x^3 \\ -1.0000240462x^2 + 1.0x + 0.0000090551$$

and

$$u_{10}(x) = 0.0000000661x^{10} - 0.0000002889x^9 + 0.00000052947x^8 - 0.00000053514x^7 \\ + 0.000000328x^6 - 0.00000012512x^5 + 0.0000000293x^4 \\ - 0.00000000370x^3 - 0.999999999656939x^2 + 1.0x + 0.00000000015$$

In Table 1, Comparison the exact solution $u(x)$ with the approximate solution $u_N(x)$ of Example 1 for $N = 3, 6$ and 10, respectively, in terms of least square error and running time.

Table 1. Compares the exact solution $u(x)$ with the approximate solution $u_N(x)$ of Example 1.

x_i	Exact Solution Example 1.	N-Approximate Solution $u_N(x)$		
		N = 3	N = 6	N = 10
0.0	0.00	$-1.9835710 \times 10^{-06}$	$9.05511829 \times 10^{-06}$	$1.53060853 \times 10^{-10}$
0.1	0.09	$8.99948769 \times 10^{-02}$	$9.00089897 \times 10^{-02}$	$9.00000002 \times 10^{-02}$
0.2	0.16	$1.59986355 \times 10^{-01}$	$1.60009167 \times 10^{-01}$	$1.60000000 \times 10^{-01}$
0.3	0.21	$2.09973797 \times 10^{-01}$	$2.10009729 \times 10^{-01}$	$2.10000000 \times 10^{-01}$
0.4	0.24	$2.39958548 \times 10^{-01}$	$2.40010676 \times 10^{-01}$	$2.40000000 \times 10^{-01}$
0.5	0.25	$2.49941953 \times 10^{-01}$	$2.50012188 \times 10^{-01}$	$2.50000000 \times 10^{-01}$
0.6	0.24	$2.39925358 \times 10^{-01}$	$2.40014679 \times 10^{-01}$	$2.40000000 \times 10^{-01}$
0.7	0.21	$2.09910109 \times 10^{-01}$	$2.10018608 \times 10^{-01}$	$2.10000000 \times 10^{-01}$
0.8	0.16	$1.59897550 \times 10^{-01}$	$1.60024025 \times 10^{-01}$	$1.60000000 \times 10^{-01}$
0.9	0.09	$8.98890288 \times 10^{-02}$	$9.00298712 \times 10^{-02}$	$9.00000005 \times 10^{-02}$
1.0	0.00	$-1.141107 \times 10^{-04}$	$3.30162295 \times 10^{-05}$	$5.74892312 \times 10^{-10}$
L.S.E.		$5.5474382 \times 10^{-08}$	$3.72530728 \times 10^{-09}$	$1.13502895 \times 10^{-18}$
Running Time/Sec		0.47589683	0.227055311	0.57032561

Example 2. Let us now consider the LF-VIFDEs on the closed bounded interval $[0, 1]$ given by

$$_0^C D_x^{1.3} u(x) + \frac{x}{2} {_0^C D_x^{0.8}} u(x) + \sqrt{x} {_0^C D_x^{0.5}} u(x) + \left(x^2 + 1\right) u(x)$$

$$= g(x) + \int_0^1 \left((\sin(x) - t) {_0^C D_t^{0.9}} u(t) + 2(e^x - t) {_0^C D_t^{1.1}} u(t)\right) dt + 3 \int_0^x t\sinh(x) {_0^C D_t^{1.9}} u(t) dt$$

where

$$g(x) = \frac{2}{\Gamma(1.7)} x^{0.7} + \frac{1}{\Gamma(2.2)} x^{2.2} + \frac{1}{\Gamma(1.2)} x^{1.2} + \frac{2}{\Gamma(2.5)} x^{1.5} \sqrt{x} + \frac{2}{\Gamma(1.5)} x + (x^2 + 1)(x + 1)^2$$

$$- \left(\frac{2}{\Gamma(3.1)} + \frac{2}{\Gamma(2.1)}\right) \sin(x) - \frac{4}{\Gamma(2.9)} e^x - \frac{6.6}{\Gamma(3.1)} x^{2.1} \sinh(x) + \left(\frac{4.2}{\Gamma(4.1)} + \frac{2.2}{\Gamma(3.1)} + \frac{7.6}{\Gamma(3.9)}\right)$$

with the boundary conditions:

$$u(0) + u^{(1)}(1) = 5 \text{ and } u(1) + u^{(1)}(0) = 6$$

The exact solution is $u(x) = (x+1)^2$.
Let us now calculate the coefficients a_∇, $\nabla = \overline{0:N}$ of approximate solution with the aid of the truncated Bessel series:

$$u(x) \cong u_N(x) = \sum_{\nabla=0}^{N} a_\nabla J_\nabla(x)$$

Here, from the considered example we have:

$\sigma_1 = 0.5, \sigma_2 = 0.8, \sigma_3 = 1.3 \to \overline{n}(\sigma_1) = \sigma_1 = 1, \overline{n}(\sigma_2) = \sigma_2 = 1, \overline{n}(\sigma_3) = \sigma_3 = 2$

$\alpha_1 = 0.9, \alpha_2 = 1.1 \to \overline{n}(\alpha_1) = \alpha_1 = 1, \overline{n}(\alpha_2) = \alpha_2 = 2$

$\beta_1 = 1.9 \to \overline{n}(\beta_1) = \beta_1 = 2$, $\mu = \max\{\lceil 1.3 \rceil, \lceil 1.1 \rceil, \lceil 1.9 \rceil\} = 2$ and $\lambda_1 = 1, \lambda_2 = 2, \overline{\lambda}_1 = 3$

$P_1(x) = \frac{x}{2}, P_2(x) = \sqrt{x}, P_3(x) = (x^2+1), F_1(x,t) = \sin(x) - t, F_2(x,t) = (e^x - t), V_1(x,t) = t\sinh(x)$

Hence $\mu = 2$ so take $N = 3$, the set of collocation points, and the fundamental matrix equation of the given (LF-VFIDEs) is derived from Equation (31), written as

$$[y(3)X\,C(\overline{n}(\sigma_3) - \sigma_3)\,(B^T)^{\overline{n}(\sigma_3)}D^T + p_1 y(2)X\,C(\overline{n}(\sigma_2) - \sigma_2)\,(B^T)^{\overline{n}(\sigma_2)}D^T$$
$$+ p_2 y(1)X\,C(\overline{n}(\sigma_1) - \sigma_1)\,(B^T)^{\overline{n}(\sigma_1)}D^T + p_3 XD^T$$
$$- \lambda_1 XF_t^1\,H_{f,1}\,C(\overline{n}(\alpha_1) - \alpha_1)\,(B^T)^{\overline{n}(\alpha_1)}D^T$$
$$- \lambda_2 XF_t^2\,H_{f,2}\,C(\overline{n}(\alpha_2) - \alpha_2)\,(B^T)^{\overline{n}(\alpha_2)}D^T - \overline{\lambda}_1 \overline{XV}^1 \overline{H}_1 \overline{C}_1 \overline{B}^1 \overline{D}\,]A = G$$

After inputting each of the parameters above by running the general python program, which are written for this purpose for $N = 3$, we obtain the approximate solution of the problem,

$$u_3(x) = 0.0029061023x^3 + 0.99568824199x^2 + 2.0015004466x + 0.9984047624$$

Similarly, the approximate solution of the problem for $N = 6$ and 11, respectively, we obtain

$$u_6(x) = 4.353138854 \times 10^{-7}x^6 - 4.54065912 \times 10^{-5}x^5 + 5.511176789 \times 10^{-5}x^4$$
$$- 1.135105524 \times 10^{-5}x^3 + 1.0000204x^2 + 1.999983583x$$
$$+ 1.000013642$$

and

$$u_{11}(x) = 3.2299188319 \times 10^{-7}x^{11} - 1.5377530537 \times 10^{-6}x^{10}$$
$$+ 3.1687057159 \times 10^{-6}x^9 - 3.716458436 \times 10^{-6}x^8$$
$$+ 2.73904171 \times 10^{-6}x^7 - 1.319310342 \times 10^{-6}x^6$$
$$+ 4.181528996 \times 10^{-7}x^5 - 8.549654914 \times 10^{-8}x^4$$
$$+ 1.082189316 \times 10^{-8}x^3 + 0.9999999996x^2 + 1.9999999996x$$
$$+ 1.0000000003.$$

In Table 2 comparison in terms of least square error and running time the exact solution $u(x)$ with the approximate solution $u_N(x)$ of example 2 for $N = 3, 6$ and 11, respectively.

Table 2. Compares the exact solution $u(x)$ with the approximate solution $u_N(x)$ of Example 2.

x_i	Exact Solution Example 2.	N-Approximate Solutions $u_N(x)$		
		$N = 3$	$N = 6$	$N = 11$
0.0	1.00	0.99840476	1.00001364	1.00
0.1	1.21	1.2085146	1.2100122	1.21
0.2	1.44	1.43855563	1.44001116	1.44
0.3	1.69	1.6885453	1.69001058	1.69
0.4	1.96	1.95850105	1.96001056	1.96
0.5	2.25	2.24844031	2.25001115	2.25
0.6	2.56	2.55838052	2.56001232	2.56
0.7	2.89	2.88833911	2.89001391	2.89
0.8	3.24	3.23833352	3.24001556	3.24
0.9	3.61	3.60838119	3.6100167	3.61
1.0	4.00	3.99849955	4.00001642	4.00
	L.S.E.	$2.66633874 \times 10^{-05}$	$1.94276005 \times 10^{-09}$	$7.33991398 \times 10^{-19}$
	Running Time/Sec		0.2821376323	0.6396300792

Figures 1 and 2 illustrate a comparison between the exact solution and approximate solution of LF-VIFDEs of Examples 1 and 2, respectively. To show the result of the proposed method to an exact solution, we present Tables 1 and 2, respectively. Each of the plots is drawn with our Python program version 3.8.8 (2021).

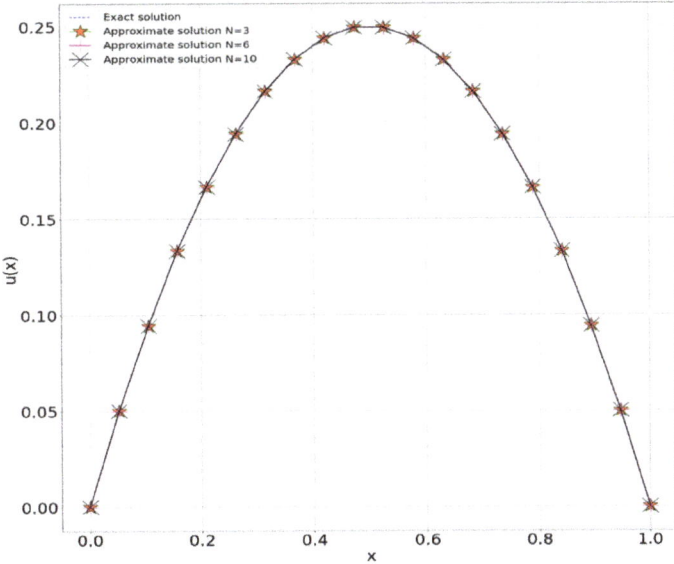

Figure 1. Comparison of the exact and approximate solution of Example 1.

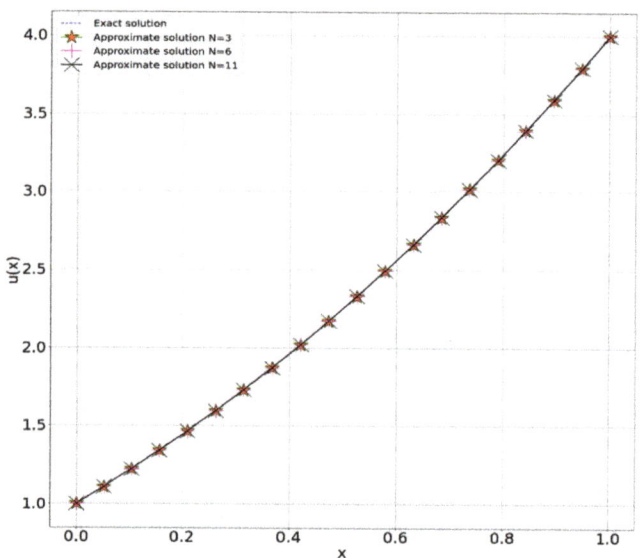

Figure 2. Comparison of the exact and approximate solution of Example 2.

Example 3. *Let us consider the linear Fredholm–Volterra fractional integro-differential equation on the closed bounded interval* $[0,1]$:

$$_0^C D_x^{0.73} u(x) = g(x) + \int_0^1 \left(t \sin\left(\frac{x}{2}\right) {_0^C D_t^{0.64}} u(t) + 2(x - \frac{t}{2}) {_0^C D_t^{1.64}} u(t) \right) dt + \int_0^x \left(-e^x {_0^C D_t^{0.5}} u(t) + (-2) t e^x {_0^C D_t^{1.5}} u(t) \right) dt$$

where

$$g(x) = \sum_{i=0}^{\infty} \left[\frac{2x}{\Gamma(i+2.36)} - \frac{x^{i+0.27}}{\Gamma(i+1.27)} + \frac{\sin(\frac{x}{2})(i+1.36)}{\Gamma(i+3.36)} - \frac{(i+1.36)}{\Gamma(i+3.36)} - \frac{e^x x^{i+1.5}}{\Gamma(i+2.5)} - \frac{2(i+1.5)e^x x^{i+2.5}}{\Gamma(i+3.5)} \right]$$

with the boundary conditions

$$u(0) + u^{(1)}(0) = -1 \text{ and } u(1) - u^{(1)}(1) = 1$$

which is the exact solution $u(x) = 1 - e^x$.

Let us now calculate the coefficients a_r, $r = \overline{0 : N}$ of approximate solution with the aid of the truncated Bessel series:

$$u(x) \cong u_N(x) = \sum_{r=0}^{N} a_r J_r(x)$$

Here, from consider example we have:

$\sigma_1 = 0.73 \to \overline{n}(\sigma_1) = \sigma_1 = 1$

$\alpha_1 = 0.64 \to \overline{n}(\alpha_1) = \alpha_1 = 1$, $\alpha_2 = 1.64 \to \overline{n}(\alpha_2) = \alpha_2 = 2$

$\beta_1 = 0.5 \to \overline{n}(\beta_1) = \beta_1 = 1$, $\beta_2 = 1.5 \to \overline{n}(\beta_1) = \beta_1 = 2$

$\mu = \max\{\lceil\sigma_1\rceil, \lceil\alpha_2\rceil, \lceil\beta_2\rceil\} = \max\{\lceil 0.73\rceil, \lceil 1.64\rceil, \lceil 1.5\rceil\} = 2$

$p_1(x) = 0$, $F_1(x,t) = t\sin(\frac{x}{2})$, $F_2(x,t) = (x - \frac{t}{2})$, $V_1(x,t) = e^x$, $V_2(x,t) = te^x$ and $\lambda_1 = 1, \lambda_2 = 1, \overline{\lambda}_1 = -1, \overline{\lambda}_2 = -2$.

Suppose that, we take $\overline{N}$ terms from the homogeneous part $g(x)$:

$$g(x) = \sum_{i=0}^{\overline{N}}\left[\frac{2x}{\Gamma(i+2.36)} - \frac{x^{i+0.27}}{\Gamma(i+1.27)} + \frac{\sin(\frac{x}{2})(i+1.36)}{\Gamma(i+3.36)} - \frac{(i+1.36)}{\Gamma(i+3.36)} - \frac{e^x x^{i+1.5}}{\Gamma(i+2.5)} - \frac{2(i+1.5)e^x x^{i+2.5}}{\Gamma(i+3.5)}\right]$$

Hence $\mu = 2$, the fundamental matrix equation of the given (LF-VFIDEs) is derived from Equation (31), written as

$$\left[y(1)X\, C(\overline{n}(\sigma_1) - \sigma_1)\,(B^T)^{\overline{n}(\sigma_1)}D^T - \lambda_1 XF_t^1\, H_{f,1}\, C(\overline{n}(\alpha_1) - \alpha_1)(B^T)^{\overline{n}(\alpha_1)}D^T \right.$$
$$\left. - \lambda_2 XF_t^2\, H_{f,2}\, C(\overline{n}(\alpha_2) - \alpha_2)\,(B^T)^{\overline{n}(\alpha_2)}D^T - \overline{\lambda}_1 \overline{XV}^1 \overline{H}_1 \overline{C}_1 B^1 D - \overline{\lambda}_2 \overline{XV}^2 \overline{H}_2 \overline{C}_2 B^2 D\right]A = G$$

We choose if $\overline{N} = 5$, the approximate solution of the problem for $N = 4, 10, 21$, respectively

$u_4(x) = -0.0569736258x^4 - 0.1665069441x^3 - 0.49426610634x^2 - 1.0017991281x$
$\qquad + 0.001799128086$

$u_{10}(x) = -0.0112791176x^{10} + 0.04361473251x^9 - 0.07167515469x^8 + 0.06498524957x^7$
$\qquad - 0.03711020557x^6 + 0.00378578712x^5 - 0.0441096211x^4$
$\qquad - 0.1664357675x^3 - 0.4999839118x^2 - 0.999999027x$
$\qquad - 9.72893359856624 \times 10^{-7}$,

and

$u_{21}(x) = 2649.390397x^{21} - 26487.723779x^{20} + 123552.0169x^{19} - 357222.928001x^{18}$
$\qquad + 717359.135397x^{17} - 1062540.52114x^{16} + 1203203.42788x^{15}$
$\qquad - 1065423.12951x^{14} + 748326.26838x^{13} - 420462.50002x^{12}$
$\qquad + 189747.93897x^{11} - 68792.817648x^{10} + 19970.45119x^9 - 4609.812041x^8$
$\qquad + 836.75503981x^7 - 117.510711061x^6 + 12.4665060412x^5$
$\qquad - 1.0098682157x^4 - 0.11440971986x^3 - 0.50180747163x^2$
$\qquad - 0.99996814632x - 3.1859671582 \times 10^{-5}$

we choose if $\overline{N} = 10$, the approximate solution of the problem for $N = 4, 10, 21$, respectively

$u_4(x) = -0.05759757087x^4 - 0.1652553696x^3 - 0.4950425756x^2 - 1.001653973x + 0.001653973$

$$u_{10}(x) = 3.399204141e - 6x^{10} - 2.0679446987e - 5x^9 + 1.254303583e - 5x^8$$
$$- 0.000242474208x^7 - 0.001356395456x^6 - 0.008348879766x^5$$
$$- 0.0416618228x^4 - 0.166667629x^3 - 0.4999998846x^2$$
$$- 1.0000000056x + 5.656145543 \times 10^{-9},$$

and

$$u_{21}(x) = 0.0366737297x^{21} - 0.36713669x^{20} + 1.714854990x^{19} - 4.9651991545x^{18}$$
$$+ 9.9858420757x^{17} - 14.8143137167x^{16} + 16.8038086x^{15}$$
$$- 14.90654178x^{14} + 10.49049965x^{13} - 5.9069139812x^{12}$$
$$+ 2.6719828271x^{11} - 0.97127108855x^{10} + 0.282792909x^9$$
$$- 0.0655249586x^8 + 0.0117379624x^7 - 0.00307302784x^6$$
$$- 0.008153528584x^5 - 0.041680721487x^4 - 0.1666659001x^3$$
$$- 0.5000000273x^2 - 0.9999999995x - 4.670435338 \times 10^{-10}$$

Similarly doing it for $\overline{N} = 16$, the approximate solution of the problem for $N = 4, 10, 21$, respectively

$$u_4(x) = -0.057597577x^4 - 0.165255357x^3 - 0.495042583x^2 - 1.001653971x + 0.0016539713$$

$$u_{10}(x) = 3.4214653141e - 6x^{10} - 2.0612514452e - 5x^9 + 1.2077128326e - 5x^8$$
$$- 0.000241586656x^7 - 0.0013572657x^6 - 0.0083483752x^5$$
$$- 0.0416620032x^4 - 0.1666675894x^3 - 0.499999889x^2$$
$$- 1.000000005x + 5.395456491 \times 10^{-9}$$

and

$$u_{21}(x) = -0.000949182285x^{21} + 0.00893091688x^{20} - 0.03880287140x^{19}$$
$$+ 0.103101802x^{18} - 0.186859983473284x^{17} + 0.24354820561x^{16}$$
$$- 0.2337645213x^{15} + 0.1652048762x^{14} - 0.08287855594x^{13}$$
$$+ 0.0253870484x^{12} - 0.0005131248x^{11} - 0.0043926365x^{10}$$
$$+ 0.00281921684x^9 - 0.0010956089x^8 + 8.623002860e - 5x^7$$
$$- 0.00144403537x^6 - 0.00832553568x^5 - 0.0416674549x^4 - 0.16666661x^3$$
$$- 0.500000002x^2 - 0.99999999995x - 4.983565034 \times 10^{-11}.$$

In Table 3 presents a comparison between the exact solution $u(x)$ and approximate solution $u_N(x)$, when we choose $\overline{N} = 5, 10$, and 16, respectively. For each of them we chose $N = 4, 10$, and 21, respectively depending on the least square error and running time.

Table 3. Comparison between the exact solution $u(x)$ and approximate solution $u_N(x)$ for Example 3.

(a)

x_i	Exact Solution Example 3.	$\overline{N}=5$ N-Approximate Solution $u_N(x)$		
		$N=4$	$N=10$	$N=21$
0.0	0.0	0.00179913	$-9.72893360 \times 10^{-7}$	$-3.1859672 \times 10^{-5}$
0.1	−0.10517092	−0.10349565	−0.105171555	−0.10520278
0.2	−0.22140276	−0.21975456	−0.221402690	−0.22143482
0.3	−0.34985881	−0.34818173	−0.349857839	−0.34989091
0.4	−0.4918247	−0.49011807	−0.491822593	−0.49185648
0.5	−0.64872127	−0.64704118	−0.648717441	−0.64875193
0.6	−0.8221188	−0.82056543	−0.822111722	−0.82214657
0.7	−1.01375271	−1.0124419	−1.01373901	−1.01377394
0.8	−1.22554093	−1.22455843	−1.22551392	−1.22554873
0.9	−1.45960311	−1.45893959	−1.45955258	−1.45958517
1.0	−1.71828183	−1.71774668	−1.71820801	−1.71823455
L.S.E.		$2.3130923 \times 10^{-05}$	$8.99107006 \times 10^{-09}$	$9.87875607 \times 10^{-09}$
Running Time/Sec		0.171678066	0.49988222	2.92118477

(b)

x_i	Exact Solution Example 3.	$\overline{N}=10$ N-Approximate Solution $u_N(x)$		
		$N=4$	$N=10$	$N=21$
0	0.0	$1.65397131 \times 10^{-03}$	$5.65614554 \times 10^{-09}$	$-4.67043534 \times 10^{-10}$
0.1	−0.10517092	$-1.03632867 \times 10^{-01}$	$-1.05170912 \times 10^{-01}$	$-1.05170919 \times 10^{-01}$
0.2	−0.22140276	$-2.19892725 \times 10^{-01}$	$-2.21402752 \times 10^{-01}$	$-2.21402758 \times 10^{-01}$
0.3	−0.34985881	$-3.48324488 \times 10^{-01}$	$-3.49858802 \times 10^{-01}$	$-3.49858808 \times 10^{-01}$
0.4	−0.4918247	$-4.90265271 \times 10^{-01}$	$-4.91824692 \times 10^{-01}$	$-4.91824698 \times 10^{-01}$
0.5	−0.64872127	$-6.47190428 \times 10^{-01}$	$-6.48721264 \times 10^{-01}$	$-6.48721271 \times 10^{-01}$
0.6	−0.8221188	$-8.20713545 \times 10^{-01}$	$-8.22118794 \times 10^{-01}$	$-8.22118801 \times 10^{-01}$
0.7	−1.01375271	−1.01258644	−1.01375270	−1.01375271
0.8	−1.22554093	−1.22469917	−1.22554092	−1.22554093
0.9	−1.45960311	−1.45908002	−1.45960311	−1.45960311
1	−1.71828183	−1.71789552	−1.71828182	−1.71828183
L.S.E.		$1.89772271 \times 10^{-05}$	$3.87412311 \times 10^{-16}$	$2.32392057 \times 10^{-18}$
Running Time/Sec		0.187295436	0.515326499	2.749377012

Table 3. Cont.

(c)

x_i	Exact Solution Example 3.	N-Approximate Solution $u_N(x)$		
		$N = 4$	$N = 10$	$N = 21$
0.0	0.	$1.65397131 \times 10^{-03}$	$5.39545649 \times 10^{-09}$	$-4.98356503 \times 10^{-11}$
0.1	-0.10517092	$-1.03632867 \times 10^{-01}$	$-1.05170913 \times 10^{-01}$	$-1.05170918 \times 10^{-01}$
0.2	-0.22140276	$-2.19892725 \times 10^{-01}$	$-2.21402753 \times 10^{-01}$	$-2.21402758 \times 10^{-01}$
0.3	-0.34985881	$-3.48324488 \times 10^{-01}$	$-3.49858802 \times 10^{-01}$	$-3.49858808 \times 10^{-01}$
0.4	-0.4918247	$-4.90265271 \times 10^{-01}$	$-4.91824692 \times 10^{-01}$	$-4.91824698 \times 10^{-01}$
0.5	-0.64872127	$-6.47190428 \times 10^{-01}$	$-6.48721265 \times 10^{-01}$	$-6.48721271 \times 10^{-01}$
0.6	-0.8221188	$-8.20713545 \times 10^{-01}$	$-8.22118794 \times 10^{-01}$	$-8.22118800 \times 10^{-01}$
0.7	-1.01375271	-1.01258644	-1.01375270	-1.01375271
0.8	-1.22554093	-1.22469917	-1.22554092	-1.22554093
0.9	-1.45960311	-1.45908002	-1.45960311	-1.45960311
1.0	-1.71828183	-1.71789552	-1.71828182	-1.71828183
	L.S.E.	$1.89771908 \times 10^{-05}$	$3.4403856 \times 10^{-16}$	$3.45079782 \times 10^{-20}$
	Running Time/Sec	0.203105688	0.5199816226	2.90862059

Figure 3a–c illustrates a comparison between the exact solution and approximate solution of (LF-VIFDEs) of equation above, respectively. To show the result of the proposed method to an exact solution, we present Table 3, respectively. Each of the plots is drawn with our Python program version 3.8.8 (2021).

Example 4. *Suppose that the following linear Fredholm–Volterra fractional integro-differential equation given by*

$$^C_0 D^{0.8}_x u(x) = g(x) + \int_0^2 \left(\tfrac{1}{2}(t-2) \,^C_0 D^{0.3}_t u(t) + (t + \cos(x)) \,^C_0 D^{1.3}_t u(t) \right) dt$$
$$+ \int_0^x \left(\tfrac{1}{4}\left(\tfrac{x-t}{2}\right) u(t) + [\tan(x)t] \,^C_0 D^{2.7}_t u(t) \right) dt, \quad 0 \leq x, t \leq 2$$

where

$$g(x) = \tfrac{2}{\Gamma(2.2)} x^{1.2} - \tfrac{1}{\Gamma(1.2)} x^{0.2} - \tfrac{(2.7)}{\Gamma(4.7)} 2^{3.7} - \tfrac{1}{\Gamma(3.7)}\left((1.7)2^{3.7} - 2^{3.7} - (1.7)2^{1.7}\right)$$
$$- \tfrac{1}{\Gamma(2.7)}\left(2^{1.7} + 2^{2.7}\cos(x)\right) - \tfrac{1}{16}\left(\tfrac{1}{6}x^4 - \tfrac{1}{3}x^3 + x^2\right)$$

with the boundary conditions

$$u(0) + u(2) = 4, \quad u^{(1)}(0) + u^{(1)}(2) = 2 \text{ and } u^{(2)}(0) + u^{(2)}(2) = 4$$

which is the exact solution $u(x) = x^2 - x + 1$.

Now let us find the approximate solution given by the N-truncated Bessel series

$$u(x) \cong u_N(x) = \sum_{r=0}^{N} a_r J_r(x)$$

Here, from consider example we have:

$$\sigma_1 = 0.8 \to \overline{n}(\sigma_1) = \sigma_1 = 1$$
$$\alpha_1 = 0.3, \alpha_2 = 1.3 \to \overline{n}(\alpha_1) = \alpha_1 = 1, \overline{n}(\alpha_2) = \alpha_2 = 2$$
$$\beta_0 = 0, \beta_1 = 2.7 \to \overline{n}(\beta_0) = \beta_0 = 0, \overline{n}(\beta_1) = \beta_1 = 3$$
$$\mu = \max\{\lceil 0.8 \rceil, \lceil 1.3 \rceil, \lceil 2.7 \rceil\} = 3, \lambda_1 = \tfrac{1}{2}, \lambda_2 = 1, \overline{\lambda}_0 = \tfrac{1}{4}, \overline{\lambda}_1 = 1$$

and $p_1(x) = 0$, $F_1(x,t) = t - 2$, $F_2(x,t) = t + \cos(x)$, $V_0(x,t) = \left(\tfrac{x-t}{2}\right)$, $V_1(x,t) = \tan(x)t$, from Equation (31), the fundamental matrix equation of the given problem is written as

$$\left[y(1) \, X \, C(\overline{n}(\sigma_1) - \sigma_1) \, (B^T)^{\overline{n}(\sigma_1)} D^T - \lambda_1 X F_t^1 \, H_{f,1} \, C(\overline{n}(\alpha_1) - \alpha_1) \, (B^T)^{\overline{n}(\alpha_1)} D^T \right.$$
$$\left. - \lambda_2 X F_t^2 \, H_{f,2} \, C(\overline{n}(\alpha_2) - \alpha_2) \, (B^T)^{\overline{n}(\alpha_2)} D^T - \overline{\lambda}_0 \overline{XV}^0 \overline{H}_0 \overline{D} - \overline{\lambda}_1 \overline{XV}^1 \overline{H}_1 \overline{C}_1 \overline{B}^1 \overline{D} \right] A = G$$

Thus, the approximate solution of the problem for $N = 5, 9, 12$, respectively

$$u_5(x) = 0.00595039974990515 x^5 - 0.0295085156834566 x^4 + 0.0436273982107945 x^3$$
$$+ 0.98520400357289 x^2 - 0.998052135471447 x + 0.99399626495166,$$

$$u_9(x) = -6.93792068979952 \times 10^{-7} x^9 + 6.12616976553793 \times 10^{-6} x^8 - 2.13390478665744 \times 10^{-5} x^7$$
$$+ 3.69079850914571 \times 10^{-5} x^6 - 3.01713615634192 \times 10^{-5} x^5$$
$$+ 1.16291867668927 \times 10^{-6} x^4 + 1.66943823538962 \times 10^{-5} x^3$$
$$+ 0.999993226980557 x^2 - 0.999998942794685 x + 0.99999725431996,$$

and

$$u_{12}(x) = -5.96684668513465 \times 10^{-10} x^{12} + 6.51569719257085 \times 10^{-9} x^{11}$$
$$- 3.10471674650442 \times 10^{-8} x^{10} + 8.49087433427835 \times 10^{-8} x^9$$
$$- 1.47156926969634 \times 10^{-7} x^8 + 1.67915537193858 \times 10^{-7} x^7$$
$$- 1.26723286902409 \times 10^{-7} x^6 + 6.23462096628266 \times 10^{-8} x^5$$
$$- 2.05277535525461 \times 10^{-8} x^4 + 5.82167120066757 \times 10^{-9} x^3$$
$$+ 0.999999998891922 x^2 - 0.999999999885384 x + 0.999999999582705.$$

In Table 4. presents a comparison between the exact solution $u(x)$ and approximate solution $u_N(x)$ for $N = 5, 9$ and 12, respectively, depending on the least square error and running time.

Table 4. Comparison between the exact solution $u(x)$ and approximate solution $u_N(x)$ for Example 4.

x_i	Exact Solution Example 4.	N-Approximate Solution $u_N(x)$		
		N = 5	N = 9	N = 12
0.0	1.00	0.99399626	0.99999725	1.00
0.1	0.91	0.90408383	0.90999731	0.91
0.2	0.84	0.83409771	0.83999732	0.84
0.3	0.79	0.78420236	0.78999737	0.79
0.4	0.76	0.75450572	0.7599975	0.76
0.5	0.75	0.74506629	0.74999774	0.75

Table 4. Cont.

x_i	Exact Solution Example 4.	N-Approximate Solution $u_N(x)$		
		N = 5	N = 9	N = 12
0.6	0.76	0.75590034	0.75999808	0.76
0.7	0.79	0.78698902	0.78999852	0.79
0.8	0.84	0.83828549	0.83999904	0.84
0.9	0.91	0.90972207	0.90999961	0.91
1.0	1.00	1.00121742	1.00000023	1.00
L.S.E.		0.0002244	$4.72053254 \times 10^{-11}$	$1.11096853 \times 10^{-18}$
Running Time/Sec		0.568155527	1.649296999	6.3344522

Figure 4 illustrates a comparison between the exact solution and approximate solution of linear (FVIFDEs). To show the result of the proposed method to an exact solution, we present Table 4. Each of the plots is drawn with our Python program version 3.8.8 (2021).

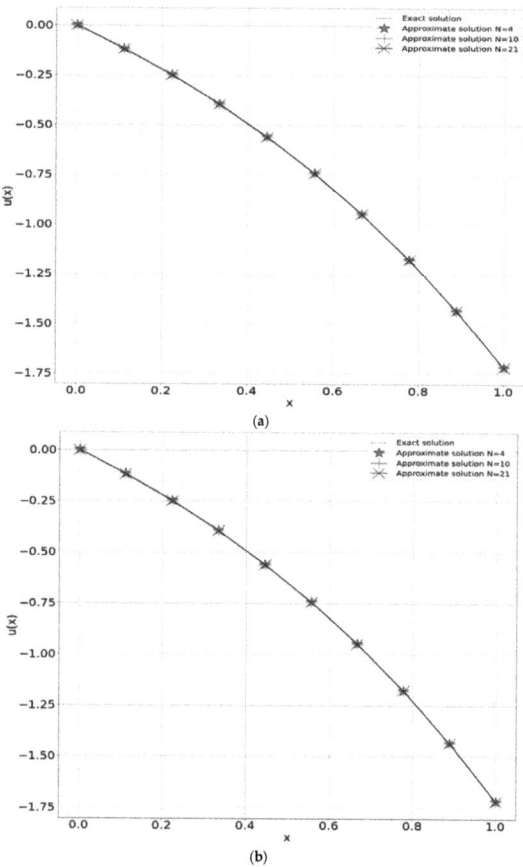

Figure 3. Cont.

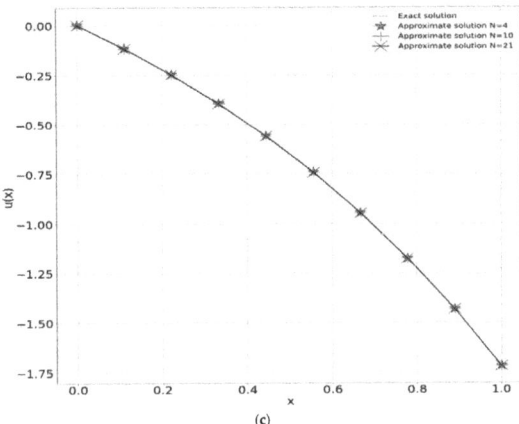

(c)

Figure 3. (a) Comparison of the exact and approximate solution, when $\overline{N} = 5$. (b) Comparison of the exact and approximate solution, when $\overline{N} = 10$. (c) Comparison of the exact and approximate solution, when $\overline{N} = 16$.

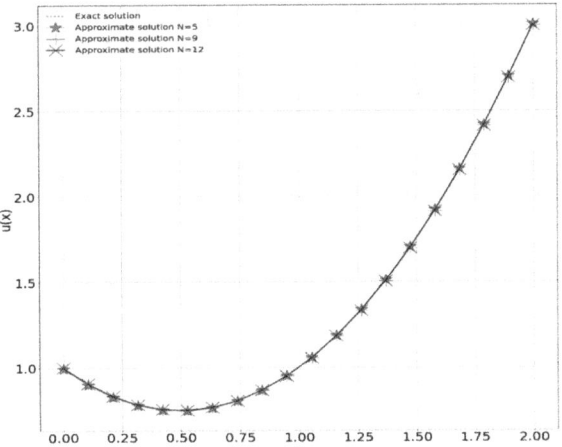

Figure 4. Comparison of the exact and approximate solution of example 4.

6. Conclusions

Multi-fractional order linear integro-differential equations are generally difficult to solve analytically. In many situations, it is necessary to approximate solutions. In this work, we present a new technique for numerically solving the linear Fredholm–Volterra integro-fractional differential equation of multi-fractional order of the Caputo sense using first-order Bessel polynomials. The comparison of the results achieved with the exact solution, the exact solution, and the other methods suggests that the procedure is very effective and convenient. We introduced this with some illustrative examples of the approach and their least square error to minimize the error terms on the specified domain and running time are also given in tabular form. It is obvious that as N rises, the error rate reduces and the answer becomes closer to the exact solution. One significant benefit of the technique is that the Bessel coefficients of the solution may be determined relatively quickly using computer code developed in Python v3.8.8 (2021). As an example, consider the Python v3.8.8 (2021).

Future directions: Using the residual error function, we can enhance the Bessel collocation method for solving the multi-high fractional-order system of Fredholm–Volterra integro-differential equations and their delay. This technique can also be used to make an accurate error estimation.

Author Contributions: Conceptualization, S.S.A. and S.J.M.; methodology, S.S.A.; software, S.J.M.; validation, S.S.A. and S.J.M.; formal analysis, S.S.A.; investigation, S.J.M.; resources, S.S.A.; data curation, S.J.M.; writing—original draft preparation, S.S.A.; writing—review and editing, S.J.M.; visualization, S.J.M.; supervision, S.S.A.; project administration, S.S.A.; funding acquisition, S.S.A. All authors have read and agreed to the published version of the manuscript.

Funding: This research received no external funding.

Institutional Review Board Statement: The study was conducted according to the guidelines of the Declaration of Helsinki, and approved by the Institutional Ethics Committee of Mathematics Department, College of Science, University of Sulaimani, Sulaymaniyah 46001, Kurdistan Region, Iraq.

Informed Consent Statement: Informed consent was obtained from all subjects involved in the study.

Data Availability Statement: The data used during the study are available from the corresponding author.

Conflicts of Interest: The authors declare no conflict of interest.

References

1. Stanković, M.S.; Anatoly, A.; Kilbas; Hari, M.S.; Juan, J. *Trujillo, Theory and Applications of Fractional Differential Equations*; Elsevier B.V.: Amsterdam, The Netherlands, 2006.
2. El-Mesiry, A.E.M.; El-Sayed, A.M.A.; El Saka, H.A.A. Numerical methods for multi-term fractional (arbitrary) orders differential equations. *Appl. Math. Comput.* **2005**, *160*, 683–699. [CrossRef]
3. Miller, K.S.; Ross, B. *An Introduction to the Fractional Calculus and Fractional Differential Equations*; John Wiley: Hoboken, NJ, USA, 1993.
4. Podlubny, I. *Fractional Differential Equations*; Academic Press: New York, NY, USA, 1999.
5. El-Wakil, S.; Elhanbaly, A.; Abdou, M. Adomian decomposition method for solving fractional nonlinear differential equations. *Appl. Math. Comput.* **2006**, *182*, 313–324. [CrossRef]
6. Das, S. Analytical solution of a fractional diffusion equation by variational iteration method. *Comput. Math. Appl.* **2009**, *57*, 483–487. [CrossRef]
7. Erturk, V.S.; Momani, S.; Odibat, Z. Application of generalized differential transform method to multi-order fractional differential equations. *Commun. Nonlinear Sci. Numer. Simul.* **2008**, *13*, 1642–1654. [CrossRef]
8. Meerschaert, M.M.; Tadjeran, C. Finite difference approximations for two-sided space-fractional partial differential equations. *Appl. Numer. Math.* **2006**, *56*, 80–90. [CrossRef]
9. Odibat, Z.M.; Shawagfeh, N.T. Generalized Taylor's formula. *Appl. Math. Comput.* **2007**, *186*, 286–293. [CrossRef]
10. Yüzbaşı, Ş. A numerical approximation based on the Bessel functions of first kind for solutions of Riccati type differential–difference equations. *Comput. Math. Appl.* **2012**, *64*, 1691–1705. [CrossRef]
11. Gülsu, M.; Gürbüz, B.; Öztürk, Y.; Sezer, M. Laguerre polynomial approach for solving linear delay difference equations. *Appl. Math. Comput.* **2011**, *217*, 6765–6776. [CrossRef]
12. Samadi, O.R.N.; Tohidi, E. Thespectral method for solving systems of Volterra integral equations. *J. Appl. Math. Comput.* **2012**, *40*, 477–497. [CrossRef]
13. Tohidi, E.; Soleymani, F.; Kilicman, A. Robustness of Operational Matrices of Differentiation for Solving State-Space Analysis and Optimal Control Problems. *Abstr. Appl. Anal.* **2013**, *2013*, 535979. [CrossRef]
14. Toutounian, F.; Tohidi, E.; Kilicman, A. Fourier Operational Matrices of Differentiation and Transmission: Introduction and Applications. *Abstr. Appl. Anal.* **2013**, *2013*, 198926. [CrossRef]
15. Yalçinbaş, S.; Aynigül, M.; Sezer, M. A collocation method using Hermite polynomials for approximate solution of pantograph equations. *J. Frankl. Inst.* **2011**, *348*, 1128–1139. [CrossRef]
16. Yüzbaşı, Ş.; Şahın, N.; Yildirim, A. A collocation approach for solving high-order linear Fredholm–Volterra integro-differential equations. *Math. Comput. Model.* **2012**, *55*, 547–563. [CrossRef]
17. Yüzbaşı, Ş.; Şahın, N.; Sezer, M. Bessel polynomial solutions of high-order linear Volterra integro-differential equations. *Comput. Math. Appl.* **2011**, *62*, 1940–1956. [CrossRef]
18. Parand, K.; Nikarya, M. Application of Bessel functions for solving differential and integro-differential equations of the fractional order. *Appl. Math. Model.* **2014**, *38*, 4137–4147. [CrossRef]
19. Izadi, M.; Cattani, C. Generalized Bessel Polynomial for Multi-Order Fractional Differential Equations. *Symmetry* **2020**, *12*, 1260. [CrossRef]
20. Ahmed, S.S. On System of Linear Volterra Integro Fractional Differential Equations. Ph.D. Thesis, University of Sulaimani, Sulaymaniyah, Iraq, July 2009.

21. Weilbeer, M. *Efficient Numerical Methods for Fractional Differential Equations and Their Analytical Background*; US Army Medical Research and Material Command: Frederick, MD, USA, 2005.
22. Ahmed, S.S.; Salih, S.A.H. Numerical treatment of the most general linear Volterra integro-fractional differential equations with Caputo derivative by Quadrature methods. *J. math. Comput. Sci.* **2012**, *2*, 1293–1311.
23. Al-Nasir, R.H. Numerical Solution of Volterra Integral Equation of the Second Kind. Master's Thesis, Technology-Iraq University, Baghdad, Iraq, September 1999.
24. Shukla, A.; Sukavanam, N.; Pandey, D.N. Approximate Controllability of Semilinear Fractional Control Systems of Order $\alpha \in (1, 2]$. In Proceedings of the SIAM Proceedings of the Conference on Control and its Applications (CT), San Diego, CA, USA, 8–10 July 2015.
25. Ordokhani, Y.; Dehestani, H. An Application of Bessel function for Solving Nonlinear Fredholm-Volterra-Hammerstein Integro-differential Equations. *J. Sci. Kharazmi Univ.* **2013**, *13*, 347–362.
26. Yüzbaşı, Ş.; Şahin, N.; Sezer, M. A Bessel polynomial approach for solving linear neutral delay differential equations with variable coefficients. *J. Adv. Res. Differ. Equ.* **2011**, *3*, 81–101.
27. Maleknejad, K.; Mahmoudi, Y. Taylor polynomial solution of high-order nonlinear Volterra–Fredholm integro-differential equations. *Appl. Math. Comput.* **2003**, *145*, 641–653. [CrossRef]
28. Yüzbaşı, Ş.; Sezer, M. A collocation approach to solve a class of Lane-Emden type equations. *J. Adv. Res. Appl. Math.* **2011**, *3*, 58–73. [CrossRef]

Article

Hermite Cubic Spline Collocation Method for Nonlinear Fractional Differential Equations with Variable-Order

Tinggang Zhao [1,*,†] and Yujiang Wu [2,†]

1   School of Mathematics, Lanzhou City University, Lanzhou 730070, China
2   School of Mathematics and Statistics, Lanzhou University, Lanzhou 730000, China; myjaw@lzu.edu.cn
*   Correspondence: zhaotg@lzcu.edu.cn; Tel.: +86-1366-939-7938
†   These authors contributed equally to this work.

Abstract: In this paper, we develop a Hermite cubic spline collocation method (HCSCM) for solving variable-order nonlinear fractional differential equations, which apply C^1-continuous nodal basis functions to an approximate problem. We also verify that the order of convergence of the HCSCM is about $O(h^{\min\{4-\alpha,p\}})$ while the interpolating function belongs to $C^p(p \geq 1)$, where h is the mesh size and α the order of the fractional derivative. Many numerical tests are performed to confirm the effectiveness of the HCSCM for fractional differential equations, which include Helmholtz equations and the fractional Burgers equation of constant-order and variable-order with Riemann-Liouville, Caputo and Patie-Simon sense as well as two-sided cases.

Keywords: collocation method; fractional calculus; hermite cubic spline; fractional burgers equation

---

Citation: Zhao, T.; Wu, Y. Hermite Cubic Spline Collocation Method for Nonlinear Fractional Differential Equations with Variable-Order. *Symmetry* **2021**, *13*, 872. https://doi.org/10.3390/sym13050872

Academic Editor: Francisco Martínez González, Mohammed KA Kaabar

Received: 8 April 2021
Accepted: 10 May 2021
Published: 13 May 2021

Publisher's Note: MDPI stays neutral with regard to jurisdictional claims in published maps and institutional affiliations.

Copyright: © 2021 by the authors. Licensee MDPI, Basel, Switzerland. This article is an open access article distributed under the terms and conditions of the Creative Commons Attribution (CC BY) license (https://creativecommons.org/licenses/by/4.0/).

1. Introduction

As a powerful tool for modeling a broad range of non-classical phenomena, fractional calculus has already gained much attention from various science and engineering fields during recent decades. For models of anomalous transport processes and diffusion, there are a lot of fractional partial differential equations proposed in publications [1,2] as well as for the modeling of frequency dependent damping behavior such as in viscoelastic, continuum and statistical mechanics, solid mechanic, economics [3,4], and so on. For modelling the energy supply-demand system, the Caputo-Fabrizio fractional derivative is applied and leads to an interesting fractional energy supply-demand equation [5]. With extensive applications of fractional calculus operators, many fractional differential equations (FDEs) are presented.

Meanwhile, there is increasing demand for a robust method to produce a high accuracy solution to FDEs. Publications on numerical methods for FDEs are largely substantial. A considerable number of them are based on finite difference, see [6–17] and the references therein. There are many works based on finite element methods, see [18–23]. Methods based on spectral/pseudo-spectral or collocation methods, even the spectral element method, can be seen in [24–33].

The main challenge of approximating FDEs is the precision deterioration caused by singularity of fractional derivatives [34]. For the spectral-type methods, which are one of the most popular numerical methods due to their high accuracy [35–38], the singularity of endpoints may damage "the spectral accuracy". Spectral or collocation methods using fractional polynomials rather than polynomials as basis functions provide a promising way to develop an efficient algorithm for numerically solving FDEs and even fractional operator-related problems. There are theoretical and practical efforts involved in publications such as [29,39–48].

The demand of flexibility may lead researchers to pursue a multi-domain method or domain decomposition method. A multi-domain spectral collocation method (MDSCM) is suggested to numerically solve FDEs [49]. Authors make use of piecewise continuous

(C^0) nodal basis functions to approximate the problem. However, a piecewise continuous function may have infinite derivatives of fractional-order. Let us introduce the following result from [49]:

Lemma 1. *Let $\alpha \in (0,2)$ be a constant, and $a, b, c \in \mathbb{R}$ such that $a < c < b$. If $u \in C^2[a,c] \cap C^2[c,b] \cap C[a,b]$ and $u'(c^-)$ and $u'(c^+)$ exist, then*

$$_{RL}D_{a,x}^{\alpha}u(x) = \sum_{k=0}^{1} \frac{(x-a)^{k-\alpha}}{\Gamma(k+1-\alpha)}u^{(k)}(a) + \frac{(x-c)^{1-\alpha}}{\Gamma(2-\alpha)}[u'(c^+) - u'(c^-)] + s(x),$$

for any $x \in (c,b)$, where

$$s(x) = \begin{cases} \frac{1}{\Gamma(3-\alpha)} \frac{d}{dx}\left[\int_a^x u''(\tau)(x-\tau)^{2-\alpha}d\tau\right], & \text{if } \alpha \in (0,1), \\ \frac{1}{\Gamma(4-\alpha)} \frac{d^2}{dx^2}\left[\int_a^x u''(\tau)(x-\tau)^{3-\alpha}d\tau\right], & \text{if } \alpha \in (1,2). \end{cases}$$

Here, the $_{RL}D_{a,x}^{\alpha}$ denotes the left Riemann-Liouville fractional derivative of α-order; we will give its definition in the following section. The above lemma shows that $\lim_{x \to c} {_{RL}D_{a,x}^{\alpha}u(x)} = \infty$ if the conditions $u'(c^+) \neq u'(c^-)$ and $\alpha > 1$ are satisfied. To overcome this drawback, C^1-continuous nodal basis functions are needed. It is well-known that spline functions are a special class of piecewise polynomials, which provide continuous differentiable solutions over the whole spatial domain with great accuracy. One promising candidate as a C^1-continuous nodal basis function is the Hermite cubic spline function.

Spline collocation methods are successfully applied to numerical approximation of differential equations (see [50–52] and references therein). However, there are a few publications devoted to the spline collocation method for FDEs. Recently, Liu et al. [53] presented an interesting result of stability and convergence of quadratic spline collocation method for time-dependent fractional diffusion equations. Majeed et al. [54] applied the cubic B-spline collocation method to solve time fractional Burgers' and Fisher's equations. Khalid et al. [55] presented a non-polynomial quintic spline collocation method to solve fourth-order fractional boundary value problems involving products terms. Emadifar et al. [56] explored exponential spline interpolation with multiple parameters to find solutions of fractional boundary value problem and conducted the convergence analysis for this technique.

In this paper, our aim is to develop a Hermite cubic spline collocation method (HCSCM) for solving variable-order nonlinear fractional differential equations, which makes use of C^1-continuous nodal basis functions to approximate a problem. In particular, the collocation fractional differentiation matrix is derived for fractional derivatives in various senses including Riemann-Liouville, Caputo, Patie-Simon. The main contributions of this work are as follows:

- A set of C^1 nodal basis functions are constructed and the corresponding collocation fractional differentiation matrix is derived for the discretization.
- Making use of the Hermite cubic spline collocation method, numerical solution could be found for variable-order nonlinear fractional differential equations. The order of convergence of the HCSCM is also analysed for the left Riemann-Liouville case.
- The effectiveness of the HCSCM is confirmed by solving fractional Helmholtz equations of constant-order and variable-order. With application the HCSCM to the fractional Burgers equation, the numerical fractional diffusion is simulated with different senses.

The paper is organized as follows: in the next Section, some definitions and properties are reviewed for later discussion. The Hermite cubic spline collocation method (HCSCM) is presented in Section 3. The key part is to set up the collocation fractional differentiation matrix. In Section 4, the order of convergence of the HCSCM approximation is analyzed for the left Riemann-Liouville case. Several numerical tests are presented in Section 5. This includes applying HCSCM to fractional Helmholtz equations and fractional Burgers equations. Finally, we conclude in Section 6.

2. Preliminaries

In this Section, some definitions of fractional calculus are reviewed for subsequent discussions. The most common-used definitions of fractional derivatives are possibly the Riemann-Liouville's and the Caputo's, found in various publications, such as ([57,58]). The following definitions are variable-order versions, which provide constant-order definitions when $\alpha(x) \equiv \alpha$ is a constant in the formulas.

Definition 1. For a function $f(x), x \in [x_L, x_R]$, the left Riemann-Liouville fractional integral of order $\alpha(x) > 0$ is defined as

$$_{x_L}I_x^{\alpha(x)} f(x) := \frac{1}{\Gamma(\alpha(x))} \int_{x_L}^{x} (x-s)^{\alpha(x)-1} f(s) ds, \qquad (1)$$

and the right Riemann-Liouville fractional integral of order $\alpha(x) > 0$ is defined as

$$_{x}I_{x_R}^{\alpha(x)} f(x) := \frac{1}{\Gamma(\alpha(x))} \int_{x}^{x_R} (s-x)^{\alpha(x)-1} f(s) ds, \qquad (2)$$

where $\Gamma(\cdot)$ is the Euler's gamma function.

Definition 2. For a function $f(x), x \in [x_L, x_R]$, the left Riemann-Liouville fractional derivative of order $\alpha(x) > 0$ is defined as

$$_{RL}D_{x_L,x}^{\alpha(x)} f(x) := \frac{1}{\Gamma(n-\alpha(x))} \left[\frac{d^n}{d\xi^n} \int_{x_L}^{\xi} (\xi-s)^{n-\alpha(x)-1} f(s) ds \right]_{\xi=x}, \qquad (3)$$

and the right Riemann-Liouville fractional derivative of order $\alpha(x) > 0$ is defined as

$$_{RL}D_{x,x_R}^{\alpha(x)} f(x) := \frac{(-1)^n}{\Gamma(n-\alpha(x))} \left[\frac{d^n}{d\xi^n} \int_{\xi}^{x_R} (s-\xi)^{n-\alpha(x)-1} f(s) ds \right]_{\xi=x}, \qquad (4)$$

where n is the positive integer such that $n-1 < \alpha(x) < n$.

Definition 3. For a function $f(x), x \in [x_L, x_R]$, the left Caputo fractional derivative of order $\alpha(x) > 0$ is defined as

$$_{C}D_{x_L,x}^{\alpha(x)} f(x) := \frac{1}{\Gamma(n-\alpha(x))} \int_{x_L}^{x} (x-s)^{n-\alpha(x)-1} f^{(n)}(s) ds, \qquad (5)$$

and the right Caputo fractional derivative of order $\alpha(x) > 0$ is defined as

$$_{C}D_{x,x_R}^{\alpha(x)} f(x) := \frac{(-1)^n}{\Gamma(n-\alpha(x))} \int_{x}^{x_R} (s-x)^{n-\alpha(x)-1} f^{(n)}(s) ds, \qquad (6)$$

where n is the positive integer such that $n-1 < \alpha(x) < n$.

The well-known relationship between Riemann-Liouville and the Caputo derivative is as follows:

Lemma 2. If $_{RL}D_{x_L,x}^{\alpha(x)} f(x), _{C}D_{x_L,x}^{\alpha(x)} f(x), _{RL}D_{x,x_R}^{\alpha(x)} f(x)$ and $_{C}D_{x,x_R}^{\alpha(x)} f(x)$ exist, then

$$_{RL}D_{x_L,x}^{\alpha(x)} f(x) = {_{C}D_{x_L,x}^{\alpha(x)}} f(x) + \sum_{k=0}^{n-1} \frac{f^{(k)}(x_L)}{\Gamma(k+1-\alpha(x))} (x-x_L)^{k-\alpha(x)}, \qquad (7)$$

and

$$_{RL}D_{x,x_R}^{\alpha(x)} f(x) = {_{C}D_{x,x_R}^{\alpha(x)}} f(x) + \sum_{k=0}^{n-1} \frac{(-1)^{n-j} f^{(k)}(x_R)}{\Gamma(k+1-\alpha(x))} (x_R-x)^{k-\alpha(x)}. \qquad (8)$$

Besides the common-used definitions above, the fractional diffusion operators which limit the order $1 < \alpha(x) \le 2$ are also considered. A definition was proposed by Patie and Simon in [59] as follows.

Definition 4. *For a function $f(x), x \in [x_L, x_R]$, the left Patie-Simon (or mixed Caputo) fractional derivative of order $1 < \alpha(x) < 2$ is defined as*

$$_{PS}D_{x_L,x}^{\alpha(x)}f(x) := \frac{1}{\Gamma(2-\alpha(x))}\left[\frac{d}{d\xi}\int_{x_L}^{\xi}(\xi-s)^{1-\alpha(x)}f'(s)ds\right]_{\xi=x}, \tag{9}$$

and the right Patie-Simon (or mixed Caputo) fractional derivative of order $1 < \alpha(x) < 2$ is defined as

$$_{PS}D_{x,x_R}^{\alpha(x)}f(x) := \frac{1}{\Gamma(2-\alpha(x))}\left[\frac{d}{d\xi}\int_{\xi}^{x_R}(s-\xi)^{1-\alpha(x)}f'(s)ds\right]_{\xi=x}. \tag{10}$$

From the above definitions and Lemma 2, hold the following relationships:

Lemma 3. *If $1 < \alpha(x) < 2$ and $_{RL}D_{x_L,x}^{\alpha(x)}f(x)$, $_CD_{x_L,x}^{\alpha(x)}f(x)$, $_{PS}D_{x_L,x}^{\alpha(x)}f(x)$, $_{RL}D_{x,x_R}^{\alpha(x)}f(x)$, $_CD_{x,x_R}^{\alpha(x)}f(x)$ and $_{PS}D_{x,x_R}^{\alpha(x)}f(x)$ exist, then*

$$_{RL}D_{x_L,x}^{\alpha(x)}f(x) = {}_{PS}D_{x_L,x}^{\alpha(x)}f(x) + \frac{f(x_L)}{\Gamma(1-\alpha(x))}(x-x_L)^{-\alpha(x)}, \tag{11}$$

and

$$_{RL}D_{x,x_R}^{\alpha(x)}f(x) = {}_{PS}D_{x,x_R}^{\alpha(x)}f(x) + \frac{f(x_R)}{\Gamma(1-\alpha(x))}(x_R-x)^{-\alpha(x)}, \tag{12}$$

and

$$_{PS}D_{x_L,x}^{\alpha(x)}f(x) = {}_CD_{x_L,x}^{\alpha(x)}f(x) + \frac{f'(x_L)}{\Gamma(2-\alpha(x))}(x-x_L)^{1-\alpha(x)}, \tag{13}$$

and

$$_{PS}D_{x,x_R}^{\alpha(x)}f(x) = {}_CD_{x,x_R}^{\alpha(x)}f(x) + \frac{f'(x_R)}{\Gamma(2-\alpha(x))}(x_R-x)^{1-\alpha(x)}. \tag{14}$$

Proof. Since

$$\int_{x_L}^{\xi}(\xi-s)^{1-\alpha(x)}f(s)ds = \frac{(\xi-x_L)^{2-\alpha(x)}}{2-\alpha(x)}f(x_L) + \frac{1}{2-\alpha(x)}\int_{x_L}^{\xi}(\xi-s)^{2-\alpha(x)}f'(s)ds.$$

Then note that $1 < \alpha(x) < 2$,

$$\frac{d^2}{d\xi^2}\left[\int_{x_L}^{\xi}(\xi-s)^{1-\alpha(x)}f(s)ds\right]$$

$$= (1-\alpha(x))(\xi-x_L)^{-\alpha(x)}f(x_L) + \frac{d}{d\xi}\int_{x_L}^{\xi}(\xi-s)^{1-\alpha(x)}f'(s)ds].$$

The equality (11) is obtained by dividing factor $\Gamma(2-\alpha(x))$. Other results can be derived by a similar argument. □

There exist the following well-known properties:

Lemma 4. Let m be an integer number, the following properties hold for $x \in [x_L, x_R]$ and Riemann-Liouville fractional calculus

$$\begin{aligned}
{}_{x_L}I_x^{\alpha(x)}(x-x_L)^m &= \frac{m!}{\Gamma(m+\alpha(x)+1)}(x-x_L)^{m+\alpha(x)}, \\
{}_xI_{x_R}^{\alpha(x)}(x_R-x)^m &= \frac{m!}{\Gamma(m+\alpha(x)+1)}(x_R-x)^{m+\alpha(x)}, \\
{}_{RL}D_{x_L,x}^{\alpha(x)}(x-x_L)^m &= \frac{m!}{\Gamma(m-\alpha(x)+1)}(x-x_L)^{m-\alpha(x)}, \\
{}_{RL}D_{x,x_R}^{\alpha(x)}(x_R-x)^m &= \frac{m!}{\Gamma(m-\alpha(x)+1)}(x_R-x)^{m-\alpha(x)},
\end{aligned} \qquad (15)$$

and for the Caputo fractional derivative,

$$\begin{aligned}
{}_C D_{x_L,x}^{\alpha(x)}(x-x_L)^m &= \begin{cases} \frac{m!}{\Gamma(m-\alpha(x)+1)}(x-x_L)^{m-\alpha(x)}, & \text{if } m > \alpha(x), \\ 0, & \text{if } m < \alpha(x), \end{cases} \\
{}_C D_{x,x_R}^{\alpha(x)}(x_R-x)^m &= \begin{cases} \frac{m!}{\Gamma(m-\alpha(x)+1)}(x_R-x)^{m-\alpha(x)}, & \text{if } m > \alpha(x), \\ 0, & \text{if } m < \alpha(x), \end{cases}
\end{aligned} \qquad (16)$$

and for the Patie-Simon fractional derivative of $1 < \alpha(x) < 2$,

$$\begin{aligned}
{}_{PS}D_{x_L,x}^{\alpha(x)}(x-x_L)^m &= \begin{cases} \frac{m!}{\Gamma(m-\alpha(x)+1)}(x-x_L)^{m-\alpha(x)}, & \text{if } m > 0, \\ 0, & \text{if } m = 0, \end{cases} \\
{}_{PS}D_{x,x_R}^{\alpha(x)}(x_R-x)^m &= \begin{cases} \frac{m!}{\Gamma(m-\alpha(x)+1)}(x_R-x)^{m-\alpha(x)}, & \text{if } m > 0, \\ 0, & \text{if } m = 0. \end{cases}
\end{aligned} \qquad (17)$$

The following operators with top-tilde are useful in HCSCM for $x > x_R$,

$$_{x_L*}\tilde{I}_{x_R}^{\alpha(x)}f(x) := \frac{1}{\Gamma(\alpha(x))}\int_{x_L}^{x_R}(x-s)^{\alpha(x)-1}f(s)ds, \qquad (18)$$

and for $x < x_L$,

$$_{x_L}\tilde{I}_{x_R*}^{\alpha(x)}f(x) := \frac{1}{\Gamma(\alpha(x))}\int_{x_L}^{x_R}(s-x)^{\alpha(x)-1}f(s)ds, \qquad (19)$$

and for $x > x_R$,

$$_{RL}\tilde{D}_{x_L*,x_R}^{\alpha(x)}f(x) := \frac{1}{\Gamma(n-\alpha(x))}\left[\frac{d^n}{d\zeta^n}\int_{x_L}^{x_R}(\zeta-s)^{n-\alpha(x)-1}f(s)ds\right]_{\zeta=x}, \qquad (20)$$

and for $x < x_L$,

$$_{RL}\tilde{D}_{x_L,x_R*}^{\alpha(x)}f(x) := \frac{(-1)^n}{\Gamma(n-\alpha(x))}\left[\frac{d^n}{d\zeta^n}\int_{x_L}^{x_R}(s-\zeta)^{n-\alpha(x)-1}f(s)ds\right]_{\zeta=x}. \qquad (21)$$

Operators ${}_C\tilde{D}_{x_L*,x_R}^{\alpha(x)}$, ${}_C\tilde{D}_{x_L,x_R*}^{\alpha(x)}$, ${}_{PS}\tilde{D}_{x_L*,x_R}^{\alpha(x)}$, ${}_{PS}\tilde{D}_{x_L,x_R*}^{\alpha(x)}$ are defined similarly.

Lemma 5. Let $x_L < x_c < x_R$ and $x \in (x_c, x_R]$, then

$$\begin{aligned}
{}_{x_L}I_x^{\alpha(x)}f(x) &= {}_{x_L*}\tilde{I}_{x_c}^{\alpha(x)}f(x) + {}_{x_c}I_x^{\alpha(x)}f(x), \\
{}_{RL}D_{x_L,x}^{\alpha(x)}f(x) &= {}_{RL}\tilde{D}_{x_L*,x_c}^{\alpha(x)}f(x) + {}_{RL}D_{x_c,x}^{\alpha(x)}f(x), \\
{}_C D_{x_L,x}^{\alpha(x)}f(x) &= {}_C\tilde{D}_{x_L*,x_c}^{\alpha(x)}f(x) + {}_C D_{x_c,x}^{\alpha(x)}f(x), \\
{}_{PS}D_{x_L,x}^{\alpha(x)}f(x) &= {}_{PS}\tilde{D}_{x_L*,x_c}^{\alpha(x)}f(x) + {}_{PS}D_{x_c,x}^{\alpha(x)}f(x),
\end{aligned} \qquad (22)$$

and when $x \in [x_L, x_c)$, we have

$$\begin{aligned}
{}_xI_{x_R}^{\alpha(x)}f(x) &= {}_xI_{x_c}^{\alpha(x)}f(x) + {}_{x_c}\tilde{I}_{x_R*}^{\alpha(x)}f(x), \\
{}_{RL}D_{x,x_R}^{\alpha(x)}f(x) &= {}_{RL}D_{x,x_c}^{\alpha(x)}f(x) + {}_{RL}\tilde{D}_{x_c,x_R*}^{\alpha(x)}f(x), \\
{}_cD_{x,x_R}^{\alpha(x)}f(x) &= {}_cD_{x,x_c}^{\alpha(x)}f(x) + {}_c\tilde{D}_{x_c,x_R*}^{\alpha(x)}f(x), \\
{}_{PS}D_{x,x_R}^{\alpha(x)}f(x) &= {}_{PS}D_{x,x_c}^{\alpha(x)}f(x) + {}_{PS}\tilde{D}_{x_c,x_R*}^{\alpha(x)}f(x).
\end{aligned} \tag{23}$$

Proof. Since

$$\int_{x_L}^x (x-s)^{\alpha(x)-1} f(s) ds = \int_{x_L}^{x_c} (x-s)^{\alpha(x)-1} f(s) ds + \int_{x_c}^x (x-s)^{\alpha(x)-1} ds,$$

Then the first equality in (22) is obtained by dividing factor $\Gamma(\alpha(x))$ and the definitions (1) and (18). Other results can be derived by a similar argument. □

If $\mathcal{D}^{*\alpha(x)}$ and $\mathcal{D}^{\alpha(x)*}$ represent all left-sided and right-sided definitions of the abovementioned, respectively, then the two-sided fractional derivative can be written as

$$\mathbb{D}_r^{\alpha(x)} := r\mathcal{D}^{*\alpha(x)} + (1-r)\mathcal{D}^{\alpha(x)*}, \quad 0 \leq r \leq 1. \tag{24}$$

3. Hermite Cubic Spline Collocation Method (HCSCM)

In the Section, the HCSCM is presented. The key role of HCSCM is the collocation fractional differentiation matrix.

3.1. Fractional Differentiation Matrix (FDM) for HCSCM

Let $\Lambda := (x_L, x_R)$, the first step is to divide the interval Λ into N elements, that is,

$$x_L = x_0 < x_1 < \cdots < x_N = x_R.$$

Denote $I_i = [x_{i-1}, x_i], i = 1, 2, \ldots, N$ the i-th element and $h_i = x_i - x_{i-1}$ the length of I_i. Let $\mathbb{P}_N^l$ be the collection of all algebraic polynomials defined on interval I with degree at most N. The piecewise Hermite cubic polynomial space is

$$\mathbb{V}_N = \{v \in C^1(\Lambda) : v|_{I_i} \in \mathbb{P}_3^{l_i}, i = 1, 2, \ldots, N\}.$$

which is defined by the following set of nodal basis functions. It contains $2N + 2$ functions as follows. The first two functions as

$$\varphi_0(x) = \begin{cases} \left(1 + 2\frac{x-x_0}{h_1}\right)\left(1 - \frac{x-x_0}{h_1}\right)^2, & \text{if } x \in I_1, \\ 0, & \text{otherwise,} \end{cases}$$

and

$$\phi_0(x) = \begin{cases} \left(\frac{x-x_0}{h_1}\right)\left(1 - \frac{x-x_0}{h_1}\right)^2 h_1, & \text{if } x \in I_1, \\ 0, & \text{otherwise.} \end{cases}$$

For $i = 1, 2, \ldots, N-1$,

$$\varphi_i(x) = \begin{cases} \left(3 - 2\frac{x-x_{i-1}}{h_i}\right)\left(\frac{x-x_{i-1}}{h_i}\right)^2, & \text{if } x \in I_i, \\ \left(1 + 2\frac{x-x_i}{h_{i+1}}\right)\left(1 - \frac{x-x_i}{h_{i+1}}\right)^2, & \text{if } x \in I_{i+1}, \\ 0, & \text{otherwise,} \end{cases}$$

and
$$\phi_i(x) = \begin{cases} -\left(1 - \frac{x-x_{i-1}}{h_i}\right)\left(\frac{x-x_{i-1}}{h_i}\right)^2 h_i, & \text{if } x \in I_i, \\ \left(\frac{x-x_i}{h_{i+1}}\right)\left(1 - \frac{x-x_i}{h_{i+1}}\right)^2 h_{i+1}, & \text{if } x \in I_{i+1}, \\ 0, & \text{otherwise}. \end{cases}$$

and the last two functions as
$$\varphi_N(x) = \begin{cases} \left(3 - 2\frac{x-x_{N-1}}{h_N}\right)\left(\frac{x-x_{N-1}}{h_N}\right)^2, & \text{if } x \in I_N, \\ 0, & \text{otherwise}, \end{cases}$$

and
$$\phi_N(x) = \begin{cases} -\left(1 - \frac{x-x_{N-1}}{h_N}\right)\left(\frac{x-x_{N-1}}{h_N}\right)^2 h_N, & \text{if } x \in I_N, \\ 0, & \text{otherwise}. \end{cases}$$

Therefore,
$$V_N = \text{span}\{\varphi_i, \phi_i, i = 0, 1, ..., N\}.$$

If $u_N \in V_N$, then can be expanded as
$$u_N(x) = \sum_{i=0}^{N}(u_N(x_i)\varphi_i(x) + u'_N(x_i)\phi_i(x)).$$

In each element I_i, $x^c_{i,1}, x^c_{i,2} \in I_i$ are the collocation points, where
$$x^c_{i,1} = x_{i-1} + \sigma_{i,1}h_i, \quad x^c_{i,2} = x_{i-1} + \sigma_{i,2}h_i, \quad i = 1, 2, ..., N,$$

and $0 \leq \sigma_{i,1} < \sigma_{i,2} \leq 1$. In fact, a choice of this points is the Gauss-type quadrature nodes, $\sigma_{i,1} = (1 - \sigma_{i,2}) = \frac{\sqrt{3}}{3}$, which is named by orthogonal spline collocation. However, the stable collocation points may not be symmetric in the view of [60,61].

As a collocation approximation to the $\alpha(x)$th-order differential operators defined in Section 2, we denote by $\mathbf{D}^\alpha$ the collocation fractional differentiation matrix, which satisfies
$$(\mathbf{D}^\alpha u_N)_l = \mathcal{D}^{\alpha(x^c_{ij})} u_N(x^c_{ij}), \quad j = 1, 2; \quad i = 1, 2, ..., N. \tag{25}$$

The structure of the collocation fractional differentiation matrix (FDM) may differ with the ordering of the collocation points and the unknowns. In natural ordering $l = 2(i-1) + j$, we have
$$\begin{aligned} \mathbf{u} &= [u'_0, u_1, u'_1, \cdots, u_{N-1}, u'_{N-1}, u'_N]^T, \\ \mathbf{x}^c &= [x^c_{11}, x^c_{12}, x^c_{21}, x^c_{22}, \cdots, x^c_{N1}, x^c_{N2}]^T. \end{aligned} \tag{26}$$

and $\mathbf{D}^\alpha$ with Dirichlet boundary conditions is
$$\mathbf{D}^\alpha = \begin{bmatrix} \mathcal{D}\phi_0(x^c_{11}) & \mathcal{D}\varphi_1(x^c_{11}) & \mathcal{D}\phi_1(x^c_{11}) & \cdots & \mathcal{D}\phi_N(x^c_{11}) \\ \mathcal{D}\phi_0(x^c_{12}) & \mathcal{D}\varphi_1(x^c_{12}) & \mathcal{D}\phi_1(x^c_{12}) & \cdots & \mathcal{D}\phi_N(x^c_{12}) \\ \vdots & \vdots & \vdots & & \vdots \\ \mathcal{D}\phi_0(x^c_{N1}) & \mathcal{D}\varphi_1(x^c_{N1}) & \mathcal{D}\phi_1(x^c_{N1}) & \cdots & \mathcal{D}\phi_N(x^c_{N1}) \\ \mathcal{D}\phi_0(x^c_{N2}) & \mathcal{D}\varphi_1(x^c_{N2}) & \mathcal{D}\phi_1(x^c_{N2}) & \cdots & \mathcal{D}\phi_N(x^c_{N2}) \end{bmatrix}, \tag{27}$$

here $\mathcal{D} = \mathcal{D}^{\alpha(x^c_{ij})}$ is one of the fractional differential operators defined in Section 2. Typically, the matrix $\mathcal{D}$ is block-triangular for left and right fractional operators.

Remark 1. *According to the Lemma 2, Lemma 3 and the special nodal basis functions, the collocation FDM $\mathbf{D}^\alpha$ of the Riemann-Liouville operators is equal to the corresponding FDM of* the Patie-Simon operators. For the Caputo operators, the corresponding FDM is different only from the first or last column.

3.2. Computing the Entries of FDM

For ease of computing, operations are shifted from an arbitrary interval $[a,b]$ to the reference interval $[-1,1]$. Let the linear transformation

$$x = \frac{h}{2}(y+1) + a, \quad \text{or} \quad y = \frac{2}{h}(x-a) - 1, \quad h := b - a \tag{28}$$

shift functions $f(x), \alpha(x)$ defined on the interval $[a,b]$ to $\hat{f}(y), \hat{\alpha}(y)$ on the reference interval $[-1,1]$. Then we have the following relations:

$$_aI_x^{\alpha(x)} f(x) = \left(\frac{h}{2}\right)^{\hat{\alpha}(y)} {}_{-1}I_y^{\hat{\alpha}(y)} \hat{f}(y),$$

$$_xI_b^{\alpha(x)} f(x) = \left(\frac{h}{2}\right)^{\hat{\alpha}(y)} {}_yI_1^{\hat{\alpha}(y)} \hat{f}(y), \tag{29}$$

$$\mathcal{D}^{\alpha(x)} f(x) = \left(\frac{2}{h}\right)^{\hat{\alpha}(y)} \mathcal{D}^{\hat{\alpha}(y)} \hat{f}(y),$$

here $\mathcal{D}^{\alpha(x)}$ be one of the seven: $_{RL}D_{a,x}^{\alpha(x)}, {}_{RL}D_{x,b}^{\alpha(x)}, {}_CD_{a,x}^{\alpha(x)}, {}_CD_{x,b}^{\alpha(x)}, {}_{PS}D_{a,x}^{\alpha(x)}, {}_{PS}D_{x,b}^{\alpha(x)}, \mathbb{D}_r^{\alpha(x)}$.

For the tilde operators, the following relations also hold:

$$_{a*}\tilde{I}_b^{\alpha(x)} f(x) = \left(\frac{h}{2}\right)^{\hat{\alpha}(y)} {}_{-1*}\tilde{I}_1^{\hat{\alpha}(y)} \hat{f}(y),$$

$$_a\tilde{I}_{b*}^{\alpha(x)} f(x) = \left(\frac{h}{2}\right)^{\hat{\alpha}(y)} {}_{-1}\tilde{I}_{1*}^{\hat{\alpha}(y)} \hat{f}(y), \tag{30}$$

$$\tilde{\mathcal{D}}^{\alpha(x)} f(x) = \left(\frac{2}{h}\right)^{\hat{\alpha}(y)} \tilde{\mathcal{D}}^{\hat{\alpha}(y)} \hat{f}(y),$$

here $\tilde{\mathcal{D}}^{\alpha(x)}$ can be one of the six: $_{RL}\tilde{D}_{a*,b}^{\alpha(x)}, {}_{RL}\tilde{D}_{a,b*}^{\alpha(x)}, {}_C\tilde{D}_{a*,b}^{\alpha(x)}, {}_C\tilde{D}_{a,b*}^{\alpha(x)}, {}_{PS}\tilde{D}_{a*,b}^{\alpha(x)}, {}_{PS}\tilde{D}_{a,b*}^{\alpha(x)}$.

The nodal basis functions presented in Section 3, are the so-called shape functions after being transferred by (28), that is,

$$\zeta_1(y) := (2+y)\left(1 - \frac{y+1}{2}\right)^2,$$

$$\zeta_2(y) := \left(\frac{y+1}{2}\right)\left(1 - \frac{y+1}{2}\right)^2, \quad \text{(except factor } h_i\text{)},$$

$$\zeta_3(y) := (2-y)\left(\frac{y+1}{2}\right)^2, \tag{31}$$

$$\zeta_4(y) := -\left(1 - \frac{y+1}{2}\right)\left(\frac{y+1}{2}\right)^2, \quad \text{(except factor } h_i\text{)},$$

and $y \in [-1,1]$. The Hermite Spline collocation method will perform all the operators mentioned above on the shape functions (31).

4. Order of Convergence of the Approximation with HCSCM

In this Section, the order of convergence of the approximation with HCSCM is analysed. Typically, the left Riemann-Liouville fractional derivative is considered. For convenience of analysis, denote $D^\alpha = {}_{RL}D^{\alpha(x)}_{x_L,x}$ and let $h_i = x_i - x_{i-1} = h$, $\sigma_{i,1} = \sigma_1$, $\sigma_{i,2} = \sigma_2$, $i = 1, 2, ..., N$. Then $x_i = x_0 + ih$, $i = 0, 1, ..., N$ and the collocation points

$$x^c_{i,1} = x_0 + (i-1+\sigma_1)h, \quad x^c_{i,2} = x_0 + (i-1+\sigma_2)h, \quad i = 1, 2, ..., N.$$

Let $\Pi_N : C^1(\Lambda) \to \mathbb{V}_N$ be the piecewise Hermite cubic interpolation operator, determined uniquely by

$$\Pi_N f(x_i) = f(x_i), \quad \frac{d}{dx}(\Pi_N f)(x_i) = f'(x_i), \quad i = 0, 1, ..., N,$$

for every $f \in C^1(\Lambda)$.

For a function $u(x) \in C(\Lambda)$, the maximum norm is defined by

$$\|u\|_\infty = \max_{x \in \Lambda} |u(x)|.$$

The following results are related to the interpolation errors [62].

Lemma 6. *Let $u(x) \in C^4(\Lambda)$. Then*

$$\left\|\frac{d^j}{dx^j}(u - \Pi_N u)\right\|_\infty \leq Ch^{4-j}\|u^{(4)}\|_\infty, \quad 0 \leq j \leq 3, \tag{32}$$

where C is a constant number which do not dependent on N.

If $u(x) \in C^p(\Lambda) (p \geq 1)$, the interpolation error holds (see [63]):

$$\left\|\frac{d^j}{dx^j}(u - \Pi_N u)\right\|_\infty \leq Ch^{s-j}\|u^{(s)}\|_\infty, \quad 0 \leq j \leq p. \tag{33}$$

where $s = \min\{p, 4\}$.

In the following, the error bound is presented for $\|D^\alpha(\Pi_N u - u)\|_\infty$ with constant-order $\alpha \in (1, 2)$. Let $\tau(x) = \Pi_N u - u$, we have

$$D^\alpha \tau(x) = \frac{1}{\Gamma(2-\alpha)} \frac{d^2}{dx^2} \int_{x_L}^{x} \frac{\tau(s)}{(x-s)^{\alpha-1}} ds.$$

Assume that $x \in [x_{j-1}, x_j]$ for some j, then the above integration can be split as

$$\int_{x_L}^{x} \frac{\tau(s)}{(x-s)^{\alpha-1}} ds = \sum_{i=1}^{j-1} \int_{x_{i-1}}^{x_i} \frac{\tau(s)}{(x-s)^{\alpha-1}} ds + \int_{x_{j-1}}^{x} \frac{\tau(s)}{(x-s)^{\alpha-1}} ds. \tag{34}$$

Let $\bar{x}_i = (x_{i-1} + x_i)/2$, $i = 1, 2, ..., N$. Under the assumption of $u(x) \in C^4(\Lambda)$, from Taylor's theorem we have

$$\frac{d^2}{dx^2} \int_{x_{i-1}}^{x_i} \frac{\tau(s)}{(x-s)^{\alpha-1}} ds = \sum_{k=0}^{3} \frac{\tau^{(k)}(\bar{x}_i)}{k!} \frac{d^2}{dx^2} \int_{x_{i-1}}^{x_i} \frac{(s-\bar{x}_i)^k}{(x-s)^{\alpha-1}} ds$$
$$+ \frac{1}{24} \frac{d^2}{dx^2} \int_{x_{i-1}}^{x_i} \frac{u^{(4)}(\zeta_i)(s-\bar{x}_i)^4}{(x-s)^{\alpha-1}} ds =: \sum_{k=0}^{3} J_k(x) + J_4(x), \tag{35}$$

where $\zeta_i \in (x_{i-1}, x_i)$. Now from the Mean Value Theorem for integrals for $k = 0, 1, 2, 3$

$$J_k(x) = \frac{\tau^{(k)}(\bar{x}_i)}{k!}(\hat{\zeta}_{i,k} - \bar{x}_i)^k(\alpha - 1)[(x-x_i)^{-\alpha} - (x-x_{i-1})^{-\alpha}], \tag{36}$$

and
$$J_4(x) = \frac{u^{(4)}(\tilde{\zeta}_i)(\hat{\zeta}_{i,4} - \overline{x}_i)^4}{24}(\alpha - 1)[(x - x_i)^{-\alpha} - (x - x_{i-1})^{-\alpha}], \tag{37}$$

where $\tilde{\zeta}_i, \hat{\zeta}_{i,k} \in (x_{i-1}, x_i)$. For the second integral in (34), we have

$$\begin{aligned}
\frac{d^2}{dx^2} \int_{x_{j-1}}^{x} \frac{\tau(s)}{(x-s)^{\alpha-1}} ds &= \tau(\overline{x}_j)(1-\alpha)(x - x_{j-1})^{-\alpha} \\
&\quad + \tau'(\overline{x}_j)\left[-\frac{h}{2}(1-\alpha)(x - x_{j-1})^{-\alpha} + (x - x_{j-1})^{1-\alpha}\right] \\
&\quad + \frac{\tau''(\overline{x}_j)}{2}\left[\frac{h^2}{4}(1-\alpha)(x - x_{j-1})^{-\alpha} - h(x - x_{j-1})^{1-\alpha} \right. \\
&\quad \left. + \frac{2}{2-\alpha}(x - x_{j-1})^{2-\alpha}\right] \\
&\quad + \frac{\tau'''(\overline{x}_j)}{6}\left[-\frac{h^3}{8}(1-\alpha)(x - x_{j-1})^{-\alpha} + \frac{3h^2}{4}(x - x_{j-1})^{1-\alpha}\right. \\
&\quad \left. + \frac{3h}{2-\alpha}(x - x_{j-1})^{2-\alpha} + \frac{6}{(2-\alpha)(3-\alpha)}(x - x_{j-1})^{3-\alpha}\right] \\
&\quad + \frac{u^{(4)}(\tilde{\zeta}_j)(\hat{\zeta}_j - \overline{x}_j)^4}{24}(1-\alpha)(x - x_{j-1})^{-\alpha} =: J(x).
\end{aligned} \tag{38}$$

Let $\sigma h = x - x_{j-1}$. Note that $\sigma \in (0,1)$, it is easy to know that

$$\sigma^{-\alpha} > \sigma^{-\alpha} - (\sigma+1)^{-\alpha} > (\sigma+1)^{-\alpha} - (\sigma+2)^{-\alpha} > \ldots \tag{39}$$

Hence, by Lemma 6, for $k = 0, 1, 2, 3$ and every $i < j$ we have

$$|J_k(x)| \leq \frac{|\tau^{(k)}(\overline{x}_i)|}{k!} h^{k-\alpha}(\alpha - 1)\left[\sigma^{-\alpha} - (\sigma+1)^{-\alpha}\right] \leq Ch^{4-\alpha}\|u^{(4)}\|_\infty, \tag{40}$$

and

$$|J_4(x)| \leq \frac{|u^{(k)}(\tilde{\zeta}_i)|}{4!} h^{4-\alpha}(\alpha - 1)\left[\sigma^{-\alpha} - (\sigma+1)^{-\alpha}\right] \leq Ch^{4-\alpha}\|u^{(4)}\|_\infty, \tag{41}$$

and

$$\begin{aligned}
|J(x)| &\leq |\tau(\overline{x}_j)h^{-\alpha}[(1-\alpha)\sigma^{-\alpha}]| \\
&\quad + \left|\tau'(\overline{x}_j)h^{1-\alpha}\left[-\frac{(1-\alpha)}{2}\sigma^{-\alpha} + \sigma^{1-\alpha}\right]\right| \\
&\quad + \left|\tau''(\overline{x}_j)h^{2-\alpha}\left[\frac{1-\alpha}{8}\sigma^{-\alpha} - \frac{1}{2}\sigma^{1-\alpha} + \frac{1}{2-\alpha}\sigma^{2-\alpha}\right]\right| \\
&\quad + \left|\tau'''(\overline{x}_j)h^{3-\alpha}\left[\frac{\alpha-1}{48}\sigma^{-\alpha} + \frac{\sigma^{1-\alpha}}{8} + \frac{\sigma^{2-\alpha}}{2(2-\alpha)} + \frac{\sigma^{3-\alpha}}{(2-\alpha)(3-\alpha)}\right]\right| \\
&\quad + \left|(\hat{\zeta}_j - \overline{x}_j)^4 h^{-\alpha}\left[\frac{u^{(4)}(\tilde{\zeta}_j)}{24}(1-\alpha)\sigma^{-\alpha}\right]\right| \\
&\leq Ch^{4-\alpha}\|u^{(4)}\|_\infty.
\end{aligned} \tag{42}$$

Now collecting the inequalities (40)–(42) gives the following result for the case of $p = 4$.

Theorem 1. *If $u(x) \in C^p(\Lambda)$ and $p \geq 1$ an integer number, then it holds the error estimate:*

$$\|D^\alpha(\Pi_N u - u)\|_\infty \leq Ch^{\min\{p, 4-\alpha\}}\|u^{(p)}\|_\infty, \tag{43}$$

where C independent on N.

Proof of Theorem 1. When $p = 1$, the Taylor's theorem gives

$$\frac{d^2}{dx^2}\int_{x_{i-1}}^{x_i}\frac{\tau(s)}{(x-s)^{\alpha-1}}dx = [\tau(\overline{x}_i)+\tau'(\tilde{\zeta}_i)(\hat{\zeta}_i-\overline{x}_i)](\alpha-1)[(x-x_i)^{-\alpha}-(x-x_{i-1})^{-\alpha}],$$

and

$$\frac{d^2}{dx^2}\int_{x_{j-1}}^{x}\frac{\tau(s)}{(x-s)^{\alpha-1}}dx = [\tau(\overline{x}_j)+\tau'(\tilde{\zeta}_j)(\hat{\zeta}_j-\overline{x}_j)](\alpha-1)[(x-x_{j-1})^{-\alpha}].$$

So, by (33), the estimate (43) follows. For the cases $p=2$ and $p=3$, the estimate (43) can be obtained by a similar argument. □

Remark 2. For a real number $p \in (0,1)$, the numerical tests show that the estimate (43) also holds.

5. Applications to Fractional Differential Equations

In this Section, some numerical examples are presented to demonstrate the efficiency of our approximation method. The following three types of meshes are used in numerical tests:

- Uniform mesh (Mesh 1):

$$x_j = x_L + \frac{(x_R-x_L)j}{N}, \quad j=0,1,...,N.$$

- Graded mesh (Mesh 2):

$$x_j = x_L + (x_R - x_L)\left(\frac{j}{N}\right)^q, \quad q>1, j=0,1,...,N.$$

Note: For the two-sided operator, two-sided graded mesh will be used with an even number N:

$$x_j = x_L + \frac{(x_R-x_L)}{2}\left(\frac{j}{N_h}\right)^{q_1}, \quad j=0,1,...,N_h,$$

$$x_j = x_R - \frac{(x_R-x_L)}{2}\left(\frac{N-j}{N_h}\right)^{q_2}, \quad j=N_h+1, N_h+2,...N,$$

where $N_h = \frac{N}{2}$ and when $q = q_1 = q_2$, the two-sided mesh is symmetric.
- Geometric mesh (Mesh 3):

$$x_0 = x_L, x_j = x_L + (x_R-x_L)*q^{N-j}, 0<q<1, j=1,2,...,N.$$

5.1. Fractional Helmholtz Equations

To measure the accuracy of the HCSCM when the exact solution is known, we define the errors by

$$E_0 = \max_{x \in \{x_1, x_2,...,x_{N-1}\}}\{|u_N(x)-u(x)|\},$$

where $u_N(x)$ and $u(x)$ are numerical and exact solution respectively. Let $\Lambda := (x_L, x_R)$ and $1 < \alpha(x) < 2$. In this subsection we apply the HCSCM to the following variable-order fractional Helmholtz equation with homogeneous boundary conditions

$$\lambda^2 u(x) - \mathcal{D}^{\alpha(x)}u(x) = f(x), \quad x \in \Lambda, \quad u(x_L) = u(x_R) = 0. \tag{44}$$

The HCSCM for (44) is to find $u_N \in \mathbb{V}_N$, such that

$$\lambda^2 u_N(x) - \mathcal{D}^{\alpha(x)}u_N(x) = f(x), \quad x \in \mathbf{x}^c, \quad u_N(x_L) = u_N(x_R) = 0. \tag{45}$$

The above equation leads to the following linear system:

$$\left(\lambda^2 \mathbf{M} - \mathbf{D}^\alpha\right)\mathbf{u} = \mathbf{f} \tag{46}$$

where $\mathbf{M} = [\phi_0(\mathbf{x}^c), \varphi_1(\mathbf{x}^c), \phi_1(\mathbf{x}^c), \cdots, \varphi_{N-1}(\mathbf{x}^c), \phi_{N-1}(\mathbf{x}^c), \phi_N(\mathbf{x}^c)]$ is the collocation matrix, $\mathbf{x}^c$ and $\mathbf{u}$ as in (26), $\mathbf{f} = f(\mathbf{x}^c)$ and $\mathbf{D}^\alpha$ is the fractional differentiation matrix with respect to the fractional operator $\mathcal{D}^{\alpha(x)}$ as in (27).

Example 1. *Our first test of HCSCM is to consider the problem (44) with exact solution $u(x) = \sin(\pi x)$ at $[x_L, x_R] = [-1, 1]$.*

The right-hand side function $f(x) = \lambda^2 u(x) - \mathcal{D}^{\alpha(x)} u(x)$ in which the fractional derivative term is approximated by when $\mathcal{D}^{\alpha(x)} = {}_{RL}D^{\alpha(x)}_{-1,x}$

$${}_{RL}D^{\alpha(x)}_{-1,x} \sin(\pi x) = \sum_{k=0}^{L}(-1)^{k+1}\frac{\pi^{2k+1}(x+1)^{2k+1-\alpha(x)}}{\Gamma(2k+2-\alpha(x))},$$

and when $\mathcal{D}^{\alpha(x)} = {}_{RL}D^{\alpha(x)}_{x,1}$

$${}_{RL}D^{\alpha(x)}_{x,1} \sin(\pi x) = \sum_{k=0}^{L}(-1)^{k}\frac{\pi^{2k+1}(1-x)^{2k+1-\alpha(x)}}{\Gamma(2k+2-\alpha(x))},$$

with $L = 50$, respectively.
For $\alpha(x)$, we consider the following two cases:
1. The constant-order $\alpha = 1.1, 1.2, 1.4, 1.5, 1.6, 1.8, 1.9$.
2. The variable-order $\alpha(x) = 1.1 + \frac{x+1}{2.5}$.

The aim of this example is to test the accuracy of the proposed method for the smooth solution. In this example, the uniform mesh is used. The error E_0 and the orders of convergence are listed in Table 1. It is shown that the order of convergence of the approximation is $4 - \alpha$.

Table 1. Error E_0 and the order of convergence (OC), for Example 1 with Mesh 1: $\alpha(x) = 1.2, 1.4, 1.6, 1.8, (\sigma_1, \sigma_2) = (0.2, 0.8), \lambda = 0$.

N	$\alpha(x) = 1.2$	OC	$\alpha(x) = 1.4$	OC	$\alpha(x) = 1.6$	OC	$\alpha(x) = 1.8$	OC
20	1.1797×10^{-4}	-	1.1326×10^{-4}	-	1.8116×10^{-4}	-	2.9010×10^{-4}	-
40	1.6776×10^{-5}	2.81	1.7641×10^{-5}	2.68	3.3277×10^{-5}	2.45	6.2240×10^{-5}	2.22
80	2.4257×10^{-6}	2.79	2.8890×10^{-6}	2.61	6.2306×10^{-6}	2.42	1.3406×10^{-5}	2.22
120	7.8179×10^{-7}	2.79	1.0058×10^{-6}	2.60	2.3442×10^{-6}	2.41	5.4714×10^{-6}	2.21
160	3.5026×10^{-7}	2.79	4.7588×10^{-7}	2.60	1.1740×10^{-6}	2.40	2.8984×10^{-6}	2.21
200	1.8588×10^{-7}	2.84	2.6617×10^{-7}	2.60	6.8516×10^{-7}	2.41	1.7712×10^{-6}	2.21
240	1.1199×10^{-7}	2.78	1.6566×10^{-7}	2.60	4.4209×10^{-7}	2.40	1.1876×10^{-6}	2.19

The error E_0 and CPU time for $\alpha = 1.4$ are listed in Table 2. Similar results can be obtained for other cases. All the computations are performed by Matlab R2020a on pc with AMD PRO A10-8770 R7, 10 COMPUTE CORES 4C+6G 3.50GHZ. The Matlab route *inv* is used to solve the linear system (46) in our numerical tests. Other faster solver such as LU decomposition, iteration-type methods and so forth might be used to improve the efficiency.

Table 2. Error E_0 and the CPU time for Example 1 with Mesh 1: $\alpha = 1.4, (\sigma_1, \sigma_2) = (0.2, 0.8), \lambda = 0$.

N	E_0	CPU Time (s)
10	3.6097×10^{-3}	0.018
50	8.8401×10^{-5}	0.165
100	1.5063×10^{-5}	0.479
150	5.2966×10^{-6}	1.046
200	2.5185×10^{-6}	1.831
250	1.4119×10^{-6}	2.721
300	8.8569×10^{-7}	3.397
500	2.3094×10^{-7}	7.363
1000	1.0710×10^{-7}	23.029

The error E_0 is given in Figures 1 and 2. In Figure 1, it is clearly shown that the orders of convergence of approximation is about $4 - \alpha$ which confirms the estimate in Theorem 1. In Figure 2, the orders of convergence of approximation is about

$$4 - \max_{-1 \leq x \leq 1} \alpha(x)$$

for the variable-order case.

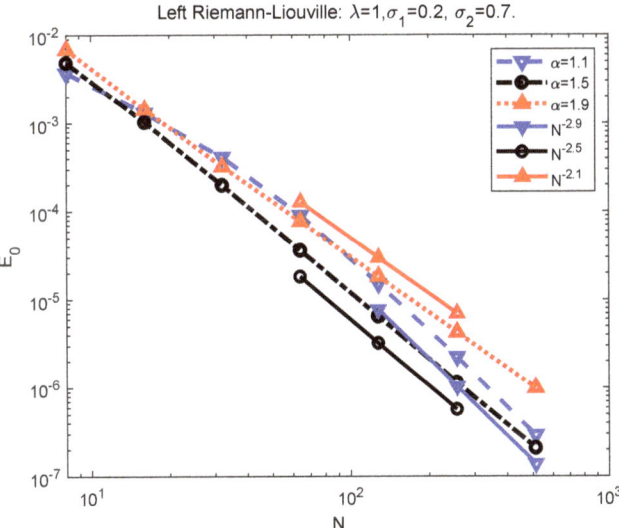

Figure 1. Error for Example 1 with Mesh 1: $\alpha(x) = 1.1, 1.5, 1.9$.

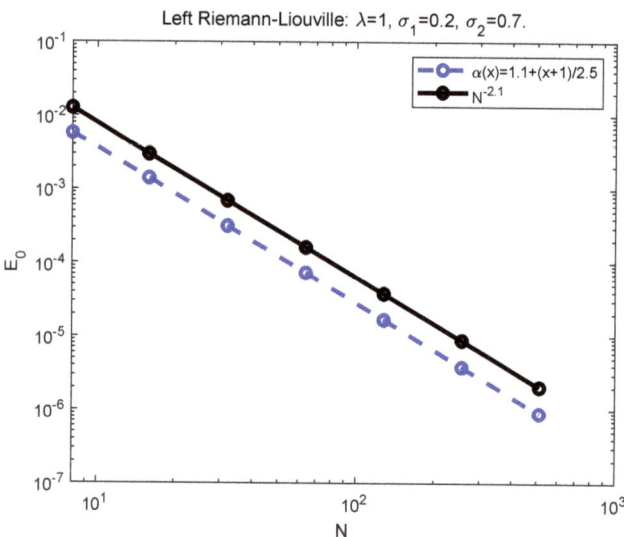

Figure 2. Error for Example 1 with Mesh 1: $\alpha(x) = 1.1 + \frac{x+1}{2.5}$.

Example 2. *Our second test of HCSCM is to consider the problem (44) with an exact solution that has low regularity.*

When we take $u(x) = (1-x)(1+x)^{\alpha(x)}$ with $x_L = -1, x_R = 1$, the right-hand functions are same for left Riemann–Liouville, left Caputo and left Patie-Simon cases, that is,

$$f(x) = \lambda^2 u(x) - 2\Gamma(1+\alpha(x)) + \Gamma(2+\alpha(x))(1+x).$$

In fact, here we have $u(x) \in C^\alpha(\Lambda)$.

The error E_0 and the orders of convergence are listed in Table 3. It is shown that the order of convergence of the approximation is α.

Table 3. Error E_0 and the order of convergence (OC) for Example 2 with Mesh 1: $\alpha(x) = 1.2, 1.4, 1.6, 1.8, (\sigma_1, \sigma_2) = (0.2, 0.8), \lambda = 0$.

N	$\alpha(x) = 1.2$	OC	$\alpha(x) = 1.4$	OC	$\alpha(x) = 1.6$	OC	$\alpha(x) = 1.8$	OC
20	3.6965×10^{-3}	-	3.5347×10^{-3}	-	1.9174×10^{-3}	-	5.9525×10^{-4}	-
40	1.6497×10^{-3}	1.16	1.3360×10^{-3}	1.40	6.3033×10^{-4}	1.60	1.6984×10^{-4}	1.81
80	7.2079×10^{-4}	1.19	5.0527×10^{-4}	1.40	2.0733×10^{-4}	1.60	4.8498×10^{-5}	1.81
120	4.4317×10^{-4}	1.20	2.8620×10^{-4}	1.40	1.0824×10^{-4}	1.60	2.3318×10^{-5}	1.81
160	3.1379×10^{-4}	1.20	1.9124×10^{-4}	1.40	6.8268×10^{-5}	1.60	1.3874×10^{-5}	1.80
200	2.4007×10^{-4}	1.20	1.3990×10^{-4}	1.40	4.7751×10^{-5}	1.60	9.2759×10^{-6}	1.80
240	1.9289×10^{-4}	1.20	1.0836×10^{-4}	1.40	3.5659×10^{-5}	1.60	6.6766×10^{-6}	1.80

The error E_0 for uniform mesh and for $\alpha = 1.1, 1.5, 1.9$ are shown in Figure 3. It is clear that the order of convergence of E_0 is α.

The error E_0 for $\alpha = 1.2$ with three types of mesh are shown in Figure 4. It is shown that the uniform mesh achieves an α order of convergence of E_0 and the graded mesh improves significantly the order of convergence. We can also observe that the geometric mesh might achieve "higher accuracy" (see the dotted line with squares Figure 3), although the precisions are damaged for large N. The errors E_0 for the Caputo case and Patie-Simon case are plotted in Figures 5 and 6.

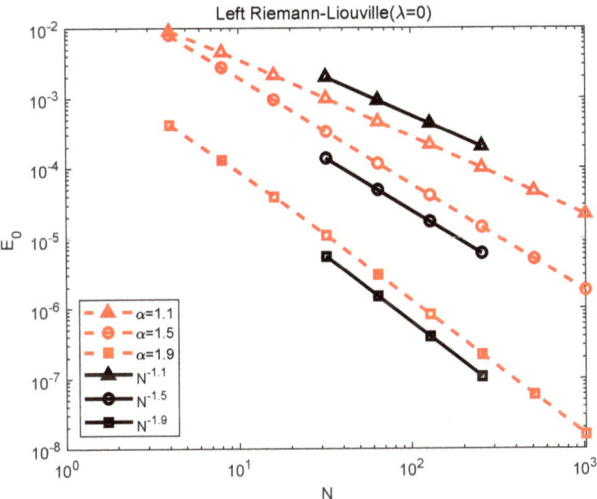

Figure 3. Error for Example 2 with exact solution: $u(x) = (1-x)(1+x)^{\alpha(x)}$ and $(\sigma_1, \sigma_2) = (\frac{\sqrt{3}}{3}, 1 - \frac{\sqrt{3}}{3})$. Mesh 1 for $\alpha = 1.1, 1.5, 1.9$.

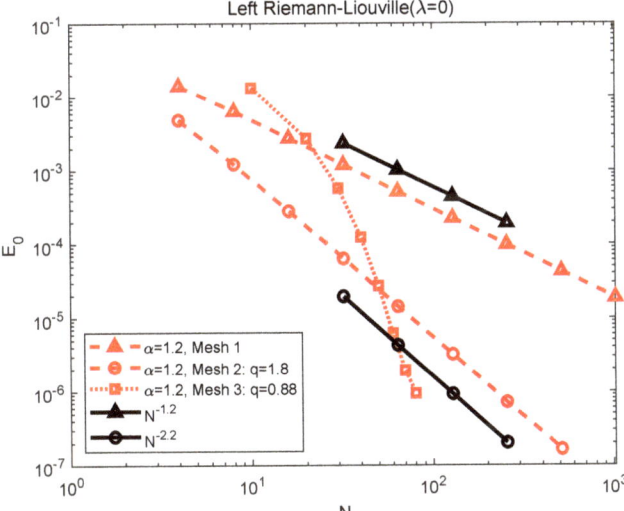

Figure 4. Error for Example 2 with exact solution: $u(x) = (1-x)(1+x)^{\alpha(x)}$ and $(\sigma_1, \sigma_2) = (\frac{\sqrt{3}}{3}, 1 - \frac{\sqrt{3}}{3})$.

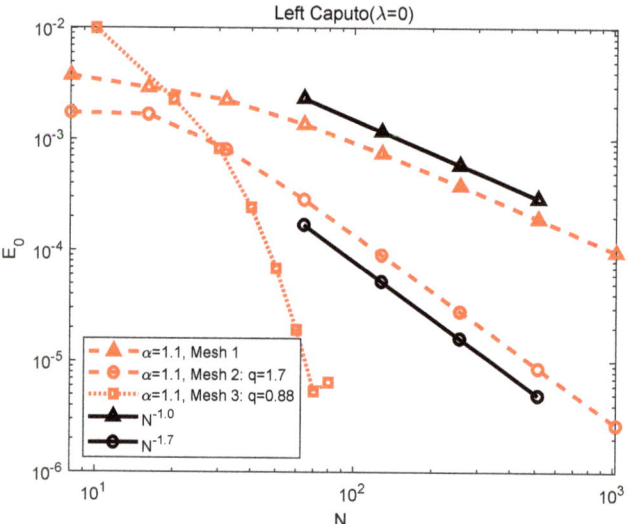

Figure 5. Error for Example 2 with exact solution: $u(x) = (1-x)(1+x)^{\alpha(x)}$ and $(\sigma_1, \sigma_2) = (\frac{\sqrt{3}}{3}, 1 - \frac{\sqrt{3}}{3})$ of left Caputo case ($\alpha = 1.1$).

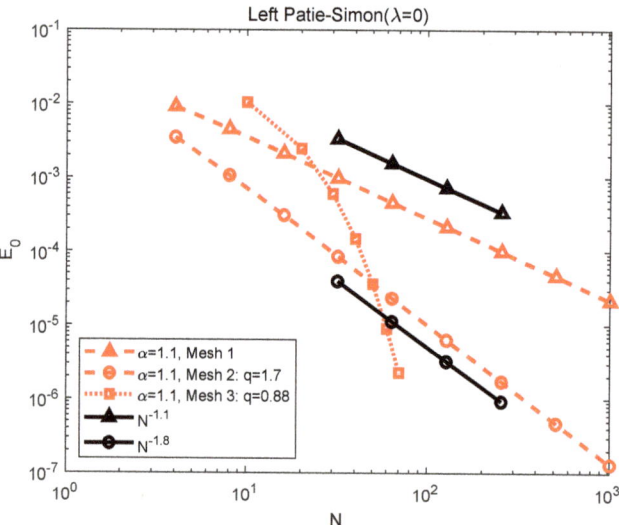

Figure 6. Error for Example 2 with exact solution: $u(x) = (1-x)(1+x)^{\alpha(x)}$ and $(\sigma_1, \sigma_2) = (\frac{\sqrt{3}}{3}, 1 - \frac{\sqrt{3}}{3})$ of left Patie-Simon($\alpha = 1.1$).

If we take $u(x) = (1-x)(1+x)^{\alpha(x)-1}$, it means that $u(x) \in C^{\alpha-1}(\Lambda)$, which has very low regularity with $x_L = -1, x_R = 1$ and then the right-hand function

$$f(x) = \lambda^2 u(x) + \Gamma(1 + \alpha(x))$$

for left Riemann-Liouville and left Patie-Simon cases (but not for left Caputo case).

The error E_0 and the orders of convergence are listed in Table 4. It is shown that the order of convergence of the approximation is $\alpha - 1$.

Table 4. Error E_0 and the order of convergence (OC) for Example 2 with Mesh 1: $\alpha(x) = 1.2, 1.4, 1.6, 1.8, (\sigma_1, \sigma_2) = (0.2, 0.8), \lambda = 0$.

N	$\alpha(x) = 1.2$	OC	$\alpha(x) = 1.4$	OC	$\alpha(x) = 1.6$	OC	$\alpha(x) = 1.8$	OC
20	$1.1624 \times 10^{+0}$	-	4.0781×10^{-1}	-	1.0764×10^{-1}	-	1.8809×10^{-2}	-
40	$1.0453 \times 10^{+0}$	0.15	3.1153×10^{-1}	0.39	7.1834×10^{-2}	0.58	1.0960×10^{-2}	0.78
80	9.1739×10^{-1}	0.19	2.3697×10^{-1}	0.39	4.7653×10^{-2}	0.59	6.3385×10^{-3}	0.79
120	8.4738×10^{-1}	0.20	2.0173×10^{-1}	0.40	3.7430×10^{-2}	0.60	4.5930×10^{-3}	0.79
160	8.0061×10^{-1}	0.20	1.7991×10^{-1}	0.40	3.1524×10^{-2}	0.60	3.6528×10^{-3}	0.80
200	7.6601×10^{-1}	0.20	1.6460×10^{-1}	0.40	2.7588×10^{-2}	0.60	3.0577×10^{-3}	0.80
240	7.3879×10^{-1}	0.20	1.5306×10^{-1}	0.40	2.4738×10^{-2}	0.60	2.6439×10^{-3}	0.80

The errors E_0 are plotted in Figures 7–10. Compared the Figure 4 with the Figure 7, we can find that the orders of convergence of E_0 are dropped to $\alpha - 1$ for the exact solution that belongs to $C^{\alpha-1}(\Lambda)$, which agree with the results in Theorem 1. It is also observed that the orders of convergence of E_0 are improved by making use of the graded mesh and the geometric mesh similarly.

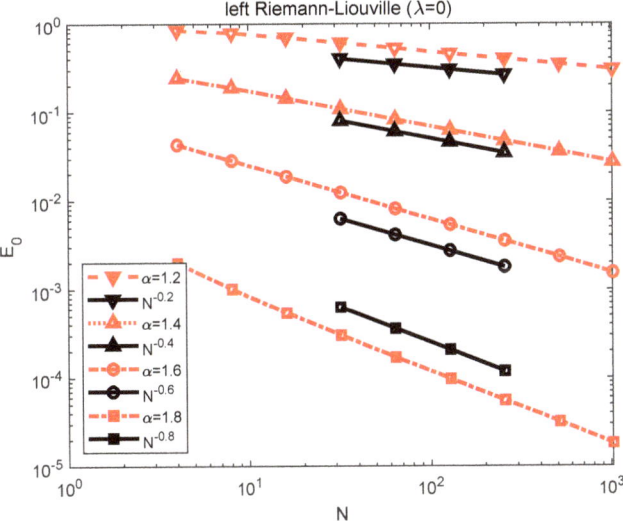

Figure 7. Error for Example 2 with exact solution $u(x) = (1-x)(1+x)^{\alpha(x)-1}$ and $(\sigma_1, \sigma_2) = (\frac{\sqrt{3}}{3}, 1 - \frac{\sqrt{3}}{3})$ by the uniform mesh.

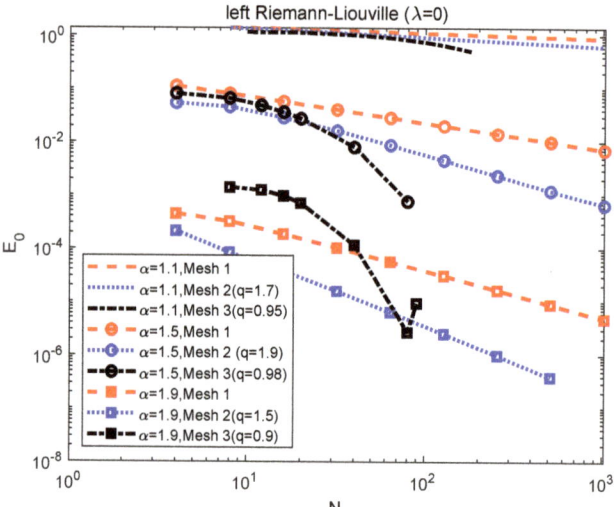

Figure 8. Error for Example 2 with exact solution: $u(x) = (1-x)(1+x)^{\alpha(x)-1}$ and $(\sigma_1, \sigma_2) = (\frac{\sqrt{3}}{3}, 1 - \frac{\sqrt{3}}{3})$.

The HCSCM is comparable to the MDSCM [49] since both of them are applied piecewise polynomial to approximation problems. The numerical errors are compared by using the HCSCM and the MDSCM. The maximum errors with the degree of freedom are plotted for the constant-order $\alpha = 1.1, 1.5$ and 1.9 and for the variable-order $\alpha(x) = 1.1 + (x+1)/2.5$ in Figures 9 and 10. The black lines are for the MDSCM with fixed $N = 3$ and the penalty parameter $\tau = 100,000$. By the choice of $N = 3$ of the MDSCM, the degree of piecewise polynomial in the HCSCM is the same as ones in the MDSCM. Both the uniform meshes are applied for two methods. It is shown that the accuracy of the HCSCM is better than those of the MDSCM [49] with h-refinement but the orders of convergence are almost same.

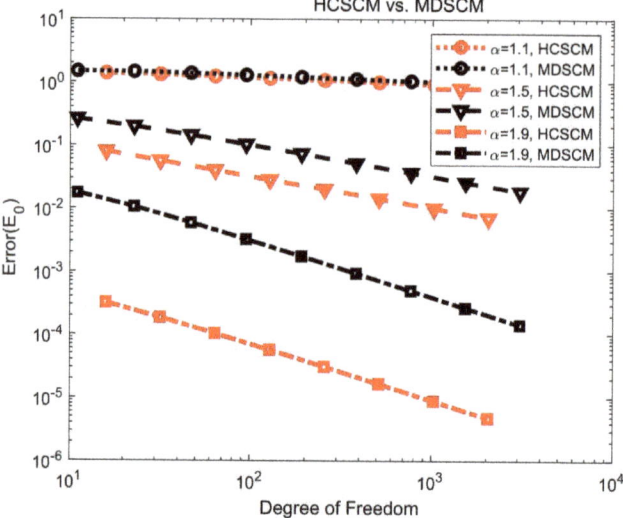

Figure 9. Error for Example 2 with exact solution $u(x) = (1-x)(1+x)^{\alpha(x)-1}$ and $\lambda = 0, (\sigma_1, \sigma_2) = (\frac{\sqrt{3}}{3}, 1 - \frac{\sqrt{3}}{3})$ of left Riemann-Liouville with constant-order case.

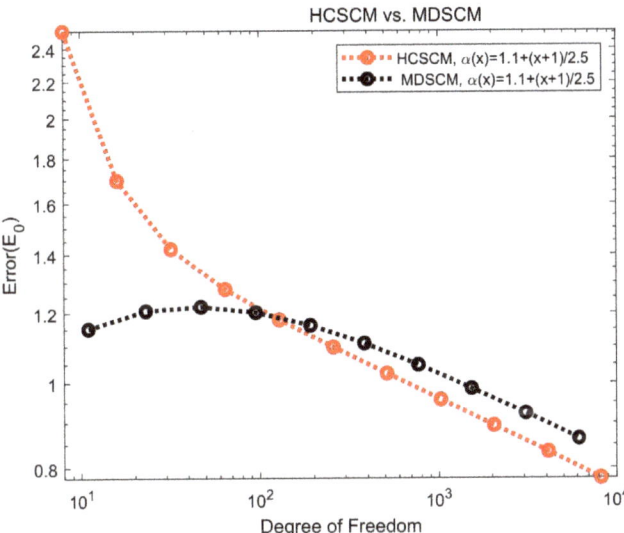

Figure 10. Error for Example 2 with exact solution $u(x) = (1-x)(1+x)^{\alpha(x)-1}$ and $\lambda = 0, (\sigma_1, \sigma_2) = (\frac{\sqrt{3}}{3}, 1 - \frac{\sqrt{3}}{3})$ of left Riemann-Liouville with variable-order case.

5.2. Fractional Burgers Equations

In this subsection, we try to solve the fractional Burgers equation as

$$\partial_t u(x,t) + u(x,t)\partial_x u(x,t) = \epsilon \mathbb{D}_r^{\alpha(x,t)} u(x,t), \quad (47)$$

subject to homogeneous Dirichlet boundary condition and initial condition $u(x,0) = u_0(x)$, where $\epsilon > 0, 1 < \alpha(x,t) < 2$ and $(x,t) \in (-1,1) \times (0,1]$.

The two-step Crank-Nicolson/leapfrog scheme is first employed for time stepping, then the HCSCM is applied to the resulting equations. Thus, the full discretization scheme reads as: for $k = 1, 2, \ldots,$

$$\begin{cases} \left(\mathbf{M} - \Delta t \epsilon \mathbf{D}^{\alpha^{k+1}}\right) \mathbf{u}^{k+1} = \mathbf{g}, \\ \mathbf{M}\mathbf{u}^1 = \left(\mathbf{M} + \Delta t \epsilon \mathbf{D}^{\alpha^0}\right) \mathbf{u}^0 - \Delta t (\mathbf{M}\mathbf{u}^0) . * (\mathbf{S}\mathbf{u}^0) \\ \mathbf{M}\mathbf{u}^0 = u_0(\mathbf{x}), \end{cases} \quad (48)$$

where

$$\mathbf{g} = \left(\mathbf{M} + \Delta t \epsilon \mathbf{D}^{\alpha^{k-1}}\right) \mathbf{u}^{k-1} - 2\Delta t \left(\mathbf{M}\mathbf{u}^k\right) . * \left(\mathbf{S}\mathbf{u}^k\right),$$

and Δt is the time stepsize, $\mathbf{M}$ the collocation matrix, $\mathbf{D}^{\alpha^k}$ the fractional differentiation matrix of order $\alpha^k = \alpha(x, k\Delta t)$ with respect to the fractional operator $\mathcal{D}^{\alpha(x,t)} = \mathbb{D}_r^{\alpha(x,t)}$ as in (25), $\mathbf{S}$ the collocation first-order differentiation matrix which defines as

$$\mathbf{S} = [\phi_0'(\mathbf{x}^c), \phi_1'(\mathbf{x}^c), \phi_1'(\mathbf{x}^c), \cdots, \phi_{N-1}'(\mathbf{x}^c), \phi_{N-1}'(\mathbf{x}^c), \phi_N'(\mathbf{x}^c)],$$

and notation .* the entry-to-entry multiplication.

Example 3. *In this example, we consider the fractional Burgers Equation (47) with the initial condition $u_0(x) = \sin(\pi x)$.*

Our first test is the numerical solutions to the Equation (47) of the left Riemann-Liouville fractional derivative. The following five cases of fractional order are considered in [40,49]:

- Case 1:(constant-order) $\alpha(x,t) = 1.1, 1.2, 1.3, 1.5, 1.8$;
- Case 2:(monotonic increasing-order) $\alpha(x,t) = 1 + \frac{5+4x}{10}$;
- Case 3:(monotonic decreasing-order) $\alpha(x,t) = 1 + \frac{5-4x}{10}$;
- Case 4:(nonsmooth order) $\alpha(x,t) = 1.1 + \frac{4}{5}|\sin(10\pi(x-t))|$;
- Case 5:(nonsmooth order) $\alpha(x,t) = 1.1 + \frac{4|xt|}{5}$.

In Figure 11, the numerical solutions at $t = 1$ is plotted for constant-order cases (Case 1). The obtained numerical result is the same as the one in Fig4.6 ([49]). We also compare some results by the multi-domain spectral collocation method(MDSCM) with those by the presented method(HCSCM) for $\alpha = 1.1, 1.5$ in Figures 12 and 13. It is shown that by the HCSCM one get smoother numerical solution near the left boundary $x = -1$ than that by the MDSCM.

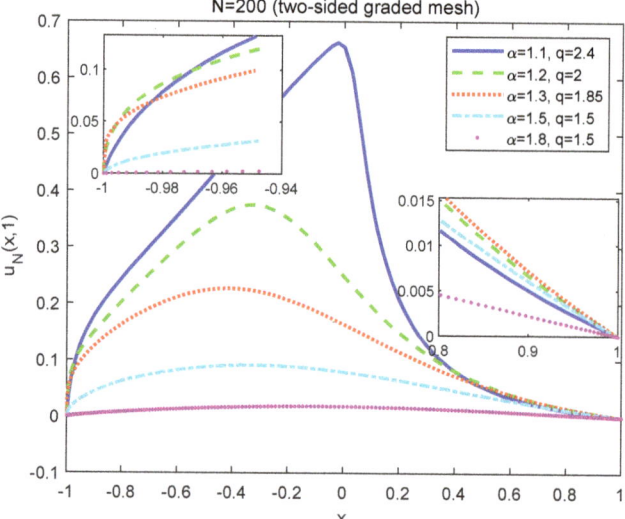

Figure 11. Numerical solutions at $t = 1$ for Example 3 (Case 1) with the graded mesh (Mesh 2). $\epsilon = 1, N = 200, \Delta t = 10^{-3}$ and $(\sigma_1, \sigma_2) = (0.09, 0.88)$ for $\alpha = 1.1, 1.3$, $(\sigma_1, \sigma_2) = (0.09, 0.85)$ for $\alpha = 1.2$, $(\sigma_1, \sigma_2) = (0.3, 0.7)$ for $\alpha = 1.5, 1.8$.

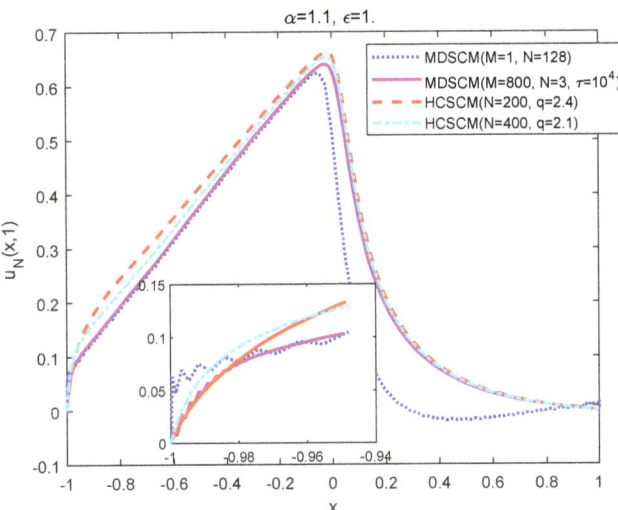

Figure 12. Numerical solutions at $t=1$ for Example 3 with left Riemann-Liouville case($\alpha=1.1, r=1$): MDSCM vs HCSCM. $(\sigma_1,\sigma_2)=(0.09,0.88)$ for $N=200,400$ and the two-sided graded mesh (Mesh 2) are used for all cases in HCSCM.

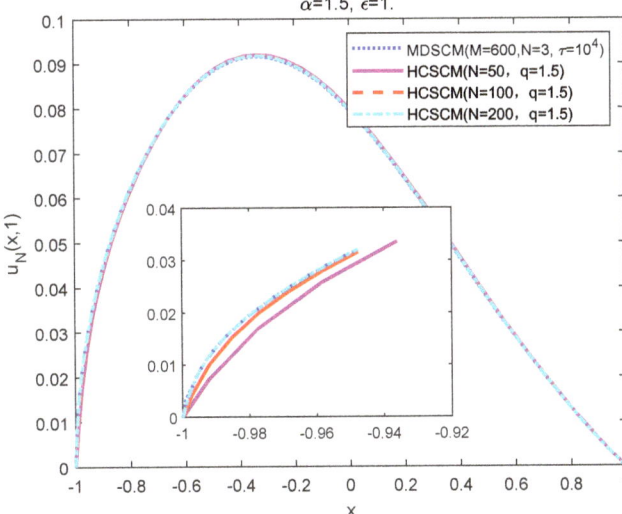

Figure 13. Numerical solutions at $t=1$ for Example 3 with left Riemann-Liouville case($r=1, \alpha=1.5$): MDSCM vs HCSCM. $(\sigma_1,\sigma_2)=(0.2,0.7)$ for $N=50,100,200$ and the two-sided graded mesh (Mesh 2) are used for all cases in HCSCM.

The numerical solutions at $t=1$ for variable-order cases(Case 2–5) are plotted in Figures 14–17, which are agree with the results in [40,49].

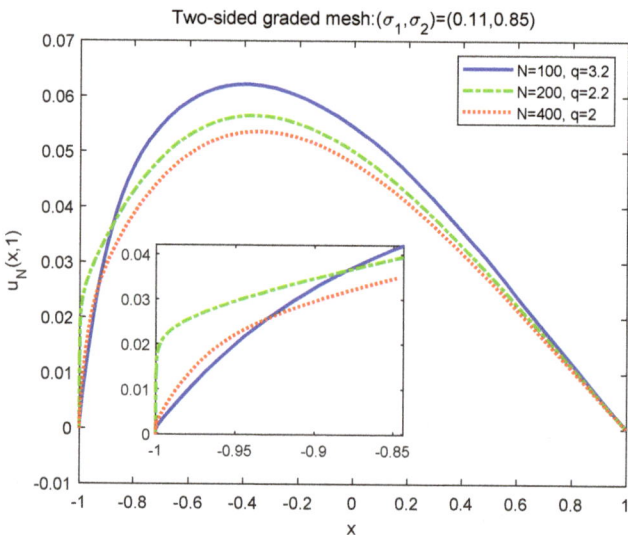

Figure 14. Numerical solutions at $t = 1$ for Example 3 with left Riemann–Liouville case ($r = 1$) and $\epsilon = 1$: $\alpha(x) = 1 + \frac{5+4x}{10}$.

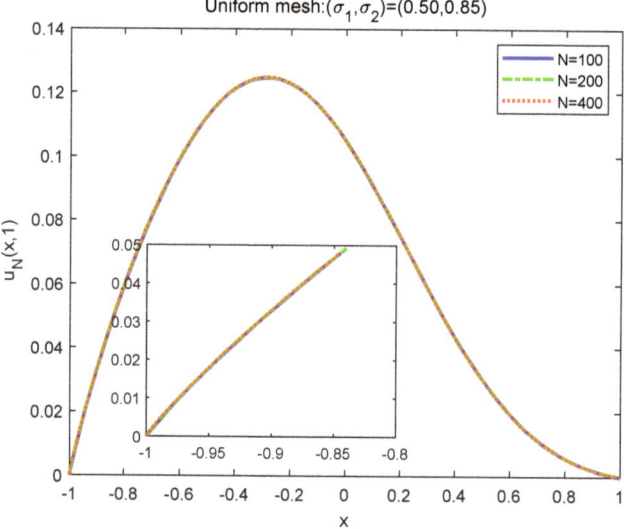

Figure 15. Numerical solutions at $t = 1$ for Example 3 with left Riemann–Liouville case ($r = 1$) and $\epsilon = 1$: $\alpha(x) = 1 + \frac{5-4x}{10}$.

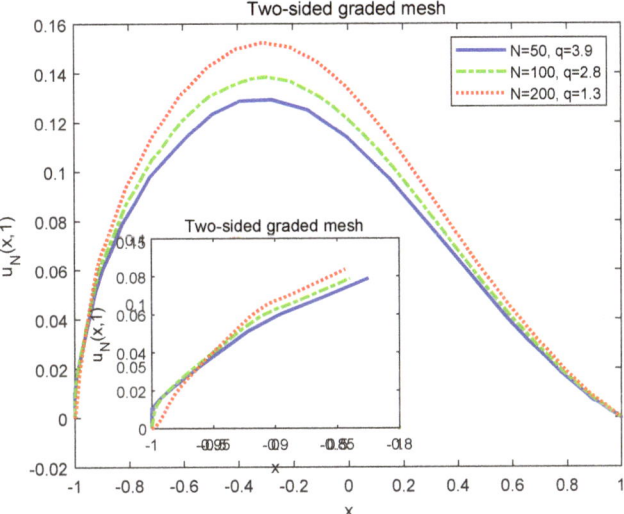

Figure 16. Numerical solutions at $t = 1$ for Example 3 with left Riemann-Liouville case ($\alpha(x,t) = 1.1 + \frac{4}{5}|\sin(10\pi(x-t))|, r = 1$) and $\epsilon = 1$. where $(\sigma_1, \sigma_2) = (0.35, 0.85)$ for $N = 50$, $(\sigma_1, \sigma_2) = (0.28, 0.75)$ for $N = 100$, $(\sigma_1, \sigma_2) = (0.2, 0.75)$ for $N = 200$.

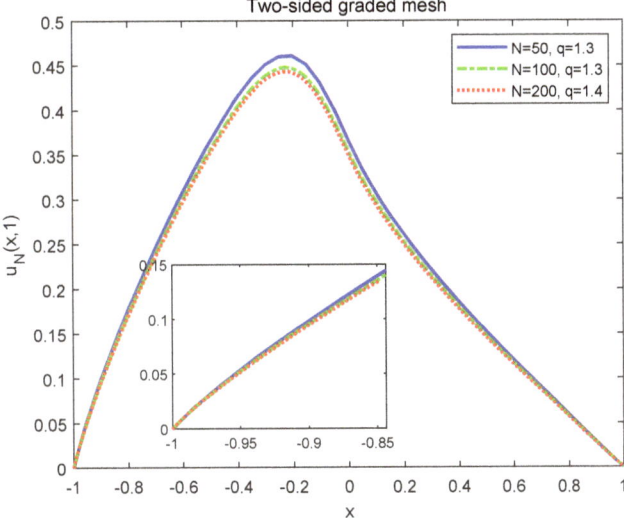

Figure 17. Numerical solutions at $t = 1$ for Example 3 with left Riemann-Liouville case ($\alpha(x,t) = 1.1 + \frac{4|xt|}{5}, r = 1$) and $\epsilon = 1$. where $(\sigma_1, \sigma_2) = (0.2, 0.75)$ for all three Ns.

The numerical solutions are also computed to the fractional Burgers equation with two-sided operators. The numerical solutions at $t = 1$ for various α's and r's are plotted in Figures 18–23.

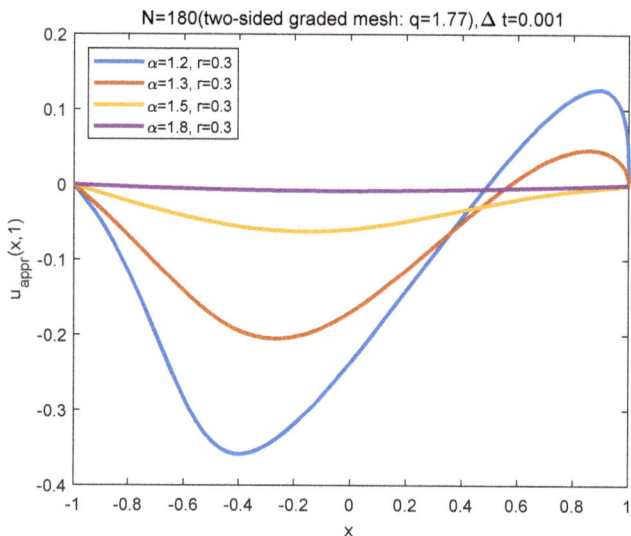

Figure 18. Numerical solutions at $t = 1$ for Example 3 of two-sided Riemann-Liouville case $(r = 0.3)$: $\epsilon = 1$, $(\sigma_1, \sigma_2) = (\frac{\sqrt{3}}{3}, 1 - \frac{\sqrt{3}}{3})$ for all cases.

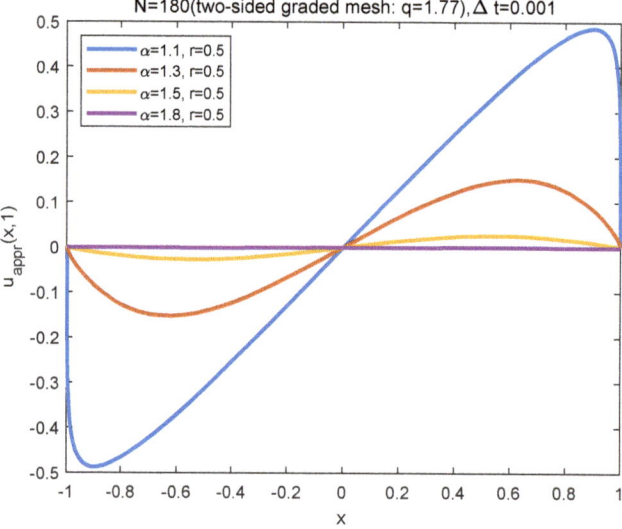

Figure 19. Numerical solutions at $t = 1$ for Example 3 of two-sided Riemann-Liouville case $(r = 0.5)$: $\epsilon = 1$, $(\sigma_1, \sigma_2) = (\frac{\sqrt{3}}{3}, 1 - \frac{\sqrt{3}}{3})$ for all cases.

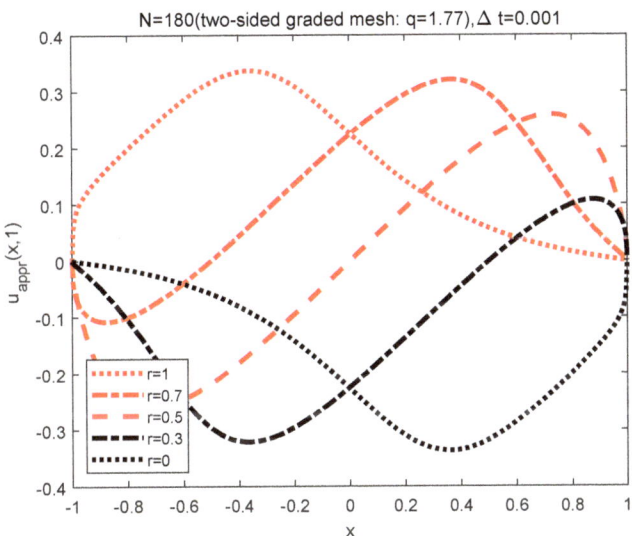

Figure 20. Numerical solutions at $t = 1$ for Example 3 of two-sided Riemann–Liouville case ($\alpha = 1.22$): $\epsilon = 1$, $(\sigma_1, \sigma_2) = (\frac{\sqrt{3}}{3}, 1 - \frac{\sqrt{3}}{3})$ for all cases.

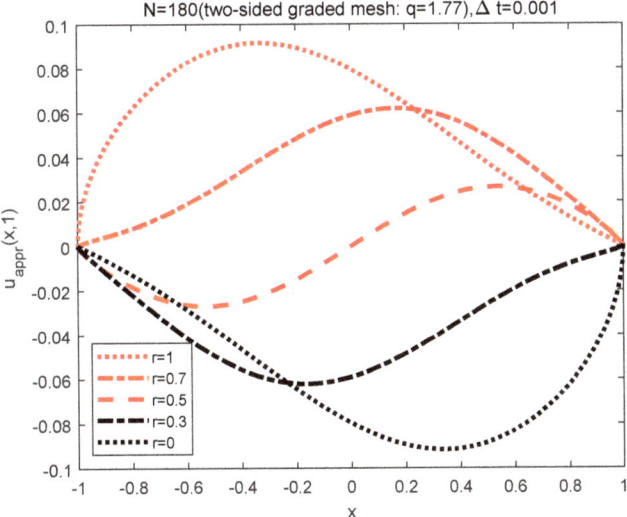

Figure 21. Numerical solutions at $t = 1$ for Example 3 of two-sided Riemann–Liouville case ($\alpha = 1.5$): $\epsilon = 1$, $(\sigma_1, \sigma_2) = (\frac{\sqrt{3}}{3}, 1 - \frac{\sqrt{3}}{3})$ for all cases.

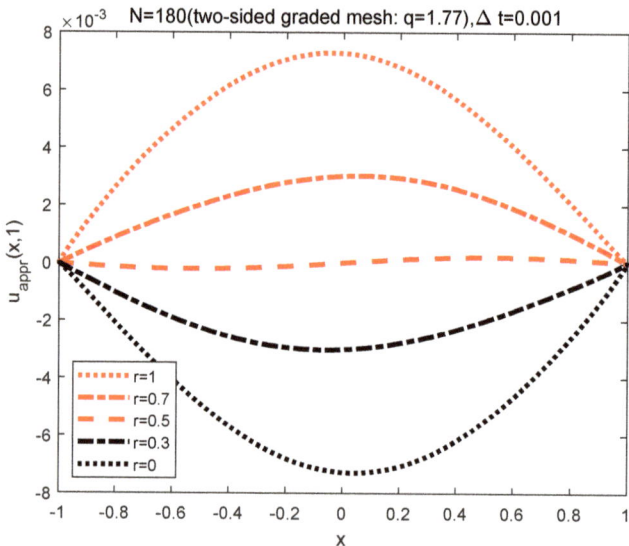

Figure 22. Numerical solutions at $t = 1$ for Example 3 of two-sided Riemann-Liouville case ($\alpha = 1.9$): $\epsilon = 1$, $(\sigma_1, \sigma_2) = (\frac{\sqrt{3}}{3}, 1 - \frac{\sqrt{3}}{3})$ for all cases.

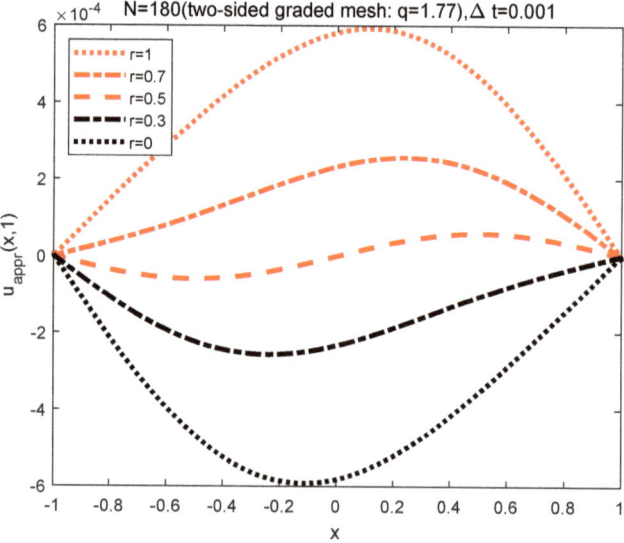

Figure 23. Numerical solutions at $t = 1$ for Example 3 of two-sided Riemann-Liouville case ($\alpha = 1.99$): $\epsilon = 1$, $(\sigma_1, \sigma_2) = (\frac{\sqrt{3}}{3}, 1 - \frac{\sqrt{3}}{3})$ for all cases.

6. Conclusions

In this paper, a Hermite cubic spline collocation method (HCSCM) are developed for solving variable-order nonlinear fractional differential equations, which apply C^1-continuous nodal basis functions to approximate problem. It is verified that the order of convergence of the HCSCM is $O(h^{\min\{4-\alpha,p\}})$, while the interpolating function belongs to $C^p(p \geq 1)$, where h is the mesh-size and α the order of the fractional derivative. The effectiveness of the HCSCM is demonstrated by solving fractional Helmholtz equations of constant-order and variable-order, and solving the fractional Burgers equation. The numerical fractional diffusions are compared with different senses.

The HCSCM can be applied to fractional-order differential equations on a two or three dimensional Descartes product domain by nodal basis tensor. Through adjusting the location of collocation points, the stability of the HCSCM can be observed numerically. Our future work will focus on the stability and error analysis of the HCSCM for some FDEs.

Author Contributions: Methodology, numerical tests and writing, T.Z.; review and editing, Y.W. All authors have read and agreed to the published version of the manuscript.

Funding: This research was funded by National Natural Science Foundation of China under Grant No. 11661048.

Institutional Review Board Statement: Not applicable.

Informed Consent Statement: Not applicable.

Data Availability Statement: Not applicable.

Conflicts of Interest: The authors declare no conflict of interest.

Abbreviations

The following abbreviations are used in this manuscript:

HCSCM	Hermite cubic spline collocation method
FDEs	Fractional differential equations
FDM	Fractional differentiation matrix
MDSCM	Multi-domain spectral collocation method

References

1. Metzler, R.; Klafter, J. The restaurant at the end of the random walk: Recent developments in the description of anomalous transport by fractional dynamics. *J. Phys. A* **2004**, *37*, 161–208. [CrossRef]
2. Metzler, R.; Klafter, J. The random walk's guide to anomalous diffusion: A fractional dynamics approach. *Phys. Rep.* **2000**, *339*, 1–77. [CrossRef]
3. Rossikhin, Y.A.; Shitikova, M.V. Applications of fractional calculus to dynamic problems of linear and nonlinear hereditary mechanics of solids. *Appl. Mech. Rev.* **1997**, *50*, 15–67. [CrossRef]
4. Mainardi, F. Fractional calculus: Some basic problems in continuum and statistical mechanics. In *Fractals and Fractional Calculus in Continuum Mechenics*; Carpinteri, A., Mainardi, F., Eds.; Springer: New York, NY, USA, 1997; pp. 291–348.
5. Noeiaghdam, S.; Sidorov, D. Caputo-Fabrizio fractional derivative to solve the fractional model of energy supply-demand system. *Math. Model. Eng. Probl.* **2020**, *7*, 359–367. [CrossRef]
6. Lin, Y.M.; Xu, C.J. Finite difference/spectral approxiamtions for the time-fractional diffusion equation. *J. Comput. Phys.* **2007**, *225*, 1533–1552. [CrossRef]
7. Liu, F.; Anh, V.; Turner, I. Numerical solution of the space fractional Fokker-Planck equation. *J. Comput. Appl. Math.* **2004**, *166*, 209–219. [CrossRef]
8. Meerschaert, M.M.; Scheffler, H.P.; Tadjeran, C. Finite difference methods for two-dimensional fractional dispersion equation. *J. Comput. Phys.* **2006**, *211*, 249–261. [CrossRef]
9. Meerschaert, M.M.; Tadjeran, C. Finite difference approximations for fractional advection-dispersion flow equations. *J. Comput. Appl. Math.* **2004**, *172*, 65–77. [CrossRef]
10. Sun, Z.Z.; Wu, X.N. A fully discrete difference scheme for a diffusion-wave system. *Appl. Numer. Math.* **2006**, *56*, 193–209. [CrossRef]
11. Tadjeran, C.; Meerschaert, M.M.; Scheffler, H.P. A second-order accurate numerical approximation for the fractional diffusion equation. *J. Comput. Phys.* **2006**, *213*, 205–213. [CrossRef]
12. Tadjeran, C.; Meerschaert, M.M. A second-order accurate numerical method for the two-dimensional fractional diffusion equation. *J. Comput. Phys.* **2007**, *220*, 813–823. [CrossRef]
13. Wang, H.; Basu, T.S. A fast finite difference method for two-dimensional space-fractional diffusion equation. *SIAM J. Sci. Comput.* **2012**, *34*, 2444–2458. [CrossRef]
14. Wang, H.; Wang, K. An $O(N\log^2 N)$ alternating-direction finite difference method for two-dimensional fractional diffusion equation. *J. Comput. Phys.* **2011**, *230*, 7830–7839. [CrossRef]
15. Zhang, Y.; Sun, Z.Z.; Wu, H.W. Error estimates of Crank-Nicolson type difference schemes for the sub-diffusion eqution. *SIAM J. Numer. Anal.* **2011**, *49*, 2302–2322. [CrossRef]

16. Zhou, H.; Tian, W.Y.; Deng, W.H. Quasi-compact finite difference schemes for space fractional diffusion equations. *J. Sci. Comput.* **2013**, *56*, 45–66. [CrossRef]
17. Zhao, X.; Sun, Z.Z.; Hao, Z.P. A fourth-order compact ADI scheme for two-dimensional nonlinear space fractional Schrödinger equation. *SIAM J. Sci. Comput.* **2014**, *36*, A2865–A2886. [CrossRef]
18. Deng, W.H. Finite element method for the space and time fractional Fokker-Planck equation. *SIAM J. Numer. Anal.* **2008**, *47*, 204–226. [CrossRef]
19. Ford, N.; Xiao, J.Y.; Yan, Y.B. A finite element method for time fractional partial differential equations. *Fract. Calc. Appl. Anal.* **2011**, *14*, 454–474. [CrossRef]
20. Jiang, Y.; Ma, J. High-order finite element methods for time-fractional partial differential equations. *J. Comput. Appl. Math.* **2011**, *235*, 3285–3290. [CrossRef]
21. Lian, Y.; Ying, Y.; Tang, S.; Lin, S.; Wagner, G.J.; Liu, W.K. A Petrev-Galerkin finite element method for the fractional advection-diffusion equation. *Comput. Methods Appl. Mech. Engrg.* **2016**, *309*, 388–410. [CrossRef]
22. Wang, H.; Yang, D.P.; Zhu, S.F. A Petrev-Galerkin finite element method for variable-coefficient fractional diffusion equations. *Comput. Methods Appl. Mech. Engrg.* **2015**, *290*, 45–56. [CrossRef]
23. Zheng, Y.Y.; Li, C.P.; Zhao, Z.G. A note on the finite element method for the space-fractional advection diffusion equation. *Comput. Math. Appl.* **2010**, *59*, 1718–1726. [CrossRef]
24. Li, C.P.; Zeng, F.H.; Liu, F. Spectral approximations to the fractional integral and derivative. *Fract. Calc. Appl. Anal.* **2012**, *15*, 383–406. [CrossRef]
25. Li, X.J.; Xu, C.J. A space-time spectral method for the time fractional diffusion equations. *SIAM J. Numer. Anal.* **2009**, *47*, 2018–2131. [CrossRef]
26. Li, X.J.; Xu, C.J. Existence and uniqueness of the weak solution of the space-time fractional diffusion equation and a spectral method approximation. *Commun. Comput. Phys.* **2010**, *8*, 1016–1051.
27. Tian, W.Y.; Deng, W.H.; Wu, Y.J. Polynomial spectral collocation method for space fractional advection-diffusion equation. *Numer. Methods Partial Differ. Eq.* **2014**, *30*, 514–535. [CrossRef]
28. Xu, Q.W.; Hesthaven, J.S. Stable multi-domian spectral penalty methods for fractional partial differential equations. *J. Comput. Phys.* **2014**, *257*, 241–258. [CrossRef]
29. Mao, Z.P.; Chen, S.; Shen, J. Efficient and accurate spectral method using generalized Jacobi functions for solving Riesz fractional differential equations. *Appl. Numer. Math.* **2016**, *106*, 165–181. [CrossRef]
30. Mao, Z.P.; Shen, J. Efficient spectral-Galerkin methods for fractional partial differential equations with variable coefficients. *J. Comput. Phys.* **2016**, *307*, 243–261. [CrossRef]
31. Zeng, F.H.; Liu, F.W.; Li, C.P.; Burrage, K.; Turner, I.; Anh V. A Crank-Nicolson ADI spectral method for a two-dimensional Riesz space fractional nonlinear reaction-diffusion equation. *SIAM J. Numer. Anal.* **2014**, *52*, 2599–2622. [CrossRef]
32. Zayernouri, M.; Karniadakis, G.E. Discontinuous spectral element methods for time- and space-fractional advection equations. *SIAM J. Sci. Comput.* **2014**, *36*, B684–B707. [CrossRef]
33. Kharazmi, E.; Zayernouri, M.; Karniadakis, G.E. A Petrev-Galerkin spectral element method for fractional elliptic problems. *Comput. Methods Appl. Mech. Engrg.* **2017**, *324*, 512–536. [CrossRef]
34. Li, C.P.; Cai, M. *Theory and Numerical Approximations of Fractional Integrals and Derivatives*; SIAM: Philadelphia, PA, USA, 2020.
35. Shen, J.; Tang, T.; Wang, L.L. *Spectral Methods: Algorithms, Analysis and Applications*; Springer: Berlin/Heidelberg, Germany, 2011.
36. Guo, B.Y. *The Spectral Methods and Its Applications*; World Scientific: Singapore, 1998.
37. Bernardi, C.; Maday, Y. Spectral methods. In *Handbook of Numerical Analysis, Vol. V, Techniques of Scientific Computing (Part 2)*; Ciarlet P.G., Lions J.L., Eds.; Elsevier: Amsterdam, The Netherlands, 1997; pp. 209–486.
38. Boyd, J.P. *Chebyshev and Fourier Spectral Methods*, 2nd ed.; Dover Publication: Mineola, NY, USA, 2000.
39. Chen, S.; Shen, J.; Wang, L.L. Generalized Jacobi functions and their applications to fractional differential equations. *Math. Comput.* **2016**, *85*, 1603–1638. [CrossRef]
40. Zeng, F.H.; Zhang, Z.Q.; Karniadakis, G.E. A generalized spectral collocation method with tunable accuracy for variable-order fractional differential equations. *SIAM J. Sci. Comput.* **2015**, *37*, A2710–A2732. [CrossRef]
41. Zeng, F.H.; Mao, Z.P.; Karniadakis, G.E. A generalized spectral collocation method with tunable accuracy for fractional differential equations with end-point singularities. *SIAM J. Sci. Comput.* **2017**, *39*, A360–A383. [CrossRef]
42. Mao, Z.P.; Karniadakis, G.E. A spectral method (of exponential convergence) for singular solutions of the diffusion equation with general two-sided fractional derivative. *SIAM J. Numer. Anal.* **2018**, *56*, 24–49. [CrossRef]
43. Zayernouri, M.; Karniadakis, G.E. Fractional Sturm-Liouville eigenproblems: Theory and numerical approximation. *J. Comput. Phys.* **2013**, *252*, 495–517. [CrossRef]
44. Zayernouri, M.; Karnidakis, G.E. Fractional spectral collocation method. *SIAM J. Sci. Comput.* **2014**, *36*, A40–A62. [CrossRef]
45. Zayernouri, M.; Karnidakis, G.E. Fractional spectral collocation methods for linear and nonlinar variable order FPDEs. *J. Comput. Phys.* **2015**, *293*, 312–338. [CrossRef]
46. Huang, C.; Jiao, Y.J.; Wang, L.L.; Zhang, Z.M. Optimal fractional integration preconditioning and error analysis of fractional collocation method using nodal generalised Jacobi functions. *SIAM J. Numer. Anal.* **2016**, *54*, 3357–3387. [CrossRef]
47. Jiao, Y.J.;Wang, L.L.; Huang, C. Well-conditioned fractional collocation methods using fractional Birkhoff interpolation basis. *J. Comput. Phys.* **2016**, *305*, 1–28. [CrossRef]

48. Zhang, Z.Q.; Zeng, F.H.; Karniadakis, G.E. Optimal error estimates of spectral petrov-Galerkin and collocation methods for initial value problems of fractional differential equations. *SIAM J. Numer. Anal.* **2015**, *53*, 2074–2096. [CrossRef]
49. Zhao, T.G.; Mao, Z.P.; Karniadakis, G.E. Multi-domain spectral collocation method for variable-order nonlinear fractional differential equations. *Comput. Methods Appl. Mech. Engrg.* **2019**, *348*, 377–395. [CrossRef]
50. Bialecki, B.; Fairweather, G. Orthogonal spline collocation methods for partial differential equations. *J. Comput. Appl. Math.* **2001**, *128*, 55–82. [CrossRef]
51. Douglas, J., Jr.; Dupont, T. Collocation methods for parabolic equations in a single space variable. In *Lecture Notes in Mathematics*; Springer: New York, NY, USA, 1974; Volume 385.
52. Greenwell-Yanik, C.E.; Fairweather, G. Analyses of spline collocation methods for parabolic and hyperbolic problems in two space variables. *SIAM J. Numer. Anal.* **1986**, *23*, 282–296. [CrossRef]
53. Liu, J.; Fu, H.F.; Chai, X.C.; Sun, Y.N.; Guo, H. Stability and convergence analysis of the quadratic spline collocation method for time-dependent fractional diffusion equations. *Appl. Math. Comput.* **2019**, *346*, 633–648. [CrossRef]
54. Majeed, A.; Kamran, M.; Iqbal, M.K.; Baleanu, D. Solving time fractional Burgers' and Fisher's equations using cubic B-spline approximation method. *Adv. Diff. Equ.* **2020**, *2020*, 175. [CrossRef]
55. Khalid, N.; Abbas, M.; Iqbal, M.K. Non-polynomial quintic spline for solving fourth-order fractional boundary value problems involving product terms. *Appl. Math. Comput.* **2019**, *349*, 393–407. [CrossRef]
56. Emadifar, H.; Jalilian, R. An exponential spline approximation for fractional Bagley-Torvik equation. *Bound. Value Probl.* **2020**, *2020*, 20. [CrossRef]
57. Kilbas, A.A.; Srivastava, H.M.; Trujillo, J.J. *Theory and Applications of Fractional Differential Equations*; Elsevier Science: Amsterdam, The Netherlands, 2006.
58. Podlubny, I. *Fractional Differential Equations: An Introduction to Fractional Derivatives, Fractional Differential Equations, to Methods of Their Solution and some of Their Applications*; Mathematics in Science and Engineering; Academic Press: San Diego, CA, USA, 1999; Volume 198.
59. Patie, P.; Simon, T. Intertwining certain fractional derivatives. *Potential Anal.* **2012**, *36*, 569–587. [CrossRef]
60. Sun, W. The spectral analysis of Hermite cubic spline collocation systems. *SIAM J. Numer. Anal.* **1999**, *36*, 1962–1975. [CrossRef]
61. Sun, W. Hermite cubic spline collocation method with upwind features. *ANZIAM J.* **2000**, *42*, C1379–C1397. [CrossRef]
62. Varma, A.K.; Howell, G. Best error bounds for derivatives in two point Birkhoff interpolation problems. *J. Approx. Theory* **1983**, *38*, 258–268. [CrossRef]
63. Birkhoff, G.; Schultz, M.H.; Varga, R.S. Piecewise Hermite interpolation in one and two variables with applications to partial differential equations. *Numer. Math.* **1968**, *11*, 232–256. [CrossRef]

Article

The Comparative Study for Solving Fractional-Order Fornberg–Whitham Equation via ρ-Laplace Transform

Pongsakorn Sunthrayuth [1], Ahmed M. Zidan [2,3], Shao-Wen Yao [4,*], Rasool Shah [5] and Mustafa Inc [6,7,8,*]

[1] Department of Mathematics and Computer Science, Faculty of Science and Technology, Rajamangala University of Technology Thanyaburi (RMUTT), Pathumthani 12110, Thailand; pongsakorn_su@rmutt.ac.th
[2] Department of Mathematics, College of Science, King Khalid University, P.O. Box 9004, Abha 61413, Saudi Arabia; ahmoahmed@kku.edu.sa
[3] Department of Mathematics, Faculty of Science, Al-Azhar University, Assuit 71511, Egypt
[4] School of Mathematics and Information Science, Henan Polytechnic University, Jiaozuo 454000, China
[5] Department of Mathematics, Abdul Wali khan University, Mardan 23200, Pakistan; rasoolshahawkum@gmail.com
[6] Department of Computer Engineering, Biruni University, Istanbul 34096, Turkey
[7] Department of Mathematics, Science Faculty, Firat University, Elazig 23119, Turkey
[8] Department of Medical Research, China Medical University Hospital, China Medical University, Taichung 40402, Taiwan
* Correspondence: yaoshaowen@hpu.edu.cn (S.-W.Y.); minc@firat.edu.tr (M.I.)

Abstract: In this article, we also introduced two well-known computational techniques for solving the time-fractional Fornberg–Whitham equations. The methods suggested are the modified form of the variational iteration and Adomian decomposition techniques by ρ-Laplace. Furthermore, an illustrative scheme is introduced to verify the accuracy of the available methods. The graphical representation of the exact and derived results is presented to show the suggested approaches reliability. The comparative solution analysis via graphs also represented the higher reliability and accuracy of the current techniques.

Keywords: ρ-Laplace variational iteration method; ρ-Laplace decomposition method; partial differential equation; caputo operator; fractional Fornberg–Whitham equation (FWE)

1. Introduction

With engineering and science development, non-linear evolution models have been analyzed as the problems to define physical phenomena in plasma waves, fluid mechanics, chemical physics, solid-state physics, etc. For the last few years, therefore, a lot of interest has been paid to the result (both numerical and analytical) of these significant models [1–4]. Different methods are available in the literature for the approximate and exact results of these models. In current years, fractional calculus (FC) applied in many phenomena in applied sciences, fluid mechanics, physics and other biology can be described as very effective using mathematical tools of FC. The fractional derivatives have occurred in many applied sciences equations such as reaction and diffusion processes, system identification, velocity signal analysis, relaxation of damping behaviour fabrics and creeping of polymer composites [5–8].

The investigation of non-linear wave models and their application is significant in different areas of engineering. Travelling wave notions are between the most attractive results for non-linear fractional-order partial differential equations (NLFPDEs). NLFPDEs are usually identified as mechanical processes and complex physical. Therefore, it is important to get exact results for non-linear time-fractional partial differential equations [9–12]. Overall, travelling wave results are between the exciting forms of products for NFPDEs. On the other hand, other NLFPDEs, such as the Camassa–Holm or the Kortewegde–Vries equations, have been well-known to have some moving wave solutions. These are non-linear multi-directional dispersive waves in shallow water design problems [13–16].

The FWE study is of crucial significance in different areas of mathematical physics. The FWEs [15,16] is defined as

$$D_\Im^\delta \mu - D_{\varphi\varphi\Im}\mu + D_\varphi\mu = \mu D_{\varphi\varphi\varphi}\mu - \mu D_\varphi\mu + 3D_\varphi\mu D_{\varphi\varphi}\mu. \tag{1}$$

The quantities performance of wave deformation, a non-linear dispersive wave model, is shown in the investigation. The FWE is presented as a mathematical model for limiting wave heights and wave breaks, allowing peakon results as a numerical model. In 1978, Fornberg and Whitham achieved a measured outcome of the form $\mu(\varphi, \eta) = Ce^{\left(\frac{-\varphi}{2} - \frac{4\Im}{3}\right)}$, where C is constant. The investigation of FWEs has been carried out by several analytical and numerical techniques, such as Adomian decomposition transform method [17], variational iteration technique [18], Lie Symmetry [19], new iterative method [20], differential transformation method [21], homotopy analysis transformation technique [22] and homotopy-perturbation technique [23].

Recently, Abdeljawad and Fahd [24] introduced the Laplace transformation of the fractional-order Caputo derivatives. We suggested a new iterative technique with ρ-Laplace transformation to investigate fractional-order ordinary and partial differential equations with fractional-order Caputo derivative. We apply this novel method for solving many fractional-order differential equations such as linear and non-linear diffusion equation, fractional-order Zakharov–Kuznetsov equation and Fokker–Planck equations. We analyzed the impact of δ and ρ in the process. The Variational iteration method (VIM) was first introduced by He [25,26] and was effectively implemented to the autonomous ordinary differential equation in [27], to non-linear polycrystalline solids [28], and other areas. Similarly, this technique is modified with ρ-Laplace transformation, so the modified method is called the ρ-Laplace variational iteration method. Many types of differential equations and partial differential equations have solved VITM. For example, this technique is analyzed for solving the time-fractional differential equation (FDEs) in [27]. In [28], this technique is applied to solve non-linear oscillator models. Compared to Adomian's decomposition process, VITM solves the problem without the need to compute Adomian's polynomials. This scheme provides a quick result to the equation, whereas the [29] mesh point techniques provide an analytical solution. This method can also be used to get a close approximation of the exact result. G. Adomian, an American mathematician, developed the Adomian decomposition technique. It focuses on finding series-like results and decomposing the non-linear operator into a sequence, with the terms presently computed using Adomian polynomials [30]. This method is modified with ρ-Laplace transform, so the modified approach is the ρ-Laplace decomposition method. This technique is used for the non-homogeneous FDEs [31–36].

This paper has implemented the ρ-Laplace variational iteration method and ρ-Laplace decomposition method to solve the time-fractional Fornberg–Whitham equations with the Caputo fractional derivative operator. The ρ-LDM and ρ-LVIM achieve the approximate results in the form of series results.

2. Basic Definitions

In this section, the fractional generalized derivative, the fractional generalized integral, the Mittag-Leffler function the ρ-Laplace transform have been discussed.

Definition 1. *The generalized fractional-order integral δ of a continuous function $f : [0, +\infty] \to R$ is expressed as [24]*

$$(I^{\delta,\rho}f)(\zeta) = \frac{1}{\Gamma(\delta)} \int_0^\zeta \left(\frac{\zeta^\rho - s^\rho}{\rho}\right)^{\delta-1} \frac{f(s)ds}{s^{1-\rho}},$$

the gamma function denote by Γ, $\rho > 0$, $\zeta > 0$ and $0 < \delta < 1$.

Definition 2. *The generalized fractional-order derivative of δ of a continuous function $f : [0, +\infty] \to R$ is given as [24].*

$$(D^{\delta,\rho}f)(\zeta) = (I^{1-\delta,\rho}f)(\zeta) = \frac{1}{\Gamma(1-\delta)} \left(\frac{d}{d\zeta}\right) \int_0^\zeta \left(\frac{\zeta^\rho - s^\rho}{\rho}\right)^{-\delta} \frac{f(s)ds}{s^{1-\rho}}.$$

where define the gamma function Γ, $\rho > 0$, $\zeta > 0$ and $0 < \delta < 1$.

Definition 3. *The Caputo fractional-order derivative δ of a continuous function $f : [0, +\infty] \to R$ is expressed as [24]*

$$(D^{\delta,\rho}f)(\zeta) = \frac{1}{\Gamma(1-\delta)} \int_0^\zeta \left(\frac{\zeta^\rho - s^\rho}{\rho}\right)^{-\delta} \beta^n \frac{f(s)ds}{s^{1-\rho}}.$$

where $n = 1$, $\rho > 0$, $\zeta > 0$, $\beta = \zeta^{1-\rho}\frac{d}{d\zeta}$ and $0 < \delta < 1$.

Definition 4. *The ρ-Laplace transformation of a continuous function $f : [0, +\infty] \to R$ is given as [24]*

$$L_\rho\{f(\zeta)\} = \int_0^\infty e^{-s\frac{\zeta^\rho}{\rho}} f(\zeta) \frac{d\zeta}{\zeta^{1-\rho}}.$$

The Caputo generalized fractional-order ρ-Laplace transform derivative of a continuous function f is defined by [24].

$$L_\rho\{D^{\delta,\rho}f(\zeta)\} = s^\delta L_\rho\{f(\zeta)\} - \sum_{k=0}^{n-1} s^{\delta-k-1}(I^{\delta,\rho}\beta^n f)(0) \quad n = 1.$$

3. The General Methodology of ρ-LDM

The ρ-LDM is a combination of the Laplace decomposition method and the ρ-Laplace transformation. In this section, we solve the ρ-LDM solution of fractional partial differential equation. The main steps of this method are described as follows:

$$D_\Im^{\delta,\rho}\omega(\varphi, \Im) + \tilde{L}(\varphi, \Im) + \mathcal{N}(\varphi, \Im) - \mathcal{H}(\varphi, \Im) = 0, \quad 0 < \delta \le 1, \tag{2}$$

where $\tilde{L}$ and $\mathcal{N}$ are linear and nonlinear functions, $\mathcal{H}$ is the sources function.
The initial condition is

$$\omega(\varphi, 0) = f(\varphi), \tag{3}$$

Apply ρ-Laplace transform to Equation (2),

$$L_\rho[D_\Im^{\delta,\rho}\omega(\varphi, \Im)] + L_\rho[\tilde{L}(\varphi, \Im) + \mathcal{N}(\varphi, \Im) - \mathcal{H}(\varphi, \Im)] = 0. \tag{4}$$

Applying the ρ-Laplace transformation differentiation property, we get

$$L_\rho[\omega(\varphi, \Im)] = \frac{1}{s}\omega(\varphi, 0) + \frac{1}{s^\delta}L_\rho[\mathcal{H}(\varphi, \Im)] - \frac{1}{s^\delta}L_\rho\{\tilde{L}(\varphi, \Im) + \mathcal{N}(\varphi, \Im)\}]. \tag{5}$$

ρ-LDM solution of infinite series $\omega(\varphi, \Im)$,

$$\omega(\varphi, \Im) = \sum_{j=0}^\infty \omega_m(\varphi, \Im). \tag{6}$$

The $\mathcal{N}$ is the nonlinear term defined as

$$\mathcal{N}(\varphi, \Im) = \sum_{j=0}^\infty \mathcal{A}_m. \tag{7}$$

So the with the help of Adomian polynomial we can define the nonlinear terms

$$\mathcal{A}_m = \frac{1}{j!}\left[\frac{\partial^m}{\partial \lambda^m}\left\{\mathcal{N}\left(\sum_{k=0}^{\infty}\lambda^k \omega_k\right)\right\}\right]_{\lambda=0}. \quad (8)$$

Putting Equations (6) and (7) into (5), we get

$$L_\rho\left[\sum_{j=0}^{\infty}\omega_m(\varphi,\Im)\right] = \frac{1}{s}\omega(\varphi,0) + \frac{1}{s^\delta}S\{\mathcal{H}(\varphi,\Im)\} - \frac{1}{s^\delta}L_\rho\left\{\tilde{\mathcal{L}}\left(\sum_{j=0}^{\infty}\omega_m\right) + \sum_{j=0}^{\infty}\mathcal{A}_m\right\}. \quad (9)$$

Using the inverse ρ-Laplace transform with Equation (9),

$$\sum_{j=0}^{\infty}\omega_m(\varphi,\Im) = L_\rho^{-1}\left[\frac{1}{s}\omega(\varphi,0) + \frac{1}{s^\delta}L_\rho\{\mathcal{H}(\varphi,\Im)\} - \frac{1}{s^\delta}L_\rho\left\{\tilde{\mathcal{L}}\left(\sum_{j=0}^{\infty}\omega_m\right) + \sum_{j=0}^{\infty}\mathcal{A}_m\right\}\right]. \quad (10)$$

we define the next terms,

$$\omega_0(\varphi,\Im) = L_\rho^{-1}\left[\frac{1}{s}\omega(\varphi,0) + \frac{1}{s^\delta}L_\rho\{\mathcal{H}(\varphi,\Im)\}\right], \quad (11)$$

$$\omega_1(\varphi,\Im) = -L_\rho^{-1}\left[\frac{1}{s^\delta}L_\rho\{\tilde{\mathcal{L}}_1(\omega_0) + \mathcal{A}_0\}\right].$$

For $m \geq 1$, is expressed as

$$\omega_{j+1}(\varphi,\Im) = -L_\rho^{-1}\left[\frac{1}{s^\delta}L_\rho\{\tilde{\mathcal{L}}(\omega_m) + \mathcal{A}_m\}\right].$$

4. Convergence Analysis

Theorem 1. [37] (Uniqueness theorem) Equation has a unique solution whenever $0 < \varepsilon < 1$ where $\varepsilon = \frac{(h_1+h_2+h_3)\Im^{\delta+1}}{(\delta-1)!}$.

Theorem 2. [37] (Convergence Theorem) The series solution (11) and (12) of the problem (3) using ρ-LTADM and ρ-LTVIM converges if $0 < \varepsilon < 1$.

Proof. Let S_ℓ be the mth partial sum, i.e., $S_\ell = \sum_{j=0}^{m}\omega_\ell(\varphi,\Im)$. We shall prove that S_ℓ is a Cauchy sequence in Banach space E. By using a new formulation of Adomian polynomials we get [37]

$$R(S_\ell) = \widehat{A_\ell} + \sum_{j=0}^{m-1}\widehat{A_j}$$

$$\aleph(S_\ell) = \widehat{A_\ell} + \sum_{n=0}^{m-1}\widehat{A_n}$$

$$||S_\ell - S_{m-1}|| = \max_{\Im \in I} |S_\ell - S_{m-1}| = \max_{\Im \in I} \left| \sum_{j=n+1}^{m} \widehat{\omega}_j(\varphi, \Im) \right|, \quad j = 0, 1, 2 \cdots$$

$$\leq \max_{\Im \in I} \left| \begin{array}{l} L_\rho^{-1}\{\frac{1}{s^\delta} L_\rho\{[\sum_{j=n+1}^{m} k[\omega_{j-1}(\varphi, \Im)]]\} \\ +L_\rho^{-1}\{\frac{1}{s^\delta} L_\rho\{[\sum_{j=n+1}^{m} M[\omega_{j-1}(\varphi, \Im)]]\} \\ +L_\rho^{-1}\{\frac{1}{s^\delta} L_\rho\{[\sum_{j=n+1}^{m} [A_{j-1}(\varphi, \Im)]]\} \end{array} \right|,$$

$$\leq \max_{\Im \in I} \left| \begin{array}{l} L_\rho^{-1}\{\frac{1}{s^\delta} L_\rho\{[\sum_{j=n+1}^{m} k[\omega_j(\varphi, \Im)]]\} \\ +L_\rho^{-1}\{\frac{1}{s^\delta} L_\rho\{[\sum_{j=n+1}^{m} M[\omega_j(\varphi, \Im)]]\} \\ +L_\rho^{-1}\{\frac{1}{s^\delta} L_\rho\{[\sum_{j=n+1}^{m} [A_j(\varphi, \Im)]]\} \end{array} \right|,$$

$$\leq \max_{\Im \in I} \left| \begin{array}{l} L_\rho^{-1}\{\frac{1}{s^\delta} L_\rho\{[\sum_{j=n+1}^{m} k[S_{m_1-1} - S_{m_2-1}]]\} \\ +L_\rho^{-1}\{\frac{1}{s^\delta} L_\rho\{[\sum_{j=n+1}^{m} M[S_{m_1-1} - S_{m_2-1}]]\} \\ +L_\rho^{-1}\{\frac{1}{s^\delta} L_\rho\{[\sum_{j=n+1}^{m} [S_{m_1-1} - S_{m_2-1}]]\} \end{array} \right|,$$

$$\leq \max_{\Im \in I} \left| \begin{array}{l} L_\rho^{-1}\{\frac{1}{s^\delta} L_\rho\{[k[S_{m_1-1} - S_{m_2-1}]]\} \\ +L_\rho^{-1}\{\frac{1}{s^\delta} L_\rho\{[M[S_{m_1-1} - S_{m_2-1}]]\} \\ +L_\rho^{-1}\{\frac{1}{s^\delta} L_\rho\{[[S_{m_1-1} - S_{m_2-1}]]\} \end{array} \right|,$$

$$\leq k_1 \max_{\Im \in I} \left| L_\rho^{-1}\{\tfrac{1}{s^\delta} L_\rho\{[S_{m_1-1} - S_{m_2-1}]\}\right|,$$

$$+ k_2 \max_{\Im \in I} \left| L_\rho^{-1}\{\tfrac{1}{s^\delta} L_\rho\{[S_{m_1-1} - S_{m_2-1}]\}\right|,$$

$$+ k_3 \max_{\Im \in I} \left| L_\rho^{-1} \tfrac{1}{s^\delta} L_\rho\{[S_{m_1-1} - S_{m_2-1}]\}\right|,$$

$$= \frac{(k_1 + k_2 + k_3)\Im^{\delta-1}}{(\delta - 1)!} ||S_{m_1-1} - S_{m_2-1}||.$$

Letting $m_1 = m_2 + 1$, we get

$$||S_{m_2+1} - S_{m_2}|| \leq \varepsilon ||S_{m_2} - S_{m_2-1}|| \leq \varepsilon^2 ||S_{m_2-1} - S_{m_2-2}|| \leq \cdots \leq \varepsilon^{m_2} ||S_1 - S_0||,$$

where $\varepsilon = \frac{(k_1+k_2+k_3)\Im^{\delta-1}}{(\delta-1)!}$ similarly, we have from the triangle inequality we get

$$||S_{m_1-1} - S_{m_2-1}|| \leq ||S_{m_1+1} - S_{m_2}|| + ||S_{m_1+2} - S_{m_2+1}|| + \cdots + ||S_{m_1} - S_{m_1-1}||,$$

$$\leq \left[\varepsilon^{m_2} + \varepsilon^{m_2+1} + \cdots + \varepsilon^{m_1-1}\right] \leq ||S_1 + S_0||,$$

$$\leq \varepsilon^{m_2}(\frac{1 - \varepsilon^{m_1-m_2}}{\varepsilon}) ||\omega_1||.$$

Since $0 < \varepsilon < 1$ we have $1 - \varepsilon^{m_1-m_2} < 1$

$$||S_{m_1} + S_{m_2}|| \leq \frac{\varepsilon^{m_2}}{1 - \varepsilon} \leq \max_{\Im \in I} ||\omega||.$$

However $|\omega| < \infty$ so, as $m_2 \to \infty$ then $||S_{m_1} - S_{m_2}|| \to 0$, hence S_{m_1} is a Cauchy sequence, the series $\sum_{m_1=0}^{\infty} \omega_{m_1}$ converges and the proof is complete. □

Theorem 3. *[37] (Error estimate) The maximum absolute error of the series solution can be given the following formula*

$$\max_{\Im \in I} |\omega(\varphi, \Im) - \sum_{\ell=1}^{\infty} \omega_\ell(\varphi, \Im)| \leq \frac{\varepsilon^{m_2}}{1 - \varepsilon} \max_{\Im \in I} ||\omega_1||.$$

5. The General Methodology of ρ-Laplace Variational Iteration Method

In this section we show the general methodology of the ρ-Laplace variational iteration method solution for fractional partial differential equations.

$$D_\Im^{\delta,\rho} w(\varphi,\Im) + \tilde{\mathcal{L}}(\varphi,\Im) + \mathcal{N}(\varphi,\Im) - \mathcal{H}(\varphi,\Im) = 0, \quad 0 < \delta \leq 1, \tag{12}$$

with the initial condition

$$w(\varphi,0) = f(\varphi), \tag{13}$$

The using ρ-Laplace transformation to Equation (12),

$$L_\rho[D_\Im^{\delta,\rho} w(\varphi,\Im)] + L_\rho[\tilde{\mathcal{L}}(\varphi,\Im) + \mathcal{N}(\varphi,\Im) - \mathcal{H}(\varphi,\Im)] = 0. \tag{14}$$

Applying the differentiation property of ρ-Laplace transform, we get

$$s^\delta L_\rho[w(\varphi,\Im)] - s^{\delta-1} w(\varphi,0) = -L_\rho[\tilde{\mathcal{L}}(\varphi,\Im) + \mathcal{N}(\varphi,\Im) - \mathcal{H}(\varphi,\Im)]. \tag{15}$$

The Lagrange multiplier is used in the iterative method

$$L_\rho[w_{j+1}(\varphi,\Im)] = L_\rho[w_j(\varphi,\Im)] + \lambda(s)\left[s^\delta L_\rho[w_j(\varphi,\Im)] - s^{\delta-1} w_j(\varphi,0)\right. \\ \left. - L_\rho\{\tilde{\mathcal{L}}(\varphi,\Im) + \mathcal{N}(\varphi,\Im)\} - L_\rho[\mathcal{H}(\varphi,\Im)]\right]. \tag{16}$$

The Lagrange multiplier is

$$\lambda(s) = -\frac{1}{s^\delta}, \tag{17}$$

using inverse ρ-Laplace transform L^{-1}, Equation (16), we get

$$w_{j+1}(\varphi,\Im) = w_j(\varphi,\Im) - L_\rho^{-1}\left[\frac{1}{s^\delta}[-L_\rho\{\tilde{\mathcal{L}}(\varphi,\Im) + \mathcal{N}(\varphi,\Im)\}] - L_\rho[\mathcal{H}(\varphi,\Im)]\right], \tag{18}$$

the initial value can be defined as

$$w_0(\varphi,\Im) = L_\rho^{-1}\left[\frac{1}{s^\delta}\{s^{\delta-1} w(\varphi,0)\}\right]. \tag{19}$$

6. Implementation of Techniques

We now proceed to derive an approximate solution to the time-fractional nonlinear FW equations using suggested techniques with generalized Caputo fractional derivative.

6.1. Problem

Consider the time-fractional nonlinear FWE is given as

$$D_\Im^{\delta,\rho} w - D_{\varphi\varphi\Im} w + D_\varphi w = w D_{\varphi\varphi\varphi} w - w D_\varphi w + 3 D_\varphi w D_{\varphi\varphi} w, \quad 0 < \delta \leq 1, \tag{20}$$

the initial condition is

$$w(\varphi,0) = e^{\left(\frac{\varphi}{2}\right)}. \tag{21}$$

Taking ρ-Laplace transform of (20),

$$s^\delta L_\rho[w(\varphi,\Im)] - s^{\delta-1} w(\varphi,0) = L_\rho[D_{\varphi\varphi\Im} w - D_\varphi w + w D_{\varphi\varphi\varphi} w - w D_\varphi w + 3 D_\varphi w D_{\varphi\varphi} w].$$

Applying inverse ρ-Laplace transform

$$w(\varphi,\Im) = L_\rho^{-1}\left[\frac{w(\varphi,0)}{s} - \frac{1}{s^\delta} L_\rho[D_{\varphi\varphi\Im} w - D_\varphi w + w D_{\varphi\varphi\varphi} w - w D_\varphi w + 3 D_\varphi w D_{\varphi\varphi} w]\right].$$

Using ADM procedure, we get

$$w_0(\varphi,\Im) = L_\rho^{-1}\left[\frac{w(\varphi,0)}{s}\right] = L_\rho^{-1}\left[\frac{e^{\left(\frac{\varphi}{2}\right)}}{s}\right],$$

$$w_0(\varphi,t) = e^{\left(\frac{\varphi}{2}\right)}, \qquad (22)$$

$$\sum_{\ell=0}^{\infty} w_{\ell+1}(\varphi,\Im) = L_\rho^{-1}\left[\frac{1}{s^\delta}L_\rho\left[\sum_{\ell=0}^{\infty}(D_{\varphi\varphi\Im}w)_\ell - \sum_{\ell=0}^{\infty}(D_\varphi w)_\ell + \sum_{\ell=0}^{\infty}A_\ell - \sum_{\ell=0}^{\infty}B_\ell + 3\sum_{\ell=0}^{\infty}C_\ell\right]\right], \quad \ell=0,1,2,\cdots$$

$$A_0(wD_{\varphi\varphi\varphi}w) = w_0 D_{\varphi\varphi\varphi}w_0,$$
$$A_1(wD_{\varphi\varphi\varphi}w) = w_0 D_{\varphi\varphi\varphi}w_1 + w_1 D_{\varphi\varphi\varphi}w_0,$$
$$A_2(wD_{\varphi\varphi\varphi}w) = w_1 D_{\varphi\varphi\varphi}w_2 + w_1 D_{\varphi\varphi\varphi}w_1 + w_2 D_{\varphi\varphi\varphi}w_0,$$

$$B_0(wD_\varphi w) = w_0 D_\varphi w_0,$$
$$B_1(wD_\varphi w) = w_0 D_\varphi w_1 + w_1 D_\varphi w_0,$$
$$B_2(wD_\varphi w) = w_1 D_\varphi w_2 + w_1 D_\varphi w_1 + w_2 D_\varphi w_0,$$

$$C_0(D_\varphi w D_{\varphi\varphi}w) = D_\varphi w_0 D_{\varphi\varphi}w_0,$$
$$C_1(D_\varphi w D_{\varphi\varphi}w) = D_\varphi w_0 D_{\varphi\varphi}w_1 + D_\varphi w_1 D_{\varphi\varphi}w_0,$$
$$C_2(D_\varphi w D_{\varphi\varphi}w) = D_\varphi w_1 D_{\varphi\varphi}w_2 + D_\varphi w_1 D_{\varphi\varphi}w_1 + D_\varphi w_2 D_{\varphi\varphi}w_0,$$

for $\ell = 1$

$$w_1(\varphi,\Im) = L_\rho^{-1}\left[\frac{1}{s^\delta}L_\rho[D_{\varphi\varphi\Im}w_0 - D_\varphi w_0 + A_0 - B_0 + 3C_0]\right],$$

$$w_1(\varphi,t) = -\frac{1}{2}L_\rho^{-1}\left[\frac{e^{\left(\frac{\varphi}{2}\right)}}{s^{\delta+1}}\right] = -\frac{1}{2}e^{\left(\frac{\varphi}{2}\right)}\frac{\left(\frac{\Im\rho}{\rho}\right)^\delta}{\Gamma(\delta+1)}. \qquad (23)$$

for $\ell = 2$

$$w_2(\varphi,\Im) = L_\rho^{-1}\left[\frac{1}{s^\delta}L_\rho[D_{\varphi\varphi\Im}w_1 - D_\varphi w_1 + A_1 - B_1 + 3C_1]\right],$$

$$w_2(\varphi,\Im) = -\frac{1}{8}e^{\left(\frac{\varphi}{2}\right)}\frac{\left(\frac{\Im\rho}{\rho}\right)^{2\delta-1}}{\Gamma(2\delta)} + \frac{1}{4}e^{\left(\frac{\varphi}{2}\right)}\frac{\left(\frac{\Im\rho}{\rho}\right)^{2\delta}}{\Gamma(2\delta+1)}, \qquad (24)$$

for $\ell = 3$

$$w_3(\varphi,\Im) = L_\rho^{-1}\left[\frac{1}{s^\delta}L_\rho[D_{\varphi\varphi\Im}w_2 - D_\varphi w_2 + A_2 - B_2 + 3C_2]\right],$$

$$w_3(\varphi,\Im) = -\frac{1}{32}e^{\left(\frac{\varphi}{2}\right)}\frac{\left(\frac{\Im\rho}{\rho}\right)^{3\delta-2}}{\Gamma(3\delta-1)} + \frac{1}{8}e^{\left(\frac{\varphi}{2}\right)}\frac{\Im^{3\delta-1}}{\Gamma(3\delta)} - \frac{1}{8}e^{\left(\frac{\varphi}{2}\right)}\frac{\left(\frac{\Im\rho}{\rho}\right)^{3\delta}}{\Gamma(3\delta+1)}, \qquad (25)$$

The ρ-LDM result of Example 1 is

$$w(\varphi,\Im) = w_0(\varphi,\Im) + w_1(\varphi,\Im) + w_2(\varphi,\Im) + w_3(\varphi,\Im) + w_4(\varphi,\Im) + \cdots,$$

$$\omega(\varphi,\Im) = e^{(\frac{\varphi}{2})} - \frac{1}{2}e^{(\frac{\varphi}{2})}\frac{\left(\frac{\Im^{\rho}}{\rho}\right)^{\delta}}{\Gamma(\delta+1)} - \frac{1}{8}e^{(\frac{\varphi}{2})}\frac{\left(\frac{\Im^{\rho}}{\rho}\right)^{2\delta-1}}{\Gamma(2\delta)} + \frac{1}{4}e^{(\frac{\varphi}{2})}\frac{\left(\frac{\Im^{\rho}}{\rho}\right)^{2\delta}}{\Gamma(2\delta+1)} - \frac{1}{32}e^{(\frac{\varphi}{2})}\frac{\left(\frac{\Im^{\rho}}{\rho}\right)^{3\delta-2}}{\Gamma(3\delta-1)}$$
$$+ \frac{1}{8}e^{(\frac{\varphi}{2})}\frac{\Im^{3\delta-1}}{\Gamma(3\delta)} - \frac{1}{8}e^{(\frac{\varphi}{2})}\frac{\left(\frac{\Im^{\rho}}{\rho}\right)^{3\delta}}{\Gamma(3\delta+1)} - \cdots. \quad (26)$$

The simplify we can write Equation (26), we get

$$\omega(\varphi,\Im) = e^{(\frac{\varphi}{2})}\left[1 - \frac{\left(\frac{\Im^{\rho}}{\rho}\right)^{\delta}}{2\Gamma(\delta+1)} - \frac{1}{8}\frac{\left(\frac{\Im^{\rho}}{\rho}\right)^{2\delta-1}}{\Gamma(2\delta)} + \frac{1}{4}\frac{\left(\frac{\Im^{\rho}}{\rho}\right)^{2\delta}}{\Gamma(2\delta+1)} - \frac{1}{32}\frac{\left(\frac{\Im^{\rho}}{\rho}\right)^{3\delta-2}}{\Gamma(3\delta-1)} + \frac{1}{8}\frac{\left(\frac{\Im^{\rho}}{\rho}\right)^{3\delta-1}}{\Gamma(3\delta)} - \frac{1}{8}\frac{\left(\frac{\Im^{\rho}}{\rho}\right)^{3\delta}}{\Gamma(3\delta+1)} + \cdots\right]. \quad (27)$$

The analytical result by ρ-LVIM.
The iteration method apply for Equation (20), we get

$$\omega_{\ell+1}(\varphi,\Im) = \omega_{\ell}(\varphi,\Im) - L_{\rho}^{-1}\left[\frac{1}{s^{\delta}}L_{\rho}\left\{s^{\delta}D_{\Im}\omega_{\ell} - D_{\varphi\varphi\Im}\omega_{\ell} + D_{\varphi}\omega_{\ell} - \omega_{\ell}D_{\varphi\varphi\varphi}\omega_{\ell} + \omega_{\ell}D_{\varphi}\omega_{\ell} - 3D_{\varphi}\omega_{\ell}D_{\varphi\varphi}\omega_{\ell}\right\}\right], \quad (28)$$

where

$$\omega_0(\varphi,\Im) = e^{(\frac{\varphi}{2})}. \quad (29)$$

For $\ell = 0, 1, 2, \cdots$

$$\omega_1(\varphi,\Im) = \omega_0(\varphi,\Im) - L_{\rho}^{-1}\left[\frac{1}{s^{\delta}}L_{\rho}\left\{s^{\delta}D_{\Im}\omega_0 - D_{\varphi\varphi\Im}\omega_0 + D_{\varphi}\omega_0 - \omega_0 D_{\varphi\varphi\varphi}\omega_0\right.\right.$$
$$\left.\left.+\omega_0 D_{\varphi}\omega_0 - 3D_{\varphi}\omega_0 D_{\varphi\varphi}\omega_0\right\}\right], \quad (30)$$
$$\omega_1(\varphi,\Im) = -\frac{1}{2}e^{(\frac{\varphi}{2})}\frac{\left(\frac{\Im^{\rho}}{\rho}\right)^{\delta}}{\Gamma(\delta+1)},$$

$$\omega_2(\varphi,\Im) = \omega_1(\varphi,\Im) - L_{\rho}^{-1}\left[\frac{1}{s^{\delta}}L_{\rho}\left\{s^{\delta}D_{\Im}\omega_1 - D_{\varphi\varphi\Im}\omega_1 + D_{\varphi}\omega_1 - \omega_1 D_{\varphi\varphi\varphi}\omega_1\right.\right.$$
$$\left.\left.+\omega_1 D_{\varphi}\omega_1 - 3D_{\varphi}\omega_1 D_{\varphi\varphi}\omega_1\right\}\right], \quad (31)$$
$$\omega_2(\varphi,\Im) = -\frac{1}{8}e^{(\frac{\varphi}{2})}\frac{\left(\frac{\Im^{\rho}}{\rho}\right)^{2\delta-1}}{\Gamma(2\delta)} + \frac{1}{4}e^{(\frac{\varphi}{2})}\frac{\left(\frac{\Im^{\rho}}{\rho}\right)^{2\delta}}{\Gamma(2\delta+1)},$$

$$\omega_3(\varphi,\Im) = \omega_2(\varphi,\Im) - L_{\rho}^{-1}\left[\frac{1}{s^{\delta}}L_{\rho}\left\{s^{\delta}D_{\Im}\omega_2 - D_{\varphi\varphi\Im}\omega_2 + D_{\varphi}\omega_2 - \omega_2 D_{\varphi\varphi\varphi}\omega_2\right.\right.$$
$$\left.\left.+\omega_2 D_{\varphi}\omega_2 - 3D_{\varphi}\omega_2 D_{\varphi\varphi}\omega_2\right\}\right], \quad (32)$$
$$\omega_3(\varphi,\Im) = -\frac{1}{32}e^{(\frac{\varphi}{2})}\frac{\left(\frac{\Im^{\rho}}{\rho}\right)^{3\delta-2}}{\Gamma(3\delta-1)} + \frac{1}{8}e^{(\frac{\varphi}{2})}\frac{\left(\frac{\Im^{\rho}}{\rho}\right)^{3\delta-1}}{\Gamma(3\delta)} - \frac{1}{8}e^{(\frac{\varphi}{2})}\frac{\left(\frac{\Im^{\rho}}{\rho}\right)^{3\delta}}{\Gamma(3\delta+1)},$$

$$\omega(\varphi,\Im) = \sum_{m=0}^{\infty}\omega_m(\varphi) = e^{(\frac{\varphi}{2})} - \frac{1}{2}e^{(\frac{\varphi}{2})}\frac{\left(\frac{\Im^{\rho}}{\rho}\right)^{\delta}}{\Gamma(\delta+1)} - \frac{1}{8}e^{(\frac{\varphi}{2})}\frac{\left(\frac{\Im^{\rho}}{\rho}\right)^{2\delta-1}}{\Gamma(2\delta)} + \frac{1}{4}e^{(\frac{\varphi}{2})}\frac{\left(\frac{\Im^{\rho}}{\rho}\right)^{2\delta}}{\Gamma(2\delta+1)}$$
$$- \frac{1}{32}e^{(\frac{\varphi}{2})}\frac{\left(\frac{\Im^{\rho}}{\rho}\right)^{3\delta-2}}{\Gamma(3\delta-1)} + \frac{1}{8}e^{(\frac{\varphi}{2})}\frac{\Im^{3\delta-1}}{\Gamma(3\delta)} - \frac{1}{8}e^{(\frac{\varphi}{2})}\frac{\left(\frac{\Im^{\rho}}{\rho}\right)^{3\delta}}{\Gamma(3\delta+1)} - \cdots. \quad (33)$$

The exact result of Equation (20) at $\delta = 1$,

$$\omega(\varphi,\Im) = e^{(\frac{\varphi}{2} - \frac{2\Im}{3})}. \quad (34)$$

Figure 1 shows the ρ-LDM and ρ-LVIM solution of the fractional Fornberg–Whitham defined by generalized fractional-order Caputo derivative in the space coordinate and time $0 < \Im \leq 0.5$, $\rho = 1$ and $\delta = 1$. Figure 2, the 3D graph shows approximate and exact solutions graph at $\delta = 1$ and $\rho = 0.9$; the figure shows that different fractional-order at δ. Similarly, in Figure 3, the 2D graph of exact and approximate solutions plot at $\delta = 1$ and $\rho = 0.9$ the figure shows that different fractional-order at δ.

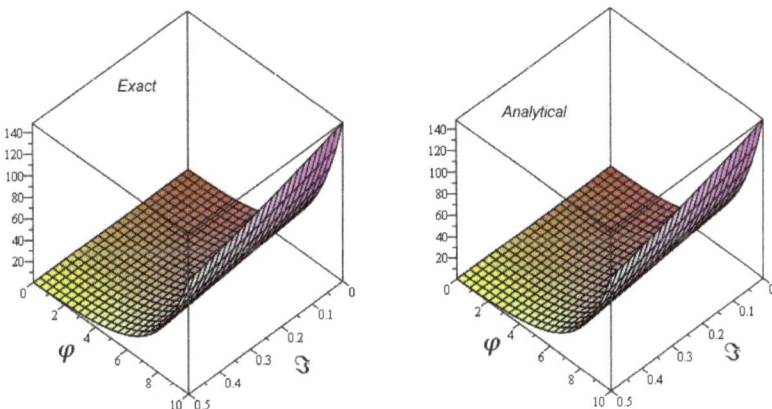

Figure 1. The graph of Exact and analytical solutions of $\delta = 1$ and $\rho = 1$ of problem 1.

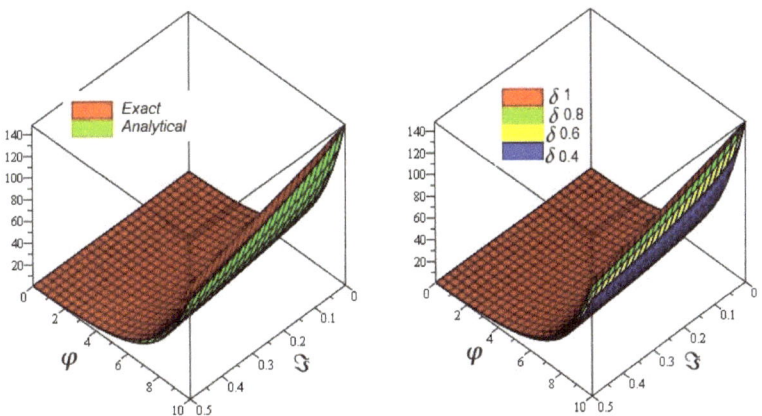

Figure 2. The first 3D graph of Exact and analytical solutions graph at $\delta = 1$ and $\rho = 0.9$ and second plot of the approximate different fractional-order of $\delta = 1$ of problem 1.

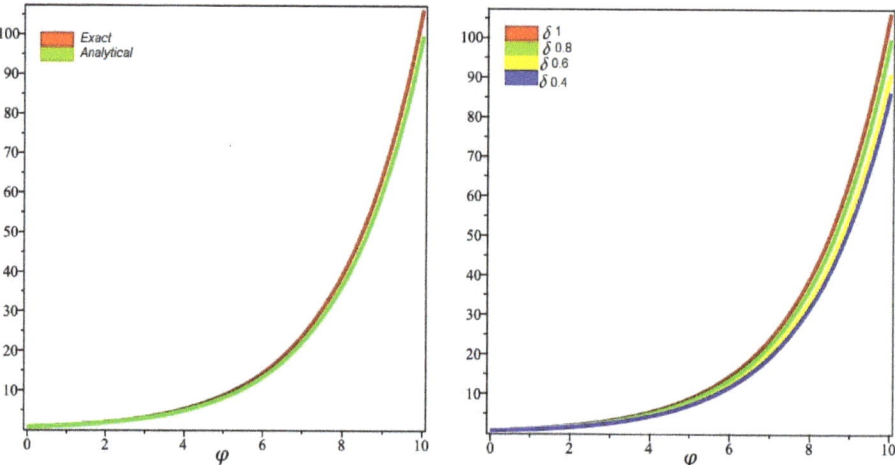

Figure 3. The first 2D graph of Exact and analytical solutions graph at $\delta = 1$ and $\rho = 0.9$ and second plot of the approximate different fractional-order of $\delta = 1$ of problem 1.

6.2. Problem

Consider the time-fractional non-linear FWE given as

$$D_\Im^{\delta,\rho}\omega - D_{\varphi\varphi\Im}\omega + D_\varphi\omega = \omega D_{\varphi\varphi\varphi}\omega - \omega D_\varphi\omega + 3D_\varphi\omega D_{\varphi\varphi}\omega, \quad \Im > 0, \quad 0 < \delta \leq 1, \quad (35)$$

with the initial condition

$$\omega(\varphi, 0) = \cosh^2\left(\frac{\varphi}{4}\right). \quad (36)$$

Taking ρ-Laplace transform of (35),

$$s^\delta L_\rho[\omega(\varphi, \Im)] - s^{\delta-1}\omega(\varphi, 0) = L_\rho\left[D_{\varphi\varphi\Im}\omega - D_\varphi\omega + \omega D_{\varphi\varphi\varphi}\omega - \omega D_\varphi\omega + 3D_\varphi\omega D_{\varphi\varphi}\omega\right].$$

Applying inverse ρ-Laplace transform

$$\omega(\varphi, \Im) = L_\rho^{-1}\left[\frac{\omega(\varphi, 0)}{s} - \frac{1}{s^\delta}L_\rho\{D_{\varphi\varphi\Im}\omega - D_\varphi\omega + \omega D_{\varphi\varphi\varphi}\omega - \omega D_\varphi\omega + 3D_\varphi\omega D_{\varphi\varphi}\omega\}\right].$$

Using ADM procedure, we get

$$\omega_0(\varphi, \Im) = L_\rho^{-1}\left[\frac{\omega(\varphi, 0)}{s}\right] = L_\rho^{-1}\left[\frac{\cosh^2\left(\frac{\varphi}{4}\right)}{s}\right],$$

$$\omega_0(\varphi, \Im) = \cosh^2\left(\frac{\varphi}{4}\right), \quad (37)$$

$$\sum_{\ell=0}^{\infty}\omega_{\ell+1}(\varphi, \Im) = L_\rho^{-1}\left[\frac{1}{s^\delta}L_\rho\left[\sum_{\ell=0}^{\infty}(D_{\varphi\varphi\Im}\omega)_\ell - \sum_{\ell=0}^{\infty}(D_\varphi\omega)_\ell + \sum_{\ell=0}^{\infty}A_\ell - \sum_{\ell=0}^{\infty}B_\ell + 3\sum_{\ell=0}^{\infty}C_\ell\right]\right], \quad \ell = 0, 1, 2, \cdots$$

for $\ell = 0$

$$\omega_1(\varphi, \Im) = L_\rho^{-1}\left[\frac{1}{s^\delta} L_\rho [D_{\varphi\varphi\Im}\omega_0 - D_\varphi\omega_0 + A_0 - B_0 + 3C_0]\right],$$

$$\omega_1(\varphi, \Im) = -\frac{11}{32} L_\rho^{-1}\left[\frac{\sinh\left(\frac{x}{2}\right)}{s^{\delta+1}}\right] = -\frac{11}{32} \sinh\left(\frac{\varphi}{4}\right)\frac{\left(\frac{\Im^\rho}{\rho}\right)^\delta}{\Gamma(\delta+1)}. \tag{38}$$

for $\ell = 1$

$$\omega_2(\varphi, \Im) = L_\rho^{-1}\left[\frac{1}{s^\delta} L_\rho [D_{\varphi\varphi\Im}\omega_1 - D_\varphi\omega_1 + A_1 - B_1 + 3C_1]\right],$$

$$\omega_2(\varphi, \Im) = -\frac{11}{28} \sinh\left(\frac{\varphi}{4}\right)\frac{\left(\frac{\Im^\rho}{\rho}\right)^\delta}{\Gamma(\delta+1)} + \frac{121}{1024} \cosh\left(\frac{\varphi}{4}\right)\frac{\left(\frac{\Im^\rho}{\rho}\right)^{2\delta}}{\Gamma(2\delta+1)}, \tag{39}$$

for $\ell = 2$

$$\omega_3(\varphi, \Im) = L_\rho^{-1}\left[\frac{1}{s^\delta} L_\rho [D_{\varphi\varphi\Im}\omega_2 - D_\varphi\omega_2 + A_2 - B_2 + 3C_2]\right],$$

$$\omega_3(\varphi, \Im) = -\frac{11}{512} \sinh\left(\frac{\varphi}{4}\right)\frac{\left(\frac{\Im^\rho}{\rho}\right)^\delta}{\Gamma(\delta+1)} + \frac{121}{2048} \cosh\left(\frac{\varphi}{4}\right)\frac{\left(\frac{\Im^\rho}{\rho}\right)^{2\delta}}{\Gamma(2\delta+1)} - \frac{1331}{49152} \sinh\left(\frac{\varphi}{4}\right)\frac{\left(\frac{\Im^\rho}{\rho}\right)^{3\delta}}{\Gamma(3\delta+1)}, \tag{40}$$

The ρ-LDM result for problem 2 is

$$\omega(\varphi, \Im) = \omega_0(\varphi, \Im) + \omega_1(\varphi, \Im) + \omega_2(\varphi, \Im) + \omega_3(\varphi, \Im) + \omega_4(\varphi, \Im) + \cdots,$$

$$\omega(\varphi, \Im) = \cosh^2\left(\frac{\varphi}{4}\right) - \frac{11}{32}\sinh\left(\frac{\varphi}{4}\right)\frac{\left(\frac{\Im^\rho}{\rho}\right)^\delta}{\Gamma(\delta+1)} - \frac{11}{28}\sinh\left(\frac{\varphi}{4}\right)\frac{\left(\frac{\Im^\rho}{\rho}\right)^\delta}{\Gamma(\delta+1)} + \frac{121}{1024}\cosh\left(\frac{\varphi}{4}\right)\frac{\left(\frac{\Im^\rho}{\rho}\right)^{2\delta}}{\Gamma(2\delta+1)}$$

$$- \frac{11}{512}\sinh\left(\frac{\varphi}{4}\right)\frac{\left(\frac{\Im^\rho}{\rho}\right)^\delta}{\Gamma(\delta+1)} + \frac{121}{2048}\cosh\left(\frac{\varphi}{4}\right)\frac{\left(\frac{\Im^\rho}{\rho}\right)^{2\delta}}{\Gamma(2\delta+1)} - \frac{1331}{49152}\sinh\left(\frac{\varphi}{4}\right)\frac{\left(\frac{\Im^\rho}{\rho}\right)^{3\delta}}{\Gamma(3\delta+1)}\cdots. \tag{41}$$

The analytical solution by ρ-LVIM.
The iteration method is apply by Equation (35), we get

$$\omega_{\ell+1}(\varphi, \Im) = \omega_\ell(\varphi, \Im) - L_\rho^{-1}\left[\frac{1}{s^\delta} L_\rho \left\{s^\delta D_\Im \omega_\ell - D_{\varphi\varphi\Im}\omega_\ell + D_\varphi\omega_\ell - \omega_\ell D_{\varphi\varphi\varphi}\omega_\ell + \omega_\ell D_\varphi\omega_\ell - 3D_\varphi\omega_\ell D_{\varphi\varphi}\omega_\ell\right\}\right], \tag{42}$$

where

$$\omega_0(\varphi, t) = \cosh^2\left(\frac{\varphi}{4}\right). \tag{43}$$

For $\ell = 0, 1, 2, \cdots$

$$\omega_1(\varphi, \Im) = \omega_0(\varphi, \Im) - L_\rho^{-1}\left[\frac{1}{s^\delta} L_\rho \left\{s^\delta D_\Im \omega_0 - D_{\varphi\varphi\Im}\omega_0 + D_\varphi\omega_0 - \omega_0 D_{\varphi\varphi\varphi}\omega_0 + \omega_0 D_\varphi\omega_0 - 3D_\varphi\omega_0 D_{\varphi\varphi}\omega_0\right\}\right],$$

$$\omega_1(\varphi, \Im) = \cosh^2\left(\frac{\varphi}{4}\right) - \frac{11}{32}\sinh\left(\frac{\varphi}{4}\right)\frac{\left(\frac{\Im^\rho}{\rho}\right)^\delta}{\Gamma(\delta+1)}, \tag{44}$$

$$w_2(\varphi,\Im) = w_1(\varphi,\Im) - L_\rho^{-1}\left[\frac{1}{s^\delta}L_\rho\left\{s^\delta D_\Im w_1 - D_{\varphi\varphi\Im}w_1 + D_\varphi w_1 - w_1 D_{\varphi\varphi\varphi}w_1 + w_1 D_\varphi w_1 - 3D_\varphi w_1 D_{\varphi\varphi}w_1\right\}\right],$$

$$w_2(\varphi,\Im) = \cosh^2\left(\frac{\varphi}{4}\right) - \frac{11}{32}\sinh\left(\frac{\varphi}{4}\right)\frac{\left(\frac{\Im^\rho}{\rho}\right)^\delta}{\Gamma(\delta+1)} - \frac{11}{28}\sinh\left(\frac{\varphi}{4}\right)\frac{\left(\frac{\Im^\rho}{\rho}\right)^\delta}{\Gamma(\delta+1)} + \frac{121}{1024}\cosh\left(\frac{\varphi}{4}\right)\frac{\left(\frac{\Im^\rho}{\rho}\right)^{2\delta}}{\Gamma(2\delta+1)}, \qquad (45)$$

$$w_3(\varphi,\Im) = w_2(\varphi,\Im) - L_\rho^{-1}\left[\frac{1}{s^\delta}L_\rho\left\{s^\delta D_\Im w_2 - D_{\varphi\varphi\Im}w_2 + D_\varphi w_2 - w_2 D_{\varphi\varphi\varphi}w_2 + w_2 D_\varphi w_2 - 3D_\varphi w_2 D_{\varphi\varphi}w_2\right\}\right],$$

$$w_3(\varphi,\Im) = \cosh^2\left(\frac{\varphi}{4}\right) - \frac{11}{32}\sinh\left(\frac{\varphi}{4}\right)\frac{\left(\frac{\Im^\rho}{\rho}\right)^\delta}{\Gamma(\delta+1)} - \frac{11}{28}\sinh\left(\frac{\varphi}{4}\right)\frac{\left(\frac{\Im^\rho}{\rho}\right)^\delta}{\Gamma(\delta+1)} + \frac{121}{1024}\cosh\left(\frac{\varphi}{4}\right)\frac{\left(\frac{\Im^\rho}{\rho}\right)^{2\delta}}{\Gamma(2\delta+1)}, \qquad (46)$$

$$-\frac{11}{512}\sinh\left(\frac{\varphi}{4}\right)\frac{\left(\frac{\Im^\rho}{\rho}\right)^\delta}{\Gamma(\delta+1)} + \frac{121}{2048}\cosh\left(\frac{\varphi}{4}\right)\frac{\left(\frac{\Im^\rho}{\rho}\right)^{2\delta}}{\Gamma(2\delta+1)} - \frac{1331}{49152}\sinh\left(\frac{\varphi}{4}\right)\frac{\left(\frac{\Im^\rho}{\rho}\right)^{3\delta}}{\Gamma(3\delta+1)},$$

$$w(\varphi,\Im) = \sum_{m=0}^\infty w_m(\varphi) = \cosh^2\left(\frac{\varphi}{4}\right) - \frac{11}{32}\sinh\left(\frac{\varphi}{4}\right)\frac{\left(\frac{\Im^\rho}{\rho}\right)^\delta}{\Gamma(\delta+1)} - \frac{11}{28}\sinh\left(\frac{\varphi}{4}\right)\frac{\left(\frac{\Im^\rho}{\rho}\right)^\delta}{\Gamma(\delta+1)} + \frac{121}{1024}\cosh\left(\frac{\varphi}{4}\right)\frac{\left(\frac{\Im^\rho}{\rho}\right)^{2\delta}}{\Gamma(2\delta+1)}, \qquad (47)$$

$$-\frac{11}{512}\sinh\left(\frac{\varphi}{4}\right)\frac{\left(\frac{\Im^\rho}{\rho}\right)^\delta}{\Gamma(\delta+1)} + \frac{121}{2048}\cosh\left(\frac{\varphi}{4}\right)\frac{\left(\frac{\Im^\rho}{\rho}\right)^{2\delta}}{\Gamma(2\delta+1)} - \frac{1331}{49152}\sinh\left(\frac{\varphi}{4}\right)\frac{\left(\frac{\Im^\rho}{\rho}\right)^{3\delta}}{\Gamma(3\delta+1)} - \cdots.$$

The exact result of Equation (35) at $\delta = 1$,

$$w(\varphi,\Im) = \cosh^2\left(\frac{\varphi}{4} - \frac{11\Im}{24}\right). \qquad (48)$$

Figure 4 shows the ρ-LDM and ρ-LVIM solution of the fractional Fornberg–Whitham defined by generalized Caputo fractional-order derivative in the space coordinate and time $0 < \Im \leq 0.5$, $\rho = 1$ and $\delta = 1$. Figure 5, the 3D graph shows exact and approximate solutions plot at $\delta = 1$ and $\rho = 0.9$; the figure shows that different fractional-order at δ. Similarly, in Figure 6, the 2D graph of exact and approximate solutions plot at $\delta = 1$ and $\rho = 0.9$ the figure shows that different fractional-order at δ.

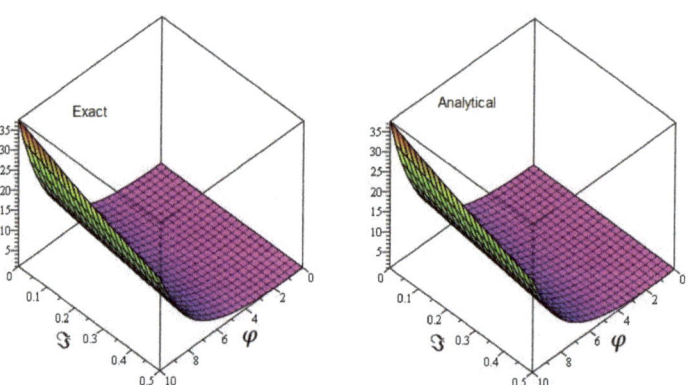

Figure 4. The graph of Exact and approximate solutions of $\delta = 1$ and $\rho = 1$ of Example 2.

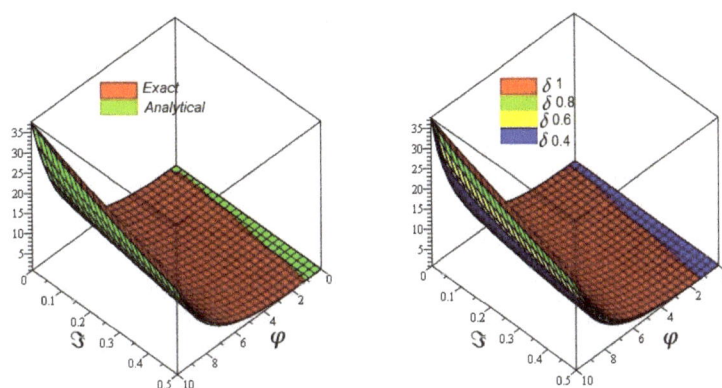

Figure 5. The first 3D graph of Exact and approximate solutions plot at $\delta = 1$ and $\rho = 0.9$ and second plot of the approximate different fractional-order of $\delta = 1$ of Example 2.

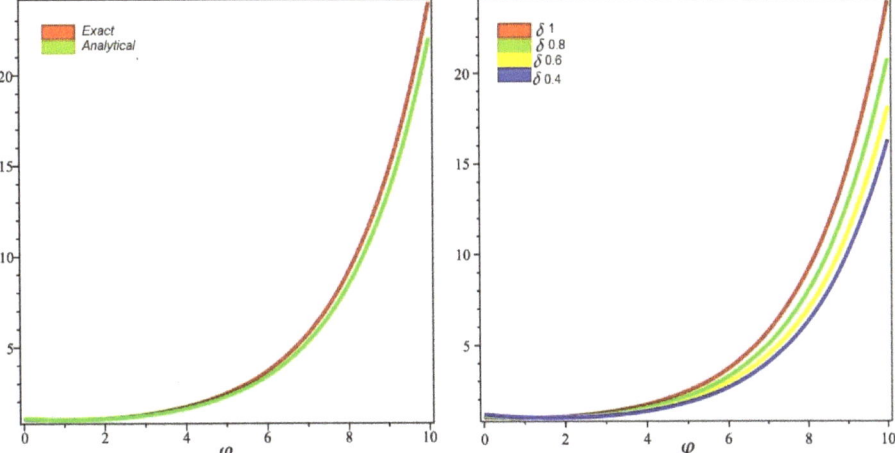

Figure 6. The first 2D graph of Exact and approximate solutions plot at $\delta = 1$ and $\rho = 0.9$ and second plot of the approximate different fractional-order of $\delta = 1$ of Example 2.

7. Conclusions

In this article, different semi-analytical techniques are implemented to solve time-fractional Fornberg–Whitham equation. The approximate solution of the equations is evaluated to confirm the validity and reliability of the proposed methods. Graphs of the solutions are plotted to display the closed relation between the obtained and exact results. In addition, the suggested techniques provide easily computable components for the series-form tests. It is investigated that the results achieved in the series form have a higher convergence rate towards the exact results. The proposed methods have a small number of calculations to achieve the approximate solution. In conclusion, it is found that the proposed technique is a sophisticated method for solving other NLFPDEs. In the future, the analytical result of non-linear fractional-order boundary values problems achieved using this technique is in the form of uniform convergence series.

Author Contributions: onceptualization, R.S. and M.I.; methodology, R.S.; software, P.S. and A.M.Z.; validation, R.S.; formal analysis, M.I. and P.S.; data curation, R.S.; writing—original draft preparation R.S.; writing—review and editing, A.M.Z.; supervision, M.I.; project administration, A.M.Z.; funding acquisition, S.-W.Y. All authors have read and agreed to the published version of the manuscript.

Funding: National Natural Science Foundation of China (No. 71601072), Key Scientific Research Project of Higher Education Institutions in Henan Province of China (No. 20B110006) and the Fundamental Research Funds for the Universities of Henan Province (No. NSFRF210314).

Data Availability Statement: Not applicable.

Acknowledgments: One of the Co-authors (A. M. Zidan) extend his appreciation to the Deanship of Scientific Research at King Khalid University, Abha 61413, Saudi Arabia for funding this work through research groups program under grant number R.G.P.1/30/42.

Conflicts of Interest: The authors declare no conflict of interest.

References

1. Qureshi, S.; Yusuf, A.; Shaikh, A.A.; Inc, M.; Baleanu, D. Fractional modeling of blood ethanol concentration system with real data application. *Chaos Interdiscip. J. Nonlinear Sci.* **2019**, *29*, 013143. [CrossRef]
2. Yusuf, A.; Qureshi, S.; Inc, M.; Aliyu, A.I.; Baleanu, D.; Shaikh, A.A. Two-strain epidemic model involving fractional derivative with Mittag-Leffler kernel. *Chaos Interdiscip. J. Nonlinear Sci.* **2018**, *28*, 123121. [CrossRef] [PubMed]
3. Kilic, B.; Inc, M. The first integral method for the time fractional Kaup-Boussinesq system with time dependent coefficient. *Appl. Math. Comput.* **2015**, *254*, 70–74. [CrossRef]
4. Inc, M.; Akgul, A.; Kilicman, A. Numerical solutions of the second-order one-dimensional telegraph equation based on reproducing kernel Hilbert space method. *Abstr. Appl. Anal.* **2013**, *2013*, 768963. [CrossRef]
5. Hilfer, R. *Applications of Fractional Calculus in Physics*; World Scientific: Singapore, 2000; Volume 35, pp. 87–130.
6. Cesarano, C.; Pierpaolo, N.; Paolo, E.R. Pseudo-Lucas Functions of Fractional Degree and Applications. *Axioms* **2021**, *10*, 51. [CrossRef]
7. Sabatier, J.A.T.M.J.; Agrawal, O.P.; Machado, J.T. *Advances in Fractional Calculus*; Springer: Dordrecht, The Netherlands, 2007; Volume 4.
8. Al-luhaibi, M.S. An analytical treatment to fractional Fornberg-Whitham equation. *Math. Sci.* **2017**, *11*, 1–6. [CrossRef]
9. Agarwal, P.; Ahs, S.; Akbare, M.; Nawaz, R.; Cesarano, C. A Reliable Algorithm for solution of Higher Dimensional Nonlinear (1+1) and (2+1) Dimensional Volterra-Fredholm Integral Equations. *Dolomites Res. Notes Approx.* **2021**, *14*, 18–25.
10. Zakarya, M.; Altanji, M.; AlNemer, G.; Abd, El-Hamid, H.A.; Cesarano, C.; Rezk, H.M. Fractional Reverse Coposn's Inequalities via Conformable Calculus on Time Scales. *Symmetry* **2021**, *13*, 542. [CrossRef]
11. Ahmad, I.; Ahmad, H.; Thounthong, P.; Chu, Y.-M.; Cesarano, C. Solution of Multi-Term Time-Fractional PDE Models Arising in Mathematical Biology and Physics by Local Meshless Method. *Symmetry* **2020**, *12*, 1195. [CrossRef]
12. Bazighifan, O.; Cesarano, C. A Philos-type oscillation criteria for fourth-order neutral differential equations. *Symmetry* **2020**, *12*, 379. [CrossRef]
13. Camacho, J.C.; Rosa, M.; Garias, M.L.; Bruzon, M.S. Classical symmetries, travelling wave solutions and conservation laws of a generalized Fornberg-Whitham equation. *J. Comput. Appl. Math.* **2017**, *318*, 149–155. [CrossRef]
14. Bruzon, M.S.; Marquez, A.P.; Garrido, T.M.; Recio, E.; de la Rosa, R. Conservation laws for a generalized seventh order KdV equation. *J. Comput. Appl. Math.* **2019**, *354*, 682–688. [CrossRef]
15. Whitham, G.B. Variational methods and applications to water waves. *Proc. R. Soc. Lond. Ser. A Math. Phys. Sci.* **1967**, *299*, 6–25.
16. Fornberg, B.; Whitham, G.B. A numerical and theoretical study of certain nonlinear wave phenomena. *Philos. Trans. R. Soc. Lond. Ser. A Math. Phys. Sci.* **1978**, *289*, 373–404.
17. Kumar, D.; Singh, J.; Baleanu, D. A new analysis of the Fornberg-Whitham equation pertaining to a fractional derivative with Mittag-Leffler-type kernel. *Eur. Phys. J. Plus* **2018**, *133*, 1–10. [CrossRef]
18. Lu, J. An analytical approach to the Fornberg-Whitham type equations by using the variational iteration method. *Comput. Math. Appl.* **2011**, *61*, 2010–2013. [CrossRef]
19. Hashemi, M.S.; Haji-Badali, A.; Vafadar, P. Group invariant solutions and conservation laws of the Fornberg-Whitham equation. *Z. Naturforschung A* **2014**, *69*, 489–496. [CrossRef]
20. Ramadan, M.A.; Al-luhaibi, M.S. New iterative method for solving the fornberg-whitham equation and comparison with homotopy perturbation transform method. *J. Adv. Math. Comput. Sci.* **2014**, *4*, 1213–1227. [CrossRef]
21. Merdan, M.; Gokdogan, A.; Yildirim, A.; Mohyud-Din, S.T. Numerical simulation of fractional Fornberg-Whitham equation by differential transformation method. *Abstr. Appl. Anal.* **2012**, *2012*, 965367. [CrossRef]
22. Wang, K.; Liu, S. Application of new iterative transform method and modified fractional homotopy analysis transform method for fractional Fornberg-Whitham equation. *J. Nonlinear Sci. Appl.* **2016**, *9*, 2419–2433. [CrossRef]
23. Abidi, F.; Omrani, K. Numerical solutions for the nonlinear Fornberg-Whitham equation by He's methods. *Int. J. Mod. Phys. B* **2011**, *25*, 4721–4732. [CrossRef]

24. Jarad, F.; Abdeljawad, T. A modified Laplace transform for certain generalized fractional operators. *Results Nonlinear Anal.* **2018**, *1*, 88–98.
25. He, J.H. Approximate solution of nonlinear differential equations with convolution product nonlinearities. *Comput. Methods Appl. Mech. Eng.* **1998**, *167*, 69–73. [CrossRef]
26. He, J.H. Variational iteration method for autonomous ordinary differential systems. *Appl. Math. Comput.* **2000**, *114*, 115–123. [CrossRef]
27. Wu, G.C.; Baleanu, D. Variational iteration method for fractional calculus-a universal approach by Laplace transform. *Adv. Differ. Equ.* **2013**, *2013*, 18. [CrossRef]
28. Anjum, N.; He, J.H. Laplace transform: Making the variational iteration method easier. *Appl. Math. Lett.* **2019**, *92*, 134–138. [CrossRef]
29. Dehghan, M. Finite difference procedures for solving a problem arising in modeling and design of certain optoelectronic devices. *Math. Comput. Simul.* **2006**, *71*, 16–30. [CrossRef]
30. Adomian, G. *Solving Frontier Problems of Physics: The Decomposition Method*; Kluwer Academic Publishers: Boston, MA, USA, 1994.
31. Khalouta, A.; Kadem, A. A New Method to Solve Fractional Differential Equations: Inverse Fractional Laplace Transform Method. *Appl. Appl. Math.* **2019**, *14*, 926–941.
32. Bokhari, A.; Baleanu, D.; Belgacem, R. Application of Laplace transform to Atangana-Baleanu derivatives. *J. Math. Comput. Sci.* **2019**, *20*, 101–107. [CrossRef]
33. Belgacem, R.; Baleanu, D.; Bokhari, A. Laplace Transform and Applications to Caputo-Fractional Differential Equations. *Int. J. Anal. Appl.* **2019**, *17*, 917–927.
34. Machado, J.; Baleanu, D.; Chen, W.; Sabatier, J. New trends in fractional dynamics. *J. Vib. Control* **2014**, *20*, 963. [CrossRef]
35. Baleanu, D.; Guvenc, Z.; Machado, J. *New Trends in Nanotechnology and Fractional Calculus Applications*; Springer: Dordrecht, The Netherlands, 2010.
36. Maitama, S.; Zhao, W. New integral transform: Laplace transform a generalization of Sumudu and Laplace transform for solving differential equations. *arXiv* **2019**, arXiv:1904.11370.
37. El-Kalla, I.L. Convergence of the Adomian method applied to a class of nonlinear integral equations. *Appl. Math. Lett.* **2008**, *21*, 372–376. [CrossRef]

Article

New Variational Problems with an Action Depending on Generalized Fractional Derivatives, the Free Endpoint Conditions, and a Real Parameter

Ricardo Almeida *[image] and Natália Martins [image]

Center for Research and Development in Mathematics and Applications (CIDMA), Department of Mathematics, University of Aveiro, 3810-193 Aveiro, Portugal; natalia@ua.pt
* Correspondence: ricardo.almeida@ua.pt

Abstract: This work presents optimality conditions for several fractional variational problems where the Lagrange function depends on fractional order operators, the initial and final state values, and a free parameter. The fractional derivatives considered in this paper are the Riemann–Liouville and the Caputo derivatives with respect to an arbitrary kernel. The new variational problems studied here are generalizations of several types of variational problems, and therefore, our results generalize well-known results from the fractional calculus of variations. Namely, we prove conditions useful to determine the optimal orders of the fractional derivatives and necessary optimality conditions involving time delays and arbitrary real positive fractional orders. Sufficient conditions for such problems are also studied. Illustrative examples are provided.

Keywords: fractional calculus; Euler–Lagrange equation; natural boundary conditions; time delay

MSC: 26A33; 49K05; 34A08

1. Introduction

Fractional calculus refers to the integration and differentiation of a non-integer order and is as old as the classical (integer order) calculus [1]. It is a subject that has gained much popularity and importance in the last few decades and has been applied in several fields of knowledge, such as mechanics [2,3], bioengineering [4], signal and image processing [5], physics [6,7], viscoelasticity [8], electrical engineering [9], economics [10], epidemiology [11,12], control theory [13,14], energy supply-demand systems [15], and fuzzy problems [16].

One of the specificities of fractional calculus is that there are many definitions of fractional derivatives that allow the researcher to choose the one that best corresponds to a given problem. Some of the most commonly used fractional derivatives are the Riemann–Liouville, the Erdélyi–Kober, the Caputo, the Hadamard, and the Grünwald–Letnikov derivatives. For a detailed study on this subject, see [1,17]. In this present work, we consider fractional operators with respect to an arbitrary kernel (see [17] for the Riemann–Liouville sense and [18] for the Caputo sense).

Fractional calculus of variations is a recent field that consists of minimizing or maximizing functionals that depend on fractional operators. The first works in this scientific area are due to Riewe [3,19]. Since then, many papers were published on different topics of the fractional calculus of variations for different types of fractional operators (see [2,20–29] and the references therein). For more details, we recommend the works [30–32].

By considering a more general form of the fractional derivative, like the Caputo fractional derivative with respect to an arbitrary kernel (see [18]), we can generalize different fractional variational problems. In [23,33], necessary and sufficient optimality conditions were proven for different variational problems depending on the Caputo fractional derivative with respect to an arbitrary kernel.

In [33], the following problem was studied: determine $x \in C^1([a,b], \mathbb{R})$ and $\zeta \in \mathbb{R}$ that extremize:

$$\mathcal{J}(x,\zeta) := \int_a^b L\left(t, x(t), (^C D_{a+}^{\gamma,g} x)(t), (^C D_{b-}^{\delta,g} x)(t), x(a), x(b), \zeta\right) dt, \tag{1}$$

where $L \in C^1([a,b] \times \mathbb{R}^6, \mathbb{R})$ and $^C D_{a+}^{\gamma,g} x$ and $^C D_{b-}^{\delta,g} x$ denote, respectively, the left and right g-Caputo fractional derivatives of x of order γ and δ, with $\gamma, \delta \in]0,1[$ (see Definition 3). The main results of [33] are optimality conditions for variational problems with or without isoperimetric and holonomic constraints. The aim of this paper is to generalize these previous results. It is important to mention that this type of generalized fractional variational problem cannot be solved using the classical theory. Moreover, since the g-Caputo fractional derivatives are generalizations of several fractional derivatives and the variational problem (1) is a generalization of several kinds of the calculus of variation problems, the results obtained in [33] not only generalize some known results, but also give new contributions to the theory of the fractional calculus of variations.

In this paper, we prove optimality conditions for different fractional variational problems that are generalizations of the one introduced in [33]. Namely, we prove the generalized fractional variational principle for problems with optimal orders, with time delay and with arbitrary real order fractional derivatives. In addition, we prove sufficient optimality conditions for all of the problems considered in the paper.

This main structure of this paper is as follows. In Section 2, we present some preliminaries on fractional calculus. In Section 3, we exhibit the main results. We finish the paper with two examples and some conclusions.

2. Preliminaries

We begin with a brief review of some important concepts and results that will be used in this paper. In what follows, Γ represents the well-known Gamma function, and the integer part of $\gamma \in \mathbb{R}$ is denoted by $[\gamma]$.

Definition 1. [17] Let γ be a positive real, $g : [a,b] \to \mathbb{R}$ a C^1 function with positive derivative, and $x \in L^1([a,b], \mathbb{R})$. The left Riemann–Liouville fractional integral of x of order γ, with respect to the kernel g, is defined as:

$$(I_{a+}^{\gamma,g} x)(t) := \frac{1}{\Gamma(\gamma)} \int_a^t g'(\tau)(g(t) - g(\tau))^{\gamma-1} x(\tau) \, d\tau, \quad t > a,$$

and the right derivative is given by:

$$(I_{b-}^{\gamma,g} x)(t) := \frac{1}{\Gamma(\gamma)} \int_t^b g'(\tau)(g(\tau) - g(t))^{\gamma-1} x(\tau) \, d\tau, \quad t < b.$$

Next, we present the definitions of the g-Riemann–Liouville fractional derivatives of a function x of order γ.

Definition 2. [17] Let γ be a positive real, $g : [a,b] \to \mathbb{R}$ a C^n function with positive derivative, and $x \in L^1([a,b], \mathbb{R})$. The left Riemann–Liouville fractional derivative of x of order γ, with respect to the kernel g, is given by:

$$(D_{a+}^{\gamma,g} x)(t) := \left(\frac{1}{g'(t)} \frac{d}{dt}\right)^n (I_{a+}^{n-\gamma,g} x)(t), \quad t > a,$$

and the right derivative by:

$$(D_{b-}^{\gamma,g} x)(t) := \left(-\frac{1}{g'(t)} \frac{d}{dt}\right)^n (I_{b-}^{n-\gamma,g} x)(t), \quad t < b,$$

where $n = [\gamma] + 1$.

Remark 1. *It is easily seen that:*

1. *for certain choices of the kernel g, we recover well-known fractional derivatives, such as Riemann–Liouville ($g(t) = t$), Hadamard ($g(t) = \ln(t)$, $a > 0$), and Erdélyi–Kober fractional derivatives ($g(t) = t^\sigma$, $\sigma > 0$);*
2. *if $\gamma = m \in \mathbb{N}$, then:*

$$(D_{a^+}^{\gamma,g} x)(t) = \left(\frac{1}{g'(t)}\frac{d}{dt}\right)^m x(t) \quad \text{and} \quad (D_{b^-}^{\gamma,g} x)(t) = \left(-\frac{1}{g'(t)}\frac{d}{dt}\right)^m x(t).$$

Next, the concept of g-Caputo fractional derivatives of x of order γ is presented, which is fundamental for the formulation of our problem.

Definition 3. *[18] Let γ be a positive real and:*

$$n = \begin{cases} [\gamma] + 1 & \text{if } \gamma \notin \mathbb{N} \\ \gamma & \text{if } \gamma \in \mathbb{N} \end{cases}$$

Let x, g be two real C^n functions defined on $[a,b]$, where g satisfies $g'(t) > 0$. The left Caputo fractional derivative of x of order γ, with respect to the kernel g, is defined as:

$$({}^C D_{a^+}^{\gamma,g} x)(t) := \left(I_{a^+}^{n-\gamma,g} \left(\frac{1}{g'(t)}\frac{d}{dt}\right)^n x \right)(t), \quad t > a,$$

and the right derivative as:

$$({}^C D_{b^-}^{\gamma,g} x)(t) := \left(I_{b^-}^{n-\gamma,g} \left(-\frac{1}{g'(t)}\frac{d}{dt}\right)^n x \right)(t), \quad t < b.$$

Remark 2. *It is clear that if g is the identity, then ${}^C D_{a^+}^{\gamma,g} x$ and ${}^C D_{b^-}^{\gamma,g} x$ are the usual Caputo fractional derivatives of x. Notice that if $\gamma = m \in \mathbb{N}$, then:*

$$({}^C D_{a^+}^{\gamma,g} x)(t) = \left(\frac{1}{g'(t)}\frac{d}{dt}\right)^m x(t) \quad \text{and} \quad ({}^C D_{b^-}^{\gamma,g} x)(t) = \left(-\frac{1}{g'(t)}\frac{d}{dt}\right)^m x(t).$$

Otherwise,

$$({}^C D_{a^+}^{\gamma,g} x)(t) = \frac{1}{\Gamma(n-\gamma)} \int_a^t g'(\tau)(g(t)-g(\tau))^{n-\gamma-1} \left(\frac{1}{g'(\tau)}\frac{d}{d\tau}\right)^n x(\tau)\, d\tau$$

and:

$$({}^C D_{b^-}^{\gamma,g} x)(t) = \frac{1}{\Gamma(n-\gamma)} \int_t^b g'(\tau)(g(\tau)-g(t))^{n-\gamma-1} \left(-\frac{1}{g'(\tau)}\frac{d}{d\tau}\right)^n x(\tau)\, d\tau.$$

Since the integration by parts formula is of great importance in the calculus of variations, we state here this basic result.

Theorem 1. *[18] Let x be a continuous function and y, g two C^n functions, with domain $[a,b]$. Then,*

$$\int_a^b x(t) \cdot ({}^C D_{a^+}^{\gamma,g} y)(t)\, dt = \int_a^b y(t) \cdot \left(D_{b^-}^{\gamma,g} \frac{x}{g'}\right)(t) g'(t)\, dt$$

$$+ \left[\sum_{k=0}^{n-1} \left(-\frac{1}{g'(t)}\frac{d}{dt}\right)^k \left(I_{b^-}^{n-\gamma,g} \frac{x}{g'}\right)(t) \cdot \left(\frac{1}{g'(t)}\frac{d}{dt}\right)^{n-k-1} y(t) \right]_{t=a}^{t=b}$$

and:

$$\int_a^b x(t) \cdot ({}^C D_{b-}^{\gamma,g} y)(t)\, dt = \int_a^b y(t) \cdot \left(D_{a+}^{\gamma,g} \frac{x}{g'}\right)(t) g'(t)\, dt$$
$$+ \left[\sum_{k=0}^{n-1} (-1)^{n-k}\left(\frac{1}{g'(t)}\frac{d}{dt}\right)^k \left(I_{a+}^{n-\gamma,g}\frac{x}{g'}\right)(t) \cdot \left(\frac{1}{g'(t)}\frac{d}{dt}\right)^{n-k-1} y(t)\right]_{t=a}^{t=b}.$$

Remark 3. *In particular, if $0 < \gamma < 1$, Theorem 1 reduces to:*

$$\int_a^b x(t) \cdot ({}^C D_{a+}^{\gamma,g} y)(t)\, dt = \int_a^b y(t) \cdot \left(D_{b-}^{\gamma,g} \frac{x}{g'}\right)(t) g'(t)\, dt + \left[\left(I_{b-}^{1-\gamma,g}\frac{x}{g'}\right)(t) \cdot y(t)\right]_{t=a}^{t=b}$$

and:

$$\int_a^b x(t) \cdot ({}^C D_{b-}^{\gamma,g} y)(t)\, dt = \int_a^b y(t) \cdot \left(D_{a+}^{\gamma,g} \frac{x}{g'}\right)(t) g'(t)\, dt - \left[\left(I_{a+}^{1-\gamma,g}\frac{x}{g'}\right)(t) \cdot y(t)\right]_{t=a}^{t=b}.$$

Next, we present the following result, which is useful in applications. For a more detailed study of the g-Caputo fractional derivatives, we refer to [18].

Lemma 1. *[18] If $n < \sigma \in \mathbb{R}$, then:*

$$^C D_{a+}^{\gamma,g}(g(t) - g(a))^{\sigma-1} = \frac{\Gamma(\sigma)}{\Gamma(\sigma - \gamma)}(g(t) - g(a))^{\sigma-\gamma-1}$$

and:

$$^C D_{b-}^{\gamma,g}(g(b) - g(t))^{\sigma-1} = \frac{\Gamma(\sigma)}{\Gamma(\sigma - \gamma)}(g(b) - g(t))^{\sigma-\gamma-1}.$$

Throughout the text, the partial derivative of L with respect to its i-th argument is denoted by $\partial_i L$.

3. Main Results

Now, we are ready to present the main contributions of this work, by proving some generalizations of the fractional variational problem studied in [33]. The results of the paper are trivially generalized for the case of vector functions x.

3.1. Generalized Fractional Variational Principle with Optimal Orders

One of the advantages of fractional derivatives is that, in many real problems, they better describe the dynamics of the problems compared to the classical derivative. With this in mind, a natural issue is to include the order of the fractional derivatives in the optimization process, that is, the variational problem under study consists of finding a curve x, a parameter ζ, and the order of the fractional derivatives γ and δ that extremize the variational functional.

Consider the following problem:

Problem 1. *Determine the functions $x : [a,b] \to \mathbb{R}$ of class C^1, the parameters $\zeta \in \mathbb{R}$, and fractional orders $\gamma, \delta \in\,]0,1[$ that minimize or maximize:*

$$\mathcal{J}_1(x,\zeta,\gamma,\delta) := \int_a^b L\left(t, x(t), ({}^C D_{a+}^{\gamma,g} x)(t), ({}^C D_{b-}^{\delta,g} x)(t), x(a), x(b), \zeta\right) dt, \qquad (2)$$

where $L \in C^1([a,b] \times \mathbb{R}^6, \mathbb{R})$ and $x(a)$ and $x(b)$ can be fixed or free.

For simplification, we use the notation:

$$[x, \zeta, \gamma, \delta](t) := \left(t, x(t), (^C D_{a+}^{\gamma,g} x)(t), (^C D_{b-}^{\delta,g} x)(t), x(a), x(b), \zeta\right).$$

The next result is the optimal fractional order variational principle for Problem 1.

Theorem 2. *If $(x^\star, \zeta^\star, \gamma^\star, \delta^\star)$ is an extremizer of functional $\mathcal{J}_1$ defined by (2) and if the maps:*

$$t \mapsto \left(D_{b-}^{\gamma^\star, g} \frac{\partial_3 L[x^\star, \zeta^\star, \gamma^\star, \delta^\star]}{g'}\right)(t) \quad \text{and} \quad t \mapsto \left(D_{a+}^{\delta^\star, g} \frac{\partial_4 L[x^\star, \zeta^\star, \gamma^\star, \delta^\star]}{g'}\right)(t)$$

are continuous, then, for all t,

$$\partial_2 L[x^\star, \zeta^\star, \gamma^\star, \delta^\star](t) + \left(D_{b-}^{\gamma^\star, g} \frac{\partial_3 L[x^\star, \zeta^\star, \gamma^\star, \delta^\star]}{g'}\right)(t) g'(t)$$

$$+ \left(D_{a+}^{\delta^\star, g} \frac{\partial_4 L[x^\star, \zeta^\star, \gamma^\star, \delta^\star]}{g'}\right)(t) g'(t) = 0. \quad (3)$$

Furthermore, the following conditions hold:

$$\int_a^b \partial_7 L[x^\star, \zeta^\star, \gamma^\star, \delta^\star](t)\, dt = 0, \quad (4)$$

$$\int_a^b f_t'(\gamma^\star) \cdot \partial_3 L[x^\star, \zeta^\star, \gamma^\star, \delta^\star](t)\, dt = 0, \quad (5)$$

$$\int_a^b g_t'(\delta^\star) \cdot \partial_4 L[x^\star, \zeta^\star, \gamma^\star, \delta^\star](t)\, dt = 0, \quad (6)$$

where, for each $t \in [a,b]$, $f_t :]0,1[\to \mathbb{R}$ and $g_t :]0,1[\to \mathbb{R}$ are the functions defined as follows:

$$f_t(\gamma) = (^C D_{a+}^{\gamma,g} x^\star)(t) \quad \text{and} \quad g_t(\delta) = (^C D_{b-}^{\delta,g} x^\star)(t).$$

If $x(a)$ is not fixed, the following is verified:

$$\int_a^b \partial_5 L[x^\star, \zeta^\star, \gamma^\star, \delta^\star](t)\, dt$$

$$= \left(I_{b-}^{1-\gamma^\star, g} \frac{\partial_3 L[x^\star, \zeta^\star, \gamma^\star, \delta^\star]}{g'}\right)(a) - \left(I_{a+}^{1-\delta^\star, g} \frac{\partial_4 L[x^\star, \zeta^\star, \gamma^\star, \delta^\star]}{g'}\right)(a); \quad (7)$$

Furthermore, if $x(b)$ is not fixed,

$$\int_a^b \partial_6 L[x^\star, \zeta^\star, \gamma^\star, \delta^\star](t)\, dt$$

$$= \left(I_{a+}^{1-\delta^\star, g} \frac{\partial_4 L[x^\star, \zeta^\star, \gamma^\star, \delta^\star]}{g'}\right)(b) - \left(I_{b-}^{1-\gamma^\star, g} \frac{\partial_3 L[x^\star, \zeta^\star, \gamma^\star, \delta^\star]}{g'}\right)(b). \quad (8)$$

Proof. Suppose that $(x^\star, \zeta^\star, \gamma^\star, \delta^\star)$ is an extremizer for functional $\mathcal{J}_1$. Hence, for any (fixed) $\eta \in C^1([a,b], \mathbb{R})$, $\Delta\zeta, \Delta\gamma, \Delta\delta \in \mathbb{R}$ such that $0 < \gamma^\star + \epsilon\Delta\gamma < 1$ and $0 < \delta^\star + \epsilon\Delta\delta < 1$, with ϵ in a neighborhood of zero, we conclude that:

$$\frac{d}{d\epsilon} \mathcal{J}_1(x^\star + \epsilon\eta, \zeta^\star + \epsilon\Delta\zeta, \gamma^\star + \epsilon\Delta\gamma, \delta^\star + \epsilon\Delta\delta)\Big|_{\epsilon=0} = 0.$$

Therefore, the following condition holds:

$$\int_a^b \Big(\partial_2 L[x^\star,\zeta^\star,\gamma^\star,\delta^\star](t)\cdot \eta(t)+\partial_3 L[x^\star,\zeta^\star,\gamma^\star,\delta^\star](t)\cdot \left(f'_t(\gamma^\star)\Delta\gamma+({}^C D_{a+}^{\gamma^\star,g}\eta)(t)\right)$$
$$+\partial_4 L[x^\star,\zeta^\star,\gamma^\star,\delta^\star](t)\cdot \left(g'_t(\delta^\star)\Delta\delta+({}^C D_{b-}^{\delta^\star,g}\eta)(t)\right)$$
$$+\partial_5 L[x^\star,\zeta^\star,\gamma^\star,\delta^\star](t)\cdot \eta(a)+\partial_6 L[x^\star,\zeta^\star,\gamma^\star,\delta^\star](t)\cdot \eta(b)+\partial_7 L[x^\star,\zeta^\star,\gamma^\star,\delta^\star](t)\cdot \Delta\zeta\Big)\,dt = 0.$$

Integration by parts gives (see Remark 3):

$$\int_a^b \left(\partial_2 L[x^\star,\zeta^\star,\gamma^\star,\delta^\star](t)+\left(D_{b-}^{\gamma^\star,g}\frac{\partial_3 L[x^\star,\zeta^\star,\gamma^\star,\delta^\star]}{g'}\right)(t)g'(t)\right.$$
$$+\left(D_{a+}^{\delta^\star,g}\frac{\partial_4 L[x^\star,\zeta^\star,\gamma^\star,\delta^\star]}{g'}\right)(t)g'(t)\Bigg)\cdot \eta(t)\,dt + \left[\left(I_{b-}^{1-\gamma^\star,g}\frac{\partial_3 L[x^\star,\zeta^\star,\gamma^\star,\delta^\star]}{g'}\right)(t)\cdot \eta(t)\right]_{t=a}^{t=b}$$
$$-\left[\left(I_{a+}^{1-\delta^\star,g}\frac{\partial_4 L[x^\star,\zeta^\star,\gamma^\star,\delta^\star]}{g'}\right)(t)\cdot \eta(t)\right]_{t=a}^{t=b}$$
$$+\Delta\gamma\int_a^b f'_t(\gamma^\star)\cdot \partial_3 L[x^\star,\zeta^\star,\gamma^\star,\delta^\star](t)\,dt + \Delta\delta\int_a^b g'_t(\delta^\star)\cdot \partial_4 L[x^\star,\zeta^\star,\gamma^\star,\delta^\star](t)\,dt$$
$$+\int_a^b \Big(\partial_5 L[x^\star,\zeta^\star,\gamma^\star,\delta^\star](t)\cdot \eta(a)+\partial_6 L[x^\star,\zeta^\star,\gamma^\star,\delta^\star](t)\cdot \eta(b)$$
$$+\partial_7 L[x^\star,\zeta^\star,\gamma^\star,\delta^\star](t)\cdot \Delta\zeta\Big)\,dt = 0. \quad (9)$$

We first consider functions η such that $\eta(a)=\eta(b)=0$. In this case, Equation (9) becomes:

$$\int_a^b \left(\partial_2 L[x^\star,\zeta^\star,\gamma^\star,\delta^\star](t)+\left(D_{b-}^{\gamma^\star,g}\frac{\partial_3 L[x^\star,\zeta^\star,\gamma^\star,\delta^\star]}{g'}\right)(t)g'(t)\right.$$
$$+\left(D_{a+}^{\delta^\star,g}\frac{\partial_4 L[x^\star,\zeta^\star,\gamma^\star,\delta^\star]}{g'}\right)(t)g'(t)\Bigg)\cdot \eta(t)\,dt + \Delta\gamma\int_a^b f'_t(\gamma^\star)\cdot \partial_3 L[x^\star,\zeta^\star,\gamma^\star,\delta^\star](t)\,dt$$
$$+\Delta\delta\int_a^b g'_t(\delta^\star)\cdot \partial_4 L[x^\star,\zeta^\star,\gamma^\star,\delta^\star](t)\,dt + \Delta\zeta\int_a^b \partial_7 L[x^\star,\zeta^\star,\gamma^\star,\delta^\star](t)\,dt = 0. \quad (10)$$

By the arbitrariness of $\Delta\gamma$, $\Delta\delta$, and $\Delta\zeta$, if we consider that all of them are null, using Lemma 2.2.2 in [34], we get:

$$\partial_2 L[x^\star,\zeta^\star,\gamma^\star,\delta^\star](t)+\left(D_{b-}^{\gamma^\star,g}\frac{\partial_3 L[x^\star,\zeta^\star,\gamma^\star,\delta^\star]}{g'}\right)(t)g'(t)$$
$$+\left(D_{a+}^{\delta^\star,g}\frac{\partial_4 L[x^\star,\zeta^\star,\gamma^\star,\delta^\star]}{g'}\right)(t)g'(t) = 0,$$

for all t, proving the Euler–Lagrange Equation (3). Since $(x^\star,\zeta^\star,\gamma^\star,\delta^\star)$ satisfies Equality (3) for all $t \in [a,b]$, the first integral in (10) vanishes, and then, it takes the form:

$$\Delta\gamma\int_a^b f'_t(\gamma^\star)\cdot \partial_3 L[x^\star,\zeta^\star,\gamma^\star,\delta^\star](t)\,dt + \Delta\delta\int_a^b g'_t(\delta^\star)\cdot \partial_4 L[x^\star,\zeta^\star,\gamma^\star,\delta^\star](t)\,dt$$
$$+\Delta\zeta\int_a^b \partial_7 L[x^\star,\zeta^\star,\gamma^\star,\delta^\star](t)\,dt = 0. \quad (11)$$

By the arbitrariness of $\Delta\gamma$, $\Delta\delta$, and $\Delta\zeta$, we deduce from (11) the necessary conditions (4)–(6). We now seek the natural boundary conditions.

1. If $x(a)$ is not fixed in the formulation of the problem, then η need not to be null at $t = a$. Restricting η to be null at $t = b$ and substituting the necessary conditions (3)–(6) into (9), it follows that:

$$\left(\left(I_{a^+}^{1-\delta^*,g} \frac{\partial_4 L[x^*,\zeta^*,\gamma^*,\delta^*]}{g'} \right)(a) - \left(I_{b^-}^{1-\gamma^*,g} \frac{\partial_3 L[x^*,\zeta^*,\gamma^*,\delta^*]}{g'} \right)(a) \right.$$
$$\left. + \int_a^b \partial_5 L[x^*,\zeta^*,\gamma^*,\delta^*](t)\,dt \right) \cdot \eta(a) = 0.$$

Since $\eta(a)$ is an arbitrary real, we prove (7).
2. Suppose now that $x(b)$ is not fixed. Restricting η to be null at $t = a$ and using similar arguments as previously, we get Equation (8).

□

Remark 4. We note that if L does not depend on $(^C D_{b^-}^{\delta,g} x)(t)$, $x(a)$, $x(b)$, and ζ, then Theorem 2 reduces to Theorem 2.9 from [23] if the final time is fixed.

3.2. Generalized Variational Problems with Time Delay

It is known that a delay is inherent in many problems, such as in control theory, bioengineering, electrochemistry, and social sciences. Differential equations with time delays have been used to model complex systems and have led to an intense topic of research for many years. Although fractional derivatives are not local in nature and are capable of modeling memory effects, delays are also very important because they take into account the system's history from a previous state. For these reasons, many real-world problems can be modeled more precisely, including fractional derivatives and time delays. In recent years, delayed fractional differential equations have started to attract the attention of many researchers [35–37]. Few works are yet devoted to fractional variational problems with time delay so far [38–40].

Encouraged by the importance of considering a delay in many real-world problems, we study here the following fractional problem with a time delay τ, where $\tau \in \mathbb{R}$ satisfies $0 \le \tau < b - a$.

Problem 2. Determine a C^1 function $x : [a - \tau, b] \to \mathbb{R}$, subject to $x(t) = X(t)$, for all $t \in [a - \tau, a]$, where X is a given initial function of class C^1 and $\zeta \in \mathbb{R}$ that minimize or maximize:

$$\mathcal{J}_2(x,\zeta) := \int_a^b L\left(t, x(t), x(t-\tau), (^C D_{a^+}^{\gamma,g} x)(t), (^C D_{b^-}^{\delta,g} x)(t), x(a), x(b), \zeta\right) dt, \quad (12)$$

where $L \in C^1([a,b] \times \mathbb{R}^7, \mathbb{R})$.

Define:

$$[x,\zeta]_\tau(t) := \left(t, x(t), x(t-\tau), (^C D_{a^+}^{\gamma,g} x)(t), (^C D_{b^-}^{\delta,g} x)(t), x(a), x(b), \zeta \right).$$

Theorem 3. Suppose that (x^*, ζ^*) is an extremizer of $\mathcal{J}_2$ defined by (12) and that the functions exist and are continuous:

$$t \mapsto \left(D_{(b-\tau)^-}^{\gamma,g} \frac{\partial_4 L[x^*,\zeta^*]_\tau}{g'} \right)(t) \quad \text{and} \quad t \mapsto \left(D_{a^+}^{\delta,g} \frac{\partial_5 L[x^*,\zeta^*]_\tau}{g'} \right)(t) \quad \text{on } [a, b-\tau]$$

and:

$$t \mapsto \left(D_{b^-}^{\gamma,g} \frac{\partial_4 L[x^*,\zeta^*]_\tau}{g'} \right)(t) \quad \text{and} \quad t \mapsto \left(D_{(b-\tau)^+}^{\delta,g} \frac{\partial_5 L[x^*,\zeta^*]_\tau}{g'} \right)(t) \quad \text{on } [b-\tau, b].$$

Then, for all $t \in [a, b - \tau]$,

$$\partial_2 L[x^*, \zeta^*]_\tau(t) + \partial_3 L[x^*, \zeta^*]_\tau(t + \tau) + \left(D_{(b-\tau)^-}^{\gamma,g} \frac{\partial_4 L[x^*, \zeta^*]_\tau}{g'}\right)(t) \cdot g'(t)$$
$$+ \left(D_{a^+}^{\delta,g} \frac{\partial_5 L[x^*, \zeta^*]_\tau}{g'}\right)(t) \cdot g'(t) - \frac{1}{\Gamma(1-\gamma)} \frac{d}{dt} \int_{b-\tau}^{b} (g(s) - g(t))^{-\gamma} \partial_4 L[x^*, \zeta^*]_\tau(s)\, ds = 0, \quad (13)$$

and for all $t \in [b - \tau, b]$,

$$\partial_2 L[x^*, \zeta^*]_\tau(t) + \left(D_{b^-}^{\gamma,g} \frac{\partial_4 L[x^*, \zeta^*]_\tau}{g'}\right)(t) \cdot g'(t) + \left(D_{(b-\tau)^+}^{\delta,g} \frac{\partial_5 L[x^*, \zeta^*]_\tau}{g'}\right)(t) \cdot g'(t)$$
$$+ \frac{1}{\Gamma(1-\delta)} \frac{d}{dt} \int_a^{b-\tau} (g(t) - g(s))^{-\delta} \partial_5 L[x^*, \zeta^*]_\tau(s)\, ds = 0. \quad (14)$$

Moreover,

$$\int_a^b \partial_8 L[x^*, \zeta^*]_\tau(t)\, dt = 0 \quad (15)$$

and if $x(b)$ is not fixed, then:

$$\int_a^b \partial_7 L[x^*, \zeta^*]_\tau(t)\, dt = \left(I_{a^+}^{1-\delta,g} \frac{\partial_5 L[x^*, \zeta^*]_\tau}{g'}\right)(b) - \left(I_{b^-}^{1-\gamma,g} \frac{\partial_4 L[x^*, \zeta^*]_\tau}{g'}\right)(b). \quad (16)$$

Proof. Let $\eta : [a - \tau, b] \to \mathbb{R}$ be a C^1 function vanishing on $[a - \tau, a]$, and let $\Delta \zeta$ be a real. Consider:

$$\omega(\epsilon) = \mathcal{J}_2(x^* + \epsilon \eta, \zeta^* + \epsilon \Delta \zeta)$$

defined on an open interval containing zero. Since (x^*, ζ^*) is an extremizer of $\mathcal{J}_2$, then $\omega'(0) = 0$, and therefore:

$$\int_a^b \Big(\partial_2 L[x^*, \zeta^*]_\tau(t) \cdot \eta(t) + \partial_3 L[x^*, \zeta^*]_\tau(t) \cdot \eta(t - \tau) + \partial_4 L[x^*, \zeta^*]_\tau(t) \cdot ({}^C D_{a^+}^{\gamma,g} \eta)(t)$$
$$+ \partial_5 L[x^*, \zeta^*]_\tau(t) \cdot ({}^C D_{b^-}^{\delta,g} \eta)(t) + \partial_6 L[x^*, \zeta^*]_\tau(t) \cdot \eta(a) + \partial_7 L[x, \zeta](t) \cdot \eta(b)$$
$$+ \partial_8 L[x^*, \zeta^*]_\tau(t) \cdot \Delta \zeta \Big) dt = 0. \quad (17)$$

Considering $t = u + \tau$, we obtain:

$$\int_a^b \partial_3 L[x^*, \zeta^*]_\tau(t) \cdot \eta(t - \tau)\, dt = \int_a^{b-\tau} \partial_3 L[x^*, \zeta^*]_\tau(t + \tau) \cdot \eta(t)\, dt. \quad (18)$$

Observe that, for $a \leq t \leq b - \tau$,

$$\left(D_{b^-}^{\gamma,g} \frac{\partial_4 L[x^*, \zeta^*]_\tau}{g'}\right)(t) = \left(D_{(b-\tau)^-}^{\gamma,g} \frac{\partial_4 L[x^*, \zeta^*]_\tau}{g'}\right)(t)$$
$$- \frac{1}{\Gamma(1-\gamma)} \left(\frac{1}{g'(t)} \frac{d}{dt}\right) \int_{b-\tau}^b (g(s) - g(t))^{-\gamma} \partial_4 L[x^*, \zeta^*]_\tau(s)\, ds \quad (19)$$

and for $b - \tau \leq t \leq b$,

$$\left(D_{a^+}^{\delta,g} \frac{\partial_5 L[x^*, \zeta^*]_\tau}{g'}\right)(t) = \left(D_{(b-\tau)^+}^{\delta,g} \frac{\partial_5 L[x^*, \zeta^*]_\tau}{g'}\right)(t)$$
$$+ \frac{1}{\Gamma(1-\delta)} \left(\frac{1}{g'(t)} \frac{d}{dt}\right) \int_a^{b-\tau} (g(t) - g(s))^{-\delta} \partial_5 L[x^*, \zeta^*]_\tau(s)\, ds. \quad (20)$$

By Theorem 1 and (19), we obtain:

$$\int_a^b \partial_4 L[x^\star,\zeta^\star]_\tau(t) \cdot (^C D_{a^+}^{\gamma,g}\eta)(t)\,dt = \int_a^{b-\tau} \left(\left(D_{(b-\tau)^-}^{\gamma,g}\frac{\partial_4 L[x^\star,\zeta^\star]_\tau}{g'}\right)(t)\cdot g'(t)\right.$$
$$\left.-\frac{1}{\Gamma(1-\gamma)}\frac{d}{dt}\int_{b-\tau}^b (g(s)-g(t))^{-\gamma}\partial_4 L[x^\star,\zeta^\star]_\tau(s)\,ds\right)\cdot\eta(t)\,dt$$
$$+\int_{b-\tau}^b \left(D_{b^-}^{\gamma,g}\frac{\partial_4 L[x^\star,\zeta^\star]_\tau}{g'}\right)(t)g'(t)\cdot\eta(t)\,dt + \left[\left(I_{b^-}^{1-\gamma,g}\frac{\partial_4 L[x^\star,\zeta^\star]_\tau}{g'}\right)(t)\cdot\eta(t)\right]_{t=a}^{t=b}. \quad (21)$$

Again, by Theorem 1 and (20), we obtain:

$$\int_a^b \partial_5 L[x^\star,\zeta^\star]_\tau(t)\cdot (^C D_{b^-}^{\delta,g}\eta)(t)\,dt = \int_a^{b-\tau}\left(D_{a^+}^{\delta,g}\frac{\partial_5 L[x^\star,\zeta^\star]_\tau}{g'}\right)(t)g'(t)\cdot\eta(t)\,dt$$
$$-\left[\left(I_{a^+}^{1-\delta,g}\frac{\partial_5 L[x^\star,\zeta^\star]_\tau}{g'}\right)(t)\cdot\eta(t)\right]_{t=a}^{t=b} + \int_{b-\tau}^b\left(\left(D_{(b-\tau)^+}^{\delta,g}\frac{\partial_5 L[x^\star,\zeta^\star]_\tau}{g'}\right)(t)\cdot g'(t)\right.$$
$$\left.+\frac{1}{\Gamma(1-\delta)}\frac{d}{dt}\int_a^{b-\tau}(g(t)-g(s))^{-\delta}\partial_5 L[x^\star,\zeta^\star]_\tau(s)\,ds\right)\cdot\eta(t)\,dt. \quad (22)$$

Introducing (18), (21), and (22) into Equation (17), we can conclude that:

$$\int_a^{b-\tau}\left(\partial_2 L[x^\star,\zeta^\star]_\tau(t)+\partial_3 L[x^\star,\zeta^\star]_\tau(t+\tau)+\left(D_{(b-\tau)^-}^{\gamma,g}\frac{\partial_4 L[x^\star,\zeta^\star]_\tau}{g'}\right)(t)g'(t)\right.$$
$$\left.-\frac{1}{\Gamma(1-\gamma)}\frac{d}{dt}\int_{b-\tau}^b(g(s)-g(t))^{-\gamma}\partial_4 L[x^\star,\zeta^\star]_\tau(s)\,ds+\left(D_{a^+}^{\delta,g}\frac{\partial_5 L[x^\star,\zeta^\star]_\tau}{g'}\right)(t)g'(t)\right)\cdot\eta(t)\,dt$$
$$+\int_{b-\tau}^b\left(\partial_2 L[x^\star,\zeta^\star]_\tau(t)+\left(D_{b^-}^{\gamma,g}\frac{\partial_4 L[x^\star,\zeta^\star]_\tau}{g'}\right)(t)g'(t)+\left(D_{(b-\tau)^+}^{\delta,g}\frac{\partial_5 L[x^\star,\zeta^\star]_\tau}{g'}\right)(t)g'(t)\right.$$
$$\left.+\frac{1}{\Gamma(1-\delta)}\frac{d}{dt}\int_a^{b-\tau}(g(t)-g(s))^{-\delta}\partial_5 L[x^\star,\zeta^\star]_\tau(s)\,ds\right)\cdot\eta(t)\,dt$$
$$+\left[\left(I_{b^-}^{1-\gamma,g}\frac{\partial_4 L[x^\star,\zeta^\star]_\tau}{g'}\right)(t)\cdot\eta(t)\right]_{t=a}^{t=b}-\left[\left(I_{a^+}^{1-\delta,g}\frac{\partial_5 L[x^\star,\zeta^\star]_\tau}{g'}\right)(t)\cdot\eta(t)\right]_{t=a}^{t=b}$$
$$+\int_a^b\left(\partial_6 L[x^\star,\zeta^\star]_\tau(t)\cdot\eta(a)+\partial_7 L[x^\star,\zeta^\star]_\tau(t)\cdot\eta(b)+\partial_8 L[x^\star,\zeta^\star]_\tau(t)\cdot\Delta\zeta\right)dt = 0. \quad (23)$$

Since Equation (23) is valid for any variations η and all $\Delta\zeta$, assuming that η vanishes on the interval $[b-\tau,b]$ and taking $\Delta\zeta = 0$, from Lemma 2.2.2 in [34], we prove that Condition (13) holds on $[a,b-\tau]$. Restricting the variations η to those functions that satisfy $\eta(b) = 0$ and introducing Condition (13) into (23), we obtain:

$$\int_{b-\tau}^b\left(\partial_2 L[x^\star,\zeta^\star]_\tau(t)+\left(D_{b^-}^{\gamma,g}\frac{\partial_4 L[x^\star,\zeta^\star]_\tau}{g'}\right)(t)g'(t)+\left(D_{(b-\tau)^+}^{\delta,g}\frac{\partial_5 L[x^\star,\zeta^\star]_\tau}{g'}\right)(t)g'(t)\right.$$
$$\left.+\frac{1}{\Gamma(1-\delta)}\frac{d}{dt}\int_a^{b-\tau}(g(t)-g(s))^{-\delta}\partial_5 L[x^\star,\zeta^\star]_\tau(s)\,ds\right)\cdot\eta(t)\,dt$$
$$+\int_a^b \partial_8 L[x^\star,\zeta^\star]_\tau(t)\cdot\Delta\zeta\,dt = 0. \quad (24)$$

Since the last equality holds for all $\Delta\zeta$, then, in particular, it holds for $\Delta\zeta = 0$; hence, from Lemma 2.2.2 in [34], Condition (14) holds on the interval $[b - \tau, b]$. Introducing (14) into (24), we conclude, from the arbitrariness of $\Delta\zeta$, that $\int_a^b \partial_8 L[x^*, \zeta^*]_\tau(t)\, dt = 0$, proving the necessary condition (15). If $x(b)$ is free, $\eta(b)$ need not to be null; in this case, we get from (23) that:

$$\int_a^b \partial_7 L[x^*,\zeta^*]_\tau(t) \cdot \eta(b)\, dt + \left(I_{b-}^{1-\gamma,g} \frac{\partial_4 L[x^*,\zeta^*]_\tau}{g'}\right)(b) \cdot \eta(b)$$
$$- \left(I_{a+}^{1-\delta,g} \frac{\partial_5 L[x^*,\zeta^*]_\tau}{g'}\right)(b) \cdot \eta(b) = 0.$$

From the arbitrariness of $\eta(b)$, we prove Condition (16), as desired. $\square$

Remark 5. *We remark that:*

1. *if the delay is removed ($\tau = 0$), then Problem 2 coincides with the problem given by (1) if we consider $x(a)$ fixed, and therefore, the fractional variational principle given by Theorem 3 in [33] can be obtained from Theorem 3;*
2. *when the final time is fixed, Theorem 2.7 in [23] can be obtained from Theorem 3.*

3.3. Generalized Higher Order Fractional Variational Principle

In this subsection, we consider an extension of the generalized variational problem given by (1), by including in the Lagrangian function arbitrary real fractional orders $\gamma, \delta > 0$. With this, we obtain what is known as a fractional variational problem with arbitrary higher order fractional derivatives. The problem formulation is the following.

Problem 3. *Find functions $x : [a,b] \to \mathbb{R}$ of class C^n and $\zeta \in \mathbb{R}$ that minimize or maximize the functional:*

$$\mathcal{J}_3(x,\zeta) := \int_a^b L\Big(t, x(t), ({}^C D_{a+}^{\gamma_1,g} x)(t), ({}^C D_{b-}^{\delta_1,g} x)(t), \ldots, ({}^C D_{a+}^{\gamma_n,g} x)(t),$$
$$({}^C D_{b-}^{\delta_n,g} x)(t), x(a), x(b), \zeta\Big)\, dt, \quad (25)$$

where $L \in C^1([a,b] \times \mathbb{R}^{2n+4}, \mathbb{R})$, and $k - 1 < \gamma_k, \delta_k < k$, for $k = 1, \ldots, n$. Furthermore, the boundary conditions:

$$x^{(k)}(a) = x_a^k \quad \text{and} \quad x^{(k)}(b) = x_b^k, \; k = 1, \ldots, n-1, \qquad (26)$$

are assumed to hold, where $x_a^k, x_b^k \in \mathbb{R}$ are fixed, for all k.

To abbreviate, define:

$$[x,\zeta]_n(t) := \Big(t, x(t), ({}^C D_{a+}^{\gamma_1,g} x)(t), ({}^C D_{b-}^{\delta_1,g} x)(t), \ldots, ({}^C D_{a+}^{\gamma_n,g} x)(t), ({}^C D_{b-}^{\delta_n,g} x)(t), x(a), x(b), \zeta\Big).$$

Theorem 4. *If (x^*, ζ^*) is an extremizer of functional $\mathcal{J}_3$ defined by (25) and the functions exist and are continuous:*

$$t \mapsto \left(D_{b-}^{\gamma_i,g} \frac{\partial_{2i+1} L[x^*,\zeta^*]_n}{g'}\right)(t) \quad \text{and} \quad t \mapsto \left(D_{a+}^{\delta_i,g} \frac{\partial_{2i+2} L[x^*,\zeta^*]_n}{g'}\right)(t),$$

for all $i = 1, \ldots, n$, then:

$$\partial_2 L[x^\star, \zeta^\star]_n(t) + \sum_{i=1}^{n}\left[\left(D_{b^-}^{\gamma_i g}\frac{\partial_{2i+1}L[x^\star,\zeta^\star]_n}{g'}\right)(t)\cdot g'(t)\right.$$
$$\left.+\left(D_{a^+}^{\delta_i g}\frac{\partial_{2i+2}L[x^\star,\zeta^\star]_n}{g'}\right)(t)\cdot g'(t)\right] = 0 \quad (27)$$

and:

$$\int_a^b \partial_{2n+5}L[x^\star,\zeta^\star]_n(t)\,dt = 0. \quad (28)$$

If $x(a)$ is not fixed, then:

$$\int_a^b \partial_{2n+3}L[x^\star,\zeta^\star]_n(t)\,dt = \left[\sum_{i=1}^{n}\left(\left(-\frac{1}{g'(t)}\frac{d}{dt}\right)^{i-1}\left(I_{b^-}^{i-\gamma_i g}\frac{\partial_{2i+1}L[x^\star,\zeta^\star]_n}{g'}\right)(t)\right.\right. \quad (29)$$
$$\left.\left.-\left(\frac{1}{g'(t)}\frac{d}{dt}\right)^{i-1}\left(I_{a^+}^{i-\delta_i g}\frac{\partial_{2i+2}L[x^\star,\zeta^\star]_n}{g'}\right)(t)\right)\right]_{t=a},$$

and if $x(b)$ is not fixed, then:

$$\int_a^b \partial_{2n+4}L[x^\star,\zeta^\star]_n(t)\,dt = \left[\sum_{i=1}^{n}\left(\left(\frac{1}{g'(t)}\frac{d}{dt}\right)^{i-1}\left(I_{a^+}^{i-\delta_i g}\frac{\partial_{2i+2}L[x^\star,\zeta^\star]_n}{g'}\right)(t)\right.\right. \quad (30)$$
$$\left.\left.-\left(-\frac{1}{g'(t)}\frac{d}{dt}\right)^{i-1}\left(I_{b^-}^{i-\gamma_i g}\frac{\partial_{2i+1}L[x^\star,\zeta^\star]_n}{g'}\right)(t)\right)\right]_{t=b}.$$

Proof. Consider the pair given by $(x^\star + \epsilon\eta, \zeta^\star + \epsilon\Delta\zeta)$, where $\eta \in C^n([a,b],\mathbb{R})$ satisfies $\eta^{(i)}(a) = 0$ and $\eta^{(i)}(b) = 0$, for all $i \in \{1,\ldots,n-1\}$, and $\Delta\zeta, \epsilon$ are two arbitrary real numbers. Observe that:

$$\left(\frac{1}{g'(t)}\frac{d}{dt}\right)^i \eta(t) = 0 \quad \text{at } t \in \{a,b\}, \quad \forall i \in \{1,\ldots,n-1\}.$$

Defining:

$$v(\epsilon) = \mathcal{J}_3(x^\star + \epsilon\eta, \zeta^\star + \epsilon\Delta\zeta),$$

the condition $v'(0) = 0$ implies that:

$$\int_a^b \left(\partial_2 L[x^\star,\zeta^\star]_n(t)\cdot\eta(t) + \sum_{i=1}^{n}\left[\partial_{2i+1}L[x^\star,\zeta^\star]_n(t)\cdot({}^C D_{a^+}^{\gamma_i g}\eta)(t)\right.\right.$$
$$\left.\left.+\partial_{2i+2}L[x^\star,\zeta^\star]_n(t)\cdot({}^C D_{b^-}^{\delta_i g}\eta)(t)\right] + \partial_{2n+3}L[x^\star,\zeta^\star]_n(t)\cdot\eta(a)\right.$$
$$\left.+\partial_{2n+4}L[x^\star,\zeta^\star]_n(t)\cdot\eta(b) + \partial_{2n+5}L[x^\star,\zeta^\star]_n(t)\cdot\Delta\zeta\right)dt = 0.$$

Applying Theorem 1, we get, for each $i \in \{1,\ldots,n\}$,

$$\int_a^b \partial_{2i+1}L[x^\star,\zeta^\star]_n(t)\cdot({}^C D_{a^+}^{\gamma_i g}\eta)(t)\,dt = \int_a^b \left(D_{b^-}^{\gamma_i g}\frac{\partial_{2i+1}L[x^\star,\zeta^\star]_n}{g'}\right)(t)g'(t)\cdot\eta(t)\,dt$$
$$+\left[\left(-\frac{1}{g'(t)}\frac{d}{dt}\right)^{i-1}\left(I_{b^-}^{i-\gamma_i g}\frac{\partial_{2i+1}L[x^\star,\zeta^\star]_n}{g'}\right)(t)\cdot\eta(t)\right]_{t=a}^{t=b}$$

and:

$$\int_a^b \partial_{2i+2}L[x^\star,\zeta^\star]_n(t)\cdot({}^C D_{b^-}^{\delta_i,g}\eta)(t)\,dt = \int_a^b \left(D_{a^+}^{\delta_i,g}\frac{\partial_{2i+2}L[x^\star,\zeta^\star]_n}{g'}\right)(t)g'(t)\cdot\eta(t)\,dt$$

$$-\left[\left(\frac{1}{g'(t)}\frac{d}{dt}\right)^{i-1}\left(I_{a^+}^{i-\delta_i,g}\frac{\partial_{2i+2}L[x^\star,\zeta^\star]_n}{g'}\right)(t)\cdot\eta(t)\right]_{t=a}^{t=b}.$$

Thus,

$$\int_a^b\Bigg(\partial_2 L[x^\star,\zeta^\star]_n(t) + \sum_{i=1}^n\left[\left(D_{b^-}^{\gamma_i,g}\frac{\partial_{2i+1}L[x^\star,\zeta^\star]_n}{g'}\right)(t)g'(t)\right.$$

$$\left.+\left(D_{a^+}^{\delta_i,g}\frac{\partial_{2i+2}L[x^\star,\zeta^\star]_n}{g'}\right)(t)g'(t)\right]\Bigg)\cdot\eta(t)\,dt$$

$$+\sum_{i=1}^n\left[\left(\left(-\frac{1}{g'(t)}\frac{d}{dt}\right)^{i-1}\left(I_{b^-}^{i-\gamma_i,g}\frac{\partial_{2i+1}L[x^\star,\zeta^\star]_n}{g'}\right)(t)\right.\right.$$

$$\left.\left.-\left(\frac{1}{g'(t)}\frac{d}{dt}\right)^{i-1}\left(I_{a^+}^{i-\delta_i,g}\frac{\partial_{2i+2}L[x^\star,\zeta^\star]_n}{g'}\right)(t)\right)\cdot\eta(t)\right]_{t=a}^{t=b}$$

$$+\int_a^b\Big(\partial_{2n+3}L[x^\star,\zeta^\star]_n(t)\cdot\eta(a)+\partial_{2n+4}L[x^\star,\zeta^\star]_n(t)\cdot\eta(b)+\partial_{2n+5}L[x^\star,\zeta^\star]_n(t)\cdot\Delta\zeta\Big)dt=0.$$

Since η and $\Delta\zeta$ are arbitrary, we prove Equations (27)–(30). □

Remark 6. *We remark that:*

1. *we considered the constraints (26) for the simplicity of presentation; of course, we could consider the case when $x^{(k)}(a)$ and $x^{(k)}(b)$, $k = 1, \ldots, n-1$, are free, and at the end deduce the respective natural boundary conditions;*
2. *Theorem 2.8 in [23] with the final time fixed is a corollary of Theorem 4.*

3.4. Sufficient Optimality Conditions

In this subsection, we give sufficient conditions of optimization for all the problems considered previously, first for Problem 1.

Theorem 5. *Suppose that L satisfies the inequality:*

$$L\Big(t, x_1+\Delta x_1, {}^C D_{a^+}^{\gamma+\Delta\gamma,g}(x_1+\Delta x_1), {}^C D_{b^-}^{\delta+\Delta\delta,g}(x_1+\Delta x_1), x_4+\Delta x_4, x_5+\Delta x_5, x_6+\Delta x_6\Big)$$

$$-L\Big(t, x_1, {}^C D_{a^+}^{\gamma,g}x_1, {}^C D_{b^-}^{\delta,g}x_1, x_4, x_5, x_6\Big)\geq (\text{resp. }\leq)\ \partial_2 L[\bullet]\Delta x_1+\sum_{i=4}^{6}\partial_{i+1}L[\bullet]\Delta x_i$$

$$+\partial_3 L[\bullet]\Big(f_t'(\gamma).\Delta\gamma+{}^C D_{a^+}^{\gamma,g}\Delta x_1\Big)+\partial_4 L[\bullet]\Big(g_t'(\delta).\Delta\delta+{}^C D_{b^-}^{\delta,g}\Delta x_1\Big) \quad (31)$$

for all $x_1, \Delta x_1 \in C^1([a,b],\mathbb{R})$, $x_4, x_5, x_6, \Delta x_4, \Delta x_5, \Delta x_6 \in \mathbb{R}$, and $\Delta\gamma, \Delta\delta \in \mathbb{R}$ such that $0 < \gamma + \Delta\gamma < 1$ and $0 < \delta + \Delta\delta < 1$, where $[\bullet] := (t, x_1, {}^C D_{a^+}^{\gamma,g} x_1, {}^C D_{b^-}^{\delta,g} x_1, x_4, x_5, x_6)$ and f_t, g_t as defined in Theorem 2. If $(x^\star, \zeta^\star, \gamma^\star, \delta^\star)$ satisfies the necessary conditions (3)–(8), then $(x^\star, \zeta^\star, \gamma^\star, \delta^\star)$ is a minimizer (resp. maximizer) of functional $\mathcal{J}_1$.

Proof. We present the proof only when the inequality (31) holds for $\geq$; the other case is similar. Let $\eta \in C^1([a,b],\mathbb{R})$, $\Delta\zeta \in \mathbb{R}$, and $\Delta\gamma, \Delta\delta \in \mathbb{R}$ such that $0 < \gamma^\star + \Delta\gamma < 1$ and $0 < \delta^\star + \Delta\delta < 1$. In what follows, we denote:

$$[\star](t) := \Big(t, x^\star(t), ({}^C D_{a^+}^{\gamma^\star,g} x^\star)(t), ({}^C D_{b^-}^{\delta^\star,g} x^\star)(t), x^\star(a), x^\star(b), \zeta^\star\Big).$$

Observe that:

$$\mathcal{J}_1(x^* + \eta, \zeta^* + \Delta\zeta, \gamma^* + \Delta\gamma, \delta^* + \Delta\delta) - \mathcal{J}_1(x^*, \zeta^*, \gamma^*, \delta^*)$$

$$= \int_a^b \left(L\left(t, x^*(t) + \eta(t), (^C D_{a+}^{\gamma^* + \Delta\gamma, g}(x^* + \eta))(t), (^C D_{b-}^{\delta^* + \Delta\delta, g}(x^* + \eta))(t), x^*(a) + \eta(a), \right.\right.$$

$$\left.\left. x^*(b) + \eta(b), \zeta^* + \Delta\zeta\right) - L\left(t, x^*(t), (^C D_{a+}^{\gamma^*, g} x^*)(t), (^C D_{b-}^{\delta^*, g} x^*)(t), x^*(a), x^*(b), \zeta^*\right) \right) dt$$

$$\geq \int_a^b \left(\partial_2 L[\star](t) \cdot \eta(t) + \partial_3 L[\star](t) \left(f_t'(\gamma^*) . \Delta\gamma + (^C D_{a+}^{\gamma^*, g} \eta)(t) \right) \right.$$

$$\left. + \partial_4 L[\star](t) \left(g_t'(\delta^*) . \Delta\delta + (^C D_{b-}^{\delta^*, g} \eta)(t) \right) + \partial_5 L[\star](t) \cdot \eta(a) \right.$$

$$\left. + \partial_6 L[\star](t) \cdot \eta(b) + \partial_7 L[\star](t) \cdot \Delta\zeta \right) dt$$

$$= \int_a^b \left(\partial_2 L[\star](t) + \left(D_{b-}^{\gamma^*, g} \frac{\partial_3 L[\star]}{g'} \right)(t) g'(t) + \left(D_{a+}^{\delta^*, g} \frac{\partial_4 L[\star]}{g'} \right)(t) g'(t) \right) \cdot \eta(t) dt$$

$$+ \Delta\gamma \int_a^b f_t'(\gamma^*) \cdot \partial_3 L[\star](t) dt + \Delta\delta \int_a^b g_t'(\delta^*) \cdot \partial_4 L[\star](t) dt$$

$$+ \eta(a) \left(\int_a^b \partial_5 L[\star](t) dt + \left(I_{a+}^{1-\delta^*, g} \frac{\partial_4 L[\star]}{g'} \right)(a) - \left(I_{b-}^{1-\gamma^*, g} \frac{\partial_3 L[\star]}{g'} \right)(a) \right)$$

$$+ \eta(b) \left(\int_a^b \partial_6 L[\star](t) dt + \left(I_{b-}^{1-\gamma^*, g} \frac{\partial_3 L[\star]}{g'} \right)(b) - \left(I_{a+}^{1-\delta^*, g} \frac{\partial_4 L[\star]}{g'} \right)(b) \right) + \Delta\zeta \int_a^b \partial_7 L[\star](t) dt.$$

Using Conditions (3)–(8), we conclude that:

$$\mathcal{J}_1(x^* + \eta, \zeta^* + \Delta\zeta, \gamma^* + \Delta\gamma, \delta^* + \Delta\delta) - \mathcal{J}_1(x^*, \zeta^*, \gamma^*, \delta^*) \geq 0$$

proving the desired result. □

Definition 4. *Let $m \in \mathbb{N}$ and $c, d \in \mathbb{R}$ such that $c < d$. Function $L(t, x_1, \ldots, x_m)$ is said to be jointly convex in $S \subseteq [c, d] \times \mathbb{R}^m$ if, for all $i = 2, 3, \ldots, m+1$, $\partial_i L$ are continuous and satisfy:*

$$L(t, x_1 + \Delta x_1, \ldots, x_m + \Delta x_m) - L(t, x_1, \ldots, x_m) \geq \sum_{i=1}^m \partial_{i+1} L(t, x_1, \ldots, x_m) \Delta x_i,$$

for all $(t, x + \Delta x_1, \ldots, x_m + \Delta x_m), (t, x_1, \ldots, x_m) \in S$. We say that L is jointly concave in $S \subseteq [c, d] \times \mathbb{R}^m$ if the previous inequality holds, replacing $\geq$ by $\leq$.

Next, we present a sufficient optimality condition for the problem considered in Section 3.2.

Theorem 6. *Let L be jointly convex (respectively jointly concave) in $[a - \tau, b] \times \mathbb{R}^7$. If (x^*, ζ^*) satisfies the necessary conditions (13)–(16), then (x^*, ζ^*) is a minimizer (respectively maximizer) of functional $\mathcal{J}_2$.*

Proof. Consider a function $\eta : [a - \tau, b] \to \mathbb{R}$, of class C^1, vanishing at $[a - \tau, a]$, and let $\Delta\zeta$ be an arbitrary real number. Using the same ideas used to prove Theorem 3, one gets:

$$\mathcal{J}_2(x^* + \eta, \zeta^* + \Delta\zeta) - \mathcal{J}_2(x^*, \zeta^*) \geq H(x^*, \zeta^*),$$

where $H(x^\star, \zeta^\star)$ denotes the left-hand side of Equation (23). Introducing (13)–(16) into the last expression, we get that $\mathcal{J}_2(x^\star + \eta, \zeta^\star + \Delta\zeta) - \mathcal{J}_2(x^\star, \zeta^\star) \geq 0$, as desired. □

The following result can be proven using the same methods as before.

Theorem 7. *Suppose that L is jointly convex (respectively jointly concave) in $[a,b] \times \mathbb{R}^{2n+4}$. If $(x^\star, \zeta^\star)$ satisfies the necessary conditions (27)–(30), then $(x^\star, \zeta^\star)$ is a minimizer (respectively maximizer) of functional $\mathcal{J}_3$.*

4. Illustrative Examples

In this section, we provide two examples that show the applicability of some of our results.

Example 1. *Suppose we want to find a minimizer of the following functional:*

$$\mathcal{J}(x,\zeta,\gamma) = \int_0^1 \left((^C D_{0+}^{\gamma,g} x)^2(t) \frac{(g(t) - g(0))^\gamma}{\Gamma(\gamma+1)} - 2 (^C D_{0+}^{\gamma,g} x)(t)(g(t) - g(0))^\gamma \right.$$

$$\left. + \frac{(x(0))^2}{2} + \frac{(\zeta - 1)^4}{2} \right)^2 dt,$$

subject to the boundary condition $x(1) = (g(1) - g(0))^\gamma$, for the case $0 < \gamma < 1$. Let $x^\star(t) = (g(t) - g(0))^\gamma$, $\zeta^\star = 1$, and $\gamma^\star$ be given later. Using Lemma 1, we get:

$$(^C D_{0+}^{\gamma,g} x^\star)(t) = \Gamma(\gamma+1),$$

and therefore,

$$\partial_3 L[x^\star, \zeta^\star, \gamma^\star] = 2\left((^C D_{0+}^{\gamma^\star,g} x^\star)^2(t) \frac{(g(t) - g(0))^{\gamma^\star}}{\Gamma(\gamma^\star+1)} - 2 (^C D_{0+}^{\gamma^\star,g} x)(t)(g(t) - g(0))^{\gamma^\star} \right.$$

$$\left. + \frac{(x^\star(0))^2}{2} + \frac{(\zeta^\star - 1)^4}{2} \right) \left(2(^C D_{0+}^{\gamma^\star,g} x^\star)(t) \frac{(g(t) - g(0))^{\gamma^\star}}{\Gamma(\gamma^\star+1)} - 2(g(t) - g(0))^{\gamma^\star} \right) = 0.$$

Following Theorem 2, we observe that $x^\star$ and $\zeta^\star$ solve Equation (3), Equation (4), and the natural boundary condition (7). Moreover,

$$\int_0^1 f_t'(\gamma^\star) \cdot \partial_3 L[x^\star, \zeta^\star, \gamma^\star](t) \, dt = 0,$$

where $f_t(\gamma) = \Gamma(\gamma+1)$, and so, $f_t'(\gamma) = g_0(\gamma+1)\Gamma(\gamma+1)$, where g_0 denotes the Digamma function, proving that Equation (5) holds. Let:

$$\Psi(\gamma) := \mathcal{J}(x^\star, \zeta^\star, \gamma) = \int_0^1 \left(\Gamma(\gamma+1)(g(t) - g(0))^\gamma \right)^2 dt.$$

In Figure 1, we present the graphs of function Ψ, with respect to three different kernels $g(t) = t$ (Figure 1a), $g(t) = 2\sin(t)$ (Figure 1b), and $g(t) = (t+1)^{3/2}$ (Figure 1c). The optimal values are $\gamma^\star \approx 0.9010$, $\gamma^\star \approx 0.4139$, and $\gamma^\star \approx 0.4335$, respectively.

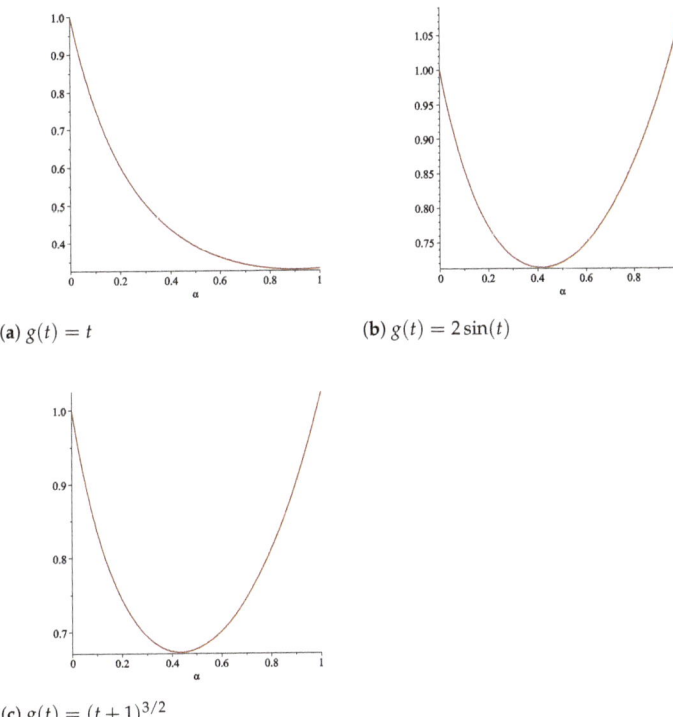

(a) $g(t) = t$

(b) $g(t) = 2\sin(t)$

(c) $g(t) = (t+1)^{3/2}$

Figure 1. Plots of function Ψ.

As we can observe, the value of the functional depends on the value of the fractional order when we evaluated it at the optimal solution (x^*, ζ^*). Thus, it is also an important question to determine the optimal value γ^* in these types of variational problems.

Example 2. *We now consider an example containing higher order derivatives. Let $\gamma \in [1,2]$ and $\delta \in [0,1]$. Suppose we want to find a minimizer of:*

$$\mathcal{J}(x,\zeta) = \int_0^1 \left(\left((^C D_{0+}^{\gamma,g} x)(t) - \frac{2(g(t)-g(0))^{2-\gamma}}{\Gamma(3-\gamma)} \right)^2 \right.$$
$$+ \left((^C D_{1-}^{\delta,g} x)(t) - \frac{2(g(1)-g(t))^{2-\delta}}{\Gamma(3-\delta)} - \frac{2(g(0)-g(1))(g(1)-g(t))^{1-\delta}}{\Gamma(2-\delta)} \right)^2$$
$$\left. + (x(1) - (g(1)-g(0))^2)^2 + \zeta^2 \right) dt,$$

under the constraints $x(0) = x'(0) = 0$ and $x'(1) = 2(g(1)-g(0))$. Let $x^*(t) = (g(t)-g(0))^2$ and $\zeta^* = 0$. Using Lemma 1, we get:

$$(^C D_{0+}^{\gamma,g} x^*)(t) = \frac{2(g(t)-g(0))^{2-\gamma}}{\Gamma(3-\gamma)},$$

and:

$$({}^C D_{1^-}^{\delta,g} x^\star)(t) = \left({}^C D_{1^-}^{\delta,g}((g(1)-g)+(g(0)-g(1)))^2\right)(t)$$
$$= \left({}^C D_{1^-}^{\delta,g}(g(1)-g)^2\right)(t) + 2(g(0)-g(1))\left({}^C D_{1^-}^{\delta,g}(g(1)-g)\right)(t)$$
$$= \frac{2(g(1)-g(t))^{2-\delta}}{\Gamma(3-\delta)} + \frac{2(g(0)-g(1))(g(1)-g(t))^{1-\delta}}{\Gamma(2-\delta)}.$$

Clearly, $(x^\star, \zeta^\star)$ satisfies (27), (28), and (30), proving that the pair $(x^\star, \zeta^\star)$ is a candidate to be a solution of the problem. Since L is jointly convex, we can conclude by Theorem 7 that $(x^\star, \zeta^\star)$ is a solution of the proposed problem.

5. Concluding Remarks

Optimization problems are an important issue in several fields of research. In particular, variational problems are useful in Newton's laws of motion, geometric optics, mathematical economics, hydrodynamics, minimal surfaces, Noether's theorems, etc. For centuries, the considered problems involved integer order derivatives only, but in the last few years, generalizations of such a rich theory were considered, by including fractional derivatives in the formulation of the variational problems. However, due to the large number of choices for such fractional derivatives, we considered here a general form of the fractional derivative. We continued our study initiated in [33], by considering three new questions: first, how to find the best order of the fractional derivatives that extremizes the functional, secondly to determine the necessary conditions of optimization with time delay, and finally, when the Lagrangian function contains higher order derivatives. To end, sufficient conditions were proven and some examples were given.

For the future, one important problem is to develop numerical methods to deal directly with the variational problems of these types, without the use of necessary conditions, for example: using discretizations of the fractional derivatives and of the integral, reduce each problem to a finite dimensional one or, using appropriate approximations of the derivatives, depending only on the first order derivative, convert the fractional variational system as an ordinary optimal control problem. Other possibilities can be studied to enrich this theory.

Author Contributions: Conceptualization, R.A. and N.M.; methodology, R.A. and N.M.; formal analysis, R.A. and N.M.; investigation, R.A. and N.M.; writing—original draft preparation, R.A. and N.M.; writing—review and editing, R.A. and N.M. All authors have read and agreed to the published version of the manuscript.

Funding: Work supported by Portuguese funds through the CIDMA (Center for Research and Development in Mathematics and Applications) and the Portuguese Foundation for Science and Technology (FCT-Fundação para a Ciência e a Tecnologia), within Project UIDB/04106/2020.

Conflicts of Interest: The authors declare no conflict of interest.

References

1. Kilbas, A.A.; Srivastava, H.M.; Trujillo, J.J. *Theory and Applications of Fractional Differential Equations*; North-Holland Mathematics Studies 204; Elsevier Science B.V.: Amsterdam, The Netherlands, 2006.
2. Klimek, M. Lagrangean and Hamiltonian fractional sequential mechanics. *Czechoslovak J. Phys.* **2002**, *52*, 1247–1253. [CrossRef]
3. Riewe, F. Mechanics with fractional derivatives. *Phys. Rev. E* **1997**, *55*, 3581–3592. [CrossRef]
4. Magin, R.L. Fractional calculus in bioengineering. *Crit. Rev. Biomed. Eng.* **2004**, *32*, 1–104.
5. Pu, Y.F. Fractional differential analysis for texture of digital image. *J. Alg. Comput. Technol.* **2007**, *1*, 357–380.
6. Hilfer, R. *Applications of Fractional Calculus in Physics*; World Scientific: Singapore, 2000.
7. Holm, S.; Sinkus, R. A unifying fractional wave equation for compressional and shear waves. *J. Acoust. Soc. Am.* **2010**, *127*, 542–548. [CrossRef]
8. Fang, C.Q.; Sun, H.Y.; Gu, J.P. Application of fractional calculus methods to viscoelastic response of amorphous shape memory polymers. *J. Mech.* **2015**, *31*, 427–432. [CrossRef]
9. Alsaedi, A.; Nieto, J.J.; Venktesh, V. Fractional electrical circuits. *Adv. Mech. Eng.* **2015**. [CrossRef]

10. Škovránek, T.; Podlubny, I.; Petrxaxš, I. Modeling of the national economies in state-space: A fractional calculus approach. *Econ. Model.* **2012**, *29*, 1322–1327. [CrossRef]
11. Pinto, C.M.A.; Carvalho, A.R.M. Fractional order model for HIV dynamics. *J. Comput. Appl. Math.* **2017**, *312*, 240–256. [CrossRef]
12. Saeedian, M.; Khalighi, M.; Azimi–Tafreshi, N.; Jafari, G.R.; Ausloos, M. Memory effects on epidemic evolution: The susceptible-infected-recovered epidemic model. *Phys. Rev. E* **2017**, *95*, 022409. [CrossRef]
13. Machado, J.A.T. Discrete-time fractional-order controllers. *Fract. Calc. Appl. Anal.* **2001**, *4*, 47–66.
14. Podlubny, I. Fractional-order systems and $PI^\lambda D^\mu$-controllers. *IEEE Trans. Autom. Control* **1999**, *44*, 208–214. [CrossRef]
15. Noeiaghdam, S.; Sidorov, D. Caputo–Fabrizio Fractional Derivative to Solve the Fractional Model of Energy Supply-Demand System. *Math. Model. Eng. Probl.* **2020**, *7*, 359–367. [CrossRef]
16. Allahviranloo, T.; Noeiaghdam, Z.; Noeiaghdam, S.; Nieto, J.J. A Fuzzy Method for Solving Fuzzy Fractional Differential Equations Based on the Generalized Fuzzy Taylor Expansion. *Mathematics* **2020**, *8*, 2166. [CrossRef]
17. Samko, S.G.; Kilbas, A.A.; Marichev, O.I. *Fractional Integrals and Derivatives, Translated from the 1987 Russian Original*; Gordon and Breach: Yverdon, Switzerland, 1993.
18. Almeida, R. A Caputo fractional derivative of a function with respect to another function, Commun. *Nonlinear Sci. Numer. Simul.* **2017**, *44*, 460–481. [CrossRef]
19. Riewe, F. Nonconservative Lagrangian and Hamiltonian mechanics. *Phys. Rev. E* **1996**, *53*, 1890–1899. [CrossRef] [PubMed]
20. Agrawal, O.P. Formulation of Euler-Lagrange equations for fractional variational problems. *J. Math. Anal. Appl.* **2002**, *272*, 368–379. [CrossRef]
21. Agrawal, O.P. Fractional variational calculus and the transversality conditions. *J. Phys. A* **2006**, *39*, 10375–10384. [CrossRef]
22. Agrawal, O.P. Generalized Euler-Lagrange equations and transversality conditions for FVPs in terms of the Caputo derivative. *J. Vib. Control* **2007**, *13*, 1217–1237. [CrossRef]
23. Almeida, R. Optimality conditions for fractional variational problems with free terminal time, Discrete Contin. *Dyn. Syst. Ser. S* **2018**, *11*, 1–19.
24. Atanacković, T.M.; Konjik, S.; Pilipović, Š. Variational problems with fractional derivatives: Euler-Lagrange equations. *J. Phys. A* **2008**, *41*, 095201. [CrossRef]
25. Baleanu, D.; Muslih, S.I.; Rabei, E.M. On fractional Euler–Lagrange and Hamilton equations and the fractional generalization of total time derivative. *Nonlinear Dynam.* **2008**, *53*, 67–74. [CrossRef]
26. Bourdin, L.; Odzijewicz T.; Torres, D.F.M. Existence of minimizers for fractional variational problems containing Caputo derivatives. *Adv. Dyn. Syst. Appl.* **2013**, *8*, 3–12.
27. Herzallah M.A.E.; Baleanu, D. Fractional-order Euler-Lagrange equations and formulation of Hamiltonian equations. *Nonlinear Dynam.* **2009**, *58*, 385–391. [CrossRef]
28. Hoffman, K.A. Stability results for constrained calculus of variations problems: An analysis of the twisted elastic loop. *Proc. Roy. Soc. A Math. Phy.* **2005**, *461*, 1357–1381. [CrossRef]
29. Malinowska, A.B.; Torres, D.F.M. Generalized natural boundary conditions for fractional variational problems in terms of the Caputo derivative. *Comput. Math. Appl.* **2010**, *59*, 3110–3116. [CrossRef]
30. Almeida, R.; Pooseh, S.; Torres, D.F.M. *Computational Methods in the Fractional Calculus of Variations*; Imp. Coll. Press: London, UK, 2015.
31. Malinowska, A.B.; Odzijewicz, T.; Torres, D.F.M. *Advanced Methods in the Fractional Calculus of Variations*; Springer Briefs in Applied Sciences and Technology; Springer: Cham, Germany, 2015.
32. Malinowska, A.B.; Torres, D.F.M. *Introduction to the Fractional Calculus of Variations*; Imp. Coll. Press: London, UK, 2012.
33. Almeida, R.; Martins, N. A generalization of a fractional variational problem with dependence on the boundaries and a real parameter. *Fractal Fract.* **2021**, *5*, 24. [CrossRef]
34. Van Brunt, B. *The Calculus of Variations*; Universitext; Springer: New York, NY, USA, 2004.
35. Machado, J.A.T. Time-Delay and Fractional Derivatives. *Adv. Differ. Equ.* **2011**, *12*, 934094.
36. Morgado, M.L.; Ford, N.J.; Lima, P.M. Analysis and numerical methods for fractional differential equations with delay. *J. Comput. Appl. Math.* **2013**, *252*, 159–168. [CrossRef]
37. Wang, Z. A Numerical Method for Delayed Fractional–Order Differential Equations. *J. Appl. Math.* **2013**, *7*, 256071. [CrossRef]
38. Almeida, R. Fractional Variational Problems Depending on Indefinite Integrals and with Delay. *Bull. Malays. Math. Sci. Soc.* **2016**, *39*, 1515–1528. [CrossRef]
39. Jarad, F.; Abdeljawad, T.; Baleanu, D. Fractional variational principles with delay within Caputo derivatives. *Rep. Math. Phys.* **2010**, *65*, 17–28. [CrossRef]
40. Sayevand, K.; Rostami, M.R.; Attari, H.S. A new study on delay fractional variational problems. *Int. J. Comput. Math.* **2018**, *95*, 1170–1194. [CrossRef]

1. Introduction

In the field of fractional order differential equations, prevalent advancement is currently speculated. The dominant use of multifarious projects which are masked by fractional differential equations (FDEs), lies in the field of nano-technology, bio-informatics, control system, chemical engineering, heat conduction, ion-acoustic wave, mechanical engineering, diffusion equations and, additionally, several other sciences. Because of its prodigious scope and applications in the various area of science and technology, congruent consideration has been given to the exact solutions of FDEs. There are many techniques that can be used to analyze NLFPDEs [1–14]. The exact solution provides a proper understanding of the physical phenomena modeled by NLFPDEs. Finding exact solutions to NLFPDEs are quite difficult as compared to approximate solutions. The Lie symmetry method is one of the most powerful methods used to find the exact solution of NLFPDEs [15–23]. This technique is used to reduce the NLFPDEs into a lower dimension. The conservation laws can be investigated for nonlinear FPDEs, which are very important tool for the study of differential equations. Noether's theorem involves a methodology for constructing conservation laws, using symmetries associated with Noether's operator [19–22,24–29]. In general, there is no technique that provides specific solutions for the system. In recent years, many researchers have concentrated on the approximate analytical solutions to the FDE system and some methods have been developed. One of the most useful techniques for solving the linear system and non-linear system of fractional differential equations with a quick convergence rate and small calculation error is the fractional power series method. Another major benefit is that this approach can be used directly, without requiring linearization, discretization, Adomian polynomials, etc., to the non-linear fractional PDE system. The power series method is applied to finding an exact solution in the form of a power series of a fractional differential equation. The (2 + 1) dimensional Kadomtsev-Petviashvili (KP) system [30,31] is given by

$$u_{tx} - uu_{xx} - u_x^2 - u_{xxxx} = u_{yy},$$

which can also be written as the system

$$u_t - uu_x - u_{xxx} - w_y = 0,$$
$$w_x - u_y = 0. \qquad (1)$$

In nonlinear wave theory, the KP system is one of the most universal models which arises as a reduction in the system with quadratic nonlinearity. This system has been broadly studied in terms of its mathematical association in recent years. The KP equation was originated by the two Soviet physicists, Boris Kadomtsev and Vladimir Petviashvili in [32]. The KP equation has been studied by many authors for integer-order or fraction-order derivatives by different methods in recent years. Exact traveling wave solutions have been analyzed in [31]. In [30], KP equation is studied for symmetry reduction using a loop algebra. In [33], KP solitary waves has been studied. Symmetries of the integer order KP equation have been studied in [34]. In [35], the Cauchy problem for the fractional KP equations has been discussed.

The main goal of this work is to analyze the fractional order KP system with arbitrary constant coefficients as

$$\partial_t^\alpha u - A_1 u \partial_x^\beta u - A_2 \partial_y^\gamma w - A_3 u_{xxx} = 0,$$
$$\partial_x^\beta w - A_4 \partial_y^\gamma u = 0. \tag{2}$$

This is a system of NLPDEs of fractional order, which depicts the evolution of nonlinear long waves with small amplitude. Here, u and w are dependent functions of x, y, t, and A_1, A_2, A_3, A_4 are arbitrary constants. x and y are the longitudinal and transverse spatial coordinates, respectively.

In this work, the KP system (2) is considered for symmetry reduction. The exact solutions, in the form of power series, are obtained, and the conservation laws are investigated.

To find some new exact solutions to the system (2), we apply the Lie symmetry method to reduce the system into lower dimensions. The system is also studied for conservation laws by using the new conservation theorem [27]. The preliminary material is given in Section 2. In Section 3, the symmetry of system (2) is obtained via the classical Lie method. Through the corresponding generators, we reduce system (2) to lower-dimensional NLFPDEs. Some exact solutions are obtained, corresponding to the reduced equation, by using the power series method in Section 4. In Section 5 the obtained power series solutions are analyzed for convergence. Some conservation laws are investigated in Section 6. In the last section, the conclusion to the study is presented.

2. Preliminaries

In this section, we will discuss basic definitions and theories for Lie symmetry analysis.

Definition 1. *Riemann-Liouville fractional derivative [36,37]*

Let $f : [a, b] \subseteq \mathcal{R} \longrightarrow \mathcal{R}$, such that $\frac{\partial^n f}{\partial t^n}$ is continuous and integrable for all $n \in \mathbb{N} \cup \{0\}$ and $n - 1 < \alpha < n$, then the Riemann-Liouville fractional derivative of order $\alpha > 0$ is defined by

$$_0 D_t^\alpha f(x, y, t) = \frac{\partial^\alpha f(x, y, t)}{\partial t^\alpha} = \begin{cases} \frac{1}{\Gamma(n-\alpha)} \frac{\partial^n}{\partial t^n} \int_0^t (t-s)^{n-\alpha-1} f(x, y, s) ds, & t > 0, \ n-1 < \alpha < n, \\ \frac{\partial^n f(x,y,t)}{\partial t^n}, & \alpha = n \in \mathbb{N}, \end{cases} \tag{3}$$

where $\Gamma(\alpha)$ is the Euler's gamma function.

Definition 2. *Erdèlyi-Kober operator*

The left-hand-side Erdèlyi-Kober fractional differential operator $(\mathcal{P}_{\varrho_1, \varrho_2}^{\vartheta, \alpha})$ is defined as

$$(\mathcal{P}_{\varrho_1, \varrho_2}^{\vartheta, \alpha} g)(y_1, y_2) = \prod_{k=0}^{r-1} \left(\vartheta + k - \frac{1}{\varrho_1} y_1 \frac{d}{dy_1} - \frac{1}{\varrho_2} y_2 \frac{d}{dy_2} \right) (\mathcal{M}_{\varrho_1, \varrho_2}^{\vartheta+\alpha, r-\alpha} g)(y_1, y_2), \quad y_i > 0, \ \varrho_i > 0, \ \alpha > 0,$$
$$i = 1, 2, \tag{4}$$

$$r = \begin{cases} [\alpha] + 1 & \text{if } \alpha \notin \mathbb{N}, \\ \alpha & \text{if } \alpha \in \mathbb{N}, \end{cases}$$

where

$$(\mathcal{M}_{\varrho_1, \varrho_2}^{\vartheta, \alpha} g)(y_1, y_2) = \begin{cases} \frac{1}{\Gamma(\alpha)} \int_1^\infty (\rho - 1)^{\alpha-1} \rho^{-(\vartheta+\alpha)} g(y_1 \rho^{\frac{1}{\varrho_1}}, y_2 \rho^{\frac{1}{\varrho_2}}) d\rho & \text{if } \alpha > 0, \\ g(y_1, y_2) & \text{if } \alpha = 0, \end{cases} \tag{5}$$

is the left-hand-side Erdèlyi-Kober fractional integral operator.

The right-hand-side Erdèlyi-Kober fractional differential operator $(\mathcal{D}_{\varrho_1, \varrho_2}^{\vartheta, \beta})$ is defined as

$$(\mathcal{D}^{\vartheta,\beta}_{\varrho_1,\varrho_2}g)(y_1,y_2) = \prod_{k=1}^{r}\left(\vartheta + k + \frac{1}{\varrho_1}y_1\frac{d}{dy_1} + \frac{1}{\varrho_2}y_2\frac{d}{dy_2}\right)(\mathcal{I}^{\vartheta+\beta,r-\beta}_\varrho g)(y_1,y_2), \quad y_i > 0, \varrho_i > 0, \beta > 0,$$
$$i = 1,2, \quad (6)$$

$$r = \begin{cases} [\beta]+1 & \text{if } \beta \notin \mathbb{N}, \\ \beta & \text{if } \beta \in \mathbb{N}, \end{cases}$$

where

$$(\mathcal{I}^{\vartheta,\beta}_{\varrho_1,\varrho_2}g)(y_1,y_2) = \begin{cases} \frac{1}{\Gamma(\beta)}\int_0^1 (1-\rho)^{\beta-1}\rho^\vartheta g(y_1\rho^{\frac{1}{\varrho_1}},y_2\rho^{\frac{1}{\varrho_2}})d\rho & \text{if } \beta > 0, \\ g(y_1,y_2) & \text{if } \beta = 0, \end{cases} \quad (7)$$

is the right-hand-side Erdèlyi-Kober fractional integral operator.

Symmetry Analysis

Consider the system of NLFPDEs as follows

$$\Delta_h = F_h\left(x,y,t,\mathbf{v},\frac{\partial^\alpha \mathbf{v}}{\partial t^\alpha},\frac{\partial^\beta \mathbf{v}}{\partial x^\beta},\frac{\partial^\gamma \mathbf{v}}{\partial y^\gamma},\frac{\partial \mathbf{v}}{\partial x},\frac{\partial^2 \mathbf{v}}{\partial x^2},\cdots\right), \quad h = 1,2,\cdots, \quad (8)$$

where $\frac{\partial^\alpha \mathbf{v}}{\partial t^\alpha}$, $\frac{\partial^\beta \mathbf{v}}{\partial x^\beta}$ and $\frac{\partial^\gamma \mathbf{v}}{\partial x^\gamma}$ are the fractional derivatives of Riemann-Liouville (RL) type. Suppose that the Lie group of transformations are given by

$$\begin{aligned}
x^* &= x + \varepsilon\xi(x,y,t,\mathbf{v}) + O(\varepsilon^2), \\
t^* &= t + \varepsilon\tau(x,y,t,\mathbf{v}) + O(\varepsilon^2), \\
y^* &= t + \varepsilon\mu(x,y,t,\mathbf{v}) + O(\varepsilon^2), \\
v^{r*} &= v^r + \varepsilon\eta^{(r)}(x,t,v^r) + O(\varepsilon^2), \\
\frac{\partial^\alpha v^{r*}}{\partial t^\alpha} &= \frac{\partial^\alpha v^r}{\partial t^\alpha} + \varepsilon\eta^{(r)\alpha,t} + O(\varepsilon^2), \\
\frac{\partial^\beta v^{r*}}{\partial x^\beta} &= \frac{\partial^\beta v^r}{\partial x^\beta} + \varepsilon\eta^{(r)\beta,x} + O(\varepsilon^2), \\
\frac{\partial^\gamma v^{r*}}{\partial y^\gamma} &= \frac{\partial^\gamma v^r}{\partial y^\gamma} + \varepsilon\eta^{(r)\gamma,y} + O(\varepsilon^2), \\
\frac{\partial v^{r*}}{\partial x} &= \frac{\partial v^r}{\partial x} + \varepsilon\eta^{(r)x} + O(\varepsilon^2), \\
\frac{\partial^2 v^{r*}}{\partial x^2} &= \frac{\partial^2 v^r}{\partial x^2} + \varepsilon\eta^{(r)xx} + O(\varepsilon^2), \\
&\vdots
\end{aligned} \quad (9)$$

where ϵ being the group parameter and $\xi, \tau, \mu, \eta^{(r)}$ are the infinitesimals,

$$\begin{aligned}
\eta^{(k),x} &= D_x(\eta^{(k)}) - v_x^k D_x(\xi) - v_t^k(\tau) - v_y^k D_y(\mu), \\
\eta^{(k)\alpha,t} &= D_t^\alpha(\eta^{(k)}) + \xi D_t^\alpha(v_x^k) - D_t^\alpha(\xi v_x^k) + \tau D_t^\alpha(v_t^k) - D_t^\alpha(\tau v_t^k) + \mu D_t^\alpha(v_y^k) \\
&\quad - D_t^\alpha(\mu v_y^k), \\
\eta^{(k)\beta,x} &= D_x^\beta(\eta^{(k)}) + \xi D_x^\beta(v_x^k) - D_x^\beta(\xi v_x^k) + \tau D_x^\beta(v_t^k) - D_x^\beta(\tau v_t^k) + \mu D_x^\beta(v_y^k) \\
&\quad - D_x^v(\mu v_y^k), \\
\eta^{(k)\gamma,y} &= D_y^\gamma(\eta^{(k)}) + \xi D_y^\gamma(v_x^k) - D_y^\gamma(\xi v_x^k) + \tau D_y^\gamma(v_t^k) - D_y^\gamma(\tau v_t^k) + \mu D_y^\gamma(v_y^k) \\
&\quad - D_y^\gamma(\mu v_y^k), \quad (10)
\end{aligned}$$

are extended infinitesimals. In (10), D_x and D_t are total derivative operators. The αth, β^{th} and γ^{th} extended infinitesimals related to the RL fractional derivative are given in [38].
The associated vector field is

$$X = \xi(x,y,t,\mathbf{v})\frac{\partial}{\partial x} + \mu(x,y,t,\mathbf{v})\frac{\partial}{\partial y} + \tau(x,y,t,\mathbf{v})\frac{\partial}{\partial t} + \sum_{r=1}^{p}\eta^{(r)}(x,y,t,\mathbf{v})\frac{\partial}{\partial v^r}. \quad (11)$$

The corresponding extended symmetry generator is as follows

$$\begin{aligned}
pr^{(\alpha,\beta,\gamma)}X &= X + \sum_r \eta^{(r)\alpha,t}\partial_{\partial_t^\alpha v^r} + \sum_r \eta^{(r)\beta,x}\partial_{\partial_x^\beta v^r} + \sum_r \eta^{(r)\gamma,y}\partial_{\partial_y^\gamma v^r} + \sum_r \eta^{(r)x}\partial_{v_x^r} \\
&\quad + \sum_r \eta^{(r)xx}\partial_{v_{xx}^r} + \ldots, \quad (12)
\end{aligned}$$

As the lower limit of RL fractional derivative [36,37,39] is fixed, we have

$$\xi(x,y,t,u,w)|_{x=0} = 0, \quad \tau(x,y,t,u,w)|_{t=0} = 0, \quad \mu(x,y,t,u,w)|_{y=0} = 0. \quad (13)$$

3. Symmetry Analysis of (2 + 1)-Dimensional Fractional Kadomtsev-Petviashvili System

Let us assume that the system (2) is invariant under group of transformations (9), then we have

$$\begin{aligned}
\partial_{t^*}^\alpha u^* - A_1 u^* \partial_{x^*}^\beta u^* - A_2 \partial_{y^*}^\gamma w^* - A_3 u_{x^*x^*x^*}^* &= 0, \\
\partial_{x^*}^\beta w^* - A_4 \partial_{y^*}^\gamma u^* &= 0. \quad (14)
\end{aligned}$$

Therefore, using (9) in (14) the invariance criteria for (2) are obtained as

$$\begin{aligned}
\eta^{\alpha,t} - A_1 \eta \partial_x^\beta u - A_1 u \eta^{\beta,x} - A_2 \phi^{\gamma,y} - A_3 \eta^{xxx} &= 0, \\
\phi^{\beta,x} - A_4 \eta^{\gamma,y} &= 0. \quad (15)
\end{aligned}$$

Using the value of extended infinitesimals and collecting the coefficients of various powers of u and partial derivatives of u and w, we have

$$\xi_t = \xi_u = \xi_w = 0,$$
$$\tau_x = \tau_u = \tau_v = 0,$$
$$\eta_w = \phi_u = 0,$$
$$\eta_{uu} = \phi_{ww} = 0,$$
$$\eta_u - \phi_w - \alpha D_t \tau + \gamma D_y \mu = 0,$$
$$3\xi_x - \alpha D_t \tau = 0,$$
$$\eta_u - \phi_w + \beta D_x \xi - 9\gamma D_y \mu = 0,$$
$$\eta - u\beta D_x \xi + u\alpha D_t \tau = 0,$$
$$\partial_t^\alpha \eta - A_1 u \partial_t^\alpha \eta_u - A_1 u (\partial_x^\beta \eta - u \partial_x^\beta \eta_u) - A_2 (\partial_y^\gamma \phi - w \partial_y^\gamma \phi_w) - A_3 \eta_{xxx} = 0,$$
$$\partial_x^\beta \phi - w \partial_x^\beta \phi_w - A_4 \partial_y^\gamma \eta + A_4 y \partial_y^\gamma \eta_u = 0,$$
$$\binom{\alpha}{n} \partial_t^n \eta_u - \binom{\alpha}{n+1} D_t^{n+1} \tau = 0,$$
$$\binom{\beta}{n} \partial_x^n \eta_u - \binom{\beta}{n+1} D_x^{n+1} \xi = 0,$$
$$\binom{\gamma}{n} \partial_y^n \eta_u - \binom{\gamma}{n+1} D_y^{n+1} \mu = 0,$$
$$\binom{\beta}{n} \partial_x^n \phi_w - \binom{\beta}{n+1} D_x^{n+1} \xi = 0,$$
$$\binom{\gamma}{n} \partial_y^n \phi_w - \binom{\gamma}{n+1} D_y^{n+1} \xi = 0, \tag{16}$$

where $n \in \mathbb{N}$.

Solving these equations simultaneously, we get the infinitesimals

$$\xi = C_1 x, \quad \tau = \frac{3}{\alpha} C_1 t, \quad \mu = \frac{3+\beta}{2\gamma} C_1 y,$$
$$\eta = (\beta - 3) C_1 u, \quad \phi = \frac{3(\beta - 3)}{2} C_1 w, \tag{17}$$

where C_1 is thw arbitrary constant.

Thus, the corresponding vector field is

$$V = x\partial_x + \frac{3}{\alpha} t\partial_t + \frac{3+\beta}{2\gamma} y\partial_y + (\beta - 3) u\partial_u + \frac{3(\beta-3)}{2} w\partial_w. \tag{18}$$

Corresponding to vector field V, the characterisitc equation is written as

$$\frac{dx}{x} = \frac{dt}{\frac{3}{\alpha}t} = \frac{dy}{\frac{3+\beta}{2\gamma}y} = \frac{du}{(\beta-3)u} = \frac{dw}{\frac{3(\beta-3)}{2}w}.$$

After solving these equations, we get the symmetry variables

$$z_1 = xt^{-\frac{\alpha}{3}}, \quad z_2 = yt^{\frac{-\alpha(\beta+3)}{6\gamma}}, \tag{19}$$

and symmetry transformations

$$u = t^{\frac{\alpha(\beta-3)}{3}} f(z_1, z_2), \quad w = t^{\frac{\alpha(\beta-3)}{2}} g(z_1, z_2), \tag{20}$$

where f and g are arbitrary functions.

The one-parameter Lie group of infinitesimal transformation of fractional order operator of the form (22) using symmetry of Eqs. (25), (39) and (20) two theorems of given in (19) then

$$\frac{\partial^\alpha u}{\partial t^\alpha} = t^{-\alpha}z_1^{-\beta}(D_{1,\infty}^{-\beta,\beta}f)(z_1,z_2),$$

In similar manner, the RL fractional derivatives of order $\beta, \gamma > 0$ are obtained as

Thus, Equation (22) becomes

$$\frac{\partial^\beta u}{\partial x^\beta} = t^{-\alpha}z_1^{-\beta}(D_{1,\infty}^{-\beta,\beta}f)(z_1,z_2),$$

$$\frac{\partial^\beta w}{\partial x^\beta} = t^{-\alpha}z_1^{-\beta}(D_{1,\infty}^{-\beta,\beta}g)(z_1,z_2),$$

$$\frac{\partial^\gamma v}{\partial y^\beta} = t^{-\alpha}z_2^{-\gamma}(D_{\infty,1}^{-\gamma,\gamma}f)(z_1,z_2),$$

$$\frac{\partial^\gamma w}{\partial y^\beta} = t^{-\alpha}z_2^{-\gamma}(D_{\infty,1}^{-\gamma,\gamma}g)(z_1,z_2),$$

where $(D_{1,\infty}^{-\beta,\beta})$ and $(D_{\infty,1}^{-\gamma,\gamma})$ are the differential operators defined in [36,37,39].

By using (24) and (25), the symmetry reduction fractional KP system (2) is obtained as

Proof. Let us assume $n-1 < \alpha \leq n$, where $n \in \mathbb{N}$, then, using the definition of RL fractional differentiation, we have

$$\frac{\partial^\alpha u}{\partial t^\alpha} = \frac{\partial^n}{\partial t^n}\left(\frac{1}{\Gamma(n-\alpha)}\int_1^\infty (\rho-1)^{n-\alpha-1}\rho^{-(n-\alpha+\frac{\alpha(\beta-3)}{3})}f(z_1\rho^{\frac{\alpha}{3}}, z_2\rho^{\frac{\alpha(\beta+3)}{6\gamma}})d\rho\right).$$

Let $s = \frac{t}{\rho}$; then, we get

$$\frac{\partial^\alpha u}{\partial t^\alpha} = \frac{\partial^n}{\partial t^n}\left(\frac{t^{n+\frac{\alpha(\beta-6)}{3}}}{\Gamma(n-\alpha)}\int_1^\infty (\rho-1)^{n-\alpha-1}\rho^{-(n-\alpha+\frac{\alpha(\beta-3)}{3})}f(z_1\rho^{\frac{\alpha}{3}}, z_2\rho^{\frac{\alpha(\beta+3)}{6\gamma}})d\rho\right).$$

$\square$

By using the definition of the left-hand-side EK fractional integral operator $(\mathcal{M}_\rho^{\theta,\alpha}f)$ (z_1,z_2), defined in [36,37,39], we have

$$\frac{\partial^\alpha u}{\partial t^\alpha} = \frac{\partial^n}{\partial t^n}\left(t^{n+\frac{\alpha(\beta-6)}{3}}(\mathcal{M}_{\frac{3}{\alpha},\frac{6\gamma}{\alpha(\beta+3)}}^{1+\frac{\alpha(\beta-3)}{3},n-\alpha}f)(z_1,z_2)\right), \tag{22}$$

where

$$\left(\mathcal{M}_{\frac{3}{\alpha},\frac{6\gamma}{\alpha(\beta+3)}}^{1+\frac{\alpha(\beta-3)}{3},n-\alpha}f\right)(z_1,z_2) = \frac{1}{\Gamma(n-\alpha)}\int_1^\infty (\rho-1)^{n-\alpha-1}\rho^{-(n-\alpha+\frac{\alpha(\beta-3)}{3})}f(z_1\rho^{\frac{\alpha}{3}}, z_2\rho^{\frac{\alpha(\beta+3)}{6\gamma}})d\rho.$$

$$\begin{aligned}\frac{\partial^\alpha u}{\partial t^\alpha} &= \frac{\partial^n}{\partial t^n}\left(t^{n+\frac{\alpha(\beta-6)}{3}}(\mathcal{M}_{\frac{3}{\alpha},\frac{6\gamma}{\alpha(\beta+3)}}^{1+\frac{\alpha(\beta-3)}{3},n-\alpha}f)(z_1,z_2)\right)\\ &= \frac{\partial^{n-1}}{\partial t^{n-1}}\left(t^{n+\frac{\alpha(\beta-6)}{3}}(\mathcal{M}_{\frac{3}{\alpha},\frac{6\gamma}{\alpha(\beta+3)}}^{1+\frac{\alpha(\beta-3)}{3},n-\alpha}f)(z_1,z_2)\right)\\ &= \frac{\partial^{n-1}}{\partial t^{n-1}}\left(t^{n+\frac{\alpha(\beta-6)}{3}-1}\left(n-\alpha+\frac{\alpha(\beta-3)}{3}-\frac{\alpha}{3}z_1\frac{\partial}{\partial z_1}-\frac{\alpha(\beta+3)}{6\gamma}z_2\frac{\partial}{\partial z_2}\right)\left(\mathcal{M}_{\frac{3}{\alpha},\frac{6\gamma}{\alpha(\beta+3)}}^{1+\frac{\alpha(\beta-3)}{3},n-\alpha}f\right)(z_1,z_2)\right).\end{aligned}$$

$$\frac{\partial^\gamma u}{\partial t^\gamma} = \left(t^{\frac{\alpha(\beta-6)}{3}}\prod_{j=0}^{n-1}\left(1+j+\frac{\alpha(\beta-6)}{3}-\frac{\alpha}{3}z_1\frac{\partial}{\partial z_1}-\frac{\alpha(\beta+3)}{6\gamma}z_2\frac{\partial}{\partial z_2}\right)\left(\mathcal{M}_{\frac{3}{\alpha},\frac{6\gamma}{\alpha(\beta+3)}}^{1+\frac{\alpha(\beta-3)}{3},n-\alpha}f\right)(z_1,z_2)\right). \tag{23}$$

$$\left(\mathcal{P}_{\frac{3}{\alpha},\frac{6\gamma}{\alpha(\beta+3)}}^{\frac{\alpha(\beta-6)}{3}+1,\alpha}f\right)(z_1,z_2) - A_1z_1^{-\beta}f(z_1,z_2)\left(D_{1,\infty}^{-\beta,\beta}f\right)(z_1,z_2) - A_2z_2^{-\gamma}\left(D_{\infty,1}^{-\gamma,\gamma}g\right)(z_1,z_2) - A_3\frac{\partial^3 f(z_1,z_2)}{\partial z_1^3} = 0,$$

$$z_1^{-\beta}\left(D_{1,\infty}^{-\beta,\beta}g\right)(z_1,z_2) - A_4z_2^{-\gamma}\left(D_{\infty,1}^{-\gamma,\gamma}f\right)(z_1,z_2) = 0. \tag{26}$$

4. Power Series Solutions of the NLFPDEs

In this section, we will obtain the power series solutions of NLFPDEs (26) [18,28,40]. Let us consider two double power series

$$f(z_1, z_2) = \sum_{n,m=0}^{\infty} a_{n,m} z_1^n z_2^m, \quad g(z_1,z_2) = \sum_{n,m=0}^{\infty} b_{n,m} z_1^n z_2^m. \tag{27}$$

Suppose that (30) helps to obtain a solution of the form (26) and $b_{n,m}$ $(n, m \geq 0)$ of the power series (27) for arbitrarily chosen series $\sum a_{i,m}$ $(i = 0, 1, 2)$. Therefore, the system (26) has the exact power series solution, and the coefficient of the series depends on (31). Hence, we can write the power series (27) as

$$f(z_1, z_2) = \sum_{n,m=0}^{\infty} a_{n,m} z_1^n z_2^m, \quad g(z_1, z_2) = \sum_{n,m=0}^{\infty} b_{n,m} z_1^n z_2^m. \tag{30}$$

Therefore, from (29), but both are not simultaneously zero, we have

$$\frac{\partial f}{\partial z_1} = \sum_{n,m=0}^{\infty} (n+1) a_{n+1,m} z_1^n z_2^m,$$

$$\frac{\partial^2 f}{\partial z_1^2} = \sum_{n,m=0}^{\infty} (n+2)(n+1) a_{n+2,m} z_1^n z_2^m,$$

$$\frac{\partial^3 f}{\partial z_1^3} = \sum_{n,m=0}^{\infty} (n+3)(n+2)(n+1) a_{n+3,m} z_1^n z_2^m \tag{28}$$

Inserting (27) and (28) into (26), we have

$$\sum_{n,m=0}^{\infty} \frac{\Gamma\left(\frac{\alpha(\beta-3)}{3} + 1 - \frac{n\alpha}{3} - \frac{m\alpha(\beta+3)}{6\gamma}\right)}{\Gamma\left(\frac{\alpha(\beta-6)}{3} + 1 - \frac{n\alpha}{3} - \frac{m\alpha(\beta+3)}{6\gamma}\right)} a_{n,m} z_1^n z_2^m - A_1 z_1^{-\beta} \sum_{n,m=0}^{\infty} \sum_{k=0}^{n} \sum_{j=0}^{m} \left(\frac{\Gamma(1+k)}{\Gamma(1+k-\beta)} a_{n-k,m-j} a_{k,j}\right)$$

$$- A_2 z_2^{-\gamma} \sum_{n,m=0}^{\infty} \frac{\Gamma(1+m)}{\Gamma(1+m-\gamma)} b_{n,m} z_1^n z_2^m - A_3 \sum_{n,m=0}^{\infty} \left((n+3)(n+2)(n+1) a_{n+3,m} z_1^n z_2^m\right) = 0,$$

$$z_1^{-\beta} \sum_{n,m=0}^{\infty} \frac{\Gamma(1+n)}{\Gamma(1+n-\beta)} b_{n,m} z_1^n z_2^m - A_4 z_2^{-\gamma} \sum_{n,m=0}^{\infty} \frac{\Gamma(1+m)}{\Gamma(1+m-\gamma)} a_{n,m} z_1^n z_2^m = 0. \tag{29}$$

$$a_{n+3,m} = \frac{1}{A_3(n+3)(n+2)(n+1)} \left\{ \frac{\Gamma\left(\frac{\alpha(\beta-3)}{3} + 1 - \frac{n\alpha}{3} - \frac{m\alpha(\beta+3)}{6\gamma}\right)}{\Gamma\left(\frac{\alpha(\beta-6)}{3} + 1 - \frac{n\alpha}{3} - \frac{m\alpha(\beta+3)}{6\gamma}\right)} a_{n,m} \right.$$

$$\left. - A_1 z_1^{-\beta} \sum_{k=0}^{n} \sum_{j=0}^{m} \frac{\Gamma(1+k)}{\Gamma(1+k-\beta)} a_{n-k,m-j} a_{k,j} - A_2 z_2^{-\gamma} \frac{\Gamma(1+m)}{\Gamma(1+m-\gamma)} b_{n,m} \right\},$$

$$b_{n,m} = A_4 z_1^\beta z_2^{-\gamma} \frac{\Gamma(1+n-\beta)\Gamma(1+m)}{\Gamma(1+n)\Gamma(1+m-\gamma)} a_{n,m}. \tag{31}$$

$$f(z_1, z_2) = a_{0,0} + a_{1,0} z_1 + a_{2,0} z_1^2 + \frac{1}{6A_3} \left(\frac{\Gamma(\frac{\alpha(\beta-3)}{3}+1)}{\Gamma(\frac{\alpha(\beta-6)}{3}+1)} a_{0,0} - A_1 z_1^{-\beta} \frac{1}{\Gamma(1-\beta)} a_{0,0}^2 - A_2 z_2^{-\gamma} \frac{1}{\Gamma(1-\gamma)} b_{0,0} \right) z_1^3$$

$$+ \sum_{n=1,m=0}^{\infty} \frac{1}{A_3(n+3)(n+2)(n+1)} \left(\frac{\Gamma\left(\frac{\alpha(\beta-3)}{3}+1-\frac{n\alpha}{3}-\frac{m\alpha(\beta+3)}{6\gamma}\right)}{\Gamma\left(\frac{\alpha(\beta-6)}{3}+1-\frac{n\alpha}{3}-\frac{m\alpha(\beta+3)}{6\gamma}\right)} a_{n,m} \right.$$

$$\left. - A_1 z_1^{-\beta} \sum_{k=0}^{n} \sum_{j=0}^{m} \frac{\Gamma(1+k)}{\Gamma(1+k-\beta)} a_{n-k,m-j} a_{k,j} - A_2 z_2^{-\gamma} \frac{\Gamma(1+m)}{\Gamma(1+m-\gamma)} b_{n,m} \right) z_1^n z_2^m,$$

$$g(z_1, z_2) = \sum_{n,m=0}^{\infty} A_4 z_1^\beta z_2^{-\gamma} \frac{\Gamma(1+n-\beta)\Gamma(1+m)}{\Gamma(1+n)\Gamma(1+m-\gamma)} a_{n,m} z_1^n z_2^m. \tag{32}$$

$$u(x,t) = t^{\frac{\alpha(\beta-3)}{3}} a_{0,0} + xt^{\frac{\alpha(\beta-4)}{3}} a_{1,0} + x^2 t^{\frac{\alpha(\beta-5)}{3}} a_{2,0} + \frac{1}{6A_3} \left(\frac{\Gamma(\frac{\alpha(\beta-3)}{3}+1)}{\Gamma(\frac{\alpha(\beta-6)}{3}+1)} a_{0,0} - A_1 x^{-\beta} t^{\frac{\alpha\beta}{3}} \frac{1}{\Gamma(1-\beta)} a_{0,0}^2 \right)$$

$$- A_2 y^{-\gamma} t^{\frac{\alpha(\beta+3)}{6}} \frac{1}{\Gamma(1-\gamma)} b_{0,0} \right) x^3 t^{\frac{\alpha(\beta-6)}{3}} + \sum_{n=1,m=0}^{\infty} \frac{1}{A_3(n+3)(n+2)(n+1)}$$

$$\times \left(\frac{\Gamma\left(\frac{\alpha(\beta-3)}{3}+1-\frac{n\alpha}{3}-\frac{m\alpha(\beta+3)}{6\gamma}\right)}{\Gamma\left(\frac{\alpha(\beta-6)}{3}+1-\frac{n\alpha}{3}-\frac{m\alpha(\beta+3)}{6\gamma}\right)} a_{n,m} - A_1 x^{-\beta} t^{\frac{\alpha\beta}{3}} \sum_{k=0}^{n} \sum_{j=0}^{m} \frac{\Gamma(1+k)}{\Gamma(1+k-\beta)} a_{n-k,m-j} a_{k,j} \right.$$

5. Analysis of the Convergence

In this section, we will analyze the convergence of the power series solution (33) and (34).

Theorem 2. *The power series of the solutions (33) and (34) converges.*

Proof. From (31), we have

$$|a_{n+3,m}| \leq M\left\{|a_{n,m}| + \sum_{k=0}^{n}\sum_{j=0}^{m}|a_{n-k,m-j}||a_{k,j}| + |b_{n,m}|\right\}, \quad (35)$$

where $M = \max\left(\frac{1}{A_3}\left\{\frac{\Gamma\left(\frac{\alpha(\beta-3)}{3}+1-\frac{n\alpha}{3}-\frac{m\alpha(\beta+3)}{6\gamma}\right)}{\Gamma\left(\frac{\alpha(\beta-6)}{3}+1-\frac{n\alpha}{3}-\frac{m\alpha(\beta+3)}{6\gamma}\right)}\right\}, \sum_{k=0}^{n}\frac{\Gamma(1+k)}{\Gamma(1+k-\beta)}\left(\frac{A_1 z_1^{-\beta}}{A_3}, \frac{A_2 z_2^{-\gamma}}{A_3}\right)\right).$

and

$$|b_{n,m}| \leq N(|a_{n,m}|), \quad (36)$$

where $N = \max\left(1, A_4 z_1^{\beta} z_2^{-\gamma}\frac{\Gamma(1+n-\beta)\Gamma(1+m)}{\Gamma(1+n)\Gamma(1+m-\gamma)}\right).$

Let us consider two double power series

$$P = P(z_1, z_2) = \sum_{n,m=0}^{\infty} p_{n,m} z_1^n z_2^m,$$

$$R = R(z_1, z_2) = \sum_{n,m=0}^{\infty} r_{n,m} z_1^n z_2^m, \quad (37)$$

by

$$p_{n,m} = |a_{n,m}|, \quad r_{i,j} = |b_{i,j}|, \quad n = 0,1,2, \quad m = 0, \quad i,j = 0, \quad (38)$$

and

$$p_{n+3,m} = M\left(p_{n,m} + \sum_{k=0}^{n}\sum_{j=0}^{m} p_{n-k,m-j}p_{k,j} + r_{n,m}\right),$$

$$r_{n,m} = N(p_{n,m}), \quad (39)$$

where $n = 0,1,2,3,\cdots$. Therefore, one can easily check that

$$|a_{n,m}| \leq p_{n,m}, \quad |b_{n,m}| \leq r_{n,m}, \quad n = 0,1,2,\cdots. \quad (40)$$

Therefore, the series

$$P = P(z_1, z_2) = \sum_{n,m=0}^{\infty} p_{n,m} z_1^n z_2^m$$

and

$$R = R(z_1, z_2) = \sum_{n,m=0}^{\infty} r_{n,m} z_1^n z_2^m$$

are the majorant series of the series $f(z_1, z_2)$ and $g(z_1, z_2)$, respectively. □

Let us consider one particular case,

$$A_i = \sum_{m=0}^{\infty} p_{i,m} z_1^i z_2^m, \quad i = 0,1,2.$$

F, H are analytics in the neighbourhood of $(0,0,A_0,NA_0)$. $F(0,0,A_0,NA_0) = 0$, $G(0,0,A_0,NA_0) = 0$, and the Jacobian determinant is

$$\frac{\partial(F,H)}{\partial(P,R)}\bigg|_{(0,0,A_0,NA_0)} = 1 \neq 0.$$

Then, by the implicit function theorem [41], both power series are convergent. Hence, an exact solution of KP system (2) exists.

6. Conservation Laws

In this section, conservation laws of (2) will be constructed by using the new conservation theorem and the nonlinear self adjointness [27,29].

The conservation laws for (2) are introduced as

$$D_t(C^t) + D_x(C^x) + D_y(C^y) = 0, \tag{44}$$

where $C^t(x,y,t,u,w)$, $C^x(x,y,t,u,w)$ and $C^y(x,y,t,u,w)$ are conserved vectors of (2).

The Euler–Lagrange operators given by

$$\frac{\delta}{\delta u^j} = \frac{\partial}{\partial u^j} + (D_t^\alpha)^* \frac{\partial}{\partial (D_t^\alpha u^j)} + (D_x^\beta)^* \frac{\partial}{\partial (D_x^\beta u^j)} + (D_y^\gamma)^* \frac{\partial}{\partial (D_y^\gamma u^j)}$$

$$+ \sum_{k=1}^{\infty} (-1)^k D_{i_1} D_{i_2}, \ldots, D_{i_k} \frac{\partial}{\partial (u^j)_{i_1,i_2,\ldots,i_k}}, \tag{45}$$

where D_{i_k} represents the total derivative operator. $(D_t^\alpha)^*$, $(D_x^\beta)^*$ and $(D_y^\gamma)^*$ are also the adjoint operators of the RL derivative operators [36,39] D_t^γ and D_x^β, respectively, given as follows

$$\begin{aligned}
(D_t^\alpha)^* &= (-1)^n I_p^{n-\alpha}(D_t^n) =_t^C D_p^\alpha, \\
(D_x^\beta)^* &= (-1)^m I_q^{m-\beta}(D_x^m) =_x^C D_q^\beta, \\
(D_y^\gamma)^* &= (-1)^k I_r^{k-\gamma}(D_y^r) =_y^C D_r^\gamma,
\end{aligned} \tag{46}$$

where $I_p^{n-\alpha}$, $I_q^{m-\beta}$ and $I_r^{k-\gamma}$ are the right-hand-side fractional integral operators of order $n-\alpha$, $m-\beta$ and $k-\gamma$, respectively, defined as follows

$$I_p^{n-\alpha} f(x,t) = \frac{1}{\Gamma(n-\alpha)} \int_t^p \frac{f(x,y,s)}{(s-t)^{1+\alpha-n}} ds, \tag{47}$$

where $n = [\alpha] + 1$

$$I_q^{m-\beta} f(x,t) = \frac{1}{\Gamma(m-\beta)} \int_x^q \frac{f(s,y,t)}{(s-x)^{1+\beta-m}} ds, \tag{48}$$

where $m = [\beta] + 1$

$$I_r^{k-\gamma} f(x,t) = \frac{1}{\Gamma(k-\gamma)} \int_y^r \frac{f(x,s,t)}{(s-y)^{1+\gamma-r}} ds, \tag{49}$$

where $k = [\gamma] + 1$

The formal Lagrangian of the system (2) is given by

$$\mathcal{L} = T(\partial_t^\alpha u - A_1 u \partial_x^\beta u - A_2 \partial_y^\gamma w - A_3 u_{xxx}) + Q(\partial_x^\beta w - A_4 \partial_y^\gamma u), \tag{50}$$

where T and Q are new dependent variables.

where $m = [\alpha] + 1$, and W^j, $(j = 1, 2)$ are defined in (58) and u_1, u_2 are dependent variables. Additionally, $\mathcal{J}_1(h_1, h_2)$ is the integral

$$\mathcal{J}_1(h_1, h_2) = \frac{1}{\Gamma(m-\alpha)} \int_0^t \int_t^q \frac{h_1(x,y,s) h_2(x,y,r)}{(r-s)^{\alpha+1-m}} dr ds,$$

for any two functions $h_1(x,y,t)$ and $h_2(x,y,t)$.

In a similar way, other fractional Noether's operators C^x and C^y are defined. By using (58) and vector field (18), the characteristic functions are

$$W^1 = (\beta - 3)u - xu_x - \frac{3+\beta}{2\gamma} y u_y - \frac{3}{\alpha} t, \quad W^2 = \frac{3(\beta-3)}{2} w - xu_x - \frac{3+\beta}{2\gamma} y u_y - \frac{3}{\alpha} t. \quad (60)$$

Now, we will obtain the conserved vectors of the system (2) as follows.

Case 1. For $0 < \alpha < 1$, we have

$$C^t = I_t^{1-\alpha}(W^1)\varphi + \mathcal{J}_1(W^1, \varphi_t).$$

Case 2. For $1 < \alpha < 2$, we have

$$C^t = D_t^{\alpha-1}(W^1)\varphi - I_t^{2-\alpha}(W^1)\varphi + \mathcal{J}_1(W^1, \varphi_{tt}).$$

Case 3. Similarly, for $0 < \beta < 1$, we have

$$C^x = -I_x^{1-\beta}(W^1)(A_1 u \varphi) + I_x^{1-\beta}(W^2)(\phi) + \mathcal{J}_2(W^1, D_x(-A_1 u \varphi)) + \mathcal{J}_2(W^2, D_x(\phi)).$$

Case 4. For $1 < \beta < 2$, we have

$$\begin{aligned} C^x &= -D_x^{\beta-1}(W^1)(A_1 u \varphi)) + D_x^{\beta-1}(W^2)(\phi) + I_x^{2-\beta}(W^1) D_x(A_1 u \varphi) - I_x^{2-\beta}(W^2) D_x(\phi) \\ &\quad - \mathcal{J}_2(W^1, D_x^2(-A_1 u \varphi)) - \mathcal{J}_2(W^2, D_x^2(\phi)). \end{aligned}$$

Case 5. For $0 < \gamma < 1$, we have

$$C^y = -I_y^{1-\beta}(W^1)(A_4 \phi) - I_y^{1-\beta}(W^2)(A_2 \varphi) + \mathcal{J}_3(W^1, D_x(-A_4 \phi)) + \mathcal{J}_3(W^2, D_x(-A_2 \varphi)).$$

Case 6. For $1 < \gamma < 2$, we have

$$\begin{aligned} C^y &= -D_x^{\beta-1}(W^1)(A_4 \phi)) - D_x^{\beta-1}(W^2)(A_2 \varphi) + I_x^{2-\beta}(W^1) D_x(A_4 \phi) + I_x^{2-\beta}(W^2) D_x(A_2 \varphi) \\ &\quad - \mathcal{J}_3(W^1, D_x^2(-A_4 \phi)) - \mathcal{J}_2(W^2, D_x^2(-A_2 \varphi)). \end{aligned}$$

7. Concluding Remarks

In this work, we have studied a (2 + 1)-dimensional fractional Kadomtsev-Petviashvili system (2) by Lie symmetry analysis and power series expansion techniques, via. an RL fractional derivative. First, we obtained the Lie point symmetries, and then the similarity transformations were successfully presented. Using the similarity transformations, we were able to reduce the system of NLFPDEs (2) of three dimensions into a system of NLFPDEs of two dimensions. Further, the explicit exact solution for the reduced NLFPDEs was obtained using the power series expansion method. The analysis of convergence for the power series solution was also performed. Using the new conservation theorem [27], the conservation laws of the system are successfully obtained. The obtained solutions might be of substantial consequence in the corresponding physical phenomena of science and applied mathematics.

Author Contributions: S.K. and B.K., methodology; S.-W.Y., funding acquisition; M.I., formal analysis; M.S.O., writing—review and editing. All authors have read and agreed to the published version of the manuscript.

$$P = \sum_{n=1}^{\infty} \left(p_{0,m} a_1^t + p_{1,m} z_1^{2i} + b \bar{p}_{2,m} \bar{z}_1^2 x + z_{3,m} \bar{z}_1^3 \right) + \sum p_{n,m}(3e^{n} g_1 t, u_1^t \bar{\omega}_1^t), \tag{51}$$

From (50) and (51), the adjoint equations are

$$E = \sum \left(p_{0,m} z_2^m + \sum p_{1,m} z_1 z_2^m + \sum p_{2,m} z_1^2 z_2^m + \sum p_{3,m} z_1^3 z_2^m \right)$$

$$\frac{\delta \mathcal{L}}{\delta T} = D_t^{\alpha *} T - A_4 (D_y^{\gamma})^* Q + A_3 D_x^{\gamma} T = 0,$$

$$\frac{\delta \mathcal{L}}{\delta w} = \sum M \left(p_{n,m} + \sum \sum p_{n-k,m-j} p_{k,j} + r_{n,m} \right) z_1^{n+3} z_2^m \tag{41}$$

By the fractional Noether operator [2] Ψ_i defined as

$$H_1 = \left[(P^2 - A_0^2) + N(P - A_0) \right], \tag{42}$$

and

$$T = \varphi(x, y, t, u, w), \qquad Q = \psi(x, y, t, u, w), \tag{53}$$

Equation (52) satisfies, with at least one of T, Q variable being non-zero, the system (2) is called the nonlinear self adjoint. Now, the derivative(s) of $T = \varphi(x, y, t, u, w)$ with respect to x, are

$$R = \sum_{n,m=0}^{\infty} r_{n,m} z_1^n z_2^m = \sum_{n,m=0}^{\infty} N p_{n,m} z_1^n z_2^m$$

$$= N \left[A_0 + A_1 z_1 + A_2 z_1^2 + A_3 z_1^3 + M z_1^3 \left((P - A_0) + (P^2 - A_0^2) + N(P - A_0) \right) \right] \tag{43}$$

$$T_x = \varphi_x + \varphi_u u_x + \varphi_w w_x,$$

$$T_{xx} = \varphi_{xx} + 2\varphi_{xu} u_x + 2\varphi_{xw} w_x + \varphi_{uu} u_x^2 + 2\varphi_{uw} u_x w_x + \varphi_{ww} w_x^2 + \varphi_u u_{xx} + \varphi_w w_{xx},$$

$$T_{xxx} = \varphi_{xxx} + 6\varphi_{xuw} u_x w_x + 3\varphi_{uuw} u_x^2 w_x + 3\varphi_{uuw} u_x w_{xx} + 3\varphi_{uuw} u_x w_x^2 + 3\varphi_{xu} u_{xx}$$

$$F(z_1, z_2, P, R) = P - A_0 - A_1 z_1 - A_2 z_1^2 - A_3 z_1^3 - M z_1^3 \left[(P - A_0) + (P^2 - A_0^2) + N(P - A_0) \right]$$

$$+ 3\varphi_{uw}(u_x w_{xx} + w_x u_{xx}) + 3\varphi_{ww} w_x w_{xx} + 3\varphi_{xxw} w_x + 3\varphi_{xxu} u_x + 3\varphi_{xw} w_{xx}$$

$$H(z_1, z_2, P, R) = R - N \left[A_0 + A_1 z_1 + A_2 z_1^2 + A_3 z_1^3 + M z_1^3 \left((P - A_0) + (P^2 - A_0^2) + N(P - A_0) \right) \right]$$

$$+ 3\varphi_{uu} u_{xxx} + \varphi_w w_{xxx} + 3\varphi_{xu} u_{xx} + 3\varphi_{uww} u_x w_x^2 + \varphi_{uuu} u_x^3 + \varphi_{www} w_x^3. \tag{54}$$

Thus, the nonlinear self adjointness conditions are

$$\frac{\delta \mathcal{L}}{\delta u} = \lambda_1 (\partial_t^{\alpha} u - A_1 u \partial_x^{\beta} u - A_2 \partial_y^{\gamma} w - A_3 u_{xxx}) + \lambda_2 (\partial_x^{\beta} w - A_4 \partial_y^{\gamma} u)$$

$$\frac{\delta \mathcal{L}}{\delta w} = \lambda_3 (\partial_t^{\alpha} u - A_1 u \partial_x^{\beta} u - A_2 \partial_y^{\gamma} w - A_3 u_{xxx}) + \lambda_4 (\partial_x^{\beta} w - A_4 \partial_y^{\gamma} u), \tag{55}$$

where λ_i ($i = 1, 2, 3, 4$) are to be determined.

Therefore, we have

$$(D_t^{\alpha})^* \varphi + -A_1 u (D_x^{\beta})^* \varphi - A_4 (D_y^{\gamma}) * \psi + A_3 \bigg(\varphi_{xxx} + 6\varphi_{xuw} u_x w_x + 3\varphi_{uuw} u_x^2 w_x + 3\varphi_{uu} u_x w_{xx}$$

$$+ 3\varphi_{ww} w_x w_{xx} + 3\varphi_{uw} (u_x w_{xx} + w_x u_{xx}) + 3\varphi_{xxw} w_x + 3\varphi_{xxu} u_x + 3\varphi_{xw} w_{xx} + \varphi_u u_{xxx} + 3\varphi_{xuu} u_x^2$$

$$+ 3\varphi_{xww} w_x^2 + \varphi_w w_{xxx} + 3\varphi_{xu} u_{xx} + 3\varphi_{uww} u_x w_x^2 + \varphi_{uuu} u_x^3 + \varphi_{www} w_x^3 \bigg)$$

$$= \lambda_1 (\partial_t^{\alpha} u - A_1 u \partial_x^{\beta} u - A_2 \partial_y^{\gamma} w - A_3 u_{xxx}) + \lambda_2 (\partial_x^{\beta} w - A_4 \partial_y^{\gamma} u),$$

$$(D_x^{\beta})^* \psi - A_2 (D_y^{\gamma})^* \varphi$$

$$= \lambda_3 (\partial_t^{\alpha} u - A_1 u \partial_x^{\beta} u - A_2 \partial_y^{\gamma} w - A_3 u_{xxx}) + \lambda_4 (\partial_x^{\beta} w - A_4 \partial_y^{\gamma} u). \tag{56}$$

$$C^t = \sum_{j=1}^{2} \left[\sum_{k=0}^{m-1} (-1)^k D_t^{\alpha-1-k}(W^j) D_t^k \left(\frac{\partial \mathcal{L}}{\partial (D_t^{\alpha} u_j)} \right) - (-1)^m \mathcal{J}_1 \left(W^j, D_t^m \left(\frac{\partial \mathcal{L}}{\partial (D_t^{\alpha} u_j)} \right) \right) \right] \tag{59}$$

Funding: This research was funded by the Key Scientific Research Project of Higher Education of Henan Province of China (No.1, 2)3] 40000 and the Fundamental Research Funds for the Universities of Henan Province. (57)

Institutional Review Board Statement: Not applicable.

Informed Consent Statement: Not applicable.

Data Availability Statement: Data sharing not applicable to this article as no datasets were generated or analyzed during the current study.

Acknowledgments: The author Baljinder Kour Dhiman fully acknowledges the support of CSIR SRF G(58).

Conflicts of Interest: The authors declare no conflict of interest.

References

1. Akbulut, A.; Kaplan, M. Symmetries, Auxiliary equation method for time-fractional differential equations with conformable derivative. *Comput. Math. Appl.* **2018**, *75*, 876–882. [CrossRef]
2. Jia, J.; Wang, H. Symmetries, A fast finite difference method for distributed-order space-fractional partial differential equations on convex domains. *Comput. Math. Appl.* **2018**, *75*, 2031–2043. [CrossRef]
3. Chen, C.; Jiang, Y.L. Symmetries, Simplest equation method for some time-fractional partial differential equations with conformable derivative. *Comput. Math. Appl.* **2018**, *75*, 2978–2988. [CrossRef]
4. Höök, L.J.; Ludvigsson, G.; von Sydow, L. The Kolmogorov forward fractional partial differential equation for the CGMY-process with applications in option pricing. *Comput. Math. Appl.* **2018**, *76*, 2330–2344. [CrossRef]
5. Osman, M.S.; Rezazadeh, H.; Eslami, M. Traveling wave solutions for (3 + 1) dimensional conformable fractional Zakharov-Kuznetsov equation with power law nonlinearity. *Nonlinear Eng.* **2019**, *8*, 559–567. [CrossRef]
6. Abdel-Gawad, H.I.; Osman, M.S. On the variational approach for analyzing the stability of solutions of evolution equations. *Kyungpook Math. J.* **2013**, *53*, 661–680. [CrossRef]
7. Akinyemi, L.; Iyiola, O.S. A reliable technique to study nonlinear time-fractional coupled Korteweg-de Vries equations. *Adv. Differ. Equ.* **2020**, *169*, 1–27. [CrossRef]
8. Senol, M.; Iyiola, O.S.; Kasmaei, H.D.; Akinyemi, L. Efficient analytical techniques for solving time-fractional nonlinear coupled Jaulent-Miodek system with energy-dependent Schrödinger potential. *Adv. Differ. Equ.* **2019**, *1*, 462. [CrossRef]
9. Kumar, S.; Kumar, A.; Samet, B.; Gómez-Aguilar, J.F.; Osman, M.S. A chaos study of tumor and effector cells in fractional tumor-immune model for cancer treatment. *Chaos Solitons Fractals* **2020**, *141*, 110321. [CrossRef]
10. Arqub, O.A.; Osman, M.S.; Abdel-Aty, A.H.; Mohamed, A.B.A.; Momani, S. A numerical algorithm for the solutions of ABC singular Lane-Emden type models arising in astrophysics using reproducing kernel discretization method. *Mathematics* **2020**, *8*, 923. [CrossRef]
11. Senol, M. New analytical solutions of fractional symmetric regularized-long-wave equation. *Rev. Mex. Fis.* **2020**, *66*, 297–307. [CrossRef]
12. Kour, B.; Kumar, S. Time fractional Biswas Milovic equation: Group analysis, soliton solutions, conservation laws and residual power series solution. *Optik* **2019**, *183*, 1085–1098. [CrossRef]
13. Kumar, S.; Kumar, R.; Osman, M.S.; Samet, B. A wavelet based numerical scheme for fractional order SEIR epidemic of measles by using Genocchi polynomials. *Numer. Method Partial Differ. Equ.* **2021**, *37*, 1250–1268. [CrossRef]
14. Kumar, D.; Kaplan, M.; Haque, M.; Osman, M.S.; Baleanu, D. A Variety of Novel Exact Solutions for Different Models With the Conformable Derivative in Shallow Water. *Front. Phys.* **2020**, *8*, 177. [CrossRef]
15. Kour, B.; Kumar, S. Space time fractional Drinfel'd-Sokolov-Wilson system with time-dependent variable coefficients: Symmetry analysis, power series solutions and conservation laws. *Eur. Phys. J. Plus* **2019**, *134*, 467. [CrossRef]
16. Kumari, P.; Gupta, R.K.; Kumar, S. The time fractional D (m, n) system: Invariant analysis, explicit solution, conservation laws and optical soliton. *Waves Random Complex Media* **2020**. [CrossRef]
17. Feng, L.L.; Tian, S.F.; Wang, X.B.; Zhang, T.T. Lie symmetry analysis, conservation laws and exact power series solutions for time-fractional Fordy-Gibbons equation. *Commun. Theor. Phys.* **2016**, *66*, 321–329. [CrossRef]
18. Zhang, Y. Lie symmetry analysis and exact solutions of general time fractional fifth-order Korteweg-de Vries equation. *Int. J. Appl. Math.* **2017**, *47*, 66–74.
19. Rui, W. Zhang, X. Lie symmetries and conservation laws for the time fractional Derrida-Lebowitz-Speer-Spohn equation. *Commun. Nonlinear Sci. Numer. Simul.* **2016**, *34*, 38–44. [CrossRef]
20. Sil, S.; Sekhar, T.R.; Zeidan, D. Nonlocal conservation laws, nonlocal symmetries and exact solutions of an integrable soliton equation. *Chaos Soliton Fractals* **2020**, *139*, 110010. [CrossRef]
21. Du, X.X.; Tian, B.; Qu, Q.X.; Yuan, Y.-Q.; Zhao, X.-H. Lie group analysis, solitons, self-adjointness and conservation laws of the modified Zakharov-Kuznetsov equation in an electron-positron-ion magnetoplasma. *Chaos Soliton Fractals* **2020**, *134*, 109709. [CrossRef]

Article

A Comparative Analysis of Fractional-Order Kaup–Kupershmidt Equation within Different Operators

Nehad Ali Shah [1], Yasser S. Hamed [2], Khadijah M. Abualnaja [2], Jae-Dong Chung [1,*], Rasool Shah [3] and Adnan Khan [3]

[1] Department of Mechanical Engineering, Sejong University, Seoul 05006, Korea; nehadali199@yahoo.com
[2] Department of Mathematics and Statistics, College of Science, Taif University, P.O. Box 11099, Taif 21944, Saudi Arabia; yasersalah@tu.edu.sa (Y.S.H.); kh.abualnaja@tu.edu.sa (K.M.A.)
[3] Department of Mathematics, Abdul Wali Khan University, Mardan 23200, Pakistan; rasoolshah@awkum.edu.pk (R.S.); adnanmummand@gmail.com (A.K.)
* Correspondence: jdchung@sejong.ac.kr

Abstract: In this paper, we find the solution of the fractional-order Kaup–Kupershmidt (KK) equation by implementing the natural decomposition method with the aid of two different fractional derivatives, namely the Atangana–Baleanu derivative in Caputo manner (ABC) and Caputo–Fabrizio (CF). When investigating capillary gravity waves and nonlinear dispersive waves, the KK equation is extremely important. To demonstrate the accuracy and efficiency of the proposed technique, we study the nonlinear fractional KK equation in three distinct cases. The results are given in the form of a series, which converges quickly. The numerical simulations are presented through tables to illustrate the validity of the suggested technique. Numerical simulations in terms of absolute error are performed to ensure that the proposed methodologies are trustworthy and accurate. The resulting solutions are graphically shown to ensure the applicability and validity of the algorithms under consideration. The results that we obtain confirm that the proposed method is the best tool for handling any nonlinear problems arising in science and technology.

Keywords: Caputo–Fabrizio and Atangana-Baleanu operators; time-fractional Kaup–Kupershmidt equation; natural transform; Adomian decomposition method

1. Introduction

Fractional calculus has grown in popularity over the last three decades, owing to its well-established applications in a wide range of scientific and engineering areas. Many pioneers have demonstrated that fractional-order models can effectively describe complicated phenomena when modified by integer-order models [1,2]. The integer-order derivatives are local in nature, whereas the Caputo fractional derivatives are nonlocal. That is, we can investigate changes in the neighbourhood of a point with the integer-order derivative, but we can analyse changes in the entire interval with the Caputo fractional derivative. Senior mathematicians worked together to establish the basic framework for fractional-order derivatives and integrals, such as Caputo [3], Riemann [4], Liouville [5], Podlubny [6], Miller and Ross [7] and others. Fractional-order calculus theory has been linked to practical projects and it has been applied to signal processing [8], chaos theory [9], human diseases [10,11], electrodynamics [12] and other areas.

Fractional differential equations are becoming more well known nowadays as a result of their numerous applications in science and engineering, such as electrodynamics [13], chaos theory [14], finance [15], fluid and continuum mechanics [16], signal processing [17], biological population models [18] and some others, which are well described by fractional differential equations. The elegance of symmetry analysis is most evident in the study of partial differential equations—more precisely, those derived from finance mathematics. The secret of nature is symmetry, but most observations in nature do not exhibit symmetry.

The phenomenon of spontaneous symmetry breaking is an effective approach to conceal symmetry. Symmetries are classified into two types: finite and infinitesimal. Discrete or continuous symmetries can exist for finite symmetries. Symmetry and time reverse are discrete natural symmetries, whereas space is a continuous transformation. Patterns have captivated mathematicians for centuries. In the nineteenth century, systematic classifications of planar and spatial patterns emerged. Regrettably, solving nonlinear fractional differential equations accurately has proven to be rather challenging [19]. Effective tools are required to solve such problems. As a result, in this article, we will try to use an effective analytic method to obtain a more accurate solution for nonlinear arbitrary-order differential equations. Fractional differential equations can pleasantly and even more precisely analyse a variety of schemes in collaborative areas. In this connection, different techniques have been developed, among which some are as follows: the reduced differential transform method (RDTM) [20], the fractional Adomian decomposition method (FADM) [21], the fractional variational iteration method (FVIM) [22], the Elzaki transform decomposition method (ETDM) [23,24], the iterative Laplace transform method (ILTM) [25], the fractional natural decomposition method (FNDM) [26], the fractional homotopy perturbation method (FHPM) [27] and the Yang transform decomposition method (YTDM) [28]. The main goal of the present paper is to implement the natural decomposition method with the help of two different fractional derivatives to study the fractional-order Kaup–Kupershmidt (KK) equation. Natural decomposition methods avoid round-off errors by not requiring prescriptive assumptions, linearization, discretization or perturbation.

Kaup presented the famous dispersive classical Kaup–Kupershmidt equation [29] in 1980, and Kupershmidt modified it in 1994 [30]. The purpose of this paper is to look at the time-fractional modified Kaup–Kupershmidt (KK) equation. The study of nonlinear dispersive waves and the behaviour of capillary gravity waves is examined using the fractional-order Kaup-Kupershmidt equation. The nonlinear fifth-order evolution equation is of the form:

$$D_\kappa^\gamma \zeta(\varphi,\kappa) + j\zeta\zeta_{\varphi\varphi\varphi} + kp\zeta_\varphi\zeta_{\varphi\varphi} + l\zeta^2\zeta_\varphi + \zeta_{\varphi\varphi\varphi\varphi\varphi} = 0, \qquad (1)$$

where j, k and l are constants, and $0 < \gamma \leq 1$ represents the order time-fractional derivative. The above fifth-order nonlinear evolution equation can be transformed into the fifth-order time-fractional Kaup–Kupershmidt equation by changing the values of j, k and l. Thus, by taking $j = -15, k = -15$ and $l = 45$, the given equation reduces to

$$D_\kappa^\gamma \zeta(\varphi,\kappa) - 15\zeta\zeta_{\varphi\varphi\varphi} - 15p\zeta_\varphi\zeta_{\varphi\varphi} + 45\zeta^2\zeta_\varphi + \zeta_{\varphi\varphi\varphi\varphi\varphi} = 0, \qquad (2)$$

Extensive research has been dedicated in recent years to the investigation of the classical Kaup–Kupershmidt equation. At $p = \frac{5}{2}$, the classical KK equation is integrable [31] and has bilinear representations [32]. For general nonlinear evolution equations, solitary and soliton wave solutions can be obtained by independently applying four different approaches. Ablowitz and Clarkson used the inverse scattering approach in the creation of soliton solutions to investigate nonlinear equations having physical implications [33]. Tam and Hu employed Hirota's approach and used Mathematica to determine the equivalent answer [34]. Musette and Verhoeven reported the fifth-order Kaup–Kupershmidt equation, which was one of the integrable examples of the Henon–Heiles system.

The rest of the paper is organized as follows: in Section 2, some of the suitable definitions related to fractional derivatives and used in our present work are given. For the fractional-order Kaup–Kupershmidt equation, the basic idea of the natural decomposition method with the aid of two different fractional derivatives is presented in Section 3. The convergence phenomenon for the proposed method is presented in Section 4. Section 5 is concerned with the implementation of the suggested technique for the solution of various problems of the fractional-order Kaup–Kupershmidt equation. At the end, a brief conclusion of the whole paper is given.

2. Basic Preliminaries

In this part of the article, we present some basic definitions related to fractional calculus that are further used in our work too.

Definition 1. *For a function $j \in C_v, v \geq -1$, the Riemann–Liouville integral for non-integer order is given as [35]*

$$I^\gamma j(\vartheta) = \frac{1}{\Gamma(\gamma)} \int_0^\vartheta (\vartheta - \mu)^{\gamma-1} j(\mu) d\mu, \quad \gamma > 0, \ \vartheta > 0. \tag{3}$$

and $I^0 j(\vartheta) = j(\vartheta)$

Definition 2. *For a function $j(\vartheta)$, the fractional Caputo derivative is defined as [35]*

$$^C D_\vartheta^\gamma j(\vartheta) = I^{n-\gamma} D^n j(\vartheta) = \frac{1}{n-\gamma} \int_\vartheta^0 (\vartheta - \mu)^{n-\gamma-1} j^n(\mu) d\mu \tag{4}$$

for $n - 1 < \gamma \leq n$, $n \in N$, $\vartheta > 0$, $j \in C_v^n$, $v \geq -1$.

Definition 3. *For a function $j(\vartheta)$, the fractional Caputo–Fabrizio derivative is given as [35]*

$$^{CF} D_\vartheta^\gamma j(\vartheta) = \frac{F(\gamma)}{1-\gamma} \int_0^\vartheta \exp\left(\frac{-\gamma(\vartheta - \mu)}{1-\gamma}\right) D(j(\mu)) d\mu, \tag{5}$$

where $0 < \gamma < 1$ and the normalization function is represented by $F(\gamma)$ with $F(0) = F(1) = 1$.

Definition 4. *For a function $j(\vartheta)$, the fractional Atangana–Baleanu Caputo derivative is defined as [35]*

$$^{ABC} D_\vartheta^\gamma j(\vartheta) = \frac{B(\gamma)}{1-\gamma} \int_0^\vartheta E_\gamma\left(\frac{-\gamma(\vartheta - \mu)}{1-\gamma}\right) D(j(\mu)) d\mu, \tag{6}$$

where $0 < \gamma < 1$, $B(\gamma)$ represents the normalization function with a similar property as $F(\gamma)$ and $E_\gamma(z) = \sum_{m=0}^\infty \frac{z^m}{\Gamma(m\gamma+1)}$ represents the Mittag–Leffler function.

Definition 5. *By applying the natural transform, the function $\zeta(\kappa)$ can be rewritten as*

$$\mathcal{N}(\zeta(\kappa)) = \mathcal{V}(\omega, v) = \int_{-\infty}^\infty e^{-\omega\kappa} \zeta(v\kappa) d\kappa, \quad \omega, v \in (-\infty, \infty). \tag{7}$$

Natural transformation of $\zeta(\kappa)$ for $\kappa \in (0, \infty)$ is given as

$$\mathcal{N}(\zeta(\kappa) H(\kappa)) = \mathcal{N}^+ \zeta(\kappa) = \mathcal{V}^+(\omega, v) = \int_{-\infty}^\infty e^{-\omega\kappa} \zeta(v\kappa) d\kappa, \quad \omega, v \in (0, \infty). \tag{8}$$

where $H(\kappa)$ is the Heaviside function.

Definition 6. *On applying the natural inverse transform, the function $\mathcal{V}(\omega, v)$ can be written as*

$$\mathcal{N}^{-1}[\mathcal{V}(\omega, v)] = \zeta(\kappa), \quad \forall \kappa \geq 0 \tag{9}$$

Lemma 1. *If the linearity property having natural transformation for $\zeta_1(\kappa)$ is $\zeta_1(\omega, v)$ and $\zeta_2(\kappa)$ is $\zeta_2(\omega, v)$, then*

$$\mathcal{N}[c_1\zeta_1(\kappa) + c_2\zeta_2(\kappa)] = c_1 \mathcal{N}[\zeta_1(\kappa)] + c_2 \mathcal{N}[\zeta_2(\kappa)] = c_1 \mathcal{V}_1(\omega, v) + c_2 \mathcal{V}_2(\omega, v), \tag{10}$$

where c_1 and c_2 are constants.

Lemma 2. *If the inverse natural transforms of $\mathcal{V}_1(\varpi, v)$ and $\mathcal{V}_2(\varpi, v)$ are $\zeta_1(\kappa)$ and $\zeta_2(\kappa)$, respectively, then*

$$\mathcal{N}^{-1}[c_1\mathcal{V}_1(\varpi, v) + c_2\mathcal{V}_2(\varpi, v)] = c_1\mathcal{N}^{-1}[\mathcal{V}_1(\varpi, v)] + c_2\mathcal{N}^{-1}[\mathcal{V}_2(\varpi, v)] = c_1\zeta_1(\kappa) + c_2\zeta_2(\kappa), \quad (11)$$

where c_1 and c_2 are constants.

Definition 7. *The natural transformation of $D_\kappa^\gamma \zeta(\kappa)$ in the Caputo sense is defined as [35]*

$$\mathcal{N}[^C D_\kappa^\gamma] = \left(\frac{\varpi}{v}\right)^\gamma \left(\mathcal{N}[\zeta(\kappa)] - \left(\frac{1}{\varpi}\right)\zeta(0)\right) \quad (12)$$

Definition 8. *The natural transformation of $D_\kappa^\gamma \zeta(\kappa)$ in the Caputo–Fabrizio sense is defined as [35]*

$$\mathcal{N}[^{CF} D_\kappa^\gamma] = \frac{1}{1 - \gamma + \gamma(\frac{v}{\varpi})} \left(\mathcal{N}[\zeta(\kappa)] - \left(\frac{1}{\varpi}\right)\zeta(0)\right) \quad (13)$$

Definition 9. *The natural transformation of $D_\kappa^\gamma \zeta(\kappa)$ in the Atangana–Baleanu Caputo sense is defined as [35]*

$$\mathcal{N}[^{ABC} D_\kappa^\gamma] = \frac{B(\gamma)}{1 - \gamma + \gamma(\frac{v}{\varpi})^\gamma} \left(\mathcal{N}[\zeta(\kappa)] - \left(\frac{1}{\varpi}\right)\zeta(0)\right) \quad (14)$$

3. Methodology

In this section, we give the general implementation of the natural transform decomposition method with the aid of two different derivatives for solving the given equation [36,37].

$$D_\kappa^\gamma \zeta(\varphi, \kappa) = \mathcal{L}(\zeta(\varphi, \kappa)) + \mathbb{N}(\zeta(\varphi, \kappa)) + h(\varphi, \kappa), \quad (15)$$

with initial condition

$$\zeta(\varphi, 0) = \phi(\varphi), \quad (16)$$

having $\mathcal{L}$ linear term, $\mathbb{N}$ nonlinear term and the source term $h(\varphi, \kappa)$.

3.1. Case I ($NTDM_{CF}$)

By applying the natural transform with the aid of the fractional Caputo–Fabrizio derivative, Equation (1) can be rewritten as

$$\frac{1}{p(\gamma, v, \varpi)} \left(\mathcal{N}[\zeta(\varphi, \kappa)] - \frac{\phi(\varphi)}{\varpi}\right) = \mathcal{N}\left[\mathcal{L}(\zeta(\varphi, \kappa)) + \mathbb{N}(\zeta(\varphi, \kappa)) + h(\varphi, \kappa)\right], \quad (17)$$

with

$$p(\gamma, v, \varpi) = 1 - \gamma + \gamma(\frac{v}{\varpi}). \quad (18)$$

On applying natural inverse transformation, Equation (3) can be presented as

$$\zeta(\varphi, \kappa) = \mathcal{N}^{-1}\left[\frac{\phi(\varphi)}{\varpi} + p(\gamma, v, \varpi)\mathcal{N}[h(\varphi, \kappa)]\right] + \mathcal{N}^{-1}\left[p(\gamma, v, \varpi)\mathcal{N}\left(\mathcal{L}(\zeta(\varphi, \kappa)) + \mathbb{N}(\zeta(\varphi, \kappa))\right)\right]. \quad (19)$$

$\mathbb{N}(\zeta(\varphi,\kappa))$ can be decomposed into

$$\mathbb{N}(\zeta(\varphi,\kappa)) = \sum_{i=0}^{\infty} A_i, \qquad (20)$$

The series form solution for $\zeta^{CF}(\varphi,\kappa)$ is given as

$$\zeta^{CF}(\varphi,\kappa) = \sum_{i=0}^{\infty} \zeta_i^{CF}(\varphi,\kappa). \qquad (21)$$

Substituting Equations (6) and (7) into (5), we get

$$\sum_{i=0}^{\infty} \zeta_i(\varphi,\kappa) = \mathcal{N}^{-1}\left(\frac{\phi(\varphi)}{\varpi} + p(\gamma,v,\varpi)\mathcal{N}[h(\varphi,\kappa)]\right)$$
$$+ \mathcal{N}^{-1}\left(p(\gamma,v,\varpi)\mathcal{N}\left[\sum_{i=0}^{\infty}\mathcal{L}(\zeta_i(\varphi,\kappa)) + A_\kappa\right]\right) \qquad (22)$$

From (8), we have

$$\zeta_0^{CF}(\varphi,\kappa) = \mathcal{N}^{-1}\left(\frac{\phi(\varphi)}{\varpi} + p(\gamma,v,\varpi)\mathcal{N}[h(\varphi,\kappa)]\right),$$
$$\zeta_1^{CF}(\varphi,\kappa) = \mathcal{N}^{-1}(p(\gamma,v,\varpi)\mathcal{N}[\mathcal{L}(\zeta_0(\varphi,\kappa)) + A_0]),$$
$$\vdots \qquad (23)$$
$$\zeta_{l+1}^{CF}(\varphi,\kappa) = \mathcal{N}^{-1}(p(\gamma,v,\varpi)\mathcal{N}[\mathcal{L}(\zeta_l(\varphi,\kappa)) + A_l]), \; l=1,2,3,\cdots$$

Finally, we obtain the $NTDM_{CF}$ solution to (1) by putting (23) into (7),

$$\zeta^{CF}(\varphi,\kappa) = \zeta_0^{CF}(\varphi,\kappa) + \zeta_1^{CF}(\varphi,\kappa) + \zeta_2^{CF}(\varphi,\kappa) + \cdots \qquad (24)$$

3.2. Case II ($NTDM_{ABC}$)

By applying the natural transform with the aid of the fractional Atangana–Baleanu Caputo derivative, Equation (1) can be rewritten as

$$\frac{1}{q(\gamma,v,\varpi)}\left(\mathcal{N}[\zeta(\varphi,\kappa)] - \frac{\phi(\varphi)}{\varpi}\right) = \mathcal{N}\left[\mathcal{L}(\zeta(\varphi,\kappa)) + \mathbb{N}(\zeta(\varphi,\kappa)) + h(\varphi,\kappa)\right], \qquad (25)$$

with

$$q(\gamma,v,\varpi) = \frac{1-\gamma+\gamma(\frac{v}{\varpi})^\gamma}{B(\gamma)}. \qquad (26)$$

On applying the natural inverse transform, Equation (25) can be presented as

$$\zeta(\varphi,\kappa) = \mathcal{N}^{-1}\left(\frac{\phi(\varphi)}{\varpi} + q(\gamma,v,\varpi)\mathcal{N}[h(\varphi,\kappa)]\right) + \mathcal{N}^{-1}\left[q(\gamma,v,\varpi)\mathcal{N}\left(\mathcal{L}(\zeta(\varphi,\kappa)) + \mathbb{N}(\zeta(\varphi,\kappa))\right)\right]. \qquad (27)$$

$\mathbb{N}(\zeta(\varphi,\kappa))$ can be decomposed into

$$\mathbb{N}(\zeta(\varphi,\kappa)) = \sum_{i=0}^{\infty} A_i, \qquad (28)$$

The series form solution for $\zeta^{ABC}(\varphi,\kappa)$ is given as

$$\zeta^{ABC}(\varphi,\kappa) = \sum_{i=0}^{\infty} \zeta_i^{ABC}(\varphi,\kappa). \tag{29}$$

Substituting Equations (28) and (29) into (27), we get

$$\sum_{i=0}^{\infty} \zeta_i(\varphi,\kappa) = \mathcal{N}^{-1}\left(\frac{\phi(\varphi)}{\varpi} + q(\gamma,v,\varpi)\mathcal{N}[h(\varphi,\kappa)]\right)$$
$$+ \mathcal{N}^{-1}\left(q(\gamma,v,\varpi)\mathcal{N}\left[\sum_{i=0}^{\infty} \mathcal{L}(\zeta_i(\varphi,\kappa)) + A_\kappa\right]\right) \tag{30}$$

From (8), we have

$$\zeta_0^{ABC}(\varphi,\kappa) = \mathcal{N}^{-1}\left(\frac{\phi(\varphi)}{\varpi} + q(\gamma,v,\varpi)\mathcal{N}[h(\varphi,\kappa)]\right),$$
$$\zeta_1^{ABC}(\varphi,\kappa) = \mathcal{N}^{-1}(q(\gamma,v,\varpi)\mathcal{N}[\mathcal{L}(\zeta_0(\varphi,\kappa)) + A_0]),$$
$$\vdots \tag{31}$$
$$\zeta_{l+1}^{ABC}(\varphi,\kappa) = \mathcal{N}^{-1}(q(\gamma,v,\varpi)\mathcal{N}[\mathcal{L}(\zeta_l(\varphi,\kappa)) + A_l]), \quad l = 1,2,3,\cdots$$

Finally, we obtain the $NTDM_{ABC}$ solution to (1) by putting (31) into (29):

$$\zeta^{ABC}(\varphi,\kappa) = \zeta_0^{ABC}(\varphi,\kappa) + \zeta_1^{ABC}(\varphi,\kappa) + \zeta_2^{ABC}(\varphi,\kappa) + \cdots \tag{32}$$

4. Convergence Analysis

The convergence and uniqueness analysis of the $NTDM_{CF}$ and $NTDM_{ABC}$ is discussed here.

Theorem 1. *The result of (1) is unique for $NTDM_{CF}$ when $0 < (\Im_1 + \Im_2)(1 - \gamma + \gamma\kappa) < 1$.*

Proof. Let $H = (C[J], ||.||)$ with the norm $||\phi(\kappa)|| = max_{\kappa \in J}|\phi(\kappa)|$ as Banach space, with $\forall$ continuous function on J. Let $I : H \to H$ be a nonlinear mapping, where

$$\zeta_{l+1}^C = \zeta_0^C + \mathcal{N}^{-1}[p(\gamma,v,\varpi)\mathcal{N}[\mathcal{L}(\zeta_l(\mu,\kappa)) + \mathbb{N}(\zeta_l(\mu,\kappa))]], \quad l \geq 0.$$

Suppose that $|\mathcal{L}(\zeta) - \mathcal{L}(\zeta^*)| < \Im_1|\zeta - \zeta^*|$ and $|\mathbb{N}(\zeta) - \mathbb{N}(\zeta^*)| < \Im_2|\zeta - \zeta^*|$, where $\zeta := \zeta(\mu,\kappa)$ and $\zeta^* := \zeta^*(\mu,\kappa)$ are two different function values and $\Im_1, \Im_2$ are Lipschitz constants.

$$||I\zeta - I\zeta^*|| \leq max_{t \in J}|\mathcal{N}^{-1}\left[p(\gamma,v,\varpi)\mathcal{N}[\mathcal{L}(\zeta) - \mathcal{L}(\zeta^*)]\right.$$
$$+ p(\gamma,v,\varpi)\mathcal{N}[\mathbb{N}(\zeta) - \mathbb{N}(\zeta^*)]|\right]$$
$$\leq max_{\kappa \in J}\left[\Im_1\mathcal{N}^{-1}[p(\gamma,v,\varpi)\mathcal{N}[|\zeta - \zeta^*|]]\right.$$
$$+ \Im_2\mathcal{N}^{-1}[p(\gamma,v,\varpi)\mathcal{N}[|\zeta - \zeta^*|]]\right] \tag{33}$$
$$\leq max_{t \in J}(\Im_1 + \Im_2)\left[\mathcal{N}^{-1}[p(\gamma,v,\varpi)\mathcal{N}|\zeta - \zeta^*|]\right]$$
$$\leq (\Im_1 + \Im_2)\left[\mathcal{N}^{-1}[p(\gamma,v,\varpi)\mathcal{N}||\zeta - \zeta^*||]\right]$$
$$= (\Im_1 + \Im_2)(1 - \gamma + \gamma\kappa)||\zeta - \zeta^*||.$$

I is a contraction as $0 < (\Im_1 + \Im_2)(1 - \gamma + \gamma\kappa) < 1$. From Banach fixed point theorem, the result of (1) is unique. □

Theorem 2. *The result of (1) is unique for* $NTDM_{ABC}$ *when* $0 < (\Im_1 + \Im_2)(1 - \gamma + \gamma \frac{\kappa^\nu}{\Gamma(\nu+1)}) < 1$.

Proof. Let $H = (C[J], ||.||)$ with the norm $||\phi(\kappa)|| = max_{\kappa \in J}|\phi(\kappa)|$ be the Banach space, with $\forall$ continuous function on J. Let $I : H \to H$ be a nonlinear mapping, where

$$\zeta_{l+1}^C = \zeta_0^C + \mathcal{N}^{-1}[p(\gamma, v, \omega)\mathcal{N}[\mathcal{L}(\zeta_l(\varphi, \kappa)) + \mathbb{N}(\zeta_l(\varphi, \kappa))]], \ l \geq 0.$$

Suppose that $|\mathcal{L}(\zeta) - \mathcal{L}(\zeta^*)| < \Im_1|\zeta - \zeta^*|$ and $|\mathbb{N}(\zeta) - \mathbb{N}(\zeta^*)| < \Im_2|\zeta - \zeta^*|$, where $\zeta := \zeta(\mu, \kappa)$ and $\zeta^* := \zeta^*(\mu, \kappa)$ are two different function values and $\Im_1, \Im_2$ are Lipschitz constants.

$$\begin{aligned}
||I\zeta - I\zeta^*|| &\leq max_{t \in J}|\mathcal{N}^{-1}\Big[q(\gamma, v, \omega)\mathcal{N}[\mathcal{L}(\zeta) - \mathcal{L}(\zeta^*)] \\
&\quad + q(\gamma, v, \omega)\mathcal{N}[\mathbb{N}(\zeta) - \mathbb{N}(\zeta^*)]|\Big] \\
&\leq max_{t \in J}\Big[\Im_1\mathcal{N}^{-1}[q(\gamma, v, \omega)\mathcal{N}[|\zeta - \zeta^*|]] \\
&\quad + \Im_2\mathcal{N}^{-1}[q(\gamma, v, \omega)\mathcal{N}[|\zeta - \zeta^*|]]\Big] \\
&\leq max_{t \in J}(\Im_1 + \Im_2)\Big[\mathcal{N}^{-1}[q(\gamma, v, \omega)\mathcal{N}|\zeta - \zeta^*|]\Big] \\
&\leq (\Im_1 + \Im_2)\Big[\mathcal{N}^{-1}[q(\gamma, v, \omega)\mathcal{N}||\zeta - \zeta^*||]\Big] \\
&= (\Im_1 + \Im_2)(1 - \gamma + \gamma\frac{\kappa^\gamma}{\Gamma\gamma + 1})||\zeta - \zeta^*||.
\end{aligned} \quad (34)$$

I is a contraction as $0 < (\Im_1 + \Im_2)(1 - \gamma + \gamma\frac{\kappa^\gamma}{\Gamma\gamma+1}) < 1$. From Banach fixed point theorem, the result of (1) is unique. □

Theorem 3. *The* $NTDM_{CF}$ *result of (1) is convergent.*

Proof. Let $\zeta_m = \sum_{r=0}^m \zeta_r(\varphi, \kappa)$. To show that ζ_m is a Cauchy sequence in H, let

$$\begin{aligned}
||\zeta_m - \zeta_n|| &= max_{\kappa \in J}|\sum_{r=n+1}^m \zeta_r|, \ n = 1, 2, 3, \cdots \\
&\leq max_{\kappa \in J}\Big|\mathcal{N}^{-1}\Big[p(\gamma, v, \omega)\mathcal{N}\Big[\sum_{r=n+1}^m (\mathcal{L}(\zeta_{r-1}) + \mathbb{N}(\zeta_{r-1}))\Big]\Big]\Big| \\
&= max_{\kappa \in J}\Big|\mathcal{N}^{-1}\Big[p(\gamma, v, \omega)\mathcal{N}\Big[\sum_{r=n+1}^{m-1} (\mathcal{L}(\zeta_r) + \mathbb{N}(\zeta_r))\Big]\Big]\Big| \\
&\leq max_{\kappa \in J}|\mathcal{N}^{-1}[p(\gamma, v, \omega)\mathcal{N}[(\mathcal{L}(\zeta_{m-1}) - \mathcal{L}(\zeta_{n-1}) + \mathbb{N}(\zeta_{m-1}) - \mathbb{N}(\zeta_{n-1}))]]| \\
&\leq \Im_1 max_{\kappa \in J}|\mathcal{N}^{-1}[p(\gamma, v, \omega)\mathcal{N}[(\mathcal{L}(\zeta_{m-1}) - \mathcal{L}(\zeta_{n-1}))]]| \\
&\quad + \Im_2 max_{\kappa \in J}|\mathcal{N}^{-1}[p(\gamma, v, \omega)\mathcal{N}[(\mathbb{N}(\zeta_{m-1}) - \mathbb{N}(\zeta_{n-1}))]]| \\
&= (\Im_1 + \Im_2)(1 - \gamma + \gamma\kappa)||\zeta_{m-1} - \zeta_{n-1}||
\end{aligned} \quad (35)$$

Let $m = n + 1$, then

$$||\zeta_{n+1} - \zeta_n|| \leq \Im||\zeta_n - \zeta_{n-1}|| \leq \Im^2||\zeta_{n-1}\zeta_{n-2}|| \leq \cdots \leq \Im^n||\zeta_1 - \zeta_0||, \quad (36)$$

where $\Im = (\Im_1 + \Im_2)(1 - \gamma + \gamma\kappa)$. Similarly, we have

$$\begin{aligned}
||\zeta_m - \zeta_n|| &\leq ||\zeta_{n+1} - \zeta_n|| + ||\zeta_{n+2}\zeta_{n+1}|| + \cdots + ||\zeta_m - \zeta_{m-1}||, \\
&\quad (\Im^n + \Im^{n+1} + \cdots + \Im^{m-1})||\zeta_1 - \zeta_0|| \\
&\leq \Im^n\left(\frac{1 - \Im^{m-n}}{1 - \Im}\right)||\zeta_1||,
\end{aligned} \quad (37)$$

As $0 < \Im < 1$, we get $1 - \Im^{m-n} < 1$. Therefore,

$$||\zeta_m - \zeta_n|| \leq \frac{\Im^n}{1-\Im} \max_{\kappa \in J} ||\zeta_1||. \quad (38)$$

Since $||\zeta_1|| < \infty$, $||\zeta_m - \zeta_n|| \to 0$ when $n \to \infty$. As a result, ζ_m is a Cauchy sequence in H, implying that the series ζ_m is convergent. □

Theorem 4. *The NTDM$_{ABC}$ result of (1) is convergent.*

Proof. Let $\zeta_m = \sum_{r=0}^{m} \zeta_r(\varphi, \kappa)$. To show that ζ_m is a Cauchy sequence in H, let

$$||\zeta_m - \zeta_n|| = \max_{\kappa \in J} \left| \sum_{r=n+1}^{m} \zeta_r \right|, \quad n = 1, 2, 3, \cdots$$

$$\leq \max_{\kappa \in J} \left| \mathcal{N}^{-1} \left[q(\gamma, v, \omega) \mathcal{N} \left[\sum_{r=n+1}^{m} (\mathcal{L}(\zeta_{r-1}) + \mathbb{N}(\zeta_{r-1})) \right] \right] \right|$$

$$= \max_{\kappa \in J} \left| \mathcal{N}^{-1} \left[q(\gamma, v, \omega) \mathcal{N} \left[\sum_{r=n+1}^{m-1} (\mathcal{L}(\zeta_r) + \mathbb{N}(u_r)) \right] \right] \right| \quad (39)$$

$$\leq \max_{\kappa \in J} |\mathcal{N}^{-1}[q(\gamma, v, \omega) \mathcal{N}[(\mathcal{L}(\zeta_{m-1}) - \mathcal{L}(\zeta_{n-1}) + \mathbb{N}(\zeta_{m-1}) - \mathbb{N}(\zeta_{n-1}))]]|$$

$$\leq \Im_1 \max_{\kappa \in J} |\mathcal{N}^{-1}[q(\gamma, v, \omega) \mathcal{N}[(\mathcal{L}(\zeta_{m-1}) - \mathcal{L}(\zeta_{n-1}))]]|$$

$$+ \Im_2 \max_{\kappa \in J} |\mathcal{N}^{-1}[q(\gamma, v, \omega) \mathcal{N}[(\mathbb{N}(\zeta_{m-1}) - \mathbb{N}(\zeta_{n-1}))]]|$$

$$= (\Im_1 + \Im_2)(1 - \gamma + \gamma \frac{\kappa^\gamma}{\Gamma(\gamma+1)}) ||\zeta_{m-1} - \zeta_{n-1}||$$

Let $m = n + 1$, then

$$||\zeta_{n+1} - \zeta_n|| \leq \Im ||\zeta_n - \zeta_{n-1}|| \leq \Im^2 ||\zeta_{n-1} \zeta_{n-2}|| \leq \cdots \leq \Im^n ||\zeta_1 - \zeta_0||, \quad (40)$$

where $\Im = (\Im_1 + \Im_2)(1 - \gamma + \gamma \frac{\kappa^\gamma}{\Gamma(\gamma+1)})$. Similarly, we have

$$||\zeta_m - \zeta_n|| \leq ||\zeta_{n+1} - \zeta_n|| + ||\zeta_{n+2}\zeta_{n+1}|| + \cdots + ||\zeta_m - \zeta_{m-1}||,$$
$$(\Im^n + \Im^{n+1} + \cdots + \Im^{m-1}) ||\zeta_1 - \zeta_0|| \quad (41)$$
$$\leq \Im^n \left(\frac{1 - \Im^{m-n}}{1 - \Im} \right) ||\zeta_1||,$$

As $0 < \Im < 1$, we get $1 - \Im^{m-n} < 1$. Therefore,

$$||\zeta_m - \zeta_n|| \leq \frac{\Im^n}{1-\Im} \max_{t \in J} ||\zeta_1||. \quad (42)$$

Since $||\zeta_1|| < \infty$, $||\zeta_m - \zeta_n|| \to 0$ when $n \to \infty$. As a result, ζ_m is a Cauchy sequence in H, implying that the series ζ_m is convergent. □

5. Numerical Examples

In this section, we find the analytical solution of the time-fractional Kaup–Kupershmidt equation.

Example 1. *Consider the time-fractional Kaup–Kupershmidt equation [38]*

$$D_\kappa^\gamma \zeta(\varphi, \kappa) - 15\zeta\zeta_{\varphi\varphi\varphi} - 15p\zeta_\varphi \zeta_{\varphi\varphi} + 45\zeta^2 \zeta_\varphi + \zeta_{\varphi\varphi\varphi\varphi\varphi} = 0, \quad 0 < \gamma \leq 1, \quad (43)$$

with initial condition

$$\zeta(\varphi, 0) = \frac{1}{4} w^2 Y^2 \operatorname{sech}^2\left(\frac{w\varphi Y}{2}\right) + \frac{w^2 Y^2}{12}, \quad (44)$$

Equation (43) can be expressed as follows with the use of the natural transform:

$$\mathcal{N}[D_\kappa^\gamma \zeta(\varphi,\kappa)] = \mathcal{N}\left\{15\zeta\zeta_{\varphi\varphi\varphi}\right\} + \mathcal{N}\left\{15p\zeta_\varphi\zeta_{\varphi\varphi}\right\} - \mathcal{N}\left\{45\zeta^2\zeta_\varphi\right\} - \mathcal{N}\left\{\zeta_{\varphi\varphi\varphi\varphi\varphi}\right\}, \quad (45)$$

Characterize the non-linear operator as

$$\frac{1}{\omega^\gamma}\mathcal{N}[\zeta(\varphi,\kappa)] - \omega^{2-\gamma}\zeta(\varphi,0) = \mathcal{N}\left[15\zeta\zeta_{\varphi\varphi\varphi} + 15p\zeta_\varphi\zeta_{\varphi\varphi} - 45\zeta^2\zeta_\varphi - \zeta_{\varphi\varphi\varphi\varphi\varphi}\right], \quad (46)$$

We obtain the following when it comes to simplification:

$$\mathcal{N}[\zeta(\varphi,\kappa)] = \omega^2\left[\frac{1}{4}w^2Y^2\,\text{sech}^2(\frac{w\varphi Y}{2}) + \frac{w^2Y^2}{12}\right] + \frac{\gamma(\omega-\gamma(\omega-\gamma))}{\omega^2}\mathcal{N}\left[15\zeta\zeta_{\varphi\varphi\varphi} + 15p\zeta_\varphi\zeta_{\varphi\varphi} - 45\zeta^2\zeta_\varphi - \zeta_{\varphi\varphi\varphi\varphi\varphi}\right], \quad (47)$$

Equation (47) can be written as follows with inverse NT:

$$\zeta(\varphi,\kappa) = \left[\frac{1}{4}w^2Y^2\,\text{sech}^2(\frac{w\varphi Y}{2}) + \frac{w^2Y^2}{12}\right]$$
$$+ \mathcal{N}^{-1}\left[\frac{\gamma(\omega-\gamma(\omega-\gamma))}{\omega^2}\mathcal{N}\left\{15\zeta\zeta_{\varphi\varphi\varphi} + 15p\zeta_\varphi\zeta_{\varphi\varphi} - 45\zeta^2\zeta_\varphi - \zeta_{\varphi\varphi\varphi\varphi\varphi}\right\}\right], \quad (48)$$

5.1. Implementing NDM$_{CF}$

The unknown function $\zeta(\varphi,\kappa)$ has a series form solution, which is stated as

$$\zeta(\varphi,\kappa) = \sum_{l=0}^{\infty}\zeta_l(\varphi,\kappa) \quad (49)$$

The nonlinear terms are illustrated by using Adomian polynomials $\zeta\zeta_{\varphi\varphi\varphi} = \sum_{l=0}^{\infty}\mathcal{A}_l$, $\zeta_\varphi\zeta_{\varphi\varphi} = \sum_{l=0}^{\infty}\mathcal{B}_l$ and $\zeta^2\zeta_\varphi = \sum_{l=0}^{\infty}\mathcal{C}_l$ Thus, Equation (48) can be expressed with the help of the following terms

$$\sum_{l=0}^{\infty}\zeta_{l+1}(\varphi,\kappa) = \frac{1}{4}w^2Y^2\,\text{sech}^2(\frac{w\varphi Y}{2}) + \frac{w^2Y^2}{12}$$
$$+ \mathcal{N}^{-1}\left[\frac{\gamma(\omega-\gamma(\omega-\gamma))}{\omega^2}\mathcal{N}\left\{15\sum_{l=0}^{\infty}\mathcal{A}_l + 15\sum_{l=0}^{\infty}\mathcal{B}_l - 45\sum_{l=0}^{\infty}\mathcal{C}_l - \sum_{l=0}^{\infty}\zeta_{l\varphi\varphi\varphi\varphi\varphi}\right\}\right], \quad (50)$$

When both sides of Equation (50) are compared, we obtain

$$\zeta_0(\varphi,\kappa) = \frac{1}{4}w^2Y^2\,\text{sech}^2(\frac{w\varphi Y}{2}) + \frac{w^2Y^2}{12},$$

$$\zeta_1(\varphi,\kappa) = -\left(-\frac{1}{512}w^7Y^7(3843 + 480p - 4(209 + 60p)\cosh(w\varphi Y) + \cosh(2w\varphi Y))\,\text{sech}^6\left(\frac{w\varphi Y}{2}\right)\right.$$
$$\left.\tanh\left(\frac{w\varphi Y}{2}\right)\right)(\gamma(\kappa-1)+1), \quad (51)$$

$$\zeta_2(\varphi,\kappa) = \frac{w^{12}Y^{12}}{524288}(-3947228724 - 733469760p - 20736000p^2 + 6(777305099 + 148082560p + 4358400p^2)$$
$$\cosh(w\varphi Y) - 48(18859301 + 3850520p + 124800p^2)\cosh(2w\varphi Y) + 46313277\cosh(3w\varphi Y) + 10287360p$$
$$\cosh(3w\varphi Y) + 345600p^2\cosh(3w\varphi Y) - 305756\cosh(4w\varphi Y) - 87360p\cosh(4w\varphi Y) + \cosh(5w\varphi Y)) \quad (52)$$
$$\mathrm{sech}^{12}\left(\frac{w\varphi Y}{2}\right)\left((1-\gamma)^2 + 2\gamma(1-\gamma)\kappa + \frac{\gamma^2\kappa^2}{2}\right),$$

Using the same procedure, we can easily find the remaining ζ_l components for $(l \geq 3)$. Following this, we define series form solutions as

$$\zeta(\varphi,\kappa) = \sum_{l=0}^{\infty} \zeta_l(\varphi,\kappa) = \zeta_0(\varphi,\kappa) + \zeta_1(\varphi,\kappa) + \zeta_2(\varphi,\kappa) + \cdots,$$

$$\zeta(\varphi,\kappa) = \frac{1}{4}w^2Y^2\,\mathrm{sech}^2(\frac{w\varphi Y}{2}) + \frac{w^2Y^2}{12} - \left(-\frac{1}{512}w^7Y^7(3843 + 480p - 4(209 + 60p)\cosh(w\varphi Y) + \cosh(2w\varphi Y))\right.$$

$$\mathrm{sech}^6\left(\frac{w\varphi Y}{2}\right)\tanh\left(\frac{w\varphi Y}{2}\right)\right)(\gamma(\kappa-1)+1) + \frac{w^{12}Y^{12}}{524288}(-3947228724 - 733469760p - 20736000p^2+ \quad (53)$$

$$6(777305099 + 148082560p + 4358400p^2)\cosh(w\varphi Y) - 48(18859301 + 3850520p + 124800p^2)\cosh(2w\varphi Y)$$
$$+ 46313277\cosh(3w\varphi Y) + 10287360p\cosh(3w\varphi Y) + 345600p^2\cosh(3w\varphi Y) - 305756\cosh(4w\varphi Y)$$
$$- 87360p\cosh(4w\varphi Y) + \cosh(5w\varphi Y))\mathrm{sech}^{12}\left(\frac{w\varphi Y}{2}\right)\left((1-\gamma)^2 + 2\gamma(1-\gamma)\kappa + \frac{\gamma^2\kappa^2}{2}\right) + \cdots.$$

5.2. Implementing NDM$_{ABC}$

The unknown function $\zeta(\varphi,\kappa)$ has a series form solution, which is stated as

$$\zeta(\varphi,\kappa) = \sum_{l=0}^{\infty} \zeta_l(\varphi,\kappa) \quad (54)$$

The nonlinear terms are illustrated by using Adomian polynomials $\zeta\zeta_{\varphi\varphi\varphi} = \sum_{l=0}^{\infty} A_l$, $\zeta_\varphi\zeta_{\varphi\varphi} = \sum_{l=0}^{\infty} B_l$ and $\zeta^2\zeta_\varphi = \sum_{l=0}^{\infty} C_l$. Thus, Equation (48) can be expressed with the help of the following terms:

$$\sum_{l=0}^{\infty} \zeta_{l+1}(\varphi,\kappa) = \frac{1}{2} + \frac{1}{2}\tanh\left(\frac{\varphi}{2}\right)$$
$$+ \mathcal{N}^{-1}\left[\frac{v^\gamma(\omega^\gamma + \gamma(v^\gamma - \omega^\gamma))}{\omega^{2\gamma}}\mathcal{N}\left\{15\sum_{l=0}^{\infty} A_l + 15\sum_{l=0}^{\infty} B_l - 45\sum_{l=0}^{\infty} C_l - \sum_{l=0}^{\infty} \zeta_{l\varphi\varphi\varphi\varphi}\right\}\right], \quad (55)$$

When both sides of Equation (55) are compared, we obtain

$$\zeta_0(\varphi,\kappa) = \frac{1}{4}w^2Y^2\,\mathrm{sech}^2(\frac{w\varphi Y}{2}) + \frac{w^2Y^2}{12},$$

$$\zeta_1(\varphi,\kappa) = -\left(-\frac{1}{512}w^7Y^7(3843 + 480p - 4(209 + 60p)\cosh(w\varphi Y) + \cosh(2w\varphi Y))\mathrm{sech}^6\left(\frac{w\varphi Y}{2}\right)\right.$$
$$\tanh\left(\frac{w\varphi Y}{2}\right)\right)\left(1 - \gamma + \frac{\gamma\kappa^\gamma}{\Gamma(\gamma+1)}\right), \quad (56)$$

$$\zeta_2(\varphi,\kappa) = \frac{w^{12}Y^{12}}{524288}(-3947228724 - 733469760p - 20736000p^2 + 6(777305099 + 148082560p + 4358400p^2)$$
$$\cosh(w\varphi Y) - 48(18859301 + 3850520p + 124800p^2)\cosh(2w\varphi Y) + 46313277\cosh(3w\varphi Y) + 10287360p$$
$$\cosh(3w\varphi Y) + 345600p^2\cosh(3w\varphi Y) - 305756\cosh(4w\varphi Y) - 87360p\cosh(4w\varphi Y) + \cosh(5w\varphi Y)) \quad (57)$$
$$\mathrm{sech}^{12}\left(\frac{w\varphi Y}{2}\right)\left[\frac{\gamma^2\kappa^{2\gamma}}{\Gamma(2\gamma+1)} + 2\gamma(1-\gamma)\frac{\kappa^\gamma}{\Gamma(\gamma+1)} + (1-\gamma)^2\right],$$

Using the same procedure, we can easily find the remaining ζ_l components for $(l \geq 3)$. Following this, we define series form solutions as

$$\zeta(\varphi,\kappa) = \sum_{l=0}^{\infty} \zeta_l(\varphi,\kappa) = \zeta_0(\varphi,\kappa) + \zeta_1(\varphi,\kappa) + \zeta_2(\varphi,\kappa) + \cdots,$$

$$\zeta(\varphi,\kappa) = \frac{1}{4}w^2Y^2\mathrm{sech}^2(\frac{w\varphi Y}{2}) + \frac{w^2Y^2}{12} - \left(-\frac{1}{512}w^7Y^7(3843 + 480p - 4(209+60p)\cosh(w\varphi Y) + \cosh(2w\varphi Y))\right)$$

$$\mathrm{sech}^6\left(\frac{w\varphi Y}{2}\right)\tanh\left(\frac{w\varphi Y}{2}\right)\left(1 - \gamma + \frac{\gamma\kappa^\gamma}{\Gamma(\gamma+1)}\right) + \frac{w^{12}Y^{12}}{524288}(-3947228724 - 733469760p - 20736000p^2 + \quad (58)$$

$$6(777305099 + 148082560p + 4358400p^2)\cosh(w\varphi Y) - 48(18859301 + 3850520p + 124800p^2)\cosh(2w\varphi Y)$$
$$+ 46313277\cosh(3w\varphi Y) + 10287360p\cosh(3w\varphi Y) + 345600p^2\cosh(3w\varphi Y) - 305756\cosh(4w\varphi Y)$$
$$- 87360p\cosh(4w\varphi Y) + \cosh(5w\varphi Y))\mathrm{sech}^{12}\left(\frac{w\varphi Y}{2}\right)\left[\frac{\gamma^2\kappa^{2\gamma}}{\Gamma(2\gamma+1)} + 2\gamma(1-\gamma)\frac{\kappa^\gamma}{\Gamma(\gamma+1)} + (1-\gamma)^2\right] + \cdots.$$

We obtain the exact solution if we set $\gamma = 1$

$$\zeta(\varphi,\kappa) = \frac{1}{4}w^2Y^2\mathrm{sech}^2\left(\frac{Y}{2}\left(\frac{-w^5(-8Y^2\ell + 16\ell^2 + Y^4)}{16} + w\varphi\right)\frac{w^2Y^2}{12}\right), \quad (59)$$

Example 2. *Consider the nonlinear time-fractional Kaup–Kupershmidt equation [38]*

$$D_\kappa^\gamma \zeta(\varphi,\kappa) - 15\zeta\zeta_{\varphi\varphi\varphi} - 15p\zeta_\varphi\zeta_{\varphi\varphi} + 45\zeta^2\zeta_\varphi + \zeta_{\varphi\varphi\varphi\varphi\varphi} = 0, \quad 0 < \gamma \leq 1, \quad (60)$$

with initial condition

$$\zeta(\varphi,0) = \frac{4}{3}c - \frac{4}{p}c\,\mathrm{sech}^2(\sqrt{c}\varphi) \quad (61)$$

Equation (60) can be expressed as follows with the use of the natural transform:

$$\mathcal{N}[D_\kappa^\gamma \zeta(\varphi,\kappa)] = \mathcal{N}\left\{15\zeta\zeta_{\varphi\varphi\varphi}\right\} + \mathcal{N}\left\{15p\zeta_\varphi\zeta_{\varphi\varphi}\right\} - \mathcal{N}\left\{45\zeta^2\zeta_\varphi\right\} - \mathcal{N}\left\{\zeta_{\varphi\varphi\varphi\varphi\varphi}\right\}, \quad (62)$$

Characterize the nonlinear operator as

$$\frac{1}{\omega^\gamma}\mathcal{N}[\zeta(\varphi,\kappa)] - \omega^{2-\gamma}\zeta(\varphi,0) = \mathcal{N}\left[15\zeta\zeta_{\varphi\varphi\varphi} + 15p\zeta_\varphi\zeta_{\varphi\varphi} - 45\zeta^2\zeta_\varphi - \zeta_{\varphi\varphi\varphi\varphi\varphi}\right], \quad (63)$$

We obtain the following when it comes to simplification:

$$\mathcal{N}[\zeta(\varphi,\kappa)] = \omega^2\left[\frac{4}{3}c - \frac{4}{p}c\,\mathrm{sech}^2(\sqrt{c}\varphi)\right] + \frac{\gamma(\omega - \gamma(\omega - \gamma))}{\omega^2}\mathcal{N}\left[15\zeta\zeta_{\varphi\varphi\varphi} + 15p\zeta_\varphi\zeta_{\varphi\varphi} - 45\zeta^2\zeta_\varphi - \zeta_{\varphi\varphi\varphi\varphi\varphi}\right], \quad (64)$$

Equation (64) can be written as follows with inverse NT:

$$\zeta(\varphi,\kappa) = \left[\frac{4}{3}c - \frac{4}{p}c\,\text{sech}^2(\sqrt{c\varphi})\right]$$
$$+ \mathcal{N}^{-1}\left[\frac{\gamma(\omega - \gamma(\omega - \gamma))}{\omega^2}\mathcal{N}\left\{15\zeta\zeta_{\varphi\varphi\varphi} + 15p\zeta_{\varphi}\zeta_{\varphi\varphi} - 45\zeta^2\zeta_{\varphi} - \zeta_{\varphi\varphi\varphi\varphi\varphi}\right\}\right], \quad (65)$$

5.3. Applying NDM$_{CF}$

The unknown function $\zeta(\varphi,\kappa)$ has a series form solution, which is stated as

$$\zeta(\varphi,\kappa) = \sum_{l=0}^{\infty} \zeta_l(\varphi,\kappa) \quad (66)$$

The nonlinear terms are illustrated by using Adomian polynomials $\zeta\zeta_{\varphi\varphi\varphi} = \sum_{l=0}^{\infty} A_l$, $\zeta_{\varphi}\zeta_{\varphi\varphi} = \sum_{l=0}^{\infty} B_l$ and $\zeta^2\zeta_{\varphi} = \sum_{l=0}^{\infty} C_l$. Thus, Equation (65) can be expressed with the help of the following terms:

$$\sum_{l=0}^{\infty} \zeta_{l+1}(\varphi,\kappa) = \frac{4}{3}c - \frac{4}{p}c\,\text{sech}^2(\sqrt{c\varphi})$$
$$+ \mathcal{N}^{-1}\left[\frac{\gamma(\omega - \gamma(\omega - \gamma))}{\omega^2}\mathcal{N}\left\{15\sum_{l=0}^{\infty} A_l + 15\sum_{l=0}^{\infty} B_l - 45\sum_{l=0}^{\infty} C_l - \sum_{l=0}^{\infty} \zeta_{l\varphi\varphi\varphi\varphi}\right\}\right], \quad (67)$$

When both sides of Equation (67) are compared, we obtain

$$\zeta_0(\varphi,\kappa) = \frac{4}{3}c - \frac{4}{p}\,\text{csech}^2(\sqrt{c\varphi}),$$

$$\zeta_1(\varphi,\kappa) = -\frac{16c^{\frac{7}{2}}}{p^3}(360 - 420p + 63p^2 + 4p(-15 + 16p)\cosh(2\sqrt{cx}) + p^2\cosh(4\sqrt{cx}))\,\text{sech}^6(\sqrt{cx})$$
$$\tanh(\sqrt{cx})(\gamma(\kappa - 1) + 1)$$

$$\zeta_2(\varphi,\kappa) = \frac{16c^6\,\text{sech}^{12}(\sqrt{c\varphi})}{p^5}(-3110400 + 14515200p - 26369280p^2 + 15270480p^3 - 306084p^4 - 6$$
$$(-432000 + 2217600p - 4451160p^2 + 2656400p^3 + 9181p^4)\cosh(2\sqrt{c\varphi}) + 48p(14400 - 60780p + 41590p^2 +$$
$$4789p^3)\cosh(4\sqrt{c\varphi}) + 79920p^2\cosh(6\sqrt{c\varphi}) - 59040p^3\cosh(6\sqrt{c\varphi}) - 20883p^4\cosh(6\sqrt{c\varphi}) -$$
$$240p^3\cosh(8\sqrt{c\varphi}) + 244p^4\cosh(8\sqrt{c\varphi}) + p^4\cosh(10\sqrt{c\varphi}))\left((1-\gamma)^2 + 2\gamma(1-\gamma)\kappa + \frac{\gamma^2\kappa^2}{2}\right),$$

Using the same procedure, we can easily find the remaining ζ_l components for ($l \geq 3$). Following this, we define series form solutions as

$$\zeta(\varphi,\kappa) = \sum_{l=0}^{\infty} \zeta_l(\varphi,\kappa) = \zeta_0(\varphi,\kappa) + \zeta_1(\varphi,\kappa) + \zeta_2(\varphi,\kappa) + \cdots,$$

$$\zeta(\varphi,\kappa) = \frac{4}{3}c - \frac{4}{p}c\operatorname{sech}^2(\sqrt{c\varphi}) - \frac{16c^{\frac{7}{2}}}{p^3}(360 - 420p + 63p^2 + 4p(-15 + 16p)\cosh(2\sqrt{cx}) + p^2\cosh(4\sqrt{cx}))$$

$$\operatorname{sech}^6(\sqrt{cx})\tanh(\sqrt{cx})(\gamma(\kappa - 1) + 1)\frac{16c^6\operatorname{sech}^{12}(\sqrt{c\varphi})}{p^5}(-3110400 + 14515200p - 26369280p^2 + \qquad (68)$$

$$15270480p^3 - 306084p^4 - 6(-432000 + 2217600p - 4451160p^2 + 2656400p^3 + 9181p^4)\cosh(2\sqrt{c\varphi})$$
$$+ 48p(14400 - 60780p + 41590p^2 + 4789p^3)\cosh(4\sqrt{c\varphi}) + 79920p^2\cosh(6\sqrt{c\varphi}) - 59040p^3$$
$$\cosh(6\sqrt{c\varphi}) - 20883p^4\cosh(6\sqrt{c\varphi}) - 240p^3\cosh(8\sqrt{c\varphi}) + 244p^4\cosh(8\sqrt{c\varphi})$$
$$+ p^4\cosh(10\sqrt{c\varphi}))\left((1-\gamma)^2 + 2\gamma(1-\gamma)\kappa + \frac{\gamma^2\kappa^2}{2}\right) + \cdots,$$

5.4. Applying NDM$_{ABC}$

The unknown function $\zeta(\varphi,\kappa)$ has a series form solution, which is stated as

$$\zeta(\varphi,\kappa) = \sum_{l=0}^{\infty} \zeta_l(\varphi,\kappa) \qquad (69)$$

The nonlinear terms are illustrated by using Adomian polynomials $\zeta\zeta_{\varphi\varphi\varphi} = \sum_{l=0}^{\infty} \mathcal{A}_l$, $\zeta_\varphi\zeta_{\varphi\varphi} = \sum_{l=0}^{\infty} \mathcal{B}_l$ and $\zeta^2\zeta_\varphi = \sum_{l=0}^{\infty} \mathcal{C}_l$. Thus, Equation (65) can be expressed with the help of the following terms:

$$\sum_{l=0}^{\infty} \zeta_l(\varphi,\kappa) = \frac{4}{3}c - \frac{4}{p}c\operatorname{sech}^2(\sqrt{c\varphi})$$

$$+ \mathcal{N}^{-1}\left[\frac{v^\gamma(\omega^\gamma + \gamma(v^\gamma - \omega^\gamma))}{\omega^{2\gamma}}\mathcal{N}\left\{15\sum_{l=0}^{\infty}\mathcal{A}_l + 15\sum_{l=0}^{\infty}\mathcal{B}_l - 45\sum_{l=0}^{\infty}\mathcal{C}_l - \sum_{l=0}^{\infty}\zeta_{l\varphi\varphi\varphi\varphi}\right\}\right], \qquad (70)$$

When both sides of Equation (70) are compared, we obtain

$$\zeta_0(\varphi,\kappa) = \frac{4}{3}c - \frac{4}{p}c\operatorname{sech}^2(\sqrt{c\varphi}),$$

$$\zeta_1(\varphi,\kappa) = -\frac{16c^{\frac{7}{2}}}{p^3}(360 - 420p + 63p^2 + 4p(-15 + 16p)\cosh(2\sqrt{cx}) + p^2\cosh(4\sqrt{cx}))\operatorname{sech}^6(\sqrt{cx})$$

$$\tanh(\sqrt{cx})\left(1 - \gamma + \frac{\gamma\kappa^\gamma}{\Gamma(\gamma+1)}\right),$$

$$\zeta_2(\varphi,\kappa) = \frac{16c^6\operatorname{sech}^{12}(\sqrt{c\varphi})}{p^5}(-3110400 + 14515200p - 26369280p^2 +$$

$$15270480p^3 - 306084p^4 - 6(-432000 + 2217600p - 4451160p^2 + 2656400p^3 + 9181p^4)\cosh(2\sqrt{c\varphi})$$
$$+ 48p(14400 - 60780p + 41590p^2 + 4789p^3)\cosh(4\sqrt{c\varphi}) + 79920p^2\cosh(6\sqrt{c\varphi}) - 59040p^3$$
$$\cosh(6\sqrt{c\varphi}) - 20883p^4\cosh(6\sqrt{c\varphi}) - 240p^3\cosh(8\sqrt{c\varphi}) + 244p^4\cosh(8\sqrt{c\varphi})$$
$$+ p^4\cosh(10\sqrt{c\varphi}))\left[\frac{\gamma^2\kappa^{2\gamma}}{\Gamma(2\gamma+1)} + 2\gamma(1-\gamma)\frac{\kappa^\gamma}{\Gamma(\gamma+1)} + (1-\gamma)^2\right]$$

Using the same procedure, we can easily find the remaining ζ_l components for $(l \geq 3)$. Following this, we define series form solutions as

$$\zeta(\varphi,\kappa) = \sum_{l=0}^{\infty} \zeta_l(\varphi,\kappa) = \zeta_0(\varphi,\kappa) + \zeta_1(\varphi,\kappa) + \zeta_2(\varphi,\kappa) + \cdots,$$

$$\zeta(\varphi,\kappa) = \frac{4}{3}c - \frac{4}{p}c\,\text{sech}^2(\sqrt{c\varphi}) - \frac{16c^{\frac{7}{2}}}{p^3}(360 - 420p + 63p^2 + 4p(-15 + 16p)\cosh(2\sqrt{c}x) + p^2\cosh(4\sqrt{c}x))$$

$$\text{sech}^6(\sqrt{c}x)\tanh(\sqrt{c}x)\left(1 - \gamma + \frac{\gamma\kappa^\gamma}{\Gamma(\gamma+1)}\right) \frac{16c^6\,\text{sech}^{12}(\sqrt{c\varphi})}{p^5}(-3110400 + 14515200p - 26369280p^2 +$$

$$15270480p^3 - 306084p^4 - 6(-432000 + 2217600p - 4451160p^2 + 2656400p^3 + 9181p^4)\cosh(2\sqrt{c\varphi}) \quad (71)$$

$$+ 48p(14400 - 60780p + 41590p^2 + 4789p^3)\cosh(4\sqrt{c\varphi}) + 79920p^2\cosh(6\sqrt{c\varphi}) - 59040p^3$$

$$\cosh(6\sqrt{c\varphi}) - 20883p^4\cosh(6\sqrt{c\varphi}) - 240p^3\cosh(8\sqrt{c\varphi}) + 244p^4\cosh(8\sqrt{c\varphi})$$

$$+ p^4\cosh(10\sqrt{c\varphi}))\left[\frac{\gamma^2\kappa^{2\gamma}}{\Gamma(2\gamma+1)} + 2\gamma(1-\gamma)\frac{\kappa^\gamma}{\Gamma(\gamma+1)} + (1-\gamma)^2\right] + \cdots,$$

We achieve the exact solution if we set $\gamma = 1$

$$\zeta(\varphi,\kappa) = \frac{4}{3}c - \frac{4}{p}c\,\text{sech}^2(\sqrt{c} + (\varphi + 8(3c^2 - 5pc)\kappa)). \quad (72)$$

Example 3. *Consider the nonlinear time-fractional Kaup–Kupershmidt equation [38]*

$$D_\kappa^\gamma \zeta(\varphi,\kappa) = 5\zeta\zeta_{\varphi\varphi\varphi} + \frac{25}{2}\zeta_\varphi\zeta_{\varphi\varphi} + 5\zeta^2\zeta_\varphi + \zeta_{\varphi\varphi\varphi\varphi\varphi}, \quad 0 < \gamma \leq 1, \quad (73)$$

with initial condition

$$\zeta(\varphi,0) = -2k^2 + \frac{24k^2}{1+e^{k\varphi}} - \frac{24k^2}{(1+e^{k\varphi})^2} \quad (74)$$

Equation (73) can be expressed as follows with the use of the natural transform:

$$\mathcal{N}[D_\kappa^\gamma \zeta(\varphi,\kappa)] = \mathcal{N}\left\{5\zeta\zeta_{\varphi\varphi\varphi}\right\} + \mathcal{N}\left\{\frac{25}{2}\zeta_\varphi\zeta_{\varphi\varphi}\right\} + \mathcal{N}\left\{5\zeta^2\zeta_\varphi\right\} + \mathcal{N}\left\{\zeta_{\varphi\varphi\varphi\varphi\varphi}\right\}, \quad (75)$$

Characterize the nonlinear operator as

$$\frac{1}{\omega^\gamma}\mathcal{N}[\zeta(\varphi,\kappa)] - \omega^{2-\gamma}\zeta(\varphi,0) = \mathcal{N}\left[5\zeta\zeta_{\varphi\varphi\varphi} + \frac{25}{2}\zeta_\varphi\zeta_{\varphi\varphi} + 5\zeta^2\zeta_\varphi + \zeta_{\varphi\varphi\varphi\varphi\varphi}\right], \quad (76)$$

We obtain the following when it comes to simplification:

$$\mathcal{N}[\zeta(\varphi,\kappa)] = \omega^2\left[-2k^2 + \frac{24k^2}{1+e^{k\varphi}} - \frac{24k^2}{(1+e^{k\varphi})^2}\right] + \frac{\gamma(\omega - \gamma(\omega-\gamma))}{\omega^2}\mathcal{N}\left[5\zeta\zeta_{\varphi\varphi\varphi} + \frac{25}{2}\zeta_\varphi\zeta_{\varphi\varphi} + 5\zeta^2\zeta_\varphi + \zeta_{\varphi\varphi\varphi\varphi\varphi}\right], \quad (77)$$

Equation (77) can be written as follows with inverse NT

$$\zeta(\varphi,\kappa) = \left[-2k^2 + \frac{24k^2}{1+e^{k\varphi}} - \frac{24k^2}{(1+e^{k\varphi})^2}\right]$$

$$+ \mathcal{N}^{-1}\left[\frac{\gamma(\omega - \gamma(\omega-\gamma))}{\omega^2}\mathcal{N}\left\{5\zeta\zeta_{\varphi\varphi\varphi} + \frac{25}{2}\zeta_\varphi\zeta_{\varphi\varphi} + 5\zeta^2\zeta_\varphi + \zeta_{\varphi\varphi\varphi\varphi\varphi}\right\}\right], \quad (78)$$

5.5. Applying NDM$_{CF}$

The unknown function $\zeta(\varphi,\kappa)$ has a series form solution, which is stated as

$$\zeta(\varphi,\kappa) = \sum_{l=0}^{\infty} \zeta_l(\varphi,\kappa) \qquad (79)$$

The nonlinear terms are illustrated by using Adomian polynomials $\zeta\zeta_{\varphi\varphi\varphi} = \sum_{l=0}^{\infty} A_l$, $\zeta_\varphi \zeta_{\varphi\varphi} = \sum_{l=0}^{\infty} B_l$ and $\zeta^2 \zeta_\varphi = \sum_{l=0}^{\infty} C_l$. Thus, Equation (78) can be expressed with the help of the following terms:

$$\sum_{l=0}^{\infty} \zeta_{l+1}(\varphi,\kappa) = -2k^2 + \frac{24k^2}{1+e^{k\varphi}} - \frac{24k^2}{(1+e^{k\varphi})^2}$$

$$+ \mathcal{N}^{-1}\left[\frac{\gamma(\omega - \gamma(\omega - \gamma))}{\omega^2}\mathcal{N}\left\{5\sum_{l=0}^{\infty} A_l + \frac{25}{2}\sum_{l=0}^{\infty} B_l + 5\sum_{l=0}^{\infty} C_l + \sum_{l=0}^{\infty} \zeta_{l\varphi\varphi\varphi\varphi}\right\}\right], \qquad (80)$$

When both sides of Equation (80) are compared, we obtain

$$\zeta_0(\varphi,\kappa) = -2k^2 + \frac{24k^2}{1+e^{k\varphi}} - \frac{24k^2}{(1+e^{k\varphi})^2},$$

$$\zeta_1(\varphi,\kappa) = -\left(\frac{264e^{k\varphi}(-1+e^{k\varphi})k^7}{(1+e^{k\varphi})^3}\right)(\gamma(\kappa-1)+1)$$

$$\zeta_2(\varphi,\kappa) = 2904e^{k\varphi}\left(\frac{264e^{k\varphi}(1 - 4e^{k\varphi} + e^{2k\varphi})k^{12}}{(1+e^{k\varphi})^4}\right)\left((1-\gamma)^2 + 2\gamma(1-\gamma)\kappa + \frac{\gamma^2\kappa^2}{2}\right),$$

Using the same procedure, we can easily find the remaining ζ_l components for $(l \geq 3)$. Following this, we define series form solutions as

$$\zeta(\varphi,\kappa) = \sum_{l=0}^{\infty} \zeta_l(\varphi,\kappa) = \zeta_0(\varphi,\kappa) + \zeta_1(\varphi,\kappa) + \zeta_2(\varphi,\kappa) + \cdots,$$

$$\zeta(\varphi,\kappa) = -2k^2 + \frac{24k^2}{1+e^{k\varphi}} - \frac{24k^2}{(1+e^{k\varphi})^2} - \left(\frac{264e^{k\varphi}(-1+e^{k\varphi})k^7}{(1+e^{k\varphi})^3}\right)(\gamma(\kappa-1)+1)+ \qquad (81)$$

$$2904e^{k\varphi}\left(\frac{264e^{k\varphi}(1-4e^{k\varphi}+e^{2k\varphi})k^{12}}{(1+e^{k\varphi})^4}\right)\left((1-\gamma)^2 + 2\gamma(1-\gamma)\kappa + \frac{\gamma^2\kappa^2}{2}\right) + \cdots,$$

5.6. Applying NDM$_{ABC}$

The unknown function $\zeta(\varphi,\kappa)$ has a series form solution, which is stated as

$$\zeta(\varphi,\kappa) = \sum_{l=0}^{\infty} \zeta_l(\varphi,\kappa) \qquad (82)$$

The nonlinear terms are illustrated by using Adomian polynomials $\zeta\zeta_{\varphi\varphi\varphi} = \sum_{l=0}^{\infty} A_l$, $\zeta_\varphi \zeta_{\varphi\varphi} = \sum_{l=0}^{\infty} B_l$ and $\zeta^2 \zeta_\varphi = \sum_{l=0}^{\infty} C_l$. Thus, Equation (78) can be expressed with the help of the following terms:

$$\sum_{l=0}^{\infty} \zeta_l(\varphi,\kappa) = -2k^2 + \frac{24k^2}{1+e^{k\varphi}} - \frac{24k^2}{(1+e^{k\varphi})^2}$$

$$+ \mathcal{N}^{-1}\left[\frac{v^\gamma(\omega^\gamma + \gamma(v^\gamma - \omega^\gamma))}{\omega^{2\gamma}}\mathcal{N}\left\{5\sum_{l=0}^{\infty} A_l + \frac{25}{2}\sum_{l=0}^{\infty} B_l + 5\sum_{l=0}^{\infty} C_l + \sum_{l=0}^{\infty} \zeta_{l\varphi\varphi\varphi\varphi}\right\}\right], \qquad (83)$$

When both sides of Equation (83) are compared, we obtain

$$\zeta_0(\varphi,\kappa) = -2k^2 + \frac{24k^2}{1+e^{k\varphi}} - \frac{24k^2}{(1+e^{k\varphi})^2},$$

$$\zeta_1(\varphi,\kappa) = -\left(\frac{264e^{k\varphi}(-1+e^{k\varphi})k^7}{(1+e^{k\varphi})^3}\right)\left(1-\gamma+\frac{\gamma\kappa^\gamma}{\Gamma(\gamma+1)}\right),$$

$$\zeta_2(\varphi,\kappa) = 2904e^{k\varphi}\left(\frac{264e^{k\varphi}(1-4e^{k\varphi}+e^{2k\varphi})k^{12}}{(1+e^{k\varphi})^4}\right)\left[\frac{\gamma^2\kappa^{2\gamma}}{\Gamma(2\gamma+1)}+2\gamma(1-\gamma)\frac{\kappa^\gamma}{\Gamma(\gamma+1)}+(1-\gamma)^2\right]$$

Using the same procedure, we can easily find the remaining ζ_l components for $(l \geq 3)$. Following this, we define series form solutions as

$$\zeta(\varphi,\kappa) = \sum_{l=0}^{\infty}\zeta_l(\varphi,\kappa) = \zeta_0(\varphi,\kappa) + \zeta_1(\varphi,\kappa) + \zeta_2(\varphi,\kappa) + \cdots,$$

$$\zeta(\varphi,\kappa) = -2k^2 + \frac{24k^2}{1+e^{k\varphi}} - \frac{24k^2}{(1+e^{k\varphi})^2} - \left(\frac{264e^{k\varphi}(-1+e^{k\varphi})k^7}{(1+e^{k\varphi})^3}\right)\left(1-\gamma+\frac{\gamma\kappa^\gamma}{\Gamma(\gamma+1)}\right)+ \quad (84)$$

$$2904e^{k\varphi}\left(\frac{264e^{k\varphi}(1-4e^{k\varphi}+e^{2k\varphi})k^{12}}{(1+e^{k\varphi})^4}\right)\left[\frac{\gamma^2\kappa^{2\gamma}}{\Gamma(2\gamma+1)}+2\gamma(1-\gamma)\frac{\kappa^\gamma}{\Gamma(\gamma+1)}+(1-\gamma)^2\right]+\cdots,$$

We achieve the exact solution if we set $\gamma = 1$

$$\zeta(\varphi,\kappa) = -2k^2 + \frac{24k^2}{1+e^{k\varphi+11k^5t}} - \frac{24k^2}{(1+e^{k\varphi+11k^5t})^2}. \quad (85)$$

6. Results and Discussion

We find the solution of fractional-order Kaup-Kupershmidt (KK) equation by implementing Natural decomposition method with the aid of two different fractional derivatives. Figure 1 exhibits the nature of the exact and proposed method solution while Figure 2 shows nature of the absolute error of example 1 at $\gamma = 1$. Figure 3 exhibits the nature of the exact and proposed method solution whereas Figures 4 and 5 shows the nature of the proposed method solution at different fractional orders. Figure 6 exhibits the nature of the exact and proposed method solution whereas Figures 7 and 8 shows the nature of the proposed method solution at different fractional orders.

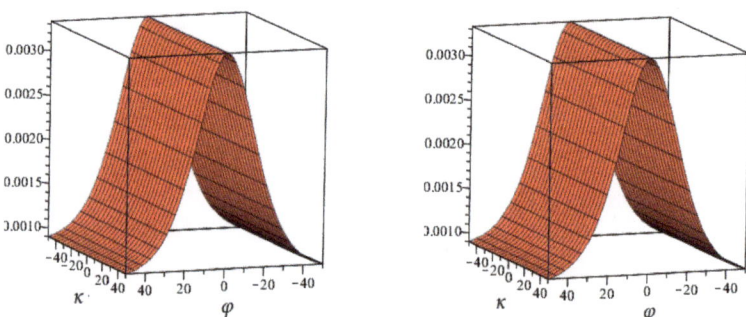

Figure 1. Nature of the exact and proposed method solution of example 1 at $\gamma = 1$.

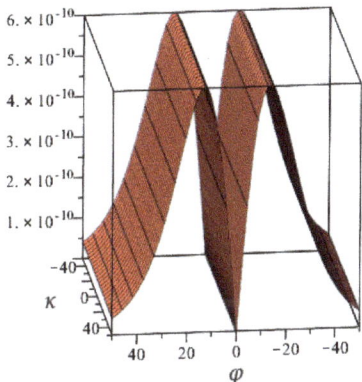

Figure 2. Nature of the absolute error of example 1.

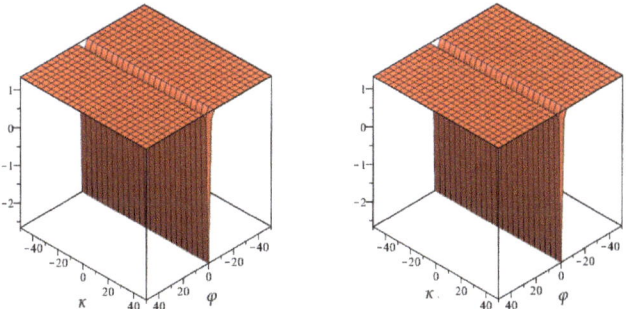

Figure 3. Nature of the exact and proposed method solution of example 2 at at $\gamma = 1$.

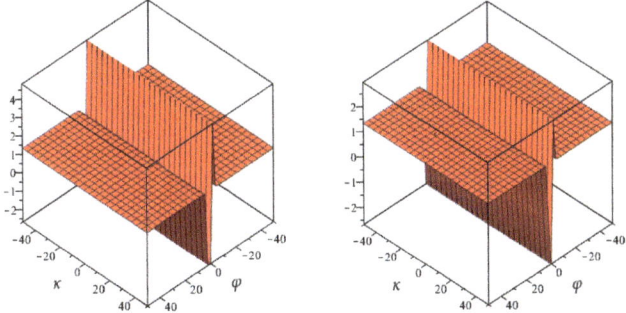

Figure 4. Nature of the proposed method solution of example 2 at $\gamma = 0.8, 0.6$.

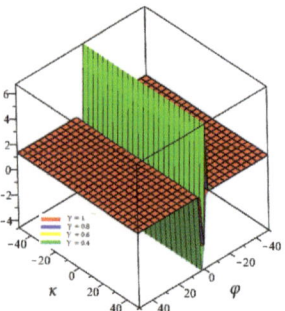

Figure 5. Nature of the proposed method solution at various orders of γ for example 2.

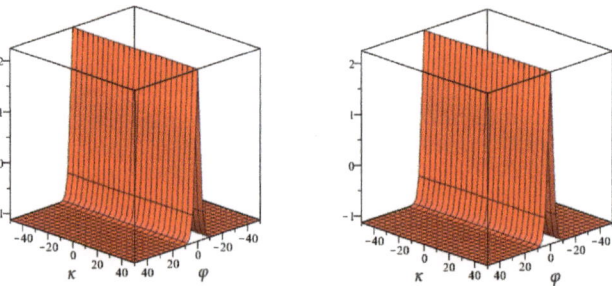

Figure 6. Nature of the exact and proposed method solution of example 3 at $\gamma = 1$.

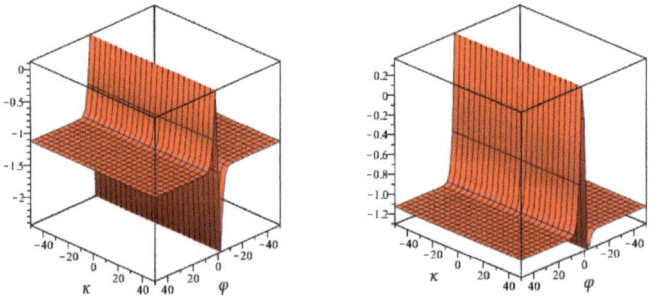

Figure 7. Nature of the proposed method solution of example 3 at $\gamma = 0.8, 0.6$.

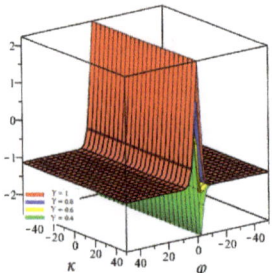

Figure 8. Nature of the proposed method solution at various orders of γ for example 3.

7. Conclusions

In this paper, we find the solution of the time-fractional Kaup–Kupershmidt equation by means of the natural decomposition method with the aid of two different fractional derivatives. To demonstrate the validity of the proposed method, we study the time-fractional KK equation in three different cases. The results that we obtain by implementing the proposed methods show that our results are in good agreement with the exact solution. The results shown in Tables 1–7 are suitable when compared with other techniques such as the two-dimensional Legendre multiwavelet method, optimal homotopy analysis transform method (OHAM) and q-homotopy analysis transform method (q-HATM). Finally, we can conclude that the suggested method is sufficiently consistent and can be used to examine a wide range of fractional-order nonlinear mathematical models that enable us to understand the behaviour of highly nonlinear complicated phenomena in related fields of science and engineering.

Table 1. Comparison at different fractional order of γ on the basis of error for example 1.

κ	φ	$\gamma = 0.4$	$\gamma = 0.6$	$\gamma = 0.8$	$\gamma = 1(NTDM_{CF})$	$\gamma = 1(NTDM_{ABC})$
0.1	0.2	$7.7794000000 \times 10^{-8}$	$5.9046000000 \times 10^{-8}$	$3.1881000000 \times 10^{-8}$	$1.5379000000 \times 10^{-8}$	$1.5379000000 \times 10^{-8}$
	0.4	$1.5668400000 \times 10^{-7}$	$1.1893800000 \times 10^{-7}$	$6.4232000000 \times 10^{-8}$	$3.0990000000 \times 10^{-8}$	$3.0990000000 \times 10^{-8}$
	0.6	$2.3529200000 \times 10^{-7}$	$1.7864200000 \times 10^{-7}$	$9.6509000000 \times 10^{-8}$	$4.6575000000 \times 10^{-8}$	$4.6575000000 \times 10^{-8}$
	0.8	$3.1347800000 \times 10^{-7}$	$2.3807000000 \times 10^{-7}$	$1.2867500000 \times 10^{-7}$	$6.2123000000 \times 10^{-8}$	$6.2123000000 \times 10^{-8}$
	1	$3.9110400000 \times 10^{-7}$	$2.9712500000 \times 10^{-7}$	$1.6069300000 \times 10^{-7}$	$7.7622000000 \times 10^{-8}$	$7.7622000000 \times 10^{-8}$
0.2	0.2	$1.5316700000 \times 10^{-7}$	$1.1624400000 \times 10^{-7}$	$6.2757000000 \times 10^{-8}$	$3.0269000000 \times 10^{-8}$	$3.0269000000 \times 10^{-8}$
	0.4	$3.1100100000 \times 10^{-7}$	$2.3605500000 \times 10^{-7}$	$1.2746600000 \times 10^{-7}$	$6.1491000000 \times 10^{-8}$	$6.1491000000 \times 10^{-8}$
	0.6	$4.6827700000 \times 10^{-7}$	$3.5549900000 \times 10^{-7}$	$1.9202900000 \times 10^{-7}$	$9.2664000000 \times 10^{-8}$	$9.2664000000 \times 10^{-8}$
	0.8	$6.2471400000 \times 10^{-7}$	$4.7438600000 \times 10^{-7}$	$2.5637000000 \times 10^{-7}$	$1.2376100000 \times 10^{-7}$	$1.2376100000 \times 10^{-7}$
	1	$7.8003300000 \times 10^{-7}$	$5.9253400000 \times 10^{-7}$	$3.2041800000 \times 10^{-7}$	$1.5476200000 \times 10^{-7}$	$1.5476200000 \times 10^{-7}$
0.3	0.2	$2.2606900000 \times 10^{-7}$	$1.7156700000 \times 10^{-7}$	$9.2620000000 \times 10^{-8}$	$4.4671000000 \times 10^{-8}$	$4.4671000000 \times 10^{-8}$
	0.4	$4.6285600000 \times 10^{-7}$	$3.5130300000 \times 10^{-7}$	$1.8968900000 \times 10^{-7}$	$9.1506000000 \times 10^{-8}$	$9.1506000000 \times 10^{-8}$
	0.6	$6.9881100000 \times 10^{-7}$	$5.3049100000 \times 10^{-7}$	$2.8654200000 \times 10^{-7}$	$1.3826500000 \times 10^{-7}$	$1.3826500000 \times 10^{-7}$
	0.8	$9.3351200000 \times 10^{-7}$	$7.0884900000 \times 10^{-7}$	$3.8306300000 \times 10^{-7}$	$1.8491500000 \times 10^{-7}$	$1.8491500000 \times 10^{-7}$
	1	$1.1665440000 \times 10^{-6}$	$8.8610300000 \times 10^{-7}$	$4.7914600000 \times 10^{-7}$	$2.3141700000 \times 10^{-7}$	$2.3141700000 \times 10^{-7}$
0.4	0.2	$2.9650200000 \times 10^{-7}$	$2.2501400000 \times 10^{-7}$	$1.2147100000 \times 10^{-7}$	$5.8586000000 \times 10^{-8}$	$5.8586000000 \times 10^{-8}$
	0.4	$6.1224700000 \times 10^{-7}$	$4.6468100000 \times 10^{-7}$	$2.5090400000 \times 10^{-7}$	$1.2103200000 \times 10^{-7}$	$1.2103200000 \times 10^{-7}$
	0.6	$9.2689100000 \times 10^{-7}$	$7.0362200000 \times 10^{-7}$	$3.8004700000 \times 10^{-7}$	$1.8338100000 \times 10^{-7}$	$1.8338100000 \times 10^{-7}$
	0.8	$1.2398730000 \times 10^{-6}$	$9.4146100000 \times 10^{-7}$	$5.0875300000 \times 10^{-7}$	$2.4558200000 \times 10^{-7}$	$2.4558200000 \times 10^{-7}$
	1	$1.5506360000 \times 10^{-6}$	$1.1778340000 \times 10^{-6}$	$6.3687500000 \times 10^{-7}$	$3.0758800000 \times 10^{-7}$	$3.0758800000 \times 10^{-7}$
0.5	0.2	$3.6446500000 \times 10^{-7}$	$2.7658900000 \times 10^{-7}$	$1.4931100000 \times 10^{-7}$	$7.2012000000 \times 10^{-8}$	$7.2012000000 \times 10^{-8}$
	0.4	$7.5917400000 \times 10^{-7}$	$5.7618900000 \times 10^{-7}$	$3.1110700000 \times 10^{-7}$	$1.5007100000 \times 10^{-7}$	$1.5007100000 \times 10^{-7}$
	0.6	$1.1525180000 \times 10^{-6}$	$8.7488900000 \times 10^{-7}$	$4.7254400000 \times 10^{-7}$	$2.2800900000 \times 10^{-7}$	$2.2800900000 \times 10^{-7}$
	0.8	$1.5437950000 \times 10^{-6}$	$1.1722200000 \times 10^{-6}$	$6.3343900000 \times 10^{-7}$	$3.0576500000 \times 10^{-7}$	$3.0576500000 \times 10^{-7}$
	1	$1.9323090000 \times 10^{-6}$	$1.4677220000 \times 10^{-6}$	$7.9360600000 \times 10^{-7}$	$3.8327700000 \times 10^{-7}$	$3.8327700000 \times 10^{-7}$

Table 2. Comparison of absolute error among Legendre Multiwavelet [39], OHAM [39], $q-HATM$ [38], NDM_{CF} and NDM_{ABC} for example 1 at $w = 1, \ell = 0, Y = 0.1, \gamma = 1$ and $\kappa = 0.1$.

φ	$\|Legendre\ Multiwelet\|$	$\|OHAM\|$	$\|q-HATM\|$	$\|NTDM_{CF}\|$	$\|NTDM_{ABC}\|$
0.1	3.5268×10^{-10}	3.4968×10^{-10}	3.1482×10^{-10}	$7.5000000000 \times 10^{-13}$	$7.5000000000 \times 10^{-13}$
0.2	7.0308×10^{-10}	7.2934×10^{-6}	6.3101×10^{-10}	$1.5400000000 \times 10^{-12}$	$1.5400000000 \times 10^{-12}$
0.3	1.0532×10^{-9}	2.6793×10^{-5}	9.4682×10^{-10}	$2.3200000000 \times 10^{-12}$	$2.3200000000 \times 10^{-12}$
0.4	1.4028×10^{-9}	5.8103×10^{-5}	1.2620×10^{-9}	$3.1000000000 \times 10^{-12}$	$3.1000000000 \times 10^{-12}$
0.5	1.7520×10^{-9}	1.0061×10^{-4}	1.5765×10^{-9}	$3.8800000000 \times 10^{-12}$	$3.8800000000 \times 10^{-12}$

Table 3. Comparison of absolute error among Legendre Multiwavelet [39], OHAM [39], $q-HATM$ [38], NDM_{CF} and NDM_{ABC} for example 1 at $w = 1, \ell = 0, Y = 0.1, \gamma = 0.75$ and $\kappa = 0.1$.

φ	$|Legendre\ Multiwavelet|$	$|OHAM|$	$|q-HATM|$	$|NTDM_{CF}|$	$|NTDM_{ABC}|$
0.1	6.7734×10^{-10}	6.7141×10^{-10}	6.0478×10^{-10}	$1.4700000000 \times 10^{-12}$	$1.4700000000 \times 10^{-12}$
0.2	1.3533×10^{-9}	7.2899×10^{-6}	1.2165×10^{-10}	$3.0200000000 \times 10^{-12}$	$3.0200000000 \times 10^{-12}$
0.3	2.0287×10^{-9}	2.6785×10^{-5}	1.8276×10^{-10}	$4.5900000000 \times 10^{-12}$	$4.5900000000 \times 10^{-12}$
0.4	2.7033×10^{-9}	5.8094×10^{-5}	2.4376×10^{-9}	$6.1500000000 \times 10^{-12}$	$6.1500000000 \times 10^{-12}$
0.5	3.3768×10^{-9}	1.0060×10^{-4}	3.0461×10^{-9}	$7.7100000000 \times 10^{-12}$	$7.7100000000 \times 10^{-12}$

Table 4. Comparison of absolute error among Legendre Multiwavelet [39], OHAM [39], $q-HATM$ [38], NDM_{CF} and NDM_{ABC} for example 1 at $w = 1, \ell = 0, Y = 0.1, \gamma = 0.5$ and $\kappa = 0.1$.

φ	$|Legendre\ Multiwavelet|$	$|OHAM|$	$|q-HATM|$	$|NTDM_{CF}|$	$|NTDM_{ABC}|$
0.1	1.2348×10^{-9}	1.2175×10^{-9}	1.0979×10^{-9}	$2.1300000000 \times 10^{-12}$	$2.1300000000 \times 10^{-12}$
0.2	2.4789×10^{-9}	7.2836×10^{-6}	2.2262×10^{-9}	$1.5400000000 \times 10^{-12}$	$4.4700000000 \times 10^{-12}$
0.3	3.7221×10^{-9}	2.6773×10^{-5}	3.3531×10^{-9}	$6.8100000000 \times 10^{-12}$	$6.8100000000 \times 10^{-12}$
0.4	4.9638×10^{-9}	5.8078×10^{-5}	4.4781×10^{-9}	$9.1600000000 \times 10^{-12}$	$9.1600000000 \times 10^{-12}$
0.5	6.2035×10^{-9}	1.0058×10^{-4}	5.6004×10^{-9}	$1.1500000000 \times 10^{-11}$	$1.1500000000 \times 10^{-11}$

Table 5. Comparison at different fractional order of γ on the basis of error for example 2.

κ	φ	$\gamma = 0.4$	$\gamma = 0.6$	$\gamma = 0.8$	$\gamma = 1(NTDM_{CF})$	$\gamma = 1(NTDM_{ABC})$
0.1	0.2	$5.2120000000 \times 10^{-7}$	$3.6017600000 \times 10^{-7}$	$2.0943200000 \times 10^{-7}$	$6.4513000000 \times 10^{-8}$	$6.4513000000 \times 10^{-8}$
	0.4	$1.0384330000 \times 10^{-6}$	$7.1776700000 \times 10^{-7}$	$4.1757400000 \times 10^{-7}$	$1.2893800000 \times 10^{-7}$	$1.2893800000 \times 10^{-7}$
	0.6	$1.5474640000 \times 10^{-6}$	$1.0698980000 \times 10^{-6}$	$6.2282300000 \times 10^{-7}$	$1.9293900000 \times 10^{-7}$	$1.9293900000 \times 10^{-7}$
	0.8	$2.0443500000 \times 10^{-6}$	$1.4139470000 \times 10^{-6}$	$8.2379300000 \times 10^{-7}$	$2.5631800000 \times 10^{-7}$	$2.5631800000 \times 10^{-7}$
	1	$2.5253100000 \times 10^{-6}$	$1.7473930000 \times 10^{-6}$	$1.0191440000 \times 10^{-6}$	$3.1886900000 \times 10^{-7}$	$3.1886900000 \times 10^{-7}$
0.2	0.2	$5.8984400000 \times 10^{-7}$	$4.2773700000 \times 10^{-7}$	$2.7455400000 \times 10^{-7}$	$1.2876600000 \times 10^{-7}$	$1.2876600000 \times 10^{-7}$
	0.4	$1.1759660000 \times 10^{-6}$	$8.5314400000 \times 10^{-7}$	$5.4809200000 \times 10^{-7}$	$2.5761600000 \times 10^{-7}$	$2.5761600000 \times 10^{-7}$
	0.6	$1.7533840000 \times 10^{-6}$	$1.2726080000 \times 10^{-6}$	$8.1829600000 \times 10^{-7}$	$3.8562800000 \times 10^{-7}$	$3.8562800000 \times 10^{-7}$
	0.8	$2.3179040000 \times 10^{-6}$	$1.6832640000 \times 10^{-6}$	$1.0835570000 \times 10^{-6}$	$5.1239700000 \times 10^{-7}$	$5.1239700000 \times 10^{-7}$
	1	$2.8655420000 \times 10^{-6}$	$2.0823970000 \times 10^{-6}$	$1.3423590000 \times 10^{-6}$	$6.3748800000 \times 10^{-7}$	$6.3748800000 \times 10^{-7}$
0.3	0.2	$6.5642700000 \times 10^{-7}$	$4.9400400000 \times 10^{-7}$	$3.3914500000 \times 10^{-7}$	$1.9276800000 \times 10^{-7}$	$1.9276800000 \times 10^{-7}$
	0.4	$1.3096920000 \times 10^{-6}$	$9.8624000000 \times 10^{-7}$	$6.7785000000 \times 10^{-7}$	$3.8605500000 \times 10^{-7}$	$3.8605500000 \times 10^{-7}$
	0.6	$1.9537820000 \times 10^{-6}$	$1.4720680000 \times 10^{-6}$	$1.0127840000 \times 10^{-6}$	$5.7806700000 \times 10^{-7}$	$5.7806700000 \times 10^{-7}$
	0.8	$2.5842940000 \times 10^{-6}$	$1.9484160000 \times 10^{-6}$	$1.3421460000 \times 10^{-6}$	$7.6821500000 \times 10^{-7}$	$7.6821500000 \times 10^{-7}$
	1	$3.1969960000 \times 10^{-6}$	$2.4123230000 \times 10^{-6}$	$1.6641870000 \times 10^{-6}$	$9.5586800000 \times 10^{-7}$	$9.5586800000 \times 10^{-7}$
0.4	0.2	$7.2188300000 \times 10^{-7}$	$5.5944200000 \times 10^{-7}$	$4.0328700000 \times 10^{-7}$	$2.5652100000 \times 10^{-7}$	$2.5652100000 \times 10^{-7}$
	0.4	$1.4414910000 \times 10^{-6}$	$1.1180020000 \times 10^{-6}$	$8.0703200000 \times 10^{-7}$	$5.1423300000 \times 10^{-7}$	$5.1423300000 \times 10^{-7}$
	0.6	$2.1514870000 \times 10^{-6}$	$1.6697170000 \times 10^{-6}$	$1.2065920000 \times 10^{-6}$	$7.7025600000 \times 10^{-7}$	$7.7025600000 \times 10^{-7}$
	0.8	$2.8472250000 \times 10^{-6}$	$2.2112730000 \times 10^{-6}$	$1.5999330000 \times 10^{-6}$	$1.0237930000 \times 10^{-6}$	$1.0237930000 \times 10^{-6}$
	1	$3.5242520000 \times 10^{-6}$	$2.7394880000 \times 10^{-6}$	$1.9850950000 \times 10^{-6}$	$1.2740070000 \times 10^{-6}$	$1.2740070000 \times 10^{-6}$
0.5	0.2	$7.8654200000 \times 10^{-7}$	$6.2423400000 \times 10^{-7}$	$4.6702100000 \times 10^{-7}$	$3.2001400000 \times 10^{-7}$	$3.2001400000 \times 10^{-7}$
	0.4	$1.5720280000 \times 10^{-6}$	$1.2488040000 \times 10^{-6}$	$9.3572800000 \times 10^{-7}$	$6.4215100000 \times 10^{-7}$	$6.4215100000 \times 10^{-7}$
	0.6	$2.3474550000 \times 10^{-6}$	$1.8660800000 \times 10^{-6}$	$1.3998180000 \times 10^{-6}$	$9.6219500000 \times 10^{-7}$	$9.6219500000 \times 10^{-7}$
	0.8	$3.1079660000 \times 10^{-6}$	$2.4725350000 \times 10^{-6}$	$1.8570540000 \times 10^{-6}$	$1.2791210000 \times 10^{-6}$	$1.2791210000 \times 10^{-6}$
	1	$3.8489010000 \times 10^{-6}$	$3.0647790000 \times 10^{-6}$	$2.3052760000 \times 10^{-6}$	$1.5918960000 \times 10^{-6}$	$1.5918960000 \times 10^{-6}$

Table 6. Comparison at different fractional order of γ on the basis of error for example 3.

κ	φ	$\gamma = 0.4$	$\gamma = 0.6$	$\gamma = 0.8$	$\gamma = 1(NTDM_{CF})$	$\gamma = 1(NTDM_{ABC})$
0.1	0.2	$6.4600000000 \times 10^{-10}$	$4.8300000000 \times 10^{-10}$	$2.7900000000 \times 10^{-10}$	$6.7000000000 \times 10^{-11}$	$6.7000000000 \times 10^{-11}$
	0.4	$6.4300000000 \times 10^{-10}$	$4.8100000000 \times 10^{-10}$	$2.7700000000 \times 10^{-10}$	$6.6000000000 \times 10^{-11}$	$6.6000000000 \times 10^{-11}$
	0.6	$6.4400000000 \times 10^{-10}$	$4.8200000000 \times 10^{-10}$	$2.8000000000 \times 10^{-10}$	$6.9000000000 \times 10^{-11}$	$6.9000000000 \times 10^{-11}$
	0.8	$6.4500000000 \times 10^{-10}$	$4.8400000000 \times 10^{-10}$	$2.8200000000 \times 10^{-10}$	$7.1000000000 \times 10^{-11}$	$7.1000000000 \times 10^{-11}$
	1	$6.4100000000 \times 10^{-10}$	$4.8000000000 \times 10^{-10}$	$2.7900000000 \times 10^{-10}$	$6.9000000000 \times 10^{-11}$	$6.9000000000 \times 10^{-11}$
0.2	0.2	$6.7100000000 \times 10^{-10}$	$5.4000000000 \times 10^{-10}$	$3.5400000000 \times 10^{-10}$	$1.4300000000 \times 10^{-10}$	$1.4300000000 \times 10^{-10}$
	0.4	$6.6700000000 \times 10^{-10}$	$5.3700000000 \times 10^{-10}$	$3.5100000000 \times 10^{-10}$	$1.4000000000 \times 10^{-10}$	$1.4000000000 \times 10^{-10}$
	0.6	$6.5800000000 \times 10^{-10}$	$5.2800000000 \times 10^{-10}$	$3.4300000000 \times 10^{-10}$	$1.3300000000 \times 10^{-10}$	$1.3300000000 \times 10^{-10}$
	0.8	$6.6300000000 \times 10^{-10}$	$5.3300000000 \times 10^{-10}$	$3.4900000000 \times 10^{-10}$	$1.4000000000 \times 10^{-10}$	$1.4000000000 \times 10^{-10}$
	1	$6.5700000000 \times 10^{-10}$	$5.2800000000 \times 10^{-10}$	$3.4400000000 \times 10^{-10}$	$1.3500000000 \times 10^{-10}$	$1.3500000000 \times 10^{-10}$
0.3	0.2	$6.8100000000 \times 10^{-10}$	$5.7600000000 \times 10^{-10}$	$4.1200000000 \times 10^{-10}$	$2.1000000000 \times 10^{-10}$	$2.1000000000 \times 10^{-10}$
	0.4	$6.8400000000 \times 10^{-10}$	$5.8000000000 \times 10^{-10}$	$4.1700000000 \times 10^{-10}$	$2.1500000000 \times 10^{-10}$	$2.1500000000 \times 10^{-10}$
	0.6	$6.7400000000 \times 10^{-10}$	$5.7100000000 \times 10^{-10}$	$4.0700000000 \times 10^{-10}$	$2.0700000000 \times 10^{-10}$	$2.0700000000 \times 10^{-10}$
	0.8	$6.7700000000 \times 10^{-10}$	$5.7400000000 \times 10^{-10}$	$4.1100000000 \times 10^{-10}$	$2.1100000000 \times 10^{-10}$	$2.1100000000 \times 10^{-10}$
	1	$6.6500000000 \times 10^{-10}$	$5.6200000000 \times 10^{-10}$	$4.0000000000 \times 10^{-10}$	$2.0000000000 \times 10^{-10}$	$2.0000000000 \times 10^{-10}$
0.4	0.2	$6.9600000000 \times 10^{-10}$	$6.1600000000 \times 10^{-10}$	$4.7400000000 \times 10^{-10}$	$2.8700000000 \times 10^{-10}$	$2.8700000000 \times 10^{-10}$
	0.4	$6.8900000000 \times 10^{-10}$	$6.0900000000 \times 10^{-10}$	$4.6800000000 \times 10^{-10}$	$2.8100000000 \times 10^{-10}$	$2.8100000000 \times 10^{-10}$
	0.6	$6.8600000000 \times 10^{-10}$	$6.0600000000 \times 10^{-10}$	$4.6500000000 \times 10^{-10}$	$2.7900000000 \times 10^{-10}$	$2.7900000000 \times 10^{-10}$
	0.8	$6.8900000000 \times 10^{-10}$	$6.0900000000 \times 10^{-10}$	$4.6900000000 \times 10^{-10}$	$2.8300000000 \times 10^{-10}$	$2.8300000000 \times 10^{-10}$
	1	$6.7700000000 \times 10^{-10}$	$5.9700000000 \times 10^{-10}$	$4.5700000000 \times 10^{-10}$	$2.7200000000 \times 10^{-10}$	$2.7200000000 \times 10^{-10}$
0.5	0.2	$6.9800000000 \times 10^{-10}$	$6.3800000000 \times 10^{-10}$	$5.2000000000 \times 10^{-10}$	$3.5100000000 \times 10^{-10}$	$3.5100000000 \times 10^{-10}$
	0.4	$7.0100000000 \times 10^{-10}$	$6.4200000000 \times 10^{-10}$	$5.2400000000 \times 10^{-10}$	$3.5600000000 \times 10^{-10}$	$3.5600000000 \times 10^{-10}$
	0.6	$6.9700000000 \times 10^{-10}$	$6.3800000000 \times 10^{-10}$	$5.2100000000 \times 10^{-10}$	$3.5300000000 \times 10^{-10}$	$3.5300000000 \times 10^{-10}$
	0.8	$6.9300000000 \times 10^{-10}$	$6.3400000000 \times 10^{-10}$	$5.1700000000 \times 10^{-10}$	$3.4900000000 \times 10^{-10}$	$3.4900000000 \times 10^{-10}$
	1	$6.9000000000 \times 10^{-10}$	$6.3100000000 \times 10^{-10}$	$5.1500000000 \times 10^{-10}$	$3.4700000000 \times 10^{-10}$	$3.4700000000 \times 10^{-10}$

Table 7. Comparison of absolute error among $q-HATM$ [38], NDM_{CF} and NDM_{ABC} for example 3 at $k = 0.25$.

κ	φ	$|q-HATM|$	$|NTDM_{CF}|$	$|NTDM_{ABC}|$
0.25	1	7.0832×10^{-13}	$2.0000000000 \times 10^{-13}$	$2.0000000000 \times 10^{-13}$
	2	4.4031×10^{-13}	$1.0000000000 \times 10^{-13}$	$1.0000000000 \times 10^{-13}$
	3	1.1304×10^{-13}	$1.0000000000 \times 10^{-13}$	$1.0000000000 \times 10^{-13}$
	4	1.6642×10^{-13}	$1.0000000000 \times 10^{-13}$	$1.0000000000 \times 10^{-13}$
	5	3.3639×10^{-13}	$1.0000000000 \times 10^{-13}$	$1.0000000000 \times 10^{-13}$

Author Contributions: Formal analysis, N.A.S.; Methodology, R.S.; Project administration, Y.S.H.; Software, A.K.; Supervision, J.-D.C.; Validation, K.M.A.; Writing—original draft, R.S., and A.K.; Writing—review & editing, Y.S.H. and K.M.A. Also, N.A.S. and R.S. have equal contribution. All authors have read and agreed to the published version of the manuscript.

Funding: This research received no external funding.

Institutional Review Board Statement: Not applicable.

Informed Consent Statement: Not applicable.

Data Availability Statement: No data were used to support this study.

Acknowledgments: This research was supported by Taif University Researchers Supporting Project Number (TURSP-2020/217), Taif University, Taif, Saudi Arabia. This research was supported by Basic Science Research Program through the National Research Foundation of Korea (NRF) funded by the Ministry of Education (No. 2017R1D1A1B05030422).

Conflicts of Interest: The authors declare that they have no competing interest.

References

1. Kilbas, A.A.; Srivastava, H.M.; Trujillo, J.J. *Theory and Applications of Fractional Differential Equations*; Elsevier: Amsterdam, The Netherlands, 2006.
2. Baleanu, D.; Guvenc, Z.B.; Tenreiro Machado, J.A. *New Trends in Nanotechnology and Fractional Calculus Applications*; Springer: New York, NY, USA, 2010.
3. Caputo, M. *Elasticita e Dissipazione*; Zanichelli: Bologna, Italy, 1969.
4. Riemann, G.F.B. *Versuch Einer Allgemeinen Auffassung der Integration und Differentiation*; Gesammelte Mathematische Werke: Leipzig, Germany, 1896.
5. Liouville, J. Memoire sur quelques questions de geometrie et de mecanique, et sur un nouveaugenre de calcul pour resoudre ces questions. *J. Ecole Polytech.* **1832**, *13*, 1–69.
6. Podlubny, I. *Fractional Differential Equations*; Academic Press: New York, NY, USA, 1999.
7. Miller, K.S.; Ross, B. *An Introduction to Fractional Calculus and Fractional Differential Equations*; Wiley: New York, NY, USA, 1993.
8. Li, Y.; Liub, F.; Turner, I.W.; Li, T. Time-fractional diffusion equation for signal smoothing. *Appl. Math Comput.* **2018**, *326*, 108–116. [CrossRef]
9. Lin, W. Global existence theory and chaos control of fractional differential equations. *J. Math Anal. Appl.* **2007**, *332*, 709–726. [CrossRef]
10. Veeresha, P.; Prakasha, D.G.; Baskonus, H.M. Solving smoking epidemic model of fractional order using a modified homotopy analysis transform method. *Math. Sci.* **2019**, *13*, 115–128. [CrossRef]
11. Prakasha, D.G.; Veeresha, P.; Baskonus, H.M. Analysis of the dynamics of hepatitis E virus using the Atangana-Baleanu fractional derivative. *Eur. Phys. J. Plus* **2019**, *134*, 1–11. [CrossRef]
12. Shah, N.; Alyousef, H.; El-Tantawy, S.; Shah, R.; Chung, J. Analytical Investigation of Fractional-Order Korteweg–De-Vries-Type Equations under Atangana–Baleanu–Caputo Operator: Modeling Nonlinear Waves in a Plasma and Fluid. *Symmetry* **2022**, *14*, 739. [CrossRef]
13. Sunthrayuth, P.; Zidan, A.; Yao, S.; Shah, R.; Inc, M. The Comparative Study for Solving Fractional-Order Fornberg–Whitham Equation via ρ-Laplace Transform. *Symmetry* **2021**, *13*, 784. [CrossRef]
14. Baleanu, D.; Wu, G.C.; Zeng, S.D. Chaos analysis and asymptotic stability of generalized Caputo fractional differential equations. *Chaos Solitons Fractals* **2017**, *102*, 99–105. [CrossRef]
15. Scalas, E.; Gorenflo, R.; Mainardi, F. Fractional calculus and continuous-time finance. *Physica A* **2000**, *284*, 376–384. [CrossRef]
16. Drapaka, C.S.; Sivaloganathan, S. A fractional model of continuum mechanics. *J. Elst.* **2012**, *107*, 105–123. [CrossRef]
17. Cruz-Duarte, J.M.; Rosales-Garcia, J.; Correa-Cely, C.R.; Garcia-Perez, A.; Avina-Cervantes, J.G. A closed form expression for the Gaussian-based Caputo-Fabrizio fractional derivative for signal processing applications. *Commun. Nonlinear Sci. Numer. Simul.* **2018**, *61*, 138–148. [CrossRef]
18. Singh, B.K. A novel approach for numeric study of 2D biological population model. *Cogent Math.* **2016**, *3*, 1–15. [CrossRef]
19. El-Sayed, A.; Hamdallah, E.; Ba-Ali, M. Qualitative Study for a Delay Quadratic Functional Integro-Differential Equation of Arbitrary (Fractional) Orders. *Symmetry* **2022**, *14*, 784. [CrossRef]
20. Keskin, Y.; Oturanc, G. Reduced differential transform method for partial differential equations. *Int. J. Nonlinear Sci. Numer. Simul.* **2009**, *10*, 741–750. [CrossRef]
21. Momani, S.; Odibat, Z. Analytical solution of a time-fractional Navier-Stokes equation by Adomian decomposition method. *Appl. Math. Comput.* **2006**, *177*, 488–494. [CrossRef]
22. Wu, G.C. A fractional variational iteration method for solving fractional nonlinear differential equations. *Comput. Math. Appl.* **2011**, *61*, 2186–2190. [CrossRef]
23. Khan, H.; Khan, A.; Kumam, P.; Baleanu, D.; Arif, M. An approximate analytical solution of the Navier-Stokes equations within Caputo operator and Elzaki transform decomposition method. *Adv. Differ. Equ.* **2020**, *2011*, 1–23. [CrossRef]
24. Nonlaopon, K.; Alsharif, A.M.; Zidan, A.M.; Khan, A.; Hamed, Y.S.; Shah, R. Numerical investigation of fractional-order Swif-Hohenberg equations via a Novel transform. *Symmetry* **2021**, *13*, 1263. [CrossRef]
25. Khan, H.; Khan, A.; Al-Qurashi, M.; Shah, R.; Baleanu, D. Modified modelling for heat like equations within Caputo operator. *Energies* **2020**, *13*, 2002. [CrossRef]
26. Aljahdaly, N.; Agarwal, R.; Shah, R.; Botmart, T. Analysis of the Time Fractional-Order Coupled Burgers Equations with Non-Singular Kernel Operators. *Mathematics* **2021**, *9*, 2326. [CrossRef]
27. Qin, Y.; Khan, A.; Ali, I.; Al Qurashi, M.; Khan, H.; Shah, R.; Baleanu, D. An efficient analytical approach for the solution of certain fractional-order dynamical systems. *Energies* **2020**, *13*, 2725. [CrossRef]
28. Alaoui, M.K.; Fayyaz, R.; Khan, A.; Shah, R.; Abdo, M.S. Analytical Investigation of Noyes-Field Model for Time-Fractional Belousov-Zhabotinsky Reaction. *Complexity* **2021**, *2021*, 3248376. [CrossRef]
29. Kaup, D.J. On the inverse scattering for cubic eigenvalue problems of the class equations. *Stud. Appl. Math.* **1980**, *62*, 183–195. [CrossRef]
30. Kupershmidt, B.A. A super Korteweg-de-Vries equations: An integrable system. *Phys. Lett. A* **1994**, *102*, 213–218. [CrossRef]
31. Fan, E. Uniformly constructing a series of explicit exact solutions to nonlinear equations in mathematical physics. *Chaos Solitons Fractals* **2003**, *16*, 819–839. [CrossRef]

32. Inc, M. On numerical soliton solution of the Kaup-Kupershmidt equation and convergence analysis of the decomposition method. *Appl. Math. Comput.* **2006**, *172*, 72–85.
33. Rashid, S.; Khalid, A.; Sultana, S.; Hammouch, Z.; Shah, R.; Alsharif, A. A Novel Analytical View of Time-Fractional Korteweg-De Vries Equations via a New Integral Transform. *Symmetry* **2021**, *13*, 1254. [CrossRef]
34. Tam, H.W.; Hu, X.B. Two integrable differential-difference equations exhibiting soliton solutions of the Kaup-Kupershimdt equation type. *Phys. Lett. A* **2000**, *272*, 174–183. [CrossRef]
35. Zhou, M.X.; Kanth, A.S.V.; Aruna, K.; Raghavendar, K.; Rezazadeh, H.; Inc, M.; Aly, A.A. Numerical Solutions of Time Fractional Zakharov-Kuznetsov Equation via Natural Transform Decomposition Method with Nonsingular Kernel Derivatives. *J. Funct. Spaces* **2021**, *2021*, 9884027. [CrossRef]
36. Adomian, G. A new approach to nonlinear partial differential equations. *J. Math. Anal. Appl.* **1984**, *102*, 420–434. [CrossRef]
37. Shah, R.; Khan, H.; Baleanu, D.; Kumam, P.; Arif, M. The analytical investigation of time-fractional multi-dimensional Navier–Stokes equation. *Alex. Eng. J.* **2020**, *59*, 2941–2956. [CrossRef]
38. Prakasha, D.G.; Malagi, N.S.; Veeresha, P.; Prasannakumara, B.C. An efficient computational technique for time-fractional Kaup-Kupershmidt equation. *Numer. Methods Partial. Differ. Equ.* **2021**, *37*, 1299–1316. [CrossRef]
39. Gupta, A.K.; Ray, S.S. The comparison of two reliable methods for accurate solution of time-fractional Kaup-Kupershmidt equation arising in capillary gravity waves. *Math. Methods Appl. Sci.* **2016**, *39*, 583–592. [CrossRef]

Article

Midpoint Inequalities in Fractional Calculus Defined Using Positive Weighted Symmetry Function Kernels

Pshtiwan Othman Mohammed [1,*], **Hassen Aydi** [2,3,4,*], **Artion Kashuri** [5], **Y. S. Hamed** [6] **and Khadijah M. Abualnaja** [6]

[1] Department of Mathematics, College of Education, University of Sulaimani, Sulaimani 46001, Kurdistan Region, Iraq
[2] Institut Supérieur d'Informatique et des Techniques de Communication, Université de Sousse, Hammam Sousse 4000, Tunisia
[3] Department of Mathematics and Applied Mathematics, Sefako Makgatho Health Sciences University, Ga-Rankuwa, South Africa
[4] China Medical University Hospital, China Medical University, Taichung 40402, Taiwan
[5] Department of Mathematics, Faculty of Technical Science, University Ismail Qemali, Vlora 9401, Albania; artionkashuri@gmail.com
[6] Department of Mathematics and Statistics, College of Science, Taif University, P.O. Box 11099, Taif 21944, Saudi Arabia; yasersalah@tu.edu.sa (Y.S.H.); Kh.abualnaja@tu.edu.sa (K.M.A.)
* Correspondence: pshtiwansangawi@gmail.com (P.O.M.); hassen.aydi@isima.rnu.tn (H.A.)

Citation: Mohammed, P.O.; Aydi, H.; Kashuri, A.; Hamed, Y.S.; Abualnaja, K.M. Midpoint Inequalities in Fractional Calculus Defined Using Positive Weighted Symmetry Function Kernels. *Symmetry* **2021**, *13*, 550. https://doi.org/10.3390/sym13040550

Academic Editor: Aviv Gibali

Received: 25 February 2021
Accepted: 24 March 2021
Published: 26 March 2021

Publisher's Note: MDPI stays neutral with regard to jurisdictional claims in published maps and institutional affiliations.

Copyright: © 2021 by the authors. Licensee MDPI, Basel, Switzerland. This article is an open access article distributed under the terms and conditions of the Creative Commons Attribution (CC BY) license (https://creativecommons.org/licenses/by/4.0/).

Abstract: The aim of our study is to establish, for convex functions on an interval, a midpoint version of the fractional HHF type inequality. The corresponding fractional integral has a symmetric weight function composed with an increasing function as integral kernel. We also consider a midpoint identity and establish some related inequalities based on this identity. Some special cases can be considered from our main results. These results confirm the generality of our attempt.

Keywords: symmetry; weighted fractional operators; convex functions; HHF type inequality

1. Introduction

Let $\mathcal{J} \subset \mathcal{R}$ be an interval and let $u : \mathcal{J} \to \mathcal{R}$ be a continuous function. Then, the function u is called convex if it satisfies

$$u(\kappa c_1 + (1-\kappa)c_2) \leq \kappa u(c_1) + (1-\kappa)u(c_2), \quad \forall c_1, c_2 \in \mathcal{J} \text{ and } \kappa \in [0,1]. \tag{1}$$

The function u is called concave whenever $-u$ is convex.

For convex functions $u : \mathcal{J} \to \mathcal{R}$, there is an important integral inequality in the literature, namely the Hermite–Hadamard or, briefly, the HH integral inequality, which is given by [1]:

$$u\left(\frac{c_1 + c_2}{2}\right) \leq \frac{1}{c_2 - c_1} \int_{c_1}^{c_2} u(x)dx \leq \frac{u(c_1) + u(c_2)}{2}, \tag{2}$$

where $c_1 < c_2$ belong to $\mathcal{J}$. In the literature, one can observe that the HH integral inequality (2) has been applied to different classes of convexity such as GA–convexity [2], quasi-convexity [3,4], s–convexity [5], (α, m)–convexity [6], exponentially convexity [7,8], MT–convexity [9], and the readers can consult [10,11] to find other types.

As we know, fractional calculus is a generalized form of integer order calculus. Various forms of fractional derivatives including RL, Hadamard, Caputo, Caputo–Hadamard, Riesz, ψ–RL, Prabhakar, and weighted versions [12–16] have been developed to date. Most of these versions are described in the RL sense based on the corresponding fractional integral. Many integer-order integral inequalities such as Ostrowski [17], Simpson [18],

Hardy [19], Olsen [20], Gagliardo–Nirenberg [21], Opial [22,23] and Rozanova [24] have been generalized and reformulated from the fractional point of view.

In addition, in 2013, the HH integral inequality (2) was generalized and reformulated by Sarikaya et al. [25] in terms of RL fractional integrals. Their result is given by:

$$u\left(\frac{c_1+c_2}{2}\right) \leq \frac{\Gamma(\nu+1)}{2(c_2-c_1)^\nu}\left[{}^{RL}\mathcal{I}^\nu_{c_1+}u(c_2) + {}^{RL}\mathcal{I}^\nu_{c_2-}u(c_1)\right] \leq \frac{u(c_1)+u(c_2)}{2}, \quad (3)$$

where $u : \mathcal{J} \to \mathcal{R}$ is assumed to be a positive convex function, continuous on the closed interval $[c_1, c_2]$, and for Lebesgue, almost all $x \in [c_1, c_2]$ when $u(x) \in L^1[c_1, c_2]$ with $c_1 < c_2$, where ${}^{RL}\mathcal{I}^\nu_{c_1+}$ and ${}^{RL}\mathcal{I}^\nu_{c_2-}$ are the left- and right-sided RL fractional integrals of order $\nu > 0$, defined by [12]:

$$\begin{aligned}{}^{RL}\mathcal{I}^\nu_{c_1+}u(x) &= \frac{1}{\Gamma(\nu)}\int_{c_1}^x (x-\kappa)^{\nu-1}u(\kappa)d\kappa, \quad x > c_1; \\ {}^{RL}\mathcal{I}^\nu_{c_2-}u(x) &= \frac{1}{\Gamma(\nu)}\int_x^{c_2} (\kappa-x)^{\nu-1}u(\kappa)d\kappa, \quad x < c_2, \end{aligned} \quad (4)$$

respectively.

The inequality (3) is also known as the endpoint HH inequality due to using the ends c_1, c_2 of the interval.

On the other hand, the endpoint HH inequality (3) has been applied for various classes of convexity such as λ_ψ-convexity [26], F-convexity [27], (α, m)-convexity [28], MT-convexity [29]. The reader can find other types of convexity in the literature, which in particular, is true for [30]. In the mean time, applying the end-point HH inequality to other models of fractional calculus has received a huge amount of attention. For example, this is true for RL fractional models [31], conformable fractional models [32,33], generalized fractional models [34], ψ RL fractional models [35,36], tempered fractional models [37], and AB- and Prabhakar fractional models [38].

After extending the important field of the integral inequalities in (2) and (3), a new version of the endpoint HH inequality (3) was found by Sarikaya and Yildirim [39], namely the midpoint HH inequality due to using the midpoint $\frac{c_1+c_2}{2}$ of the interval, which is given by

$$u\left(\frac{c_1+c_2}{2}\right) \leq \frac{2^{\nu-1}\Gamma(\nu+1)}{(c_2-c_1)^\nu}\left[{}^{RL}\mathcal{I}^\nu_{\left(\frac{c_1+c_2}{2}\right)+}u(c_2) + {}^{RL}\mathcal{I}^\nu_{\left(\frac{c_1+c_2}{2}\right)-}u(c_1)\right] \leq \frac{u(c_1)+u(c_2)}{2}, \quad (5)$$

where the function $u : [c_1, c_2] \to \mathcal{R}$ is convex and continuous.

Definition 1 ([40]). *Let $g : [c_1, c_2] \to [0, \infty)$ be a function. Then, we say g is symmetric with respect to $(c_1 + c_2)/2$ if*

$$g(c_1 + c_2 - x) = g(x), \quad \forall x \in [c_1, c_2] \quad (6)$$

Based on above definition, in [41], Fejér found a new extension of the HH type inequality (2), namely the HHF type inequality, and the result is as follows:

$$u\left(\frac{c_1+c_2}{2}\right)\int_{c_1}^{c_2} g(x)dx \leq \int_{c_1}^{c_2} u(x)g(x)dx \leq \frac{u(c_1)+u(c_2)}{2}\int_{c_1}^{c_2} g(x)dx, \quad (7)$$

where g is the integrable function, and Işcan [42] found the endpoint version of (7) in the sense of RL fractional integrals, which is also the extension of (3). The result is as follows:

$$u\left(\frac{c_1+c_2}{2}\right)\left[{}^{RL}\mathcal{I}_{c_1+}^\nu g(c_2) + {}^{RL}\mathcal{I}_{c_2-}^\nu g(c_1)\right] \leq \left[{}^{RL}\mathcal{I}_{c_1+}^\nu (ug)(c_2) + {}^{RL}\mathcal{I}_{c_2-}^\nu (ug)(c_1)\right]$$

$$\leq \frac{u(c_1)+u(c_2)}{2}\left[{}^{RL}\mathcal{I}_{c_1+}^\nu g(c_2) + {}^{RL}\mathcal{I}_{c_2-}^\nu g(c_1)\right], \quad (8)$$

where u is convex and continuous and the function g belongs to $L^1[c_1, c_2]$ and is symmetric (see Definition 1).

It is worth mentioning that the midpoint version of (8) has not been found yet, even though many related inequalities of midpoint type were obtained in [43].

Recently, Mohammed et al. [44] found a new endpoint HHF-inequality in terms of weighted fractional integrals with positive weighted symmetric function in a kernel, and their result is as follows:

$$u\left(\frac{c_1+c_2}{2}\right)\left[\left({}_{\varrho^{-1}(c_1)+}\mathcal{I}^{\nu:\varrho}(w\circ\varrho)\right)\left(\varrho^{-1}(c_2)\right) + \left(\mathcal{I}_{\varrho^{-1}(c_2)-}^{\nu:\varrho}(w\circ\varrho)\right)\left(\varrho^{-1}(c_1)\right)\right]$$

$$\leq w(c_2)\left({}_{\varrho^{-1}(c_1)+}\mathcal{I}_{w\circ\varrho}^{\nu:\varrho}(u\circ\varrho)\right)\left(\varrho^{-1}(c_2)\right) + w(c_1)\left({}_{w\circ\varrho}\mathcal{I}_{\varrho^{-1}(c_2)-}^{\nu:\varrho}(u\circ\varrho)\right)\left(\varrho^{-1}(c_1)\right)$$

$$\leq \frac{u(c_1)+u(c_2)}{2}\left[\left({}_{\varrho^{-1}(c_1)+}\mathcal{I}^{\nu:\varrho}(w\circ\varrho)\right)\left(\varrho^{-1}(c_2)\right)\right.$$

$$\left. + \left(\mathcal{I}_{\varrho^{-1}(c_2)-}^{\nu:\varrho}(w\circ\varrho)\right)\left(\varrho^{-1}(c_1)\right)\right]. \quad (9)$$

Here, u is a convex and continuous function, $\varrho(x)$ a monotone increasing function from the interval $(c_1, c_2]$ onto itself with a continuous derivative $\varrho'(x)$ on the open interval (c_1, c_2), and $w : [c_1, c_2] \to (0, \infty)$ is an integrable function, which is symmetric with respect to $(c_1 + c_2)/2$, where $c_1 < c_2$.

Definition 2. *Let $(c_1, c_2) \subseteq \mathcal{R}$ and $\varrho(x)$ be an increasing positive and monotone function on the interval $(c_1, c_2]$ with a continuous derivative $\varrho'(x)$ on the open interval (c_1, c_2). Then, the left-sided and right-sided the weighted fractional integrals of a function u according to another function $\varrho(x)$ on $[c_1, c_2]$ are defined by [15]:*

$$\left({}_{c_1+}\mathcal{I}_w^{\nu:\varrho} u\right)(x) = \frac{[w(x)]^{-1}}{\Gamma(\nu)} \int_{c_1}^x \varrho'(\kappa)(\varrho(x) - \varrho(\kappa))^{\nu-1} u(\kappa) w(\kappa) d\kappa,$$

$$\left({}_w\mathcal{I}_{c_2-}^{\nu:\varrho} u\right)(x) = \frac{[w(x)]^{-1}}{\Gamma(\nu)} \int_x^{c_2} \varrho'(\kappa)(\varrho(\kappa) - \varrho(x))^{\nu-1} u(\kappa) w(\kappa) d\kappa, \quad \nu > 0, \quad (10)$$

for $[w(x)]^{-1} := \frac{1}{w(x)}$ such that $w(x) \neq 0$.

Remark 1. *From Definition 2, we can obtain the following special cases.*

- *If $\varrho(x) = x$ and $w(x) = 1$, then the weighted fractional integrals (10) reduce to the classical RL fractional integrals (4).*

- *If $w(x) = 1$, we obtain the fractional integrals of the function u with respect to the function $\varrho(x)$, which is defined by [13,14]:*

$$\left({}_{c_1+}\mathcal{I}^{\nu:\varrho} u\right)(x) = \frac{1}{\Gamma(\nu)} \int_{c_1}^x \varrho'(\kappa)(\varrho(x) - \varrho(\kappa))^{\nu-1} u(\kappa) d\kappa,$$

$$\left(\mathcal{I}_{c_2-}^{\nu:\varrho} u\right)(x) = \frac{1}{\Gamma(\nu)} \int_x^{c_2} \varrho'(\kappa)(\varrho(\kappa) - \varrho(x))^{\nu-1} u(\kappa) d\kappa, \quad \nu > 0. \quad (11)$$

In this article, we will investigate the midpoint version of (9) and some related HHF inequalities by using the weighted fractional integrals (10) with positive weighted symmetric functions in the kernel.

The rest of our article is structured in the following way: In Section 2, we will prove the necessary and auxiliary lemmas, including the midpoint version of (9). In Section 3, we will prove our main results, including new midpoint fractional HHF integral inequalities with some related results. We will present some concluding remarks in Section 4.

2. Auxiliary Results

In this section, we prove analogues of the fractional HH inequalities (2)–(3) and HHF inequalities (7)–(8) for weighted fractional integral operators with positive weighted symmetric function kernels. Here, the main results are as follows: Theorem 1 (it is a generalisation of HH inequalities (2)–(3) and HHF inequality (7), and a reformulation of HHF inequality (8)) and Lemma 2 (it is a consequence of Theorem 1).

At first, we need the following lemma.

Lemma 1. *Assume that* $w : [c_1, c_2] \to (0, \infty)$ *is an integrable function and symmetric with respect to* $(c_1 + c_2)/2$, $c_1 < c_2$. *Then,*

(i) for each $\kappa \in [0, 1]$, *we have*

$$w\left(\frac{\kappa}{2}c_1 + \frac{2-\kappa}{2}c_2\right) = w\left(\frac{2-\kappa}{2}c_1 + \frac{\kappa}{2}c_2\right). \quad (12)$$

(ii) For $\nu > 0$, *we have*

$$\left({}_{\varrho^{-1}\left(\frac{c_1+c_2}{2}\right)+}\mathcal{I}^{\nu;\varrho}(w \circ \varrho)\right)\left(\varrho^{-1}(c_2)\right) = \left({}_{\varrho^{-1}\left(\frac{c_1+c_2}{2}\right)-}\mathcal{I}^{\nu;\varrho}(w \circ \varrho)\right)\left(\varrho^{-1}(c_1)\right)$$

$$= \frac{1}{2}\left[\left({}_{\varrho^{-1}\left(\frac{c_1+c_2}{2}\right)+}\mathcal{I}^{\nu;\varrho}(w \circ \varrho)\right)\left(\varrho^{-1}(c_2)\right) + \left({}_{\varrho^{-1}\left(\frac{c_1+c_2}{2}\right)-}\mathcal{I}^{\nu;\varrho}(w \circ \varrho)\right)\left(\varrho^{-1}(c_1)\right)\right]. \quad (13)$$

Proof.

(i) Let $x = \frac{\kappa}{2}c_1 + \frac{2-\kappa}{2}c_2$. It is clear that $x \in [c_1, c_2]$ for each $\kappa \in [0, 1]$ and that $c_1 + c_2 - x = \frac{2-\kappa}{2}c_1 + \frac{\kappa}{2}c_2$. Then, by making use of the assumptions and Definition 1, we can obtain (12).

(ii) The symmetry property of w leads to

$$(w \circ \varrho)(\kappa) = w(\varrho(\kappa)) = w(c_1 + c_2 - \varrho(\kappa)), \quad \forall \kappa \in \left[\varrho^{-1}(c_1), \varrho^{-1}(c_2)\right].$$

From above and setting $\varrho(x) := c_1 + c_2 - \varrho(\kappa)$, it follows that

$$\left({}_{\varrho^{-1}\left(\frac{c_1+c_2}{2}\right)+}\mathcal{I}^{\nu;\varrho}(w \circ \varrho)\right)\left(\varrho^{-1}(c_2)\right)$$

$$= \frac{1}{\Gamma(\nu)}\int_{\varrho^{-1}\left(\frac{c_1+c_2}{2}\right)}^{\varrho^{-1}(c_2)}(c_2 - \varrho(x))^{\nu-1}(w \circ \varrho)(x)\varrho'(x)dx$$

$$= \frac{1}{\Gamma(\nu)}\int_{\varrho^{-1}(c_1)}^{\varrho^{-1}\left(\frac{c_1+c_2}{2}\right)}(\varrho(\kappa) - c_1)^{\nu-1}w(c_1 + c_2 - \varrho(\kappa))\varrho'(\kappa)d\kappa$$

$$= \frac{1}{\Gamma(\nu)}\int_{\varrho^{-1}(c_1)}^{\varrho^{-1}\left(\frac{c_1+c_2}{2}\right)}(\varrho(\kappa) - c_1)^{\nu-1}(w \circ \varrho)(\kappa)\varrho'(\kappa)d\kappa$$

$$= \left({}_{\varrho^{-1}\left(\frac{c_1+c_2}{2}\right)-}\mathcal{I}^{\nu;\varrho}(w \circ \varrho)\right)\left(\varrho^{-1}(c_1)\right),$$

which completes the desired equality (13). □

Remark 2. *Throughout the present article, we denote* $[w(x)]^{-1} = \frac{1}{w(x)}$ *and* $\varrho^{-1}(x)$ *the inverse of the function* $\varrho(x)$.

Theorem 1. *Let* $0 \leq c_1 < c_2$, *let* $u : [c_1, c_2] \to \mathcal{R}$ *be an* L^1 *convex function and* $w : [c_1, c_2] \to \mathcal{R}$ *be an integrable, positive and weighted symmetric function with respect to* $\frac{c_1+c_2}{2}$. *If, in addition,* ϱ *is an increasing and positive function from* $[c_1, c_2)$ *onto itself such that its derivative* $\varrho'(x)$ *is continuous on* (c_1, c_2), *then for* $\nu > 0$, *the following inequalities are valid:*

$$u\left(\frac{c_1+c_2}{2}\right)\left[\left(\mathcal{I}^{\nu;\varrho}_{\varrho^{-1}\left(\frac{c_1+c_2}{2}\right)_+}(w \circ \varrho)\right)\left(\varrho^{-1}(c_2)\right)\right.$$
$$\left.+\left(\mathcal{I}^{\nu;\varrho}_{\varrho^{-1}\left(\frac{c_1+c_2}{2}\right)_-}(w \circ \varrho)\right)\left(\varrho^{-1}(c_1)\right)\right] \leq w(c_2)\left(\mathcal{I}^{\nu;\varrho}_{\varrho^{-1}\left(\frac{c_1+c_2}{2}\right)_+}\mathcal{I}^{\nu;\varrho}_{w \circ \varrho}(u \circ \varrho)\right)\left(\varrho^{-1}(c_2)\right)$$
$$+ w(c_1)\left(w \circ \varrho \mathcal{I}^{\nu;\varrho}_{\varrho^{-1}\left(\frac{c_1+c_2}{2}\right)_-}(u \circ \varrho)\right)\left(\varrho^{-1}(c_1)\right)$$
$$\leq \frac{u(c_1)+u(c_2)}{2}\left[\left(\mathcal{I}^{\nu;\varrho}_{\varrho^{-1}\left(\frac{c_1+c_2}{2}\right)_+}(w \circ \varrho)\right)\left(\varrho^{-1}(c_2)\right)\right.$$
$$\left.+\left(\mathcal{I}^{\nu;\varrho}_{\varrho^{-1}\left(\frac{c_1+c_2}{2}\right)_-}(w \circ \varrho)\right)\left(\varrho^{-1}(c_1)\right)\right]. \quad (14)$$

Proof. The convexity of u on $[c_1, c_2]$ gives

$$u\left(\frac{x+y}{2}\right) \leq \frac{u(x)+u(y)}{2} \quad \text{for all } x, y \in [c_1, c_2].$$

So, for $x = \frac{\kappa}{2}c_1 + \frac{2-\kappa}{2}c_2$ and $y = \frac{2-\kappa}{2}c_1 + \frac{\kappa}{2}c_2$, $\kappa \in [0,1]$, it follows that

$$2u\left(\frac{c_1+c_2}{2}\right) \leq u\left(\frac{\kappa}{2}c_1 + \frac{2-\kappa}{2}c_2\right) + u\left(\frac{2-\kappa}{2}c_1 + \frac{\kappa}{2}c_2\right). \quad (15)$$

Multiplying both sides of (15) by $\kappa^{\nu-1}w\left(\frac{\kappa}{2}c_1 + \frac{2-\kappa}{2}c_2\right)$ and integrating the resulting inequality with respect to κ over $[0,1]$, we obtain

$$2u\left(\frac{c_1+c_2}{2}\right)\int_0^1 \kappa^{\nu-1}w\left(\frac{\kappa}{2}c_1 + \frac{2-\kappa}{2}c_2\right)d\kappa$$
$$\leq \int_0^1 \kappa^{\nu-1}u\left(\frac{\kappa}{2}c_1 + \frac{2-\kappa}{2}c_2\right)w\left(\frac{\kappa}{2}c_1 + \frac{2-\kappa}{2}c_2\right)d\kappa$$
$$+ \int_0^1 \kappa^{\nu-1}u\left(\frac{2-\kappa}{2}c_1 + \frac{\kappa}{2}c_2\right)w\left(\frac{\kappa}{2}c_1 + \frac{2-\kappa}{2}c_2\right)d\kappa. \quad (16)$$

From the left-hand side of the inequality in (16), we use (13) to obtain

$$\frac{2^{\nu-1}\Gamma(\nu)}{(c_2-c_1)^\nu}\left[\left({}_{\varrho^{-1}\left(\frac{c_1+c_2}{2}\right)+}\mathcal{I}^{\nu;\varrho}(w\circ\varrho)\right)\left(\varrho^{-1}(c_2)\right)+\left(\mathcal{I}^{\nu;\varrho}_{\varrho^{-1}\left(\frac{c_1+c_2}{2}\right)-}(w\circ\varrho)\right)\left(\varrho^{-1}(c_1)\right)\right]$$

$$=\frac{2^\nu\Gamma(\nu)}{(c_2-c_1)^\nu}\left({}_{\varrho^{-1}\left(\frac{c_1+c_2}{2}\right)+}\mathcal{I}^{\nu;\varrho}(w\circ\varrho)\right)\left(\varrho^{-1}(c_2)\right)$$

$$=\frac{2^\nu}{(c_2-c_1)^\nu}\int_{\varrho^{-1}\left(\frac{c_1+c_2}{2}\right)}^{\varrho^{-1}(c_2)}(c_2-\varrho(x))^{\nu-1}(w\circ\varrho)(x)\varrho'(x)dx$$

$$=\int_{\varrho^{-1}\left(\frac{c_1+c_2}{2}\right)}^{\varrho^{-1}(c_2)}\left(\frac{2(c_2-\varrho(x))}{c_2-c_1}\right)^{\nu-1}(w\circ\varrho)(x)\varrho'(x)\frac{2dx}{c_2-c_1}$$

$$=\int_0^1 \kappa^{\nu-1}w\left(\frac{\kappa}{2}c_1+\frac{2-\kappa}{2}c_2\right)d\kappa,\quad\left[\text{denoting }\kappa:=\frac{2(c_2-\varrho(x))}{c_2-c_1}\right].$$

It follows that

$$2u\left(\frac{c_1+c_2}{2}\right)\int_0^1\kappa^{\nu-1}w\left(\frac{\kappa}{2}c_1+\frac{2-\kappa}{2}c_2\right)d\kappa=\frac{2^\nu\Gamma(\nu)}{(c_2-c_1)^\nu}u\left(\frac{c_1+c_2}{2}\right)$$

$$\times\left[\left({}_{\varrho^{-1}\left(\frac{c_1+c_2}{2}\right)+}\mathcal{I}^{\nu;\varrho}(w\circ\varrho)\right)\left(\varrho^{-1}(c_2)\right)+\left(\mathcal{I}^{\nu;\varrho}_{\varrho^{-1}\left(\frac{c_1+c_2}{2}\right)-}(w\circ\varrho)\right)\left(\varrho^{-1}(c_1)\right)\right].\quad(17)$$

By evaluating the weighted fractional operators, we see that

$$w(c_2)\left({}_{\varrho^{-1}\left(\frac{c_1+c_2}{2}\right)+}\mathcal{I}^{\nu;\varrho}_{w\circ\varrho}(u\circ\varrho)\right)\left(\varrho^{-1}(c_2)\right)+w(c_1)\left({}_{w\circ\varrho}\mathcal{I}^{\nu;\varrho}_{\varrho^{-1}\left(\frac{c_1+c_2}{2}\right)-}(u\circ\varrho)\right)\left(\varrho^{-1}(c_1)\right)$$

$$=w(c_2)\frac{(w\circ\varrho)^{-1}\left(\varrho^{-1}(c_2)\right)}{\Gamma(\nu)}\int_{\varrho^{-1}\left(\frac{c_1+c_2}{2}\right)}^{\varrho^{-1}(c_2)}(c_2-\varrho(x))^{\nu-1}(u\circ\varrho)(x)(w\circ\varrho)(x)\varrho'(x)dx$$

$$+w(c_1)\frac{(w\circ\varrho)^{-1}\left(\varrho^{-1}(c_1)\right)}{\Gamma(\nu)}\int_{\varrho^{-1}(c_1)}^{\varrho^{-1}\left(\frac{c_1+c_2}{2}\right)}(\varrho(x)-c_1)^{\nu-1}(u\circ\varrho)(x)(w\circ\varrho)(x)\varrho'(x)dx$$

$$=\frac{(c_2-c_1)^\nu}{2^\nu\Gamma(\nu)}\int_{\varrho^{-1}\left(\frac{c_1+c_2}{2}\right)}^{\varrho^{-1}(c_2)}\left(\frac{2(c_2-\varrho(x))}{c_2-c_1}\right)^{\nu-1}(u\circ\varrho)(x)(w\circ\varrho)(x)\varrho'(x)\frac{2dx}{c_2-c_1}$$

$$+\frac{(c_2-c_1)^\nu}{2^\nu\Gamma(\nu)}\int_{\varrho^{-1}(c_1)}^{\varrho^{-1}\left(\frac{c_1+c_2}{2}\right)}\left(\frac{2(\varrho(x)-c_1)}{c_2-c_1}\right)^{\nu-1}(u\circ\varrho)(x)(w\circ\varrho)(x)\varrho'(x)\frac{2dx}{c_2-c_1},$$

where we used

$$\left[(w\circ\varrho)(\varrho^{-1}(y))\right]^{-1}=\frac{1}{(w\circ\varrho)(\varrho^{-1}(y))}=\frac{1}{w(y)}\quad\text{for } y=c_1,c_2.\quad(18)$$

Setting $t_1 = \frac{2(c_2 - \varrho(x))}{c_2 - c_1}$ and $t_2 = \frac{2(\varrho(x) - c_1)}{c_2 - c_1}$, one can deduce that

$$w(c_2)\left({}_{\varrho^{-1}\left(\frac{c_1+c_2}{2}\right)+}\mathcal{I}^{v;\varrho}_{w \circ \varrho}(u \circ \varrho)\right)\left(\varrho^{-1}(c_2)\right) + w(c_1)\left({}_{w \circ \varrho}\mathcal{I}^{v;\varrho}_{\varrho^{-1}\left(\frac{c_1+c_2}{2}\right)-}(u \circ \varrho)\right)\left(\varrho^{-1}(c_1)\right)$$

$$= \frac{(c_2 - c_1)^v}{2^v \Gamma(v)} \left[\int_0^1 t_1^{v-1} u\left(\frac{t_1}{2}c_1 + \frac{2 - t_1}{2}c_2\right) w\left(\frac{t_1}{2}c_1 + \frac{2 - t_1}{2}c_2\right) dt_1 \right.$$

$$+ \int_0^1 t_2^{v-1} u\left(\frac{2 - t_2}{2}c_1 + \frac{t_2}{2}c_2\right) w\left(\frac{2 - t_2}{2}c_1 + \frac{t_2}{2}c_2\right) dt_2 \right]$$

$$= \frac{(c_2 - c_1)^v}{2^v \Gamma(v)} \left[\int_0^1 \kappa^{v-1} u\left(\frac{\kappa}{2}c_1 + \frac{2 - \kappa}{2}c_2\right) w\left(\frac{\kappa}{2}c_1 + \frac{2 - \kappa}{2}c_2\right) d\kappa \right.$$

$$+ \underbrace{\int_0^1 \kappa^{v-1} u\left(\frac{2 - \kappa}{2}c_1 + \frac{\kappa}{2}c_2\right) w\left(\frac{\kappa}{2}c_1 + \frac{2 - \kappa}{2}c_2\right) d\kappa}_{\text{by using (12)}} \right].$$

It follows that

$$\int_0^1 \kappa^{v-1} u\left(\frac{\kappa}{2}c_1 + \frac{2 - \kappa}{2}c_2\right) w\left(\frac{\kappa}{2}c_1 + \frac{2 - \kappa}{2}c_2\right) d\kappa$$

$$+ \int_0^1 \kappa^{v-1} u\left(\frac{2 - \kappa}{2}c_1 + \frac{\kappa}{2}c_2\right) w\left(\frac{\kappa}{2}c_1 + \frac{2 - \kappa}{2}c_2\right) d\kappa$$

$$= \frac{2^v \Gamma(v)}{(c_2 - c_1)^v} \left[w(c_2) \left({}_{\varrho^{-1}\left(\frac{c_1+c_2}{2}\right)+} \mathcal{I}^{v;\varrho}_{w \circ \varrho}(u \circ \varrho)\right)\left(\varrho^{-1}(c_2)\right) \right.$$

$$+ w(c_1) \left({}_{w \circ \varrho}\mathcal{I}^{v;\varrho}_{\varrho^{-1}\left(\frac{c_1+c_2}{2}\right)-}(u \circ \varrho)\right)\left(\varrho^{-1}(c_1)\right) \right]. \quad (19)$$

By making use of (17) and (19) in (16), we get

$$u\left(\frac{c_1 + c_2}{2}\right) \left[\left({}_{\varrho^{-1}\left(\frac{c_1+c_2}{2}\right)+} \mathcal{I}^{v;\varrho}(w \circ \varrho)\right)\left(\varrho^{-1}(c_2)\right) \right.$$

$$+ \left. \left(\mathcal{I}^{v;\varrho}_{\varrho^{-1}\left(\frac{c_1+c_2}{2}\right)-}(w \circ \varrho)\right)\left(\varrho^{-1}(c_1)\right) \right] \leq w(c_2) \left({}_{\varrho^{-1}\left(\frac{c_1+c_2}{2}\right)+} \mathcal{I}^{v;\varrho}_{w \circ \varrho}(u \circ \varrho)\right)\left(\varrho^{-1}(c_2)\right)$$

$$+ w(c_1) \left({}_{w \circ \varrho}\mathcal{I}^{v;\varrho}_{\varrho^{-1}\left(\frac{c_1+c_2}{2}\right)-}(u \circ \varrho)\right)\left(\varrho^{-1}(c_1)\right). \quad (20)$$

Thus, the proof of the first inequality of (14) is completed.
On the other hand, we can prove the second inequality of (14) by making use of the convexity of u to get

$$u\left(\frac{\kappa}{2}c_1 + \frac{2 - \kappa}{2}c_2\right) + u\left(\frac{2 - \kappa}{2}c_1 + \frac{\kappa}{2}c_2\right) \leq u(c_1) + u(c_2). \quad (21)$$

Multiplying both sides of (21) by $\kappa^{\nu-1} w\left(\frac{\kappa}{2}c_1 + \frac{2-\kappa}{2}c_2\right)$ and integrating with respect to κ over $[0,1]$ to get

$$\int_0^1 \kappa^{\nu-1} u\left(\frac{\kappa}{2}c_1 + \frac{2-\kappa}{2}c_2\right) w\left(\frac{\kappa}{2}c_1 + \frac{2-\kappa}{2}c_2\right) d\kappa$$
$$+ \int_0^1 \kappa^{\nu-1} u\left(\frac{2-\kappa}{2}c_1 + \frac{\kappa}{2}c_2\right) w\left(\frac{\kappa}{2}c_1 + \frac{2-\kappa}{2}c_2\right) d\kappa$$
$$\leq (u(c_1) + u(c_2)) \int_0^1 \kappa^{\nu-1} w\left(\frac{\kappa}{2}c_1 + \frac{2-\kappa}{2}c_2\right) d\kappa. \quad (22)$$

Then, by using (12) and (19) in (22), we get

$$w(c_2) \left(\mathcal{I}^{\nu;\varrho}_{\varrho^{-1}\left(\frac{c_1+c_2}{2}\right)+} (u \circ \varrho) \right) \left(\varrho^{-1}(c_2)\right)$$
$$+ w(c_1) \left({}_{w \circ \varrho}\mathcal{I}^{\nu;\varrho}_{\varrho^{-1}\left(\frac{c_1+c_2}{2}\right)-} (u \circ \varrho) \right) \left(\varrho^{-1}(c_1)\right)$$
$$\leq \frac{u(c_1) + u(c_2)}{2} \left[\left(\mathcal{I}^{\nu;\varrho}_{\varrho^{-1}\left(\frac{c_1+c_2}{2}\right)+} (w \circ \varrho) \right) \left(\varrho^{-1}(c_2)\right) \right.$$
$$\left. + \left(\mathcal{I}^{\nu;\varrho}_{\varrho^{-1}\left(\frac{c_1+c_2}{2}\right)-} (w \circ \varrho) \right) \left(\varrho^{-1}(c_1)\right) \right].$$

This ends our proof. □

Remark 3. *From Theorem 1, we can obtain some special cases as follows:*

(i) *If $\varrho(x) = x$, then inequality (14) becomes*

$$u\left(\frac{c_1+c_2}{2}\right) \left[{}^{RL}\mathcal{I}^{\nu}_{\left(\frac{c_1+c_2}{2}\right)+} w(c_2) + {}^{RL}\mathcal{I}^{\nu}_{\left(\frac{c_1+c_2}{2}\right)-} w(c_1) \right]$$
$$\leq w(c_2) \left({}^{RL}\mathcal{I}^{\nu}_{\left(\frac{c_1+c_2}{2}\right)+w} u \right)(c_2) + w(c_1) \left({}^{RL}_{w}\mathcal{I}^{\nu}_{\left(\frac{c_1+c_2}{2}\right)-} u \right)(c_1)$$
$$\leq \frac{u(c_1) + u(c_2)}{2} \left[{}^{RL}\mathcal{I}^{\nu}_{\left(\frac{c_1+c_2}{2}\right)+} w(c_2) + {}^{RL}\mathcal{I}^{\nu}_{\left(\frac{c_1+c_2}{2}\right)-} w(c_1) \right], \quad (23)$$

where ${}^{RL}\mathcal{I}^{\nu}_{c_1+w}$ and ${}^{RL}_{w}\mathcal{I}^{\nu}_{c_2-}$ are the left- and right-weighted RL fractional integrals, respectively, given by

$$\left({}^{RL}_{c_1+}\mathcal{I}^{\nu}_{w} u \right)(x) = \frac{w^{-1}(x)}{\Gamma(\nu)} \int_{c_1}^{x} (x - \kappa)^{\nu-1} u(\kappa) w(\kappa) d\kappa,$$
$$\left({}^{RL}_{w}\mathcal{I}^{\nu}_{c_2-} u \right)(x) = \frac{w^{-1}(x)}{\Gamma(\nu)} \int_{x}^{c_2} (\kappa - x)^{\nu-1} u(\kappa) w(\kappa) d\kappa, \quad \nu > 0.$$

(ii) *If $\varrho(x) = x$ and $\nu = 1$, then inequality (14) becomes the inequality in (7).*

(iii) *If $\varrho(x) = x$ and $w(x) = 1$, then inequality (14) becomes the inequality in (5).*

(iv) *If $\varrho(x) = x$, $w(x) = 1$ and $\nu = 1$, then inequality (14) becomes the inequality in (2).*

Lemma 2. *Let $0 \leq c_1 < c_2$, let $u : [c_1, c_2] \to \mathbb{R}$ be a continuous with a derivative $u' \in L^1[c_1, c_2]$ such that $u(x) = u(c_1) + \int_{c_1}^{x} u'(\kappa) d\kappa$ and let $w : [c_1, c_2] \to \mathbb{R}$ be an integrable, positive and weighted symmetric function with respect to $\frac{c_1+c_2}{2}$. If ϱ is a continuous increasing mapping from* the interval $[c_1, c_2]$ onto itself with a derivative $\varrho'(x)$ which is continuous on (c_1, c_2), then for $\nu > 0$, the following equality is valid:

$$u\left(\frac{c_1+c_2}{2}\right)\left[\left(\mathcal{I}^{\nu:\varrho}_{\varrho^{-1}\left(\frac{c_1+c_2}{2}\right)+}(w\circ\varrho)\right)\left(\varrho^{-1}(c_2)\right)\right.$$

$$\left.+\left(\mathcal{I}^{\nu:\varrho}_{\varrho^{-1}\left(\frac{c_1+c_2}{2}\right)-}(w\circ\varrho)\right)\left(\varrho^{-1}(c_1)\right)\right]$$

$$-\left[w(c_2)\left(\mathcal{I}^{\nu:\varrho}_{\varrho^{-1}\left(\frac{c_1+c_2}{2}\right)+}(u\circ\varrho)\right)\left(\varrho^{-1}(c_2)\right)\right.$$

$$\left.+w(c_1)\left(w\circ\mathcal{I}^{\nu:\varrho}_{\varrho^{-1}\left(\frac{c_1+c_2}{2}\right)-}(u\circ\varrho)\right)\left(\varrho^{-1}(c_1)\right)\right]$$

$$=\frac{1}{\Gamma(\nu)}\int_{\varrho^{-1}(c_1)}^{\varrho^{-1}\left(\frac{c_1+c_2}{2}\right)}\left[\int_{\varrho^{-1}(c_1)}^{\kappa}\varrho'(x)(\varrho(x)-c_1)^{\nu-1}(w\circ\varrho)(x)dx\right](u'\circ\varrho)(\kappa)\varrho'(\kappa)d\kappa$$

$$-\frac{1}{\Gamma(\nu)}\int_{\varrho^{-1}\left(\frac{c_1+c_2}{2}\right)}^{\varrho^{-1}(c_2)}\left[\int_{\kappa}^{\varrho^{-1}(c_2)}\varrho'(x)(c_2-\varrho(x))^{\nu-1}(w\circ\varrho)(x)dx\right](u'\circ\varrho)(\kappa)\varrho'(\kappa)d\kappa. \quad (24)$$

Proof. Let us set

$$\frac{1}{\Gamma(\nu)}\int_{\varrho^{-1}(c_1)}^{\varrho^{-1}\left(\frac{c_1+c_2}{2}\right)}\left[\int_{\varrho^{-1}(c_1)}^{\kappa}\varrho'(x)(\varrho(x)-c_1)^{\nu-1}(w\circ\varrho)(x)dx\right](u'\circ\varrho)(\kappa)\varrho'(\kappa)d\kappa$$

$$-\frac{1}{\Gamma(\nu)}\int_{\varrho^{-1}\left(\frac{c_1+c_2}{2}\right)}^{\varrho^{-1}(c_2)}\left[\int_{\kappa}^{\varrho^{-1}(c_2)}\varrho'(x)(c_2-\varrho(x))^{\nu-1}(w\circ\varrho)(x)dx\right](u'\circ\varrho)(\kappa)\varrho'(\kappa)d\kappa$$

$$=\frac{1}{\Gamma(\nu)}\int_{\varrho^{-1}(c_1)}^{\varrho^{-1}\left(\frac{c_1+c_2}{2}\right)}\left[\int_{\varrho^{-1}(c_1)}^{\kappa}\varrho'(x)(\varrho(x)-c_1)^{\nu-1}(w\circ\varrho)(x)dx\right](u'\circ\varrho)(\kappa)\varrho'(\kappa)d\kappa$$

$$+\frac{-1}{\Gamma(\nu)}\int_{\varrho^{-1}\left(\frac{c_1+c_2}{2}\right)}^{\varrho^{-1}(c_2)}\left[\int_{\kappa}^{\varrho^{-1}(c_2)}\varrho'(x)(c_2-\varrho(x))^{\nu-1}(w\circ\varrho)(x)dx\right](u'\circ\varrho)(\kappa)\varrho'(\kappa)d\kappa$$

$$:= \Xi_1 + \Xi_2.$$

By integrating by parts, using Lemma 1, and (10) and (11), we obtain

$$\Xi_1 = \frac{1}{\Gamma(\nu)}\left(\int_{\varrho^{-1}(c_1)}^{\kappa}\varrho'(x)(\varrho(x)-c_1)^{\nu-1}(w\circ\varrho)(x)dx\right)(u\circ\varrho)(\kappa)d\kappa\bigg|_{\kappa=\varrho^{-1}(c_1)}^{\varrho^{-1}\left(\frac{c_1+c_2}{2}\right)}$$

$$-\frac{1}{\Gamma(\nu)}\int_{\varrho^{-1}(c_1)}^{\varrho^{-1}\left(\frac{c_1+c_2}{2}\right)}\varrho'(\kappa)(\varrho(\kappa)-c_1)^{\nu-1}(w\circ\varrho)(\kappa)(u\circ\varrho)(\kappa)d\kappa$$

$$=\left(\frac{1}{\Gamma(\nu)}\int_{\varrho^{-1}(c_1)}^{\varrho^{-1}\left(\frac{c_1+c_2}{2}\right)}\varrho'(x)(\varrho(x)-c_1)^{\nu-1}(w\circ\varrho)(x)dx\right)u\left(\frac{c_1+c_2}{2}\right)$$

$$-\underbrace{w(c_1)\frac{(w\circ\varrho)^{-1}(\varrho^{-1}(c_1))}{\Gamma(\nu)}\int_{\varrho^{-1}(c_1)}^{\varrho^{-1}\left(\frac{c_1+c_2}{2}\right)}\varrho'(\kappa)(\varrho(\kappa)-c_1)^{\nu-1}(w\circ\varrho)(\kappa)(u\circ\varrho)(\kappa)d\kappa}_{\text{by using (18)}}$$

$$= u\left(\frac{c_1+c_2}{2}\right)\left(\mathcal{I}^{\nu;\varrho}_{\varrho^{-1}\left(\frac{c_1+c_2}{2}\right)-}(w\circ\varrho)\right)\left(\varrho^{-1}(c_1)\right)$$

$$- w(c_1)\left({}_{w\circ\varrho}\mathcal{I}^{\nu;\varrho}_{\varrho^{-1}\left(\frac{c_1+c_2}{2}\right)-}(u\circ\varrho)\right)\left(\varrho^{-1}(c_1)\right)$$

$$= \frac{1}{2}u\left(\frac{c_1+c_2}{2}\right)\left[\left(\mathcal{I}^{\nu;\varrho}_{\varrho^{-1}\left(\frac{c_1+c_2}{2}\right)+}(w\circ\varrho)\right)\left(\varrho^{-1}(c_2)\right)\right.$$

$$+ \left(\mathcal{I}^{\nu;\varrho}_{\varrho^{-1}\left(\frac{c_1+c_2}{2}\right)-}(w\circ\varrho)\right)\left(\varrho^{-1}(c_1)\right)\right] - w(c_1)\left({}_{w\circ\varrho}\mathcal{I}^{\nu;\varrho}_{\varrho^{-1}\left(\frac{c_1+c_2}{2}\right)-}(u\circ\varrho)\right)\left(\varrho^{-1}(c_1)\right).$$

Analogously, we get

$$\Xi_2 = \frac{-1}{\Gamma(\nu)}\left(\int_\kappa^{\varrho^{-1}(c_2)} \varrho'(x)(c_2-\varrho(x))^{\nu-1}(w\circ\varrho)(x)dx\right)(u\circ\varrho)(\kappa)d\kappa\Bigg|_{t=\varrho^{-1}\left(\frac{c_1+c_2}{2}\right)}^{\varrho^{-1}(c_2)}$$

$$- \frac{1}{\Gamma(\nu)}\int_{\varrho^{-1}\left(\frac{c_1+c_2}{2}\right)}^{\varrho^{-1}(c_2)} \varrho'(\kappa)(c_2-\varrho(\kappa))^{\nu-1}(w\circ\varrho)(\kappa)(u\circ\varrho)(\kappa)d\kappa$$

$$= \left(\frac{1}{\Gamma(\nu)}\int_{\varrho^{-1}\left(\frac{c_1+c_2}{2}\right)}^{\varrho^{-1}(c_2)} \varrho'(x)(c_2-\varrho(x))^{\nu-1}(w\circ\varrho)(x)dx\right)u\left(\frac{c_1+c_2}{2}\right)$$

$$- w(c_2)\underbrace{\frac{(w\circ\varrho)^{-1}(\varrho^{-1}(c_2))}{\Gamma(\nu)}\int_{\varrho^{-1}\left(\frac{c_1+c_2}{2}\right)}^{\varrho^{-1}(c_2)} \varrho'(\kappa)(c_2-\varrho(\kappa))^{\nu-1}(w\circ\varrho)(\kappa)(u\circ\varrho)(\kappa)d\kappa}_{\text{by using (18)}}$$

$$= u\left(\frac{c_1+c_2}{2}\right)\left(\mathcal{I}^{\nu;\varrho}_{\varrho^{-1}\left(\frac{c_1+c_2}{2}\right)+}(w\circ\varrho)\right)\left(\varrho^{-1}(c_2)\right)$$

$$- w(c_2)\left(\mathcal{I}^{\nu;\varrho}_{\varrho^{-1}\left(\frac{c_1+c_2}{2}\right)+\,w\circ\varrho}(u\circ\varrho)\right)\left(\varrho^{-1}(c_2)\right)$$

$$= \frac{1}{2}u\left(\frac{c_1+c_2}{2}\right)\left[\left(\mathcal{I}^{\nu;\varrho}_{\varrho^{-1}\left(\frac{c_1+c_2}{2}\right)+}(w\circ\varrho)\right)\left(\varrho^{-1}(c_2)\right)\right.$$

$$+ \left(\mathcal{I}^{\nu;\varrho}_{\varrho^{-1}(c_2)-}(w\circ\varrho)\right)\left(\varrho^{-1}(c_1)\right)\right] - w(c_2)\left(\mathcal{I}^{\nu;\varrho}_{\varrho^{-1}\left(\frac{c_1+c_2}{2}\right)+\,w\circ\varrho}(u\circ\varrho)\right)\left(\varrho^{-1}(c_2)\right).$$

Thus, we deduce:

$$\Xi_1 + \Xi_2 = u\left(\frac{c_1+c_2}{2}\right)\left[\left(\mathcal{I}^{\nu;\varrho}_{\varrho^{-1}\left(\frac{c_1+c_2}{2}\right)+}(w\circ\varrho)\right)\left(\varrho^{-1}(c_2)\right)\right.$$

$$+ \left(\mathcal{I}^{\nu;\varrho}_{\varrho^{-1}\left(\frac{c_1+c_2}{2}\right)-}(w\circ\varrho)\right)\left(\varrho^{-1}(c_1)\right)\right] - \left[w(c_2)\left(\mathcal{I}^{\nu;\varrho}_{\varrho^{-1}\left(\frac{c_1+c_2}{2}\right)+\,w\circ\varrho}(u\circ\varrho)\right)\left(\varrho^{-1}(c_2)\right)\right.$$

$$+ w(c_1)\left({}_{w\circ\varrho}\mathcal{I}^{\nu;\varrho}_{\varrho^{-1}\left(\frac{c_1+c_2}{2}\right)-}(u\circ\varrho)\right)\left(\varrho^{-1}(c_1)\right)\right],$$

which completes the proof of Lemma 2. □

Remark 4. *From Lemma 2, we can obtain some special cases as follows:*

(i) If $\varrho(x) = x$, then equality (24) becomes

$$u\left(\frac{c_1+c_2}{2}\right)\left[{}^{RL}_{\left(\frac{c_1+c_2}{2}\right)_+}\mathcal{I}^\nu w(c_2) + {}^{RL}\mathcal{I}^\nu_{\left(\frac{c_1+c_2}{2}\right)_-}w(c_1)\right]$$

$$- \left[w(c_2)\left({}^{RL}_{\left(\frac{c_1+c_2}{2}\right)_+}\mathcal{I}^\nu_w u\right)(c_2) + w(c_1)\left({}^{RL}\mathcal{I}^\nu_w{}_{\left(\frac{c_1+c_2}{2}\right)_-}u\right)(c_1)\right]$$

$$= \frac{1}{\Gamma(\nu)}\int_{c_1}^{\frac{c_1+c_2}{2}}\left[\int_{c_1}^{\kappa}(x-c_1)^{\nu-1}w(x)dx\right]u'(\kappa)d\kappa$$

$$- \frac{1}{\Gamma(\nu)}\int_{\frac{c_1+c_2}{2}}^{c_2}\left[\int_{\kappa}^{c_2}(c_2-x)^{\nu-1}w(x)dx\right]u'(\kappa)d\kappa, \quad (25)$$

where ${}^{RL}_{\left(\frac{c_1+c_2}{2}\right)_+}\mathcal{I}^\nu_w$ and ${}^{RL}\mathcal{I}^\nu_w{}_{\left(\frac{c_1+c_2}{2}\right)_-}$ are as defined in Remark 3.

(ii) If $\varrho(x) = x$ and $w(x) = 1$, then equality (24) becomes

$$\frac{2^{\nu-1}\Gamma(\nu+1)}{(c_2-c_1)^\nu}\left[{}^{RL}_{\left(\frac{c_1+c_2}{2}\right)_+}\mathcal{I}^\nu u(c_2) + {}^{RL}\mathcal{I}^\nu_{\left(\frac{c_1+c_2}{2}\right)_-}u(c_1)\right] - u\left(\frac{c_1+c_2}{2}\right) = \frac{c_2-c_1}{4}$$

$$\times \left[\int_0^1 \kappa^\nu u'\left(\frac{\kappa}{2}c_1 + \frac{2-\kappa}{2}c_2\right)d\kappa - \int_0^1 \kappa^\nu u'\left(\frac{2-\kappa}{2}c_1 + \frac{\kappa}{2}c_2\right)d\kappa\right],$$

which is already obtained in ([39] [Lemma 3]).

(iii) If $\varrho(x) = x$, $w(x) = 1$ and $\nu = 1$, then equality (24) becomes

$$\frac{1}{c_2-c_1}\int_{c_1}^{c_2}u(x)dx - u\left(\frac{c_1+c_2}{2}\right) = \frac{c_2-c_1}{4}\left[\int_0^1 \kappa u'\left(\frac{\kappa}{2}c_1 + \frac{2-\kappa}{2}c_2\right)d\kappa\right.$$

$$\left. - \int_0^1 \kappa u'\left(\frac{2-\kappa}{2}c_1 + \frac{\kappa}{2}c_2\right)d\kappa\right], \quad (26)$$

which is already obtained in ([39] [Corollary 1]).

3. Main Results

By the help of Lemma 2, we can deduce the following HHF inequalities.

Theorem 2. Let $0 \leq c_1 < c_2$, let $u : [c_1, c_2] \subseteq [0, \infty) \to \mathcal{R}$ be a (continuously) differentiable function on the interval $[c_1, c_2]$ such that $u(x) = u(c_1) + \int_{c_1}^x u'(\kappa)d\kappa$, and let $w : [c_1, c_2] \to \mathcal{R}$ be an integrable, positive and weighted symmetric function with respect to $\frac{c_1+c_2}{2}$. If, in addition, $|u'|$ is convex on $[c_1, c_2]$, and ϱ is an increasing and positive function from $[c_1, c_2)$ onto itself such that its derivative $\varrho'(x)$ is continuous on (c_1, c_2), then for $\nu > 0$ the following inequalities are valid:

$$|\Xi_1 + \Xi_2| = \left|\frac{1}{\Gamma(\nu)}\int_{\varrho^{-1}(c_1)}^{\varrho^{-1}\left(\frac{c_1+c_2}{2}\right)}\left[\int_{\varrho^{-1}(c_1)}^{\kappa}\varrho'(x)(\varrho(x)-c_1)^{\nu-1}(w \circ \varrho)(x)dx\right]\right.$$

$$\times (u' \circ \varrho)(\kappa)\varrho'(\kappa)d\kappa$$

$$\left. - \frac{1}{\Gamma(\nu)}\int_{\varrho^{-1}\left(\frac{c_1+c_2}{2}\right)}^{\varrho^{-1}(c_2)}\left[\int_{\kappa}^{\varrho^{-1}(c_2)}\varrho'(x)(c_2-\varrho(x))^{\nu-1}(w \circ \varrho)(x)dx\right](u' \circ \varrho)(\kappa)\varrho'(\kappa)d\kappa\right|$$

$$\leq \frac{(c_2 - c_1)^{\nu+1}}{2^{\nu+2}\Gamma(\nu+3)} \left\{ \|w\|_{\left[c_1, \frac{c_1+c_2}{2}\right], \infty} [(\nu+3)|u'(c_1)| + (\nu+1)|u'(c_2)|] \right.$$
$$\left. + \|w\|_{\left[\frac{c_1+c_2}{2}, c_2\right], \infty} [(\nu+1)|u'(c_1)| + (\nu+3)|u'(c_2)|] \right\}$$
$$\leq \frac{(c_2 - c_1)^{\nu+1} \|w\|_{[c_1, c_2], \infty}}{2^{\nu+1}\Gamma(\nu+2)} [|u'(c_1)| + |u'(c_2)|]. \quad (27)$$

Proof. By making use of Lemma 2 and properties of the modulus, we obtain

$$|\Xi_1 + \Xi_2|$$
$$= \left| \frac{1}{\Gamma(\nu)} \int_{\varrho^{-1}(c_1)}^{\varrho^{-1}\left(\frac{c_1+c_2}{2}\right)} \left[\int_{\varrho^{-1}(c_1)}^{\kappa} \varrho'(x)(\varrho(x) - c_1)^{\nu-1}(w \circ \varrho)(x) dx \right] (u' \circ \varrho)(\kappa) \varrho'(\kappa) d\kappa \right.$$
$$\left. - \frac{1}{\Gamma(\nu)} \int_{\varrho^{-1}\left(\frac{c_1+c_2}{2}\right)}^{\varrho^{-1}(c_2)} \left[\int_{\kappa}^{\varrho^{-1}(c_2)} \varrho'(x)(c_2 - \varrho(x))^{\nu-1}(w \circ \varrho)(x) dx \right] (u' \circ \varrho)(\kappa) \varrho'(\kappa) d\kappa \right|$$
$$\leq \frac{1}{\Gamma(\nu)} \int_{\varrho^{-1}(c_1)}^{\varrho^{-1}\left(\frac{c_1+c_2}{2}\right)} \left| \int_{\varrho^{-1}(c_1)}^{\kappa} \varrho'(x)(\varrho(x) - c_1)^{\nu-1}(w \circ \varrho)(x) dx \right| |(u' \circ \varrho)(\kappa)| \varrho'(\kappa) d\kappa$$
$$+ \frac{1}{\Gamma(\nu)} \int_{\varrho^{-1}\left(\frac{c_1+c_2}{2}\right)}^{\varrho^{-1}(c_2)} \left| \int_{\kappa}^{\varrho^{-1}(c_2)} \varrho'(x)(c_2 - \varrho(x))^{\nu-1}(w \circ \varrho)(x) dx \right|$$
$$\times |(u' \circ \varrho)(\kappa)| \varrho'(\kappa) d\kappa. \quad (28)$$

Since $|u'|$ is convex on $[c_1, c_2]$, we get for $\kappa \in [\varrho^{-1}(c_1), \varrho^{-1}(c_2)]$:

$$|(u' \circ \varrho)(\kappa)| = \left| u'\left(\frac{c_2 - \varrho(\kappa)}{c_2 - c_1} c_1 + \frac{\varrho(\kappa) - c_1}{c_2 - c_1} c_2 \right) \right|$$
$$\leq \frac{c_2 - \varrho(\kappa)}{c_2 - c_1} |u'(c_1)| + \frac{\varrho(\kappa) - c_1}{c_2 - c_1} |u'(c_2)|. \quad (29)$$

Hence, we obtain

$$|\Xi_1 + \Xi_2| \leq \frac{\|w\|_{\left[c_1, \frac{c_1+c_2}{2}\right], \infty}}{(c_2 - c_1)\Gamma(\nu)} \int_{\varrho^{-1}(c_1)}^{\varrho^{-1}\left(\frac{c_1+c_2}{2}\right)} \left| \int_{\varrho^{-1}(c_1)}^{\kappa} \varrho'(x)(\varrho(x) - c_1)^{\nu-1} dx \right|$$
$$\times [(c_2 - \varrho(\kappa))|u'(c_1)| + (\varrho(\kappa) - c_1)|u'(c_2)|] \varrho'(\kappa) d\kappa$$
$$+ \frac{\|w\|_{\left[\frac{c_1+c_2}{2}, c_2\right], \infty}}{(c_2 - c_1)\Gamma(\nu)} \int_{\varrho^{-1}\left(\frac{c_1+c_2}{2}\right)}^{\varrho^{-1}(c_2)} \left| \int_{\kappa}^{\varrho^{-1}(c_2)} \varrho'(x)(c_2 - \varrho(x))^{\nu-1} dx \right|$$
$$\times [(c_2 - \varrho(\kappa))|u'(c_1)| + (\varrho(\kappa) - c_1)|u'(c_2)|] \varrho'(\kappa) d\kappa$$
$$= \frac{(c_2 - c_1)^{\nu+1}}{2^{\nu+2}\Gamma(\nu+3)} \left\{ \|w\|_{\left[c_1, \frac{c_1+c_2}{2}\right], \infty} [(\nu+3)|u'(c_1)| + (\nu+1)|u'(c_2)|] \right.$$
$$\left. + \|w\|_{\left[\frac{c_1+c_2}{2}, c_2\right], \infty} [(\nu+1)|u'(c_1)| + (\nu+3)|u'(c_2)|] \right\}$$
$$\leq \frac{(c_2 - c_1)^{\nu+1} \|w\|_{[c_1, c_2], \infty}}{2^{\nu+1}\Gamma(\nu+2)} [|u'(c_1)| + |u'(c_2)|], \quad (30)$$

where

$$\int_{\varrho^{-1}(c_1)}^{\kappa} \varrho'(x)(\varrho(x) - c_1)^{\nu-1} dx = \frac{(\varrho(\kappa) - c_1)^{\nu}}{\nu};$$

$$\int_{\kappa}^{\varrho^{-1}(c_2)} \varrho'(x)(c_2 - \varrho(x))^{\nu-1} dx = \frac{(c_2 - \varrho(\kappa))^{\nu}}{\nu};$$

$$\int_{\varrho^{-1}(c_1)}^{\varrho^{-1}\left(\frac{c_1+c_2}{2}\right)} (\varrho(\kappa) - c_1)^{\nu+1} \varrho'(\kappa) d\kappa = \int_{\varrho^{-1}\left(\frac{c_1+c_2}{2}\right)}^{\varrho^{-1}(c_2)} (c_2 - \varrho(\kappa))^{\nu+1} \varrho'(\kappa) d\kappa = \frac{(c_2 - c_1)^{\nu+2}}{2^{\nu+2}(\nu + 2)};$$

$$\int_{\varrho^{-1}(c_1)}^{\varrho^{-1}\left(\frac{c_1+c_2}{2}\right)} (\varrho(\kappa) - c_1)^{\nu}(c_2 - \varrho(\kappa))\varrho'(\kappa) d\kappa$$

$$= \int_{\varrho^{-1}\left(\frac{c_1+c_2}{2}\right)}^{\varrho^{-1}(c_2)} (c_2 - \varrho(\kappa))^{\nu}(\varrho(\kappa) - c_1)\varrho'(\kappa) d\kappa = \frac{(c_2 - c_1)^{\nu+2}(\nu + 3)}{2^{\nu+2}(\nu + 1)(\nu + 2)}.$$

This completes our proof. □

Remark 5. *From Theorem 2, we can obtain some special cases as follows:*

(i) If $\varrho(x) = x$, then inequality (27) becomes

$$\left| u\left(\frac{c_1 + c_2}{2}\right)\left[{}^{RL}\mathcal{I}^{\nu}_{\left(\frac{c_1+c_2}{2}\right)^+} w(c_2) + {}^{RL}\mathcal{I}^{\nu}_{\left(\frac{c_1+c_2}{2}\right)^-} w(c_1) \right] \right.$$

$$\left. - \left[w(c_2)\left({}^{RL}\mathcal{I}^{\nu}_{\left(\frac{c_1+c_2}{2}\right)^+} u\right)(c_2) + w(c_1)\left({}^{RL}\mathcal{I}^{\nu}_{\left(\frac{c_1+c_2}{2}\right)^-} u\right)(c_1) \right] \right|$$

$$\leq \frac{(c_2 - c_1)^{\nu+1}}{2^{\nu+2}\Gamma(\nu + 3)} \left\{ \|w\|_{[c_1, \frac{c_1+c_2}{2}], \infty} [(\nu + 3)|u'(c_1)| + (\nu + 1)|u'(c_2)|] \right.$$

$$\left. + \|w\|_{[\frac{c_1+c_2}{2}, c_2], \infty} [(\nu + 1)|u'(c_1)| + (\nu + 3)|u'(c_2)|] \right\}$$

$$\leq \frac{(c_2 - c_1)^{\nu+1} \|w\|_{[c_1, c_2], \infty}}{2^{\nu+1}\Gamma(\nu + 2)} [|u'(c_1)| + |u'(c_2)|]. \quad (31)$$

(ii) If $\varrho(x) = x$ and $w(x) = 1$, then inequality (27) becomes

$$\left| \frac{2^{\nu-1}\Gamma(\nu + 1)}{(c_2 - c_1)^{\nu}} \left[{}^{RL}\mathcal{I}^{\nu}_{\left(\frac{c_1+c_2}{2}\right)^+} u(c_2) + {}^{RL}\mathcal{I}^{\nu}_{\left(\frac{c_1+c_2}{2}\right)^-} u(c_1) \right] - u\left(\frac{c_1 + c_2}{2}\right) \right|$$

$$\leq \frac{(c_2 - c_1)^{\nu+1}}{2^{\nu+2}\Gamma(\nu + 3)} \left\{ [(\nu + 3)|u'(c_1)| + (\nu + 1)|u'(c_2)|] \right.$$

$$\left. + [(\nu + 1)|u'(c_1)| + (\nu + 3)|u'(c_2)|] \right\} \leq \frac{(c_2 - c_1)^{\nu+1}}{2^{\nu+1}\Gamma(\nu + 2)} [|u'(c_1)| + |u'(c_2)|], \quad (32)$$

which is already obtained in ([39] [Theorem 5]).

(iii) If $\varrho(x) = x$, $w(x) = 1$ and $\nu = 1$, then inequality (27) becomes

$$\left| \frac{1}{c_2 - c_1} \int_{c_1}^{c_2} u(x) dx - u\left(\frac{c_1 + c_2}{2}\right) \right| \leq \frac{c_2 - c_1}{8} [|u'(c_1)| + |u'(c_2)|], \quad (33)$$

which is already obtained in ([45] [Theorem 2.2]).

Theorem 3. Let $0 \leq c_1 < c_2$, let $u : [c_1, c_2] \subseteq [0, \infty) \to \mathcal{R}$ be a (continuously) differentiable function on the interval $[c_1, c_2]$ such that $u(x) = u(c_1) + \int_{c_1}^{x} u'(\kappa) d\kappa$, and let $w : [c_1, c_2] \to \mathcal{R}$ be an integrable, positive and weighted symmetric function with respect to $\frac{c_1+c_2}{2}$. If, in addition, $|u'|^q$ is convex on $[c_1, c_2]$ with $q \geq 1$, and ϱ is an increasing and positive function from $[c_1, c_2]$ onto itself such that its derivative $\varrho'(x)$ is continuous on (c_1, c_2), then for $\nu > 0$, the following inequalities are valid:

$$|\Xi_1 + \Xi_2| \leq \frac{(c_2 - c_1)^{\nu+1}}{2^{\nu+1+\frac{1}{q}}(\nu+2)^{\frac{1}{q}}\Gamma(\nu+2)}$$

$$\times \left\{ \|w\|_{[c_1, \frac{c_1+c_2}{2}], \infty} \left[(\nu+3)|u'(c_1)|^q + (\nu+1)|u'(c_2)|^q \right]^{\frac{1}{q}} \right.$$

$$\left. + \|w\|_{[\frac{c_1+c_2}{2}, c_2], \infty} \left[(\nu+1)|u'(c_1)|^q + (\nu+3)|u'(c_2)|^q \right]^{\frac{1}{q}} \right\}$$

$$\leq \frac{(c_2 - c_1)^{\nu+1}\|w\|_{[c_1, c_2], \infty}}{2^{\nu+1+\frac{1}{q}}(\nu+2)^{\frac{1}{q}}\Gamma(\nu+2)} \left\{ \left[(\nu+3)|u'(c_1)|^q + (\nu+1)|u'(c_2)|^q \right]^{\frac{1}{q}} \right.$$

$$\left. + \left[(\nu+1)|u'(c_1)|^q + (\nu+3)|u'(c_2)|^q \right]^{\frac{1}{q}} \right\}. \quad (34)$$

Proof. Since $|u'|^q$ is convex on $[c_1, c_2]$, we get for $\kappa \in [\varrho^{-1}(c_1), \varrho^{-1}(c_2)]$:

$$|(u' \circ \varrho)(\kappa)|^q = \left| u'\left(\frac{c_2 - \varrho(\kappa)}{c_2 - c_1} c_1 + \frac{\varrho(\kappa) - c_1}{c_2 - c_1} c_2 \right) \right|^q$$

$$\leq \frac{c_2 - \varrho(\kappa)}{c_2 - c_1} |u'(c_1)|^q + \frac{\varrho(\kappa) - c_1}{c_2 - c_1} |u'(c_2)|^q. \quad (35)$$

By making use of Lemma 2, power mean inequality and convexity of $|u'|^q$, we get

$$|\Xi_1 + \Xi_2|$$

$$\leq \frac{1}{\Gamma(\nu)} \int_{\varrho^{-1}(c_1)}^{\varrho^{-1}\left(\frac{c_1+c_2}{2}\right)} \left| \int_{\varrho^{-1}(c_1)}^{\kappa} \varrho'(x)(\varrho(x) - c_1)^{\nu-1}(w \circ \varrho)(x)dx \right| |(u' \circ \varrho)(\kappa)|\varrho'(\kappa)d\kappa$$

$$+ \frac{1}{\Gamma(\nu)} \int_{\varrho^{-1}\left(\frac{c_1+c_2}{2}\right)}^{\varrho^{-1}(c_2)} \left| \int_{\kappa}^{\varrho^{-1}(c_2)} \varrho'(x)(c_2 - \varrho(x))^{\nu-1}(w \circ \varrho)(x)dx \right| |(u' \circ \varrho)(\kappa)|\varrho'(\kappa)d\kappa$$

$$\leq \frac{1}{\Gamma(\nu)} \left(\int_{\varrho^{-1}(c_1)}^{\varrho^{-1}\left(\frac{c_1+c_2}{2}\right)} \left| \int_{\varrho^{-1}(c_1)}^{\kappa} \varrho'(x)(\varrho(x) - c_1)^{\nu-1}(w \circ \varrho)(x)dx \right| \varrho'(\kappa)d\kappa \right)^{1-\frac{1}{q}}$$

$$\times \left(\int_{\varrho^{-1}(c_1)}^{\varrho^{-1}\left(\frac{c_1+c_2}{2}\right)} \left| \int_{\varrho^{-1}(c_1)}^{\kappa} \varrho'(x)(\varrho(x) - c_1)^{\nu-1}(w \circ \varrho)(x)dx \right| |(u' \circ \varrho)(\kappa)|^q \varrho'(\kappa)d\kappa \right)^{\frac{1}{q}}$$

$$+ \frac{1}{\Gamma(\nu)} \left(\int_{\varrho^{-1}\left(\frac{c_1+c_2}{2}\right)}^{\varrho^{-1}(c_2)} \left| \int_{\kappa}^{\varrho^{-1}(c_2)} \varrho'(x)(c_2 - \varrho(x))^{\nu-1}(w \circ \varrho)(x)dx \right| \varrho'(\kappa)d\kappa \right)^{1-\frac{1}{q}}$$

$$\times \left(\int_{\varrho^{-1}\left(\frac{c_1+c_2}{2}\right)}^{\varrho^{-1}(c_2)} \left| \int_{\kappa}^{\varrho^{-1}(c_2)} \varrho'(x)(c_2 - \varrho(x))^{\nu-1}(w \circ \varrho)(x)dx \right| |(u' \circ \varrho)(\kappa)|^q \varrho'(\kappa)d\kappa \right)^{\frac{1}{q}}$$

$$
\begin{aligned}
&\leq \frac{\|w\|_{\left[c_1,\frac{c_1+c_2}{2}\right],\infty}}{\Gamma(\nu)}\\
&\quad\times \left(\int_{\varrho^{-1}(c_1)}^{\varrho^{-1}\left(\frac{c_1+c_2}{2}\right)}\left|\int_{\varrho^{-1}(c_1)}^{\kappa}\varrho'(x)(\varrho(x)-c_1)^{\nu-1}dx\right|\varrho'(\kappa)d\kappa\right)^{1-\frac{1}{q}}\\
&\quad\times \left(\int_{\varrho^{-1}(c_1)}^{\varrho^{-1}\left(\frac{c_1+c_2}{2}\right)}\left|\int_{\varrho^{-1}(c_1)}^{\kappa}\varrho'(x)(\varrho(x)-c_1)^{\nu-1}dx\right||(u'\circ\varrho)(\kappa)|^q\varrho'(\kappa)d\kappa\right)^{\frac{1}{q}}\\
&+\frac{\|w\|_{\left[\frac{c_1+c_2}{2},c_2\right],\infty}}{\Gamma(\nu)}\\
&\quad\times \left(\int_{\varrho^{-1}\left(\frac{c_1+c_2}{2}\right)}^{\varrho^{-1}(c_2)}\left|\int_{\kappa}^{\varrho^{-1}(c_2)}\varrho'(x)(c_2-\varrho(x))^{\nu-1}dx\right|\varrho'(\kappa)d\kappa\right)^{1-\frac{1}{q}}\\
&\quad\times \left(\int_{\varrho^{-1}\left(\frac{c_1+c_2}{2}\right)}^{\varrho^{-1}(c_2)}\left|\int_{\kappa}^{\varrho^{-1}(c_2)}\varrho'(x)(c_2-\varrho(x))^{\nu-1}dx\right||(u'\circ\varrho)(\kappa)|^q\varrho'(\kappa)d\kappa\right)^{\frac{1}{q}}\\
&\leq \frac{\|w\|_{\left[c_1,\frac{c_1+c_2}{2}\right],\infty}}{\Gamma(\nu)}\\
&\quad \left(\int_{\varrho^{-1}(c_1)}^{\varrho^{-1}\left(\frac{c_1+c_2}{2}\right)}\left|\int_{\varrho^{-1}(c_1)}^{\kappa}\varrho'(x)(\varrho(x)-c_1)^{\nu-1}dx\right|\varrho'(\kappa)d\kappa\right)^{1-\frac{1}{q}}\\
&\quad\times \left[\int_{\varrho^{-1}(c_1)}^{\varrho^{-1}\left(\frac{c_1+c_2}{2}\right)}\left|\int_{\varrho^{-1}(c_1)}^{\kappa}\varrho'(x)(\varrho(x)-c_1)^{\nu-1}dx\right|\right.\\
&\quad\left.\times \left(\frac{c_2-\varrho(\kappa)}{c_2-c_1}|u'(c_1)|^q+\frac{\varrho(\kappa)-c_1}{c_2-c_1}|u'(c_2)|^q\right)\varrho'(\kappa)d\kappa\right]^{\frac{1}{q}}\\
&+\frac{\|w\|_{\left[\frac{c_1+c_2}{2},c_2\right],\infty}}{\Gamma(\nu)}\left(\int_{\varrho^{-1}\left(\frac{c_1+c_2}{2}\right)}^{\varrho^{-1}(c_2)}\left|\int_{\kappa}^{\varrho^{-1}(c_2)}\varrho'(x)(c_2-\varrho(x))^{\nu-1}dx\right|\varrho'(\kappa)d\kappa\right)^{1-\frac{1}{q}}\\
&\quad\times \left[\int_{\varrho^{-1}\left(\frac{c_1+c_2}{2}\right)}^{\varrho^{-1}(c_2)}\left|\int_{\kappa}^{\varrho^{-1}(c_2)}\varrho'(x)(c_2-\varrho(x))^{\nu-1}dx\right|\right.\\
&\quad\left.\times \left(\frac{c_2-\varrho(\kappa)}{c_2-c_1}|u'(c_1)|^q+\frac{\varrho(\kappa)-c_1}{c_2-c_1}|u'(c_2)|^q\right)\varrho'(\kappa)d\kappa\right]^{\frac{1}{q}}\\
&=\frac{(c_2-c_1)^{\nu+1}}{2^{\nu+1+\frac{1}{q}}(\nu+2)^{\frac{1}{q}}\Gamma(\nu+2)}\\
&\quad\times \left\{\|w\|_{\left[c_1,\frac{c_1+c_2}{2}\right],\infty}\left[(\nu+3)|u'(c_1)|^q+(\nu+1)|u'(c_2)|^q\right]^{\frac{1}{q}}\right.\\
&\quad\left.+\|w\|_{\left[\frac{c_1+c_2}{2},c_2\right],\infty}\left[(\nu+1)|u'(c_1)|^q+(\nu+3)|u'(c_2)|^q\right]^{\frac{1}{q}}\right\}
\end{aligned}
$$

$$\leq \frac{(c_2-c_1)^{\nu+1}\|w\|_{[c_1,c_2],\infty}}{2^{\nu+1+\frac{1}{q}}(\nu+2)^{\frac{1}{q}}\Gamma(\nu+2)}\left\{\left[(\nu+3)|u'(c_1)|^q+(\nu+1)|u'(c_2)|^q\right]^{\frac{1}{q}}\right.$$
$$\left.+\left[(\nu+1)|u'(c_1)|^q+(\nu+3)|u'(c_2)|^q\right]^{\frac{1}{q}}\right\}, \quad (36)$$

where it is easily seen that

$$\int_{\varrho^{-1}(c_1)}^{\varrho^{-1}\left(\frac{c_1+c_2}{2}\right)}\left|\int_{\varrho^{-1}(c_1)}^{\kappa}\varrho'(x)(\varrho(x)-c_1)^{\nu-1}dx\right|\varrho'(\kappa)d\kappa$$
$$=\int_{\varrho^{-1}\left(\frac{c_1+c_2}{2}\right)}^{\varrho^{-1}(c_2)}\left|\int_{\kappa}^{\varrho^{-1}(c_2)}\varrho'(x)(c_2-\varrho(x))^{\nu-1}dx\right|\varrho'(\kappa)d\kappa = \frac{(c_2-c_1)^{\nu+1}}{2^{\nu+1}\nu(\nu+1)}.$$

Hence, the proof is completed. □

Remark 6. *From Theorem 3, we can obtain some special cases as follows:*

(i) *If $\varrho(x) = x$, then inequality (34) becomes*

$$\left|u\left(\frac{c_1+c_2}{2}\right)\left[{}^{RL}_{\left(\frac{c_1+c_2}{2}\right)^+}\mathcal{I}^\nu w(c_2)+{}^{RL}\mathcal{I}^\nu_{\left(\frac{c_1+c_2}{2}\right)^-}w(c_1)\right]\right.$$
$$\left.-\left[w(c_2)\left({}^{RL}_{\left(\frac{c_1+c_2}{2}\right)^+}\mathcal{I}^\nu_w u\right)(c_2)+w(c_1)\left({}^{RL}\mathcal{I}^\nu_{w\left(\frac{c_1+c_2}{2}\right)^-}u\right)(c_1)\right]\right|$$

$$\leq \frac{(c_2-c_1)^{\nu+1}}{2^{\nu+1+\frac{1}{q}}(\nu+2)^{\frac{1}{q}}\Gamma(\nu+2)}$$
$$\times\left\{\|w\|_{\left[c_1,\frac{c_1+c_2}{2}\right],\infty}\left[(\nu+3)|u'(c_1)|^q+(\nu+1)|u'(c_2)|^q\right]^{\frac{1}{q}}\right.$$
$$\left.+\|w\|_{\left[\frac{c_1+c_2}{2},c_2\right],\infty}\left[(\nu+1)|u'(c_1)|^q+(\nu+3)|u'(c_2)|^q\right]^{\frac{1}{q}}\right\}$$
$$\leq \frac{(c_2-c_1)^{\nu+1}\|w\|_{[c_1,c_2],\infty}}{2^{\nu+1+\frac{1}{q}}(\nu+2)^{\frac{1}{q}}\Gamma(\nu+2)}\left\{\left[(\nu+3)|u'(c_1)|^q+(\nu+1)|u'(c_2)|^q\right]^{\frac{1}{q}}\right.$$
$$\left.+\left[(\nu+1)|u'(c_1)|^q+(\nu+3)|u'(c_2)|^q\right]^{\frac{1}{q}}\right\}. \quad (37)$$

(ii) *If $\varrho(x) = x$ and $w(x) = 1$, then inequality (34) becomes*

$$\left|\frac{2^{\nu-1}\Gamma(\nu+1)}{(c_2-c_1)^\nu}\left[{}^{RL}_{\left(\frac{c_1+c_2}{2}\right)^+}\mathcal{I}^\nu u(c_2)+{}^{RL}\mathcal{I}^\nu_{\left(\frac{c_1+c_2}{2}\right)^-}u(c_1)\right]-u\left(\frac{c_1+c_2}{2}\right)\right|$$
$$\leq \frac{(c_2-c_1)^{\nu+1}}{2^{\nu+1+\frac{1}{q}}(\nu+2)^{\frac{1}{q}}\Gamma(\nu+2)}\left\{\left[(\nu+3)|u'(c_1)|^q+(\nu+1)|u'(c_2)|^q\right]^{\frac{1}{q}}\right.$$
$$\left.+\left[(\nu+1)|u'(c_1)|^q+(\nu+3)|u'(c_2)|^q\right]^{\frac{1}{q}}\right\}, \quad (38)$$

which is already obtained in ([39] [Theorem 5]).

(iii) If $\varrho(x) = x$, $w(x) = 1$ and $\nu = 1$, then inequality (34) becomes

$$\left| \frac{1}{c_2 - c_1} \int_{c_1}^{c_2} u(x)dx - u\left(\frac{c_1 + c_2}{2}\right) \right|$$
$$\leq \frac{c_2 - c_1}{8\sqrt[q]{3}} \left\{ \left[|u'(c_1)|^q + 2|u'(c_2)|^q \right]^{\frac{1}{q}} + \left[2|u'(c_1)|^q + |u'(c_2)|^q \right]^{\frac{1}{q}} \right\}. \quad (39)$$

Theorem 4. *Let $0 \leq c_1 < c_2$, let $u : [c_1, c_2] \subseteq [0, \infty) \to \mathcal{R}$ be a (continuously) differentiable function on the interval $[c_1, c_2]$ such that $u(x) = u(c_1) + \int_{c_1}^{x} u'(\kappa) d\kappa$, and let $w : [c_1, c_2] \to \mathcal{R}$ be an integrable, positive and weighted symmetric function with respect to $\frac{c_1 + c_2}{2}$. If, in addition, $|u'|^q$ is convex on $[c_1, c_2]$ with $\frac{1}{p} + \frac{1}{q} = 1$ and $q > 1$, and ϱ is an increasing and positive function from $[c_1, c_2)$ onto itself such that its derivative $\varrho'(x)$ is continuous on (c_1, c_2), then for $\nu > 0$ the following inequalities are valid:*

$$|\Xi_1 + \Xi_2| \leq \frac{(c_2 - c_1)^{\nu+1}}{2^{\nu+1+\frac{2}{q}} (p\nu + 1)^{\frac{1}{p}} \Gamma(\nu + 1)} \left\{ \|w\|_{[c_1, \frac{c_1+c_2}{2}], \infty} \right.$$
$$\times \left[3|u'(c_1)|^q + |u'(c_2)|^q \right]^{\frac{1}{q}} + \|w\|_{[\frac{c_1+c_2}{2}, c_2], \infty} \left[|u'(c_1)|^q + 3|u'(c_2)|^q \right]^{\frac{1}{q}} \right\}$$
$$\leq \frac{(c_2 - c_1)^{\nu+1} \|w\|_{[c_1, c_2], \infty}}{2^{\nu+1+\frac{2}{q}} (p\nu + 1)^{\frac{1}{p}} \Gamma(\nu + 1)}$$
$$\times \left\{ \left[3|u'(c_1)|^q + |u'(c_2)|^q \right]^{\frac{1}{q}} + \left[|u'(c_1)|^q + 3|u'(c_2)|^q \right]^{\frac{1}{q}} \right\}. \quad (40)$$

Proof. Since $|u'|^q$ is convex on $[c_1, c_2]$, we get for $\kappa \in [\varrho^{-1}(c_1), \varrho^{-1}(c_2)]$:

$$|(u' \circ \varrho)(\kappa)|^q = \left| u'\left(\frac{c_2 - \varrho(\kappa)}{c_2 - c_1} c_1 + \frac{\varrho(\kappa) - c_1}{c_2 - c_1} c_2 \right) \right|^q$$
$$\leq \frac{c_2 - \varrho(\kappa)}{c_2 - c_1} |u'(c_1)|^q + \frac{\varrho(\kappa) - c_1}{c_2 - c_1} |u'(c_2)|^q.$$

By using Lemma 2, Hölder's inequality, convexity of $|u'|^q$ and properties of modulus, we have

$$|\Xi_1 + \Xi_2| \leq \frac{1}{\Gamma(\nu)} \int_{\varrho^{-1}(c_1)}^{\varrho^{-1}\left(\frac{c_1+c_2}{2}\right)} \left| \int_{\varrho^{-1}(c_1)}^{\kappa} \varrho'(x)(\varrho(x) - c_1)^{\nu-1} (w \circ \varrho)(x) dx \right|$$
$$\times |(u' \circ \varrho)(\kappa)| \varrho'(\kappa) d\kappa$$
$$+ \frac{1}{\Gamma(\nu)} \int_{\varrho^{-1}\left(\frac{c_1+c_2}{2}\right)}^{\varrho^{-1}(c_2)} \left| \int_{\kappa}^{\varrho^{-1}(c_2)} \varrho'(x)(c_2 - \varrho(x))^{\nu-1} (w \circ \varrho)(x) dx \right| |(u' \circ \varrho)(\kappa)| \varrho'(\kappa) d\kappa$$
$$\leq \frac{1}{\Gamma(\nu)} \left(\int_{\varrho^{-1}(c_1)}^{\varrho^{-1}\left(\frac{c_1+c_2}{2}\right)} \left| \int_{\varrho^{-1}(c_1)}^{\kappa} \varrho'(x)(\varrho(x) - c_1)^{\nu-1} (w \circ \varrho)(x) dx \right|^p \varrho'(\kappa) d\kappa \right)^{\frac{1}{p}}$$

$$\times \left(\int_{\varrho^{-1}(c_1)}^{\varrho^{-1}\left(\frac{c_1+c_2}{2}\right)} |(u' \circ \varrho)(\kappa)|^q \varrho'(\kappa) d\kappa \right)^{\frac{1}{q}}$$

$$+ \frac{1}{\Gamma(\nu)} \left(\int_{\varrho^{-1}\left(\frac{c_1+c_2}{2}\right)}^{\varrho^{-1}(c_2)} \left| \int_{\kappa}^{\varrho^{-1}(c_2)} \varrho'(x)(c_2 - \varrho(x))^{\nu-1} (w \circ \varrho)(x) dx \right|^p \varrho'(\kappa) d\kappa \right)^{\frac{1}{p}}$$

$$\times \left(\int_{\varrho^{-1}\left(\frac{c_1+c_2}{2}\right)}^{\varrho^{-1}(c_2)} |(u' \circ \varrho)(\kappa)|^q \varrho'(\kappa) d\kappa \right)^{\frac{1}{q}}$$

$$\leq \frac{\|w\|_{\left[c_1, \frac{c_1+c_2}{2}\right],\infty}}{\Gamma(\nu)}$$

$$\times \left(\int_{\varrho^{-1}(c_1)}^{\varrho^{-1}\left(\frac{c_1+c_2}{2}\right)} \left| \int_{\varrho^{-1}(c_1)}^{\kappa} \varrho'(x)(\varrho(x) - c_1)^{\nu-1} dx \right|^p \varrho'(\kappa) d\kappa \right)^{\frac{1}{p}}$$

$$\times \left(\int_{\varrho^{-1}(c_1)}^{\varrho^{-1}\left(\frac{c_1+c_2}{2}\right)} |(u' \circ \varrho)(\kappa)|^q \varrho'(\kappa) d\kappa \right)^{\frac{1}{q}} + \frac{\|w\|_{\left[\frac{c_1+c_2}{2}, c_2\right],\infty}}{\Gamma(\nu)}$$

$$\times \left(\int_{\varrho^{-1}\left(\frac{c_1+c_2}{2}\right)}^{\varrho^{-1}(c_2)} \left| \int_{\kappa}^{\varrho^{-1}(c_2)} \varrho'(x)(c_2 - \varrho(x))^{\nu-1} dx \right|^p \varrho'(\kappa) d\kappa \right)^{\frac{1}{p}}$$

$$\times \left(\int_{\varrho^{-1}\left(\frac{c_1+c_2}{2}\right)}^{\varrho^{-1}(c_2)} |(u' \circ \varrho)(\kappa)|^q \varrho'(\kappa) d\kappa \right)^{\frac{1}{q}}$$

$$\leq \frac{\|w\|_{\left[c_1, \frac{c_1+c_2}{2}\right],\infty}}{\Gamma(\nu)}$$

$$\times \left(\int_{\varrho^{-1}(c_1)}^{\varrho^{-1}\left(\frac{c_1+c_2}{2}\right)} \left| \int_{\varrho^{-1}(c_1)}^{\kappa} \varrho'(x)(\varrho(x) - c_1)^{\nu-1} dx \right|^p \varrho'(\kappa) d\kappa \right)^{\frac{1}{p}}$$

$$\times \left[\int_{\varrho^{-1}(c_1)}^{\varrho^{-1}\left(\frac{c_1+c_2}{2}\right)} \left(\frac{c_2 - \varrho(\kappa)}{c_2 - c_1} |u'(c_1)|^q + \frac{\varrho(\kappa) - c_1}{c_2 - c_1} |u'(c_2)|^q \right) \varrho'(\kappa) d\kappa \right]^{\frac{1}{q}}$$

$$+ \frac{\|w\|_{\left[\frac{c_1+c_2}{2}, c_2\right],\infty}}{\Gamma(\nu)}$$

$$\times \left(\int_{\varrho^{-1}\left(\frac{c_1+c_2}{2}\right)}^{\varrho^{-1}(c_2)} \left| \int_{\kappa}^{\varrho^{-1}(c_2)} \varrho'(x)(c_2 - \varrho(x))^{\nu-1} dx \right|^p \varrho'(\kappa) d\kappa \right)^{\frac{1}{p}}$$

$$\times \left[\int_{\varrho^{-1}\left(\frac{c_1+c_2}{2}\right)}^{\varrho^{-1}(c_2)} \left(\frac{c_2 - \varrho(\kappa)}{c_2 - c_1} |u'(c_1)|^q + \frac{\varrho(\kappa) - c_1}{c_2 - c_1} |u'(c_2)|^q \right) \varrho'(\kappa) d\kappa \right]^{\frac{1}{q}}$$

$$= \frac{(c_2 - c_1)^{\nu+1}}{2^{\nu+1+\frac{2}{q}}(p\nu+1)^{\frac{1}{p}}\Gamma(\nu+1)} \left\{ \|w\|_{[c_1, \frac{c_1+c_2}{2}], \infty} [3|u'(c_1)|^q + |u'(c_2)|^q]^{\frac{1}{q}} \right.$$

$$+ \|w\|_{[\frac{c_1+c_2}{2}, c_2], \infty} [|u'(c_1)|^q + 3|u'(c_2)|^q]^{\frac{1}{q}}$$

$$\leq \frac{(c_2 - c_1)^{\nu+1} \|w\|_{[c_1, c_2], \infty}}{2^{\nu+1+\frac{2}{q}}(p\nu+1)^{\frac{1}{p}}\Gamma(\nu+1)}$$

$$\times \left\{ [3|u'(c_1)|^q + |u'(c_2)|^q]^{\frac{1}{q}} + [|u'(c_1)|^q + 3|u'(c_2)|^q]^{\frac{1}{q}} \right\},$$

where we used the identity

$$\int_{\varrho^{-1}(c_1)}^{\varrho^{-1}\left(\frac{c_1+c_2}{2}\right)} \left| \int_{\varrho^{-1}(c_1)}^{\kappa} \varrho'(x)(\varrho(x) - c_1)^{\nu-1} dx \right|^p \varrho'(\kappa) d\kappa$$

$$= \int_{\varrho^{-1}\left(\frac{c_1+c_2}{2}\right)}^{\varrho^{-1}(c_2)} \left| \int_{\kappa}^{\varrho^{-1}(c_2)} \varrho'(x)(c_2 - \varrho(x))^{\nu-1} dx \right|^p \varrho'(\kappa) d\kappa = \frac{(c_2 - c_1)^{p\nu+1}}{2^{p\nu+1}(p\nu+1)\nu^p}.$$

This ends our proof. □

Remark 7. *From Theorem 4, we can obtain some special cases as follows:*

(i) If $\varrho(x) = x$, then inequality (40) becomes

$$\left| u\left(\frac{c_1+c_2}{2}\right) \left[{}^{RL}_{\left(\frac{c_1+c_2}{2}\right)^+} \mathcal{I}^\nu w(c_2) + {}^{RL}\mathcal{I}^\nu_{\left(\frac{c_1+c_2}{2}\right)^-} w(c_1) \right] \right.$$

$$\left. - \left[w(c_2) \left({}^{RL}_{\left(\frac{c_1+c_2}{2}\right)^+} \mathcal{I}^\nu_w u \right)(c_2) + w(c_1) \left({}^{RL}\mathcal{I}^\nu_{w \left(\frac{c_1+c_2}{2}\right)^-} u \right)(c_1) \right] \right|$$

$$\leq \frac{(c_2 - c_1)^{\nu+1}}{2^{\nu+1+\frac{2}{q}}(p\nu+1)^{\frac{1}{p}}\Gamma(\nu+1)} \left\{ \|w\|_{[c_1, \frac{c_1+c_2}{2}], \infty} [3|u'(c_1)|^q + |u'(c_2)|^q]^{\frac{1}{q}} \right.$$

$$+ \|w\|_{[\frac{c_1+c_2}{2}, c_2], \infty} [|u'(c_1)|^q + 3|u'(c_2)|^q]^{\frac{1}{q}} \right\} \leq \frac{(c_2 - c_1)^{\nu+1} \|w\|_{[c_1, c_2], \infty}}{2^{\nu+1+\frac{2}{q}}(p\nu+1)^{\frac{1}{p}}\Gamma(\nu+1)}$$

$$\times \left\{ [3|u'(c_1)|^q + |u'(c_2)|^q]^{\frac{1}{q}} + [|u'(c_1)|^q + 3|u'(c_2)|^q]^{\frac{1}{q}} \right\}.$$

(ii) If $\varrho(x) = x$ and $w(x) = 1$, then inequality (40) becomes

$$\left| \frac{2^{\nu-1}\Gamma(\nu+1)}{(c_2 - c_1)^\nu} \left[{}^{RL}_{\left(\frac{c_1+c_2}{2}\right)^+}\mathcal{I}^\nu u(c_2) + {}^{RL}\mathcal{I}^\nu_{\left(\frac{c_1+c_2}{2}\right)^-} u(c_1) \right] - u\left(\frac{c_1+c_2}{2}\right) \right|$$

$$\leq \frac{(c_2 - c_1)^{\nu+1}\|w\|_{[c_1, c_2], \infty}}{2^{\nu+1+\frac{2}{q}}(p\nu+1)^{\frac{1}{p}}\Gamma(\nu+1)}$$

$$\times \left\{ [3|u'(c_1)|^q + |u'(c_2)|^q]^{\frac{1}{q}} + [|u'(c_1)|^q + 3|u'(c_2)|^q]^{\frac{1}{q}} \right\},$$

which is already obtained in ([39] [Theorem 6]).

(iii) If $\varrho(x) = x$, $w(x) = 1$ and $\nu = 1$, then inequality (40) becomes

$$\left| \frac{1}{c_2 - c_1} \int_{c_1}^{c_2} u(x)dx - u\left(\frac{c_1 + c_2}{2}\right) \right| \leq \frac{c_2 - c_1}{16} \left(\frac{4}{p+1}\right)^{\frac{1}{p}}$$
$$\times \left\{ \left[3|u'(c_1)|^q + |u'(c_2)|^q\right]^{\frac{1}{q}} + \left[|u'(c_1)|^q + 3|u'(c_2)|^q\right]^{\frac{1}{q}} \right\},$$

which is already obtained in ([45] [Theorem 2.3]).

4. Concluding Remarks

In the present article, we have investigated a midpoint fractional HHF integral inequality by using the weighted fractional integrals with positive weighted symmetric function kernels, which is also the midpoint version of (9). Moreover, we have investigated some related results.

The existing versions of HHF integral inequalities (7) and (8) have been successfully applied to other classes of convex functions, see [46–48]. Therefore, our present results can be applied to those classes of convex functions as well.

Furthermore, one can observe that our results in this article are very generic and can be extended to give further potentially useful and interesting HHF integral inequalities of end-midpoint version, like the following one

$$u\left(\frac{c_1 + c_2}{2}\right) \leq \frac{2^{\nu - 1} \Gamma(\nu + 1)}{(c_2 - c_1)^\nu} \left[{}^{RL}\mathcal{I}^\nu_{c_1^+} u\left(\frac{c_1 + c_2}{2}\right) + {}^{RL}\mathcal{I}^\nu_{c_2^-} u\left(\frac{c_1 + c_2}{2}\right) \right]$$
$$\leq \frac{u(c_1) + u(c_2)}{2},$$

which was already established by Mohammed and Brevik in [49].

Author Contributions: Conceptualization, P.O.M., H.A., Y.S.H.; methodology, P.O.M., A.K., H.A.; software, P.O.M., A.K., Y.S.H.; validation, P.O.M., A.K., K.M.A., H.A.; formal analysis, P.O.M., A.K., K.M.A.; investigation, P.O.M.; resources, P.O.M., H.A., Y.S.H.; data curation, P.O.M., A.K.; writing—original draft preparation, A.K.; writing—review and editing, A.K., P.O.M., H.A.; visualization, A.K., H.A., K.M.A.; supervision, P.O.M., A.K., H.A., Y.S.H. All authors have read and agreed to the final version of the manuscript.

Funding: This research received no external funding.

Institutional Review Board Statement: Not applicable.

Informed Consent Statement: Not applicable.

Data Availability Statement: Not applicable.

Acknowledgments: This work was supported by the Taif University Researchers Supporting Project (No. TURSP-2020/217), Taif University, Taif, Saudi Arabia.

Conflicts of Interest: The authors declare no conflict of interest.

Abbreviations

The following abbreviations are used in our manuscript:

HH   Hermite–Hadamard
HHF Hermite–Hadamard–Fejér
RL   Riemann–Liouville

References

1. Hadamard, J. Essay on the study of functions given by their Taylor expansion: Study on the properties of integer functions and in particular of a function considered by Riemann. *J. Math. Pures Appl.* **1893**, *58*, 171–215.
2. Zhang, T.-Y.; Ji, A.-P.; Qi, F. Some inequalities of Hermite–Hadamard type for GA-convex functions with applications to means. *LeMat* **2013**, *68*, 229–239.
3. Mohammed, P.O.; Vivas-Cortez, M.; Abdeljawad, T.; Rangel-Oliveros, Y. Integral inequalities of Hermite–Hadamard type for quasi-convex functions with applications. *AIMS Math.* **2020**, *5*, 7316–7331. [CrossRef]
4. Latif, M.A.; Rashid, S.; Dragomir, S.S.; Chu, Y.M. Hermite—Hadamard type inequalities for co-ordinated convex and quasi-convex functions and their applications. *J. Inequal. Appl.* **2019**, *2019*, 317. [CrossRef]
5. Kashuri, A.; Meftah, B.; Mohammed, P.O. Some weighted Simpson type inequalities for differentiable s-convex functions and their applications. *J. Fract. Calc. Nonlinear Syst.* **2021**, *1*, 75–94. [CrossRef]
6. Shi, D.-P.; Xi, B.-Y.; Qi, F. Hermite–Hadamard type inequalities for Riemann–Liouville fractional integrals of (α, m)–convex functions. *Fract. Differ. Calc.* **2014**, *4*, 31–43. [CrossRef]
7. Zhou, S.S.; Rashid, S.; Noor, M.A.; Noor, K.I.; Safdar, F.; Chu, Y.M. New Hermite–Hadamard type inequalities for exponentially convex functions and applications. *AIMS Math.* **2020**, *5*, 6874–6901. [CrossRef]
8. Rashid, S.; Noor, M.A.; Noor K.I. Fractional exponentially m–convex functions and inequalities. *Int. J. Anal. Appl.* **2019**, *17*, 464–478.
9. Mohammed, P.O. Some new Hermite–Hadamard type inequalities for MT–convex functions on differentiable coordinates. *J. King Saud Univ. Sci.* **2018**, *30*, 258–262. [CrossRef]
10. Dragomir, S.S.; Pearce, C.E.M. *Selected Topics on Hermite–Hadamard Inequalities and Applications*; RGMIA Monographs; Victoria University: Footscray, VIC, Australia, 2000.
11. Rashid, S.; Noor, M.A.; Noor, K.I.; Akdemir, A.O. Some new generalizations for exponentially s–convex functions and inequalities via fractional operators. *Fractal Fract.* **2019**, *3*, 24. [CrossRef]
12. Kilbas, A.A.; Srivastava, H.M.; Trujillo, J.J. *Theory and Applications of Fractional Differential Equations*; North-Holland Mathematics Studies; Elsevier Science B.V.: Amsterdam, The Netherlands, 2006; Volume 204.
13. Vanterler, J.; Sousa, C.; Capelas de Oliveira, E. On the Ψ–Hilfer fractional derivative. *Commun. Nonlinear Sci. Numer. Simul.* **2018**, *60*, 72–91. [CrossRef]
14. Osler, T.J. The fractional derivative of a composite function. *SIAM J. Math. Anal.* **1970**, *1*, 288–293. [CrossRef]
15. Jarad, F.; Abdeljawad, T.; Shah, K. On the weighted fractional operators of a function with respect to another function. *Fractals* **2020**, *28*, 12. [CrossRef]
16. Fernandez, A.; Abdeljawad, T.; Baleanu, D. Relations between fractional models with three-parameter Mittag–Leffler kernels. *Adv. Differ. Equ.* **2020**, *2020*, 186. [CrossRef]
17. Gavrea, B.; Gavrea, I. On some Ostrowski type inequalities. *Gen. Math.* **2010**, *18*, 33–44.
18. Vivas-Cortez, M.; Abdeljawad, T.; Mohammed, P.O.; Rangel-Oliveros, Y. Simpson's integral inequalities for twice differentiable convex functions. *Math. Probl. Eng.* **2020**, *2020*, 1936461. [CrossRef]
19. Kaijser, S.; Nikolova, L.; Persson, L.-E.; Wedestig, A. Hardy type inequalities via convexity. *Math. Inequal. Appl.* **2005**, *8*, 403–417. [CrossRef]
20. Gunawan, H. Fractional integrals and generalized Olsen inequalities. *Kyungpook Math. J.* **2009**, *49*, 31–39. [CrossRef]
21. Sawano, Y.; Wadade, H. On the Gagliardo–Nirenberg type inequality in the critical Sobolev–Morrey space. *J. Fourier Anal. Appl.* **2013**, *19*, 20–47. [CrossRef]
22. Mohammed, P.O.; Abdeljawad, T. Opial integral inequalities for generalized fractional operators with nonsingular kernel. *J. Inequal. Appl.* **2020**, *2020*, 148. [CrossRef]
23. Sarikaya, M.Z.; Bilisik, C.C.; Mohammed, P.O. Some generalizations of Opial type inequalities. *Appl. Math. Inf. Sci.* **2020**, *14*, 809–816.
24. Zhao, C.-J.; Cheung, W.-S. On improvements of the Rozanova's inequality. *J. Inequal. Appl.* **2011**, *2011*, 33. [CrossRef]
25. Sarikaya, M.Z.; Set, E.; Yaldiz, H.; Başak, N. Hermite–Hadamard's inequalities for fractional integrals and related fractional inequalities. *Math. Comput. Model.* **2013**, *57*, 2403–2407. [CrossRef]
26. Baleanu, D.; Mohammed, P.O.; Zeng, S. Inequalities of trapezoidal type involving generalized fractional integrals. *Alex. Eng. J.* **2020**. [CrossRef]
27. Mohammed, P.O.; Sarikaya, M.Z. Hermite–Hadamard type inequalities for F–convex function involving fractional integrals. *J. Inequal. Appl.* **2018**, *2018*, 359. [CrossRef]
28. Qi, F.; Mohammed P.O.; Yao, J.C.; Yao, Y.H. Generalized fractional integral inequalities of Hermite–Hadamard type for (α, m)–convex functions. *J. Inequal. Appl.* **2019**, *2019*, 135. [CrossRef]
29. Han, J.; Mohammed, P.O.; Zeng, H. Generalized fractional integral inequalities of Hermite–Hadamard-type for a convex function. *Open Math.* **2020**, *18*, 794–806. [CrossRef]
30. Mohammed, P.O.; Abdeljawad, T.; Zeng, S.; Kashuri, A. Fractional Hermite–Hadamard integral inequalities for a new class of convex functions. *Symmetry* **2020**, *12*, 1485. [CrossRef]

31. Mohammed, P.O.; Abdeljawad, T.; Alqudah, M.A.; Jarad, F. New discrete inequalities of Hermite-Hadamard type for convex functions. *Adv. Differ. Equ.* **2021**, *2021*, 122. [CrossRef]
32. Baleanu, D.; Mohammed, P.O.; Vivas-Cortez, M.; Rangel-Oliveros, Y. Some modifications in conformable fractional integral inequalities. *Adv. Differ. Equ.* **2020**, *2020*, 374. [CrossRef]
33. Abdeljawad, T.; Mohammed, P.O.; Kashuri, A. New modified conformable fractional integral inequalities of Hermite–Hadamard type with applications. *J. Funct. Spaces* **2020**, *2020*, 4352357. [CrossRef]
34. Baleanu, D.; Kashuri, A.; Mohammed, P.O.; Meftah, B. General Raina fractional integral inequalities on coordinates of convex functions. *Adv. Differ. Equ.* **2021**, *2021*, 82. [CrossRef]
35. Mohammed, P.O.; Abdeljawad, T. Integral inequalities for a fractional operator of a function with respect to another function with nonsingular kernel. *Adv. Differ. Equ.* **2020**, *2020*, 363. [CrossRef]
36. Mohammed, P.O. Hermite–Hadamard inequalities for Riemann–Liouville fractional integrals of a convex function with respect to a monotone function. *Math. Meth. Appl. Sci.* **2019**, 1–11. [CrossRef]
37. Mohammed, P.O.; Sarikaya, M.Z.; Baleanu, D. On the generalized Hermite–Hadamard inequalities via the tempered fractional integrals. *Symmetry* **2020**, *12*, 595. [CrossRef]
38. Fernandez, A.; Mohammed, P. Hermite–Hadamard inequalities in fractional calculus defined using Mittag–Leffler kernels. *Math. Meth. Appl. Sci.* **2020**, 1–18. [CrossRef]
39. Sarikaya, M.Z.; Yildirim, H. On Hermite–Hadamard type inequalities for Riemann–Liouville fractional integrals. *Miskolc Math. Notes* **2017**, *17*, 1049–1059. [CrossRef]
40. Macdonald, I.G. *Symmetric Functions and Orthogonal Polynomials*; American Mathematical Society: New York, NY, USA, 1997.
41. Fejér, L. Uberdie Fourierreihen, II, Math. *Naturwise Anz Ungar. Akad. Wiss.* **1906**, *24*, 369–390.
42. Işcan, İ. Hermite–Hadamard–Fejér type inequalities for convex functions via fractional integrals. *Stud. Univ. Babeş Bolyai Math.* **2015**, *60*, 355–366.
43. Set, E; Işcan, İ.; Sarikaya, M.Z.; Özdemir, M. E. On new inequalities of Hermite-Hadamard-Fejér type for convex functions via fractional integrals. *Appl. Math. Comput.* **2015**, *259*, 875–881. [CrossRef]
44. Mohammed, P.O.; Abdeljawad, T.; Kashuri, A. Fractional Hermite–Hadamard–Fejér inequalities for a convex function with respect to an increasing function involving a positive weighted symmetric function. *Symmetry* **2020**, *12*, 1503. [CrossRef]
45. Kırmacı, U. Inequalities for differentiable mappings and applications to special means of real numbers to midpoint formula. *Appl. Math. Comput.* **2004**, *147*, 137–146.
46. Kunt, M.; Işcan, İ. On new Hermite–Hadamard–Fejér type inequalities for p–convex functions via fractional integrals. *Commun. Math. Model. Appl.* **2017**, *2*, 1–15. [CrossRef]
47. Delavar, M.R.; Aslani, M.; De La Sen, M. Hermite–Hadamard–Fejér inequality related to generalized convex functions via fractional integrals. *Sci. Asia* **2018**, *2018*, 5864091.
48. Mehmood, S.; Zafar, F.; Asmin, N. New Hermite–Hadamard–Fejér type inequalities for (η_1, η_2)–convex functions via fractional calculus. *Sci. Asia* **2020**, *46*, 102–108. [CrossRef]
49. Mohammed, P.O.; Brevik, I. A new version of the Hermite–Hadamard inequality for Riemann–Liouville fractional integrals. *Symmetry* **2020**, *12*, 610. [CrossRef]

Article

A Novel Analytical View of Time-Fractional Korteweg-De Vries Equations via a New Integral Transform

Saima Rashid [1], Aasma Khalid [2], Sobia Sultana [3], Zakia Hammouch [4,5,*], Rasool Shah [6] and Abdullah M. Alsharif [7]

1. Department of Mathematics, Government College University, Faisalabad 38000, Pakistan; saimarashid@gcuf.edu.pk
2. Department of Mathematics, Government College Women University, Faisalabad 38000, Pakistan; asmakhalid@gcwuf.edu.pk
3. Department of Mathematics, Imam Mohammad Ibn Saud Islamic University, Riyadh 12211, Saudi Arabia; smahmood@imamu.edu.sa
4. Division of Applied Mathematics, Thu Dau Mot University, Thu Dau Mot City 75000, Vietnam
5. Department of Sciences, École Normale Superieure de Meknes, Moulay Ismail University, Meknes 50000, Morocco
6. Department of Mathematics, Abdul Wali Khan University Mardan (AWKUM), Mardan 23200, Pakistan; rasoolshahawkum@gmail.com
7. Department of Mathematics, Faculty of Science, Taif University, P.O. Box 11099, Taif 21944, Saudi Arabia; a.alshrif@tu.edu.sa
* Correspondence: hammouch_zakia@tdmu.edu.vn

Abstract: We put into practice relatively new analytical techniques, the Shehu decomposition method and the Shehu iterative transform method, for solving the nonlinear fractional coupled Korteweg-de Vries (KdV) equation. The KdV equation has been developed to represent a broad spectrum of physics behaviors of the evolution and association of nonlinear waves. Approximate-analytical solutions are presented in the form of a series with simple and straightforward components, and some aspects show an appropriate dependence on the values of the fractional-order derivatives that are, in a certain sense, symmetric. The fractional derivative is proposed in the Caputo sense. The uniqueness and convergence analysis is carried out. To comprehend the analytical procedure of both methods, three test examples are provided for the analytical results of the time-fractional KdV equation. Additionally, the efficiency of the mentioned procedures and the reduction in calculations provide broader applicability. It is also illustrated that the findings of the current methodology are in close harmony with the exact solutions. It is worth mentioning that the proposed methods are powerful and are some of the best procedures to tackle nonlinear fractional PDEs.

Keywords: Shehu transform; Caputo fractional derivative; Shehu decomposition method; new iterative transform method; fractional KdV equation

1. Introduction

The formulation of exact and explicit PDE solutions is essential for a good perspective on the mechanisms of diverse physical processes. Hirota and Satsuma proposed a coupled KdV framework to address the effects of two long waves with independent dispersion correlations. It was developed as an evolution equation regulating the propagation of a one-dimensional, small-amplitude, long-surface gravity wave in a shallow water channel. The non-linear coupled system of partial differential equations (PDEs) has several applications in physical systems such as fluid mechanics, aquifers, chaos, thermodynamics, plasma physics and many more. By examining a spectral 4×4 equation with three possibilities, Wu et al. [1] established a unique hierarchy of nonlinear equations of evolution. Therefore, the action of the KdV solitons acknowledges the impact of the former's existence. It is demonstrated that it determines the velocity [2,3] of the KdV subsystem. The fractional-order paired KdV equations are written as follows:

$$\frac{\partial^\delta \Phi}{\partial \bar{t}^\delta} = -\sigma \frac{\partial^3 \Phi}{\partial x^3} - 6\sigma \Phi \frac{\partial \Phi}{\partial x} + 6\Psi \frac{\partial \Psi}{\partial x},$$
$$\frac{\partial^\delta \Psi}{\partial \bar{t}^\delta} = -\sigma \frac{\partial^3 \Psi}{\partial x^3} - 3\zeta \Phi \frac{\partial \Psi}{\partial x}, \quad \bar{t} > 0, \ 0 < \delta \leq 1, \qquad (1)$$

where σ and ζ are constants and δ is the fractional order derivative of $\Phi(x, \bar{t})$ and $\Psi(x, \bar{t})$, respectively. The functions $\Phi(x, \bar{t})$ and $\Psi(x, \bar{t})$ are regarded as essential functions of space and time, vanishing for $\bar{t}$ and x, respectively. The latter technique reduces to the conventional paired KdV equations since $\sigma = \zeta = 1$ is utilized.

An exemplary equation in this scheme is the modified coupled KdV system (MCKdV). This equation is governed by the non-linear PDEs listed in [4]:

$$\frac{\partial^\delta \Phi}{\partial \bar{t}^\delta} = \frac{1}{2}\frac{\partial^3 \Phi}{\partial \bar{t}^3} - 3\Phi^2 \frac{\partial \Phi}{\partial x} + \frac{3}{2}Y \frac{\partial^2 \Psi}{\partial x^2} + 3\frac{\partial \Psi}{\partial x}\frac{\partial Y}{\partial x} + \frac{3}{2}\Psi \frac{\partial^2 Y}{\partial x^2} + 3\Psi Y \frac{\partial \Phi}{\partial x} + 3\Phi Y \frac{\partial \Psi}{\partial x} + 3\Phi \Psi \frac{\partial Y}{\partial x},$$
$$\frac{\partial^\delta \Psi}{\partial \bar{t}^\delta} = -\frac{\partial^3 \Psi}{\partial x^3} - 3\frac{\partial \Phi}{\partial x}\frac{\partial Y}{\partial x} - 3\Psi \frac{\partial^2 \Phi}{\partial x^2} - 3Y^2 \frac{\partial Y}{\partial x} + 6\Phi \Psi \frac{\partial \Phi}{\partial x} + 3\Phi^2 \frac{\partial \Psi}{\partial x},$$
$$\frac{\partial^\delta Y}{\partial \bar{t}^\delta} = -\frac{\partial^3 Y}{\partial x^3} - 3\frac{\partial \Phi}{\partial x}\frac{\partial Y}{\partial x} - 3Y \frac{\partial^2 \Phi}{\partial x^2} - 3Y^2 \frac{\partial \Psi}{\partial x} + 6\Phi Y \frac{\partial \Phi}{\partial x} + 3\Phi^2 \frac{\partial Y}{\partial x}, \quad \bar{t} > 0, \ 0 < \delta \leq 1. \qquad (2)$$

The modified KdV equation in its usual form is simplified by the MCKdV Equation (2), with $\Psi = Y = 0$. KdV equations are a source of non evolution equations that have a variety of applications in technology and physical sciences. The KdV equations, for example, produce ion acoustic solutions in plasma physics [5,6]. Geophysical fluid dynamics in shallow waters and deep oceans are characterized by long waves [7,8].

Numerous researchers have proposed several schemes to solve the time-fractional KdV equation using different methods, such as the Adomian decomposition method (ADM) [9], differential transform method (DTM) [10], homotopy analysis method (HAM) [11], Natural decomposition method (NDM) [12], variational iteration method [13], Elzaki projected differential transform method (EPDTM) [14], modified tanh technique (MTT) [15], new iterative method (NIM) [16], Lie symmetry analysis (LSA) [17], spectral volume method (SVM) [18,19] and so on. Analogously, similar results for (2) have been proposed by Fan [20], Cavlak and Inc [21], Inc et al. [22], Lin et al. [23], Karczewska and Szczeciński [24] and Ghoreishi et al. [25].

In recent years, the modeling of dynamical processes has progressed by incorporating notions acquired from fractional-order differential equations (FDEs). Fractional calculus resulted in the emergence of the generalization of derivatives and integrals. However, fractional calculus has a long history. Recently, it has become popular in applied sciences such as viscoelastoplastic materials, random walks, optical fibers, solid state physics, plasma physics, chaos, bifurcation, condensed matter, electromagnetic flux, image processing, virology, and biological models; memory operators called fractional derivatives are used to describe damping impacts or deterioration. Several formulations and notions of fractional derivatives were introduced by Coimbra, Davison and Essex, Riesz, Riemann–Liouville, Hadamard, Weyl, Jumarie, Grünwald–Letnikov, and Liouville–Caputo [26–29], and the characteristics of these derivatives are investigated in [30–33]. Because of their prominent features and direct physical interpretation, the implementation of the Caputo fractional derivative is gaining popularity in physics, whereas the Liouville–Caputo has a singularity in its kernel.

Maitama and Zhao [34] recently identified the Shehu transformation as an important integral transformation. The Shehu transform (ST) is a modification of the Laplace transformation. Alternately, by inserting $\omega = 1$ in ST, then we recapture the Laplace transform. Complex non-linear PDEs can be converted to simpler equations via this procedure.

Despite the tremendous boost that was provided by Gorge Adomian in 1980 is known as the Adomian decomposition method (ADM). It has been successfully applied to a variety of physical models of PDEs, such as Burger's equation, a nonlinear second-order PDE with numerous applications in applied sciences. The ADM correlated with several integral transforms, such as Laplace, modified Laplace, Mohand, Aboodh, Elzaki and many more. Recently, modified Laplace ADM [35] for solving nonlinear Volterra integral and integro-differential equations based on the Newton–Raphson formula, Discrete ADM [36] used for solving time-fractional Navier–Stokes equation, Laplace ADM [37] for finding the numerical solution of a fractional order epidemic model of a vector-born disease and hence forth.

Daftardar-Gejji and Jafari [38] proposed a new recursive approach for solving functional equations with asymptotic solutions. The novel recursive process is framed on the basis of decaying the nonlinear terms is known as the iterative Laplace transform method [39]. This process is fast and precise, and it avoids the use of an unconditioned matrix, complicated integrals, and infinite series forms. This method does not necessitate any explicit settings for the problem. Several studies have considered NITM to solve PDEs, such as the KdV Equation [16], Fornberg–Whitham equation [40], and Klein–Gordon equations [41].

Despite the significant body of work on fractional PDEs models, estimating the approximate-analytical solutions of the corresponding governing PDE is not a trivial task. In this context, we aim to develop two efficient algorithms for estimating the approximate-analytical solutions of KdV and MCKdV equations that model the dynamics of the process under investigation. The ADM and NITM are modified with the ST, and the new method is known to be the Shehu decomposition method and Shehu iterative transform method. The novel methods are applied to examining the fractional-order of the system of KDV equations. The outcome of some test examples was examined in order to demonstrate the practicality of the proposed strategy. Innovative techniques are used to derive the outcomes of the fractional-order and integral-order models. The convergence and uniqueness analysis for SDM is also presented. Using synthetic trajectories derived from the KdV and MCKdV models, we demonstrate the validity and feasibility of the suggested algorithmic approaches to deriving the approximate-analytical solutions in a simulation study. The proposed method can be used to solve other fractional orders of linear and non-linear PDEs.

2. Preliminaries

Several definitions and axiom outcomes from the literature are prerequisites in our analysis.

Definition 1 ([34]). *Shehu transform (ST) for a function $\Phi(\bar{t})$ having exponential order over the set of functions is stated as*

$$\mathbb{S} = \left\{\Phi(\bar{t}) | \exists \mathcal{K}, k_1, k_2 > 0, |\Phi(\bar{t})| < \mathcal{K}\exp(|\bar{t}|/k_j), \text{ if } \bar{t} \in (-1)^j \times [0, \infty), j = 1,2; (\mathcal{K}, k_1, k_2 > 0)\right\}, \tag{3}$$

where $\Phi(\bar{t})$ is represented by $\mathbb{S}[\Phi(\bar{t})] = \mathcal{S}(\xi, \omega)$, is described as

$$\mathbb{S}[\Phi(\bar{t})] = \int_0^\infty \Phi(\bar{t}) \exp\left(-\frac{\xi}{\omega}\bar{t}\right) d\bar{t} = \mathcal{S}(\xi, \omega), \bar{t} \leq 0, \omega \in [\kappa_1, \kappa_2]. \tag{4}$$

A useful result of the ST is stated as:

$$\mathbb{S}[\bar{t}^\delta] = \int_0^\infty \exp\left(-\frac{\xi}{\omega}\bar{t}\right)\bar{t}^\delta d\bar{t} = \Gamma(\delta+1)\left(\frac{\omega}{\xi}\right)^{\delta+1}. \tag{5}$$

Definition 2 ([34]). *The inverse ST of a mapping $\Phi(\bar{t})$ is stated as*

$$\mathbb{S}^{-1}\left[\left(\frac{\omega}{\xi}\right)^{m\delta+1}\right] = \frac{\bar{t}^{m\delta}}{\Gamma(m\delta+1)}, \quad \Re(\delta) > 0, \text{ and } m > 0. \tag{6}$$

Lemma 1. *(Linearity property of ST) Let ST of $\Phi_1(\bar{t})$ and $\Phi_2(\bar{t})$ are $\mathcal{P}(\xi,\omega)$ and $\mathcal{Q}(\xi,\omega)$, respectively,* [34],

$$\begin{aligned}\mathbb{S}[\gamma_1\Phi_1(\bar{t}) + \gamma_2\Phi_2(\bar{t})] &= \mathbb{S}[\gamma_1\Phi_1(\bar{t})] + \mathbb{S}[\gamma_2\Phi_2(\bar{t})] \\ &= \gamma_1\mathcal{P}(\xi,\omega) + \gamma_2\mathcal{Q}(\xi,\omega),\end{aligned} \tag{7}$$

where γ_1 and γ_2 are arbitrary constants.

Lemma 2 ([34]). *ST of Caputo fractional derivative of order δ is stated as*

$$\mathbb{S}[\mathcal{D}_{\bar{t}}^\delta \Phi(\bar{t})] = \left(\frac{\xi}{\omega}\right)^\delta \mathbb{S}[\Phi(\mathbf{x},\bar{t})] - \sum_{\kappa=0}^{m-1} \left(\frac{\xi}{\omega}\right)^{\delta-\kappa-1} \Phi^{(\kappa)}(\mathbf{x},0), \quad m-1 \leq \delta \leq m, \, m \in \mathbb{N}. \tag{8}$$

3. Configuration of the SDM

Assume the nonlinear fractional PDE:

$$\mathcal{D}_{\bar{t}}^\delta \Phi(\mathbf{x},\bar{t}) + \mathcal{L}\Phi(\mathbf{x},\bar{t}) + \mathcal{N}\Phi(\mathbf{x},\bar{t}) = \mathcal{F}(\mathbf{x},\bar{t}), \quad \bar{t} > 0, \, 0 < \delta \leq 1 \tag{9}$$

subject to the condition

$$\Phi(\mathbf{x},0) = \mathcal{G}(\mathbf{x}), \tag{10}$$

where $\mathcal{D}_{\bar{t}}^\delta = \frac{\partial^\delta \Phi(\mathbf{x},\bar{t})}{\partial \bar{t}^\delta}$ denotes the fractional-order Caputo derivative operator with $0 < \delta \leq 1$ while $\mathcal{L}$ and $\mathcal{N}$ are linear and nonlinear terms and $\mathcal{F}(\mathbf{x},\bar{t})$ indicates the source term.

Employing the Shehu transform to (9), and we acquire

$$\mathbb{S}[\mathcal{D}_{\bar{t}}^\delta \Phi(\mathbf{x},\bar{t}) + \mathcal{L}\Phi(\mathbf{x},\bar{t}) + \mathcal{N}\Phi(\mathbf{x},\bar{t})] = \mathbb{S}[\mathcal{F}(\mathbf{x},\bar{t})].$$

Taking differentiation property of Shehu transform, we find

$$\frac{\xi^\delta}{\omega^\delta}\mathcal{U}(\xi,\omega) = \sum_{\kappa=0}^{m-1}\left(\frac{\xi}{\omega}\right)^{\delta-\kappa-1}\Phi^{(\kappa)}(0) + \mathbb{S}[\mathcal{L}\Phi(\mathbf{x},\bar{t}) + \mathcal{N}\Phi(\mathbf{x},\bar{t})] + \mathbb{S}[\mathcal{F}(\mathbf{x},\bar{t})]. \tag{11}$$

Th inverse Shehu transform of (11) gives

$$\Phi(\mathbf{x},\bar{t}) = \mathbb{S}^{-1}\left[\sum_{\kappa=0}^{m-1}\left(\frac{\xi}{\omega}\right)^{\delta-\kappa-1}\Phi^{(\kappa)}(0) + \frac{\omega^\delta}{\xi^\delta}\mathbb{S}[\mathcal{F}(\mathbf{x},\bar{t})]\right] - \mathbb{S}^{-1}\left[\frac{\omega^\delta}{\xi^\delta}\mathbb{S}[\mathcal{L}\Phi(\mathbf{x},\bar{t}) + \mathcal{N}\Phi(\mathbf{x},\bar{t})]\right]. \tag{12}$$

The Shehu decomposition method solution $\Phi(\mathbf{x},\bar{t})$ is represented by the following infinite series

$$\Phi(\mathbf{x},\bar{t}) = \sum_{m=0}^{\infty} \Phi_m(\mathbf{x},\bar{t}). \tag{13}$$

Thus, the nonlinear term $\mathcal{N}(\mathbf{x},\bar{t})$ can be evaluated by the Adomian decomposition method prescribed as

$$\mathcal{N}\Phi(\mathbf{x},\bar{t}) = \sum_{m=0}^{\infty} \tilde{A}_m(\Phi_0,\Phi_1,...), \quad m = 0,1,..., \tag{14}$$

where
$$\tilde{A}_m(\Phi_0, \Phi_1, \ldots) = \frac{1}{m!}\left[\frac{d^m}{d\lambda^m}\mathcal{N}\left(\sum_{j=0}^{\infty}\lambda^j\Phi_j\right)\right]_{\lambda=0}, \quad m > 0.$$

Substituting (13) and (14) into (12), we have

$$\sum_{m=0}^{\infty}\Phi_m(\mathbf{x},\bar{t}) = \mathcal{G}(\mathbf{x}) + \tilde{\mathcal{G}}(\mathbf{x}) - \mathbb{S}^{-1}\left[\frac{\omega^\delta}{\zeta^\delta}\mathbb{S}\left[\mathcal{L}\Phi(\mathbf{x},\bar{t}) + \sum_{m=0}^{\infty}\tilde{A}_m\right]\right]. \quad (15)$$

Finally, the iterative procedure for (15) is obtained as follows:

$$\Phi_0(\mathbf{x},\bar{t}) = \mathcal{G}(\mathbf{x}) + \tilde{\mathcal{G}}(\mathbf{x}), \quad m = 0,$$
$$\Phi_{m+1}(\mathbf{x},\bar{t}) = -\mathbb{S}^{-1}\left[\frac{\omega^\delta}{\zeta^\delta}\mathbb{S}\left[\mathcal{L}(\Phi_m(\mathbf{x},\bar{t})) + \sum_{m=0}^{\infty}\tilde{A}_m\right]\right], \quad m \geq 1. \quad (16)$$

4. Basic Formulation of the SITM

Let us suppose the following general fractional PDE:

$$D_{\bar{t}}^\delta \Phi(\mathbf{x},\bar{t}) + \mathcal{L}\Phi(\mathbf{x},\bar{t}) + \mathcal{N}\Phi(\mathbf{x},\bar{t}) = \mathcal{F}(\mathbf{x},\bar{t}), \quad \bar{t} > 0, \ m-1 < \delta \leq m, \ m \in \mathbb{N} \quad (17)$$

subject to the condition

$$\Phi^{(\kappa)}(\mathbf{x},0) = \mathcal{G}_\kappa(\mathbf{x}), \quad \kappa = 0,1,2,\ldots,m-1, \quad (18)$$

where $\mathcal{L}$ and $\mathcal{N}$ are linear and nonlinear terms and $\mathcal{F}(\mathbf{x},\bar{t})$ indicates the source term. Utilizing the Shehu transform to (17), we obtain

$$\mathbb{S}[D_{\bar{t}}^\delta \Phi(\mathbf{x},\bar{t}) + \mathcal{L}\Phi(\mathbf{x},\bar{t}) + \mathcal{N}\Phi(\mathbf{x},\bar{t})] = \mathbb{S}[\mathcal{F}(\mathbf{x},\bar{t})].$$

Taking differentiation property of Shehu transform, we find

$$\frac{\zeta^\delta}{\omega^\delta}\mathcal{U}(\zeta,\omega) = \sum_{\kappa=0}^{m-1}\left(\frac{\zeta}{\omega}\right)^{\delta-\kappa-1}\Phi^{(\kappa)}(0) + \mathbb{S}[\mathcal{L}\Phi(\mathbf{x},\bar{t}) + \mathcal{N}\Phi(\mathbf{x},\bar{t})] + \mathbb{S}[\mathcal{F}(\mathbf{x},\bar{t})]. \quad (19)$$

Th inverse Shehu transform of (19) gives

$$\Phi(\mathbf{x},\bar{t}) = \mathbb{S}^{-1}\left[\sum_{\kappa=0}^{m-1}\left(\frac{\zeta}{\omega}\right)^{\delta-\kappa-1}\Phi^{(\kappa)}(0) + \frac{\omega^\delta}{\zeta^\delta}\mathbb{S}[\mathcal{F}(\mathbf{x},\bar{t})]\right] - \mathbb{S}^{-1}\left[\frac{\omega^\delta}{\zeta^\delta}\mathbb{S}[\mathcal{L}\Phi(\mathbf{x},\bar{t}) + \mathcal{N}\Phi(\mathbf{x},\bar{t})]\right]. \quad (20)$$

From the recursive relation, we obtain

$$\Phi(\mathbf{x},\bar{t}) = \sum_{m=0}^{\infty}\Phi_m(\mathbf{x},\bar{t}). \quad (21)$$

Furthermore, the operator $\mathcal{L}$ is linear, therefore

$$\mathcal{L}\left(\sum_{m=0}^{\infty}\Phi_m(\mathbf{x},\bar{t})\right) = \sum_{m=0}^{\infty}\mathcal{L}[\Phi_m(\mathbf{x},\bar{t})], \quad (22)$$

and we decomposed the nonlinear operator $\mathcal{N}$ as in [38]

$$\mathcal{N}\left(\sum_{m=0}^{\infty}\Phi_m(\mathbf{x},\bar{t})\right) = \mathcal{N}(\Phi_0(\mathbf{x},\bar{t})) + \sum_{m=0}^{\infty}\left[\mathcal{N}\left(\sum_{\kappa=0}^{\infty}\Phi_\kappa(\mathbf{x},\bar{t})\right) - \mathcal{N}\left(\sum_{\kappa=0}^{m-1}\Phi_\kappa(\mathbf{x},\bar{t})\right)\right]$$
$$= \mathcal{N}(\Phi_0) + \sum_{\kappa=1}^{\infty}D_m, \quad (23)$$

where $D_m = \mathcal{N}\left(\sum_{\kappa=0}^{m} \Phi_\kappa\right) - \mathcal{N}\left(\sum_{\kappa=0}^{m-1} \Phi_\kappa\right)$.

By putting (21), (22) and (23) into (24), we obtain

$$\sum_{m=0}^{\infty} \Phi_m(\mathbf{x}, \bar{t}) = \mathbb{S}^{-1}\left[\sum_{\kappa=0}^{m-1}\left(\frac{\xi}{\omega}\right)^{\delta-\kappa-1} \Phi^{(\kappa)}(0) + \frac{\omega^\delta}{\xi^\delta}\mathbb{S}[\mathcal{F}(\mathbf{x}, \bar{t})]\right]$$
$$-\mathbb{S}^{-1}\left\{\frac{\omega^\delta}{\xi^\delta}\mathbb{S}\left[\mathcal{L}\left(\sum_{\kappa=0}^{m} \Phi_\kappa(\mathbf{x}, \bar{t})\right) + \mathcal{N}(\Phi_0) + \sum_{\kappa=1}^{m} D_m\right]\right\}. \quad (24)$$

Thus, we establish the subsequent iteration

$$\Phi_0(\mathbf{x}, \bar{t}) = \mathbb{S}^{-1}\left[\sum_{\kappa=0}^{m-1}\left(\frac{\xi}{\omega}\right)^{\delta-\kappa-1} \Phi^{(\kappa)}(0) + \frac{\omega^\delta}{\xi^\delta}\mathbb{S}[\mathcal{F}(\mathbf{x}, \bar{t})]\right],$$

$$\Phi_1(\mathbf{x}, \bar{t}) = -\mathbb{S}^{-1}\left\{\frac{\omega^\delta}{\xi^\delta}\mathbb{S}\left[\mathcal{L}(\Phi_0(\mathbf{x}, \bar{t})) + \mathcal{N}(\Phi_0(\mathbf{x}, \bar{t}))\right]\right\},$$

$$\vdots$$

$$\Phi_{m+1}(\mathbf{x}, \bar{t}) = -\mathbb{S}^{-1}\left\{\frac{\omega^\delta}{\xi^\delta}\mathbb{S}\left[\mathcal{L}(\Phi_m(\mathbf{x}, \bar{t})) + D_m\right]\right\}, \; m \geq 1. \quad (25)$$

Finally, (17) and (18) yield the m-term solution in series form, described as

$$\Phi(\mathbf{x}, \bar{t}) \approx \Phi_0(\mathbf{x}, \bar{t}) + \Phi_1(\mathbf{x}, \bar{t}) + \Phi_2(\mathbf{x}, \bar{t}) + \ldots + \Phi_m(\mathbf{x}, \bar{t}), \; m \in \mathbb{N}. \quad (26)$$

5. Existence and Uniqueness Results for Shehu Decomposition Method

In what follows, we will demonstrate that the sufficient conditions assure the existence of a unique solution. Our desired existence of solutions in the case of SDM follows by [42].

Theorem 1. *(Uniqueness theorem): Equation (16) has a unique solution whenever $0 < \epsilon < 1$, where $\epsilon = \frac{((L_1+L_2+L_3))\bar{t}^{(\delta-1)}}{\delta!}$.*

Proof. Assume that $M = (\mathcal{C}[I], \|.\|)$ represents all continuous mappings on the Banach space, defined on $I = [0, \mathbb{T}]$ having the norm $\|.\|$. For this we introduce a mapping $W : M \mapsto M$, we have

$$\Phi_{n+1}(\mathbf{x}, \bar{t}) = \Phi(\mathbf{x}, \bar{t}) + \mathbb{S}^{-1}\left[\left(\frac{\omega}{\xi}\right)^\delta \mathbb{S}[\mathcal{L}[\Phi_n(\mathbf{x}, \bar{t})] + \mathcal{R}[\Phi_n(\mathbf{x}, \bar{t})] + \mathcal{N}[\Phi_n(\mathbf{x}, \bar{t})]]\right], \; n \geq 0, \quad (27)$$

where $\mathcal{L}[\Phi(\mathbf{x}, \bar{t})] \equiv \frac{\partial^3 \Phi(\mathbf{x}, \bar{t})}{\partial x^2}$ and $\mathcal{R}[\Phi(\mathbf{x}, \bar{t})] \equiv \frac{\partial \Phi(\mathbf{x}, \bar{t})}{\partial x}$. Now assume that $\mathcal{L}[\Phi(\mathbf{x}, \bar{t})]$ and $\mathcal{M}[\Phi(\mathbf{x}, \bar{t})]$ are also Lipschitzian with $|\mathcal{R}\Phi - \mathcal{R}\check{\Phi}| < \check{L}_1|\Phi - \check{\Phi}|$ and $|\mathcal{L}\Phi - \mathcal{L}\check{\Phi}| < \check{L}_2|\Phi - \check{\Phi}|$, where $\check{L}_1$ and $\check{L}_2$ are Lipschitz constant, respectively, and $\Phi, \check{\Phi}$ are various values of the mapping.

$$\|W\Phi - W\check{\Phi}\| = \max_{\bar{t} \in I} \left| \begin{array}{l} \mathbb{S}^{-1}\left[\left(\frac{\omega}{\xi}\right)^\delta \mathbb{S}[\mathcal{L}[\Phi(\mathbf{x}, \bar{t})] + \mathcal{R}[\Phi(\mathbf{x}, \bar{t})] + \mathcal{N}[\Phi(\mathbf{x}, \bar{t})]]\right] \\ - \mathbb{S}^{-1}\left[\left(\frac{\omega}{\xi}\right)^\delta \mathbb{S}[\mathcal{L}[\check{\Phi}(\mathbf{x}, \bar{t})] + \mathcal{R}[\check{\Phi}(\mathbf{x}, \bar{t})] + \mathcal{N}[\check{\Phi}(\mathbf{x}, \bar{t})]]\right] \end{array} \right|$$

$$\leq \max_{\bar{t}\in I} \left| \begin{array}{l} \mathbb{S}^{-1}\left[\left(\frac{\omega}{\zeta}\right)^\delta \mathbb{S}\left[\mathcal{L}[\Phi(\mathbf{x},\bar{t})] - \mathcal{L}[\check{\Phi}(\mathbf{x},\bar{t})]\right]\right] \\ + \mathbb{S}^{-1}\left[\left(\frac{\omega}{\zeta}\right)^\delta \mathbb{S}\left[\mathcal{R}[\Phi(\mathbf{x},\bar{t})] - \mathcal{R}[\check{\Phi}(\mathbf{x},\bar{t})]\right]\right] \\ + \mathbb{S}^{-1}\left[\left(\frac{\omega}{\zeta}\right)^\delta \mathbb{S}\left[\mathcal{N}[\Phi(\mathbf{x},\bar{t})] - \mathcal{N}[\check{\Phi}(\mathbf{x},\bar{t})]\right]\right] \end{array} \right|$$

$$\leq \max_{\bar{t}\in I} \left[\begin{array}{l} \check{L}_1 \mathbb{S}^{-1}\left[\left(\frac{\omega}{\zeta}\right)^\delta \mathbb{S}\left|\Phi(\mathbf{x},\bar{t}) - \check{\Phi}(\mathbf{x},\bar{t})\right|\right] \\ + \check{L}_2 \mathbb{S}^{-1}\left[\left(\frac{\omega}{\zeta}\right)^\delta \mathbb{S}\left|\Phi(\mathbf{x},\bar{t}) - \check{\Phi}(\mathbf{x},\bar{t})\right|\right] \\ + \check{L}_3 \mathbb{S}^{-1}\left[\left(\frac{\omega}{\zeta}\right)^\delta \mathbb{S}\left|\Phi(\mathbf{x},\bar{t}) - \check{\Phi}(\mathbf{x},\bar{t})\right|\right] \end{array} \right]$$

$$\leq \max_{\bar{t}\in I} (\check{L}_1 + \check{L}_2 + \check{L}_3) \mathbb{S}^{-1}\left[\left(\frac{\omega}{\zeta}\right)^\delta \mathbb{S}\left|\Phi(\mathbf{x},\bar{t}) - \check{\Phi}(\mathbf{x},\bar{t})\right|\right]$$

$$\leq (\check{L}_1 + \check{L}_2 + \check{L}_3) \mathbb{S}^{-1}\left[\left(\frac{\omega}{\zeta}\right)^\delta \mathbb{S}\left\|\Phi(\mathbf{x},\bar{t}) - \check{\Phi}(\mathbf{x},\bar{t})\right\|\right]$$

$$= \frac{((\check{L}_1 + \check{L}_2 + \check{L}_3))\bar{t}^{(\delta-1)}}{(\delta)!}\left\|\Phi(\mathbf{x},\bar{t}) - \check{\Phi}(\mathbf{x},\bar{t})\right\|.$$

Under the assumption $0 < \epsilon < 1$, the mapping is contraction. Thus, by Banach contraction fixed point theorem, there exists a unique solution to (9). Hence, this completes the proof. □

Theorem 2. *(Convergence Analysis) The general form solution of (9) will be convergent.*

Proof. Suppose $\widehat{S}_n$ be the *n*th partial sum, that is $\widehat{S}_n = \sum_{m=0}^{n} \Phi_m(\mathbf{x},\bar{t})$. Firstly, we show that $\{\widehat{S}_n\}$ is a Cauchy sequence in Banach space in M. Taking into consideration a new representation of Adomian polynomials we obtain

$$\bar{R}(\widehat{S}_n) = \check{H}_n + \sum_{p=0}^{n-1} \check{H}_p,$$

$$\bar{N}(\widehat{S}_n) = \check{H}_n + \sum_{c=0}^{n-1} \check{H}_c. \tag{28}$$

Now

$$\|\widehat{S}_n - \widehat{S}_q\| = \max_{\bar{t}\in I} |\widehat{S}_n - \widehat{S}_q|$$

$$= \max_{\bar{t}\in I} \left|\sum_{m=q+1}^{n} \check{\Phi}(\mathbf{x},\bar{t})\right|, (m = 1,2,3,\ldots) \tag{29}$$

$$\leq \max_{\bar{t}\in I} \left| \begin{array}{l} \mathbb{S}^{-1}\left[\left(\frac{\omega}{\zeta}\right)^\delta \mathbb{S}\left[\sum_{m=q+1}^{n} \mathcal{L}[\Phi_{n-1}(\mathbf{x},\bar{t})]\right]\right] \\ + \mathbb{S}^{-1}\left[\left(\frac{\omega}{\zeta}\right)^\delta \mathbb{S}\left[\sum_{m=q+1}^{n} \mathcal{R}[\Phi_{n-1}(\mathbf{x},\bar{t})]\right]\right] \\ + \mathbb{S}^{-1}\left[\left(\frac{\omega}{\zeta}\right)^\delta \mathbb{S}\left[\sum_{m=q+1}^{n} \check{H}_{n-1}(\mathbf{x},\bar{t})\right]\right] \end{array} \right|$$

$$= \max_{\bar{t} \in I} \left| \begin{array}{l} \mathbb{S}^{-1}\left[\left(\frac{\omega}{\zeta}\right)^{\delta} \mathbb{S}\left[\sum_{m=q}^{n-1} \mathcal{L}[\Phi_n(\mathbf{x},\bar{t})]\right]\right] \\ + \mathbb{S}^{-1}\left[\left(\frac{\omega}{\zeta}\right)^{\delta} \mathbb{S}\left[\sum_{m=q}^{n-1} \mathcal{R}[\Phi_n(\mathbf{x},\bar{t})]\right]\right] \\ + \mathbb{S}^{-1}\left[\left(\frac{\omega}{\zeta}\right)^{\delta} \mathbb{S}\left[\sum_{m=q}^{n-1} \check{H}_n(\mathbf{x},\bar{t})\right]\right] \end{array} \right|$$

$$\leq \max_{\bar{t} \in I} \left| \begin{array}{l} \mathbb{S}^{-1}\left[\left(\frac{\omega}{\zeta}\right)^{\delta} \mathbb{S}\left[\sum_{m=q}^{n-1} \mathcal{L}(\hat{S}_{n-1}) - \mathcal{L}(\hat{S}_{q-1})\right]\right] \\ + \mathbb{S}^{-1}\left[\left(\frac{\omega}{\zeta}\right)^{\delta} \mathbb{S}\left[\sum_{m=q}^{n-1} \mathcal{R}(\hat{S}_{n-1}) - \mathcal{R}(\hat{S}_{q-1})\right]\right] \\ + \mathbb{S}^{-1}\left[\left(\frac{\omega}{\zeta}\right)^{\delta} \mathbb{S}\left[\sum_{m=q}^{n-1} \mathcal{N}(\hat{S}_{n-1}) - \mathcal{N}(\hat{S}_{q-1})\right]\right] \end{array} \right|$$

$$\leq \max_{\bar{t} \in I} \left| \begin{array}{l} \mathbb{S}^{-1}\left[\left(\frac{\omega}{\zeta}\right)^{\delta} \mathbb{S}\left[\mathcal{L}(\hat{S}_{n-1}) - \mathcal{L}(\hat{S}_{q-1})\right]\right] \\ + \mathbb{S}^{-1}\left[\left(\frac{\omega}{\zeta}\right)^{\delta} \mathbb{S}\left[\mathcal{R}(\hat{S}_{n-1}) - \mathcal{R}(\hat{S}_{q-1})\right]\right] \\ + \mathbb{S}^{-1}\left[\left(\frac{\omega}{\zeta}\right)^{\delta} \mathbb{S}\left[\mathcal{N}(\hat{S}_{n-1}) - \mathcal{N}(\hat{S}_{q-1})\right]\right] \end{array} \right|$$

$$\leq \check{L}_1 \max_{\bar{t} \in I} \left| \mathbb{S}^{-1}\left[\left(\frac{\omega}{\zeta}\right)^{\delta} \mathbb{S}\left[(\hat{S}_{n-1}) - (\hat{S}_{q-1})\right]\right] \right|$$

$$+ \check{L}_2 \max_{\bar{t} \in I} \left| \mathbb{S}^{-1}\left[\left(\frac{\omega}{\zeta}\right)^{\delta} \mathbb{S}\left[(\hat{S}_{n-1}) - (\hat{S}_{q-1})\right]\right] \right|$$

$$+ \check{L}_3 \max_{\bar{t} \in I} \left| \mathbb{S}^{-1}\left[\left(\frac{\omega}{\zeta}\right)^{\delta} \mathbb{S}\left[(\hat{S}_{n-1}) - (\hat{S}_{q-1})\right]\right] \right|$$

$$= \frac{(\check{L}_1 + \check{L}_2 + \check{L}_3)\bar{t}^{(\delta-1)}}{\delta!} \|\hat{S}_{n-1} - \hat{S}_{q-1}\|.$$

Consider $n = q + 1$; then

$$\|\hat{S}_{q+1} - \hat{S}_q\| \leq \epsilon \|\hat{S}_q - \hat{S}_{q-1}\| \leq \epsilon^2 \|\hat{S}_{q-1} - \hat{S}_{q-2}\| \leq \ldots \leq \epsilon^q \|\hat{S}_1 - \hat{S}_0\|,$$

where $\frac{(\check{L}_1+\check{L}_2+\check{L}_3)\bar{t}^{(\delta-1)}}{\delta!}$. Analogously, from the triangular inequality we have

$$\begin{aligned} \|\hat{S}_n - \hat{S}_q\| &\leq \|\hat{S}_{q+1} - \hat{S}_q\| + \|\hat{S}_{q+2} - \hat{S}_{q+1}\| + \ldots + \|\hat{S}_n - \hat{S}_{n-1}\| \\ &\leq \left[\epsilon^q + \epsilon^{q+1} + \ldots + \epsilon^{n-1}\right] \|\hat{S}_1 - \hat{S}_0\| \\ &\leq \epsilon^q \left(\frac{1 - \epsilon^{n-q}}{\epsilon}\right) \|\Phi_1\|, \end{aligned}$$

since $0 < \epsilon < 1$, we have $(1 - \epsilon^{n-q}) < 1$, then

$$\|\hat{S}_n - \hat{S}_q\| \leq \frac{\epsilon^q}{1-\epsilon} \max_{\bar{t} \in I} \|\Phi_1\|.$$

However, $|\Phi_1| < \infty$ (since $\Phi(\mathbf{x}, \bar{t})$ is bounded). Thus, as $q \mapsto \infty$, then $\|\hat{S}_n - \hat{S}_q\| \mapsto 0$. Hence, $\{\hat{S}_1\}$ is a Cauchy sequence in K. As a result, the series $\sum_{n=0}^{\infty} \Phi_n$ is convergent and this completes the proof. □

Theorem 3 ([42]). *(Error estimate) The maximum absolute truncation error of the series solution (9)–(16) is computed as*

$$\max_{\bar{t}\in I}\left|\Phi(x,\bar{t}) - \sum_{n=1}^{q}\Phi_n(x,\bar{t})\right| \leq \frac{\epsilon^q}{1-\epsilon}\max_{\bar{t}\in I}\|\Phi_1\|. \tag{30}$$

6. Evaluation of the Fractional KdV Model

This section represents some test examples by employing two novel methods, SDM and SITM via the Caputo derivative operator. Furthermore, the convergence and stability of the method are elaborated on.

Problem 1 ([16]). *Assume the time-fractional coupled nonlinear KdV Equation (1) with $\sigma = \zeta = 1$, subject to the condition*

$$\Phi(x,0) = \ell^2 \operatorname{sech}^2\left(\frac{\beta}{2} + \frac{\ell x}{2}\right), \quad \Psi(x,0) = \sqrt{\frac{\sigma}{2}}\ell^2 \operatorname{sech}^2\left(\frac{\beta}{2} + \frac{\ell x}{2}\right). \tag{31}$$

Case I. First, we surmise the Shehu decomposition method for Problem 1. Employing the Shehu transformation to (1), we find

$$\frac{\zeta^\delta}{\omega^\delta}\mathcal{U}(\zeta,\omega) - \sum_{\kappa=0}^{m-1}\left(\frac{\zeta}{\omega}\right)^{\delta-\kappa-1}\Phi^{(\kappa)}(0) = \mathbb{S}\left[-\sigma\frac{\partial^3\Phi}{\partial x^3} - 6\sigma\Phi\frac{\partial\Phi}{\partial x} + 6\Psi\frac{\partial\Psi}{\partial x}\right],$$

$$\frac{\zeta^\delta}{\omega^\delta}\mathcal{V}(\zeta,\omega) - \sum_{\kappa=0}^{m-1}\left(\frac{\zeta}{\omega}\right)^{\delta-\kappa-1}\Psi^{(\kappa)}(0) = \mathbb{S}\left[-\sigma\frac{\partial^3\Psi}{\partial x^3} - 3\sigma\Phi\frac{\partial\Psi}{\partial x}\right]. \tag{32}$$

In view of (31) and simple computations yield

$$\mathcal{U}(\zeta,\omega) = \frac{\omega}{\zeta}\Phi^{(0)}(x,0) + \frac{\omega^\delta}{\zeta^\delta}\mathbb{S}\left[-\sigma\frac{\partial^3\Phi}{\partial x^3} - 6\sigma\Phi\frac{\partial\Phi}{\partial x} + 6\Psi\frac{\partial\Psi}{\partial x}\right],$$

$$\mathcal{V}(\zeta,\omega) = \frac{\omega}{\zeta}\Psi^{(0)}(x,0) + \frac{\omega^\delta}{\zeta^\delta}\mathbb{S}\left[-\sigma\frac{\partial^3\Psi}{\partial x^3} - 3\sigma\Phi\frac{\partial\Psi}{\partial x}\right].$$

Applying the inverse Shehu transform, we have

$$\Phi(x,\bar{t}) = \mathbb{S}^{-1}\left[\frac{\omega}{\zeta}\Phi(x,0)\right] + \mathbb{S}^{-1}\left[\frac{\omega^\delta}{\zeta^\delta}\mathbb{S}\left[-\sigma\frac{\partial^3\Phi}{\partial x^3} - 6\sigma\Phi\frac{\partial\Phi}{\partial x} + 6\Psi\frac{\partial\Psi}{\partial x}\right]\right],$$

$$\Psi(x,\bar{t}) = \mathbb{S}^{-1}\left[\frac{\omega}{\zeta}\Psi(x,0)\right] + \mathbb{S}^{-1}\left[\frac{\omega^\delta}{\zeta^\delta}\mathbb{S}\left[-\sigma\frac{\partial^3\Psi}{\partial x^3} - 3\sigma\Phi\frac{\partial\Psi}{\partial x}\right]\right]. \tag{33}$$

By virtue of the Shehu decomposition method, we have

$$\Phi_0(x,\bar{t}) = \mathbb{S}^{-1}\left[\frac{\omega}{\zeta}\Phi(x,0)\right] = \mathbb{S}^{-1}\left[\frac{\zeta}{\omega}\ell^2\operatorname{sech}^2\left(\frac{\beta}{2} + \frac{\ell x}{2}\right)\right]$$

$$= \ell^2\operatorname{sech}^2\left(\frac{\beta}{2} + \frac{\ell x}{2}\right),$$

$$\Psi_0(x,\bar{t}) = \mathbb{S}^{-1}\left[\frac{\omega}{\zeta}\Psi(x,0)\right]$$

$$= \sqrt{\frac{\sigma}{2}}\ell^2\operatorname{sech}^2\left(\frac{\beta}{2} + \frac{\ell x}{2}\right).$$

$$\sum_{m=0}^{\infty} \Phi_{m+1}(\mathbf{x}, \bar{t}) = \mathbb{S}^{-1}\left[\frac{\omega^{\delta}}{\xi^{\delta}}\mathbb{S}\left[-\sigma\sum_{m=0}^{\infty}(\Phi_{xxx})_m - 6\sigma\sum_{m=0}^{\infty}\mathcal{A}_m + 6\sum_{m=0}^{\infty}\mathcal{B}_m\right]\right],$$

$$\sum_{m=0}^{\infty} \Psi_{m+1}(\mathbf{x}, \bar{t}) = \mathbb{S}^{-1}\left[\frac{\omega^{\delta}}{\xi^{\delta}}\mathbb{S}\left[-\sigma\sum_{m=0}^{\infty}(\Psi_{xxx})_m - 3\sigma\sum_{m=0}^{\infty}\mathcal{C}_m\right]\right], \quad m = 0, 1, 2, \ldots.$$

The first few Adomian polynomials are presented as follows:

$$\begin{aligned}
\mathcal{A}_0(\Phi\Phi_x) &= \Phi_0\Phi_{0x}, \\
\mathcal{A}_1(\Phi\Phi_x) &= \Phi_0\Phi_{1x} + \Phi_1\Phi_{0x}, \\
\mathcal{A}_2(\Phi\Phi_x) &= \Phi_1\Phi_{2x} + \Phi_1\Phi_{1x} + \Phi_2\Phi_{0x}, \\
\mathcal{B}_0(\Psi\Psi_x) &= \Psi_0\Psi_{0x}, \\
\mathcal{B}_1(\Psi\Psi_x) &= \Psi_0\Psi_{1x} + \Psi_1\Psi_{0x}, \\
\mathcal{B}_2(\Psi\Psi_x) &= \Psi_1\Psi_{2x} + \Psi_1\Psi_{1x} + \Psi_2\Psi_{0x}, \\
\mathcal{C}_0(\Phi\Psi_x) &= \Phi_0\Psi_{0x}, \\
\mathcal{C}_1(\Phi\Psi_x) &= \Phi_0\Psi_{1x} + \Phi_1\Psi_{0x}, \\
\mathcal{C}_2(\Phi\Psi_x) &= \Phi_1\Psi_{2x} + \Phi_1\Psi_{1x} + \Phi_2\Psi_{0x}.
\end{aligned}$$

For $m = 0, 1, 2, 3, \ldots$

$$\begin{aligned}
\Phi_1(\mathbf{x}, \bar{t}) &= \mathbb{S}^{-1}\left[\frac{\omega^{\delta}}{\xi^{\delta}}\mathbb{S}\left[-\sigma(\Phi_{xxx})_0 - 6\sigma\mathcal{A}_0 + 6\mathcal{B}_0\right]\right] \\
&= \mathbb{S}^{-1}\left[\frac{\omega^{\delta+2}}{\xi^{\delta+2}}\ell^5\sigma\tanh\left(\frac{\beta}{2} + \frac{\ell x}{2}\right)\mathrm{sech}^2\left(\frac{\beta}{2} + \frac{\ell x}{2}\right)\right] \\
&= \ell^5\sigma\tanh\left(\frac{\beta}{2} + \frac{\ell x}{2}\right)\mathrm{sech}^2\left(\frac{\beta}{2} + \frac{\ell x}{2}\right)\frac{\bar{t}^{\delta}}{\Gamma(\delta+1)},
\end{aligned}$$

$$\begin{aligned}
\Psi_1(\mathbf{x}, \bar{t}) &= \mathbb{S}^{-1}\left[\frac{\omega^{\delta}}{\xi^{\delta}}\mathbb{S}\left[-\sigma(\Psi_{xxx})_0 - 3\sigma\mathcal{C}_0\right]\right] \\
&= \frac{\ell^5\sigma^{3/2}}{\sqrt{2}}\tanh\left(\frac{\beta}{2} + \frac{\ell x}{2}\right)\mathrm{sech}^2\left(\frac{\beta}{2} + \frac{\ell x}{2}\right)\frac{\bar{t}^{\delta}}{\Gamma(\delta+1)}.
\end{aligned}$$

$$\begin{aligned}
\Phi_2(\mathbf{x}, \bar{t}) &= \mathbb{S}^{-1}\left[\frac{\omega^{\delta}}{\xi^{\delta}}\mathbb{S}\left[-\sigma(\Phi_{xxx})_1 - 6\sigma\mathcal{A}_1 + 6\mathcal{B}_1\right]\right] \\
&= \mathbb{S}^{-1}\left[\frac{\omega^{2\delta+2}}{\xi^{2\delta+2}}\frac{\ell^8\sigma^2}{2}\left[2\cosh^2\left(\frac{\sigma}{2} + \frac{\ell x}{2}\right) - 3\right]\mathrm{sech}^4\left(\frac{\sigma}{2} + \frac{\ell x}{2}\right)\right] \\
&= \frac{\ell^8\sigma^2}{2}\left[2\cosh^2\left(\frac{\sigma}{2} + \frac{\ell x}{2}\right) - 3\right]\mathrm{sech}^4\left(\frac{\sigma}{2} + \frac{\ell x}{2}\right)\frac{\bar{t}^{2\delta}}{\Gamma(2\delta+1)},
\end{aligned}$$

$$\begin{aligned}
\Psi_2(\mathbf{x}, \bar{t}) &= \mathbb{S}^{-1}\left[\frac{\omega^{\delta}}{\xi^{\delta}}\mathbb{S}\left[-\sigma(\Psi_{xxx})_1 - 3\sigma\mathcal{C}_1\right]\right] \\
&= \frac{\ell^5\sigma^{5/2}}{2\sqrt{2}}\left[2\cosh^2\left(\frac{\sigma}{2} + \frac{\ell x}{2}\right) - 3\right]\mathrm{sech}^4\left(\frac{\sigma}{2} + \frac{\ell x}{2}\right)\frac{\bar{t}^{2\delta}}{\Gamma(2\delta+1)},
\end{aligned}$$

$$\Phi_3(\mathbf{x},\bar{t}) = \mathbb{S}^{-1}\left[\frac{\omega^\delta}{\zeta^\delta}\mathbb{S}\left[-\sigma(\Phi_{xxx})_2 - 6\sigma A_2 + 6B_2\right]\right]$$

$$= \frac{\sigma^3 \ell^4 \sin h(\frac{\sigma}{2}+\frac{\ell x}{2})}{2\Gamma^2(\delta+1)\Gamma(3\delta+1)\cosh^7(\frac{\beta}{2}+\frac{\ell x}{2})}\left[2\Gamma^2(\delta+1)\cos h^4\left(\frac{\sigma}{2}+\frac{\ell x}{2}\right) - 18\Gamma^2(\delta+1)\cos h^2\left(\frac{\sigma}{2}+\frac{\ell x}{2}\right)\right.$$

$$\left. + 6\Gamma(2\delta+1)\cos h^2\left(\frac{\sigma}{2}+\frac{\ell x}{2}\right) + 18\Gamma^2(\delta+1) - 9\Gamma(2\delta+1)\right],$$

$$\Psi_3(\mathbf{x},\bar{t}) = \mathbb{S}^{-1}\left[\frac{\omega^\delta}{\zeta^\delta}\mathbb{S}\left[-\sigma(\Psi_{xxx})_2 - 3\sigma C_2\right]\right]$$

$$= \frac{\sigma^{7/2}\ell^{11}\sin h(\frac{\sigma}{2}+\frac{\ell x}{2})}{2\sqrt{2}\Gamma^2(\delta+1)\Gamma(3\delta+1)\cosh^7(\frac{\beta}{2}+\frac{\ell x}{2})}\left[2\Gamma^2(\delta+1)\cos h^4\left(\frac{\sigma}{2}+\frac{\ell x}{2}\right) - 18\Gamma^2(\delta+1)\cos h^2\left(\frac{\sigma}{2}+\frac{\ell x}{2}\right)\right.$$

$$\left. + 6\Gamma(2\delta+1)\cos h^2\left(\frac{\sigma}{2}+\frac{\ell x}{2}\right) + 18\Gamma^2(\delta+1) - 9\Gamma(2\delta+1)\right],$$

$$\vdots$$

The Shehu decomposition method solution for Problem 1 is presented as:

$$\Phi(\mathbf{x},\bar{t}) = \Phi_0(\mathbf{x},\bar{t}) + \Phi_1(\mathbf{x},\bar{t}) + \Phi_2(\mathbf{x},\bar{t}) + \Phi_3(\mathbf{x},\bar{t}) + \ldots,$$

$$= \ell^2 \sec h^2\left(\frac{\beta}{2}+\frac{\ell x}{2}\right) + \ell^5 \sigma \tan h\left(\frac{\beta}{2}+\frac{\ell x}{2}\right)\sec h^2\left(\frac{\beta}{2}+\frac{\ell x}{2}\right)\frac{\bar{t}^\delta}{\Gamma(\delta+1)}$$

$$+ \frac{\ell^8 \sigma^2}{2}\left[2\cosh^2\left(\frac{\sigma}{2}+\frac{\ell x}{2}\right) - 3\right]\sec h^4\left(\frac{\sigma}{2}+\frac{\ell x}{2}\right)\frac{\bar{t}^{2\delta}}{\Gamma(2\delta+1)}$$

$$+ \frac{\sigma^3 \ell^4 \sin h(\frac{\sigma}{2}+\frac{\ell x}{2})}{2\Gamma^2(\delta+1)\Gamma(3\delta+1)\cosh^7(\frac{\beta}{2}+\frac{\ell x}{2})}\left[2\Gamma^2(\delta+1)\cos h^4\left(\frac{\sigma}{2}+\frac{\ell x}{2}\right) - 18\Gamma^2(\delta+1)\cos h^2\left(\frac{\sigma}{2}+\frac{\ell x}{2}\right)\right.$$

$$\left. + 6\Gamma(2\delta+1)\cos h^2\left(\frac{\sigma}{2}+\frac{\ell x}{2}\right) + 18\Gamma^2(\delta+1) - 9\Gamma(2\delta+1)\right] + \ldots.$$

Analogously, we have

$$\Psi(\mathbf{x},\bar{t}) = \sqrt{\frac{\sigma}{2}}\ell^2 \sec h^2\left(\frac{\beta}{2}+\frac{\ell x}{2}\right) + \frac{\ell^5 \sigma^{3/2}}{\sqrt{2}}\tan h\left(\frac{\beta}{2}+\frac{\ell x}{2}\right)\sec h^2\left(\frac{\beta}{2}+\frac{\ell x}{2}\right)\frac{\bar{t}^\delta}{\Gamma(\delta+1)}$$

$$+ \frac{\ell^5 \sigma^{5/2}}{2\sqrt{2}}\left[2\cosh^2\left(\frac{\sigma}{2}+\frac{\ell x}{2}\right) - 3\right]\sec h^4\left(\frac{\sigma}{2}+\frac{\ell x}{2}\right)\frac{\bar{t}^{2\delta}}{\Gamma(2\delta+1)}$$

$$+ \frac{\sigma^{7/2}\ell^{11}\sin h(\frac{\sigma}{2}+\frac{\ell x}{2})}{2\sqrt{2}\Gamma^2(\delta+1)\Gamma(3\delta+1)\cosh^7(\frac{\beta}{2}+\frac{\ell x}{2})}\left[2\Gamma^2(\delta+1)\cos h^4\left(\frac{\sigma}{2}+\frac{\ell x}{2}\right) - 18\Gamma^2(\delta+1)\cos h^2\left(\frac{\sigma}{2}+\frac{\ell x}{2}\right)\right.$$

$$\left. + 6\Gamma(2\delta+1)\cos h^2\left(\frac{\sigma}{2}+\frac{\ell x}{2}\right) + 18\Gamma^2(\delta+1) - 9\Gamma(2\delta+1)\right] + \ldots.$$

By setting $\delta = 1$, we then obtain the exact solution of coupled KdV Equation (1)

$$\Phi(\mathbf{x},\bar{t}) = \ell^2 \sec h^2\left(\frac{\beta}{2} + \frac{\ell x}{2} - \frac{\sigma \ell^3 \bar{t}}{2}\right),$$

$$\Psi(\mathbf{x},\bar{t}) = \sqrt{\frac{\sigma}{2}}\ell^2 \sec h^2\left(\frac{\beta}{2} + \frac{\ell x}{2} - \frac{\sigma \ell^3 \bar{t}}{2}\right).$$

Case II. Now, we surmise the Shehu iterative transform method on Problem 1.

Applying the proposed analytical approach to (33), yields

$$\Phi_0(\mathbf{x},\bar{t}) = \mathbb{S}^{-1}\left[\frac{\omega}{\xi}\Phi(\mathbf{x},0)\right] = \mathbb{S}^{-1}\left[\frac{\xi}{\omega}\ell^2 \sec h^2\left(\frac{\beta}{2}+\frac{\ell\mathbf{x}}{2}\right)\right]$$

$$= \ell^2 \sec h^2\left(\frac{\beta}{2}+\frac{\ell\mathbf{x}}{2}\right),$$

$$\Psi_0(\mathbf{x},\bar{t}) = \mathbb{S}^{-1}\left[\frac{\omega}{\xi}\Psi(\mathbf{x},0)\right]$$

$$= \sqrt{\frac{\sigma}{2}}\ell^2 \sec h^2\left(\frac{\beta}{2}+\frac{\ell\mathbf{x}}{2}\right).$$

$$\Phi_1(\mathbf{x},\bar{t}) = \mathbb{S}^{-1}\left[\frac{\omega^\delta}{\xi^\delta}\mathbb{S}\left[-\sigma\frac{\partial^3\Phi_0}{\partial x^3}-6\sigma\Phi_0\frac{\partial\Phi_0}{\partial x}+6\Psi_0\frac{\partial\Psi_0}{\partial x}\right]\right]$$

$$= \mathbb{S}^{-1}\left[\frac{\omega^{\delta+2}}{\xi^{\delta+2}}\ell^5\sigma\tan h\left(\frac{\beta}{2}+\frac{\ell\mathbf{x}}{2}\right)\sec h^2\left(\frac{\beta}{2}+\frac{\ell\mathbf{x}}{2}\right)\right]$$

$$= \ell^5\sigma\tan h\left(\frac{\beta}{2}+\frac{\ell\mathbf{x}}{2}\right)\sec h^2\left(\frac{\beta}{2}+\frac{\ell\mathbf{x}}{2}\right)\frac{\bar{t}^\delta}{\Gamma(\delta+1)},$$

$$\Psi_1(\mathbf{x},\bar{t}) = \mathbb{S}^{-1}\left[\frac{\omega^\delta}{\xi^\delta}\mathbb{S}\left[-\sigma\frac{\partial^3\Psi_0}{\partial x^3}-3\sigma\Phi_0\frac{\partial\Psi_0}{\partial x}\right]\right]$$

$$= \frac{\ell^5\sigma^{3/2}}{\sqrt{2}}\tan h\left(\frac{\beta}{2}+\frac{\ell\mathbf{x}}{2}\right)\sec h^2\left(\frac{\beta}{2}+\frac{\ell\mathbf{x}}{2}\right)\frac{\bar{t}^\delta}{\Gamma(\delta+1)}.$$

$$\Phi_2(\mathbf{x},\bar{t}) = \mathbb{S}^{-1}\left[\frac{\omega^\delta}{\xi^\delta}\mathbb{S}\left[-\sigma\frac{\partial^3\Phi_1}{\partial x^3}-6\sigma\Phi_1\frac{\partial\Phi_1}{\partial x}+6\Psi_1\frac{\partial\Psi_1}{\partial x}\right]\right]$$

$$= \mathbb{S}^{-1}\left[\frac{\omega^{2\delta+2}}{\xi^{2\delta+2}}\frac{\ell^8\sigma^2}{2}\left[2\cosh^2\left(\frac{\sigma}{2}+\frac{\ell\mathbf{x}}{2}\right)-3\right]\sec h^4\left(\frac{\sigma}{2}+\frac{\ell\mathbf{x}}{2}\right)\right]$$

$$= \frac{\ell^8\sigma^2}{2}\left[2\cosh^2\left(\frac{\sigma}{2}+\frac{\ell\mathbf{x}}{2}\right)-3\right]\sec h^4\left(\frac{\sigma}{2}+\frac{\ell\mathbf{x}}{2}\right)\frac{\bar{t}^{2\delta}}{\Gamma(2\delta+1)},$$

$$\Psi_2(\mathbf{x},\bar{t}) = \mathbb{S}^{-1}\left[\frac{\omega^\delta}{\xi^\delta}\mathbb{S}\left[-\sigma\frac{\partial^3\Psi_1}{\partial x^3}-3\sigma\Phi_1\frac{\partial\Psi_1}{\partial x}\right]\right]$$

$$= \frac{\ell^5\sigma^{5/2}}{2\sqrt{2}}\left[2\cosh^2\left(\frac{\sigma}{2}+\frac{\ell\mathbf{x}}{2}\right)-3\right]\sec h^4\left(\frac{\sigma}{2}+\frac{\ell\mathbf{x}}{2}\right)\frac{\bar{t}^{2\delta}}{\Gamma(2\delta+1)},$$

$$\Phi_3(\mathbf{x},\bar{t}) = \mathbb{S}^{-1}\left[\frac{\omega^\delta}{\xi^\delta}\mathbb{S}\left[-\sigma\frac{\partial^3\Phi_2}{\partial x^3}-6\sigma\Phi_2\frac{\partial\Phi_2}{\partial x}+6\Psi_2\frac{\partial\Psi_2}{\partial x}\right]\right]$$

$$= \frac{\sigma^3\ell^4\sin h\left(\frac{\sigma}{2}+\frac{\ell\mathbf{x}}{2}\right)}{2\Gamma^2(\delta+1)\Gamma(3\delta+1)\cosh^7\left(\frac{\beta}{2}+\frac{\ell\mathbf{x}}{2}\right)}\left[2\Gamma^2(\delta+1)\cos h^4\left(\frac{\sigma}{2}+\frac{\ell\mathbf{x}}{2}\right)-18\Gamma^2(\delta+1)\cos h^2\left(\frac{\sigma}{2}+\frac{\ell\mathbf{x}}{2}\right)\right.$$

$$\left.+6\Gamma(2\delta+1)\cos h^2\left(\frac{\sigma}{2}+\frac{\ell\mathbf{x}}{2}\right)+18\Gamma^2(\delta+1)-9\Gamma(2\delta+1)\right],$$

$$\Psi_3(\mathbf{x},\bar{t}) = \mathbb{S}^{-1}\left[\frac{\omega^\delta}{\xi^\delta}\mathbb{S}\left[-\sigma\frac{\partial^3\Psi_2}{\partial x^3}-3\sigma\Phi_2\frac{\partial\Psi_2}{\partial x}\right]\right]$$

$$= \frac{\sigma^{7/2}\ell^{11}\sin h\left(\frac{\sigma}{2}+\frac{\ell\mathbf{x}}{2}\right)}{2\sqrt{2}\Gamma^2(\delta+1)\Gamma(3\delta+1)\cosh^7\left(\frac{\beta}{2}+\frac{\ell\mathbf{x}}{2}\right)}\left[2\Gamma^2(\delta+1)\cos h^4\left(\frac{\sigma}{2}+\frac{\ell\mathbf{x}}{2}\right)-18\Gamma^2(\delta+1)\cos h^2\left(\frac{\sigma}{2}+\frac{\ell\mathbf{x}}{2}\right)\right.$$

$$\left.+6\Gamma(2\delta+1)\cos h^2\left(\frac{\sigma}{2}+\frac{\ell\mathbf{x}}{2}\right)+18\Gamma^2(\delta+1)-9\Gamma(2\delta+1)\right],$$

$\vdots$

$$\Phi_n(\mathbf{x},\bar{t}) = \mathbb{S}^{-1}\left[\frac{\omega^\delta}{\xi^\delta}\mathbb{S}\left[-\sigma\frac{\partial^3\Phi_{m-1}}{\partial x^3} - 6\sigma\Phi_{m-1}\frac{\partial\Phi_{m-1}}{\partial x} + 6\Psi_{m-1}\frac{\partial\Psi_{m-1}}{\partial x}\right]\right],$$

$$\Psi_m(\mathbf{x},\bar{t}) = \mathbb{S}^{-1}\left[\frac{\omega^\delta}{\xi^\delta}\mathbb{S}\left[-\sigma\frac{\partial^3\Psi_{m-1}}{\partial x^3} - 3\sigma\Phi_{m-1}\frac{\partial\Psi_{m-1}}{\partial x}\right]\right].$$

The series of solutions for Problem 1 is presented as:

$$\Phi(\mathbf{x},\bar{t}) = \Phi_0(\mathbf{x},\bar{t}) + \Phi_1(\mathbf{x},\bar{t}) + \Phi_2(\mathbf{x},\bar{t}) + \Phi_3(\mathbf{x},\bar{t}) + \ldots \Phi_m(\mathbf{x},\bar{t}),$$
$$\Psi(\mathbf{x},\bar{t}) = \Psi_0(\mathbf{x},\bar{t}) + \Psi_1(\mathbf{x},\bar{t}) + \Psi_2(\mathbf{x},\bar{t}) + \Psi_3(\mathbf{x},\bar{t}) + \ldots \Psi_m(\mathbf{x},\bar{t}).$$

Consequently, we have

$$\Phi(\mathbf{x},\bar{t}) = \ell^2 \operatorname{sech}^2\left(\frac{\beta}{2} + \frac{\ell \mathbf{x}}{2}\right) + \ell^5 \sigma \tanh\left(\frac{\beta}{2} + \frac{\ell \mathbf{x}}{2}\right)\operatorname{sech}^2\left(\frac{\beta}{2} + \frac{\ell \mathbf{x}}{2}\right)\frac{\bar{t}^\delta}{\Gamma(\delta+1)}$$
$$+ \frac{\ell^8 \sigma^2}{2}\left[2\cosh^2\left(\frac{\sigma}{2} + \frac{\ell \mathbf{x}}{2}\right) - 3\right]\operatorname{sech}^4\left(\frac{\sigma}{2} + \frac{\ell \mathbf{x}}{2}\right)\frac{\bar{t}^{2\delta}}{\Gamma(2\delta+1)}$$
$$+ \frac{\sigma^3 \ell^4 \sinh\left(\frac{\sigma}{2} + \frac{\ell \mathbf{x}}{2}\right)}{2\Gamma^2(\delta+1)\Gamma(3\delta+1)\cosh^7\left(\frac{\beta}{2} + \frac{\ell \mathbf{x}}{2}\right)}\left[2\Gamma^2(\delta+1)\cos h^4\left(\frac{\sigma}{2} + \frac{\ell \mathbf{x}}{2}\right) - 18\Gamma^2(\delta+1)\cos h^2\left(\frac{\sigma}{2} + \frac{\ell \mathbf{x}}{2}\right)\right.$$
$$\left. + 6\Gamma(2\delta+1)\cos h^2\left(\frac{\sigma}{2} + \frac{\ell \mathbf{x}}{2}\right) + 18\Gamma^2(\delta+1) - 9\Gamma(2\delta+1)\right] + \ldots,$$

$$\Psi(\mathbf{x},\bar{t}) = \sqrt{\frac{\sigma}{2}}\ell^2 \operatorname{sech}^2\left(\frac{\beta}{2} + \frac{\ell \mathbf{x}}{2}\right) + \frac{\ell^5 \sigma^{3/2}}{\sqrt{2}}\tanh\left(\frac{\beta}{2} + \frac{\ell \mathbf{x}}{2}\right)\operatorname{sech}^2\left(\frac{\beta}{2} + \frac{\ell \mathbf{x}}{2}\right)\frac{\bar{t}^\delta}{\Gamma(\delta+1)}$$
$$+ \frac{\ell^5 \sigma^{5/2}}{2\sqrt{2}}\left[2\cosh^2\left(\frac{\sigma}{2} + \frac{\ell \mathbf{x}}{2}\right) - 3\right]\operatorname{sech}^4\left(\frac{\sigma}{2} + \frac{\ell \mathbf{x}}{2}\right)\frac{\bar{t}^{2\delta}}{\Gamma(2\delta+1)}$$
$$+ \frac{\sigma^{7/2}\ell^{11}\sinh\left(\frac{\sigma}{2} + \frac{\ell \mathbf{x}}{2}\right)}{2\sqrt{2}\Gamma^2(\delta+1)\Gamma(3\delta+1)\cosh^7\left(\frac{\beta}{2} + \frac{\ell \mathbf{x}}{2}\right)}\left[2\Gamma^2(\delta+1)\cos h^4\left(\frac{\sigma}{2} + \frac{\ell \mathbf{x}}{2}\right) - 18\Gamma^2(\delta+1)\cos h^2\left(\frac{\sigma}{2} + \frac{\ell \mathbf{x}}{2}\right)\right.$$
$$\left. + 6\Gamma(2\delta+1)\cos h^2\left(\frac{\sigma}{2} + \frac{\ell \mathbf{x}}{2}\right) + 18\Gamma^2(\delta+1) - 9\Gamma(2\delta+1)\right] + \ldots.$$

By setting $\delta = 1$, we then obtain the exact solution of coupled KdV Equation (1)

$$\Phi(\mathbf{x},\bar{t}) = \ell^2 \operatorname{sech}^2\left(\frac{\beta}{2} + \frac{\ell \mathbf{x}}{2} - \frac{\sigma\ell^3 \bar{t}}{2}\right),$$

$$\Psi(\mathbf{x},\bar{t}) = \sqrt{\frac{\sigma}{2}}\ell^2 \operatorname{sech}^2\left(\frac{\beta}{2} + \frac{\ell \mathbf{x}}{2} - \frac{\sigma\ell^3 \bar{t}}{2}\right).$$

In Figures 1 and 2, the exact and approximate results of $\Phi(\mathbf{x},\bar{t})$ and $\Psi(\mathbf{x},\bar{t})$ are demonstrated at $\ell = 1$, $\sigma = 0.5$ and $\beta = 2$. In Figures 3 and 4, the surface and 2D graph for $\Phi(\mathbf{x},\bar{t})$ and $\Psi(\mathbf{x},\bar{t})$ for various fractional order are presented which shows that the SDM/SITM approximated results derived are in a strong agreement with the exact and the numerical ones. This comparison represents a strong correlation between the SDM and exact findings. Therefore, the SDM/SITM are reliable novel approaches which require less computation time and is quite straightforward and more flexible than the homotopy perturbation method and homotopy analysis method.

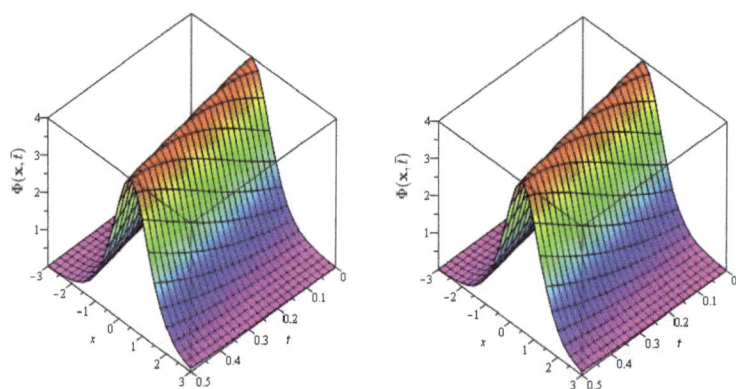

Figure 1. The exact and approximate (SDM/SITM) solution graph at $\Phi(x,\bar{t})$ of Problem 1 for $\ell = 1$, $\sigma = 0.5$ and $\beta = 2$.

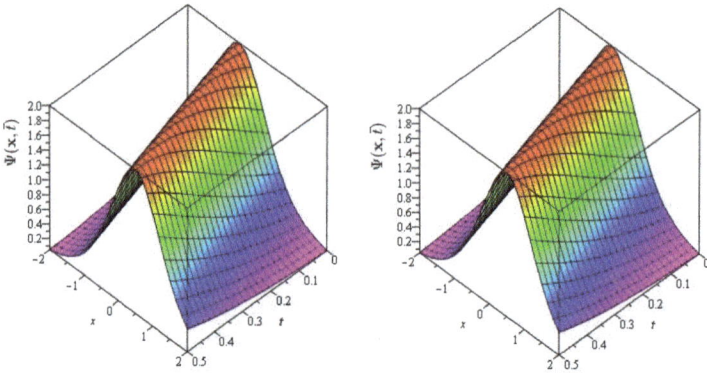

Figure 2. The exact and approximate (SDM/SITM) solution graph at $\Psi(x,\bar{t})$ of Problem 1 for $\ell = 1$, $\sigma = 0.5$ and $\beta = 2$.

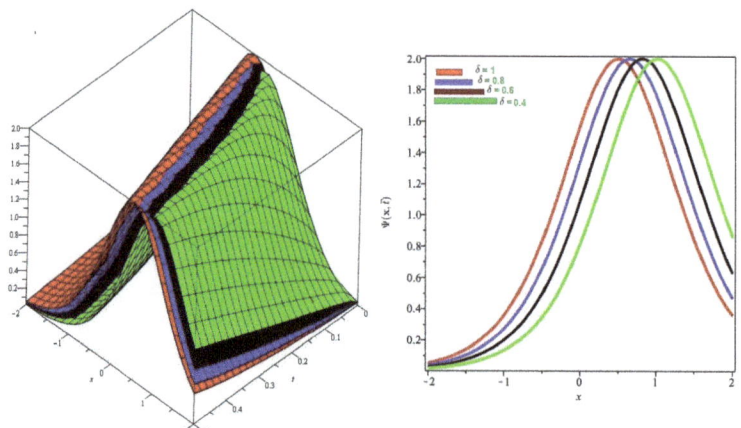

Figure 3. Numerical evaluation of graph of $\Psi(x,\bar{t})$ for Problem 1 for various fractional order $\delta = 0.4, 0.6, 0.8, 1$, $\ell = 1$, $\sigma = 0.5$ and $\beta = 2$.

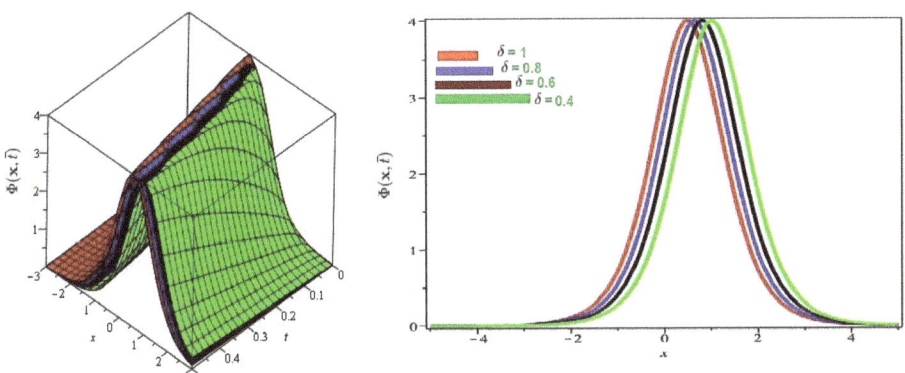

Figure 4. Numerical evaluation of graph $\Phi(x,\bar{t})$ for Problem 1 for various fractional order $\delta = 0.4, 0.6, 0.8, 1, \ell = 1, \sigma = 0.5$ and $\beta = 2$.

Problem 2 ([16]). *Assume the time-fractional coupled nonlinear KdV equation is presented as:*

$$\frac{\partial^\delta \Phi}{\partial \bar{t}^\delta} = -\frac{\partial \Psi}{\partial x} - \frac{1}{2}\frac{\partial \Phi^2}{\partial x}$$

$$\frac{\partial^\delta \Psi}{\partial \bar{t}^\delta} = -\frac{\partial \Phi}{\partial x} - \frac{\partial^3 \Phi}{\partial x^3} - \frac{\partial \Phi \Psi}{\partial x}, \quad \bar{t} > 0, \; 0 < \delta \le 1 \qquad (34)$$

subject to the condition

$$\Phi(x,0) = \sigma\left[\tan h\left(\frac{\ell}{2} + \frac{\sigma x}{2}\right) + 1\right], \quad \Psi(x,0) = \frac{\sigma^2}{2}\sec h^2\left(\frac{\ell}{2} + \frac{\sigma x}{2}\right) - 1. \qquad (35)$$

Case I. First, we surmise the Shehu decomposition method for Problem 2. Employing the Shehu transformation to (34), we find

$$\frac{\xi^\delta}{\omega^\delta}\mathcal{U}(\xi,\omega) - \sum_{\kappa=0}^{m-1}\left(\frac{\xi}{\omega}\right)^{\delta-\kappa-1}\Phi^{(\kappa)}(0) = \mathbb{S}\left[-\frac{\partial \Psi}{\partial x} - \frac{1}{2}\frac{\partial \Phi^2}{\partial x}\right],$$

$$\frac{\xi^\delta}{\omega^\delta}\mathcal{V}(\xi,\omega) - \sum_{\kappa=0}^{m-1}\left(\frac{\xi}{\omega}\right)^{\delta-\kappa-1}\Psi^{(\kappa)}(0) = \mathbb{S}\left[-\frac{\partial \Phi}{\partial x} - \frac{\partial^3 \Phi}{\partial x^3} - \frac{\partial \Phi \Psi}{\partial x}\right]. \qquad (36)$$

In view of (35) and simple computations yield

$$\mathcal{U}(\xi,\omega) = \frac{\omega}{\xi}\Phi^{(0)}(x,0) + \frac{\omega^\delta}{\xi^\delta}\mathbb{S}\left[-\frac{\partial \Psi}{\partial x} - \frac{1}{2}\frac{\partial \Phi^2}{\partial x}\right],$$

$$\mathcal{V}(\xi,\omega) = \frac{\omega}{\xi}\Psi^{(0)}(x,0) + \frac{\omega^\delta}{\xi^\delta}\mathbb{S}\left[-\frac{\partial \Phi}{\partial x} - \frac{\partial^3 \Phi}{\partial x^3} - \frac{\partial \Phi \Psi}{\partial x}\right]. \qquad (37)$$

Applying the inverse Shehu transform, we have

$$\Phi(x,\bar{t}) = \mathbb{S}^{-1}\left[\frac{\omega}{\xi}\Phi(x,0)\right] + \mathbb{S}^{-1}\left[\frac{\omega^\delta}{\xi^\delta}\mathbb{S}\left[-\frac{\partial \Psi}{\partial x} - \frac{1}{2}\frac{\partial \Phi^2}{\partial x}\right]\right],$$

$$\Psi(x,\bar{t}) = \mathbb{S}^{-1}\left[\frac{\omega}{\xi}\Psi(x,0)\right] + \mathbb{S}^{-1}\left[\frac{\omega^\delta}{\xi^\delta}\mathbb{S}\left[-\frac{\partial \Phi}{\partial x} - \frac{\partial^3 \Phi}{\partial x^3} - \frac{\partial \Phi \Psi}{\partial x}\right]\right]. \qquad (38)$$

By virtue of the Shehu decomposition method, we have

$$\Phi_0(\mathbf{x},\bar{t}) = \mathbb{S}^{-1}\left[\frac{\omega}{\xi}\Phi(x,0)\right] = \mathbb{S}^{-1}\left[\frac{\omega}{\xi}\sigma\left(\tanh\left(\frac{\ell}{2}+\frac{\sigma x}{2}\right)+1\right)\right]$$

$$= \sigma\left(\tanh\left(\frac{\ell}{2}+\frac{\sigma x}{2}\right)+1\right),$$

$$\Psi_0(\mathbf{x},\bar{t}) = \mathbb{S}^{-1}\left[\frac{\omega}{\xi}\Psi(x,0)\right]$$

$$= \frac{\sigma^2}{2}\operatorname{sech}^2\left(\frac{\ell}{2}+\frac{\sigma x}{2}\right)-1.$$

It follows that

$$\sum_{m=0}^{\infty}\Phi_{m+1}(\mathbf{x},\bar{t}) = \mathbb{S}^{-1}\left[\frac{\omega^\delta}{\xi^\delta}\mathbb{S}\left[-\sigma\sum_{m=0}^{\infty}(\Psi_x)_m - \frac{1}{2}\sum_{m=0}^{\infty}\mathcal{D}_m\right]\right],$$

$$\sum_{m=0}^{\infty}\Psi_{m+1}(\mathbf{x},\bar{t}) = \mathbb{S}^{-1}\left[\frac{\omega^\delta}{\xi^\delta}\mathbb{S}\left[-\sum_{m=0}^{\infty}(\Phi_x)_m - \sum_{m=0}^{\infty}(\Psi_{xxx})_m - \sum_{m=0}^{\infty}((\Phi\Psi)_x)_m\right]\right], \quad m=0,1,2,\ldots.$$

The first few Adomian polynomials are presented as follows:

$$\mathcal{D}_0(\Phi^2) = \Phi_0^2,$$
$$\mathcal{D}_1(\Phi^2) = 2\Phi_0\Phi_1,$$
$$\mathcal{D}_2(\Phi^2) = 2\Phi_0\Phi_2 + \Phi_1^2.$$

For $m = 0,1,2,\ldots$

$$\Phi_1(\mathbf{x},\bar{t}) = \mathbb{S}^{-1}\left[\frac{\omega^\delta}{\xi^\delta}\mathbb{S}\left[-\sigma(\Psi_x)_0 - \frac{1}{2}\mathcal{D}_0\right]\right]$$

$$= -\frac{\sigma^2}{2}\mathbb{S}^{-1}\left[\frac{\omega^{\delta+2}}{\xi^{\delta+2}}\operatorname{sech}^2\left(\frac{\ell}{2}+\frac{\sigma x}{2}\right)\right]$$

$$= -\frac{\sigma^2}{2}\operatorname{sech}^2\left(\frac{\ell}{2}+\frac{\sigma x}{2}\right)\frac{\bar{t}^\delta}{\Gamma(\delta+1)},$$

$$\Psi_1(\mathbf{x},\bar{t}) = \mathbb{S}^{-1}\left[\frac{\omega^\delta}{\xi^\delta}\mathbb{S}\left[-(\Phi_x)_0 - (\Psi_{xxx})_0 - ((\Phi\Psi)_x)_0\right]\right]$$

$$= \frac{\sigma^3}{2}\sinh\left(\frac{\ell}{2}+\frac{\sigma x}{2}\right)\operatorname{sech}^3\left(\frac{\ell}{2}+\frac{\sigma x}{2}\right)\frac{\bar{t}^\delta}{\Gamma(\delta+1)}.$$

$$\Phi_2(\mathbf{x},\bar{t}) = \mathbb{S}^{-1}\left[\frac{\omega^\delta}{\xi^\delta}\mathbb{S}\left[-\sigma(\Psi_x)_1 - \frac{1}{2}\mathcal{D}_1\right]\right]$$

$$= \mathbb{S}^{-1}\left[-\frac{\sigma^5}{4}\frac{\omega^{2\delta+2}}{\xi^{2\delta+2}}\operatorname{sech}^2\left(\frac{\ell}{2}+\frac{\sigma x}{2}\right)+\frac{3\sigma^5}{4}\frac{\omega^{2\delta+2}}{\xi^{2\delta+2}}\sinh^2\left(\frac{\ell}{2}+\frac{\sigma x}{2}\right)\operatorname{sech}^4\left(\frac{\ell}{2}+\frac{\sigma x}{2}\right)\right]$$

$$+\frac{\sigma^7}{4}\mathbb{S}^{-1}\left[\frac{\Gamma(2\delta+1)}{\Gamma^2(\delta+1)}\frac{\omega^{3\delta+2}}{\xi^{3\delta+2}}\sinh\left(\frac{\ell}{2}+\frac{\sigma x}{2}\right)\operatorname{sech}^5\left(\frac{\ell}{2}+\frac{\sigma x}{2}\right)\right]$$

$$= \left[-\frac{\sigma^5}{4}\operatorname{sech}^2\left(\frac{\ell}{2}+\frac{\sigma x}{2}\right)+\frac{3\sigma^5}{4}\sinh^2\left(\frac{\ell}{2}+\frac{\sigma x}{2}\right)\operatorname{sech}^4\left(\frac{\ell}{2}+\frac{\sigma x}{2}\right)\right]\frac{\bar{t}^{2\delta}}{\Gamma(2\delta+1)}$$

$$+\frac{\sigma^7}{4}\sinh\left(\frac{\ell}{2}+\frac{\sigma x}{2}\right)\operatorname{sech}^5\left(\frac{\ell}{2}+\frac{\sigma x}{2}\right)\frac{\Gamma(2\delta+1)\bar{t}^{3\delta}}{\Gamma^2(\delta+1)\Gamma(3\delta+1)},$$

$$\Psi_2(x,\bar{t}) = \mathbb{S}^{-1}\left[\frac{\omega^\delta}{\zeta^\delta}\left[-(\Phi_x)_1 - (\Psi_{xxx})_1 - ((\Phi\Psi)_x)_1\right]\right]$$

$$= \frac{\sigma^6}{4}\left[2\cosh^2\left(\frac{\ell}{2} + \frac{\sigma x}{2}\right) - 3\right]\sec h^4\left(\frac{\ell}{2} + \frac{\sigma x}{2}\right)\frac{\bar{t}^{2\delta}}{\Gamma(2\delta+1)},$$

$$\vdots$$

The Shehu decomposition method solution for Problem 2 is presented as:

$$\Phi(x,\bar{t}) = \Phi_0(x,\bar{t}) + \Phi_1(x,\bar{t}) + \Phi_2(x,\bar{t}) + \dots,$$

$$= \sigma\left(\tanh\left(\frac{\ell}{2} + \frac{\sigma x}{2}\right) + 1\right) - \frac{\sigma^2}{2}\sec h^2\left(\frac{\ell}{2} + \frac{\sigma x}{2}\right)\frac{\bar{t}^\delta}{\Gamma(\delta+1)}$$

$$+ \left[-\frac{\sigma^5}{4}\sec h^2\left(\frac{\ell}{2} + \frac{\sigma x}{2}\right) + \frac{3\sigma^5}{4}\sin h^2\left(\frac{\ell}{2} + \frac{\sigma x}{2}\right)\sec h^4\left(\frac{\ell}{2} + \frac{\sigma x}{2}\right)\right]\frac{\bar{t}^{2\delta}}{\Gamma(2\delta+1)}$$

$$+ \frac{\sigma^7}{4}\sin h\left(\frac{\ell}{2} + \frac{\sigma x}{2}\right)\sec h^5\left(\frac{\ell}{2} + \frac{\sigma x}{2}\right)\frac{\Gamma(2\delta+1)\bar{t}^{3\delta}}{\Gamma^2(\delta+1)\Gamma(3\delta+1)} + \dots.$$

Analogously, we have

$$\Psi(x,\bar{t}) = -1 + \frac{\sigma^2}{2}\sec h^2\left(\frac{\ell}{2} + \frac{\sigma x}{2}\right) + \frac{\sigma^3}{2}\sin h\left(\frac{\ell}{2} + \frac{\sigma x}{2}\right)\sec h^3\left(\frac{\ell}{2} + \frac{\sigma x}{2}\right)\frac{\bar{t}^\delta}{\Gamma(\delta+1)}$$

$$+ \frac{\sigma^6}{4}\left[2\cosh^2\left(\frac{\ell}{2} + \frac{\sigma x}{2}\right) - 3\right]\sec h^4\left(\frac{\ell}{2} + \frac{\sigma x}{2}\right)\frac{\bar{t}^{2\delta}}{\Gamma(2\delta+1)} + \dots.$$

By setting $\delta = 1$, we obtain the exact solution of the coupled KdV Equation (34)

$$\Phi(x,\bar{t}) = \sigma\left(\tanh\left(\frac{\ell}{2} + \frac{\sigma x}{2} - \frac{\sigma^2 \bar{t}}{2}\right) + 1\right),$$

$$\Psi(x,\bar{t}) = \frac{\sigma^2}{2}\sec h^2\left(\frac{\ell}{2} + \frac{\sigma x}{2} - \frac{\sigma^2 \bar{t}}{2}\right) - 1.$$

Case II. Now, we surmise the Shehu iterative transform method on Problem 2. Applying the proposed analytical approach to (38) yields

$$\Phi_0(x,\bar{t}) = \mathbb{S}^{-1}\left[\frac{\omega}{\zeta}\Phi(x,0)\right] = \mathbb{S}^{-1}\left[\frac{\omega}{\zeta}\sigma\left(\tanh\left(\frac{\ell}{2} + \frac{\sigma x}{2}\right) + 1\right)\right]$$

$$= \sigma\left(\tanh\left(\frac{\ell}{2} + \frac{\sigma x}{2}\right) + 1\right),$$

$$\Psi_0(x,\bar{t}) = \mathbb{S}^{-1}\left[\frac{\omega}{\zeta}\Psi(x,0)\right]$$

$$= \frac{\sigma^2}{2}\sec h^2\left(\frac{\ell}{2} + \frac{\sigma x}{2}\right) - 1$$

$$\Phi_1(x,\bar{t}) = \mathbb{S}^{-1}\left[\frac{\omega^\delta}{\zeta^\delta}\mathbb{S}\left[-\frac{\partial \Psi_0}{\partial x} - \frac{1}{2}\frac{\partial \Phi_0^2}{\partial x}\right]\right]$$

$$= -\frac{\sigma^2}{2}\mathbb{S}^{-1}\left[\frac{\omega^{\delta+2}}{\zeta^{\delta+2}}\sec h^2\left(\frac{\ell}{2} + \frac{\sigma x}{2}\right)\right]$$

$$= -\frac{\sigma^2}{2}\sec h^2\left(\frac{\ell}{2} + \frac{\sigma x}{2}\right)\frac{\bar{t}^\delta}{\Gamma(\delta+1)},$$

$$\Psi_1(\mathbf{x},\bar{t}) = \mathbb{S}^{-1}\left[\frac{\omega^\delta}{\zeta^\delta}\mathbb{S}\left[-\frac{\partial\Phi_0}{\partial x}-\frac{\partial^3\Phi_0}{\partial x^3}-\frac{\partial\Phi_0\Psi_0}{\partial x}\right]\right]$$

$$= \frac{\sigma^3}{2}\sinh\left(\frac{\ell}{2}+\frac{\sigma x}{2}\right)\operatorname{sech}^3\left(\frac{\ell}{2}+\frac{\sigma x}{2}\right)\frac{\bar{t}^\delta}{\Gamma(\delta+1)},$$

$$\Phi_2(\mathbf{x},\bar{t}) = \mathbb{S}^{-1}\left[\frac{\omega^\delta}{\zeta^\delta}\mathbb{S}\left[-\frac{\partial\Psi_1}{\partial x}-\frac{1}{2}\frac{\partial\Phi_1^2}{\partial x}\right]\right]$$

$$= \mathbb{S}^{-1}\left[-\frac{\sigma^5}{4}\frac{\omega^{2\delta+2}}{\zeta^{2\delta+2}}\operatorname{sech}^2\left(\frac{\ell}{2}+\frac{\sigma x}{2}\right)+\frac{3\sigma^5}{4}\frac{\omega^{2\delta+2}}{\zeta^{2\delta+2}}\sinh^2\left(\frac{\ell}{2}+\frac{\sigma x}{2}\right)\operatorname{sech}^4\left(\frac{\ell}{2}+\frac{\sigma x}{2}\right)\right]$$

$$+\frac{\sigma^7}{4}\mathbb{S}^{-1}\left[\frac{\Gamma(2\delta+1)}{\Gamma^2(\delta+1)}\frac{\omega^{3\delta+2}}{\zeta^{3\delta+2}}\sinh\left(\frac{\ell}{2}+\frac{\sigma x}{2}\right)\operatorname{sech}^5\left(\frac{\ell}{2}+\frac{\sigma x}{2}\right)\right]$$

$$= \left[-\frac{\sigma^5}{4}\operatorname{sech}^2\left(\frac{\ell}{2}+\frac{\sigma x}{2}\right)+\frac{3\sigma^5}{4}\sinh^2\left(\frac{\ell}{2}+\frac{\sigma x}{2}\right)\operatorname{sech}^4\left(\frac{\ell}{2}+\frac{\sigma x}{2}\right)\right]\frac{\bar{t}^{2\delta}}{\Gamma(2\delta+1)}$$

$$+\frac{\sigma^7}{4}\sinh\left(\frac{\ell}{2}+\frac{\sigma x}{2}\right)\operatorname{sech}^5\left(\frac{\ell}{2}+\frac{\sigma x}{2}\right)\frac{\Gamma(2\delta+1)\bar{t}^{3\delta}}{\Gamma^2(\delta+1)\Gamma(3\delta+1)},$$

$$\Psi_2(\mathbf{x},\bar{t}) = \mathbb{S}^{-1}\left[\frac{\omega^\delta}{\zeta^\delta}\left[-\frac{\partial\Phi_1}{\partial x}-\frac{\partial^3\Phi_1}{\partial x^3}-\frac{\partial\Phi_1\Psi_1}{\partial x}\right]\right]$$

$$= \frac{\sigma^6}{4}\left[2\cosh^2\left(\frac{\ell}{2}+\frac{\sigma x}{2}\right)-3\right]\operatorname{sech}^4\left(\frac{\ell}{2}+\frac{\sigma x}{2}\right)\frac{\bar{t}^{2\delta}}{\Gamma(2\delta+1)},$$

$$\vdots$$

$$\Phi_m(\mathbf{x},\bar{t}) = \mathbb{S}^{-1}\left[\frac{\omega^\delta}{\zeta^\delta}\mathbb{S}\left[-\frac{\partial\Psi_{m-1}}{\partial x}-\frac{1}{2}\frac{\partial\Phi_{m-1}^2}{\partial x}\right]\right],$$

$$\Psi_m(\mathbf{x},\bar{t}) = \mathbb{S}^{-1}\left[\frac{\omega^\delta}{\zeta^\delta}\mathbb{S}\left[-\frac{\partial\Phi_{m-1}}{\partial x}-\frac{\partial^3\Phi_{m-1}}{\partial x^3}-\frac{\partial\Phi_{m-1}\Psi_{m-1}}{\partial x}\right]\right].$$

The series of solution for Problem 2 is presented as:

$$\begin{aligned}\Phi(\mathbf{x},\bar{t}) &= \Phi_0(\mathbf{x},\bar{t})+\Phi_1(\mathbf{x},\bar{t})+\Phi_2(\mathbf{x},\bar{t})+\ldots\Phi_m(\mathbf{x},\bar{t}),\\ \Psi(\mathbf{x},\bar{t}) &= \Psi_0(\mathbf{x},\bar{t})+\Psi_1(\mathbf{x},\bar{t})+\Psi_2(\mathbf{x},\bar{t})+\ldots\Psi_m(\mathbf{x},\bar{t}).\end{aligned}$$

Consequently, we have

$$\Phi(\mathbf{x},\bar{t}) = \sigma\left(\tanh\left(\frac{\ell}{2}+\frac{\sigma x}{2}\right)+1\right)-\frac{\sigma^2}{2}\operatorname{sech}^2\left(\frac{\ell}{2}+\frac{\sigma x}{2}\right)\frac{\bar{t}^\delta}{\Gamma(\delta+1)}$$

$$+\left[-\frac{\sigma^5}{4}\operatorname{sech}^2\left(\frac{\ell}{2}+\frac{\sigma x}{2}\right)+\frac{3\sigma^5}{4}\sinh^2\left(\frac{\ell}{2}+\frac{\sigma x}{2}\right)\operatorname{sech}^4\left(\frac{\ell}{2}+\frac{\sigma x}{2}\right)\right]\frac{\bar{t}^{2\delta}}{\Gamma(2\delta+1)}$$

$$+\frac{\sigma^7}{4}\sinh\left(\frac{\ell}{2}+\frac{\sigma x}{2}\right)\operatorname{sech}^5\left(\frac{\ell}{2}+\frac{\sigma x}{2}\right)\frac{\Gamma(2\delta+1)\bar{t}^{3\delta}}{\Gamma^2(\delta+1)\Gamma(3\delta+1)}+\ldots,$$

$$\Psi(\mathbf{x},\bar{t}) = -1+\frac{\sigma^2}{2}\operatorname{sech}^2\left(\frac{\ell}{2}+\frac{\sigma x}{2}\right)+\frac{\sigma^3}{2}\sinh\left(\frac{\ell}{2}+\frac{\sigma x}{2}\right)\operatorname{sech}^3\left(\frac{\ell}{2}+\frac{\sigma x}{2}\right)\frac{\bar{t}^\delta}{\Gamma(\delta+1)}$$

$$+\frac{\sigma^6}{4}\left[2\cosh^2\left(\frac{\ell}{2}+\frac{\sigma x}{2}\right)-3\right]\operatorname{sech}^4\left(\frac{\ell}{2}+\frac{\sigma x}{2}\right)\frac{\bar{t}^{2\delta}}{\Gamma(2\delta+1)}+\ldots.$$

By setting $\delta = 1$, we then obtain the exact solution of coupled KdV Equation (34)

$$\Phi(x, \bar{t}) = \sigma\left(\tanh\left(\frac{\ell}{2} + \frac{\sigma x}{2} - \frac{\sigma^2 \bar{t}}{2}\right) + 1\right),$$

$$\Psi(x, \bar{t}) = \frac{\sigma^2}{2}\operatorname{sech}^2\left(\frac{\ell}{2} + \frac{\sigma x}{2} - \frac{\sigma^2 \bar{t}}{2}\right) - 1.$$

In Figures 5 and 6, the exact and approximate results of $\Phi(x, \bar{t})$ and $\Psi(x, \bar{t})$ are demonstrated at $\ell = 1$, $\sigma = 0.5$ and $\beta = 2$. In Figures 7 and 8, the surface and 2D graph for $\Phi(x, \bar{t})$ and $\Psi(x, \bar{t})$ for various fractional order are presented which shows that the SDM/SITM approximated results derived are in a strong agreement with the exact and the numerical ones. This comparison represents a strong correlation between the SDM and exact findings. Therefore, the SDM/SITM are reliable novel approaches which require less computation time and are quite straightforward and more flexible than the homotopy perturbation method and the homotopy analysis method.

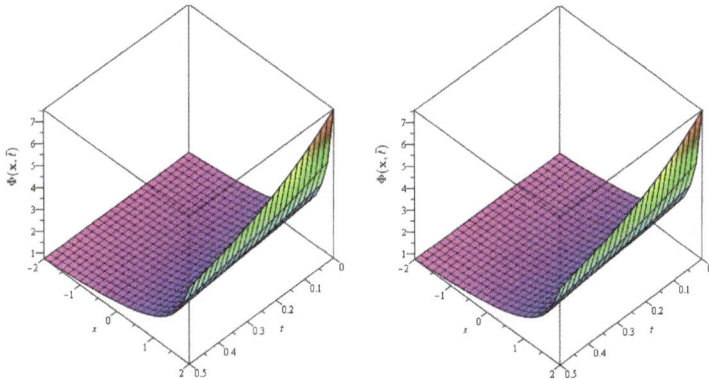

Figure 5. The exact and approximate (SDM/SITM) solution graph at $\Phi(x, \bar{t})$ of Problem 2 for $\ell = 1$, $\sigma = 0.5$ and $\beta = 2$.

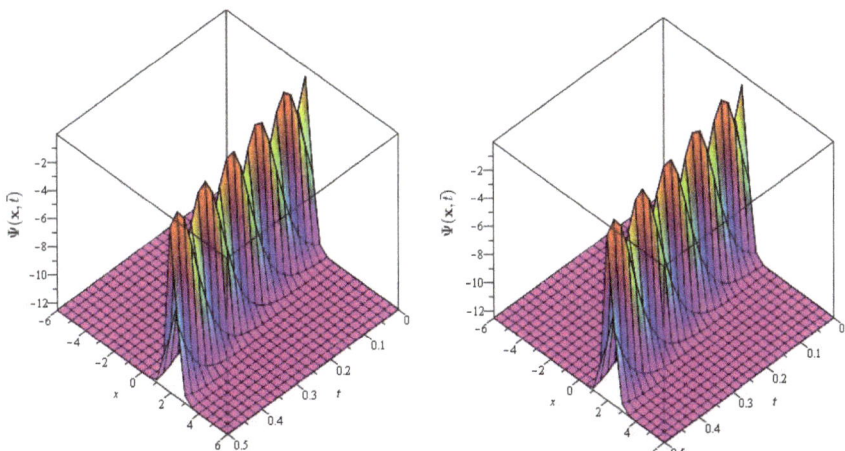

Figure 6. The exact and approximate (SDM/SITM) solution graph at $\Psi(x, \bar{t})$ of Problem 2 for $\ell = 1$, $\sigma = 0.5$ and $\beta = 2$.

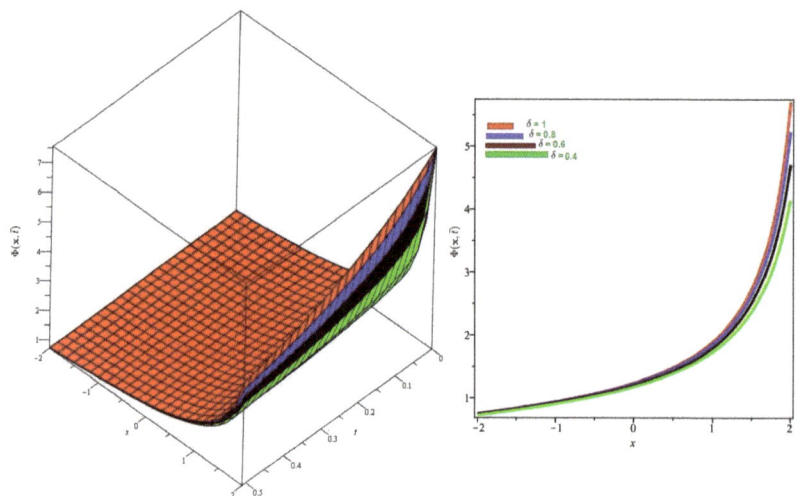

Figure 7. Numerical evaluation of graph of $\Psi(x, \bar{t})$ for Problem 2 for various fractional order $\delta = 0.4, 0.6, 0.8, 1$, $\ell = 1$, $\sigma = 0.5$ and $\beta = 2$.

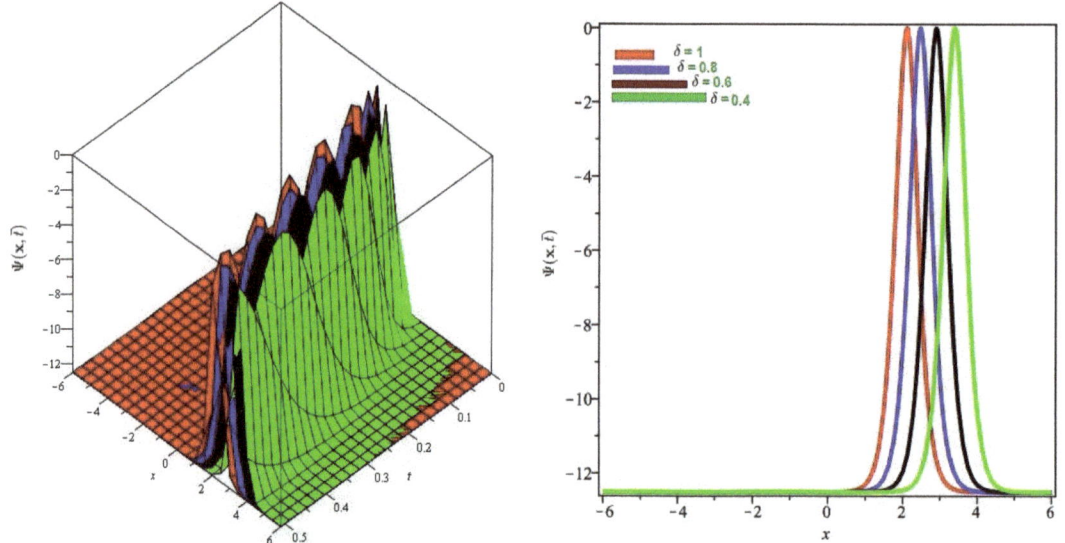

Figure 8. Numerical evaluation of graph of $\Psi(x, \bar{t})$ for Problem 2 for various fractional order $\delta = 0.4, 0.6, 0.8, 1$, $\ell = 1$, $\sigma = 0.5$ and $\beta = 2$.

Problem 3 ([16]). *Assume the time-fractional coupled nonlinear MCKdV equations is presented as (2) subject to the condition*

$$\Phi(x,0) = \frac{2 + \tanh x}{2}, \quad \Psi(x,0) = \frac{2 - \tanh x}{4}, \quad Y(x,0) = 2 - \tanh x. \qquad (39)$$

Case I. First, we surmise the Shehu decomposition method for Problem 3. Employing the Shehu transformation to (2), we find

$$\frac{\zeta^\delta}{\omega^\delta}\mathcal{U}(\xi,\omega) - \sum_{\kappa=0}^{m-1}\left(\frac{\xi}{\omega}\right)^{\delta-\kappa-1}\Phi^{(\kappa)}(0) = \mathbb{S}\left[\frac{1}{2}\frac{\partial^3\Phi}{\partial \bar{t}^3} - 3\Phi^2\frac{\partial\Phi}{\partial x} + \frac{3}{2}Y\frac{\partial^2\Psi}{\partial x^2} + 3\frac{\partial\Psi}{\partial x}\frac{\partial Y}{\partial x} + \frac{3}{2}\Psi\frac{\partial^2 Y}{\partial x^2}\right.$$

$$\left. + 3\Psi Y\frac{\partial \Phi}{\partial x} + 3\Phi Y\frac{\partial \Psi}{\partial x} + 3\Phi\Psi\frac{\partial Y}{\partial x}\right],$$

$$\frac{\zeta^\delta}{\omega^\delta}\mathcal{V}(\xi,\omega) - \sum_{\kappa=0}^{m-1}\left(\frac{\xi}{\omega}\right)^{\delta-\kappa-1}\Psi^{(\kappa)}(0) = \mathbb{S}\left[-\frac{\partial^3\Psi}{\partial x^3} - 3\frac{\partial\Phi}{\partial x}\frac{\partial\Psi}{\partial x} - 3\Psi\frac{\partial^2\Phi}{\partial x^2} - 3\Psi^2\frac{\partial Y}{\partial x} + 6\Phi\Psi\frac{\partial\Phi}{\partial x} + 3\Phi^2\frac{\partial\Psi}{\partial x}\right],$$

$$\frac{\zeta^\delta}{\omega^\delta}\mathcal{W}(\xi,\omega) - \sum_{\kappa=0}^{m-1}\left(\frac{\xi}{\omega}\right)^{\delta-\kappa-1}Y^{(\kappa)}(0) = \mathbb{S}\left[-\frac{\partial^3 Y}{\partial x^3} - 3\frac{\partial\Phi}{\partial x}\frac{\partial Y}{\partial x} - 3Y\frac{\partial^2\Phi}{\partial x^2} - 3Y^2\frac{\partial\Psi}{\partial x} + 6\Phi Y\frac{\partial\Phi}{\partial x} + 3\Phi^2\frac{\partial Y}{\partial x}\right]. \quad (40)$$

In view of (39) and simple computations yield

$$\mathcal{U}(\xi,\omega) = \frac{\omega}{\zeta}\Phi^{(0)}(x,0) + \frac{\omega^\delta}{\zeta^\delta}\mathbb{S}\left[\frac{1}{2}\frac{\partial^3\Phi}{\partial \bar{t}^3} - 3\Phi^2\frac{\partial\Phi}{\partial x} + \frac{3}{2}Y\frac{\partial^2\Psi}{\partial x^2} + 3\frac{\partial\Psi}{\partial x}\frac{\partial Y}{\partial x} + \frac{3}{2}\Psi\frac{\partial^2 Y}{\partial x^2}\right.$$

$$\left. + 3\Psi Y\frac{\partial \Phi}{\partial x} + 3\Phi Y\frac{\partial \Psi}{\partial x} + 3\Phi\Psi\frac{\partial Y}{\partial x}\right],$$

$$\mathcal{V}(\xi,\omega) = \frac{\omega}{\zeta}\Psi^{(0)}(x,0) + \frac{\omega^\delta}{\zeta^\delta}\mathbb{S}\left[-\frac{\partial^3\Psi}{\partial x^3} - 3\frac{\partial\Phi}{\partial x}\frac{\partial\Psi}{\partial x} - 3\Psi\frac{\partial^2\Phi}{\partial x^2} - 3\Psi^2\frac{\partial Y}{\partial x} + 6\Phi\Psi\frac{\partial\Phi}{\partial x} + 3\Phi^2\frac{\partial\Psi}{\partial x}\right],$$

$$\mathcal{W}(\xi,\omega) = \frac{\omega}{\zeta}Y^{(0)}(x,0) + \frac{\omega^\delta}{\zeta^\delta}\mathbb{S}\left[-\frac{\partial^3 Y}{\partial x^3} - 3\frac{\partial\Phi}{\partial x}\frac{\partial Y}{\partial x} - 3Y\frac{\partial^2\Phi}{\partial x^2} - 3Y^2\frac{\partial\Psi}{\partial x} + 6\Phi Y\frac{\partial\Phi}{\partial x} + 3\Phi^2\frac{\partial Y}{\partial x}\right]. \quad (41)$$

Applying the inverse Shehu transform, we have

$$\Phi(x,\bar{t}) = \mathbb{S}^{-1}\left[\frac{\omega}{\zeta}\Phi(x,0)\right] + \mathbb{S}^{-1}\left[\frac{\omega^\delta}{\zeta^\delta}\mathbb{S}\left[\frac{1}{2}\frac{\partial^3\Phi}{\partial \bar{t}^3} - 3\Phi^2\frac{\partial\Phi}{\partial x} + \frac{3}{2}Y\frac{\partial^2\Psi}{\partial x^2} + 3\frac{\partial\Psi}{\partial x}\frac{\partial Y}{\partial x} + \frac{3}{2}\Psi\frac{\partial^2 Y}{\partial x^2}\right.\right.$$

$$\left.\left. + 3\Psi Y\frac{\partial \Phi}{\partial x} + 3\Phi Y\frac{\partial \Psi}{\partial x} + 3\Phi\Psi\frac{\partial Y}{\partial x}\right]\right],$$

$$\Psi(x,\bar{t}) = \mathbb{S}^{-1}\left[\frac{\omega}{\zeta}\Psi(x,0)\right] + \mathbb{S}^{-1}\left[\frac{\omega^\delta}{\zeta^\delta}\mathbb{S}\left[-\frac{\partial^3\Psi}{\partial x^3} - 3\frac{\partial\Phi}{\partial x}\frac{\partial\Psi}{\partial x} - 3\Psi\frac{\partial^2\Phi}{\partial x^2} - 3\Psi^2\frac{\partial Y}{\partial x} + 6\Phi\Psi\frac{\partial\Phi}{\partial x} + 3\Phi^2\frac{\partial\Psi}{\partial x}\right]\right],$$

$$Y(x,\bar{t}) = \mathbb{S}^{-1}\left[\frac{\omega}{\zeta}Y(x,0)\right] + \mathbb{S}^{-1}\left[\frac{\omega^\delta}{\zeta^\delta}\mathbb{S}\left[-\frac{\partial^3 Y}{\partial x^3} - 3\frac{\partial\Phi}{\partial x}\frac{\partial Y}{\partial x} - 3Y\frac{\partial^2\Phi}{\partial x^2} - 3Y^2\frac{\partial\Psi}{\partial x} + 6\Phi Y\frac{\partial\Phi}{\partial x} + 3\Phi^2\frac{\partial Y}{\partial x}\right]\right]. \quad (42)$$

By virtue of the Shehu decomposition method, we have

$$\Phi_0(x,\bar{t}) = \mathbb{S}^{-1}\left[\frac{\omega}{\zeta}\Phi(x,0)\right] = \frac{1}{2}\mathbb{S}^{-1}\left[\frac{\omega}{\zeta}(2+\tanh x)\right]$$

$$= \frac{1}{2}(2+\tanh x),$$

$$\Psi_0(x,\bar{t}) = \mathbb{S}^{-1}\left[\frac{\omega}{\zeta}\Psi(x,0)\right] = \frac{1}{4}(2-\tanh x),$$

$$Y_0(x,\bar{t}) = \mathbb{S}^{-1}\left[\frac{\omega}{\zeta}Y(x,0)\right] = (2-\tanh x).$$

It follows that

$$\sum_{m=0}^{\infty} \Phi_{m+1}(\mathbf{x},\bar{t}) = \mathbb{S}^{-1}\left[\frac{\omega^{\delta}}{\xi^{\delta}}\mathbb{S}\left[\frac{1}{2}\sum_{m=0}^{\infty}(\Phi_{xxx})_m - 3\sum_{m=0}^{\infty}\mathcal{E}_m + \frac{3}{2}\sum_{m=0}^{\infty}\mathcal{F}_m + 3\sum_{m=0}^{\infty}\mathcal{G}_m + \frac{3}{2}\sum_{m=0}^{\infty}\mathcal{H}_m\right.\right.$$
$$\left.\left. + 3\sum_{m=0}^{\infty}\mathcal{I}_m + 3\sum_{m=0}^{\infty}\mathcal{J}_m + 3\sum_{m=0}^{\infty}\mathcal{K}_m\right]\right],$$

$$\sum_{m=0}^{\infty}\Psi_{m+1}(\mathbf{x},\bar{t}) = \mathbb{S}^{-1}\left[\frac{\omega^{\delta}}{\xi^{\delta}}\mathbb{S}\left[-\sum_{m=0}^{\infty}(\Psi_{xxx})_m - 3\sum_{m=0}^{\infty}\mathcal{M}_m - 3\sum_{m=0}^{\infty}\mathcal{N}_m - 3\sum_{m=0}^{\infty}\mathcal{O}_m + 6\sum_{m=0}^{\infty}\mathcal{P}_m + 3\sum_{m=0}^{\infty}\mathcal{Q}_m\right]\right],$$

$$\sum_{m=0}^{\infty}Y_{m+1}(\mathbf{x},\bar{t}) = \mathbb{S}^{-1}\left[\frac{\omega^{\delta}}{\xi^{\delta}}\mathbb{S}\left[-\sum_{m=0}^{\infty}(Y_{xxx})_m - 3\sum_{m=0}^{\infty}\mathcal{R}_m - 3\sum_{m=0}^{\infty}\mathcal{S}_m - 3\sum_{m=0}^{\infty}\mathcal{T}_m + 6\sum_{m=0}^{\infty}\mathcal{X}_m + 3\sum_{m=0}^{\infty}\mathcal{Y}_m\right]\right], m = 0,1,2,\ldots.$$

The first few Adomian polynomials are presented as follows:

$$\mathcal{E}_j(\Phi^2\Phi_x) = \begin{cases} \Phi_0^2\Phi_{0x}, & \text{for } j=0 \\ (2\Phi_0\Phi_1)\Phi_{0x} + \Phi_0^2\Phi_{1x}, & \text{for } j=1 \\ (2\Phi_0\Phi_2 + \Phi_1^2)\Phi_{0x} + (2\Phi_0\Phi_1)\Phi_{1x} + \Phi_0^2\Phi_{2x}, & \text{for } j=2 \end{cases}$$

$$\mathcal{F}_j(Y\Psi_{xx}) = \begin{cases} Y_0\Psi_{0xx}, & \text{for } j=0 \\ Y_1\Psi_{0xx} + Y_0\Psi_{1xx}, & \text{for } j=1 \\ Y_2\Psi_{0xx} + Y_1\Psi_{1xx} + Y_0\Psi_{2xx}, & \text{for } j=2 \end{cases}$$

$$\mathcal{G}_j(\Psi_x Y_x) = \begin{cases} \Psi_{0x}Y_{0x}, & \text{for } j=0 \\ \Psi_{0x}Y_{1x} + \Psi_{1x}Y_{0x}, & \text{for } j=1 \\ \Psi_{2x}Y_{0x} + \Psi_{1x}Y_{1x} + \Psi_{0x}Y_{2x}, & \text{for } j=2 \end{cases}$$

$$\mathcal{H}_j(\Psi_x Y_{xx}) = \begin{cases} \Psi_{0x}Y_{0xx}, & \text{for } j=0 \\ \Psi_{0x}Y_{1xx} + \Psi_{1x}Y_{0xx}, & \text{for } j=1 \\ \Psi_{2x}Y_{0xx} + \Psi_{1x}Y_{1xx} + \Psi_{0x}Y_{2xx}, & \text{for } j=2 \end{cases}$$

$$\mathcal{I}_j(\Psi z \Phi_x) = \begin{cases} (\Psi Y)_0\Phi_{0x}, & \text{for } j=0 \\ (\Psi Y)_0\Phi_{1x} + (\Psi Y)_1\Phi_{0x}, & \text{for } j=1 \\ (\Psi Y)_0\Phi_{2x} + (\Psi Y)_1\Phi_{1x} + (\Psi Y)_2\Phi_{0x}, & \text{for } j=2 \end{cases}$$

$$\mathcal{J}_j(\Phi z \Psi_x) = \begin{cases} (\Phi Y)_0\Psi_{0x}, & \text{for } j=0 \\ (\Phi Y)_0\Psi_{1x} + (\Phi Y)_1\Psi_{0x}, & \text{for } j=1 \\ (\Phi Y)_0\Psi_{2x} + (\Phi Y)_1\Psi_{1x} + (\Phi Y)_2\Psi_{0x}, & \text{for } j=2 \end{cases}$$

$$\mathcal{K}_j(\Phi\Psi Y_x) = \begin{cases} (\Phi\Psi)_0 Y_{0x}, & \text{for } j=0 \\ (\Phi\Psi)_0 Y_{1x} + (\Phi\Psi)_1 Y_{0x}, & \text{for } j=1 \\ (\Phi\Psi)_0 Y_{2x} + (\Phi\Psi)_1 Y_{1x} + (\Phi\Psi)_2 Y_{0x}, & \text{for } j=2 \end{cases}$$

$$\mathcal{M}_j(\Phi_x \Psi_x) = \begin{cases} \Phi_{0x}\Psi_{0x}, & \text{for } j=0 \\ \Phi_{0x}\Psi_{1x} + \Phi_x\Psi_{0x}, & \text{for } j=1 \\ \Phi_{2x}\Psi_{0x} + \Phi_{1x}\Psi_{1x} + \Phi_{0x}\Psi_{1x}, & \text{for } j=2 \end{cases}$$

$$\mathcal{N}_j(\Psi\Phi_{xx}) = \begin{cases} \Psi_0\Phi_{0xx}, & \text{for } j=0 \\ \Psi_0\Phi_{1xx} + \Psi_1\Phi_{0xx}, & \text{for } j=1 \\ \Psi_2\Phi_{0xx} + \Psi_1\Phi_{1xx} + \Psi_0\Phi_{2xx}, & \text{for } j=2 \end{cases}$$

$$\mathcal{O}_j(\Psi^2 Y_x) = \begin{cases} \Psi_0^2 Y_{0x}, & \text{for } j=0 \\ (2\Psi_0\Psi_1)Y_{0x} + \Psi_0^2 Y_{1x}, & \text{for } j=1 \\ (2\Psi_0\Psi_2 + \Psi_1^2)Y_{0x} + (2\Psi_0\Psi_1)Y_{1x} + \Psi_0^2 Y_{2x}, & \text{for } j=2 \end{cases}$$

$$\mathcal{P}_J(\Phi\Psi\Phi_x) = \begin{cases} (\Phi\Psi)_0\Phi_{0x}, & \text{for } j = 0 \\ (\Phi\Psi)_0\Phi_{1x} + (\Phi\Psi)_0\Phi_{1x}, & \text{for } j = 1 \\ (\Phi\Psi)_0\Phi_{2x} + (\Phi\Psi)_1\Phi_{1x} + (\Phi\Psi)_2\Phi_{0x}, & \text{for } j = 2 \end{cases}$$

$$\mathcal{Q}_J(\Phi^2\Psi_x) = \begin{cases} \Phi_0^2\Psi_{0x}, & \text{for } j = 0 \\ (2\Phi_0\Phi_1)\Psi_{0x} + \Phi_0^2\Psi_{1x}, & \text{for } j = 1 \\ (2\Phi_0\Phi_2 + \Phi_1^2)\Psi_{0x} + (2\Phi_0\Phi_1)\Psi_{1x} + \Phi_0^2\Psi_{2x}, & \text{for } j = 2 \end{cases}$$

$$\mathcal{R}_0(\Phi_x\Psi_x) = \begin{cases} \Phi_{0x}\Psi_{0x}, & \text{for } j = 0 \\ \Phi_{0x}\Psi_{1x} + \Phi_{1x}\Psi_{0x}, & \text{for } j = 1 \\ \Phi_{2x}\Psi_{0x} + \Phi_{1x}\Psi_{1x} + \Phi_{0x}\Psi_{2x}, & \text{for } j = 2 \end{cases}$$

$$\mathcal{S}_J(Y\Phi_{xx}) = \begin{cases} Y_0\Phi_{0xx}, & \text{for } j = 0 \\ Y_0\Phi_{1xx} + Y_1\Phi_{0xx}, & \text{for } j = 1 \\ Y_2\Phi_{0xx} + Y_1\Phi_{1xx} + Y_0\Phi_{2xx}, & \text{for } j = 2 \end{cases}$$

$$\mathcal{T}_J(Y^2\Psi_x) = \begin{cases} Y_0^2\Psi_{0x}, & \text{for } j = 0 \\ (2Y_0Y_1)\Psi_{0x} + Y_0^2\Psi_{1x}, & \text{for } j = 1 \\ (2Y_0Y_2 + Y_1^2)\Psi_{0x} + (2Y_0Y_1)\Psi_{1x} + Y_0^2\Psi_{2x}, & \text{for } j = 2 \end{cases}$$

$$\mathcal{X}_J(\Phi Y_1\Phi_x) = \begin{cases} (\Phi Y)_0\Phi_{0x}, & \text{for } j = 0 \\ (\Phi Y)_0\Phi_{1x} + (\Phi Y)_1\Phi_{0x}, & \text{for } j = 1 \\ (\Phi Y)_2\Phi_{0x} + (\Phi Y)_1\Phi_{1x} + (\Phi Y)_2\Phi_{0x}, & \text{for } j = 2 \end{cases}$$

$$\mathcal{Y}_J(\Phi^2 Y_x) = \begin{cases} \Phi_0^2 Y_{0x}, & \text{for } j = 0 \\ (2\Phi_0\Phi_1)Y_{0x} + \Phi_0^2 Y_{1x}, & \text{for } j = 1 \\ (2\Phi_0\Phi_2 + \Phi_1^2)Y_{0x} + (2\Phi_0\Phi_1)Y_{1x} + \Phi_0^2 Y_{2x}, & \text{for } j = 2. \end{cases}$$

For $m = 0, 1, 2, 3, \ldots$

$$\Phi_1(x,\bar{t}) = \mathbb{S}^{-1}\left[\frac{\omega^\delta}{\zeta^\delta}\mathbb{S}\left[\frac{1}{2}(\Phi_{xxx})_0 - 3\mathcal{E}_0 + \frac{3}{2}\mathcal{F}_0 + 3\mathcal{G}_0 + \frac{3}{2}\mathcal{H}_0 + 3I_0 + 3\mathcal{J}_0 + 3\mathcal{K}_0\right]\right]$$

$$= \frac{11}{2}\operatorname{sech}^2(x)\mathbb{S}^{-1}\left[\frac{\omega^{\delta+2}}{\zeta^{\delta+2}}\right] = \frac{11}{2}\operatorname{sech}^2(x)\frac{\bar{t}^\delta}{\Gamma(\delta+1)},$$

$$\Psi_1(x,\bar{t}) = \mathbb{S}^{-1}\left[\frac{\omega^\delta}{\zeta^\delta}\mathbb{S}\left[(\Psi_{xxx})_0 - 3\mathcal{M}_0 - 3\mathcal{N}_0 - 3\mathcal{O}_0 + 6\mathcal{P}_0 + 3\mathcal{Q}_0\right]\right]$$

$$= -\frac{11}{8}\operatorname{sech}^2(x)\frac{\bar{t}^\delta}{\Gamma(\delta+1)},$$

$$Y_1(x,\bar{t}) = \mathbb{S}^{-1}\left[\frac{\omega^\delta}{\zeta^\delta}\mathbb{S}\left[(Y_{xxx})_0 - 3\mathcal{R}_0 - 3\widehat{\mathcal{S}}_0 - 3\mathcal{T}_0 + 6\mathcal{X}_0 + 3\mathcal{Y}_0\right]\right]$$

$$= -\frac{11}{2}\operatorname{sech}^2(x)\frac{\bar{t}^\delta}{\Gamma(\delta+1)}.$$

$$\Phi_2(x,\bar{t}) = \mathbb{S}^{-1}\left[\frac{\omega^\delta}{\xi^\delta}\mathbb{S}\left[\frac{1}{2}(\Phi_{xxx})_1 - 3\mathcal{E}_1 + \frac{3}{2}\mathcal{F}_1 + 3\mathcal{G}_1 + \frac{3}{2}\mathcal{H}_1 + 3\mathcal{I}_1 + 3\mathcal{J}_1 + 3\mathcal{K}_1\right]\right]$$

$$= \frac{-121}{8}\tan h(x)\sec h^2(x)\frac{\bar{t}^{2\delta}}{\Gamma(2\delta+1)},$$

$$\Psi_2(x,\bar{t}) = \mathbb{S}^{-1}\left[\frac{\omega^\delta}{\xi^\delta}\mathbb{S}\left[(\Psi_{xxx})_1 - 3\mathcal{M}_1 - 3\mathcal{N}_1 - 3\mathcal{O}_1 + 6\mathcal{P}_1 + 3\mathcal{Q}_1\right]\right]$$

$$= \frac{121}{8}\tan h(x)\sec h^2(x)\frac{\bar{t}^{2\delta}}{\Gamma(2\delta+1)},$$

$$Y_2(x,\bar{t}) = \mathbb{S}^{-1}\left[\frac{\omega^\delta}{\xi^\delta}\mathbb{S}\left[(Y_{xxx})_1 - 3\mathcal{R}_1 - 3\hat{\mathcal{S}}_1 - 3\mathcal{T}_1 + 6\mathcal{X}_1 + 3\mathcal{Y}_1\right]\right]$$

$$= \frac{242}{8}\tan h(x)\sec h^2(x)\frac{\bar{t}^{2\delta}}{\Gamma(2\delta+1)},$$

$$\Phi_3(x,\bar{t}) = \mathbb{S}^{-1}\left[\frac{\omega^\delta}{\xi^\delta}\mathbb{S}\left[\frac{1}{2}(\Phi_{xxx})_2 - 3\mathcal{E}_2 + \frac{3}{2}\mathcal{F}_2 + 3\mathcal{G}_2 + \frac{3}{2}\mathcal{H}_2 + 3\mathcal{I}_2 + 3\mathcal{J}_2 + 3\mathcal{K}_2\right]\right]$$

$$= \frac{1331}{48}\sec h^4(x)[\cosh(2x) - 2]\frac{\bar{t}^{3\delta}}{\Gamma(3\delta+1)},$$

$$\Psi_3(x,\bar{t}) = \mathbb{S}^{-1}\left[\frac{\omega^\delta}{\xi^\delta}\mathbb{S}\left[(\Psi_{xxx})_2 - 3\mathcal{M}_2 - 3\mathcal{N}_2 - 3\mathcal{O}_2 + 6\mathcal{P}_2 + 3\mathcal{Q}_2\right]\right]$$

$$= \frac{2662}{96}\sec h^4(x)[\cosh(2x) - 2]\frac{\bar{t}^{3\delta}}{\Gamma(3\delta+1)},$$

$$Y_3(x,\bar{t}) = \mathbb{S}^{-1}\left[\frac{\omega^\delta}{\xi^\delta}\mathbb{S}\left[(Y_{xxx})_2 - 3\mathcal{R}_2 - 3\hat{\mathcal{S}}_2 - 3\mathcal{T}_2 + 6\mathcal{X}_2 + 3\mathcal{Y}_2\right]\right]$$

$$= \frac{-2662}{48}\sec h^4(x)[\cosh(2x) - 2]\frac{\bar{t}^{3\delta}}{\Gamma(3\delta+1)},$$

$$\vdots$$

The Shehu decomposition method solution for Problem 3 is presented as:

$$\Phi(x,\bar{t}) = \Phi_0(x,\bar{t}) + \Phi_1(x,\bar{t}) + \Phi_2(x,\bar{t}) + \Phi_3(x,\bar{t})\dots,$$
$$= \frac{1}{2}(2+\tanh x) + \frac{11}{2}\sec h^2(x)\frac{\bar{t}^\delta}{\Gamma(\delta+1)} - \frac{121}{8}\tan h(x)\sec h^2(x)\frac{\bar{t}^{2\delta}}{\Gamma(2\delta+1)}$$
$$+ \frac{1331}{48}\sec h^4(x)[\cosh(2x)-2]\frac{\bar{t}^{3\delta}}{\Gamma(3\delta+1)} + \dots.$$

Analogously, we have

$$\Psi(x,\bar{t}) = \frac{1}{4}(2-\tanh x) - \frac{11}{8}\sec h^2(x)\frac{\bar{t}^\delta}{\Gamma(\delta+1)} + \frac{121}{8}\tan h(x)\sec h^2(x)\frac{\bar{t}^{2\delta}}{\Gamma(2\delta+1)}$$
$$- \frac{1331}{48}\sec h^4(x)[\cosh(2x)-2]\frac{\bar{t}^{3\delta}}{\Gamma(3\delta+1)} + \dots,$$

$$Y(x,\bar{t}) = (2-\tanh x) - \frac{11}{2}\sec h^2(x)\frac{\bar{t}^\delta}{\Gamma(\delta+1)} + \frac{121}{4}\tan h(x)\sec h^2(x)\frac{\bar{t}^{2\delta}}{\Gamma(2\delta+1)}$$
$$- \frac{2662}{48}\sec h^4(x)[\cosh(2x)-2]\frac{\bar{t}^{3\delta}}{\Gamma(3\delta+1)} + \dots.$$

By Setting $\delta = 1$, we then obtain the exact solution of coupled KdV Equation (2)

$$\Phi(x,\bar{t}) = \frac{1}{2}\left(2 + \tanh\left(x - \frac{11\bar{t}}{2}\right)\right), \quad \Psi(x,\bar{t}) = \frac{1}{4}\left(2 - \tanh\left(x - \frac{11\bar{t}}{2}\right)\right),$$

$$Y(x,\bar{t}) = \left(2 - \tanh\left(x - \frac{11\bar{t}}{2}\right)\right).$$

Case II. Now, we surmise the new iterative transform method for Problem 3. Applying the proposed analytical approach to (42) yields

$$\Phi_0(x,\bar{t}) = \frac{1}{2}(2 + \tanh x),$$

$$\Psi_0(x,\bar{t}) = \frac{1}{4}(2 - \tanh x),$$

$$Y_0(x,\bar{t}) = (2 - \tanh x),$$

$$\Phi_1(x,\bar{t}) = \mathbb{S}^{-1}\left[\frac{\omega^\delta}{\zeta^\delta}\mathbb{S}\left[\frac{1}{2}\frac{\partial^3\Phi_0}{\partial \bar{t}^3} - 3\Phi_0^2\frac{\partial\Phi_0}{\partial x} + \frac{3}{2}Y_0\frac{\partial^2\Psi_0}{\partial x^2} + 3\frac{\partial\Psi_0}{\partial x}\frac{\partial Y_0}{\partial x} + \frac{3}{2}\Psi_0\frac{\partial^2 Y_0}{\partial x^2}\right.\right.$$

$$\left.\left. + 3\Psi_0 Y_0\frac{\partial\Phi_0}{\partial x} + 3\Phi_0 Y_0\frac{\partial\Psi_0}{\partial x} + 3\Phi_0\Psi_0\frac{\partial Y_0}{\partial x}\right]\right]$$

$$= \frac{11}{2}\operatorname{sech}^2(x)\mathbb{S}^{-1}\left[\frac{\omega^{\delta+2}}{\zeta^{\delta+2}}\right] = \frac{11}{2}\operatorname{sech}^2(x)\frac{\bar{t}^\delta}{\Gamma(\delta+1)},$$

$$\Psi_1(x,\bar{t}) = \mathbb{S}^{-1}\left[\frac{\omega^\delta}{\zeta^\delta}\mathbb{S}\left[-\frac{\partial^3\Psi_0}{\partial x^3} - 3\frac{\partial\Phi_0}{\partial x}\frac{\partial\Psi_0}{\partial x} - 3\Psi_0\frac{\partial^2\Phi_0}{\partial x^2} - 3\Psi_0^2\frac{\partial Y_0}{\partial x} + 6\Phi_0\Psi_0\frac{\partial\Phi_0}{\partial x} + 3\Phi_0^2\frac{\partial\Psi_0}{\partial x}\right]\right]$$

$$= -\frac{11}{8}\operatorname{sech}^2(x)\frac{\bar{t}^\delta}{\Gamma(\delta+1)},$$

$$Y_1(x,\bar{t}) = \mathbb{S}^{-1}\left[\frac{\omega^\delta}{\zeta^\delta}\mathbb{S}\left[-\frac{\partial^3 Y_0}{\partial x^3} - 3\frac{\partial\Phi_0}{\partial x}\frac{\partial Y_0}{\partial x} - 3Y_0\frac{\partial^2\Phi_0}{\partial x^2} - 3Y_0^2\frac{\partial\Psi_0}{\partial x} + 6\Phi_0 Y_0\frac{\partial\Phi_0}{\partial x} + 3\Phi_0^2\frac{\partial Y_0}{\partial x}\right]\right]$$

$$= -\frac{11}{2}\operatorname{sech}^2(x)\frac{\bar{t}^\delta}{\Gamma(\delta+1)},$$

$$\Phi_2(x,\bar{t}) = \mathbb{S}^{-1}\left[\frac{\omega^\delta}{\zeta^\delta}\mathbb{S}\left[\frac{1}{2}\frac{\partial^3\Phi_1}{\partial \bar{t}^3} - 3\Phi_1^2\frac{\partial\Phi_1}{\partial x} + \frac{3}{2}Y_1\frac{\partial^2\Psi_1}{\partial x^2} + 3\frac{\partial\Psi_1}{\partial x}\frac{\partial Y_1}{\partial x} + \frac{3}{2}\Psi_1\frac{\partial^2 Y_1}{\partial x^2}\right.\right.$$

$$\left.\left. + 3\Psi_1 Y_1\frac{\partial\Phi_1}{\partial x} + 3\Phi_1 Y_1\frac{\partial\Psi_1}{\partial x} + 3\Phi_1\Psi_1\frac{\partial Y_1}{\partial x}\right]\right]$$

$$= \frac{-121}{8}\tan h(x)\operatorname{sech}^2(x)\frac{\bar{t}^{2\delta}}{\Gamma(2\delta+1)},$$

$$\Psi_2(x,\bar{t}) = \mathbb{S}^{-1}\left[\frac{\omega^\delta}{\zeta^\delta}\mathbb{S}\left[-\frac{\partial^3\Psi_1}{\partial x^3} - 3\frac{\partial\Phi_1}{\partial x}\frac{\partial\Psi_1}{\partial x} - 3\Psi_1\frac{\partial^2\Phi_1}{\partial x^2} - 3\Psi_1^2\frac{\partial Y_1}{\partial x} + 6\Phi_1\Psi_1\frac{\partial\Phi_1}{\partial x} + 3\Phi_1^2\frac{\partial\Psi_1}{\partial x}\right]\right]$$

$$= \frac{121}{8}\tan h(x)\operatorname{sech}^2(x)\frac{\bar{t}^{2\delta}}{\Gamma(2\delta+1)},$$

$$Y_2(x,\bar{t}) = \mathbb{S}^{-1}\left[\frac{\omega^\delta}{\zeta^\delta}\mathbb{S}\left[-\frac{\partial^3 Y_1}{\partial x^3} - 3\frac{\partial\Phi_1}{\partial x}\frac{\partial Y_1}{\partial x} - 3Y_1\frac{\partial^2\Phi_1}{\partial x^2} - 3Y_1^2\frac{\partial\Psi_1}{\partial x} + 6\Phi_1 Y_1\frac{\partial\Phi_1}{\partial x} + 3\Phi_1^2\frac{\partial Y_1}{\partial x}\right]\right]$$

$$= \frac{242}{8}\tan h(x)\operatorname{sech}^2(x)\frac{\bar{t}^{2\delta}}{\Gamma(2\delta+1)},$$

$$\Phi_3(\mathbf{x},\bar{t}) = \mathbb{S}^{-1}\left[\frac{\omega^\delta}{\xi^\delta}\mathbb{S}\left[\frac{1}{2}\frac{\partial^3\Phi_2}{\partial \bar{t}^3} - 3\Phi_2^2\frac{\partial\Phi_2}{\partial \mathbf{x}} + \frac{3}{2}Y_2\frac{\partial^2\Psi_2}{\partial \mathbf{x}^2} + 3\frac{\partial\Psi_2}{\partial \mathbf{x}}\frac{\partial Y_2}{\partial \mathbf{x}} + \frac{3}{2}\Psi_2\frac{\partial^2 Y_2}{\partial \mathbf{x}^2}\right.\right.$$
$$\left.\left. + 3\Psi_2 Y_2\frac{\partial\Phi_2}{\partial \mathbf{x}} + 3\Phi_2 Y_2\frac{\partial\Psi_2}{\partial \mathbf{x}} + 3\Phi_2 \Psi_2\frac{\partial Y_2}{\partial \mathbf{x}}\right]\right]$$
$$= \frac{1331}{48}\operatorname{sech}^4(\mathbf{x})[\cosh(2\mathbf{x}) - 2]\frac{\bar{t}^{3\delta}}{\Gamma(3\delta + 1)},$$

$$\Psi_3(\mathbf{x},\bar{t}) = \mathbb{S}^{-1}\left[\frac{\omega^\delta}{\xi^\delta}\mathbb{S}\left[-\frac{\partial^3\Psi_2}{\partial \mathbf{x}^3} - 3\frac{\partial\Phi_2}{\partial \mathbf{x}}\frac{\partial\Psi_2}{\partial \mathbf{x}} - 3\Psi_2\frac{\partial^2\Phi_2}{\partial \mathbf{x}^2} - 3\Psi_2^2\frac{\partial Y_2}{\partial \mathbf{x}} + 6\Phi_2\Psi_2\frac{\partial\Phi_2}{\partial \mathbf{x}} + 3\Phi_2^2\frac{\partial\Psi_2}{\partial \mathbf{x}}\right]\right]$$
$$= \frac{2662}{96}\operatorname{sech}^4(\mathbf{x})[\cosh(2\mathbf{x}) - 2]\frac{\bar{t}^{3\delta}}{\Gamma(3\delta + 1)},$$

$$Y_3(\mathbf{x},\bar{t}) = \mathbb{S}^{-1}\left[\frac{\omega^\delta}{\xi^\delta}\mathbb{S}\left[-\frac{\partial^3 Y_2}{\partial \mathbf{x}^3} - 3\frac{\partial\Phi_2}{\partial \mathbf{x}}\frac{\partial Y_2}{\partial \mathbf{x}} - 3Y_2\frac{\partial^2\Phi_2}{\partial \mathbf{x}^2} - 3Y_2^2\frac{\partial\Psi_2}{\partial \mathbf{x}} + 6\Phi_2 Y_2\frac{\partial\Phi_2}{\partial \mathbf{x}} + 3\Phi_2^2\frac{\partial Y_2}{\partial \mathbf{x}}\right]\right]$$
$$= \frac{-2662}{48}\operatorname{sech}^4(\mathbf{x})[\cosh(2\mathbf{x}) - 2]\frac{\bar{t}^{3\delta}}{\Gamma(3\delta + 1)},$$

$$\vdots$$

$$\Phi_m(\mathbf{x},\bar{t}) = \mathbb{S}^{-1}\left[\frac{\omega^\delta}{\xi^\delta}\mathbb{S}\left[\frac{1}{2}\frac{\partial^3\Phi_{m-1}}{\partial \bar{t}^3} - 3\Phi_{m-1}^2\frac{\partial\Phi_{m-1}}{\partial \mathbf{x}} + \frac{3}{2}Y_{m-1}\frac{\partial^2\Psi_{m-1}}{\partial \mathbf{x}^2} + 3\frac{\partial\Psi_{m-1}}{\partial \mathbf{x}}\frac{\partial Y_{m-1}}{\partial \mathbf{x}} + \frac{3}{2}\Psi_{m-1}\frac{\partial^2 Y_{m-1}}{\partial \mathbf{x}^2}\right.\right.$$
$$\left.\left. + 3\Psi_{m-1}Y_{m-1}\frac{\partial\Phi_{m-1}}{\partial \mathbf{x}} + 3\Phi_{m-1}Y_{m-1}\frac{\partial\Psi_{m-1}}{\partial \mathbf{x}} + 3\Phi_{m-1}\Psi_{m-1}\frac{\partial Y_{m-1}}{\partial \mathbf{x}}\right]\right],$$

$$\Psi_m(\mathbf{x},\bar{t}) = \mathbb{S}^{-1}\left[\frac{\omega^\delta}{\xi^\delta}\mathbb{S}\left[-\frac{\partial^3\Psi_{m-1}}{\partial \mathbf{x}^3} - 3\frac{\partial\Phi_{m-1}}{\partial \mathbf{x}}\frac{\partial\Psi_{m-1}}{\partial \mathbf{x}} - 3\Psi_{m-1}\frac{\partial^2\Phi_{m-1}}{\partial \mathbf{x}^2} - 3\Psi_{m-1}^2\frac{\partial Y_{m-1}}{\partial \mathbf{x}}\right.\right.$$
$$\left.\left. + 6\Phi_{m-1}\Psi_{m-1}\frac{\partial\Phi_{m-1}}{\partial \mathbf{x}} + 3\Phi_{m-1}^2\frac{\partial\Psi_{m-1}}{\partial \mathbf{x}}\right]\right],$$

$$Y_m(\mathbf{x},\bar{t}) = \mathbb{S}^{-1}\left[\frac{\omega^\delta}{\xi^\delta}\mathbb{S}\left[-\frac{\partial^3 Y_{m-1}}{\partial \mathbf{x}^3} - 3\frac{\partial\Phi_{m-1}}{\partial \mathbf{x}}\frac{\partial Y_{m-1}}{\partial \mathbf{x}} - 3Y_{m-1}\frac{\partial^2\Phi_{m-1}}{\partial \mathbf{x}^2} - 3Y_{m-1}^2\frac{\partial\Psi_{m-1}}{\partial \mathbf{x}}\right.\right.$$
$$\left.\left. + 6\Phi_{m-1}Y_{m-1}\frac{\partial\Phi_{m-1}}{\partial \mathbf{x}} + 3\Phi_{m-1}^2\frac{\partial Y_{m-1}}{\partial \mathbf{x}}\right]\right].$$

The series solution for Problem 3 is presented as:

$$\Phi(\mathbf{x},\bar{t}) = \Phi_0(\mathbf{x},\bar{t}) + \Phi_1(\mathbf{x},\bar{t}) + \Phi_2(\mathbf{x},\bar{t}) + \Phi_3(\mathbf{x},\bar{t}) + \ldots \Phi_m(\mathbf{x},\bar{t}),$$
$$= \frac{1}{2}(2 + \tanh \mathbf{x}) + \frac{11}{2}\operatorname{sech}^2(\mathbf{x})\frac{\bar{t}^\delta}{\Gamma(\delta + 1)} - \frac{121}{8}\tanh(\mathbf{x})\operatorname{sech}^2(\mathbf{x})\frac{\bar{t}^{2\delta}}{\Gamma(2\delta + 1)}$$
$$+ \frac{1331}{48}\operatorname{sech}^4(\mathbf{x})[\cosh(2\mathbf{x}) - 2]\frac{\bar{t}^{3\delta}}{\Gamma(3\delta + 1)} + \ldots.$$

Analogously, we have

$$\Psi(x,\bar{t}) = \frac{1}{4}(2-\tanh x) - \frac{11}{8}\sec h^2(x)\frac{\bar{t}^\delta}{\Gamma(\delta+1)} + \frac{121}{8}\tan h(x)\sec h^2(x)\frac{\bar{t}^{2\delta}}{\Gamma(2\delta+1)}$$
$$- \frac{1331}{48}\sec h^4(x)[\cosh(2x)-2]\frac{\bar{t}^{3\delta}}{\Gamma(3\delta+1)} + \ldots,$$

$$Y(x,\bar{t}) = (2-\tanh x) - \frac{11}{2}\sec h^2(x)\frac{\bar{t}^\delta}{\Gamma(\delta+1)} + \frac{121}{4}\tan h(x)\sec h^2(x)\frac{\bar{t}^{2\delta}}{\Gamma(2\delta+1)}$$
$$- \frac{2662}{48}\sec h^4(x)[\cosh(2x)-2]\frac{\bar{t}^{3\delta}}{\Gamma(3\delta+1)} + \ldots.$$

By setting $\delta = 1$, we then obtain the exact solution of MCKdV Equation (2)

$$\Phi(x,\bar{t}) = \frac{1}{2}\left(2+\tanh\left(x-\frac{11\bar{t}}{2}\right)\right),\ \Psi(x,\bar{t}) = \frac{1}{4}\left(2-\tanh\left(x-\frac{11\bar{t}}{2}\right)\right),$$
$$Y(x,\bar{t}) = \left(2-\tanh\left(x-\frac{11\bar{t}}{2}\right)\right).$$

In Figures 9–11 the exact and approximate results of $\Phi(x,\bar{t})$, $\Psi(x,\bar{t})$ and $Y(x,\bar{t})$ are demonstrated at $\ell = 1$, $\sigma = 0.5$ and $\beta = 2$, respectively. In Figures 12–14, the surface and 2D graph for $\Phi(x,\bar{t})$, $\Psi(x,\bar{t})$ and $Y(x,\bar{t})$ for various fractional orders are presented which show that the SDM/SITM approximated results derived are in a strong agreement with the exact and the numerical ones. This comparison represents a strong correlation between the SDM and exact findings. Therefore, the SDM/SITM are reliable novel approaches which require less computation time and is quite straightforward and more flexible than the homotopy perturbation method or homotopy analysis method, because the ST permits one of several scenarios to reduce the deficiency mainly occurs because of unsatisfied initial conditions that appear in other semi-analytical methods such as the SDM/SITM.

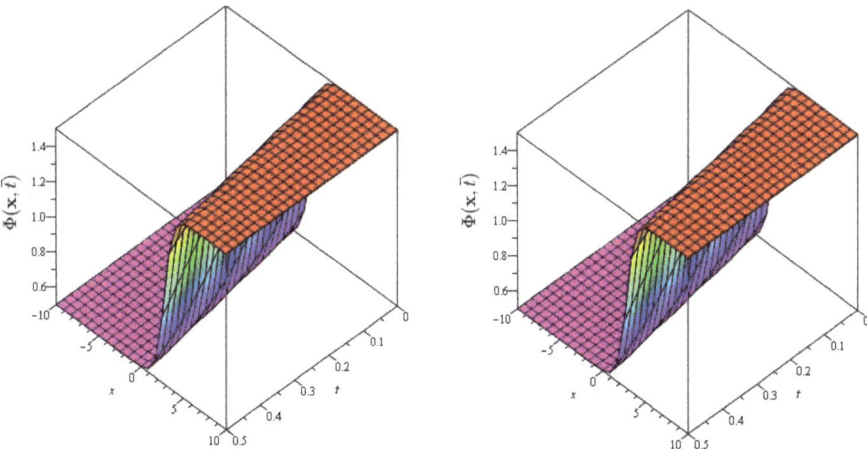

Figure 9. The exact and analytical solution graph at $\Phi(x,\bar{t})$ of Problem 3 for $\ell = 1, \sigma = 0.5$ and $\beta = 2$.

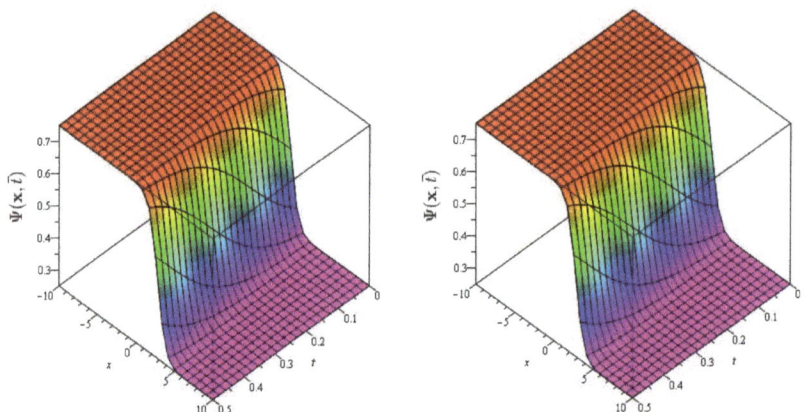

Figure 10. The exact and analytical solution graph at $\Psi(\mathbf{x}, \bar{t})$ of Problem 3 for $\ell = 1, \sigma = 0.5$ and $\beta = 2$.

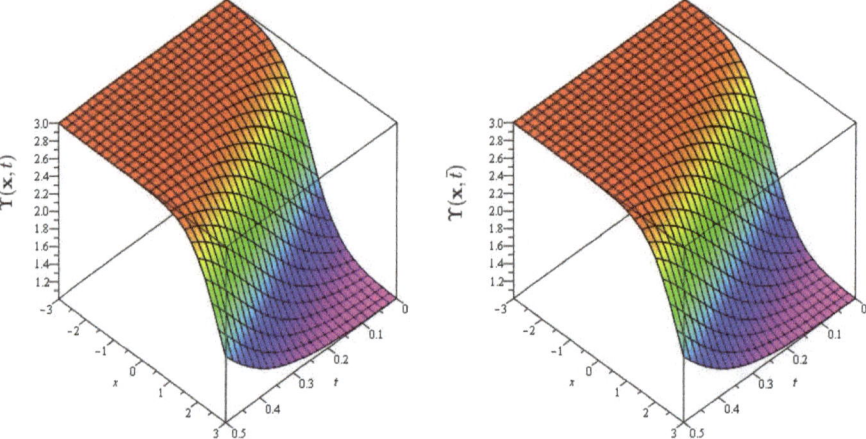

Figure 11. The exact and analytical solution graph at $\Upsilon(\mathbf{x}, \bar{t})$ of Problem 3 for $\ell = 1, \sigma = 0.5$ and $\beta = 2$.

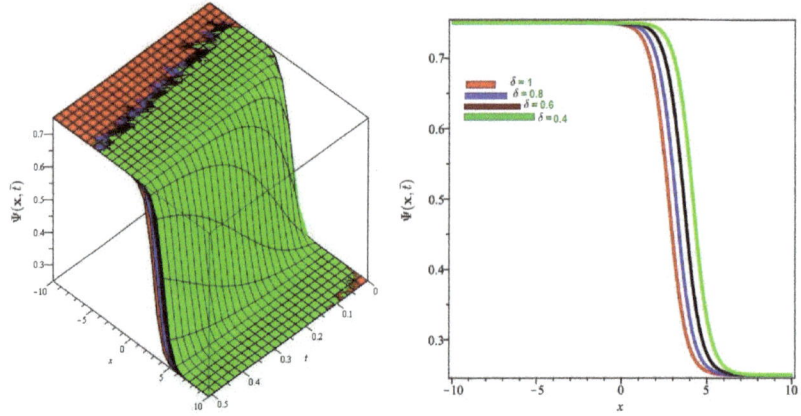

Figure 12. Numerical evaluation of graph at $\Psi(\mathbf{x}, \bar{t})$ Problem 3 for various fractional order $\delta = 0.4, 0.6, 0.8, 1, \ell = 1, \sigma = 0.5$ and $\beta = 2$.

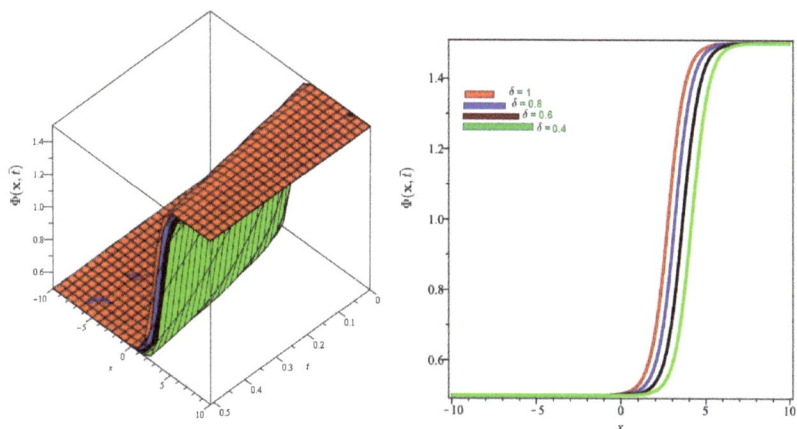

Figure 13. Numerical evaluation of graph at $\Phi(x,\bar{t})$ Problem 3 for various fractional order $\delta = 0.4, 0.6, 0.8, 1, \ell = 1, \sigma = 0.5$ and $\beta = 2$.

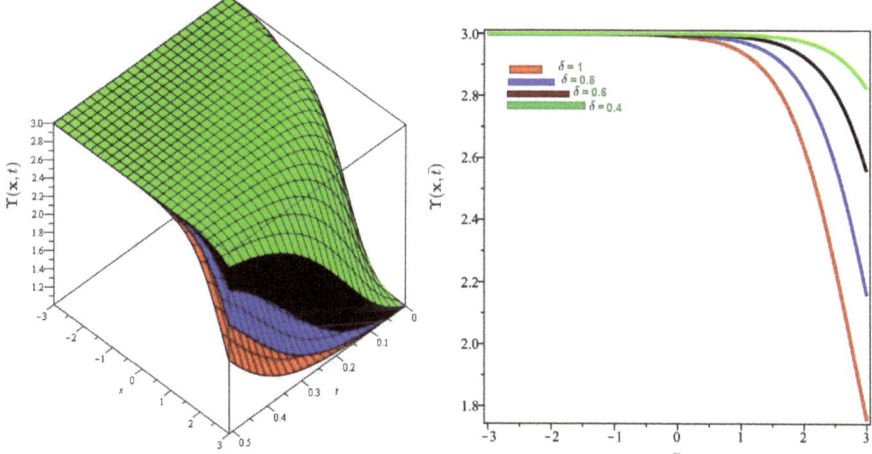

Figure 14. Numerical evaluation of graph at $Y(x,\bar{t})$ Problem 3 for various fractional order $\delta = 0.4, 0.6, 0.8, 1, \ell = 1, \sigma = 0.5$ and $\beta = 2$.

7. Conclusions

Understanding complex nonlinear PDEs remains a difficult challenge when their generative model is unknown. This challenge becomes more complex when it comes to evaluating time fractional nonlinear PDEs, surmising the model that governs their evolution. To cope with this difficulty, numerous numerical methods have been employed for dealing with nonlinear physical phenomena. Toward addressing this goal, in this paper, we have considered a time-fractional KdV equation and have developed effective, rigorous and robust algorithmic strategies (Shehu decompsition method and Shehu iterative transform method) to estimate approximate-analytical solutions and so identify the main numerical solutions appearing in the literature. In this approach, we do not need the Lagrange multiplier, correction functional, stationary conditions, or to calculate heavy integrals because the results established are noise free, which overcomes the shortcomings of existing methods. It is remarkable that the projected approaches are well-organized analytical methods for finding approximate-analytical solutions to complex nonlinear

PDEs. Finally, we conclude that this scheme will be taken into account in order to cope with other complex non-linear fractional order systems of equations.

Author Contributions: All authors contributed equally. All authors have read and agreed to the published version of the manuscript.

Funding: This research received no external funding.

Institutional Review Board Statement: Not applicable.

Informed Consent Statement: Not applicable.

Data Availability Statement: Not applicable.

Acknowledgments: This research was supported by Taif University Researchers Supporting Project Number (TURSP-2020/96), Taif University, Taif, Saudi Arabia.

Conflicts of Interest: The authors declare no conflict of interest.

References

1. Wu, Y.; Geng, X.; Hu, X.; Zhu, S. A generalized Hirota-Satsuma coupled Korteweg-de Vries equation and Miura transformations. *Phys. Lett. A* **1999**, *255*, 259–264. [CrossRef]
2. Abazari, R.; Abazari, M. Numerical simulation of generalized Hirota-Satsuma coupled KdV equation by RDTM and comparison with DTM. *Commun. Nonlinear Sci. Numer. Simul.* **2012**, *17*, 619–629. [CrossRef]
3. Ganji, D.D.; Rafei, M. Solitary wave solutions for a generalized Hirota-Satsuma coupled KdV equation by homotopy perturbation method. *Phys. Lett. A* **2006**, *356*, 131–137. [CrossRef]
4. Akinyemi, L.; Huseen, S.N. A powerful approach to study the new modified coupled Korteweg-de Vries system. *Math. Comput. Simul.* **2020**, *177*, 556–567. [CrossRef]
5. Chen, C.K.; Ho, S.H. Solving partial differential equations by two-dimensional differential transform method. *Appl. Math. Comput.* **1999**, *106*, 171–179.
6. Gao, Y.T.; Tian, B. Ion-acoustic shocks in space and laboratory dusty plasmas: Two dimensional and non-traveling-wave observable effects. *Phys. Plasmas* **2001**, *8*, 3146–3149. [CrossRef]
7. Osborne, A. The inverse scattering transform: Tools for the nonlinear fourier analysis and filtering of ocean surface waves. *Chaos Solitons Fract.* **1995**, *5*, 2623–2637. [CrossRef]
8. Ostrovsky, L.Y.; Stepanyants, A. Do internal solutions exist in the ocean. *Rev. Geophys.* **1989**, *27*, 293–310. [CrossRef]
9. Wang, M.; Zhou, Y.; Li, Z. Application of a homogeneous balance method to exact solutions of non-linear equations in mathematical physics. *Phys. Lett. A.* **1996**, *216*, 67–75. [CrossRef]
10. Gokdogan, A.; Yildirim, A.; Merdan, M. Solving coupled-KdV equations by differential transformation method. *World Appl. Sci. J.* **2012**, *19*, 1823–1828.
11. Jafari, H.; Firoozjaee, M.A. Homotopy analysis method for solving KdV equations. *Surv. Math. Appl.* **2010**, *5*, 89–98.
12. Rashid, S.; Kubra, K.T.; Rauf, A.; Chu, Y.-M.; Hamed, Y.S. New numerical approach for time-fractional partial differential equations arising in physical system involving natural decomposition method. *Phys. Sci.* **2021**, *96*, 105204. [CrossRef]
13. Lu, D.; Suleman, M.; Ramzan, M.; Ul Rahman, J. Numerical solutions of coupled nonlinear fractional KdV equations using He's fractional calculus. *Int. J. Mod. Phys. B* **2021**, *35*, 2150023. [CrossRef]
14. Mohamed, M.A.; Torky, M.S. Numerical solution of non-linear system of partial differential equations by the Laplace decomposition method and the Padé approximation. *Am. J. Comput. Math.* **2013**, *3*, 175. [CrossRef]
15. Seadawy, A.R.; El-Rashidy, K. Water wave solutions of the coupled system Zakharov-Kuznetsov and generalized coupled KdV equations. *Sci. World J.* **2014**, *2014*, 1–6. [CrossRef] [PubMed]
16. He, W.; Chen, N.; Dassios, I.; Shah, N.A.; Chung, J.E. Fractional System of Korteweg-De Vries equations via Elzaki transform. *Mathematics* **2021**, *9*, 673. [CrossRef]
17. De la Rosa, R.; Recio, E.; Garrido, T.M.; Bruzón, M.S. Lie symmetry analysis of $(2+1)$-dimensional KdV equations with variable coefficients. *Int. J. Comput. Math.* **2019**, *97*, 1–13. [CrossRef]
18. Kannan, R.; Wang, Z.J. A high order spectral volume solution to the Burgers' equation using the Hopf-Cole transformation. *Int. J. Numer. Meth. Fluids* **2012**, *69*, 781–801. [CrossRef]
19. Kannan R. A high order spectral volume formulation for solving equations containing higher spatial derivative terms II: Improving the third derivative spatial discretization using the LDG2 method. *Commun. Comput. Phy.* **2012**, *12*, 767–788. [CrossRef]
20. Fan, E. Using symbolic computation to exactly solve a new coupled MKdV system. *Phys. Lett. A* **2002**, *299*, 46–48. [CrossRef]
21. Inc, M.; Cavlak, E. On numerical solutions of a new coupled MKdV system by using the Adomian decomposition method and He's variational iteration method. *Phys. Sci.* **2008**, *78*, 1–7. [CrossRef]
22. Inc, M.; Parto-Haghighi, M.; Akinlar, M.A.; Chu, Y.-M. New numerical solutions of fractional-order Korteweg-de Vries equation. *Res. Phy.* **2020**, *19*, 103326. [CrossRef]

23. Lin, G.; Grinberg, L.; Karniadakis, G.E. Numerical studies of the stochastic Korteweg-de Vries equation. *J. Comput. Phys.* **2006**, *213*, 676–703. [CrossRef]
24. Karczewska, A.; Szczeciński, M. Martingale solution to stochastic extended Korteweg-de Vries equation. *Adv. Pure Math.* **2018**, *8*. [CrossRef]
25. Ghoreishi, M.; Ismail, A.I.; Rashid, A. The solution of coupled modifed KdV system by the homotopy analysis method. *TWMS J. Pure Appl. Math.* **2012**, *3*, 122–134.
26. Baleanu, D.; Diethelm, K.; Scalas, E.; Trujillo, J.J. *Fractional Calculus Models and Numerical Methods*; Series on Complexity, Nonlinearity and Chaos; World Scientific: Singapore, 2012.
27. Podlubny, I. *Fractional Differential Equations*; Academic Press: New York, NY, USA, 1999.
28. Atangana, A.; Secer, A. A note on fractional order derivatives and table of fractional derivatives of some special functions. *Abstr. Appl. Anal.* **2013**, *2013*, 279681. [CrossRef]
29. Caputo, M.; Fabrizio, M. A new definition of fractional derivative without singular kernel. *Prog. Fract. Differ. Appl.* **2015**, *1*, 73–85.
30. Atangana, A.; Alkahtani, B.S.T. New model of groundwater flowing within a confine aquifer: Application of Caputo-Fabrizio derivative. *Arab. J. Geosci.* **2016**, *9*, 1–6. [CrossRef]
31. Diethelm, K.; Ford, N.J.; Freed, A.D. A predictor-corrector approach for the numerical solution of fractional differential equations. *Nonlinear Dyn.* **2002**, *29*, 3–22. [CrossRef]
32. Meerschaert, M.M.; Benson, D.A.; Scheffler, H.-P.; Baeumer, B. Stochastic solution of space-time fractional diffusion equations. *Phys. Rev. E* **2002**, *65*, 041103. [CrossRef]
33. Fulger, D.; Scalas, E.; Germano, G. Monte Carlo simulation of uncoupled continuous-time random walks yielding a stochastic solution of the space-time fractional diffusion equation. *Phys. Rev. E* **2008**, *77*, 1–7. [CrossRef]
34. Maitama, S.; Zhao, W. New integral transform: Shehu transform a generalization of Sumudu and Laplace transform for solving differential equations. *Int. J. Anal. Appl.* **2019**, *17*, 167–190.
35. Rani, D.; Mishra, V. Modification of Laplace adomian decomposition method for solving nonlinear Volterra integral and integro-differential equations based on Newton–Raphson formula. *Eur. J. Pure Appl. Math.* **2018**, *11*, 202–214. [CrossRef]
36. Birajdar, G. Numerical solution of time fractional Navier–Stokes equation by discrete Adomian decomposition method. *Nonlinear Eng.* **2014**, *3*, 21–26. [CrossRef]
37. Haq, F.; Shah, K.; Khan, A.; Shahzad, M.; Rahman, G. Numerical solution of fractional order epidemic model of a vector born disease by Laplace Adomian decomposition method. *Punjab Univ. J. Math.* **2017**, *49*, 13–22.
38. Daftardar-Gejji, V.; Jafari, H. An iterative method for solving nonlinear functional equations. *J. Math. Anal. Appl.* **2006**, *316*, 753–763. [CrossRef]
39. Jafari, H.; Nazari, M.; Baleanu, D.; Khalique, C.M. A new approach for solving a system of fractional partial differential equations. *Comput. Math. Appl.* **2013**, *66*, 838–843. [CrossRef]
40. Ramadan, M.A.; Al-luhaibi, M.S. New iterative method for solving the Fornberg-Whitham equation and comparison with homotopy perturbation transform method. *J. Adv. Math. Comput. Sci.* **2014**, *4*, 1213–1227. [CrossRef]
41. Alderremy, A.A.; Elzaki, T.M.; Chamekh, M. New transform iterative method for solving some Klein-Gordon equations. *Res. Phys.* **2018**, *10*, 655–659. [CrossRef]
42. El-Kalla, I. Convergence of the Adomian method applied to a class of nonlinear integral equations. *Appl. Math. Lett.* **2008**, *21*, 372–376. [CrossRef]

MDPI
St. Alban-Anlage 66
4052 Basel
Switzerland
Tel. +41 61 683 77 34
Fax +41 61 302 89 18
www.mdpi.com

Symmetry Editorial Office
E-mail: symmetry@mdpi.com
www.mdpi.com/journal/symmetry

MDPI
St. Alban-Anlage 66
4052 Basel
Switzerland
Tel. +41 61 683 77 34
Fax +41 61 302 89 18
www.mdpi.com

Symmetry Editorial Office
E-mail: symmetry@mdpi.com
www.mdpi.com/journal/symmetry

www.ingramcontent.com/pod-product-compliance
Lightning Source LLC
LaVergne TN
LVHW070125100526
838202LV00016B/2230